HANDBOOK OF
SCALING METHODS
IN AQUATIC ECOLOGY
MEASUREMENT, ANALYSIS, SIMULATION

HANDBOOK OF
SCALING METHODS
IN AQUATIC ECOLOGY

MEASUREMENT, ANALYSIS, SIMULATION

EDITED BY

LAURENT SEURONT
PETER G. STRUTTON

CRC Press
Taylor & Francis Group
Boca Raton London New York

CRC Press is an imprint of the
Taylor & Francis Group, an **informa** business

Cover: *Mount Fuji from the Offing*, also known as *The Great Wave off Kanagawa*, from the series of block prints *36 Views of Mount Fuji* (1823–1829) by Katsushika Hokusai (1760–1849).

Senior Editor: John Sulzycki
Production Editor: Christine Andreasen
Project Coordinator: Erika Dery
Marketing Manager: Nadja English

CRC Press
Taylor & Francis Group
6000 Broken Sound Parkway NW, Suite 300
Boca Raton, FL 33487-2742

First issued in paperback 2019

© 2004 by Taylor & Francis Group, LLC
CRC Press is an imprint of Taylor & Francis Group, an Informa business

No claim to original U.S. Government works

ISBN-13: 978-0-8493-1344-8 (hbk)
ISBN-13: 978-0-367-39498-1 (pbk)
Library of Congress Card Number 2003051467

Library of Congress Cataloging-in-Publication Data

Handbook of scaling methods in aquatic ecology : measurement, analysis, simulation /
edited by Laurent Seuront and Peter G. Strutton.
      p. cm.
    Includes bibliographical references and index.
    ISBN 0-8493-1344-9
      1. Aquatic ecology—Research—Methodology. 2. Aquatic ecology—Measurement. 3.
  Aquatic ecology—Simulation methods. I. Seuront, Laurent. II. Strutton, Peter G.

QH541.5.W3H36 2003
577.6′072—dc21

2003051467

Visit the Taylor & Francis Web site at
http://www.taylorandfrancis.com

and the CRC Press Web site at
http://www.crcpress.com

Preface

Aquatic scientists have always been intrigued with concepts of scale. This interest perhaps stems from the nature of fluid dynamics in oceans and lakes — energy cascades from spatial scales of kilometers down to viscous scales at centimeters or less. Turbulent processes affect not only an organism's perception of, and response to, the physical environment, but also the interaction between species, both within and across trophic levels.

Our ability to understand processes that act across scales has traditionally been technically limited by the availability of appropriate instruments, suitable analysis, modeling and simulation techniques, and sufficient computing power. In some respects we have also lacked a theoretical framework for conducting these observations, analyses, and simulations. Since the 1970s these problems have partially been overcome and our understanding of the relationship between scale and aquatic processes has advanced accordingly. In fact, it was in the early 1970s that the first applications of spectral analysis to biological oceanographic data emerged. This initial work described the scale-dependent nature of biological–physical interactions and stimulated investigations of such interactions across a vast array of time and space scales. However, even if the increase in computer power during the last three decades has opened new perspectives in space–time complexity in aquatic ecology, it is still unfortunately not sufficient. For example, a realistic framework for turbulence simulations will still require several decades of technical improvements, simply to be able to handle high Reynolds number flows, while a theoretical framework for intermittent processes — increasingly recognized as playing a crucial role in aquatic ecosystems — still does not exist.

Only in recent years has aquatic ecology begun to incorporate new, exciting, and often interrelated observational, analysis, modeling, and simulation techniques. These include the following:

- Development of techniques to observe small-scale biological processes such as bacterial chemotaxis, zooplankton behavior, and organisms' responses to turbulence
- Increasing availability and use of satellite data to view the other end of the spatial spectrum — basin scale dynamics
- Incorporation of nonlinear analysis techniques and application of concepts from chaos theory to problems of spatial and temporal processes
- Advancement of models and simulations to mimic and hence understand complex biological processes

This volume compiles a comprehensive selection of papers, illustrating some of the recent advances that have been made toward understanding physical, biological, and chemical processes across multiple time and space scales. The chapters cover a range of ecosystems, both oceanic and freshwater, from pelagic to benthic/rocky intertidal to seagrass beds. The scale of processes considered ranges from the microscopic to almost global, spanning topics such as physiological cues in individual phytoplankton cells and mating signals in zooplankton to basin-scale primary productivity. The range of organisms studied is equally diverse, from phyto- and zooplankton to large fish dynamics. A broad range of up-to-date observational, data analysis, and simulation techniques is presented. These include (1) new bio-optical, video, acoustic, remote sensing, and synchrotron-based imaging systems, (2) different scaling methods (i.e., fractals, wavelets, rank-order relationship) to assess a broad range of spatial and temporal patterns and processes, and (3) innovative simulation techniques that allow insights into processes ranging from individual behavior to population dynamics, the structure of turbulent intermittency and its effects on swimming organisms, and the effect of large-scale physical forcings on particle distributions at small scales. Measurement, analysis, and simulation at the organismal level might be crucial to investigate the poorly understood cumulative effect of fine-scale processes on broad-scale biosphere processes. This approach may eventually link dynamic processes at several spatiotemporal scales both to understand complex ecological systems and to address old research questions from new perspectives.

It is our hope that this compilation will expose exciting new research to those already working in the field, as well as facilitate a type of cross-pollination by introducing other sections of the scientific community to recent developments. We thus believe that the combination of three potentially disparate fields — measurement, analysis, and simulation — in one volume will serve to build bridges between experimentalists and theoreticians. Only by the close collaboration of these fields will we continue to gain a solid understanding of complex aquatic ecosystems.

Laurent Seuront and Peter G. Strutton

Acknowledgments

This book arose out of a special session entitled "Dealing with Scales in Aquatic Ecology: Structure and Function in Aquatic Ecosystems" at the 2001 ASLO Aquatic Science Meeting. We gratefully acknowledge the Organizing Committee chairs, Josef Ackerman and Saran Twombly, for their enthusiastic support of what was a successful and popular session, and what we hope will be a successful and popular compilation. Specifically, we thank the contributors to that session for their quality presentations. We acknowledge, in alphabetical order, C. Avois, J.A. Barth, A.S. Cohen, E.A. Cowen, T.J. Cowles, T.L. Cucci, H. Cyr, M.M. Dekshenieks (McManus), P.L. Donaghay, K.E. Fisher, C. Greenlaw, R.E. Hecky, D.V. Holliday, Z. Johnson, P. Legendre, M. Louis, D. McGehee, C.M. O'Reilly, V. Pasour, J.E. Rines, F.G. Schmitt, C.S. Sieracki, M.E. Sieracki, J. Sullivan, E.C.U. Thier, A.K. Yamazaki, and H. Yamazaki. Finally, we owe our thanks to the reviewers of the chapters for improving the quality of the published work.

Editors

Laurent Seuront, Ph.D., is a CNRS Research Scientist at the Wimereux Marine Station, University of Lille 1 — CNRS UMR 8013 ELICO, France. His education includes a B.S. in population biology and ecology from the University of Lille 1 (1992); an M.S. in marine ecology, data analysis, and modelling from the University of Paris 6 (1995); and a Ph.D. in biological oceanography from the University of Lille 1 (1999). Prior to his present position, he was a research fellow of the Japanese Society for the Promotion of Science at the Tokyo University of Fisheries, working with Hidekatsu Yamazaki.

Dr. Seuront's research concerns biological–physical coupling in aquatic/marine systems/environments, particularly with regard to the effect of microscale (submeter) patterns and processes on large-scale processes. Aspects of his work combine field, laboratory, and numerical experiments to study the centimeter-scale distribution of biological (nutrient, bacteria, phytoplankton, microphytobenthos, and microzoobenthos) and physical parameters (temperature, salinity, light, turbulence), as well as the motile behavior of individual organisms in response to different biophysical forcings. His work to date has been the subject of more than 30 publications in international journals and contributed books, more than 30 presentations at international conferences, and invited seminars at more than 20 locations throughout the world.

Peter G. Strutton, Ph.D., is Assistant Professor of Oceanography at the Marine Sciences Research Center, State University of New York, Stony Brook. Prior to his current appointment he was Postdoctoral Scientist with Francisco Chavez at the Monterey Bay Aquarium Research Institute. He received his B.Sc. (Honors) and Ph.D. in marine science from Flinders University of South Australia, working with Jim Mitchell, who was in turn a student of Akira Okubo at Stony Brook. Professor Okubo's legacy is apparent in many of the chapters contained in this volume.

Dr. Strutton's work focuses on the interaction among physics, biology, and chemistry in the ocean at a broad range of time and space scales. Current areas of interest include the spatial and temporal variability of carbon cycling in the equatorial Pacific, the influence of phytoplankton on the heat budget of the upper ocean, and the biological–physical interactions associated with open ocean iron fertilization. Since 1996 he has authored or co-authored approximately 20 publications and has presented his work at more than 30 meetings and invited seminars.

Contributors

Neil S. Banas
School of Oceanography
University of Washington
Seattle, Washington, U.S.A.

Richard T. Barber
Nicholas School of the Environment
    and Earth Sciences
Duke University
Beaufort, North Carolina, U.S.A.

Carlos Bas
Institute of Marine Science (CSIC)
Paseo Juan de Borbón
Barcelona, Spain

Alberto Basset
Department of Biology
University of Lecce
Lecce, Italy

Mark C. Benfield
Department of Oceanography
    and Coastal Sciences
Coastal Fisheries Institute
Louisiana State University
Baton Rouge, Louisiana, U.S.A.

Uta Berger
Center for Tropical Marine Ecology
Bremen, Germany

Olivier Bernard
COMORE–INRIA
Sophia-Antipolis, France

David R. Blakeway
Department of Biological Sciences
California State University
Los Angeles, California, U.S.A.

Matthew C. Brewer
Department of Zoology
University of Florida
Gainesville, Florida, U.S.A.

Jason Brown
School of Biology
Georgia Institute of Technology
Atlanta, Georgia, U.S.A.

Maria Bunta
Department of Physics
University of Wisconsin–Milwaukee
Milwaukee, Wisconsin, U.S.A.

Janet W. Campbell
Ocean Process Analysis Laboratory
University of New Hampshire
Durham, New Hampshire, U.S.A.

Philippe Caparroy
Insect Biology Research Institute
CNRS–University of Tours
Tours, France

Jose Juan Castro
Fisheries Research Group
University of Las Palmas
Las Palmas, Canary Islands, Spain

Michael Caun
Department of Life Sciences
University of California at Santa Barbara
Santa Barbara, California, U.S.A.

Francisco P. Chavez
Monterey Bay Aquarium
    Research Institute
Moss Landing, California, U.S.A.

Andrew S. Cohen
Department of Geosciences
University of Arizona
Tucson, Arizona, U.S.A.

Timothy J. Cowles
College of Oceanic
    and Atmospheric Sciences
Oregon State University
Corvallis, Oregon, U.S.A.

Hélène Cyr
Department of Zoology
University of Toronto
Toronto, Ontario, Canada

John Davenport
Environmental Research Institute
Department of Zoology, Ecology,
    and Plant Science
University College Cork
Cork, Ireland

Dominique Davoult
Biological Station of Roscoff
University of Paris 6 and CNRS
Roscoff, France

Donald L. DeAngelis
Biological Resources Division
U.S. Geological Survey
    and Department of Biology
University of Miami
Coral Gables, Florida, U.S.A.

Robert A. Desharnais
Department of Biological Sciences
California State University
Los Angeles, California, U.S.A.

Peter J. Dillon
Department of Chemistry
Trent University
Peterborough, Ontario, Canada

Michael Doall
Functional Ecology Laboratory
State University of New York
    at Stony Brook
Stony Brook, New York, U.S.A.

Douglas D. Donalson
Department of Biological Sciences
California State University
Los Angeles, California, U.S.A.

Igor M. Dremin
Theory Department
Lebedev Physical Institute
Moscow, Russia

Karen E. Fisher
Los Alamos National Laboratory
Los Alamos, New Mexico, U.S.A.

Rodney G. Fredericks
Coastal Studies Field Support Group
Louisiana State University
Baton Rouge, Louisiana, U.S.A.

David A. Fuentes
Department of Biological Sciences
California State University
Los Angeles, California, U.S.A.

Vincent Ginot
INRA Biometry Unit
Domaine St. Paul
Avignon, France

Mario Giordano
Institute of Marine Sciences
University of Ancona
Ancona, Italy

Jarl Giske
Department of Fisheries
    and Marine Biology
University of Bergen
Bergen, Norway

Volker Grimm
Department of Ecological Modelling
UFZ Centre for Environmental Research
    Leipzig–Halle
Leipzig, Germany

Robert E. Hecky
Department of Biology
University of Waterloo
Waterloo, Ontario, Canada

Jean-Pierre Hermand
Department of Optics and Acoustics
University of Brussels (ULB)
Brussels, Belgium

Carol J. Hirschmugl
Department of Physics
University of Wisconsin–Milwaukee
Milwaukee, Wisconsin, U.S.A.

Geir Huse
Department of Fisheries
    and Marine Biology
University of Bergen
Bergen, Norway

Oleg V. Ivanov
Theory Department
Lebedev Physical Institute
Moscow, Russia

Houshuo Jiang
Department of Applied Ocean Physics
    and Engineering
Woods Hole Oceanographic Institution
Woods Hole, Massachusetts, U.S.A.

Mark P. Johnson
School of Biology and Biochemistry
The Queen's University of Belfast
Belfast, United Kingdom

Daniel Kamykowski
Department of Marine, Earth,
    and Atmospheric Sciences
North Carolina State University
Raleigh, North Carolina, U.S.A.

Sean F. Keenan
Department of Oceanography
    and Coastal Sciences
Louisiana State University
Baton Rouge, Louisiana, U.S.A.

Sophie Leterme
Wimereux Marine Station
CNRS
and
University of Lille 1
Wimereux, France

Amala Mahadevan
Department of Applied Mathematics
    and Theoretical Physics
University of Cambridge
Cambridge, United Kingdom
and
Ocean Process Analysis Laboratory
University of New Hampshire
Durham, New Hampshire, U.S.A.

Bruce D. Malamud
Environmental Monitoring
    and Modelling Group
Department of Geography
King's College London, Strand
London, United Kingdom

Horst Malchow
Department of Mathematics
    and Computer Science
Institute for Environmental
    Systems Research
University of Osnabrück
Osnabrück, Germany

Alexander B. Medvinsky
Institute for Theoretical
    and Experimental Biophysics
Russian Academy of Sciences
Pushchino, Moscow Region, Russia

Aline Migné
Laboratory of Hydrobiology
University of Paris 6 and CNRS
Paris, France

James G. Mitchell
School of Biological Sciences
Flinders University
Adelaide, Australia

Wolf M. Mooij
Centre for Limnology
The Netherlands Institute of Ecology
Nieuwersluis, the Netherlands

Vladimir A. Nechitailo
Theory Department
Lebedev Physical Institute
Moscow, Russia

Roger M. Nisbet
Department of Ecology, Evolution,
    and Marine Biology
University of California at Santa Barbara
Santa Barbara, California, U.S.A.

Catherine M. O'Reilly
Environmental Science Program
Vassar College
Poughkeepsie, New York, U.S.A.

Thomas Osborn
Department of Earth
    and Planetary Sciences
The Johns Hopkins University
Baltimore, Maryland, U.S.A.

Gary K. Ostrander
Department of Biology
The Johns Hopkins University
Baltimore, Maryland, U.S.A.

Julie E. Parker
CEH Windermere Laboratory
Ambleside, United Kingdom

Sergei V. Petrovskii
Shirshov Institute of Oceanology
Russian Academy of Sciences
Moscow, Russia

Pierre-Denis Plisnier
Royal Museum of Central Africa
Tervuren, Belgium

Anne C. Prusak
School of Biology
Georgia Institute of Technology
Atlanta, Georgia, U.S.A.

Hong-Lie Qiu
Department of Geography
  and Urban Analysis
California State University
Los Angeles, California, U.S.A.

Carlos D. Robles
Department of Biological Sciences
California State University
Los Angeles, California, U.S.A.

François G. Schmitt
Wimereux Marine Station
CNRS
and
University of Lille 1
Wimereux, France

Christopher J. Schwehm
Division of Engineering Services
Louisiana State University
Baton Rouge, Louisiana, U.S.A.

Laurent Seuront
Wimereux Marine Station
CNRS
and
University of Lille 1
Wimereux, France

Eric D. Skyllingstad
College of Oceanic
  and Atmospheric Sciences
Oregon State University
Corvallis, Oregon, U.S.A.

Aldo P. Solari
Fisheries Research Group
University of Las Palmas
Las Palmas, Canary Islands, Spain

Sami Souissi
Wimereux Marine Station
CNRS
and
University of Lille 1
Wimereux, France

Nicolas Spilmont
Laboratory of Biogeochemistry
  and Littoral Environment
CNRS
and
University of Littoral Côte d'Opale
Wimereux, France

Kyle D. Squires
Mechanical and Aerospace
  Engineering Department
Arizona State University
Tempe, Arizona, U.S.A.

Gregory Squyres
Coastal Studies Field Support Group
Louisiana State University
Baton Rouge, Louisiana, U.S.A.

J. Rudi Strickler
Great Lakes WATER Institute
University of Wisconsin–Milwaukee
Milwaukee, Wisconsin, U.S.A.

Peter G. Strutton
Marine Science Research Center
State University of New York
  at Stony Brook
Stony Brook, New York, U.S.A.

Mark V. Trevorrow
Defence Research and Development
  Canada–Atlantic
Dartmouth, Nova Scotia, Canada

Shin-Ichi Uye
Faculty of Applied Biological Science
Hiroshima University
Higashi-Hiroshima, Japan

Pieter Verburg
Department of Biology
University of Waterloo
Waterloo, Ontario, Canada

Dong-Ping Wang
Marine Sciences Research Center
State University of New York
    at Stony Brook
Stony Brook, New York, U.S.A.

Peter H. Wiebe
Department of Biology
Woods Hole Oceanographic Institution
Woods Hole, Massachusetts, U.S.A.

Fabian Wolk
Rockland Oceanographic Services, Inc.
Victoria, British Columbia, Canada

Atsuko K. Yamazaki
Department of Manufacturing
    Technologists
Monotsukuri Institute of Technologists
Gyoda-shi, Japan

Hidekatsu Yamazaki
Department of Ocean Sciences
Tokyo University of Fisheries
Tokyo, Japan

Jeannette Yen
School of Biology
Georgia Institute of Technology
Atlanta, Georgia, U.S.A.

Contents

Section I Measurements

1 Comparison of Biological Scale Resolution from CTD
and Microstructure Measurements..3
Fabian Wolk, Laurent Seuront, Hidekatsu Yamazaki, and Sophie Leterme

2 Measurement of Zooplankton Distributions
with a High-Resolution Digital Camera System..17
*Mark C. Benfield, Christopher J. Schwehm, Rodney G. Fredericks,
Gregory Squyres, Sean F. Keenan, and Mark V. Trevorrow*

3 Planktonic Layers: Physical and Biological Interactions
on the Small Scale...31
Timothy J. Cowles

4 Scales of Biological–Physical Coupling in the Equatorial Pacific 51
Peter G. Strutton and Francisco P. Chavez

5 Acoustic Remote Sensing of Photosynthetic Activity in Seagrass Beds.....65
Jean-Pierre Hermand

6 Multiscale *in Situ* Measurements
of Intertidal Benthic Production and Respiration 97
Dominique Davoult, Aline Migné, and Nicolas Spilmont

7 Spatially Extensive, High Resolution Images
of Rocky Shore Communities..109
David R. Blakeway, Carlos D. Robles, David A. Fuentes, and Hong-Lie Qiu

8 Food Web Dynamics in Stable Isotope Ecology:
Time Integration of Different Trophic Levels 125
*Catherine M. O'Reilly, Pieter Verburg, Robert E. Hecky,
Pierre-Denis Plisnier, and Andrew S. Cohen*

9 Synchrotron-Based Infrared Imaging of *Euglena gracilis* Single Cells.....135
Carol J. Hirschmugl, Maria Bunta, and Mario Giordano

10 Signaling during Mating in the Pelagic Copepod, *Temora longicornis* 149
*Jeannette Yen, Anne C. Prusak, Michael Caun, Michael Doall,
Jason Brown, and J. Rudi Strickler*

**11 Experimental Validation of an Individual-Based Model
for Zooplankton Swarming** 161
Neil S. Banas, Dong-Ping Wang, and Jeannette Yen

Section II Analysis

12 On Skipjack Tuna Dynamics: Similarity at Several Scales 183
Aldo P. Solari, Jose Juan Castro, and Carlos Bas

**13 The Temporal Scaling of Environmental Variability
in Rivers and Lakes** 201
Hélène Cyr, Peter J. Dillon, and Julie E. Parker

**14 Biogeochemical Variability at the Sea Surface:
How It Is Linked to Process Response Times** 215
Amala Mahadevan and Janet W. Campbell

**15 Challenges in the Analysis and Simulation
of Benthic Community Patterns** 229
Mark P. Johnson

**16 Fractal Dimension Estimation in Studies of Epiphytal
and Epilithic Communities: Strengths and Weaknesses** 245
John Davenport

17 Rank-Size Analysis and Vertical Phytoplankton Distribution Patterns ... 257
James G. Mitchell

18 An Introduction to Wavelets 279
Igor M. Dremin, Oleg V. Ivanov, and Vladimir A. Nechitailo

**19 Fractal Characterization of Local Hydrographic and Biological Scales
of Patchiness on Georges Bank** 297
Karen E. Fisher, Peter H. Wiebe, and Bruce D. Malamud

20 Orientation of Sea Fans Perpendicular to the Flow 321
Thomas Osborn and Gary K. Ostrander

**21 Why Are Large, Delicate, Gelatinous Organisms So Successful
in the Ocean's Interior?** 329
Thomas Osborn and Richard T. Barber

**22 Quantifying Zooplankton Swimming Behavior:
The Question of Scale** 333
Laurent Seuront, Matthew C. Brewer, and J. Rudi Strickler

23 Identification of Interactions in Copepod Populations Using
a Qualitative Study of Stage-Structured Population Models....................361
Sami Souissi and Olivier Bernard

Section III Simulation

24 The Importance of Spatial Scale in the Modeling
of Aquatic Ecosystems...383
Donald L. DeAngelis, Wolf M. Mooij, and Alberto Basset

25 Patterns in Models of Plankton Dynamics
in a Heterogeneous Environment..401
Horst Malchow, Alexander B. Medvinsky, and Sergei V. Petrovskii

26 Seeing the Forest for the Trees, and Vice Versa:
Pattern-Oriented Ecological Modeling 411
Volker Grimm and Uta Berger

27 Spatial Dynamics of a Benthic Community:
Applying Multiple Models to a Single System 429
Douglas D. Donalson, Robert A. Desharnais, Carlos D. Robles, and Roger M. Nisbet

28 The Effects of Langmuir Circulation on Buoyant Particles.......................445
Eric D. Skyllingstad

29 Modeling of Turbulent Intermittency:
Multifractal Stochastic Processes and Their Simulation 453
François G. Schmitt

30 An Application of the Lognormal Theory
to Moderate Reynolds Number Turbulent Structures 469
Hidekatsu Yamazaki and Kyle D. Squires

31 Numerical Simulation of the Flow Field at the Scale Size
of an Individual Copepod...479
Houshuo Jiang

32 Can Turbulence Reduce the Energy Costs of Hovering
for Planktonic Organisms?..493
Hidekatsu Yamazaki, Kyle D. Squires, and J. Rudi Strickler

33 Utilizing Different Levels of Adaptation
in Individual-Based Modeling 507
Geir Huse and Jarl Giske

34 Using MultiAgent Systems to Develop Individual-Based Models
for Copepods: Consequences of Individual Behavior
and Spatial Heterogeneity on the Emerging Properties
at the Population Scale..523
Sami Souissi, Vincent Ginot, Laurent Seuront, and Shin-Ichi Uye

35 Modeling Planktonic Behavior as a Complex Adaptive System.................543
Atsuko K. Yamazaki and Daniel Kamykowski

36 Discrete Events-Based Lagrangian Approach as a Tool
for Modeling Predator–Prey Interactions in the Plankton 559
Philippe Caparroy

Index..575

Section I

Measurements

1

Comparison of Biological Scale Resolution from CTD and Microstructure Measurements

Fabian Wolk, Laurent Seuront, Hidekatsu Yamazaki, and Sophie Leterme

CONTENTS

1.1 Introduction 3
1.2 Microscale Structure in Aquatic Ecosystems: Perspectives 4
    1.2.1 Aquatic Ecosystem Functioning 4
    1.2.2 Impact of the Sampling Process 5
1.3 Comparison of High-Resolution Data and Conventional Techniques 7
    1.3.1 Instrument Description...7
    1.3.2 Sensor Deployment 9
    1.3.3 Differential Structure of Standard and High-Resolution Fluorescence Signals 10
1.4 Conclusion..12
Acknowledgments..13
References...14

1.1 Introduction

The existence of small-scale (<1 m) planktonic structures and their importance to the dynamics of the aquatic ecosystem are now widely acknowledged in the oceanographic and limnology community (e.g., Hanson and Donaghay, 1998; Holliday et al., 1998; Jaffe et al., 1998; Franks and Jaffe, 2001). Despite recent advances in experimental technology (Mitchell and Fuhrman, 1989; Donaghay et al., 1992; Desiderio et al., 1993), most field work is conducted using conventional sampling methods (such as Niskin bottles or *in situ* fluorometers mounted on CTD cages), which do not resolve the small-scale biological structures.

Recently, the study of the intermittent variability of biological processes in aquatic ecosystems has benefited from the development of small-scale monitoring systems (Hanson and Donaghay, 1998; Holliday et al., 1998; Jaffe et al., 1998; Franks and Jaffe, 2001), and novel theoretical approaches to characterize intermittent patterns (Pascual et al., 1995; Strutton et al., 1996, 1997; Seuront et al., 1999). These techniques confirm the notion that any advances in understanding the response of fine-scale and microscale biological structures to physical forcing require simultaneous measurements of both physical and biological parameters over a congruent range of spatial scales. With the exception of a few studies (Donaghay et al., 1992; Desiderio et al., 1993; Cowles et al., 1998; Yamazaki et al., in revision) previous small-scale biological observations have lacked concomitant measurements of physical variables. Such information is, however, of crucial importance because physical processes at these scales (e.g., shear instabilities, convective overturns, salt fingering, etc.) lead to intermittent vertical mixing and the redistribution of biomass. Such correlated measurements are particularly important in highly dissipative environments, such as tidally mixed coastal waters, frontal structures, or turbulent patches in the seasonal thermocline, where the space and time variability of physical and biological processes and the resultant biophysical interactions are very high (Yamazaki and Osborn, 1988; Yamazaki et al., 2002).

To understand the response of phytoplankton cells to physical forcing, it is necessary to accomplish the following:

1. Describe the vertical structure of the sampled water column in terms of physical parameters
2. Identify the spatial patterns of concurrently sampled *in vivo* fluorescence (a proxy of phytoplankton biomass)
3. Investigate its potential fine and microscale relationships with the surrounding physical environment.

In this context, the purpose of this chapter is to stress the importance of fine-scale sampling methods in aquatic ecology, and to demonstrate how the use of adequate high-resolution experimental equipment, coupled with novel statistical tools for processing and analyzing the data, can increase our understanding of the structures of the aquatic ecosystem. It is shown that the chlorophyll *a* concentration inside a single Niskin bottle is far from homogeneous; concentration values can vary more than the annual distribution range from the same sampling area. Data from a high-resolution fluorometer deployed in a lake and a well-mixed tidal channel corroborate the high degree of small-scale variance found in the Niskin bottles. This small-scale structure is often overlooked in standard CTD sampling methods. Finally, it is shown that new experimental equipment and appropriate higher-order statistical tools make it possible to condense the high-resolution data and effectively characterize the experimental results. In Section 1.2.1, the current understanding of aquatic ecosystem structures and functions is outlined. An example of field samples taken from Niskin bottles demonstrates the impact of the sampling strategy on the observed microscale distribution of phytoplankton biomass (Section 1.2.2). A direct comparison of field data from a recently developed high-resolution bio-optical sensor and a conventional field fluorometer illustrates how inappropriate equipment can lead to a distorted representation of the biological structures in the water column. The results of the comparison are described in Section 1.3.

1.2   Microscale Structure in Aquatic Ecosystems: Perspectives

Accurate characterization of phytoplankton distributions, as well as their sources and scales of variability, is important for a variety of applications; e.g., basic studies of primary production (Platt et al., 1989; Seuront et al., 1999), issues related to the role of particle aggregates in the vertical flux of organic matter (Jackson and Burd, 1998), modeling the effects of thin layer reflectance on remote sensing (Petrenko et al., 1998; Zaneveld and Pegau, 1998), and studying the effects of light absorption by phytoplankton and particulate material (Sosik and Mitchell, 1995).

It has been recognized for more than two decades that physical and biological structures, identified in terms of spatial patchiness, temporal cycles, or disturbances, are a key feature of aquatic ecosystems (Denman and Powell, 1984; Mackas et al., 1985). The geometry of such structures and their effect on the aquatic ecosystem depends on both their magnitude and their spatial and temporal scales.

1.2.1   Aquatic Ecosystem Functioning

Biophysical interactions affect the ecosystem in a subtle manner because the effects depend on the coupling between physical scales of patchiness and biologically significant scales, such as generation time or ambits. The effects of a particular scale of patchiness may vary for different types of organisms with different characteristic biological scales. Biophysical disturbances, for example, have been proposed as a potential mechanism for maintenance of diversity under conditions where competitive exclusion should otherwise lead to lower diversity (Hutchinson, 1961; Scheffer, 1995; Siegel, 1998; Seuront et al., 2002; Seuront and Spilmont, 2002). The scales of disturbance that are necessary to allow coexistence in this manner for phytoplankton (with generation times of days and ambits of decameters or less) may differ from that of macrozooplankton (with generation times of weeks to months and ambits over this time of at least tens of kilometers). The outcome of other interactions, such as feeding, predation, or migration of zooplankton, may also depend on the interrelation between scales of physical structure and the biological and/or ecological scale on which the process takes place.

The ambit of planktonic organisms depends on both their movements in the water and the motion of the water. The planktonic patchiness in the ocean depends highly on mixing and stirring, as well as the size, intensity, and persistence of patches. Hence, a description of the relative patchiness of physical and biological processes, together with the extent of their spatial and temporal scales, as well as a comparison of such patterns with biologically important scales, may lead to a better understanding of the effect of biophysical patchiness on aquatic ecosystem structure and function. For example, observations of zooplankton swimming behavior have demonstrated that swimming abilities of zooplankton can in most aquatic environments overcome the effects of the root-mean-square turbulent velocities (Schmitt and Seuront, 2001; Chapter 22, this volume), which confirms previous hypotheses based on literature survey (Yamazaki and Squires, 1996; Seuront, 2001). This suggests that the effects of turbulence on planktonic contact rates could be less important than previously thought.

The key processes of the structure and function of the aquatic ecosystem take place at the microscale (i.e., at scales where molecular viscosity and diffusion become important). A salient issue in aquatic ecology is, therefore, the development of instrumentation and numerical tools to identify and characterize patchiness in both physical parameters (e.g., temperature, salinity, and shear) and phytoplankton distribution. Recent numerical investigations focus on (1) the effects of turbulence intermittency on predator–prey encounter rates, physical coagulation rates, and the flux of nutrient toward nonmotile phytoplankton cells; and (2) the effect of phytoplankton patchiness on predator–prey encounter rates via predator behavioral adaptation (Seuront, 2001; Seuront et al., 2001). To improve our understanding of aquatic ecosystem structure and function we must, therefore, integrate the microscale structure of physical and biological parameters to estimate major biochemical fluxes.

1.2.2 Impact of the Sampling Process

Conventional sampling approaches implicitly assume that biological processes are in a steady state. Measurements from different cruises or stations are compared assuming that spatial and temporal changes are minimal. However, patchiness, spatial gradients (e.g., fronts), temporal cycles (e.g., tidal, seasonal, or interannual), and both spatial and temporal microscale patchiness elevate the uncertainty of such comparisons. Generally speaking, a description of patchiness in different scales may help design effective sampling schemes. More specifically, if microscale patchiness exists, an appropriate sampling device is required to investigate the nature of patchiness.

As an example, we investigated the effect of patchiness in the sampled water in a Niskin bottle (Figure 1.1A). It is standard experimental practice to draw a number of subsamples from the Niskin bottle, according to the number of studied parameters. The unused remainder inside the Niskin bottle is often dumped. For example, if information is needed of phytoplankton biomass and production, particulate organic material (POM), dissolved organic material (DOM), and nutrient concentration, the sampling unit will be divided into five uniform subsamples (as illustrated in Figure 1.1B). Each subsample is regarded as representative of the entire bottle contents, analyzed separately, and then correlated with the other parameters to infer causality between them. However, such a sampling scheme assumes spatial homogeneity within the original sampling bottle, and thus spatial homogeneity between the subsamples. This is, however, deeply questioned, at least in the case of phytoplankton populations, where the centimeter-scale patchiness of phytoplankton has been clearly demonstrated in several studies (Yamazaki et al., 2002a; present study). Moreover, based on recent results obtained on bacteria (Seymour et al., 2000) and nutrient (Seuront et al., 2002), it is reasonable to assume comparable patchiness for a variety of components of aquatic ecosystems. Therefore, the subsamples of the original Niskin bottle samples cannot be regarded as homogeneous (Figure 1.1C), and the results of any comparison conducted between the parameters estimated from subsamples within the bottle are questionable.

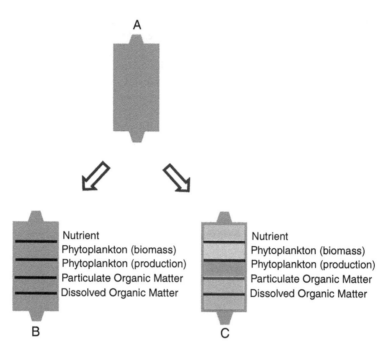

FIGURE 1.1 Schematic illustration of a standard sampling procedure using a Niskin bottle, the elementary sampling unit in aquatic ecology (A) The study of different parameters estimated from different subsamples taken from the same bottle assumes spatial homogeneity within the sampling unit (B). However, such a sampling scheme is irrelevant if there is spatial variability within the sampling volume (C); as a result, each studied parameter is taken from a different water mass. (Adapted from Leterme, 2002.)

To investigate the impact of patchiness, we took Niskin bottle samples at two stations located in the inshore (50°47′300 N, 1°33′500 E) and the offshore (50°46′950 N, 1°16′680 E) waters of the Eastern English Channel during the spring bloom on 16 and 17 April 2002, respectively. During the recovery, the Niskin bottles were handled gently to avoid stirring of the water inside the bottles. From the 5-l bottles collected at each station, we carefully drew 192 subsamples of 20 cc. (Note: The sample analysis from a comparison experiment in which Niskin bottles collected at the same site were thoroughly stirred before drawing the subsamples is still ongoing.) The volume of the subsamples is equivalent to a vertical spatial resolution of 2.4 mm, which is comparable with the resolution of the high-resolution bio-optical sensor described in Section 1.3. After determination of chlorophyll *a* concentration following Suzuki and Ishimaru (1990), and subsequent fluorometry quantity determination (Leterme, 2002), chlorophyll *a* concentrations have been plotted against the corresponding vertical position within the Niskin bottle (Figure 1.2). The phytoplankton biomass concentrations appear clearly patchy, for both inshore and offshore waters, with very sharp variations from one sample to the next. Interestingly, the patchy vertical distribution inside the bottle is reminiscent of the profiles obtained with high-resolution sensors (cf. Section 1.3.2). The inshore chlorophyll estimates range from 0.70 to 67.03 µg l^{-1} (26.68 ± 10.49; $\bar{x} \pm$ SD), and a coefficient of variation CV = 39.32%. The offshore chlorophyll estimates range from 0.94 to 12.45 µg l^{-1} (3.79 ± 1.88; $\bar{x} \pm$ SD) and a coefficient of variation CV = 49.47%. The sharpest variations observed for the inshore and offshore samples correspond to increases in chlorophyll concentration of a factor 2.13 and 8.32 over the smallest resolution reached (i.e., 2.4 mm), respectively. This corresponds to gradients of 208 and 27.54 µg l^{-1} cm^{-1}, respectively.

The ratio between maximum and minimum chlorophyll concentrations can be considered as an estimate of phytoplankton biomass variability (Seuront and Spilmont, 2002). The ratios within the primary sampling unit are very high: 96.25 and 13.22 for inshore and offshore samples, respectively. In particular, these ratios are higher than those obtained in the framework of an annual survey conducted in the Eastern English Channel from four depths in the inshore waters and five depths in the offshore waters every 2 weeks (i.e., 30 and 6 for inshore and offshore waters, respectively; Gentilhomme and Lizon, 1998).

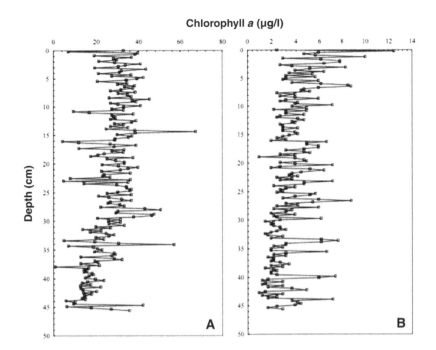

FIGURE 1.2 Spatial distribution of chlorophyll *a* concentrations (μg l⁻¹) obtained from 192 subsamples of 2.4 mm vertical resolution taken from 5-l Niskin bottles, sampled in the subsurface waters at inshore (A) and offshore (B) stations located in the Eastern English Channel. (Adapted from Leterme, 2002.)

During routine fieldwork, the size of the subsamples drawn from the primary Niskin is about 15 times larger (usually 250 to 500 ml) than the subsamples drawn here. This integration averages out the small-scale variability observed in our 20-cc samples. However, Figure 1.2 shows that even under such averaging there remains a noticeable trend of the chlorophyll concentration through the Niskin bottle. This trend is more pronounced in samples collected near the sea bottom (not shown). The microscale variability within a single Niskin bottle demonstrates that small subsamples taken from a larger sample may not represent the realistic phytoplankton distribution in the ocean. As a consequence, inferring correlations (or more generally any causality) between parameters estimated from different subsamples would lead to spurious results at best, in some cases even to utterly wrong conclusions.

1.3   Comparison of High-Resolution Data and Conventional Techniques

A high-resolution bio-optical sensor capable of resolving centimeter scales of fluorescence and turbidity was deployed in two very different environments: Lake Biwa (Japan) and in Seto Inlet (a tidally mixed channel in Hiroshima prefecture, Japan). Fluorescence data from both deployments are presented and discussed with attention to the low-frequency response of sensor as well as the small-scale resolution. Using depth averages of the high-resolution data allows us to simulate the scale resolution of conventional sampling techniques (such as conventional field fluorometers). Structure function analysis can be used to demonstrate the qualitative and quantitative differences between the original and averaged data sets.

1.3.1   Instrument Description

The high-resolution sensor is mounted on the free-fall profiler "Turbulence Ocean Microstructure Acquisition Profiler" (TurboMAP). The instrument is specifically designed to record simultaneously biological and physical properties of the water column, i.e., shear, temperature, conductivity, *in vivo*

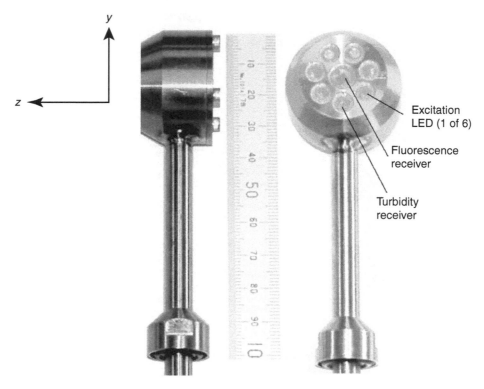

FIGURE 1.3 Side and front view of the high-resolution bio-optical probe. Numbers on the ruler are in millimeters. During operation, the sensor travels in the z direction.

fluorescence, and backscatter (Wolk et al., 2002). The operating principle of the high-resolution sensor is similar to standard backscatter fluorometers; however, instead of measuring inside a small sampling cavity, the sensor projects the sampling volume away from the sensor into the free flow where the small-scale structure of the water column is not compromised by flow distortion or mixing around the sensor housing (Figure 1.3). An array of six light-emitting diodes (LEDs) provides the blue excitation light for chlorophyll a fluorescence (400 to 480 nm). The intersection of the light beams defines a sampling volume with a center approximately 14 mm in front of the optical receiver (640 to 720 nm). A second receiver diode detects the direct light backscatter of light from suspended particles (400 to 480 nm), which is a measure of the turbidity. Details of the sensor construction are given by Wolk et al. (2001).

The size of the sampling volume, spatial resolution, and response to naturally occurring fluorescent sources, such as algae and pure chlorophyll a solutions, were investigated in several laboratory tests (Wolk et al., 2001). The sampling volume is defined by the geometry of the excitation light beams and the directivity of the receiver diode. The effective size of the sampling volume was mapped by determining both the sensitivity of the probe as a function of distance from the sensor face (z direction) and its spatial resolution in the tangential direction (y direction). Figure 1.4 shows the resulting composite sensitivity. The puck centered on the origin represents the sensor head and the conical surface represents the outline of the excitation light as deduced from the LED geometry. This shape agrees with observations when the sensor is placed in turbid water. The sensitivity decreases exponentially in the z direction; 90% of the received fluorescent light comes from the first 25 mm in front of the sensor. Over the width of the sensor face (x and y direction) the sensitivity resembles a cosine window with a width of 20 mm.

The shape of sensor housing minimizes flow distortions and eddy generation around the sensor head. During deployment on a free-fall profiler or towed instrument, the probe looks "sideways" so that the sampling volume is located in the free-flow region. This setup preserves the small-scale structures of fluorescent material in the water column. When mounted on TurboMAP, the sensor is located on the nose cone of the instrument where there is no disturbance of the flow from other parts of the instrument.

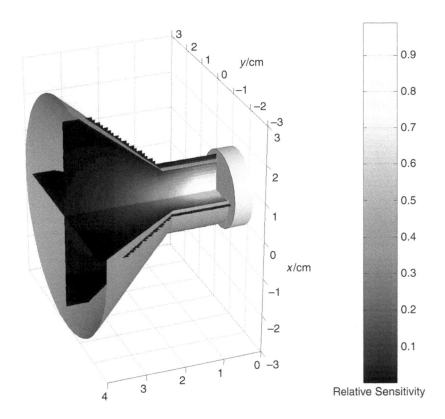

FIGURE 1.4 Composite spatial sensitivity of the high-resolution probe deduced from laboratory experiments. The gray puck in the *x–y* plane represents the sensor head and the cone shows the outline of the excitation light.

1.3.2 Sensor Deployment

The data were collected in Lake Biwa (Figure 1.5A, B) on 31 August 2001 and in Seto Inlet (Figure 1.5C, D) on 21 August 1998. During both deployments, the small-scale sensor was mounted on the nose section of the TurboMAP profiler, which descended in free-fall mode at an average fall rate of 0.64 m s^{-1} (SD = 0.01 m s^{-1}) in Lake Biwa and 0.69m s^{-1} (SD = 0.1 m s^{-1}) in Seto Inlet. The TurboMAP signal is sampled at 256 Hz, which gives one data point every 2.5 mm. The signal shown is low-pass-filtered at 30 Hz to include all spatial scales resolved by the sensor while suppressing the instrumentation noise. TurboMAP reached its terminal fall speed at about 6 m depth, so both fluorescence profiles are cropped to exclude the region above 6 m.

In the Lake Biwa data, the slowly varying part of the fluorescence signal shows a well-mixed region between 6 and 9 m depth, followed by a rapid decrease between 9 and 11 m, coinciding with the temperature drop. The temperature signals are almost constant between 11 and 14 m, and the fluorescence signal also shows homogeneous features. The signal then continues to decrease slowly between 11 and 25 m, and below 25 m it is constant within the standard deviation of the signal for that depth range. The sharp "fluorocline" is ubiquitous in all profiles collected at the 4-day Lake Biwa campaign.

The dotted line in Figure 1.5B is the 1-m depth-averaged fluorometer signal, which represents the signal we would expect from a fluorometer mounted on a typical CTD cage under the same experimental conditions. Rather than descending steadily, CTD cages often heave as a result of the ship's roll. Typical CTD cages with bottle samplers have a length and width of approximately 1 m, and thus one cannot expect to obtain a scale resolution of less than 1 m from such measurement. We note, however, that under certain conditions it is possible to obtain much higher spatial resolution. Robert C. Beardsley

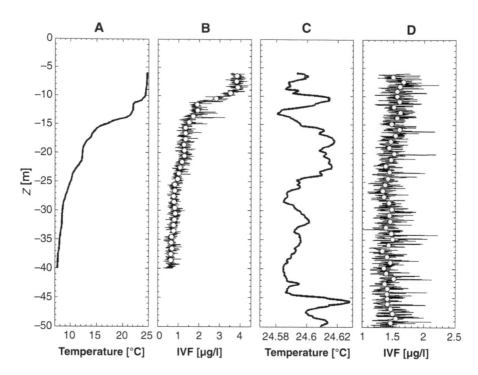

FIGURE 1.5 Temperature and *in vivo* fluorescence measured by TurboMAP in Lake Biwa (A, B) and in Seto Inlet (C, D). The high-resolution fluorescence signal from TurboMAP (black line) and after 1-m averaging (open dots). Note that the averaged signal is similar to the profiles that could have been obtained using the conventional CTD fluorescence sensor.

(personal communication, 2003) recently obtained temperature microstructure measurements with a resolution of O (10^{-2}) m from CTD rosette sampler that was lowered from an icebreaker in the ice of the Antarctic. One of the Niskin bottles of the rosette was replaced with the microstructure instrument and the CTD was lowered using the ship's CTD winch.

The numerous narrow excursions (or "spikes") seen in the high-resolution signal (e.g., at $z = -13.8$, -15.0, and -16.5 m) have a typical width between 5 and 10 cm, and they are caused by small patches of algae moving relative to the sensor. The same characteristic of the fluorescence signal is also evident in the data from the Seto Inlet tidal channel, where the temperature varied only in a narrow range of $0.05°C$ (Figure 1.5C). Even though the water column was well mixed throughout, the fluorescence signal (Figure 1.5D) shows spikes caused by particulate or aggregate phytoplankton. Clearly, in both the Biwa and Seto data sets the rich fine structure evident in the high-resolution signal is lost in the averaged data (Figure 1.5B, D).

1.3.3   Differential Structure of Standard and High-Resolution Fluorescence Signals

To investigate the structure of *in vivo* fluorescence signals, we use two related but conceptually different analysis methods. The first method is based on the study of the cumulative density function (CDF), defined as:

$$P[X > x] \propto x^{-\phi} \tag{1.1}$$

where x is a threshold value, and ϕ is the slope of a log–log plot of $P[X > x]$ vs. x. Note that Equation 1.1 can be equivalently rewritten in terms of the probability density function (PDF) as $P[X = x] \propto x^{-\gamma}$, where $\gamma (\gamma = \phi + 1)$ is the slope of a log–log plot of $P[X = x]$ vs. x, respectively. The absolute of the algebraic tail of the CDF is directly related to the moment of divergence q_D as $q_D = \phi$, and might be a signature of a multifractal behavior (Schertzer et al., 1988). The moment of divergence characterizes the highest

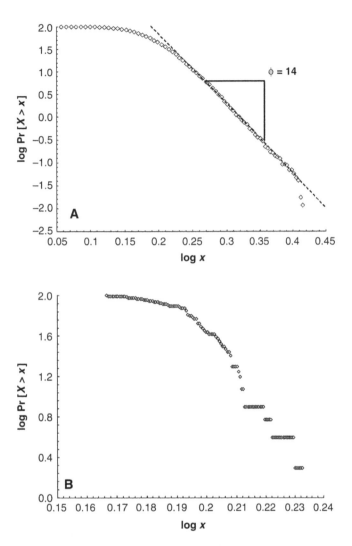

FIGURE 1.6 The CDF $P[X > x]$ vs. $x^{-\phi}$ in a log–log plot of a high-resolution fluorescence vertical profile recorded with TurboMAP in Seto Inlet (A), compared with its 1-m resolution average (B).

statistical moment that can be reliably estimated from a given data set. For moments higher than q_D, the moments cannot be defined as their values becomes intrinsically linked to the length of the data sets, and thus diverge.

The CPD of the *in vivo* fluorescence profile shown in Figure 1.5D is characterized in a log–log plot by a linear behavior (Figure 1.6A) for fluorescence values bounded between 1.74 and 2.51 with a characteristic slope $\phi = 14$ ($r^2 = 0.99$), which is in the range of ϕ values estimated for all the vertical profiles recorded in Seto Inlet, with $\phi = 13.68 \pm 1.45$ ($\bar{x} \pm \text{SD}$). The roll off observed for highest and lowest fluorescence values are related to systematic oversampling and undersampling of, respectively, the most common (i.e., the lowest) and most infrequent (i.e., the highest) fluorescence values. Similar results have been obtained from all the profiles recorded in Lake Biwa (not shown) with a slightly shallower slope $\phi = 11.25 \pm 1.50$ ($\bar{x} \pm \text{SD}$). On the other hand, the CDF estimated from the 1-m resolution fluorescence profiles (Figure 1.6B) do not exhibit the specific features observed from the high-resolution profiles (Figure 1.6A). While the roll off toward high probability related to an oversampling of the most common fluorescence values is still visible, both the linear behavior and the roll off toward low probability has disappeared, demonstrating the inability of a low-resolution sampling process to capture the micro-scale structure of fluorescence distributions.

The second method of analysis is specifically designed to quantify intermittency by adopting a generalization of the qth-order structure functions, defined by

$$\left\langle \left(\Delta IVF_l \right)^q \right\rangle \propto l^{\zeta(q)} \tag{1.2}$$

where ΔIVF_l is the fluctuation of the *in vivo* fluorescence signal at scale l and angle brackets indicating an average; see Seuront et al. (1999; Chapter 22, this volume). Equation 1.2 thus gives the scale-invariant structure function exponents $\zeta(q)$, which characterize the statistics of the whole field. The scaling exponent $\zeta(q)$ is estimated from the slope of the linear trend of $\left\langle \left(\Delta IVF_l \right)^q \right\rangle$ vs. l in a log–log plot, for different values of the statistical order of moments q. For example, the first moment ($q = 1$) gives the constant scaling exponent $\zeta(1) = H$, which describes the scale dependence of the average fluctuations. When $H \neq 1$ the fluctuations ΔIVF_l will depend on the spatial scale; H then characterizes the degree of stationarity of the process. The second moment ($q = 2$) is linked to the slope $\beta = 1 + \zeta(2)$ of the signal's power spectrum. For fractal processes, the scaling exponent $\zeta(q)$ is linear and is defined as $\zeta(q) = qH$. In particular, $\zeta(q) = q/2$ for a Brownian process and $\zeta(q) = q/3$ for nonintermittent, homogeneous turbulence. On the other hand, $\zeta(q)$ is a nonlinear convex function for multifractal (i.e., intermittent) processes.

The structure functions $\left\langle \left(\Delta IVF_l \right)^q \right\rangle$ of the high-resolution TurboMAP profiles from Lake Biwa for $q = 1,2$, and 3 are shown in Figure 1.7. The structure function scaling exponents $\zeta(q)$ were determined following Seuront and Lagadeuc (1997) by linear regression of $\log\left\langle \left(\Delta IVF_l \right)^q \right\rangle$ vs. $\log l$ over the range of scales that maximized the coefficient of determination r^2 and minimized the total sum of squared residuals for the regression. For spatial scales larger than 1 m, high-resolution and low-resolution structure functions are similar, as is seen from the identical slopes at these scales (Figure 1.8). However, the high-resolution structure functions clearly exhibit a scaling range for spatial scales bounded between 0.015 and 0.113 m for Lake Biwa (closed symbols; Figure 1.7) and between 0.050 and 0.333 m for Seto Inlet. For these two scaling ranges, the empirical exponents $\zeta(q)$ were computed for q between 0 and 5 with increments of 0.1. The resulting empirical curves (diamond symbols in Figure 1.8) are clearly nonlinear, showing that the fluorescence fluctuations for $0.015 \leq l \leq 0.113$ m in Lake Biwa and $0.050 \leq l \leq 0.333$ m in Seto Inlet can be regarded as multifractal. The TurboMAP curve clearly deviates from the straight, dotted line expected for nonintermittent turbulence ($\beta = 5/3$). In particular, the scaling of the first moment gives $H = \zeta(1) = 0.34 \pm 0.01$ (the error bars result from the separate analysis of different sections of the vertical profile). This indicates that the fluorescence distribution is far from conservative or stationary (in which case $H = 0$). The scaling of the second-order moment confirms the estimate from the slope $\beta = 1 + \zeta(2)$ of the power spectrum of the high-resolution fluorescence profile (not shown): $\zeta(2) = 0.64 \pm 0.02$, consistent within their 95% confidence intervals.

1.4   Conclusion

Data sets of biological parameters collected with conventional sampling methods were compared to records of a recently developed high-resolution fluorometer. It was shown that the conventional procedures can lead to a misrepresentation of the structure of the aquatic system. A random sample drawn from a Niskin bottle, for example, can give an erroneous representation of the average concentration of chlorophyll *a* concentration at the depth where the bottle sample was taken. Variations inside the bottle exceeded the annual variations of chlorophyll *a* at the sampling site.

The low-resolution fluorescence data, which are similar to what could be obtained from a CTD-mounted *in situ* fluorometer, showed a markedly different structure from the high-resolution data. Although the shape (i.e., slow variation) of the measured fluorescence signal (i.e., CTD-like sensor) is similar to many other fluorescence profiles published in the literature, the high-frequency part of the data shows structures that are not readily accessible with conventional instrumentation.

The importance of obtaining high-resolution biological records to our understanding of the aquatic ecosystem is undisputed. It was shown how the use of multifractal analysis gives us the necessary tools to condense the high-resolution data efficiently, which allows us to describe quantitatively the nature

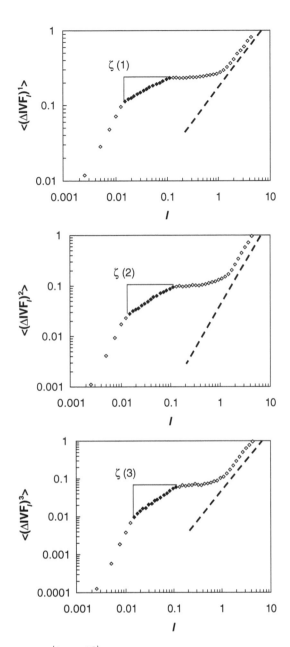

FIGURE 1.7 The structure functions $\left\langle \left(\Delta IVF_l \right)^q \right\rangle$ vs. l in log–log plots for $q = 1$, 2, and 3 (from top to bottom) for the TurboMAP fluorescence profile. The slopes of the closed symbols provide estimates of the first, second, and third scaling exponents $\zeta(1) = H, \zeta(2)$, and $\zeta(3)$. The dashed lines correspond to the slopes of the structure functions obtained from the 1-m resolution fluorescence profile.

(e.g., degree of intermittency) of the distributions. We now have at our disposal the instrumentation and data processing tools necessary for a systematic investigation of the smallest scales of aquatic ecology.

Acknowledgments

The field campaign at Lake Biwa was supported by a grant from Lake Biwa Research Institute. Daniel Kamykowski provided useful comments on an early version of this chapter.

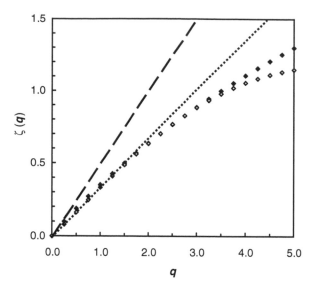

FIGURE 1.8 The empirical scaling exponents $\zeta(q)$ obtained from the TurboMAP fluorescence profile in Lake Biwa (open diamonds) and in Seto Inlet (black diamonds). The nonlinear behavior is the signature of the multifractal character of the fluorescence fluctuations. The functions for a Brownian motion process (dashed line) and for nonintermittent turbulence (dotted line) are shown for comparison.

References

Cowles, T.J., R.A. Desiderio, and M.-E. Carr, 1998: Small-scale planktonic structure: persistence and trophic consequences. *Oceanography*, 11(1), 4–9.

Denman, K.L. and T.M. Powell, 1984: Effects of physical processes on planktonic ecosystems in the coastal ocean. *Oceanogr. Mar. Biol. Annu. Rev.*, 22, 125–168.

Desiderio, R.A., T.J. Cowles, and J.N. Moum, 1993: Microstructure profiles of laser-induced chlorophyll fluorescence spectra: evaluation of backscatter and forward-scatter fiber-optic sensors. *J. Atmos. Ocean. Tech.*, 10, 209–224.

Donaghay, P., H.M. Rines, and J.M. Sieburth, 1992: Simultaneous sampling of fine scale biological, chemical, and physical structure in stratified waters. *Arch. Hydrobiol. Beih.*, 36, 97–108.

Franks, P.J.S. and J.F. Jaffe, 2001: Microscale distributions of phytoplankton: initial results from a two-dimensional imaging fluorometer, OSST. *Mar. Ecol. Prog. Ser.*, 220, 59–72.

Gentilhomme, V. and F. Lizon, 1998: Seasonal cycle of nitrogen and phytoplankton biomass in a well-mixed coastal system (Eastern English Channel). *Hydrobiologia*, 361, 191–199.

Hanson, A.K. and P.L. Donaghay, 1998: Micro- to fine-scale chemical gradients and layers in stratified coastal waters. *Oceanography*, 11, 10–17.

Holliday, D.V., R.E. Pieper, C.F. Greenlaw, and J.K. Dawson, 1998: Acoustical sensing of small-scale vertical structures in zooplankton assemblages. *Oceanography*, 11, 18–23.

Hutchinson, G.E., 1961: The paradox of the plankton. *Am. Nat.*, 95, 137–145.

Jackson, G.A. and A. Burd, 1998: Aggregation in the marine environment. *Environ. Sci. Tech.*, 32, 2805–2814.

Jaffe, J.F., P.J.S. Franks, and A.W. Leising, 1998: Simultaneous imaging of phytoplankton and zooplankton distributions. *Oceanography*, 11, 24–29.

Leterme, S., 2002: Microscale Variability of Phytoplankton Biomass in Turbulent Environment. Diplôme Supérieur de Recherche thesis, Université des Sciences et Technologies de Lille, France.

Mackas D.L., K.L. Denman, and M.R. Abbott, 1985: Plankton patchiness: biology in the physical vernacular. *Bull. Mar. Sci.*, 37, 652–674.

Mitchell J.G. and J.A. Fuhrman, 1989: Centimeter scale vertical heterogeneity in bacteria and chlorophyll *a*. *Mar. Ecol. Prog. Ser.*, 54, 141–148.

Pascual, M., F.A. Ascioti, and H. Caswell, 1995: Intermittency in the plankton: a multifractal analysis of zooplankton biomass variability. *J. Plankton Res.*, 17, 1209–1232.

Petrenko, A.A., J.R.V. Zaneveld, W.S. Pegau, A.H. Barnard, and C.D. Mobley, 1998: Effects of a thin layer on reflectance and remote-sensing reflectance. *Oceanography*, 11, 48–50.

Platt, T., W.G. Harrison, M.R. Lewis, W.K.W. Li, S. Sathyendranath, R.E. Smith, and A.F. Vezina, 1989: Biological production in the oceans: the case for a consensus. *Mar. Ecol. Prog. Ser.*, 52, 77–88.

Scheffer, M., 1995: Implications of spatial heterogeneity for the paradox of enrichment. *Ecology*, 76, 2270–2277.

Schertzer, D., S. Lovejoy, F. Schmitt, Y. Chigirinskaya, and D. Marsan, 1998: Multifractal cascade dynamics and turbulent intermittency. *Fractals*, 5, 427–471.

Schmitt, F. and L. Seuront, 2001: Multifractal random walk in copepod behavior. *Physica A*, 301, 375–396.

Seuront, L., 2001: Microscale processes in the ocean: why are they so important for ecosystem functioning? *La Mer*, 39, 1–8.

Seuront, L. and Y. Lagadeuc, 1997: Characterisation of space–time variability in stratified and mixed coastal waters (Baie des Chaleurs, Québec, Canada): application of fractal theory. *Mar. Ecol. Prog. Ser.*, 159, 81–95.

Seuront, L. and N. Spilmont, 2002: Self-organized criticality in intertidal microphytobenthos patch patterns. *Physica A*, 313, 513–539.

Seuront, L., F. Schmitt, Y. Lagadeuc, D. Schertzer, and S. Lovejoy, 1999: Universal multifractal analysis as a tool to characterize multiscale intermittent patterns: example of phytoplankton distribution in turbulent coastal waters. *J. Plankton Res.*, 21, 877–922.

Seuront, L., F. Schmitt, and Y. Lagadeuc, 2001: Turbulence intermittency, small-scale phytoplankton patchiness and encounter rates in plankton: where do we go from here? *Deep-Sea Res. I*, 48, 1199–1215.

Seuront, L., V. Gentilhomme, and Y. Lagadeuc, 2002: Small-scale nutrient patches in tidally mixed coastal waters. *Mar. Ecol. Prog. Ser.*, 232, 29–44.

Seymour, J., J.G. Mitchell, L. Pearson, and R.L. Waters, 2000: Heterogeneity in bacterioplankton abundance at centimetre scales. *Aquat. Microbial Ecol.*, 22, 143–153.

Siegel, D.A., 1998: Resource competition in a discrete environment: why are plankton distribution paradoxical? *Limnol. Oceanogr.*, 43, 1133–1146.

Sosik, H.M. and B.G. Mitchell, 1995: Light absorption by phytoplankton, photosynthetic pigments and detritus in the California Current System. *Deep Sea Res.*, 42, 1717–1748.

Strutton, P.G., J.G. Mitchell, and J.S. Parslow, 1996: Non-linear analysis of chlorophyll *a* transects as a method of quantifying spatial structure. *J. Plankton Res.*, 18, 1717–1726.

Strutton, P.G., J.G. Mitchell, and J.S. Parslow, 1997: Using non-linear analysis to compare the spatial structure of chlorophyll with passive tracers. *J. Plankton Res.*, 19, 1553–1564.

Suzuki, R. and T. Ishimaru, 1990: An improved method for the determination of phytoplankton chlorophyll using *N,N*-dimethylformamide. *J. Oceanogr. Soc. Jpn.*, 46, 190–194.

Wolk, F., L. Seuront, and H. Yamazaki, 2001: Spatial resolution of a new micro-optical probe for chlorophyll and turbidity, *J. Tokyo Univ. Fish.*, 87, 13–21. (Available through the Tokyo University of Fisheries, 4-5-7 Konan, Minato-Ku, 108-8477, Tokyo, Japan.)

Wolk, F., H. Yamazaki, L. Seuront, and R.G. Lueck, 2002: A new free-fall profiler for measuring biophysical microstructure. *J. Atmos. Oceanic Technol.*, 19, 780–793.

Yamazaki, H. and T.R. Osborn, 1988: Review of oceanic turbulence: implications for biodynamics, in Rothschild, B.J., Ed., *Toward a Theory on Biological-Physical Interactions in the World Ocean.* Kluwer, Dordrecht.

Yamazaki, H. and K.D. Squires, 1996: Comparison of oceanic turbulence and copepod swimming. *Mar. Ecol. Prog. Ser.*, 144, 299–301.

Yamazaki, H., D.L. Mackas, and K.L. Denman, 2002: Coupling small-scale turbulent processes with biology, in *The Sea*, A.R. Robinson, J.J. McCarthy, and B.J. Rothschild, Eds., *Biological–Physical Interactions in the Ocean*, Vol. 12. John Wiley & Sons, New York, 51–112.

Yamazaki, H., J.G. Mitchell, L. Seuront, F. Wolk, H. Li, J.R. Seymour, and R.L. Waters, in revision: Microscale patchiness and intermittency for phytoplankton in a turbulent ocean. *Nature*, submitted.

Zaneveld, R.J. and W.S. Pegau, 1998: A model for the reflectance of thin layers, fronts, and internal waves and its inversion. *Oceanography*, 11, 44–47.

2

Measurement of Zooplankton Distributions with a High-Resolution Digital Camera System

Mark C. Benfield, Christopher J. Schwehm, Rodney G. Fredericks, Gregory Squyres, Sean F. Keenan, and Mark V. Trevorrow

CONTENTS

2.1   Introduction 17
2.2   Methods 19
    2.2.1   System Description 19
        2.2.1.1   Camera Housing..19
        2.2.1.2   Power/Telemetry Housing...19
        2.2.1.3   Strobed Light Sheet 20
        2.2.1.4   Environmental Sensors and Frame 21
        2.2.1.5   Winch and Sea Cable...21
        2.2.1.6   Command and Control...21
        2.2.1.7   Image and Data Processing 21
    2.2.2   Knight Inlet Operations...22
2.3   Results 24
    2.3.1   November 19 Cast in Hoeya Sound 25
    2.3.2   November 21 Cast in Knight Inlet 25
2.4   Discussion..26
Acknowledgments...28
References..29

2.1   Introduction

The oceans contain many biological, biogenic, and physical phenomena distributed on fine spatial scales. With the increased use of high-frequency echosounders as survey tools, oceanographers routinely observe scattering features on vertical scales of centimeters to meters and horizontal scales of meters to kilometers (e.g., Wiebe et al., 1996; Holliday et al., 1998). Advances in bio-optical sensors (e.g., multispectral absorption and attenuation meters) and their employment in specialized sampling methodologies have led to the discovery of thin layers — vertically distinct patches of phytoplankton, ranging in thickness from a few centimeters to several meters (Donaghay et al., 1992; Cowles et al., 1998; Dekshenieks et al., 2001). Supplementary acoustic sensing suggested that thin layers may also contain zooplankton (Holliday et al., 1998). The capabilities of a variety of mesozooplankton taxa to form dense, highly localized patches has been documented for a variety of taxa (e.g., Price, 1989; Buskey et al., 1996) and the importance of localized regions of enhanced zooplankton biomass on scales of millimeters to meters was identified as a emergent new issue in zooplankton ecology (Marine Zooplankton Colloquium 2, 2001).

Oceanographers have traditionally utilized nets and pumps to sample marine mesozooplankton. Although nets are useful for quantifying zooplankton distributions and abundances on horizontal scales

0-8493-1344-9/04/$0.00+$1.50
© 2004 by CRC Press LLC

of tens to hundreds of meters and vertical scales of several meters or more, they are generally inadequate to reveal the structure of patches and layers on finer scales. Pumps have proved useful for exposing vertical structure on smaller scales than nets (Smith et al., 1976; Incze et al., 1996), however, their utility is frequently limited to microzooplankton and small mesozooplankton because of avoidance by, and damage to, larger zooplankton.

Recognition of the limitations of traditional samplers has stimulated the development of optical systems for quantifying zooplankton distributions and abundances. Optical sensors can be divided into particle detection and image-forming systems. Particle detectors such as the optical plankton counter (Herman, 1988) use the interruption of a light source by zooplankton and other objects to detect, count, and measure targets as they pass through a sampling tunnel. The resulting data consist of counts and size classes of targets as a function of time. The depth and location of targets can be estimated by relating the time of detection to pressure and GPS data, however, there is no information on the two-dimensional (2D) spatial distributions of targets at any point along the tow path. Further, there is no taxonomic information from which to match target size class with target identity.

Image-forming optics use various types of cameras to image organisms along the tow path of the instrument. Quantitative instruments in this category began with the camera-net system (Ortner et al., 1981) and currently include towed systems such as the ichthyoplankton recorder (Lenz et al., 1994), video plankton recorder (VPR) (Davis et al., 1992), *in situ* video recorder (Tiselius, 1998), and the shadowed image particle platform and evaluation recorder (SIPPER) (Samson et al., 2001). In addition, there are profiling systems, such as the underwater video profiler (UVP) (Gorsky et al., 2000); and holographic instruments (Watson et al., 1998; Malkiel et al., 1999). In general, nonholographic image-forming optical systems provide a means for estimating the spatial distributions and abundances of mesozooplankton on vertical scales of centimeters or greater. The majority of optical systems utilize video and typically image small volumes of water to achieve acceptable image resolution characteristics.

There is a trade-off between image volume and image resolution. The use of conventional video formats (NTSC, PAL, SECAM) means that a finite number of pixels (~500 to 600 horizontal scan lines depending on the format) are available to describe each image. As image dimensions and hence sample volume increase, the resolution of the image must decline. The volume imaged depends on the typical dimensions of the organisms being studied. In the case of the VPR, typical image volumes are 5 to 25 ml. Small image volumes resolve the features of individual zooplankton targets well; however, there are three problems associated with low sample volumes. First, the number of individuals present in any given image tends to be either zero or one. This limits the utility of such systems for estimation of nearest neighbor distances and quantification of patchiness on very small length scales. Second, it is normal practice to estimate the concentrations of taxa on the basis of a standard volume (liters or cubic meters). Scaling up concentrations from small sample volumes can introduce large errors into the abundance estimates. Third, the probability of detecting organisms depends on their concentration in the water column and the volume of water sensed by the video system. Optical systems that sample small volumes of water may not detect organisms unless they are very abundant.

The manner in which organisms are illuminated can complicate analysis of the images. When the camera faces the illumination source, either directly or obliquely, organisms located outside of the camera's depth of field may be imaged. Such out-of-focus organisms complicate the analysis of images unless image processing routines can be developed to detect, and reject them. This is the approach that has been adopted for the VPR during image processing steps based on machine vision. The use of collimated light has been adopted by some systems such as the UVP as a means for circumventing the out-of-focus target issue. In this case, the light forms a relatively flat sheet oriented perpendicularly to the camera, within which organisms are imaged as they pass through the light. By setting the focused depth of field of the camera to match the location of the light sheet, it is possible to illuminate only objects that are in focus.

The advent of high-resolution digital imagers using charge-coupled device (CCD) or complementary metal-oxide-silicon (CMOS) detectors providing millions of pixels or more has provided an effective means of imaging larger areas without sacrificing resolution. This chapter describes the development of a new image-forming optical instrument called ZOOVIS (ZOOplankton Visualization and Imaging System). The design of ZOOVIS was influenced by both the VPR and the UVP. The system was designed to image larger volumes of water at high resolution. A structured light sheet that is matched to the depth of field of a digital still camera restricts imaging to objects that are in focus. The use of a high-resolution digital still camera with a CCD containing 4.19 megapixels allows the image volume to be increased while maintaining acceptable image resolution. ZOOVIS is equipped with a CTD package so that both physical and biological parameters can be quantified on fine vertical scales to depths of 250 m. The system was designed to be flexible so that, at the highest sample volumes, the optics could resolve large mesozooplankton, while at higher magnifications, details of smaller mesozooplankton could be resolved. Originally designed as a profiling instrument, recent tests indicate that it is effective in a towed mode as well. The system components and operation are summarized, followed by examples of data collected in a coastal fjord system.

2.2   Methods

2.2.1   System Description

ZOOVIS consists of underwater imaging and environmental sensors, a winch, and a surface control and analysis computer. Images and data are transferred from the underwater components (Figure 2.1) to the surface over a fiber-optic/conductive cable to a surface control computer. The prototype had the underwater components of ZOOVIS mounted on a frame constructed of a tubular aluminum weldment; however, a new, more compact frame constructed of anodized aluminum was built during the spring of 2002 (Figure 2.1). This latter frame allows rapid focus and illumination adjustment because both the strobe and camera are mounted on plates that may be moved toward or away from the image volume via a worm-gear mechanism.

2.2.1.1   Camera Housing — The imaging device in ZOOVIS is a monochrome digital camera (PixelVision Bioxight). The camera is a full-frame 2048 × 2048 pixel CCD camera capable of collecting 14-bit gray scale images at sampling rates of up to 2.2 Hz. The camera is equipped with a fiber-optic interface that utilizes a pair of multimode optical fibers to carry data and commands to and from the camera. Power is provided to the camera via a multiconductor cable. The camera accepts all lenses with Nikon F-mounts and we currently use a Tamron 28-300 mm f3.5 zoom lens with macrocapability. The camera and lens are mounted in an anodized aluminum, cylindrical pressure vessel with a glass optical port on one end (Figure 2.1).

2.2.1.2   Power/Telemetry Housing — The power/telemetry housing is an anodized aluminum cylindrical pressure vessel containing power supplies for all system components, a single-board computer, and fiber-optic networking components (Figure 2.1). A bulkhead connector containing single-mode optical fibers and electrical conductors connects to the sea cable. One of the optical fibers carries multiplexed telemetry in and out of the power/telemetry housing over an Ethernet, and the other two fibers are spare. Three of the four conductors in the sea cable carry 120 VAC power from the sea cable into the housing. The system is currently being converted to operate on 220 VAC to improve power transfer efficiency to the underwater components.

Within the power/telemetry housing are DC power supplies for the underwater electronic components, a single-board computer running Windows NT 4.0, and fiber-optic telemetry hardware. The single-board computer is equipped with a PCI slot that is occupied by the camera acquisition card, which has two (transmit and receive) multi-mode optical fiber connectors. The underwater PC has several functions. It controls the camera and the CTD, receives and acts on commands from a surface PC, and sends data to the surface via an Ethernet network. The underwater computer and the surface PC are linked by a 100 Mbps Ethernet network. Commands to the computer and data from the camera enter the power/telemetry housing via optical fiber while serial data (RS232) to and from the CTD enter via a conductive cable that also carries power to the CTD from the system power supply. Power for the strobe is carried from the power/telemetry housing via a conductive cable.

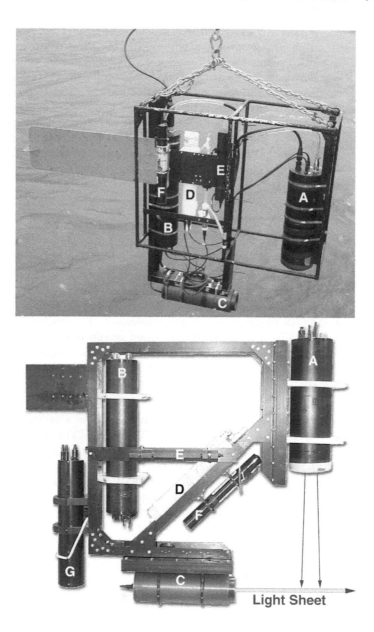

FIGURE 2.1 (Top) The underwater components of ZOOVIS being deployed from the stern of the *CCGS Vector* in Knight Inlet, British Columbia, Canada. A = camera housing; B = telemetry housing; C = strobe housing; D = CTD; E = transmissometer; and F = acoustic transponder/responder. The fluorometer is not visible in this image. In this configuration, the camera is aimed vertically into a horizontal light sheet and a stabilizing fin has been added to keep ZOOVIS oriented into the current. (Bottom) A new anodized aluminum frame with the main components identified using the same letters as in the top panel. In this frame, both the camera and strobe are mounted on travel plates that can be moved toward or away from the image volume using a worm-gear mechanism. A multifrequency acoustic system (TAPS) indicated by (G) was mounted on the new frame; however, it was not deployed in the study discussed in this chapter. The horizontal gray line from the strobe illustrates the orientation of the light sheet relative to the field of view of the camera (black arrows).

2.2.1.3   *Strobed Light Sheet* — The strobe is housed in an anodized aluminum pressure vessel fitted with a clear optical glass port on one end. A collimated sheet of light is produced by a custom-built strobe that uses a linear arc discharge tube and collimating optics. Each 20 µs pulse of the strobe produces a relatively flat sheet of light that is approximately 3.3 cm thick at the external face of the strobe housing and diverges to 3.5 and 4.5 cm at ranges of 31 and 60.5 cm, respectively. The light sheet

spreads laterally at a greater rate. It is 8.3 cm wide at the strobe housing face and spreads laterally to 18.3 and 27.9 cm wide at ranges of 31 and 60.5 cm, respectively.

2.2.1.4   Environmental Sensors and Frame — ZOOVIS is equipped with a Sea-Bird Model SBE19 CTD that incorporates a pump, fluorometer (Wetlabs Wetstar), and transmissometer (Wetlabs Seastar). The CTD is linked to the power/telemetry housing. A spare four-pin connector on the telemetry housing provides expansion capability to add additional sensors. A trackpoint-compatible acoustic multibeacon transponder/responder is attached to the frame to assist in locating the instrument should it be lost.

2.2.1.5   Winch and Sea Cable — The underwater components of ZOOVIS are connected to the surface via an electro-optical sea cable that contains three single-mode optical fibers and three copper conductors. The underwater end of the cable is connected to a mechanical termination that ends in a connector that mates to the bulkhead connector on the telemetry housing. The sea cable hangs from an instrumented sheave block that provides line tension, payout, and speed information to an LCD console. From the sheave block, the cable is spooled on to an electric-hydraulic winch fitted with a combination optical and conductive slip-ring assembly. The fiber-optic signals pass from the external pigtail on the slip-ring assembly to a box where they are demultiplexed and converted to twisted pair Ethernet. The Ethernet signal is then routed to the surface acquisition/control computer.

2.2.1.6   Command and Control — Control of ZOOVIS is achieved with two software applications consisting of a server application residing on the underwater computer and a surface client application. The client passes commands to the server, which then issues control commands to the camera or logs data from the CTD. Camera controls include the options of collection of single or multiple images. Files are normally saved to the surface hard-drive via a Microsoft network. Each image is saved as an uncompressed 16-bit TIFF file that is named using a user-specified filename that encodes the capture time. The client software also records the header and hexadecimal information that is received from the CTD. The pressure data from the underwater PC are displayed in real time in the client control window so that the user can monitor the depth of the instrument.

2.2.1.7   Image and Data Processing — There are two preprocessing operations that must be completed before examining individual images for targets: (1) correction for uneven illumination and (2) adjusting the gray scale levels and conversion to 8-bit resolution. Illumination correction is performed in Matlab; then each image is thresholded to the maximum and minimum measured values before conversion to an 8-bit file.

At present, images from ZOOVIS are processed semiautomatically using custom region-of-interest (ROI) detection, extraction, and measurement software written in Visual C++ that utilizes machine vision libraries from the Matrox Imaging Libraries. The software allows the user to specify the minimum and maximum areas and intensity thresholds of the ROIs and then locates each target meeting the ROI selection criteria. The coordinates of each ROI in the image and the area, length, and width are recorded, and each ROI is written to disk as a separate TIFF file along with a marked copy of the original image that contains rectangles around every ROI (Figure 2.2).

Once ROIs have been extracted, they are manually classified and enumerated. The volume of each image is well defined by the limits of the field of view and the thickness of the light sheet. Thus, each image is a quantitative record of the abundance of targets in a known volume of water. The concentration of each taxon is estimated for each image in a cast and plotted as a function of depth, latitude, and longitude. Measurements from the CTD package can be used to provide an estimate of the relationship between zooplankton abundance and environmental factors. The images can also provide other interesting information on zooplankton behavior and patchiness. When the camera is oriented horizontally into a vertical light sheet or vertically into a horizontal light sheet, the vertical orientations of individual targets can be determined. Nearest neighbor distances can be calculated to estimate the micropatchiness of zooplankton.

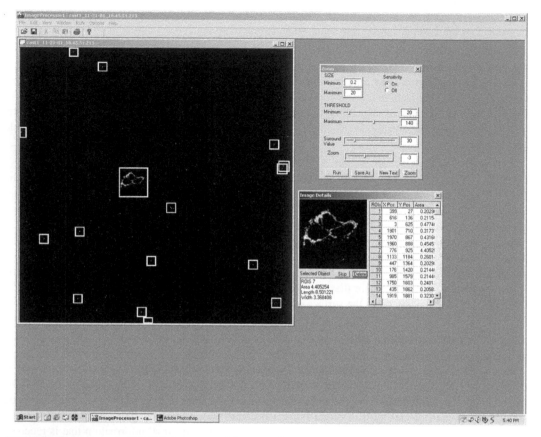

FIGURE 2.2 An example ZOOVIS image containing a marine snow particle displayed within the image processing program. White boxes surround all particles that satisfy search criteria for size and intensity. The image has been converted from 14-bit (16-bit TIFF) to 8-bit resolution.

2.2.2 Knight Inlet Operations

ZOOVIS was deployed from the *CCGS Vector* during cruise 2001-39 to Knight Inlet, British Columbia, Canada in November 2001 (Figure 2.3). The objective of this cruise was to gain a better understanding of the physical and biological processes that contribute to elevated levels of acoustic scattering within the fjord during tidal flow. During this cruise, ZOOVIS was tested in two configurations: horizontal and downward imaging. Data from two casts on November 19 and 21, 2001 that employed the downward imaging configuration are presented in this chapter.

Optical casts were collected during concurrent acoustic surveys. The echosounders (40, 100, 200 kHz) on the vessel provided a means both of guiding ZOOVIS into acoustic features of interest and of monitoring the location of ZOOVIS in relation to the bottom. One cast recorded during early evening in Hoeya Sound, a small embayment off the main inlet channel, and a second cast collected shortly after midday on November 21 (Figure 2.4) provide an illustration of the kind of data collected by ZOOVIS and its application to interpretation of echograms and zooplankton ecology. The two casts were also conducted under highly different acoustical scattering conditions (Figure 2.4).

In Hoeya Sound, dense schools of small unidentified fish were observed near the surface and were apparently actively feeding throughout much of the water column during the survey. There was also a surface lens of turbid, fresher water that extended down to approximately 15 m. The ZOOVIS image volume was 443 ml (10.95 cm × 10.95 cm × 3.69 cm) and images were acquired at approximately 0.25 Hz with the exception of brief period when images were taken singly using manual control in an unsuccessful attempt to image the fish. We lowered ZOOVIS from the stern and began acquiring images from below the turbid surface layer to near the bottom (20 to 76 m). The system was held it at depth

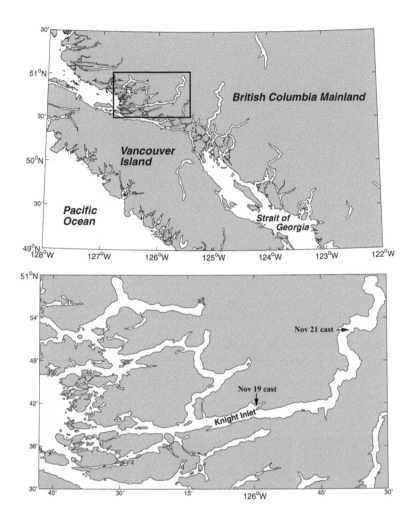

FIGURE 2.3 (Top) The location of the Knight Inlet study area on the west coast of British Columbia. (Bottom) Higher-resolution map of Knight Inlet showing the locations of the two ZOOVIS casts on November 19 and 21.

for a brief period, hauled up to 30 m, and the cast was concluded with a series of shallow stair-steps up to 20 m (Figure 2.4).

On November 21, the instrument was again deployed from the stern when the vessel was farther up the inlet and a pair of profiles were collected between the surface and 166 m (Figure 2.4). The image volume was the same as on November 19 and images were recorded at 0.25 Hz while ZOOVIS was lowered at approximately 50 cm s^{-1}. Following the profiles, with ZOOVIS still in the water, the vessel began to steam into the current at 0.5 to 1.0 knots and a series of towyos were performed at mid-depth with vertical velocities ranging from 17.4 to 57.5 cm s^{-1}. These towyos were undertaken to investigate the scattering layer at mid-depth in greater detail and to examine how well ZOOVIS performed in a towyo mode. Image acquisition was suspended at 142 m and ZOOVIS was recovered (Figure 2.4).

Images were corrected for the heterogeneous background illumination, thresholded, and converted to 8-bit images prior to image analysis. The image analysis software was used to locate and measure all particles with areas greater than 0.2 mm^2 (Figure 2.2). Recognizable targets (primarily euphausiids) were identified to the lowest taxonomic level possible. Densities of euphausiids were estimated by dividing the numbers of individuals by the volume imaged within each depth stratum or acoustical feature. The degree to which targets were aggregated or dispersed in each image during the November 21 cast was evaluated statistically using a refined nearest neighbor analysis (Boots and Getis, 1988) that tests the null hypothesis that the distribution of targets in each image was not different from a completely spatially random distribution.

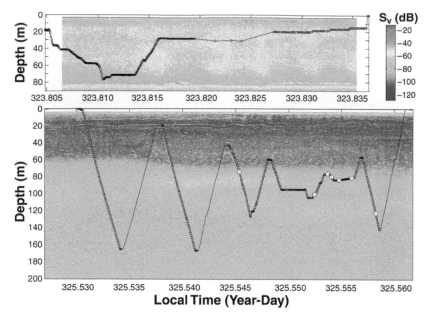

FIGURE 2.4 (Color figure follows p. 332.) Time-depth record of ZOOVIS deployments on November 19 in Hoeya Sound (top) and on November 21 farther up the Inlet (bottom). The trajectory of ZOOVIS is represented as a black line and locations where images were acquired are indicated by small white squares on the line. The trajectory is superimposed on the acoustic record (volume scattering strength) from the ship's 100 kHz echosounder. The yellow circles in the lower panel indicate locations where small euphausiids were imaged.

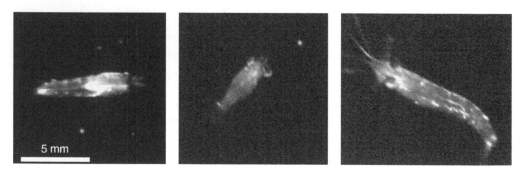

FIGURE 2.5 Examples of euphausiids (*E. pacifica*) imaged by ZOOVIS during the November 21 cast. Each image is an enlarged subsection of the complete image area (10.95 × 10.95 cm).

2.3   Results

Within the field of view selected for most Knight Inlet operations (10.95 × 10.95 cm), ZOOVIS provided clear images (Figure 2.5) of particles larger than 2 mm length from a distance of approximately 40 cm. Smaller targets were also imaged; however, anatomical details of small targets such as copepod antennae were not always visible. Sufficient resolution of sub-2-mm targets was usually available to determine their shape and likely identity (e.g., copepods, phytoplankton colony, marine snow).

As a result of parallax, the image volume was not a true rectangular volume. The area of the front of the image volume (closest to the camera) was smaller (10.7 × 10.7 cm) than at the back (11.2 × 11.2 cm). The mean width of these two measurements (10.95 cm) was used to estimate the image width and height. Our volume calculation was the product of the square of the mean width and the depth of field.

This simplified volume calculation (443.6 ml) overestimated the true image volume (443.2 ml) by 0.09%. This simplification was necessary for measurement of objects in the images because there was no way to know where an individual target to be measured was positioned relative to the front or back face of the image volume. The measurement error introduced by the use of a mean image width and height ranged from a 2.23% underestimation of true length for an object located at the back face of the image volume, to a 2.34% overestimation of the true length of an object located at the frontal face of the image volume.

The maximum image acquisition rate that we were able to achieve was 0.27 Hz, which was well below the camera's maximum rate of 2.2 Hz. This limitation was a consequence of the data transfer rate between the underwater and surface PCs. We were able to transfer images and CTD data to the surface over a single multiplexed fiber at a maximum rate of 17.3 Mbps, which was considerably lower than our network's theoretical maximum transfer rate of 100 Mbps. While lowering ZOOVIS at 0.5 m s^{-1}, we were able to achieve a vertical image separation of 2 m and finer resolution with slower payout velocities.

2.3.1 November 19 Cast in Hoeya Sound

Most of the water column was characterized by strong acoustical scattering that supported the presence of strong targets such as fish (Figure 2.4). A strong zone of scattering was present between 35 and 40 m and elevated scattering persisted to near the bottom. No large zooplanktors were imaged in Hoeya Sound and the majority of targets appeared to be small copepods and marine snow. The properties (density, mean particle area, average optical cross-sectional area) of the particles in the water column quantified by ZOOVIS were not well related to the acoustics. The number of objects ≥ 0.25 mm^2 in each image ranged from 1 to 9 (mean of 2.2) and generally increased with depth (Figure 2.6). The average area of ROIs was similar with depth and the optical cross section increased with depth. The generally low optical cross section between 25 and 35 m overlapped with much of the peak in acoustical scattering strength from 30 to 40 m (Figure 2.6).

2.3.2 November 21 Cast in Knight Inlet

The overall acoustical scattering in the channel was very different from that in Hoeya Sound and was characterized by generally low scattering to approximately 60 m, a layer of elevated scattering from 60 to 120 m, and moderately stronger scattering to 200 m (Figure 2.4). Numbers of ROIs were variable, particularly above 60 m and below 130 m (Figure 2.6). The distributions of nearest neighbors in all images were statistically completely spatially random ($p > 0.05$) throughout all depths sampled. There was no indication that particles within acoustical layers were statistically aggregated or dispersed. The size (area) distributions of particles were dominated by small objects less than 1 mm^2 (mean of 0.32 mm^2). In the example profile, the mean particle area, based on the presence of particles with areas ≥ 0.25 mm^2, showed a very gradual increase with depth (Figure 2.6). Large particles with cross-sectional areas over 1 mm^2 were frequently present between 60 and 120 m (Figure 2.6). These large objects were typically gammaridean amphipods and small euphausiids (<1 cm long) believed to be *Euphausia pacifica* (Figure 2.5). No euphausiids were detected outside the subsurface region of elevated scattering (Figure 2.4). Images of euphausiids were clear and provided unambiguous information from which to estimate their probable taxonomy, size, and orientation. Based on the volume of water imaged between 60 and 120 m (132.9 l), the density of small (<1 cm long) euphausiids within the layer was 52.7 m^{-3}. Most of these were horizontally oriented in the water column. Elevated acoustical scattering at approximately 10 m depth did not appear to be related to the presence of zooplankton-sized particles (Figure 2.6). The majority of other particles in the water column were oval or spherical in shape and appeared to be small copepods; however, details of these objects could not be resolved because the image volume that was selected for these operations was optimized to detect larger euphausiid-sized particles. Few marine snow particles were evident. The optical cross-sectional area of particles in the water column was highly variable in the upper 40 m of the water and showed a gradual increase with depth (Figure 2.6).

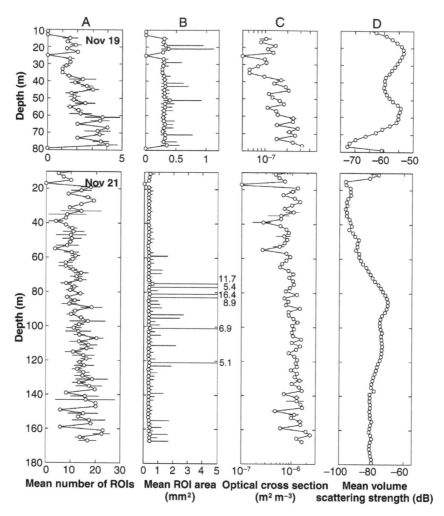

FIGURE 2.6 (A) Mean number of ROIs ≥ 0.25 mm² detected in ZOOVIS images at 2-m intervals on November 19 (upper panel) and November 21 (lower panel). Error bars indicate 1 standard error. (B) The mean areas of particles ≥ 0.25 mm² at 2 m intervals on November 19 (upper panel) and November 21 (lower panel). Error bars indicate the minimum and maximum particle areas detected in each depth interval. (C) Mean optical cross-sectional area of targets ≥ 0.25 mm² at 2-m intervals on November 19 (upper panel) and November 21 (lower panel). Error bars indicate one standard error. (D) Mean volume scattering strength at 100 kHz at 2-m intervals on November 19 (upper panel) and November 21 (lower panel).

2.4 Discussion

The images acquired by ZOOVIS contain spatially referenced information on the particle size distribution, identities of larger particles, the spatial relationships among particles, and the orientation of larger organisms at vertical intervals of at least 2 m. In some ways, ZOOVIS can provide similar information to an OPC. While an OPC provides particle size distributions, it groups them into a limited number of size categories and cannot provide information on the identities of individual particles. The particle size data from ZOOVIS can be summarized to provide the same type of information as the OPC with finer size bins and with additional information on the morphologies and likely identities of individual particles as well as their 2D spatial interrelationships. The OPC currently retains the advantage of providing a continuous record of the particle distributions along its trajectory in contrast with the discrete images of ZOOVIS; however, anticipated improvements to the sampling rate by our system should provide more complete coverage of the water column along its trajectory.

ZOOVIS was designed to collect images at vertical intervals of at least a meter. The vertical resolution of the system depends on both the rate at which the instrument descends and the sampling frequency of the camera. Our descent rate was fixed at 0.5 m s^{-1} to limit motion-induced blurring during the duration of the strobe pulse to a low value (10 μm). Our transfer rate between the underwater PC and the surface computer was well below the theoretical network maximum, which limited the sampling frequency of the camera. This meant that our actual vertical resolution was approximately 2 m. Assuming that the system could acquire images at the maximum sampling rate of the camera (2.2 Hz), a vertical resolution of 0.25 m could be achieved at the payout rates used in Knight Inlet. We plan to upgrade the network to gigabit Ethernet to address data transfer limits; however, in the short-term, our solution will be to acquire all images directly to a large-capacity hard-drive attached to the underwater PC.

The pixel density of our camera's CCD provided an image resolution of 18.7 pixels mm^{-1}. This suggests that individual objects as small as 56 μm could be detected under ideal conditions. In practice, given the illumination we were able to provide with the strobe, the sensitivity of our camera's CCD, and the water clarity, objects of approximately 75 to 100 μm and larger were clearly resolved. This was sufficient to resolve details of euphausiid antennae and the shapes of copepods, but generally did not capture information on the anatomical details of small targets such as copepod antennae. In pending future operations where the taxa of interest are copepod-sized zooplankton, we plan to reduce the field of view to 70 × 70 mm, which will increase the theoretical resolution of the system to 34 μm.

In this study, the volumetric sampling rate of ZOOVIS is 6.6 l min^{-1} at 0.25 Hz and 0.443 l image^{-1} with a potential maximum of 53.2 l min^{-1} at 2 Hz. These rates are generally comparable to the volumetric sampling rates of the VPR. For example, in a study on Georges Bank, the wide-field camera of the VPR sampled 22.08 ml image^{-1} at 60 Hz (Benfield et al., 1996), and in recent GLOBEC surveys using BIOMAPER II the camera was set to image 5.1 ml image^{-1} at 60 Hz. The volumetric sampling rates from these two settings ranged from 18.36 to 79.5 l min^{-1}. Although ZOOVIS generally samples less total water than the VPR, the number of targets within individual images was substantially greater than in VPR images providing an opportunity to assess spatial interrelationships among targets. It is important to note that ZOOVIS and the VPR are different types of instruments designed to fulfill separate sensing niches. ZOOVIS is normally deployed in a profiling mode, whereas the VPR is towed along a sawtooth trajectory called a towyo. The contiguous vertical coverage at one location provided by ZOOVIS comes at the expense of reduced spatial coverage, which the VPR can provide. The larger image volume per ZOOVIS image comes at the expense of the high sampling rate of the VPR.

The cross-sectional area of targets in ZOOVIS images acquired by a down-looking camera provides a potential optical analogue to the acoustically reflective cross section of targets scattering energy from a down-looking echosounder. One may think of the sum of the optical cross sections of targets in a ZOOVIS image as being analogous to the term $N\sigma_{bs}$ in the acoustical equation for mean volume scattering strength (S_v):

$$S_v = 10\log_{10}(\frac{N\sigma_{bs}}{V}) = 10\log_{10}(s_v) \qquad (2.1)$$

where N is the number of scatterers in an aggregation of unresolved acoustical targets, σ_{bs} is the mean differential backscattering cross section ($\sigma/4\pi$) of the targets, V is the ensonified volume, and s_v is the volume backscattering coefficient (m^2 m^{-3}) (Foote and Stanton, 2000). Although the material properties of the scatterers, and their orientation in relation to the incident acoustical beam will strongly influence their volume scattering strength, there may still be a positive relationship between the optical cross-sectional areas of targets and s_v for targets large enough to fall within the geometric scattering region of the acoustical frequency of interest.

When the mean acoustical volume scattering strength estimated for 2 m depth bins was regressed as a function of the logarithm of the corresponding optical cross section, there was a statistically significant positive relationship (adj. $R^2 = 0.11$, $p = 0.002$) for the November 21 cast. In the November 21 data set, there appeared to be an even stronger relationship between the presence of large targets in the ZOOVIS images and the acoustic scattering profile. There was a statistically significant positive relationship (adj. $R^2 = 0.27$, $p < 0.0001$) between the logarithm of cross-sectional area of the largest ROI in each depth stratum and S_v. This latter relationship suggests that it was the presence of euphausiids that dominated the scattering observed in the layer between 60 and 120 m depth. Euphausiids that were imaged by ZOOVIS had a mean length of 8 mm and would fall within the geometric scattering region at 100 kHz. Thus, a positive correlation between the optical and acoustical cross sections measured is plausible. The acoustic record also included regions of high backscatter near the surface that were not associated with high densities or cross-sectional areas of particles in ZOOVIS images. Scattering due to microturbulence may explain the poor relationship between the optical and acoustical records near the surface.

Although the broad regions of elevated acoustical scattering observed in the main channel on November 21 appeared to be a consequence of zooplankton, there was little correspondence between measured scattering and optical properties in Hoeya Sound on November 19. Visual observations suggested that small fish were abundant in Hoeya Sound. The presence of fish in the echogram is supported by high scattering at the lowest acoustical frequency (40 kHz). If these fish were feeding on zooplankton, it would explain why the optical cross sections were generally lower than those observed in the main channel and perhaps account for the very low optical cross sections in the depth stratum where the highest acoustical scattering was noted. The fish rapidly moved away from the vessel when a flashlight was shone into the water and likely avoided ZOOVIS due to its strobe or hydrodynamic signature. The zooplankton assemblage in Hoeya Sound consisted primarily of small copepods (<1 mm long) that would fall in the Rayleigh scattering range at 100 kHz. Consequently, their densities would be unlikely to be correlated with acoustical scattering.

Profiling instruments such as ZOOVIS have applications in a variety of regions that would prove problematic for towed instruments. The fjord system we examined was characterized by steep bathymetry and abrupt changes in depth that would pose a hazard for towed systems. Risks of fouling or colliding with fixed structures are also inherent problems when sampling around ice flows, from petroleum platforms, and around natural reefs. Providing that the depth of the bottom and locations of fouling structures are known, ZOOVIS can be safely operated in such areas. The ship's echosounder provided a good estimate of the safe working depth beneath the vessel in Knight Inlet and permitted ZOOVIS to be deployed to within a few meters of the bottom on several occasions. The visible presence of ZOOVIS in the echosounder also assisted us in guiding the vehicle into the appropriate scattering layers.

In summary, the performance of ZOOVIS appears promising. Further analysis of the complete Knight Inlet data set and intercomparisons with BIONESS casts and acoustic data will be required to assess potential biases of the system due to avoidance; however, the system yielded high-quality images of mesozooplankton and operated well in a fjord system. It appears to have met its design criteria for a system capable of quantifying the distributions and abundances of mesozooplankton in coastal waters. We are currently incorporating improvements in the frame (a "stealthier" smaller design) and network bandwidth (gigabit Ethernet). Ultimately, as digital camera technology evolves, improvements in the resolution of the camera will add capabilities to what is already a flexible instrument capable of quantitatively surveying a wide range of mesozooplankton in the coastal oceans.

Acknowledgments

We are grateful to the captain and crew of the *CCGS Vector* and the ship support personnel at the Institute of Ocean Sciences for their assistance with Knight Inlet operations. David Mackas, Doug Yelland, Darren Tuele, Tetjana Ross, and Robert Campbell all assisted with ZOOVIS operations. The staff of the LSU Coastal Studies Institute Field Support Shop, Malinda Sutor (Oregon State University), and Joyce Brignac (Louisiana State University) provided invaluable assistance during assembly. We appreciate technical advice and assistance from Oceaneering, Inc., Morgan City, LA. Funding for ZOOVIS was provided by the Office of Naval Research grants N00014-98-1-0563 and N00014-02-1-0012, the Louisiana Board of Regents, and the School of the Coast and Environment, Louisiana State University.

References

Benfield, M.C., C.S. Davis, P.H. Wiebe, S.M. Gallager, R.G. Lough, and N.J. Copley. 1996. Video plankton recorder estimates of copepod, pteropod and larvacean distributions from a stratified region of Georges Bank with comparative measurements from a MOCNESS sampler. *Deep-Sea Res. II*, 43, 1925–1946.

Boots, B.N. and A. Getis. 1988. *Point Pattern Analysis*, Sage Publications, Newbury Park, CA.

Buskey, E.J., J.O. Peterson, and J.W. Ambler. 1996. The swarming behavior of the copepod *Dioithona oculata*: *in situ* and laboratory studies. *Limnol. Oceanogr.*, 41, 513–521.

Cowles, T.J., R.A. Desiderio, and M.E. Carr. 1998. Small-scale planktonic structure: persistence and trophic consequences. *Oceanography*, 11, 4–9.

Davis, C.S., S.M. Gallager, M.S. Berman, L.R. Haury, and J.R. Strickler. 1992. The Video Plankton Recorder (VPR): design and initial results. *Arch. Hydrobiol. Beih. Erg. Limnol.*, 36, 67–81.

Dekshenieks, M.M., P.L. Donaghay, J.M. Sullivan, J.E.B. Rines, T.R. Osborn, and M.S. Twardowski. 2001. Temporal and spatial occurrence of thin phytoplankton layers in relation to physical processes. *Mar. Ecol. Prog. Ser.*, 223, 61–71.

Donaghay, P.L., H.M. Rines, and J.M. Sieburth. 1992. Simultaneous sampling of fine scale biological, chemical and physical structure in stratified waters. *Arch. Hydrobiol.*, 36, 97–108.

Foote, K.G. and T.K. Stanton. 2000. Acoustical methods, in Harris, R., P.H. Wiebe, J. Lenz, H.R. Skjoldal, and M. Huntley, Eds., *ICES Zooplankton Methodology Manual*, Academic Press, San Diego, 223–258.

Gorsky, G., M. Picheral, and L. Stemmann. 2000. Use of the underwater video profiler for the study of aggregate dynamics in the North Mediterranean. *Estuarine Coastal Shelf Sci.*, 50, 121–128.

Herman, A.W. 1988. Simultaneous measurement of zooplankton and light attenuance with a new optical plankton counter. *Continental Shelf Res.*, 8, 205–221.

Holliday, D.V., R.E. Pieper, R.F. Greenlaw, and J.K. Dawson. 1998. Acoustical sensing of small-scale vertical structures in zooplankton assemblages. *Oceanography*, 11, 18–23.

Incze, L.S., P. Aas, and T. Ainaire. 1996. Distributions of copepod nauplii and turbulence on the southern flank of Georges Bank: implications for feeding by larval cod. *Deep-Sea Res. II*, 43, 1855–1874.

Lenz, J., D. Schnack, D. Petersen, J. Kreikemeiser, B. Hermann, S. Mees, and K. Wieland. 1994. The ichthyoplankton recorder: a video recording system for in-situ studies of small-scale distribution patterns. *ICES J. Mar. Sci.*, 52, 409–417.

Malkiel, E., O. Alquaddoomi, and J. Katz. 1999. Measurements of plankton distribution in the ocean using submersible holography. *Measurement Sci. Technol.*, 10, 1142–1152.

Marine Zooplankton Colloquium 2. 2001. Future marine zooplankton research — a perspective. *Mar. Ecol. Prog. Ser.*, 222, 297–308.

Ortner, P.B., L.C. Hill, and H.E. Edgerton. 1981. In-situ silhouette photography of oceanic zooplankton. *Deep-Sea Res. A*, 28, 1569–1576.

Price, H.J. 1989. Swimming behavior of krill in response to algal patches: a mesocosm study. *Limnol. Oceanogr.*, 34, 649–659.

Samson, S., T. Hopkins, A. Remsen, L. Langebrake, T. Sutton, and J. Patten. 2001. A system for high-resolution zooplankton imaging. *IEEE J. Oceanic Eng.*, 26, 671–676.

Smith, L.R., C.B. Miller, and R.L. Holton. 1976. Small-scale horizontal distribution of coastal copepods. *J. Exp. Mar. Biol. Ecol.*, 23, 241–253.

Stanton, T.K., D. Chu, and P.H. Wiebe. 1998. Sound scattering by several zooplankton groups. II. Scattering models. *J. Acoust. Soc. Am.*, 103, 236–253.

Tiselius, P. 1998. An *in situ* video camera for plankton studies: design and preliminary observations. *Mar. Ecol. Prog. Ser.*, 164, 293–299.

Watson, J., V. Chalvidan, J.P. Chambard, G. Craig, A. Diard, G.L. Foresti, B. Forre, S. Gentili, P.R. Hobson, R.S. Lampitt, P. Maine, J.T. Malmo, and G. Pieroni. 1998. HOLOMAR: high resolution *in situ* HOLOgraphic recording and analysis of MARine organisms and particles, in *Proceedings of the Third European Marine Science and Technology Conference (MAST Conference)*, Lisbon, 23–27 May 1998, Vol. 3: *Generic Technologies*, 1197–1211.

Wiebe, P.H., D.G. Mountain, T.K. Stanton, C.H. Greene, G. Lough, S. Kaartvedt, J. Dawson, and N. Copley. 1996. Acoustical study of the spatial distribution of plankton on Georges Bank and the relationship between volume backscattering strength and the taxonomic composition of the plankton. *Deep-Sea Res. II,* 43, 1971–2002.

3

Planktonic Layers: Physical and Biological Interactions on the Small Scale

Timothy J. Cowles

CONTENTS

3.1 Introduction and Background...31
3.2 Small-Scale Planktonic Layers: Examples 33
    3.2.1 East Sound, Washington 33
    3.2.2 Continental Shelf, Oregon 34
3.3 Vertical Velocity Gradients: Measurements and Issues..39
    3.3.1 What the ADV Measures 39
    3.3.2 The Resolution of Vertical Shear...39
    3.3.3 The Role of Vertical Mixing...43
3.4 Dye Injections: Analogues for Plankton Layers...43
3.5 Preliminary Estimates of the Horizontal Extent of Small-Scale Planktonic Layers....................44
3.6 Trophic Implications of Sharp Vertical Gradients..44
3.7 Small-Scale Planktonic Structure: Questions to Address...45
3.8 Recommendations for Future Work 45
3.9 Conclusion..46
Acknowledgments...46
References..47

3.1 Introduction and Background

The spatial and temporal scales of plankton distribution have been evaluated and discussed for many decades (e.g., Bainbridge, 1952; Cassie, 1963; Bjørnsen and Neilsen, 1991; Denman, 1994) due to the importance of patterns of distribution in determining the rates of biological production and transfer of material through the food web. Numerous studies have shown that planktonic organisms have behavioral and physiological responses to small-scale processes (Cullen and Lewis, 1988; Cowles et al., 1988; Butler et al., 1989; Verity, 1991; Tiselius, 1992). The application of new *in situ* instrumentation in optical and acoustical oceanography has revealed vertical pattern over spatial scales of centimeters (Cowles et al., 1990, 1998; Donaghay et al., 1992; Holliday et al., 1998). On larger spatial scales, the advent of satellite remote sensing of ocean color has provided new insights into the horizontal patterns of plankton distribution and their interaction with mesoscale physical processes (Denman and Abbott, 1994; Platt and Sathyendranath, 1994; Letelier et al., 1997; Abbott and Letelier, 1998). These observations of pattern in property distributions (whether those properties are physical or biological) have stimulated the development of new approaches for quantifying and explaining differences in "pattern and process" across a wide range of spatial and temporal scales. This work in marine systems has its parallels in terrestrial ecology, where there is increasing theoretical and empirical interest in quantifying pattern (e.g., Caswell and Cohen, 1991; Gardner and O'Neill, 1991; Levin, 1992).

Although most of the research on scales of patchiness has focused on horizontal distributions (e.g., Kierstead and Slobodkin, 1953; Denman and Platt, 1976; Okubo, 1980; Bennett and Denman, 1985), steep vertical gradients in phytoplankton distributions have been noted in highly stratified habitats (e.g., Bjørnsen and Nielsen, 1991; Donaghay et al., 1992). In some cases, a two- to fourfold change in phytoplankton concentration can occur over a vertical interval of 0.25 to 1.0 m (e.g., Cowles et al., 1990, 1993; Donaghay et al., 1992; Hanson and Donaghay, 1998; Holliday et al., 1998; Dekshenieks et al., 2001). The observations suggest that these sub-1-m scale structures may persist for many hours and may have a horizontal extent of several kilometers. It is therefore essential to develop an understanding of how these structures (and the controlling processes) operate in the ocean and how they influence our ability to evaluate biological properties and rates of biological production. Until recently, estimates of the vertical distribution of many biological properties have relied on discrete samples that were often several meters apart, or lacked coincident physical measurements. Now it is appropriate to reformulate our questions about biological processes in the upper ocean to reflect our new information about sub-1-m-scale structure.

These questions about small-scale biological processes involve unraveling and quantifying the complex linkages between physical forcing and the observed small-scale pattern of plankton distribution. For example, what ranges of conditions result in the formation and persistence of small-scale features? What are the dominant processes that create and maintain small-scale structure? Over what timescales do these features form, change, or disappear? Are particular types of coastal and oceanic habitats more or less likely to possess persistent small-scale features? To what extent, and under what conditions, must we alter our sampling strategies to obtain acceptable estimates of the distribution of physical, chemical, and biological properties and rate processes in the upper ocean?

As has been suggested in the plankton literature for several decades (e.g., Bainbridge 1952, 1953; Mullin and Brooks, 1976), localized maxima in food resources are essential if zooplankton are to meet their metabolic requirements, as the local mean concentration of food often appears too low to support zooplankton metabolism. Are these local maxima distributed randomly in time and space as "patches" with a distribution of persistence times, or do these maxima represent persistent patterns that result from specific physical processes acting on phytoplankton biomass gradients?

Patchiness has been a major topic in plankton ecology for most of the past century (e.g., Hardy and Gunther, 1935; Riley, 1976; Mullin and Brooks, 1976; Haury et al., 1978; Davis et al., 1991; Grünbaum, 2002). Theoretical efforts by Kierstead and Slobodkin (1953) addressed the balance between local phytoplankton growth and horizontal diffusive processes, while observations of horizontal distributions expanded rapidly with the introduction of reliable underway fluorometric systems. The resulting long spatial records of surface fluorescence (e.g., Denman and Platt, 1976) revealed that, beyond a spatial scale of 10 to 20 km, the power spectrum of surface phytoplankton distribution has the same slope as that of temperature, suggesting that the larger-scale spatial patterns of surface phytoplankton distributions are controlled by the same physical processes that determine the patterns of surface temperature (Denman and Platt, 1976; Denman, 1994). In this chapter, however, I shift the emphasis away from patchiness in surface phytoplankton distributions, and concentrate instead on small-scale vertical patterns of distribution, and on the physical processes that appear to interact with biological processes to create extensive horizontal layers of enhanced phytoplankton concentrations (Osborn, 1998).

Extremely high concentrations of phytoplankton have been observed within narrow vertical intervals in the well-stratified waters of the Baltic (Bjørnsen and Nielsen, 1991), in some cases the layers consist of toxic phytoplankton species (Nielsen et al., 1990). Several workers developed syringe systems to collect multiple samples within narrow vertical ranges (e.g., Owen, 1989; Mitchell and Furhman, 1989) and found considerable variability on vertical scales of 1 m or less. In important related work, laboratory investigations of zooplankton feeding behavior during the 1970s and 1980s revealed that zooplankton have complex sensory structures for food detection (e.g., Strickler and Bal, 1973; Yen and Nicoll, 1990; Yen and Strickler, 1996), and these grazers display complex behavior during feeding (e.g., Koehl and Strickler, 1981; Price et al., 1983; Cowles and Strickler, 1983). The chemosensory abilities of zooplankton to detect food items were shown by Poulet and Oelluet (1982), and subsequent research revealed that zooplankton can discriminate between individual food items on the basis of chemical signals alone (e.g., Butler et al., 1989; Cowles et al., 1988). Additional behavioral experiments showed that zooplankton alter their swimming behavior to spend longer time intervals within higher concentrations of food (e.g., Tiselius, 1992). All these laboratory studies were critical in showing that zooplankton possess the adaptations to locate and exploit differences in food resources, over spatial scales from centimeters to many tens of centimeters. These adaptations may be a result of selection pressure created by nonrandom small-scale distributions of food resources in the water column, i.e., persistent pattern rather than stochastic patchiness.

These laboratory and field investigations were important precursors to the application of new technical approaches to defining small-scale vertical phytoplankton structure in the upper ocean, and the motivation for such work arose from the need to understand the food resource environment of zooplankton. As suggested by the fieldwork of Lasker (1975) on the survival of first-feeding larval fish, zooplankton make a living on the variance, not the mean, of the coarse-scale distribution of food in the water column. The challenge is to define the appropriate temporal and spatial scales for the evaluation of the variability of food resource distribution in the upper ocean.

This chapter reviews evidence for small-scale biological distributions, with particular emphasis on observations obtained by the author's research group over the continental shelf off Oregon (USA) and within a coastal fjord in the San Juan Islands in Puget Sound, Washington (USA). The chapter examines the common features displayed within the small-scale planktonic distributions at these distinct locations, and argues that much of the small-scale vertical structure observed in phytoplankton distributions is created by vertical shear over spatial scales of 1 m or less. The chapter concludes with a discussion of the observational approaches required to resolve these vertical gradients in plankton distribution in the appropriate physical context.

3.2 Small-Scale Planktonic Layers: Examples

3.2.1 East Sound, Washington

During the summer of 1998, a group of investigators focused on thin-layer dynamics in a coastal fjord of Orcas Island, WA, known as East Sound. East Sound is approximately 10 km long and 1 to 1.5 km wide, oriented with the long axis NNW to SSE. It has a relatively uniform bottom depth of 25 to 30 m, with a shallow sill extending part way across the entrance to the fjord. Several recent publications provide details of this location (Figure 3.1, and see Alldredge et al., 2002; Dekshenieks et al., 2001; Rines et al., 2002). The experiment in 1998 was the first to integrate small-scale physical observations with extensive bio-optical and bioacoustical observations on the same spatial scales, using moored acoustic doppler current profilers (ADCPs) and multifrequency bioacoustic systems as well as instruments deployed from three different vessels. I discuss a series of vertical profiles obtained during a time series of deployments from a moored vessel, the *RV Henderson*. These profiles are representative of the small-scale structure observed in East Sound.

The time series of rapidly repeated profiles (~10 h^{-1}) of the water column was obtained with a free-falling instrument package that included sensors to define the microscale and fine-scale vertical structure of hydrography, bio-optical properties, and the vertical gradients in horizontal velocity. The free-fall package consisted of a Sea-Bird™911$^+$ CTD with O$_2$ sensor, two multiwavelength absorption and attenuation meters (ac-9, WET Labs Inc.™), a multiwavelength excitation/emission fluorometer (SaFIRE, WET Labs, Inc.™), and a conventional single-wavelength fluorometer (WetStar, WET Labs, Inc.™). The instrument package also carried an Acoustic Doppler Velocimeter (OceanProbe, Sontek™) for resolution of horizontal velocity gradients on vertical scales of 0.20 to 0.25 m^{-1}. The buoyancy of the instrument package was adjusted to provide a free-fall descent rate of 0.15 to 0.20 m s^{-1}, thus resolving physical and bio-optical properties over vertical scales of a few centimeters (Cowles et al., 1998).

During a 24-h time series of repeated high-resolution vertical profiles (June 20–21, 1998), we observed a persistent layer of locally enhanced phytoplankton concentration over a 13-h interval (Figure 3.2). This small-scale feature showed a number of characteristics that are commonly found in our data sets. It was characterized by an abrupt increase in chlorophyll concentration from approximately 3 mg m^{-3} to over 12 mg m^{-3} within a vertical interval of 0.50 m or less, and was associated with a narrow density interval (21.86 to 21.88 kg m^{-3}). We obtained 93 vertical profiles in the time interval represented by the six panels in Figure 3.2, and this distinct layer of phytoplankton was present in all 93 profiles. We also observed that small-scale vertical gradients in horizontal velocity (obtained with the ADV) were associated with both the

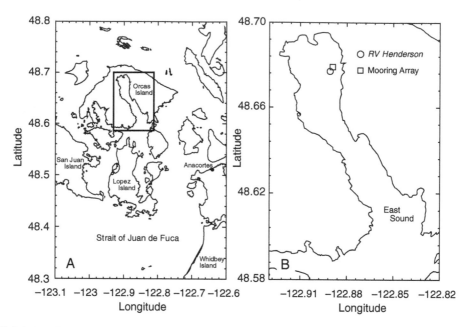

FIGURE 3.1 Location of 1998 experiment in East Sound, Orcas Island, WA. (A) Orcas Island is located within the San Juan Islands, WA, northwest of Seattle. (B) Extended time series of vertical profiles were collected from the *RV Henderson*, anchored in the northern portion of East Sound (B).

density steps and the steep vertical gradients in phytoplankton concentrations (Figure 3.3). These profiles display the same characteristics as those obtained in 1996 by Dekshenieks et al. (2001), with a majority of the narrow layers occurring within 1 m of a density gradient. Bioacoustic assessment of zooplankton concentrations also revealed small-scale structure during the East Sound experiment in 1998 (Holliday et al., in press).

3.2.2 Continental Shelf, Oregon

Our research group at Oregon State University has used the free-fall profiling system (described above) to resolve small-scale structure over the Oregon continental shelf during summer cruises in 1997, 1998, 2001, and 2002 (Figure 3.4). We have observed layers and steep vertical gradients in phytoplankton concentration (based on bio-optical indices) during every year of observation. This section presents representative examples from the over 500 vertical profiles we have obtained from this energetic, wind-driven coastal upwelling regime.

The vertical patterns of phytoplankton distribution over the shelf often exhibit steep gradients, with some of those steep gradients expressed as thin layers (Figure 3.5). As with the small-scale features observed in East Sound, the steep gradients over the continental shelf were usually associated with steps in density structure, although the steepest phytoplankton gradients may not be associated with the steepest density gradient (e.g., Figure 3.5D).

Extended time series of repeated profiles over the continental shelf have revealed the temporal evolution of steep vertical gradients in phytoplankton concentrations as well as the apparent horizontal distribution of the consequences of vertical strain on phytoplankton within specific density intervals. One example of this temporal and spatial pattern was obtained during a 15-h interval on September 12–13, 1997 at the shelf break location (see Figure 3.4), when we observed development and persistence of small-scale structure as the density structure within that depth range was strained by the internal wave field (Figure 3.6). Of particular interest is the appearance of a steep vertical phytoplankton gradient within the highlighted density interval near hour 9 of the time series (panel 3, Figure 3.6). As the density interval was strained to a thickness of less than 1 m, the chlorophyll gradient steepened, with a twofold increase in concentration over a 0.50-m interval. It is important to note that this 15-h time series at the continental

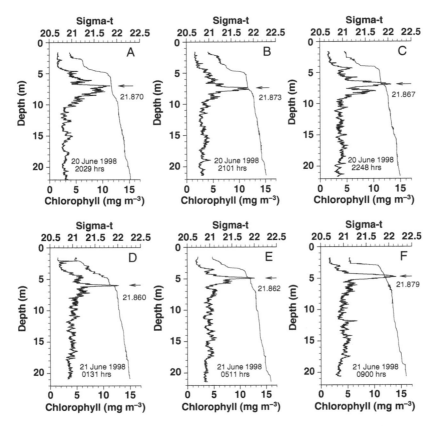

FIGURE 3.2 Six vertical profiles of phytoplankton fluorescence (mg m^{-3}) and sigma-t (kg m^{-3}) from a time series of 90 profiles obtained with a free-fall profiling package between 2000 hours on June 20, 1998 and 0900 hours on June 21, 1998 at the northern end of East Sound, WA. The time of sampling is noted on each panel. A persistent layer of phytoplankton fluorescence was observed throughout this 11-h time period. The arrow indicates the depth of the maximum fluorescence of the layer, and the density (kg m^{-3}) at that depth is shown just below the arrow. The individual profiles presented in panels A through F are characteristic of the shape, thickness, and variability of this persistent feature during the observation period. This layer occupied a consistent density interval (within 21.86 to 21.88 kg m^{-3}) throughout the time series.

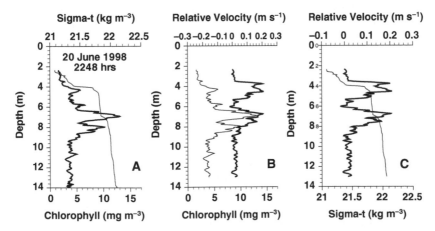

FIGURE 3.3 Vertical profile from 2248 hours on June 20, 1998 in East Sound, WA (this is the third profile in the sequence in Figure 3.2). (A) The vertical distribution of phytoplankton fluorescence and sigma-t. (B) The vertical distribution of fluorescence and the v component of relative velocity as measured from the ADV. (C) The vertical distribution of sigma-t and the v component of relative velocity from the ADV. Note that the sharp vertical gradients in phytoplankton distribution are strongly correlated with small-scale vertical gradients in horizontal velocity.

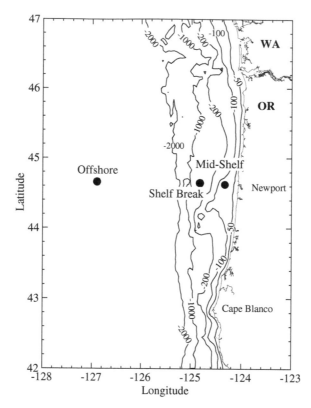

FIGURE 3.4 Locations sampled for vertical small-scale structure in 1997 and 1998 (midshelf, shelf break, and offshore) off the West Coast of North America.

shelf break displayed several of the characteristic shapes of phytoplankton layers and gradients that we have found at various locations — the "slab" (Figure 3.6, hours 1 to 3), the "notch" (Figure 3.6, hours 4.5 to 8), and the "nose" (Figure 3.6, hours 9.5 to 15).

The physical processes that create sharp vertical gradients in phytoplankton distributions (and coincident microbial and nutrient gradients) may not always create "thin layers." As seen by the time series in Figure 3.6, a range of vertical distributions occurs in conjunction with the vertical separation of stratified layers. We observe thick slabs with sharply defined upper and/or lower boundaries, we observe slabs dividing into smaller slabs, or we may find narrow layers. Each of these distributions may occur within one coherent planktonic feature, depending on the spatial differences in intermittent small-scale vertical mixing, as well as spatial differences in strain between isopycnals.

The combined effects of advection and vertical strain were also observed the following year during a time series of profiles at the midshelf location off the Oregon coast in September 1998 (Figure 3.7). During the 3-h interval shown in Figure 3.7 we observed an intensification of the vertical chlorophyll gradient associated with the 25.80 to 26.00 kg m^{-3} density interval. The steepening of the vertical chlorophyll gradient coincided with the small-scale vertical gradient in horizontal velocity, as was observed in East Sound. The temporal persistence in vertical gradients in phytoplankton concentration is matched by the persistence in the vertical shear profiles, with the velocity gradients confined to vertical intervals of 1 to 2 m extent. This point can be illustrated by examining a single profile (Figure 3.8) from the September 20, 1998 time series shown in Figure 3.7. The vertical structure of this profile displayed multiple steps in density, with strong correlations in the vertical gradient of relative velocity with those density steps (Figure 3.8A, B). The steep gradients in phytoplankton biomass corresponded with the gradients in relative velocity (Figure 3.8C, D), as had been observed for vertical distributions in East Sound (Figure 3.3). It is noteworthy that the steepest gradient and largest concentration of phytoplankton biomass (at 37 m in Figure 3.8) did not occur in conjunction with the steepest density gradient (28 m),

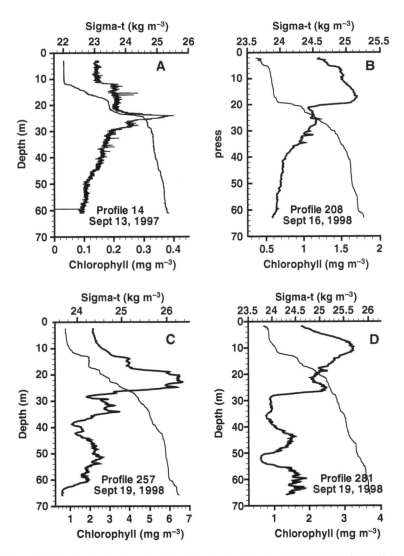

FIGURE 3.5 Characteristic steep vertical gradients in chlorophyll off the Oregon coast, associated with steps in density. The profiles in A and B were obtained over the shelf break; C and D were obtained at the midshelf station, adjacent to the ADP mooring.

yet all the local maxima in phytoplankton concentration were collocated with both density steps and vertical gradients in horizontal velocity.

The temporal coherence in phytoplankton vertical distributions over narrow vertical intervals is usually matched by temporal coherence in the small-scale velocity structure. Three vertical profiles from a 2-h time interval (profiles marked with * in Figure 3.7B) demonstrate this point (Figure 3.9). Profiles 307, 311, and 317 (as well as the intermediate profiles) each possessed a sharp transition in phytoplankton concentration at the depth of the 25.52 isopycnal (Figure 3.7B), shown at 40 m in Figure 3.9A. The steep gradient in phytoplankton concentration over the next vertical meter (seen in all profiles in this series) was matched by the vertical gradient in both the u and v components of relative velocity (Figure 3.9B, C). It is also important to note that the steep velocity gradient observed at 25 m (Figure 3.9B) did not have a corresponding steep gradient in phytoplankton concentration.

These profiles illustrate an important general feature of our observations over the continental shelf: steep gradients in phytoplankton concentration were nearly always associated with steep vertical gradients in density and horizontal velocity, whereas steep vertical gradients in density and horizontal velocity did

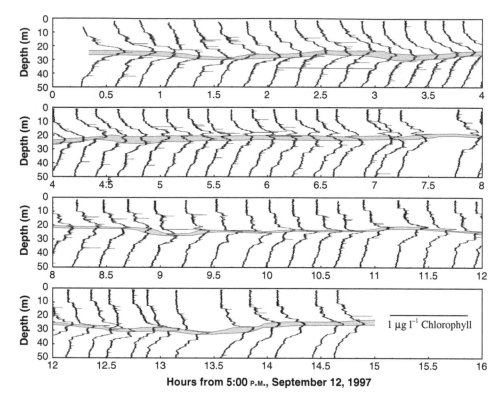

Hours from 5:00 P.M., September 12, 1997

FIGURE 3.6 Time series of 63 free-fall profiles of chlorophyll obtained between 1700 hours 12 September and 1000 hours 13 September 1997 at the shelf break station (see Figure 3.4). Average time between profiles was 12 min. The shaded interval represents the sigma-t range 24.25 to 24.50 (kg m^{-3}). Note the thickness change in this density interval during the time series, and the associated intensification of the vertical gradient in chlorophyll concentration.

not necessarily possess corresponding phytoplankton gradients. This observation points toward one possible formation process for steep vertical gradients in phytoplankton concentration, with thin layers as a special case of this process. As noted decades ago (e.g., Okubo, 1968; Woods, 1968; Kullenberg, 1974; Young et al., 1982), shear dispersion within a stratified layer results in horizontal spreading of a tracer. The thin dye layers created by Kullenberg strongly resemble the phytoplankton layers observed in East Sound and over the Oregon continental shelf, in shape as well as in temporal persistence. But such vertical gradients in tracer (phytoplankton) must be created via horizontal advection and dispersion, with little vertical mixing over the time intervals of observation. The vertical gradient in phytoplankton is a consequence of the differential motion of layers of water containing different concentrations of phytoplankton, suggesting different sources of phytoplankton "upstream" of the observation point for the waters above and below the steep vertical gradient. In addition, this mode of horizontal spreading and stretching of phytoplankton biomass implies a horizontal gradient in that biomass, as suggested by Eckart (1948).

These observations of small-scale vertical structure from East Sound and the Oregon continental shelf suggest that horizontal physical processes (such as advection and vertical shear) may be more important than vertical physical processes (such as mixing) in creating and maintaining vertical gradients in phytoplankton concentration. Key biological processes, such as phytoplankton growth and zooplankton grazing, will also influence the shape and amplitude of vertical gradients. The timescales of the impact of growth and grazing have not yet been measured in the field, but preliminary modeling studies suggest that horizontal physical processes will usually dominate.

The examples presented in this chapter supplement the results of other recent studies (e.g., Alldredge et al., 2002; Cowles et al., 1990, 1998; Dekshenieks et al., 2001; Donaghay et al., 1992; Holliday et al., 1998), and emphasize the need for better resolution of horizontal distributions and processes. In particular, over what vertical interval must we resolve changes in horizontal velocity? Conventional instrumentation

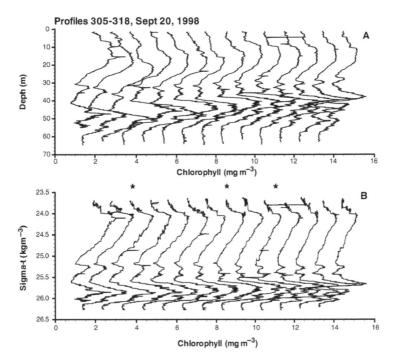

FIGURE 3.7 The 14 vertical profiles obtained during a 3-h interval over the Oregon continental shelf on September 20, 1998. (A) Time series of the vertical distribution of chlorophyll, with each profile offset horizontally by 1 mg chl m^{-3}. (B) The chlorophyll is plotted on density to illustrate the compression and straining of the water column during this time series. The small-scale velocity gradients within the three profiles denoted by ★ are featured in Figure 3.9.

(CTDs, shipboard ADCPs) are not adequate for obtaining the small-scale estimates of buoyancy and vertical shear needed for the estimation of the transition to vertical mixing (where the Richardson number, Ri, < 0.25. Ri is the ratio of squared buoyancy frequency to the squared vertical shear). The section below explores the use of shipboard and moored Acoustic Doppler systems and illustrates the relationship of those measurements with the ADV measurements obtained on the free-fall profiler.

3.3 Vertical Velocity Gradients: Measurements and Issues

3.3.1 What the ADV Measures

The ADV measures horizontal velocity relative to the sinking profiler. The profiler is also subject to the mean horizontal flow at each depth horizon, integrated over the vertical scale of the profiler (1.5 m). The relative velocity measurement is illustrated schematically in Figure 3.10, with the "mean flow" experienced by the profiler represented by the vectors on the left side of the figure, and with the relative velocity measured by the ADV represented by the vectors on the right side of the figure. The ADV relative velocity vectors are simply differences between the horizontal velocity of the profiler and the water 0.2 m below the profiler; thus, the ADV velocities will not always have same sign as the mean flow. If the profiler is advected with the mean flow with no slippage, then the relative velocity measurement is a measurement of vertical shear. If, however, the profiler does not obtain the same horizontal velocity as the mean flow, the relative velocity estimate from the ADV will be an underestimate of vertical shear.

3.3.2 The Resolution of Vertical Shear

The vertical gradients of density, phytoplankton, and relative velocity obtained with the free-fall profiler suggest that stratification and vertical shear may be the most important factors in the vertical organization

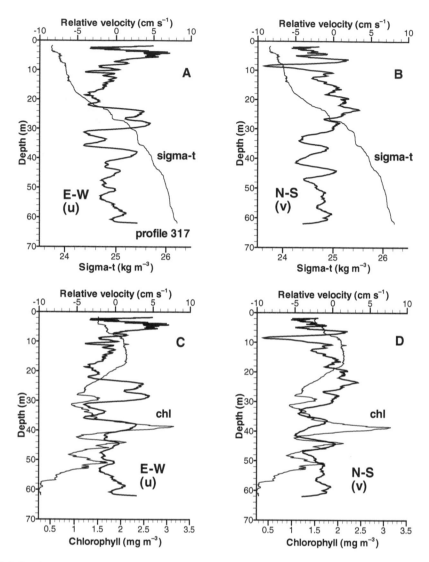

FIGURE 3.8 Vertical profile 317, September 20, 1998. (A) Sigma-t and the *u*-component of horizontal velocity, relative to the free-fall profiler. (B) Sigma-t and the *v*-component, (C) Chlorophyll and the *u*-component, (D) Chlorophyll and the *v*-component. Note the alignment of the density steps and chlorophyll gradients with the changes in relative horizontal velocity.

of the plankton biomass. The size of these vertical intervals also suggests that the vertical resolution of shear on scales of 1 to 2 m is essential for understanding the formation and persistence of planktonic structure. To what extent can shipboard or moored ADCP provide the velocity information required to understand small-scale vertical gradients in plankton distributions?

We can compare three vertical scales of velocity resolution over the Oregon continental shelf with data collected in September 1998 at the midshelf sampling location (see Figure 3.4). During our cruise, we obtained vertical profiles of velocity in 4-m depth bins from the shipboard ADCP (300 kHz RDI, narrow-band). At the midshelf location we also obtained vertical profiles of velocity in 2-m depth bins from a moored Acoustic Doppler Profiler (Sontek, 500 kHz) maintained by Dr. Michael Kosro (Oregon State University). Finally, we obtained over 100 high-resolution vertical profiles (with ADV data) within 10 km of the midshelf ADP mooring, with more than 50 profiles within 5 km of the mooring. From these three simultaneous measurements of vertical velocity gradients we can compare which vertical scale of resolution is appropriate for improving our understanding of small-scale planktonic structure.

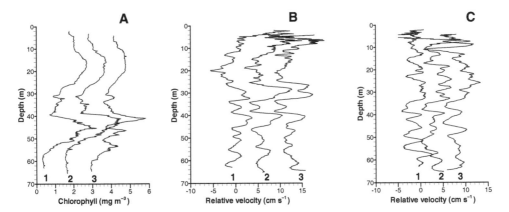

FIGURE 3.9 Three representative vertical profiles obtained within a 2-h interval on September 20, 1998, with profiles 307 (#1), 311 (#2), 317 (#3), obtained at 0604, 0647, 0808 hours GMT, respectively. Profiles are offset horizontally for plotting clarity. Profiles have been vertically aligned relative to the 25.52 sigma-t value at 40-m depth for profile 307. (A) Vertical chlorophyll distribution; (B) u-component of relative velocity obtained with the ADV; (C) v-component of relative velocity. Note the temporal coherence of the small-scale velocity at the depths of the steep gradients in chlorophyll concentration.

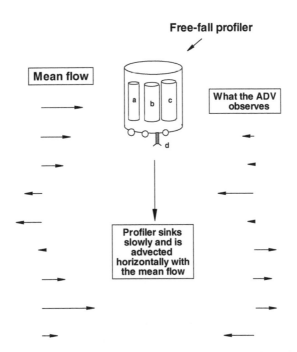

FIGURE 3.10 A schematic two-dimensional illustration of the relative velocities measured from a sinking profiler. The profiler can carry multiple instruments, such as (a) ac-9, (b) fluorometer, (c) CTD, (d) ADV. All flow-through sensor intakes are located at the leading edge of the profiler (open circles). We measure velocity 18 cm below the sinking profiler with the ADV, a velocity that is relative to the sinking profiler (the right side of this figure). The mean flow (on the left side of this figure), whatever its vertical scale, advects the profiler horizontally as it sinks. The relative velocity measurements obtained from the ADV show that significant relative velocity gradients occur over vertical scales of 0.5 to 1.0 m, but we cannot estimate the true velocity shear without knowledge of the mean flow over vertical scales of 1.0 to 2.0 m. Nor can we infer the mean flow from the relative velocity measured with the ADV, as there exist several different vertical velocity gradients in the mean flow that could produce the relative velocity gradient measured by the ADV.

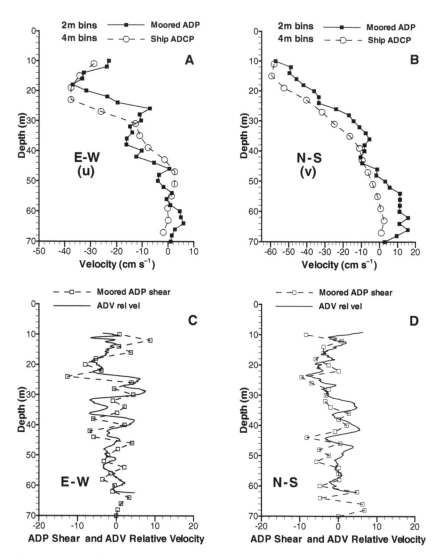

FIGURE 3.11 Comparison of three scales of resolution of vertical velocity structure at the midshelf station off the Oregon coast at 0800 September 20, 1998. The *RV Wecoma* was 4 km SW of the ADP moored in 80 m of water at the midshelf location. (A) The E–W (*u*) component of velocity obtained with the 300 kHz ADCP (open circles, dashed line) compared with the E–W component resolved with the moored 500 kHz ADP (squares, solid line). (B) The comparison of the N–S (*v*) component as measured by the two systems, as in A. (C) The shear estimated from the *u*-component from the moored ADP compared with the relative velocity obtained with the ADV on the free-fall profiler. (D) N–S component of ADV relative velocity compared with the shear obtained from the moored ADP vertical velocity gradients.

As one would expect, the moored ADP with 2-m depth resolution of velocity provided more vertical definition of the velocity field than did the shipboard ADCP with 4-m depth bins (Figure 3.11A, B). The 4-m binned velocity from the shipboard system showed the same general vertical pattern as the 2-m velocity data, but lacked the detail provided by the 2-m depth bins from the moored system. This representative profile was typical of the hundreds examined from our data set. During the same 15-min interval we obtained profile 317 with the free-fall profiling system (this profile is also highlighted in Figure 3.8). Assuming that the relative velocity measured by the ADV on the profiling system approximates a direct measurement of vertical shear (see Figure 3.10), then the vertical pattern of ADV relative velocity and the vertical shear obtained from the moored ADP should be similar. Comparison of those parameters (Figure 3.11C, D) reveals considerable similarity, with some indication that the 2-m velocity bins are missing some of the vertical structure in shear. These data provide confirmation that the profiler-based

ADV system can resolve sharp gradients in vertical shear, and indicates that the appropriate scale for resolution of vertical gradients in horizontal velocity is less than 2 m in extent, and may be as narrow as 1 m.

3.3.3   The Role of Vertical Mixing

An improved understanding of small-scale mixing processes (e.g., Denman and Gargett, 1983; Garrett, 1996; Thorpe, 1995), and their importance for biological growth in the euphotic zone (e.g., Cullen and Lewis, 1988), has led to an emphasis on the vertical processes that influence the vertical distributions of biological properties. A number of processes are involved in establishing the vertical distributions of properties in the mixed layer, including convective mixing, wind-driven overturns, linear and nonlinear internal waves, intrusions, horizontal advection, intermittent turbulent mixing, phytoplankton growth, and zooplankton grazing. Although microstructure observations of the mixed layer have demonstrated the intermittent nature of small-scale vertical mixing, the temporal spacing of intermittent mixing events can yield extended time intervals with little to no vertical mixing (Hebert and Moum, 1994). The interplay between stratification and vertical shear over the vertical gradients of small-scale phytoplankton layers will determine if the steep phytoplankton gradient persists or is mixed away.

3.4   Dye Injections: Analogues for Plankton Layers

Kullenberg (1974) conducted an innovative series of dye injections in coastal waters and lakes, and found that the dye dispersed horizontally into thin layers (usually <0.5 m in thickness) that retained their structure for 24 to 36 h. The implications of Kullenberg's results on the processes organizing the vertical distribution of phytoplankton have been overshadowed, in my opinion, by the emphasis on vertical mixing processes as the dominant physical mechanisms influencing plankton ecology in the euphotic zone. The vertical structure and temporal persistence of Kullenberg's dye layers strongly resemble the planktonic layers we have observed in East Sound and over the Oregon continental shelf, and suggest that fine-scale resolution of dye layers may reveal the key mechanisms involved in plankton layer formation and maintenance.

A number of open ocean dye studies have been conducted in recent years to quantify vertical and horizontal diffusivities (Ledwell et al., 1993, 1998). A coastal dye injection experiment over the Oregon continental shelf provided an opportunity to examine the vertical gradients of a recently injected volume of dye. In August 2001, A. Dale, J. Barth, and M. Levine injected fluorescein dye from the *R/V Elakha* onto a target isopycnal of 25.8 kg m^{-3} along the 100-m isobath over the continental shelf off Oregon. Within 1 h of the dye injection, a group of us aboard the *R/V Wecoma* began a series of CTD towyos through the dye patch to map the horizontal boundaries of the patch. The CTD was equipped with a fluorescein fluorometer. The CTD towyos were conducted during an 18-h interval of low winds (<10 knots) and calm sea state, resulting in little ship motion during the CTD profiles. The observed dye distributions had a striking similarity to the phytoplankton slabs and layers we have observed in East Sound and over the Oregon continental shelf (Figure 3.12), even after many hours within the tidally stirred waters of the continental shelf. The similarities in the shape of these dye features with our planktonic layers suggest that similar physical processes are operating in both cases. In particular, note the absence of the development of a smooth gradient between the background and the beginning of the dye layer. We were not using our free-fall system for these profiles, and therefore lack small-scale resolution of the vertical velocity gradient. The shipboard ADCP resolved vertical velocity structure within 8-m depth bins during this cruise, so the vertical resolution of shear was not sufficient for an estimation of the Richardson number over the vertical interval encompassed by the dye gradient. The persistent steep vertical gradients in fluorescein concentration imply a minimal impact of vertical mixing, while the thinning of dye concentrations into thin layers suggests an important role for shear dispersion (Young et al., 1982; Sundermeyer and Ledwell, 2001). A larger-scale dye injection experiment was conducted over the New England continental shelf by Sundermeyer and Ledwell (2001), who reported that shear dispersion alone could not account for all the horizontal dye dispersion over the shelf. Just as in the Oregon experiment, their vertical resolution of shear was based on 8-m depth bins, intervals

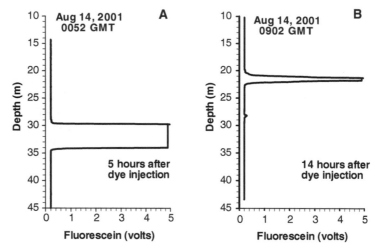

FIGURE 3.12 Examples of slab-like layer (A) and thin layer (B) of fluorescein dye at 5 and 14 h after injection on the 25.9 kg m⁻³ isopycnal, off the Oregon coast in August 2001. The high concentration of fluorescein in the thick layer (A) exceeded the 5 V maximum level of the fluorescein fluorometer. Note the similarities in the vertical gradients of fluorescein in both A and B to the phytoplankton vertical gradients observed in East Sound and off the Oregon coast (Figures 3.2 and 3.4). (Data courtesy of Andrew Dale, John Barth, and Murray Levine.)

too coarse for accurate estimates of small-scale shear. It is likely that we underestimate the potential for shear dispersion as a process if we resolve vertical velocity gradients only within 4- or 8-m depth intervals.

The absence of a gradual smoothing of the vertical gradient in dye concentrations over several hours reinforces the notion that horizontal processes dominate vertical processes in the formation and maintenance of planktonic small-scale structure. We conducted additional work on the vertical gradients of dye patches during the summer of 2003, using our free-fall profiling system.

3.5   Preliminary Estimates of the Horizontal Extent of Small-Scale Planktonic Layers

Although there is now little doubt about the presence of steep phytoplankton gradients (exemplified by "thin layers"), the horizontal extent of persistent small-scale planktonic features is still unknown. In our various time series of repeated vertical profiles, we usually permit the vessel to drift with the surface current as we conduct our profiling operations. As noted in Figure 3.11, we often experience a 20 to 30 cm s⁻¹ velocity gradient between the surface drift and the depth of a thin layer. If the layer remains distinct for 6 h, at least 5 to 8 km of the layer will have passed beneath the vessel during that time interval given a 20 to 30 cm s⁻¹ velocity difference. Our time series over the Oregon continental shelf suggest that layers typically have horizontal dimensions of at least 5 to 10 km. The aspect ratio of horizontal:vertical dimension of these persistent small-scale features thus exceeds 1000:1, and may exceed 10,000:1. Direct estimates of layer extent will be made in 2003 and 2004 with an instrumented Autonomous Underwater Vehicle (AUV).

3.6   Trophic Implications of Sharp Vertical Gradients

As mentioned in the introduction, patches of phytoplankton have long been proposed as locations for focused grazing by planktonic and nektonic consumers. Observations of persistent phytoplankton layers suggest that earlier hypotheses about stochastic patchiness (e.g., Fasham, 1978) should be reframed in terms of persistent structure within the ecosystem. It is our approach to sampling small-scale pattern that will determine if the observations resemble temporally and spatially distinct patches, or if the observations indicate a temporally and spatially coherent pattern.

The temporal and/or spatial coherence of that pattern will determine the advantage to an individual grazer of feeding within such a layer of enhanced food concentration, where the incremental increase in food consumption will depend on the shape of the nonlinear relationship between grazing rate and food concentration. As discussed by Cowles et al. (1998), the "bonus" in grazing and growth obtained by a population of grazers within a depth interval that contains one or more layers of enhanced food resources depends on the degree of aggregation by the grazer population within the layer (or layers). A uniform vertical distribution of grazers in the presence of layers will not result in a significant grazing or growth bonus, on a population basis, if the population grazes randomly within the larger depth interval. If, however, the grazer population aggregates on the layered distribution of enhanced food resources, then the population has the potential for higher grazing, growth, and reproductive rates, than would otherwise be estimated based on a mean food concentration within the entire depth interval. This estimate is dependent on the percentage of total grazing time spent within the interval with higher food concentrations. Although it is known that zooplankton can alter their swimming behavior to remain within higher food concentrations (e.g., Tiselius, 1992; Tiselius et al., 1994), it is not known to what extent grazers remain within persistent layers in the upper water column.

Layers of enhanced phytoplankton may not always be beneficial, as harmful algal species may establish toxic levels within thin layers (Nielsen et al., 1990; Bjørnsen and Nielsen, 1991; Rines et al., 2002), or layers could contain food items of low food quality and therefore become layers with lower than expected grazing activity. The fundamental trophic implication of sharp vertical gradients in resource availability is the extent to which grazers can exploit enhanced concentrations over timescales that provide significant physiological benefit to the organism. If "patches" are not ephemeral, but represent persistent layers of enhanced food resources (or unfavorable layers of harmful algae), then plankton production assessments must incorporate small-scale pattern. The trophic implications of persistent small-scale pattern in resource availability may require adjustments in sampling approaches as well as in modeling of upper ocean processes.

3.7   Small-Scale Planktonic Structure: Questions to Address

The presence of persistent small-scale planktonic structure poses a range of challenging questions for plankton ecologists as well as physical oceanographers. In addition to the questions listed in the introduction, we must ask:

To what extent is productivity enhanced within discrete layers?

To what extent is grazing localized within layers and what physiological or reproductive benefit do grazers obtain via aggregation within layers?

What fraction of the phytoplankton biomass is concentrated into fine-scale features?

What is the horizontal extent of persistent plankton layers?

How is a persistent plankton layer strained, temporally and spatially, by the internal wave field?

How long are the time intervals between intermittent vertical mixing events and what is the spatial extent of those mixing events?

Each of these questions requires field observations of small-scale structure and, in some cases, experiments on phytoplankton growth and zooplankton grazing within small-scale features. For example, there are enormous technical constraints in obtaining large water samples within narrow vertical intervals to use in experimental studies, and creative approaches are needed to address these technical problems.

3.8   Recommendations for Future Work

Small-scale planktonic gradients were difficult to detect until recent advances in *in situ* instrumentation provided higher data rates and new approaches to instrument deployment removed ship motion as a contaminating factor in the vertical resolution of properties. The use of free-fall instrument packages has permitted slow descent rates (10 to 15 cm s⁻¹) without the smearing that can occur with conventional CTD deployments on a taut cable, resulting in centimeter-scale resolution of hydrographic and bio-optical properties. Repeated deployments (several profiles per hour) permit the detection of high-frequency internal waves and changes in stratification over short timescales, processes that may have significant impact on vertical distributions of phytoplankton. Autonomous, moored profiling systems are under development and should allow the acquisition of long time series of high-resolution profiles at selected locations.

Advances in the resolution of horizontal processes and distributions are likely to emerge from the development of new AUVs that can carry CTDs, ADCPs, and bio-optical instruments, and may be programmed to track density intervals. These vehicles will provide essential information about the horizontal extent of layers, and permit evaluation of the processes influencing the spatial scales of coherence of fine-scale vertical patterns.

Controlled experiments in layer formation and maintenance can be conducted using dye injection studies. These experiments will permit the evaluation of shear dispersion as a primary mechanism for the formation and maintenance of small-scale planktonic vertical structure. Future work needs to focus on the temporal changes in the vertical gradient in dye concentration as well as the small-scale velocity shear over that vertical gradient, in the context of internal waves, strain, and intermittent vertical mixing.

All of the above approaches will benefit from improvements in the vertical resolution of horizontal currents using shipboard ADCP. As indicated by the data presented in this chapter, it is essential to close the gap between the vertical resolution of phytoplankton concentration (centimeters) and the vertical depth bins of the shipboard ADCP (4 m, 8 m).

Finally, our ability to evaluate the trophic consequences of fine-scale biological structure will depend on our ability to assess phytoplankton and zooplankton processes within small-scale features. Creative instrument development is needed to sample narrow vertical intervals (0.50 m or so) that can move up and down a few meters every 10 min through the action of the internal wave field, particularly if many liters of water must be collected from within a thin layer.

3.9   Conclusion

Any prospect for developing a predictive framework for small-scale biological structure in the upper ocean requires a mechanistic understanding of the interactions between the physical and biological processes at work on these critical spatial scales. Small-scale planktonic layers occur frequently, often associated with steep density gradients and steep gradients in vertical shear. The data suggest that vertical shear, in the presence of stratification that balances the magnitude of the shear, is a key process in the formation and maintenance of layers. Future work must expand the capabilities of shipboard ADCP to resolve velocity gradients at the same time and over the same vertical scale as the hydrographic and bio-optical data. Finally, future work must evaluate the extent to which small-scale planktonic layers (phytoplankton and zooplankton) must be accurately described if we are to assess, correctly, the magnitude of biological rates within the upper ocean.

Acknowledgments

I am indebted to Russell Desiderio, Nathan Potter, and Christopher Wingard for their technical support over the past several years of work on small-scale processes. Several former as well as current graduate students have contributed significantly to the collection and thoughtful discussion of the data summarized here — thanks to Dr. Susanne Neuer, Dr. Andrew Barnard, Dr. Lisa Eisner, Malinda Sutor, Cidney Howard, and Amanda Briggs. Dr. Mary-Elena Carr made valuable contributions to the early stages of this work during her postdoctoral work at Oregon State University. Dr. Mike Kosro has provided ADCP analysis and insights, and has generously shared his extensive data set from the ADP moored on the Oregon continental shelf. The work discussed in this chapter has been supported by the Office of Naval Research and the National Science Foundation. I am particularly grateful to Dr. James Eckman and Dr. Ronald Tipper of the Office of Naval Research for their efforts in support of research on this fascinating interface between biology and physics.

References

Abbott, M.R. and Letelier, R.M. Decorrelation scales of chlorophyll as observed from bio-optical drifters in the California Current. *Deep-Sea Res. II*, 45, 1639, 1998.

Alldredge, A.L., Cowles, T.J., MacIntyre, S., Rines, J.E.B., Donaghay, P.L., Greenlaw, C.F., Holliday, D.V., Dekshenieks, M.M., Sullivan, J.M., and Zaneveld, J.R.V. Occurrence and mechanisms of formation of a dramatic thin layer of marine snow in a shallow Pacific fjord. *Mar. Ecol. Prog. Ser.*, 233, 1, 2002.

Bainbridge, R. Underwater observations on the swimming of marine zooplankton. *J. Mar. Biol. Assoc. U.K.*, 31, 107, 1952.

Bainbridge, R. Studies on the interrelationships of zooplankton and phytoplankton. *J. Mar. Biol. Assoc. U.K.*, 32, 385, 1953.

Bennett, A.F. and Denman, K.L. Phytoplankton patchiness: inferences from particle statistics. *J. Mar. Res.*, 43, 307, 1985.

Bjørnsen, P.K. and Nielsen, T.G. Decimeter scale heterogeneity in the plankton during a pycnocline bloom of *Gyrodinium aureolum*. *Mar. Ecol. Prog. Ser.*, 73, 263, 1991.

Butler, N.M., Suttle, C.A., and Neill, W.E. Discrimination by freshwater zooplankton between single algal cells differing in nutritional status. *Oecologica*, 78, 368, 1989.

Cassie, R.M. Microdistribution of plankton. *Oceanogr. Mar. Biol. Annu. Rev.*, 1, 223, 1963.

Caswell, H. and Cohen, J.E. Communities in patchy environments: a model of disturbance, competition, and heterogeneity, in *Environmental Heterogeneity*, J. Kolasa and S.T.A. Pickett, Eds., Springer-Verlag, New York, 1991, 97.

Cowles, T.J. and Strickler, J.R. Characterization of feeding activity patterns in the planktonic copepod, *Centropages typicus* Kroyer, under various food conditions. *Limnol. Oceanogr.*, 28, 106, 1983.

Cowles, T.J., Olson, R., and Chisholm, S.W. Food selection by copepods: discrimination on the basis of food quality. *Mar. Biol.*, 100, 41, 1988.

Cowles, T.J., Desiderio, R.A., Moum, J.N., Myrick, M., Garvis, D., and Angel, S. Fluorescence microstructure using a laser/fiber optic profiler. *Proc. Soc. Photo-Opt. Instrum. Eng. Ocean Opt.*, 10, 336, 1990.

Cowles, T.J., Desiderio, R.A., and Neuer, S. *In situ* characterization of phytoplankton from vertical profiles of fluorescence emission spectra. *Mar. Biol.*, 115, 217, 1993.

Cowles, T.J., Desiderio, R.A., and Carr, M.-E. Small-scale planktonic structure: persistence and trophic consequences. *Oceanography*, 11, 4, 1998.

Cullen, J.J. and Lewis, M.R. The kinetics of algal photoadaptation in the context of vertical mixing. *J. Plankton Res.*, 10, 1039, 1988.

Davis, C.S., Flierl, G.R., Wiebe, P.H., and Franks, P.J. Micropatchiness, turbulence and recruitment in plankton. *J. Mar. Res.*, 49, 109, 1991.

Dekshenieks, M.M., Donaghay, P.L., Sullivan, J.M., Rines, J.E.B., Osborn, T.R., and Twardowski, M.S. Temporal and spatial occurrence of thin phytoplankton layers in relation to physical processes. *Mar. Ecol. Prog. Ser.*, 223, 61, 2001.

Denman, K.L. Scale-determining biological–physical interactions in oceanic food webs, in *Aquatic Ecology*, P.S. Giller, A.G. Hildrew, and D.G. Raffaelli, Eds., Blackwell, London, 1994, 377.

Denman, K.L. and Abbott, M.R. Time scales of pattern evolution from cross-spectrum analysis of advanced very high resolution radiometer and coastal zone ocean color scanner imagery. *J. Geophys. Res.*, 99, 7433, 1994.

Denman, K.L. and Gargett, A. Time and space scales of vertical mixing and advection of phytoplankton in the upper ocean. *Limnol. Oceanogr.*, 28, 801, 1983.

Denman, K.L. and Platt, T. The variance spectrum of chlorophyll and temperature in a turbulent ocean. *J. Mar. Res.*, 34, 593, 1976.

Donaghay, P.L., Rines, H.M., and Sieburth, J.McN. Simultaneous sampling of fine scale biological, chemical, and physical structure in stratified waters. *Arch. Hydrobiol. Beih.*, 36, 97, 1992.

Eckart, C. An analysis of the stirring and mixing processes in incompressible fluids. *J. Mar. Res.*, 7, 265, 1948.

Fasham, M.J.R. The statistical and mathematical analysis of plankton patchiness. *Oceanogr. Mar. Biol. Annu. Rev.*, 16, 43, 1978.

Gardner, R.H. and O'Neill, R.V. Pattern, process and predictability: the use of neutral models for landscale analysis, in *Quantitative Methods in Landscape Ecology*, M.G. Turner and R.H. Gardner, Eds., Springer-Verlag, New York, 1991, 289.

Garrett, C. Processes in the surface mixed layer of the ocean. *Dyn. Atmos. Ocean*, 23, 19, 1996.

Grünbaum, D. Predicting availability to consumers of spatially and temporally variable resources. *Hydrobiologia*, 480, 175, 2002.

Hanson, A.K., Jr. and Donaghay, P.L. Micro- to fine-scale chemical gradients and layers in vertically stratified coastal waters. *Oceanography*, 11, 10, 1998.

Hardy, A.C. and Gunther, E.R. The plankton of the South Georgia whaling grounds and adjacent waters, 1926–27. *Discovery Rep.*, 11, 1, 1935.

Haury, L., McGowan, J., and Wiebe, P. Patterns and processes in the time–space scales of plankton distributions, in *Spatial Patterns in Plankton Communities*, J.H. Steele, Ed., Plenum Press, New York, 1978, 277.

Hebert, D. and Moum, J.N. Decay of a near-inertial wave. *J. Phys. Oceanogr.*, 24, 2334, 1994.

Holliday, D.V., Pieper, R.E., Greenlaw, C.F., and Dawson, J.K. Acoustical sensing of small scale vertical structure in zooplankton assemblages. *Oceanography*, 11, 18, 1998.

Holliday, D.V., Donaghay, P.L., Greenlaw, C.F., McGehee, D.E., McManus, M.M., Sullivan, J.M., and Miksis, J.L. Advances in defining fine- and micro-scale pattern in marine plankton. *Aquat. Living Resour.*, 16(3), in press.

Kierstead, H. and Slobodkin, L.B. The size of water masses containing plankton blooms. *J. Mar. Res.*, 12, 141, 1953.

Koehl, M.A.R. and Strickler, J.R. Copepod feeding currents: food capture at low Reynolds number. *Limnol. Oceanogr.*, 26, 1062, 1981.

Kullenberg, G. Investigation of small-scale vertical mixing in relation to the temperature structure in stably stratified waters. *Adv. Geophys.*, HA, 339, 1974.

Lasker, R. Field criteria for survival of anchovy larvae: the relation between inshore chlorophyll maximum layers and successful first feeding. *Fish. Bull.*, 73, 453, 1975.

Ledwell, J.R., Watson, A.J., and Law, C.S. Evidence for slow mixing across the pycnocline from open-ocean tracer release experiments. *Nature*, 364, 701, 1993.

Ledwell, J.R., Watson, A.J., and Law, C.S. Mixing of a tracer in the pycnocline. *J. Geophys. Res.*, 103, 21, 499, 1998.

Letelier, R.M., Abbott, M.A., and Karl, D.M. Chlorophyll natural fluorescence response to upwelling events in the Southern Ocean. *Geophys. Res. Lett.*, 24, 409, 1997.

Levin, S. The problem of pattern and scale in ecology. *Ecology*, 73, 1943, 1992.

Mitchell, J.G. and Fuhrman, J.A. Centimeter scale vertical heterogeneity in bacteria and chlorophyll *a*. *Mar. Ecol. Prog. Ser.*, 54, 141, 1989.

Mullin, M.M. and Brooks, E.R. Some consequences of distributional heterogeneity of phytoplankton and zooplankton. *Limnol. Oceanogr.*, 21, 784, 1976.

Nielsen, T.G., Kiørboe, T., and Bjørnsen, P.K. The effect of a *Chrysochromulina polylepsis* subsurface bloom on the planktonic community. *Mar. Ecol. Prog. Ser.*, 62, 21, 1990.

Okubo, A. Some remarks on the importance of the "shear effect" on horizontal diffusion. *J. Oceanogr. Soc. Jpn.*, 24, 60, 1968.

Okubo, A. *Diffusion and Ecological Problems: Mathematical Models*. Springer-Verlag, New York, 1980.

Osborn, T. Finestructure, microstructure, and thin layers. *Oceanography*, 11, 36, 1998.

Owen, R.W. Microscale and finescale variations of small plankton in coastal and pelagic environments. *J. Mar. Res.*, 47, 197, 1989.

Platt, T. and Sathyendranath, S. Scale, pattern, and process in marine ecosystems, in *Aquatic Ecology*, P.S. Giller, A.G. Hildrew, and D.G. Raffaelli, Eds., Blackwell, London, 1994, 593.

Poulet, S.A. and Ouellet, G. The role of amino acids in the chemosensory swarming and feeding of marine copepods. *J. Plankton Res.*, 4, 341, 1982.

Price, H.J., Paffenhöfer, G.-A., and Strickler, J.R. Modes of cell capture in calanoid copepods. *Limnol. Oceanogr.*, 28, 116, 1983.

Riley, G. A model of plankton patchiness. *Limnol. Oceanogr.*, 21, 873, 1976.

Rines, J.E.B., Donaghay, P.L., Dekshenieks, M.M., Sullivan, J.M., and Twardowski, M.S. Thin layers and camouflage: hidden *Pseudo-nitzschia* spp. (Bacillariophyceae) populations in a fjord in the San Juan Islands, Washington, USA. *Mar. Ecol. Prog. Ser.,* 225, 123, 2002.

Strickler, J.R. and Bal, A.K. Setae of the first antennae of the copepod *Cyclops scutifer* (Sars): their structure and importance. *Proc. Natl. Acad. Sci. U.S.A.,* 70, 2656, 1973.

Sundermeyer, M.A. and Ledwell, J.R. Lateral dispersion over the continental shelf: analysis of dye release experiments. *J. Geophys. Res.,* 106, 9603, 2001.

Thorpe, S.A. Dynamical processes at the sea surface. *Prog. Oceanogr.,* 35, 315, 1995.

Tiselius, P. Behavior of *Acartia tonsa* in patchy food environments. *Limnol. Oceanogr.,* 37, 1640, 1992.

Tiselius, P., Nielsen, G., and Nielsen, T.G. Microscale patchiness of plankton within a sharp pycnocline. *J. Plankton Res.,* 16, 543, 1994.

Verity, P. Feeding in planktonic protozoans: evidence for non-random acquisition of prey. *J. Protozool.,* 38, 69, 1991.

Woods, J.D. Wave induced shear instability in the summer thermocline. *J. Fluid Mech.,* 32, 791, 1968.

Yen, J. and Nicoll, N.T. Setal array on the first antennae of a carnivorous marine copepod, *Euchaeta norvegiva. J. Crust. Biol.,* 10, 218, 1990.

Yen, J. and Strickler, J.R. Advertisement and concealment in the plankton: what makes a copepod hydrodynamically conspicuous? *Invert. Biol.,* 3, 191, 1996.

Young, W.R., Rhines, P.B., and Garrett, C.J.R. Shear-flow dispersion, internal waves, and horizontal mixing in the ocean. *J. Phys. Oceanogr.,* 12, 515, 1982.

4

Scales of Biological–Physical Coupling in the Equatorial Pacific

Peter G. Strutton and Francisco P. Chavez

CONTENTS

4.1   Introduction 51
4.2   Oceanographic Setting 52
4.3   Data..53
4.4   Analysis 54
    4.4.1   Cross-Correlation Analysis 54
    4.4.2   Decorrelation Scale Analysis..55
    4.4.3   Spectral Analysis...55
4.5   Results and Discussion...55
    4.5.1   Cross-Correlation Analysis 56
    4.5.2   Decorrelation Timescales 57
    4.5.3   Spectral Analysis...59
4.6   Conclusions 61
Acknowledgments..62
References...62

4.1   Introduction

It has become abundantly clear that understanding the controls on pelagic communities and their chemical environment cannot be accomplished without a strong foundation in ocean physics, hence the frequent use of the term *biological–physical coupling*. The physical processes of interest include upwelling, wind mixing, internal waves, and interannual climate dynamics, and are variable over a broad range of temporal and spatial scales (Harris, 1980). No single observational platform can provide the necessary data for quantifying the effect of these physical perturbations on chemical and biological dynamics. Ships provide the opportunity to conduct detailed process studies, but the spatial and temporal coverage of measurements is sparse. Satellites offer exceptional spatial and temporal coverage, but cannot quantify vertical variability in the water column beyond approximately 10 m. Moorings and drifters give excellent temporal coverage, but are limited to an Eulerian or Lagrangian spatial context, respectively. The challenge is to combine the various strengths and weaknesses of these techniques to create a view of the process of interest that is significantly clearer than the sum of the individual pieces.

In this chapter, physical measurements from the moorings that make up the Tropical Atmosphere Ocean (TAO) array are combined with surface chlorophyll data from the SeaWiFS ocean color satellite to quantify the temporal scales of variability for biological–physical coupling in the equatorial Pacific. Our purpose is to show how this combination of data — surface and subsurface, physical and biological, satellite and mooring — together with cross correlation, decorrelation, and spectral analysis can elucidate processes that each could not quantify in isolation.

0-8493-1344-9/04/$0.00+$1.50

The major physical processes of interest are the following:

- Local wind-induced mixing — important for transporting nutrients into the euphotic zone. May vary on timescales of days.

- Remote and local wind forcing of the thermocline. Remote processes that could modulate thermocline depth include Kelvin and Rossby waves, while local forcing can take the form of variability in the strength of upwelling-favorable winds.

- Tropical instability waves (TIWs): Large-scale waves generated by the shear between opposing east–west currents (the South Equatorial Current and the North Equatorial Counter Current). These are most prominent north of the equator from approximately 100°W to 150°W in the latter half of the calendar year.

The following questions are addressed in this chapter:

1. How are the physical processes such as thermocline depth, sea surface temperature, wind speed, and direction correlated with phytoplankton biomass, and does the nature of this correlation vary with location across the basin? *Hypothesis*: Thermocline variability, zonal winds, and possibly wind strength (vertical mixing) are strongly correlated with surface productivity, but meridional winds are not.

2. At what temporal scales do the physical processes most significantly impact surface productivity? *Hypothesis*: The correlation between physics and biology will change as a function of the temporal lag (physics leading biology). The timescale of maximum correlation varies spatially and between parameters, i.e., thermocline variability exerts an impact on biological productivity on a timescale different from that of wind mixing.

3. What are the decorrelation timescales of the biological and physical processes, and how do they vary in space? *Hypothesis*: Decorrelation timescales change significantly for both physical and biological parameters as a function of distance from the upwelling tongue.

4. At the timescales most relevant for the phytoplankton response to physical perturbations — 5 to 30 days — what scales of temporal variability dominate, as quantified by spectral analysis? What does this tell us about the relative importance of biological and physical processes in determining the temporal variability of productivity at a given location? *Hypothesis*: The relative importance of biological vs. physical processes in determining temporal variability of surface productivity changes with location.

4.2 Oceanographic Setting

The equatorial Pacific is a significant component of global biogeochemical cycles. Strong, persistent trade winds blow from east to west, leading to divergence at the equator. The resulting upwelling tongue stretches from the coast of South America to the International Date Line and supplies high levels of nitrate but relatively little iron to the surface waters, thus giving rise to high nutrient–low chlorophyll (HNLC) conditions (Coale et al., 1996). This lack of productivity, relative to the strong upwelling of inorganic carbon, makes the region the largest oceanic source of CO_2 to the atmosphere, at the rate of approximately 0.5 to 1.1×10^{15} gC per year, equivalent to 10% of the anthropogenic signal, or 50% of U.S. fossil fuel emissions.

The upwelling tongue and its nutrient field are perturbed by El Niño–La Niña cycles, resulting in significant fluctuations in the productivity and CO_2 dynamics of the region at interannual timescales (Chavez et al., 1999; Feely et al., 1999; Strutton and Chavez, 2000). The 1997–98 El Niño has frequently been described as the strongest ever observed. Chavez et al. (1998) showed how downwelling Kelvin waves, observed during the onset of the event, decreased primary productivity through local depressions of the thermocline. During the mature phase of the event (late 1997/early 1998) the trade winds and

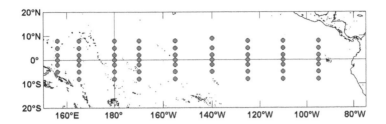

FIGURE 4.1 Map of the equatorial Pacific showing the locations of the TAO moorings from which data were obtained for this analysis.

equatorial upwelling were completely shut down, resulting in the lowest chlorophyll concentrations ever observed in the region (Chavez et al., 1999; Strutton and Chavez, 2000). When upwelling returned in April–May 1998, a period of particularly intense TIWs distorted the upwelling tongue into a wavelike structure (wavelength ~ 1000 km), and generated frontal regions of intense phytoplankton biomass (Strutton et al., 2001). These examples illustrate some of the most prominent biological–physical coupling mechanisms that influence the region.

4.3   Data

The TAO array provides a wealth of physical oceanographic and meteorological data for the equatorial Pacific (McPhaden et al., 1998). The moorings that make up the array are instrumented to measure surface air temperature, relative humidity, wind speed and direction, as well as ocean temperature at discrete depths from the surface to 500 m. Here these physical data are combined with SeaWiFS chlorophyll data to understand how the physical environment regulates phytoplankton productivity. All 60 TAO mooring locations from 8°N to 8°S, 156°E (near Papua New Guinea) to 95°W (near the Galapagos Islands) were considered. Figure 4.1 shows the location of these buoys in the context of equatorial Pacific geography. For each mooring location, the following parameters were obtained from the Pacific Marine Environmental Laboratory Web site (www.pmel.noaa.gov):

- Depth of the 20°C isotherm (m), denoted $Z_{20°C}$. A proxy for the depth of the thermocline, determined by linear interpolation of the mooring temperature profile. The term *thermocline* is used interchangeably here with $Z_{20°C}$. Thermocline depth can also be estimated using sea surface height (SSH) data from the TOPEX satellite (Wilson and Adamec, 2001; Ryan et al., 2002). Here we use $Z_{20°C}$ because the reliability of the correlation between SSH and thermocline depth decreases with distance from the equator, and because the temporal resolution of TOPEX is only 10 days (cf. daily for the data used here).
- Sea surface temperature, denoted SST (°C), obtained from a sensor at approximately 1.5 m on the buoy.
- The east–west or *zonal* component of the winds, denoted u-wind (m s^{-1}). Negative values indicate winds blowing toward the west, typical of the trades.
- The north–south or *meridional* component of the winds, denoted v-wind (m s^{-1}). Positive values indicate winds blowing toward the north.
- Total wind speed, also referred to here as the *scalar* wind (m s^{-1}). This parameter does not have a directional component. Surface wind stress data from the QuikSCAT satellite sensor could be used to quantify surface winds at greater spatial resolution than the TAO array measurements. In a relatively homogeneous environment like the equatorial Pacific, the spatial resolution of the TAO buoys is not a serious handicap to discerning basin-scale processes, but future analyses could make use of satellite data.

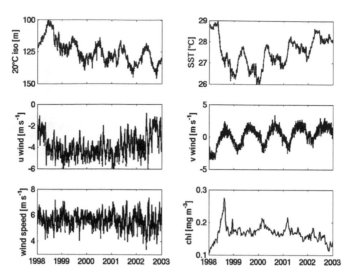

FIGURE 4.2 Time series of the physical parameters obtained from the TAO moorings and SeaWiFS chlorophyll for the region bounded by 156°E to 95°W and 8°S to 8°N, from 01 January 1998 to 31 December 2002.

SeaWiFS daily level 3, version 4 chlorophyll data were obtained from the Distributed Active Archive Center (daac.gsfc.nasa.gov) as standard mapped image (SMI) files with 9-km pixel resolution. For each TAO mooring location (Figure 4.1), time series of chlorophyll concentration were constructed by averaging all valid 9-km SeaWiFS pixels within a 1° longitude × 1° latitude box around the mooring location each day. SeaWiFS is a polar orbiting satellite that crosses the equator at local noon, but does not view the entire equatorial region in one day. Therefore, when constructing a time series of daily values, even without considering clouds, it is not possible to achieve 100% data coverage. For the time period of interest here, the mean percentage of valid daily SeaWiFS data for all mooring locations was 47% (range 26 to 82% depending on location), or one data point every 2.13 days on average. To obtain continuous time series for spectral analysis calculation of correlations, the SeaWiFS chlorophyll data were linearly interpolated to one data point per day.

The data assembled for this chapter span the period from 01 January 1998 to 31 December 2002. Figure 4.2 shows the time series of the basinwide means (8°N to 8°S, 156°E to 95°W) for the physical parameters listed above, plus SeaWiFS chlorophyll. The impact of the 1997–98 El Niño/La Niña event is clearly visible in all parameters during 1998, with the possible exception of meridional winds and wind speed. Likewise, elevated temperatures, weakened (less negative) trade winds and lower chlorophyll are also visible in 2002, indicative of a weak to moderate El Niño event. To focus on a period of relatively constant upwelling, the analysis was restricted to the time interval 01 January 1999 to 31 December 2001.

4.4　Analysis

4.4.1　Cross-Correlation Analysis

Cross-correlation analysis was performed to quantify the extent to which phytoplankton productivity (as quantified by chlorophyll concentration) is influenced by thermocline depth ($Z_{20°C}$), SST, wind speed, and wind direction. Time lags of 0, 5, 10, 20, and 30 days were considered — chlorophyll lagging physical parameters. Careful selection of data was necessary because of occasional gaps in the TAO array data due to instrument failure and vandalism. These gaps are sporadic and usually last for days to months, as opposed to the gaps in SeaWiFS data, which occur every second or third day (see above). Sections of missing TAO data were ignored, but missing SeaWiFS chlorophyll data were filled by linear interpolation.

4.4.2 Decorrelation Scale Analysis

The decorrelation scale for a given time series is defined as the timescale at which the autocorrelation function of the parameter falls below a predetermined level (e.g., the 95% confidence level or zero crossing). The calculation of decorrelation timescales has been previously applied to concurrent time series of temperature and chlorophyll (Denman and Abbott, 1994; Abbott and Letelier, 1998) to determine similarities and differences between the processes regulating physical and biological variability. Specifically, Abbott and Letelier (1998) used bio-optical drifters deployed in the California Current to show that decorrelation timescales for temperature and chlorophyll generally increased with distance from shore. It was suggested that this was due to reduced mesoscale variability farther offshore. Furthermore, the decorrelation scales for temperature and chlorophyll were very similar inshore, but diverged offshore. This suggests that physical and biological temporal variability is controlled by similar processes nearshore (intense mesoscale variability) but by differing processes offshore.

The analysis of Abbott and Letelier (1998) was performed on time series consisting of one data point per day, and the average time series length was 69 days. To be consistent with this previous study, and since the time periods of interest here are up to 30 days, the compiled TAO/SeaWiFS data files were broken into 60-day segments. For segments with more than 90% good TAO data, the missing 10% or less was filled by linear interpolation. Segments with less than 90% good data were disregarded. The seven buoy locations for which fewer than ten such good data blocks could be found were excluded from the analysis. As for the cross-correlation analysis, the SeaWiFS data were interpolated to 1-day intervals. A linear trend was removed from each 60-day segment (consistent with Abbott and Letelier, 1998), and the autocorrelation function was calculated in the standard way (Chatfield, 1989). The decorrelation scale was then calculated as the timescale for which the autocorrelation value fell below the 95% confidence interval (critical value = 0.361, $n = 30$, two-tailed; see Zar, 1999, table B17).

4.4.3 Spectral Analysis

For spectral analysis, the same 60-day data blocks selected for the decorrelation analysis were used. A linear trend was removed from each 60-day segment, and a power spectrum was calculated from each individual block using the Matlab® Finite Fast Fourier Transform function, *fft.m*. The slope of each power spectrum, in log–log space, was calculated over the range of periods from 5 to 30 days, since these intraseasonal timescales are the main focus of this work. For each buoy location, the mean slope of all valid power spectra was then calculated.

4.5 Results and Discussion

The three different analysis tools used here help to address different aspects of biological–physical coupling in the equatorial Pacific. Cross-correlation analysis quantifies the impact of physical perturbations on biological processes. It allows us to ask such questions as: What is the time lag between a change in thermocline depth or wind speed and the resulting phytoplankton response, and in what direction is this response? Decorrelation timescales are defined as the time taken for a given parameter to become decorrelated with itself. Comparing the decorrelation timescales of physical and biological parameters can show whether or not they are controlled by similar processes. Finally, spectral analysis quantifies the magnitude of variability for a given parameter across a range of timescales, in this case 5 to 30 days. By calculating the slope of the power spectrum over this range it is possible to say whether the parameter in question is dominated by variability at timescales of days or weeks. The results of each of these analyses are presented here, together with some interpretation regarding what these results mean in the context of equatorial Pacific phytoplankton dynamics. These analyses answer some important questions but also reveal interesting new research questions.

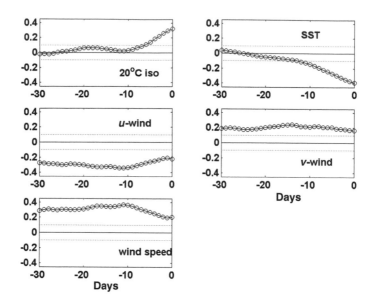

FIGURE 4.3 Example of cross-correlation plots from 2°S 125°W. A lag of −10 on the x axis indicates the physical parameter leading surface chlorophyll by 10 days. The parameters represented are the depth of the 20°C isotherm ($Z_{20°C}$ or thermocline depth, m), sea surface temperature (SST, °C), the zonal or east–west wind vector (u-wind, m s^{-1}), the meridional or north–south wind vector (v-wind, m s^{-1}), and wind speed or scalar wind (m s^{-1}).

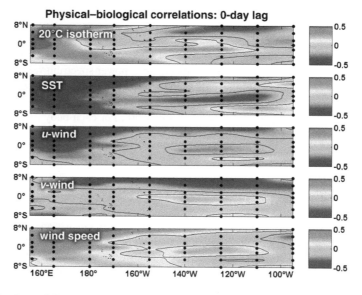

FIGURE 4.4 (Color figure follows p. 332.) Maps of the equatorial Pacific showing the correlation between SeaWiFS daily chlorophyll and mooring-derived physical parameters for a lag of 0 days. The contours indicate the 95% significance level ($r = \pm0.1$) for the correlation coefficient. The parameters represented are the depth of the 20°C isotherm ($Z_{20°C}$ or thermocline depth, m), sea surface temperature (SST, °C), the zonal or east–west wind vector (u-wind, m s^{-1}), the meridional or north–south wind vector (v-wind, m s^{-1}), and wind speed or scalar wind (m s^{-1}).

4.5.1   Cross-Correlation Analysis

The results of the cross-correlation analysis are shown in Figure 4.3 and Figure 4.4. In Figure 4.3 the full range of time lags is shown for one location (2°S 125°W) as an example, and Figure 4.4 shows the time-zero correlations for the entire basin. Although the number of valid data points varied between moorings, the sample size was always greater than 300, for which the critical value for the correlation coefficient (95% level of significance, two-tailed) is approximately ±0.1 (Zar, 1999). This critical value is plotted as horizontal lines in Figure 4.3, and as a contour in each of the maps in Figure 4.4. The time axis in Figure 4.3 ranges from –30 to 0, indicating the physical processes *leading* surface chlorophyll concentration by 30 to 0 days.

At zero lag, the generally negative correlation for chlorophyll with SST and zonal winds (*u*-wind) indicates that, for this location, higher chlorophyll (interpreted here as a proxy for increased productivity) is associated with cool sea surface temperatures and strong trade winds. The generally positive correlation of chlorophyll with $Z_{20°C}$ (thermocline depth) and wind speed suggests that higher chlorophyll is associated with a deeper thermocline and the increased upwelling and/or increased vertical mixing associated with stronger winds. The positive correlation with *v*-wind (meridional or north–south winds) implies that trade winds with a more northerly component favor enhanced productivity. Using this information we can begin to interpret the basinwide spatial variability of the correlation coefficients shown in Figure 4.4.

For a time lag of zero between physical processes and chlorophyll (Figure 4.4), there are large areas of the equatorial Pacific for which increased chlorophyll is correlated with a deeper thermocline — mostly north and south of the equator in the west and along the equator in the central region. This seems counter to the idea that a shallower thermocline brings nutrients closer to the surface, thus fueling increased productivity. This phenomenon has been described for thermocline variability induced by Kelvin waves (Chavez et al., 1998; Dunne et al., 2000), tropical instability waves (Foley et al., 1997; Strutton et al., 2001), and the El Niño/La Niña cycle (Chavez et al., 1999; Strutton and Chavez, 2000; Wilson and Adamec, 2001; Ryan et al., 2002). A possible explanation is that this correlation is part of the seasonal signal; Figure 4.2 shows that the thermocline is deepest in the boreal winter, whereas chlorophyll tends to peak at the beginning of each calendar year, particularly in 2000 and 2001. However, the time series in Figure 4.2 is a basinwide mean that does not represent the spatial variability across the basin very well.

The maps in Figure 4.4 show that the regions of high chlorophyll/deep thermocline correspond with the regions where strong winds and cool SSTs are also correlated with high chlorophyll. The likely explanation for these observations is that strong winds enhance vertical mixing of nutrients, and thus increase chlorophyll, while deepening the mixed layer and decreasing SST by surface cooling and upward mixing of deep water. Therefore, the zero-lag correlation between deep thermocline and chlorophyll occurs because the wind field drives both of these processes. It should not be interpreted as a deeper thermocline favoring increased productivity. It would be interesting to determine the relative contribution of surface cooling and deep mixing to the observed SST decrease, as this would also quantify the importance of this process for nutrient fluxes and phytoplankton growth. Such a question could be answered using emerging continuous nutrient measurement systems (Sakamoto et al., 2002) to correlate changes in nitrate flux with wind mixing events.

In the central equatorial Pacific, with increasing lag between physics and biology (Figure 4.5 shows 10 days as an example), the correlation between thermocline depth/SST and chlorophyll decreases to essentially zero. However, in this same region the relationships between chlorophyll and the wind field remain strong with increasing lag. Figure 4.3 shows that the strongest correlation between *u*-wind/wind speed and chlorophyll occurs at 10 days. That is, enhanced upwelling-favorable winds lead to increased chlorophyll in this region with lags ~10 days. This is a realistic timescale for the phytoplankton to respond to increased upwelling of nutrients; Chavez et al. (1999) observed a lag of approximately 3 weeks for a 20-fold change in chlorophyll associated with the 1998 equatorial Pacific bloom. These results are interesting in that they show a biological response to enhanced upwelling-favorable winds with no concomitant relationship between a shallower thermocline and enhanced productivity.

4.5.2 Decorrelation Timescales

The decorrelation results further emphasize the importance of atmospheric processes in regulating oceanic physics and biological variability, particularly away from the equatorial upwelling tongue. Figure 4.6 shows the decorrelation scales for the mooring at 0° 155°W as an example. Using the first crossing of the 95% confidence interval as the criterion, the decorrelation scale for $Z_{20°C}$ and SST is 3 days, and for the wind parameters it is 1 to 1.5 days. The chlorophyll decorrelation scale is 2 days,

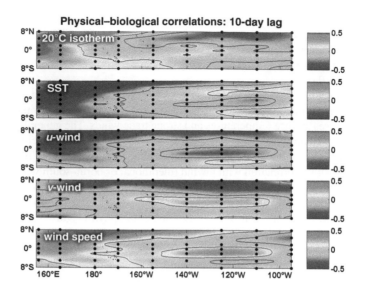

FIGURE 4.5 (Color figure follows p. 332.) Maps of the equatorial Pacific showing the correlation between SeaWiFS daily chlorophyll and mooring-derived physical parameters for a lag of 10 days. The contours indicate the 95% significance level ($r = \pm 0.1$) for the correlation coefficient. The parameters represented are the depth of the 20°C isotherm ($Z_{20°C}$ or thermocline depth, m), sea surface temperature (SST, °C), the zonal or east–west wind vector (*u*-wind, m s^{-1}), the meridional or north–south wind vector (*v*-wind, m s^{-1}), and wind speed or scalar wind (m s^{-1}).

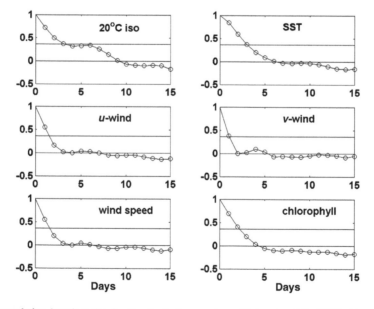

FIGURE 4.6 Decorrelation length scales for the physical parameters of interest and SeaWiFS chlorophyll; data for the TAO mooring at 0° 155°W as an example. The parameters represented are the depth of the 20°C isotherm ($Z_{20°C}$ or thermocline depth, m), sea surface temperature (SST, °C), the zonal or east–west wind vector (*u*-wind, m s^{-1}), the meridional or north–south wind vector (*v*-wind m s^{-1}), and wind speed or scalar wind (m s^{-1}).

and Figure 4.7 shows that this is a representative value for most of the central equatorial Pacific. This is less than the mean of 3.7 days calculated by Abbott and Letelier (1998) using drifters in a coastal environment, but Lagrangian drifters may experience less variability than an Eulerian platform, because drifters tend to track a parcel of water. Moorings on the other hand experience variability due to local biological and physical dynamics, as well as oceanographic features that are advected past a fixed location.

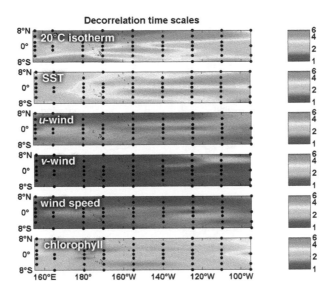

FIGURE 4.7 (Color figure follows p. 332.) Maps of the equatorial Pacific showing the decorrelation scales of mooring-derived physical parameters and SeaWiFS daily chlorophyll. The parameters represented are the depth of the 20°C isotherm ($Z_{20°C}$ or thermocline depth, m), sea surface temperature (SST, °C), the zonal or east–west wind vector (u-wind, m s^{-1}), the meridional or north–south wind vector (v-wind, m s^{-1}), wind speed or scalar wind (m s^{-1}), and chlorophyll (mg m^{-3}).

For the California Current, particularly nearshore, both Denman and Abbott (1994) and Abbott and Letelier (1998) observed equal decorrelation scales for temperature and chlorophyll, indicating that in a dynamic coastal environment phytoplankton behave as a passive tracer with spatial and temporal dynamics determined entirely by mesoscale circulation. Here we apply the same reasoning to determine the relative dominance of winds vs. oceanic physical processes in driving phytoplankton variability. That is, similarity between decorrelation scales for chlorophyll and any given physical parameter indicate similar forcing mechanisms.

The maps in Figure 4.7 show that the decorrelation scales for the wind parameters are between 1 and 2 days over the entire basin. The range for $Z_{20°C}$ and SST is 2 to 6 days, and the chlorophyll decorrelations fall between the atmospheric and oceanic data. Closer to the equator, the decorrelation scales for $Z_{20°C}$ and SST decrease (range 2 to 4 days) except for a region from about 125°W to 170°W where the $Z_{20°C}$ scale remains ~6 days. For chlorophyll, the decorrelation scale in the east is greater than for $Z_{20°C}$ and all three wind parameters, which suggests that in this region biology is not strongly influenced by any of these physical processes. Zonal currents and the Galapagos Islands plume, which is often characterized by elevated chlorophyll, may be more significant here. For vast areas of the equatorial Pacific — the central region from ~0° to 5°S and north of ~2°N across the entire basin — the chlorophyll decorrelation scale is greater than for the winds, but less than that of SST and $Z_{20°C}$. This suggests that in these areas equatorial Pacific phytoplankton are influenced by a combination of wind-induced mixing and thermocline variability. The potential importance of short-term wind variability deserves consideration, as most discussions of primary productivity north and south of the equator emphasize the dominance of processes occurring "upstream" at the equator, from where these waters are advected. The decorrelation timescales suggest that wind events may be important, and the lower panel in Figure 4.5 also shows a region to the north of the equator where productivity at a 10-day lag is correlated with wind stress but not with thermocline depth. The power spectral analysis results below add further weight to this idea.

4.5.3 Spectral Analysis

Power spectral slopes have been used since the initial days of spectral analysis in biological oceanography to discern the relative importance of physical and biological processes in determining phytoplankton distributions (Platt, 1972; Platt and Denman, 1975; Denman and Platt, 1976; Denman et al., 1977). These

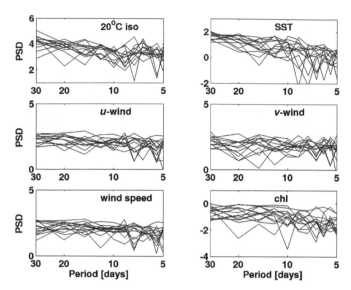

FIGURE 4.8 Example of power spectral density (PSD) as a function of period for 0° 155°W. For this location, 13 valid data segments of 60 days were found. The parameters represented are the depth of the 20°C isotherm ($Z_{20°C}$ or thermocline depth, m), sea surface temperature (SST, °C), the zonal or east–west wind vector (u-wind, m s^{-1}), the meridional or north–south wind vector (v-wind, m s^{-1}), and wind speed or scalar wind (m s^{-1}). The mean slopes of these 13 power spectra were $Z_{20°C} = -1.13$, SST $= -2.13$, u-wind $= -0.75$, v-wind $= -0.46$, wind speed $= -0.71$, and chlorophyll $= -1.21$.

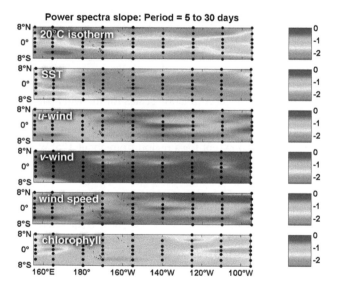

FIGURE 4.9 (Color figure follows p. 332.) Maps of the equatorial Pacific showing the mean slope of the power spectra for each buoy location for a period of 5 to 30 days. The color scale has been adjusted so that a steeper negative slope is red, indicating more variability at longer timescales. The parameters represented are the depth of the 20°C isotherm ($Z_{20°C}$ or thermocline depth, m), sea surface temperature (SST, °C), the zonal or east–west wind vector (u-wind, m s^{-1}), the meridional or north–south wind vector (v-wind, m s^{-1}), wind speed or scalar wind (m s^{-1}), and chlorophyll (mg m^{-3}).

early analyses specifically used the slopes to identify spatial scales at which physics would dominate biological processes in determining the spatial distribution of phytoplankton. Much like the comparison of decorrelation scales, where power spectra of biological and physical parameters exhibit similar shapes and/or slopes, it can be surmised that the processes controlling their dynamics are similar.

    Power spectra for all parameters (five physical variables plus chlorophyll) at all mooring locations were calculated using the procedure described in Section 4.4. The power spectra were then plotted in log-log space and slopes were calculated over the periods spanning 5 to 30 days. As described above this is the primary timescale of interest because it corresponds most closely with the response time of phytoplankton to physical perturbations. The log–log power spectra for $0°$ $155°W$ in the central Pacific are plotted as an example in Figure 4.8. The mean power spectra slope for each location across the basin is presented in Figure 4.9. A slope near zero, indicating a power spectrum that is essentially horizontal over the time periods of interest (blue in Figure 4.9), suggests that the parameter exhibits roughly equal variability over time periods of days to weeks. Increasingly, negative slopes (orange to red in Figure 4.9) indicate greater variability at timescales approaching 1 month. The atmospheric parameters (zonal, meridional, and scalar winds) have significantly flatter power spectra — a general characteristic of atmospheric vs. oceanic environments at these timescales. This is also consistent with the decorrelation scales presented in Figure 4.7. All three wind parameters do exhibit significant annual variability, as shown by Figure 4.2, but the data in Figure 4.8 and Figure 4.9 were deliberately restricted to intraseasonal timescales.

In contrast to zonal, meridional, and scalar winds, the power spectral slopes for thermocline depth, SST, and chlorophyll do exhibit some systematic spatial variability. Across the entire basin, the power spectra for thermocline depth become steeper (or "redder," i.e., more variability at longer timescales) approximately $2°$ latitude from the equator, both north and south. The gradient in SST power spectra slope is less pronounced, but shows consistently strong variability at timescales approaching 1 month in a band between $5°N$ and $5°S$. It is likely that the SST signal is due to TIWs, which generate considerable variability in SST in that same latitudinal band, at timescales of the order of 20 to 40 days. TIWs should also generate similar variability in thermocline depth, so the difference between the $Z_{20°C}$ and SST maps is curious. Near the equator, the $Z_{20°C}$ variability more closely resembles that of the wind fields. This was also visible in the decorrelation results (see upper panel of Figure 4.7), and suggests that thermocline variability at the equator is closely tied to the wind field, which drives upwelling. This upwelling is confined to a very narrow band, so farther north and south of the equator, the influence of winds rapidly diminishes, and thermocline variability appears more like that of SST.

Along the equator, the slopes of the power spectra for chlorophyll indicate that the dominant timescale of biological variability is greater than that of the winds and slightly longer than thermocline depth, but shorter than SST. This result is in contrast to the decorrelation scales, where the values for chlorophyll were intermediate between $Z_{20°C}$/SST and winds. This suggests that on the equator, at timescales less than 1 month, phytoplankton productivity is driven by a relatively complex interaction of thermocline depth and local winds. If the winds are driving thermocline depth, SST, and productivity at these timescales, then clearly the phytoplankton community is responding more slowly than the thermocline but more quickly than SST to the surface forcing.

North and south of the equator, the power spectra results are consistent with the decorrelation maps, in that the chlorophyll slopes are intermediate between the winds and thermocline/SST. This suggests that the biological processes are more strongly influenced by short-term wind events than has previously been thought. In the eastern Pacific, along the equator, a region of steeper power spectra is observed, corresponding to a similar feature in the decorrelation scales. The same mechanism could be responsible for this, i.e., influence of the Galapagos plume and zonal currents.

4.6 Conclusions

Previous work regarding physical controls of primary productivity in the equatorial Pacific has focused on signals such as Kelvin waves, TIWs, and El Niño cycles operating at temporal scales of weeks or greater. The data presented here show that variability in the wind field at timescales of days to weeks may significantly impact the phytoplankton community. For a narrow band near the equator, it seems that thermocline depth is indeed the dominant physical control. However, the decorrelation timescales suggest that even in this band of maximum upwelling, the wind field does influence biological processes at timescales shorter than that of thermocline variability. North and south of the equator, both the decorrelation timescales and the power spectra show that the temporal scales of biological variability are significantly influenced by winds. Transient wind events such as storm systems, with timescales of a week or less, may generate nutrient fluxes into the euphotic zone that stimulate surface productivity.

But if this is the mechanism driving phytoplankton dynamics, why is a corresponding signal not also seen in the SST field?

In contrast to coastal systems, where two previous studies (Denman and Abbott, 1994; Abbott and Letelier, 1998) observed phytoplankton dynamics that were dominated by circulation, the equatorial Pacific phytoplankton are not simply passive tracers in the oceanic flow. The combination of satellite and mooring data has enabled this closer look at shorter timescale variability across the whole basin, and reveals the potential importance of wind-driven processes. An interesting next step would be to compare the temporal dynamics of physics and biology at the equator with biological variability just off the equator at, say, 2° or 5° north and south, i.e., downstream of the upwelling plume. This would quantify the importance of off-equator short-term wind events in driving local productivity. The power spectral analysis could also be extended to search for significant peaks in the biological and physical data. By comparing the timescales at which these peaks occur (if they exist) it might be possible to map the distribution of characteristic timescales of variability in a more specific way than using power spectral slopes. Future analyses could also incorporate emerging satellite data products and, as longer data sets become available, compare the importance of the processes identified here during El Niño and La Niña events.

Acknowledgments

The mooring data used in this contribution were obtained from the NOAA/PMEL Tropical Atmosphere Ocean (TAO) Project, Mike McPhaden, Director. The SeaWiFS satellite chlorophyll data were obtained from the NASA Distributed Active Archive Center (daac.gsfc.nasa.gov). Comments by Zanna Chase, John Ryan, and an anonymous reviewer helped to improve this manuscript. This work was supported by the NASA SIMBIOS program, the NOAA Office of Global Programs, and the David and Lucile Packard Foundation.

References

Abbott, M.R. and Letelier, R.M., 1998. Decorrelation scales of chlorophyll as observed from bio-optical drifters in the California Current. *Deep-Sea Res. II*, 45, 1639–1667.

Chatfield, C., 1989. *The Analysis of Time Series: An Introduction.* Chapman & Hall, New York.

Chavez, F.P., Strutton, P.G., and McPhaden, M.J., 1998. Biological-physical coupling in the central equatorial Pacific during the onset of the 1997–98 El Niño. *Geophys. Res. Lett.*, 25, 3543–3546.

Chavez, F.P., Strutton, P.G., Friederich, G.E., Feely, R.A., Feldman, G.C., Foley, D.G., and McPhaden, M.J., 1999. Biological and chemical response of the equatorial Pacific Ocean to the 1997–98 El Niño. *Science*, 286, 2126–2131.

Coale, K.H., Johnson, K.S., Fitzwater, S.E., Gordon, R.M., Tanner, S., Chavez, F.P., Ferioli, L., Sakamoto, C., Rogers, P., Millero, F., Steinberg, P., Nightingale, P., Cooper, D., Cochlan, W., Landry, M., Constantinou, J., Rollwagen, G., Transvina, A., and Kudela, R., 1996. A massive phytoplankton bloom induced by an ecosystem-scale iron fertilization experiment in the equatorial Pacific Ocean. *Nature*, 383, 495–501.

Denman, K.L. and Abbott, M.R., 1994. Time scales of pattern evolution from cross-spectrum analysis of Advanced Very High Resolution Radiometer and Coastal Zone Color Scanner. *J. Geophys. Res.*, 99, 7433–7442.

Denman, K.L. and Platt, T.S., 1976. The variance spectrum of phytoplankton in a turbulent ocean. *J. Mar. Res.*, 34, 593–601.

Denman, K.L., Okubo, A., and Platt, T.S., 1977. The chlorophyll fluctuation spectrum in the sea. *Limnol. Oceanogr.*, 22, 1033–1038.

Dunne, J.P., Murray, J.W., Rodier, M., and Hansell, D.A., 2000. Export flux in the western and central equatorial Pacific: zonal and temporal variability. *Deep-Sea Res. I*, 47, 901–936.

Feely, R.A., Wanninkhof, R., Takahashi, T., and Tans, P., 1999. Influence of El Niño on the equatorial Pacific contribution to atmospheric CO_2 accumulation. *Nature*, 398, 597–601.

Foley, D.G., Dickey, T.D., McPhaden, M.J., Bidigare, R.R., Lewis, M.R., Barber, R.T., Lindley, S.T., Garside, C., Manov, D.V., and McNeil, J.D., 1997. Longwaves and primary productivity variations in the equatorial Pacific at 0°, 140°W. *Deep-Sea Res. II*, 44, 1801–1826.

Harris, G.P., 1980. The relationship between chlorophyll *a* fluorescence, diffuse attenuation changes and photosynthesis in natural phytoplankton populations. *J. Plankton Res.*, 2, 109–128.

McPhaden, M.J., Busalacchi, A.J., Cheney, R., Donguy, J.R., Gage, K.S., Halpern, D., Ji, M., Julian, P., Meyers, G., Mitchum, G.T., Niiler, P.P., Picaut, J., Reynolds, R.W., Smith, N., and Takeuchi, K., 1998. The tropical ocean global atmosphere observing system: a decade of progress. *J. Geophys. Res.*, 103, 14,169–14,240.

Platt, T.S., 1972. Local phytoplankton abundance and turbulence. *Deep Sea Res.*, 19, 183–187.

Platt, T.S. and Denman, K.L., 1975. Spectral analysis in ecology. *Annu. Rev. Ecol. Syst.*, 6, 189–210.

Ryan, J.P., Polito, P.S., Strutton, P.G., and Chavez, F.P., 2002. Unusual large-scale phytoplankton blooms in the equatorial Pacific. *Prog. Oceanogr.*, 55, 263–285.

Sakamoto, C.M., Karl, D.M., Jannasch, H.W., Bidigare, R.R., Letelier, R.M., and Johnson, K.S., 2002. Nitrate variability measured *in situ* on the HALE ALOHA open ocean mooring: influence of mesoscale eddy activity on phytoplankton community dynamics. *Eos Trans. AGU Fall Meet. Suppl.*, 83, Abstr. OS21B-0198.

Strutton, P.G. and Chavez, F.P., 2000. Primary productivity in the equatorial Pacific during the 1997–98 El Niño. *J. Geophys. Res.*, 105, 26,089–26,101.

Strutton, P.G., Ryan, J.P., and Chavez, F.P., 2001. Enhanced chlorophyll associated with tropical instability waves in the equatorial Pacific. *Geophys. Res. Lett.*, 28, 2005–2008.

Wilson, C. and Adamec, D., 2001. Correlations between surface chlorophyll and sea surface height in the tropical Pacific during the 1997–1999 El Niño–Southern Oscillation event. *J. Geophys. Res.*, 106, 31,175–31,188.

Zar, J.H., 1999. *Biostatistical Analysis*. Prentice-Hall, Upper Saddle River, NJ.

5

Acoustic Remote Sensing of Photosynthetic Activity in Seagrass Beds

Jean-Pierre Hermand

CONTENTS

5.1   Introduction 66
5.2   Influence of Photosynthesis on Acoustics 67
    5.2.1   Bubbles in Seawater..67
    5.2.2   *Posidonia* Photosynthetic Apparatus...69
    5.2.3   Oxygen Production..69
    5.2.4   Gas in Matte and Sediment 69
5.3   The USTICA 99 Experiment 69
    5.3.1   Test Site..69
    5.3.2   Experimental Configuration 71
    5.3.3   Acoustic Measurements 72
        5.3.3.1   Signal Transmission..72
        5.3.3.2   Ambient Noise Recording 72
        5.3.3.3   Transducer Calibration..72
        5.3.3.4   Equalized Matched-Filter Processing 73
    5.3.4   Oceanographic Measurements: CTD and Dissolved Oxygen Content....73
5.4   Multiscale Acoustic Effects..75
    5.4.1   Time-Varying Medium Impulse Response 75
    5.4.2   Propagation Channel Modeling 77
    5.4.3   Energy Time Distribution of Medium Response...................................80
    5.4.4   Non-Photosynthesis-Related Effects..81
        5.4.4.1   Tide...81
        5.4.4.2   Sea Surface Motion...82
        5.4.4.3   Water Temperature Profile..83
5.5   Effects of Photosynthesis on Sound Propagation..83
    5.5.1   Time Variation of Dissolved Oxygen 83
    5.5.2   Effect of Photosynthetic Bubbles on Multipaths.................................84
    5.5.3   Effect on Reverberation 88
    5.5.4   Effect on Ambient Noise 88
        5.5.4.1   Spectral Characteristics..88
        5.5.4.2   Time-Frequency Characteristics 90
        5.5.4.3   Directional Characteristics..91
        5.5.4.4   Other Observations..91
    5.5.5   Gaseous Interchange of the Leaf Blade 91
5.6   Conclusion...92
Acknowledgments...93
Appendix 5.A Comparison with Earlier Experiments 94
References..94

5.1   Introduction

To be able to prevent damage to marine and freshwater ecosystems, for example, to avert negative consequences for biodiversity, environmental surveillance and monitoring tools are required that produce data that are continuous in time and representative of extended areas of interest. In recent years, research on acoustic remote sensing of the ocean has evolved considerably, especially in studying physical and biological processes in shallow water environments.[1] Methods and systems have been developed that exploit, to different degrees, the complex nature of sound propagation to identify physical and biological markers (parameters) of the water column, its boundaries, and subbottom structures. Among these, sophisticated acoustic inversion techniques based on matched field and matched waveform processing have proved effective and reliable in determining range-average physical properties of the water column and upper sediments.[2–4]

This chapter focuses on the use of acoustics to remotely sense biological processes through an original case study: the photosynthesis by *Posidonia oceanica* (L.) Delile, an endemic marine phanerogam of the Mediterranean Sea. The organism settles most commonly on loose sediments but can develop on hard and rocky substrata and, when it encounters favorable conditions, colonizes vast areas of the sea bottom forming prairies, which extend from the surface to a depth of approximately 35 to 40 m. The prairies represent the most characteristic and, probably, most important ecosystem of the Mediterranean Sea covering an estimated surface area of 20,000 square miles. They are an important habitat for numerous fish species, marine animals, and other species of plants and algae. They create natural barriers that reduce coastal erosion. *Posidonia* is called the "green lung of the Mediterranean" for its important characteristic of producing large quantities of oxygen. Unfortunately, the plants are sensitive to environmental decay and have suffered marked regression over the last 40 years. The development of methods to assess their state of health efficiently is of considerable interest as traditional direct methods, e.g., underwater diving for inspection and sampling, and indirect methods, e.g., mechanical and high-frequency echographic soundings, require considerable time and/or costly equipment; see, e.g., References 5 through 9.

An exploratory study was started in 1995 to find ways of monitoring *in situ*, and *on the scale of a prairie*, the response of *Posidonia* plants to environmental conditions.[10] To this end, the effects of photosynthesis on long-range propagation of low frequency sound were investigated under controlled experimental conditions.[11,12] Transmission measurements in the frequency range 100 Hz to 1.6 kHz showed daily changes of frequency-dependent propagation characteristics including attenuation and dispersion (pulse spreading). The diurnal variations were attributed in part to undissolved gas present on the leaf blades during phases of photosynthesis cycle. A previously unsuspected phenomenon of waveguide propagation of sound in a bottom bubble layer was discovered, and it was shown that the phenomenon could be exploited to determine the oxygen void fraction in that layer. The proposed acoustic sampling is not invasive; i.e., it does not affect the metabolism of the plant and, in particular, the gaseous exchange with the ambient medium in ways that, for example, an incubator enclosing the plant does. The method preserves the natural life condition allowing us to obtain qualitative information about the plant response to environmental variables such as photosynthetically active radiation (light), temperature, stirring, and nutrients. Furthermore, the method alleviates the problem of space and time aliasing associated with traditional spot measurements. The acoustic propagation data "integrate" the acoustic effect of a great number of plants present along the source–receiver transect of arbitrary length within the prairie of interest. A static configuration of the transducers allows observation over long time periods (days to months) and at short time intervals (minutes to seconds), covering a great number of photosynthesis cycles.

In this chapter, we report and discuss results from a second experiment carried out under completely different conditions with respect to the first experiment in terms of the measurement geometry, acoustic transmission, and environment. The major differences were the much shorter length of acoustic path, broader frequency range of transmission, much lower oxygen productivity of the plants, lower plant density, lesser homogeneity of the prairie, and acoustically harder rocky substratum. The experiment was conducted in September 1999 over a small *Posidonia* bed off the island of Ustica. Time series of calibrated measurements of acoustic transmission were obtained during 4 days using broadband chirp signals emitted repeatedly from an underwater sound source and received on a pair of hydrophones, fixed near the sea bottom. Contemporaneous depth profiles and time series of dissolved oxygen and temperature in the water column were obtained with an oceanographic probe. Statistical analyses of the time-varying, medium impulse response allow to resolve marked changes in the propagation characteristics. Photosynthesis is seen to cause excess attenuation of multipaths, faster decay of reverberation, and lower level of ambient noise. There is a strong correlation with the release of oxygen in the water column measured with the dissolved oxygen probe. As for the first experiment, the diurnal variations are ascribed in part to undissolved gases present on the leaf blades and at the roots during phases of photosynthesis cycle. The *Posidonia* plants form a water layer where the gas void fraction varies with the time of day. The photosynthesis-driven, absorptive, scattering, and dispersive bubble layer, with a sound speed lower than bubble-free water, modifies the interaction process of waterborne acoustic energy with the substratum of volcanic basalt. Multipaths with intermediate grazing angles are shown to be the most sensitive to photosynthesis.

Section 5.2 briefly reviews the morphological features of *Posidonia* that are relevant to bubble acoustics. Section 5.3 describes the USTICA 99 experiment and data processing. In Section 5.4, the acoustical and environmental measurements are analyzed in detail. Section 5.5 focuses on the effects of photosynthesis on acoustic propagation including multipaths, reverberation, and ambient noise and provides an interpretation of the observed acoustic variations in terms of the gas transport in the seagrass. Section 5.6 concludes the chapter. In Appendix 5.A, the acoustic parameters and environmental conditions of the two experiments are compared.

In the confines of a single chapter we find it necessary to omit or pass quickly over certain notions of ocean acoustics and signal theory. Interested readers are referred to the referenced textbooks.

5.2 Influence of Photosynthesis on Acoustics

In coastal waters, the gas content in dissolved and bubble forms is determined by air–sea flux and specific environmental and biomass conditions including photosynthesis of aquatic plants, life processes of animals, and decomposition of organic materials.

5.2.1 Bubbles in Seawater

It is well established that the presence of gas bubbles in seawater influences sound propagation in a way that depends on the resonant frequency of the bubbles.[13,14]

In coastal waters, the presence of bubbles of many sizes, each with (sound) scattering and absorption cross sections,[*] cause frequency-dependent scatter, attenuation, and dispersion. Bubble radii are in the range 10 to 500 μm with a peak density typically somewhere in the range 10 to 15 μm. Recently published observations, using laser holography near the ocean surface, have shown that the densities of 10 to 15 μm radius bubbles can be as high as 10^6 m^{-3} μm^{-1} increment within 3 m of the surface of calm seas.[15] The density and distribution of bubble radius vary with depth, time of day, season, wind, and sea state (see, e.g., References 16 and 17). The bubble population is sensitive to the physical, chemical, and biological processes in the volume and on the seafloor, which are quite specific to each environment such as photosynthesis considered in this chapter.

Typically, bubbles form only a very small percentage, by volume, of the sea in which they occur. Nevertheless, because air, or more generally gas, has a markedly different density and compressibility than seawater, and because of the resonant characteristics of bubbles, the suspended gas content has a profound effect on underwater sound. At frequencies of resonance, gas bubbles pulsate radially in response to a signal frequency dependent on bubble radius. For a spherical air bubble in water a simplified[**] expression of the resonant (breathing) frequency is as follows:

---

[*] Ratio of the scattered power, referred to a unit distance, to the intensity incident on a unit area (or unit volume).
[**] i.e., no surface tension, adiabatic gas oscillation and no energy absorption.

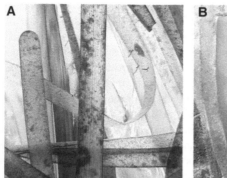 

FIGURE 5.1 *Posidonia oceanica* leaves. (A) Adult and intermediate leaves covered by epiphytes and encrustation. (B) Juvenile leaves and rhizome. (Underwater photographs taken during USTICA 99 experiments).

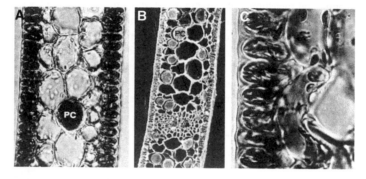

FIGURE 5.2 Photosynthesis apparatus of *P. oceanica*. Leaf blade cross sections. (A, B) Monolayered epidermidis and mesophyll with large cells and small intercellular spaces (×320 and ×200). PC: phenolic cell. (C) Detail of the porous region under the cuticle (×1100). (From P. Colombo, N. Rascio, and F. Cinelli, *Posidonia oceanica* (L.) Delile: a structural study of the photosynthetic apparatus, *Mar. Ecol.*, 4(2), 133–145, 1983. With permission.)

$$f_r \approx \frac{3.25 \times 10^6}{a}(1 + 0.1z)^{1/2} \qquad \text{for} \qquad ka \ll 1 \qquad (5.1)$$

where a is the bubble radius in μm, z is the depth in m, and k is the wavenumber.[*] For example, a bubble of radius 100 μm near the sea surface resonates at a frequency of $\approx$32.5 kHz. The extinction (scattering plus absorption) cross section has a maximum at the resonant frequency and falls off with frequency away from the resonance. Well below the resonance the cross section increases as f^4. Bubbles of near-resonant size extract a large amount of energy from the incident sound wave through scattering in all directions and conversion to heat. Also, in the vicinity of resonance, large changes in sound speed take place. Hence, over the range of resonance frequencies the medium is highly attenuative and dispersive.[18] By contrast, at high frequencies well beyond the resonant frequency of the smallest bubble present in the mixture, the effect of suspended gas content is negligible. At frequencies below resonance, the mixture of bubbles increases the compressibility of the water medium thereby reducing the sound speed below that obtained from pressure, temperature, and salinity measurements alone.[**] When gas is *dissolved*

---

[*] $k = 2\pi/\lambda$ [radians/m] where λ [m] is the wavelength. The wavenumber and angular frequency $\omega = 2\pi f$ [radians/s] are related through the equation $k = \omega/c$ where c [m/s] is the speed of sound.

[**] Sound speed is related to density and compressibility and, in the ocean, density is related to static pressure, salinity, and temperature. Sound speed is an increasing function of temperature, salinity, and pressure, with the latter a function of depth. It is customary to express sound speed c as an empirical function of three independent variables: temperature T in °C, salinity S in parts per thousand, and depth z in m. A simplified expression for this dependence is $c = 1449.2 + 4.6T - 0.055T^2 + 0.00029T^3 + (1.34 - 0.01T)(S - 35) + 0.016z$.[19]

in water the effect on sound speed is completely negligible, even when the water is completely saturated with gas.[20]

5.2.2 *Posidonia* Photosynthetic Apparatus

Posidonia oceanica (L.) Delile is an endemic phanerogam of the Mediterranean Sea. Its long ribbon-shaped leaves are grouped in shoots, which develop on various substrates in 1 to 50 m water depths (Figure 5.1). Leaf morphology allows for maximum release of photosynthetic oxygen to the ambient medium. The leaf blade consists of a monolayered epidermis and a three- to four-layered mesophyll[21] (Figure 5.2). The blade width and thickness are respectively ≈1 cm and ≈180 μm.

The major site of photosynthesis is the epidermis where chloroplasts are densely arranged in small radially elongated cells. The outer wall of epidermidal cells is formed by an outer continuous layer (cuticle) and an underlying much thicker (≈20 μm) porous region with irregularly shaped cavities. The lacunar system is constituted of connected air channels within the mesophyll. The particularly small dimensions of the lacunar system is a distinctive feature of *P. oceanica*.

Photosynthesis is the major driving force for exchange of gases among seawater, the epidermal cells, and the lacunar system. Respiratory activity is nearly an order of magnitude lower and largely involves the lacunar system. Unlike other aquatic plants, gaseous exchanges with seawater are effected by molecular diffusion as there are no stomata. The processes of oxygen uptake for respiration and release of photosynthetic oxygen are constrained by the diffusion boundary (unstirred) layer, and to a lesser extent, by the cuticle and cell wall.

5.2.3 Oxygen Production

Photosynthesis by seagrass substantially increases the quantity of oxygen in dissolved and bubble forms in the water column. For *Posidonia*, a productivity of 5 to 10 g of fixed carbon m^{-2} day^{-1} was reported.[22] Values of up to 14 l m^{-2} day^{-1} of produced oxygen have been reported for prairies of the Tuscan Arcipelago.[23] Specialized surveys showed that the time variation of oxygen concentration in the water column was determined principally by the daily cycle of oxygen productivity, with depth and seasonal dependence, including the possible occurrence of supersaturation conditions below sea surface.[11,24]

5.2.4 Gas in Matte and Sediment

The *Posidonia* matte is formed by the intertwining of various strata of rhyzomes, roots, and trapped sediments.[25] Typically, the sediments are made of poorly sorted sands, primarily organogenous.

The geoacoustic properties of the matte are virtually unknown. Attenuation is known to be high as acoustic energy of a boomer hardly penetrates the matte layer owing to scattering and absorption. Sound speed is expected to be low due to the uneven nature of water-saturated loose sediments and to the presence of slow materials (rhyzomes and roots) and gas from the decomposition of organic material. For signal frequencies below the bubble resonances the bulk material properties of the matte is expected to dominate its mechanical behavior, producing an acoustic response equivalent to a monophasic material of low sound speed. Comparable conditions are encountered with soft porous sediments with high gas content (see, e.g., Reference 26).

5.3 The USTICA 99 Experiment

5.3.1 Test Site

The experiment was conducted over a *Posidonia* bed off the island of Ustica in September 1999 (Figure 5.3A). The island lies in the southern Tyrrhenian Sea, off the northern coast of Sicily, at a distance of 65 km from Palermo (13°10′ E, 38°42′ N). It represents the relict of a vast submarine volcanic system of the Pleistocene age, which emerged 2000 m above the seabottom.[27,28] The island is characterized by

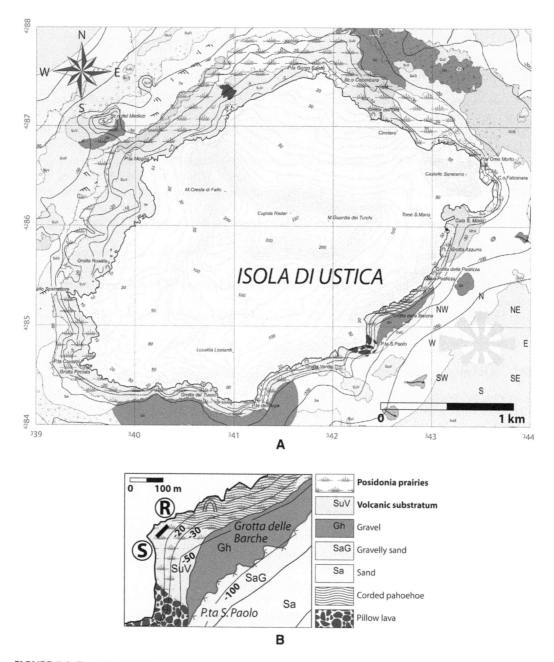

FIGURE 5.3 Test site. (A) Geological marine map of Ustica island showing the location of the investigated *Posidonia* bed. (B) Sediments, biocenosis, and stratigraphy at the test site. The thick black line shows the position of the acoustic transect. S: source. R: receivers. (Adapted from Reference 48.)

pillow-shaped outcrops of lava emerging from the sea surface. It has an area of 8 km², a coastal perimeter of 12 km, and a summit elevation of 248 m. The coast is irregular and fretted, forming little inlets like the one where the experiment was conducted (Figure 5.3B).

The island is surrounded by notably clear waters, which are subject to intense renewal, and seabeds abundant with marine flora and fauna in an ecosystem still practically intact and now protected.* The seabed is settled by benthic communities typical of hard substrata. Marine vegetation includes surface

_____
* Since 1986, a marine reserve has been established, covering an area of 3 miles from the coast.

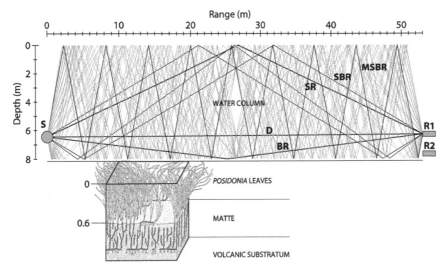

FIGURE 5.4 Experimental configuration for the acoustic remote sensing of undissolved oxygen produced by *Posidonia* photosynthesis. The positions of the underwater sound source (S) and hydrophones (R1, R2) are indicated. Eigenray diagram: The lines are the acoustic rays joining S and R1. Ray groups 1 through 10 are displayed. The black lines are the early arrivals of groups 1 and 2. The thin black line is one of the four paths belonging to group 10: nine reflections at each of the boundaries. S: surface; B: bottom; R: reflected; M: multiple. Horizontal scale 1:400. Vertical scale 1:200.

FIGURE 5.5 Acoustic instrumentation deployed on the seagrass bed. (A) Sound-source tower, rear view. (B) Two-hydrophone vertical pole, front view.

formations, hard calcareous algae, and various species of the seaweed *Cystoseira* distributed over the water depths 0 to 35 m. The most euphotic sandy and subhorizontal bottoms are carpeted by the seagrass *P. oceanica* (0 to 30 m). The deep rocky seabed, which is washed by intense currents, is capped with dense oceanic settlements of *Laminaria rodriguezi* (50 to 70 m). The marine fauna is very rich and can be deemed as representative of the Central Mediterranean basin, with a notable host of subtropical forms. The richness of the encrusting biocoenoses is the most noticeable feature of the island's seascape.

5.3.2 Experimental Configuration

A sound source (S) and a pair of receivers (R1, R2) were deployed on the seafloor (Figure 5.4). The positions were chosen to minimize bathymetric variations between S and R and the acoustical effects of nearby rock scatterers and coastal reflectors. The S–R horizontal distance was $R = 53$ m and the water depth, d, in the vertical section varied in the range 8 to 8.8 m.

The source was a broadband piezoelectric transducer mounted in a ballasted tower and positioned at a height $H_S = 1.55$ m above the seafloor (Figure 5.5). The monopole-like source has a frequency range of 200 Hz to 20 kHz and is omnidirectional up to 2 kHz. It was cable-connected through an impedance transformer to a power driving amplifier and signal generator in the laboratory, located on shore, several hundred meters uphill.

The receivers were two calibrated hydrophones mounted on a rigid pole and decoupled mechanically. A hydrophone (R1) was positioned within the *Posidonia* leaf layer and the other (R2) in the water layer, at respective heights of $H_{R1} = 0.3$ m and $H_{R2} = 1.7$ m above the seafloor. The hydrophone signals were amplified and bandpass-filtered with a high-pass RC filter and third-order Bessel filter $f_{-3dB} = 500$ Hz and a low-pass eight-order linear phase filter $f_{-3dB} = 16.7$ kHz. The signals, carried by analog symmetric lines, were recorded by a portable data acquisition unit.

The acoustic instrumentation was deployed by two divers with the support of a local fishing boat and a work boat.

5.3.3   Acoustic Measurements

5.3.3.1   *Signal Transmission* — The coded signal, $s(t)$, transmitted to measure the band-limited impulse response of the acoustic channel, $g(t)$, consisted of a low-power, long duration, linearly frequency modulated (LFM) waveform:

$$s(t) = \mathrm{Re}\left[\mathrm{rect}(t/\Delta t)\exp\left(j\pi\Delta f t^2/\Delta t\right)\exp\left(j2\pi f_0 t\right)\right] \tag{5.2}$$

where

$$f_0 = 8.1 \text{ kHz}, \Delta f = 15.8 \text{ kHz}, \text{ and } \Delta t = 15.8 \text{ s} \tag{5.3}$$

Re stands for real part, rect is the rectangular function, f_0 is the carrier frequency, Δf is the bandwidth, and Δt is the duration. The frequency range is from $f_1 = 200$ Hz to $f_2 = 16$ kHz. Pulse compression was achieved through the use of a correlation receiver or matched filter (MF) whose impulse response is the same as the waveform of the signal emitted by the source, reversed in time. The large time-bandwidth product, $\Delta t \Delta f \approx 2.5 \cdot 10^5$, permitted to resolve closely spaced multipath arrivals with a sufficient ratio of peak to ambient noise in spite of the limited power of the sound source, 180 dB μPa^{-1} re 1 m at resonance (940 Hz). The pulse repetition rate was fixed at 1 ppm to obtain sufficient statistics in sampling the physical and biological processes over the timescales of interest. About $3 \cdot 10^3$ probe signals were transmitted over a 4-day period.

The reader is referred to the original paper[29] for a conceptual description of the coded signal and its matched filter, to, e.g., References 30 through 32 for related theory of signal detection and estimation and optimum filtering, and to, e.g., References 33 and 34 for aspects of digital signal processing that are relevant to this chapter. Further details are found in References 4, 35, and 36 that deal specifically with the application of broadband, LFM-coded signals and MF receivers to inverse problems including the geoacoustic characterization of fine-grained sediments in shallow water.

5.3.3.2   *Ambient Noise Recording* — Physical and biological sounds were recorded during the "silent" intervals of the acoustic transmissions.

5.3.3.3   *Transducer Calibration* — The transducers and electronics of the S and R chains were calibrated *in situ* after the experiments. A pole-mounted hydrophone was repositioned on the source axis at a distance $R_0 = 1.93$ m and the probe signal was retransmitted.

Figure 5.6 shows that the transmitted waveform was perfectly reproducible, which was an important requirement for precise measurements of the forward acoustic propagation. The first bottom and surface bounces are recognized at time delays $\tau = 1$ ms and $\tau = 7$ ms.

The surface-reflected signal displayed variability due to surface motion. The $\approx$20-dB attenuation is somewhat larger than the spherical spreading loss calculated from the calibration geometry, i.e., $20\log(R/R_0) = 16.4$ dB where $R = 13.05$ m, due to the frequency-dependent source directivity and surface scattering loss. The bottom-reflected signal was strongly attenuated ($>>5.4$ dB geometrical loss) since the grazing angle $\theta = 58°$ was beyond the expected critical angle of the basalt interface

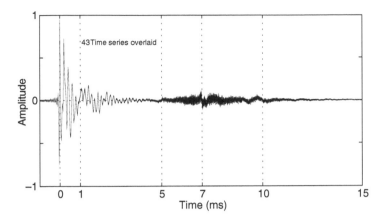

FIGURE 5.6 Matched, filtered pressure signal measured in front of the source. The overlaid, 43 signal realizations show the stability of the transmission. The hydrophone was placed at an axial distance $R_0 = 1.93$ m which largely satisfied the far field condition: $R_0 > \pi a^2/\lambda_2 = 0.46$ m, where $a = 0.12$ m is the radius of the circular piston source, $c = 1538.5$ m/s is the sound speed near the bottom and $\lambda_2 = c/f_2 = 9.6$ cm is the shortest transmitted wavelength. The dotted lines indicate the delays of the first bottom and surface echoes, calculated from the geometry.

and there was a two-way, excess attenuation due to photosynthesis in the intervening seagrass layer as discussed subsequently.

5.3.3.4 *Equalized Matched-Filter Processing* — The emitted pressure waveform (Figure 5.7A) was used to design the reference signal of an MF receiver that compensated for amplitude and phase distortion of the source.

1. Hanning windowing was applied to the MF waveform to reduce the local bottom and surface echoes.
2. The transmitting sensitivity response, measured on the radiation axis (Figure 5.7B) was equalized for flat spectrum. An inverse, finite impulse response (IFIR) filter was designed on the basis of a frequency-decimated version of the source spectrum magnitude.
3. The source waveform was convolved with the IFIR filter, which, in the frequency domain, is equivalent to multiplying by the IFIR squared magnitude with zero-phase distortion.
4. The resulting reference signal, time reversed, was convolved with the received signals.

In Figure 5.7C, raw and equalized MF (EMF) outputs are compared for one realization of the received signal. The first multipath arrivals, which were not identified in the MF output, were perfectly resolved in the EMF output, recovering the time resolution limit, $1/\Delta f = 63$ μs. The EMF output represents the convolution of the transmitted autocorrelation function (sinc function) with the actual impulse response of the medium. For the encountered conditions of limited source peak-power and high-level background noise, the achieved processing gain allowed estimation of (the coherent part of) the medium response as if the source transmitted an ideal high-energy pulse.

5.3.4 Oceanographic Measurements: CTD and Dissolved Oxygen Content

Physical and chemical conditions of seawater were monitored during the acoustic transmissions with a multiparameter, oceanographic probe Idromar IM51-201. Depth profiles and time series were alternated to obtain vertical and temporal sampling of the water column. For most time series, the probe sensors were positioned just above the *Posidonia* leaves. Figure 5.8 shows the depth profiles of temperature and oxygen concentration at different times of day. Figure 5.9 shows time series of temperature and oxygen concentration at the sea surface and seafloor. The conductivity (salinity) was nearly homogeneous over the whole water column with mild time variability, $S = 37.8 - 38$ ppt. The probe was deployed at a short

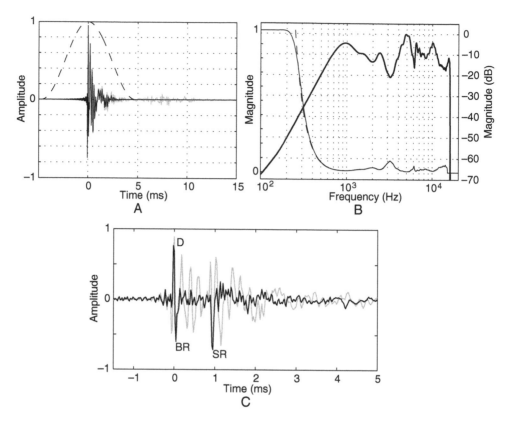

FIGURE 5.7 Equalized MF processing. (A) Raw MF transmitted signal (gray line) and Hanning (dashed) windowed version (black). (B) Spectrum magnitude in dB (thick, right scale); ideal (thin, left scale), and designed (dashed) inverse filter response. (C) Comparison of raw (gray line) and equalized (black) MF signal received on hydrophone R1. D, SR, and BR stand for direct, surface- and bottom-reflected paths, respectively.

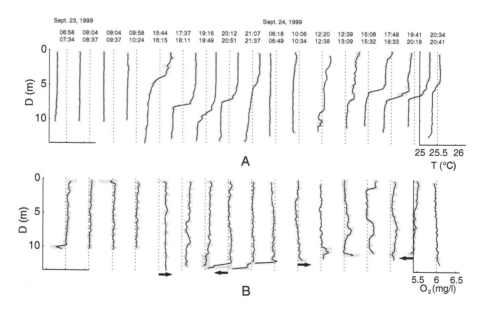

FIGURE 5.8 Depth profiles of (A) temperature and (B) oxygen concentration at different times of day. Gray circles: raw data; solid lines: smoothed data; dotted lines: references at $T = 25.5°C$ and $C_O = 6$ mg/l for visual appraisal of the depth and time variations. The arrows indicate the direction of oxygen variation near the bottom (time-series part of the profiles).

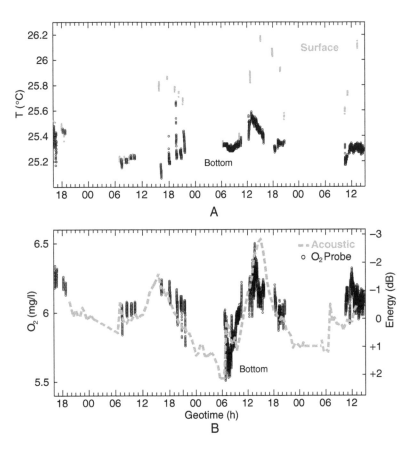

FIGURE 5.9 Time series of (A) temperature and (B) oxygen concentration. The small circles are the raw data. Gray circles: near the sea surface; dark circles: near the bottom. The surface and bottom data are in the depth ranges $z \leq 1$ m and $z \geq 8$ m, respectively. The dashed line in (B) replicates the acoustic result of Figure 5.18(A) for comparison (right scale).

distance off the transect to avoid acoustic reflections from the hull of the fishing boat, which explains the deeper depths in the displayed data.

5.4 Multiscale Acoustic Effects

In this section, the time history of acoustic transmission data is analyzed to assess their sensitivity to the time variations of all environmental parameters that are relevant to acoustic frequencies below 16 kHz.

5.4.1 Time-Varying Medium Impulse Response

Changes in the physical and biological properties of the very shallow water environment are conveniently related to the characteristics of the transmitted acoustic signals through the establishment of the time-varying impulse response of the medium.

Figure 5.10 shows the leading part of the medium response as a function of time of day[*] separated by 1-min intervals. The first 4 ms of the envelope[**] of the EMF output are displayed. The envelope representation is chosen here to highlight the temporal structure of the received energy.[***] The real-valued response was shown, for one realization, in Figure 5.7C. The leading part of the response shows

---

[*] Herewith and henceforth referred to as geotime.
[**] Magnitude of the analytical signal calculated by Hilbert transform.
[***] More precisely, the coherent component of the coded-signal energy transmitted through the medium.

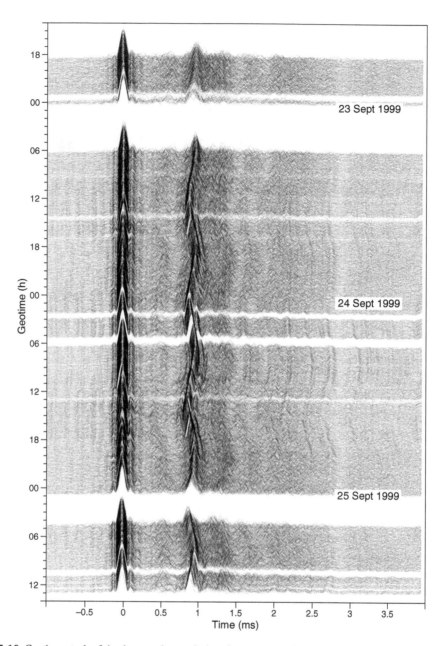

FIGURE 5.10 Geotime stack of 1-min-spaced acoustic-impulse response of the medium measured on hydrophone R1. The responses were aligned in time using the real-valued peak of the first arrival (D path) as reference. The envelopes are displayed. The white regions are missing data. Period: September 22, 1913 hours to September 25, 1301 hours, 1999.

distinct arrivals, which were fairly stable from one signal to the next. The remaining part of the response, which will be shown later, displayed much greater signal-to-signal variability and sensitivity to the time-varying environmental conditions.

The *static* properties of the acoustic propagation channel were determined by (1) the range and depths of the source and receivers, (2) the bathymetry, and (3) the acoustic properties of the basement. The *dynamic* properties were the function of (4) the tidal cycle, (5) the sea state and wind speed, and of the depth-dependent (6) temperature, (7) fish populations, and (8) gas bubbles in the water column. These processes operated on different timescales and caused both deterministic and stochastic fluctuations of the medium acoustic-impulse response.

5.4.2 Propagation Channel Modeling

For acoustic modeling purposes, the environment is described as an horizontally layered medium formed by the water layer, leaf layer (*Posidonia* plants), rhyzomes/roots and trapped sediment layer (matte), underlying sediment layers, and a semi-infinite half space (rock basement).[11]

There are different theoretical (and computational) ways of describing the propagation of sound in the ocean. Normal-mode theory[37] is particularly suited for shallow water but gives little insight, compared to ray theory, on the distribution in space and time of the energy radiating from the source. Like its analog in optics, ray acoustics provides a more intuitive description of the propagation in the form of a ray diagram.* In Figure 5.4 the acoustic rays joining the source S and receiver R1 (eigenrays) are traced for isospeed water and flat boundary conditions. Under these simplifying assumptions the rays follow straight lines in the water column and, at the boundaries, are specularly (mirror) reflected with the same angle relative to the vertical.

In the experimental area, the *Posidonia* plants grow directly on a rocky substratum. They form a real prairie, relatively dense and in good state of health. The average thickness of the leaf layer is 60 cm. In contrast to the previous experiment (see Appendix 5.A), there is not a real matte spreading. The matte is scarce and, when present, in small cavities, is thin and made of loose sediments with a predominance of bioclast. The organogenous debris come from erosion of the substratum and bioactivities. The matte-sediment composite layer is thin (a few centimeters) compared to most of the transmitted wavelengths and thus was considered acoustically transparent. The substratum is a volcanic basalt that supports both compressional and shear waves with speeds smaller than the typical values $c_p = 5250$ m/s and $c_s = 2500$ m/s,[38] due to the young age of formation and subsequent alteration of the material.

Let us first assume that the seabed vegetation is acoustically transparent. For an ideal hard bottom, i.e., a non-absorbing material with a sound speed larger than water, there is "total reflection" of the acoustic rays with grazing angles smaller than a critical value. For basalt, the apparent critical angle is approximately

$$\theta_c = \cos^{-1}\left(c_w/c_b\right) \approx 40° \tag{5.4}$$

where $c_w = 1538.5$ m/s is the sound speed of the bottom water, assumed bubble free, and $c_b = c_s = 2000$ m/s since in consolidated materials for which $c_s > c_w$ the shear speed takes on the role of compressional speed in unconsolidated sediments.[37] The shear wave provides an additional degree of freedom for the acoustic energy to penetrate into the bottom so that the interface appears "softer" than the equivalent liquid layer (compressional wave only).

The curves in Figure 5.11A show the complex-valued, elastic reflection coefficient R_C vs. grazing angle, in terms of the reflection loss and phase shift:

$$R_{dB} = -20\log\left|R_C(\theta)\right| \text{ dB and } \phi = \tan^{-1}\left[\text{Im}(R_C) / \text{Re}(R_C)\right] \tag{5.5}$$

calculated with a seismo-acoustic propagation model.[39] For a homogeneous bottom the reflection loss and phase shift curves are independent of frequency. Referring to Figure 5.4, energy that propagates near the horizontal within an aperture of $2\theta_c$ suffers only a moderate loss increasing with angle due to the lossy properties of the basalt. For an ideal, nonlossy material, there would be a perfect reflection of the rays with subcritical incidence, i.e., no bottom loss. Outside this aperture a substantial part of the energy is transmitted into the bottom at each bounce, which results in a strong decay with range of the reflected component (Figure 5.12). The bottom loss at the intermediate angles $40° < \theta < 70°$ is larger than at the steeper angles because of the elastic properties of the basalt. The small-scale roughness of the actual bottom, not accounted for in the predicted mirrorlike reflection loss of Figure 5.11A, causes an additional loss due to scattering, increasing with frequency and extending to lower and higher grazing angles.

Figure 5.12 shows the 4-day average of the medium impulse response measured between S and R1. The logarithm of the envelope is displayed because of the large dynamic range. The leading part of the response

---

*Ray-based models lack accuracy in predicting the low-frequency part of the propagation because of the inherent (high frequency) approximation.

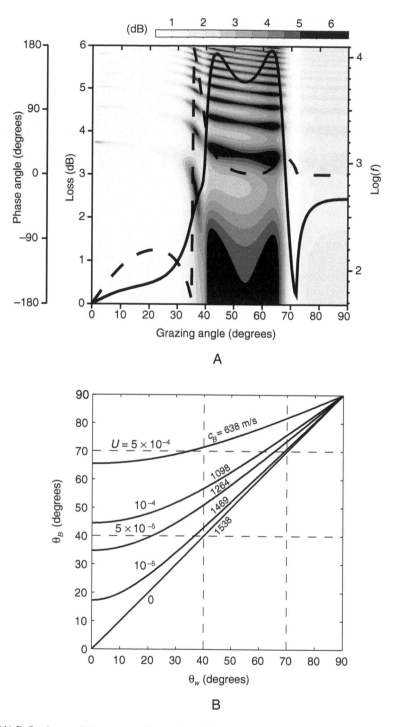

FIGURE 5.11 (A) Reflection coefficient vs. grazing angle and frequency. solid and dotted lines: Frequency-independent loss in dB and phase shift of a half-space basalt bottom (left scales). Plot: Frequency-dependent loss of a composite seagrass-basalt bottom during active photosynthesis (right and top scales). The basalt geoacoustic parameters used for the calculation are compressional speed $c_p = 5000$ m/s, shear speed $c_s = 2000$ m/s, compressional attenuation $\alpha_p = 0.2$ dB/λ, shear attenuation $\alpha_s = 0.5$ dB/λ and relative density $\rho_b/\rho_w = 2.2$. The water sound speed is $c_w = 1538.5$ m/s. The sound speed of the bubble seagrass layer is $c_B = 1264$ m/s corresponding to a gas void fraction $U = 5 \cdot 10^{-5}$. The bottom roughness and attenuation in the seagrass layer are not accounted for. (B) Modified grazing angle at the basalt bottom for different values of the gas void fraction U in the seagrass layer. Corresponding sound speeds c_B are indicated.

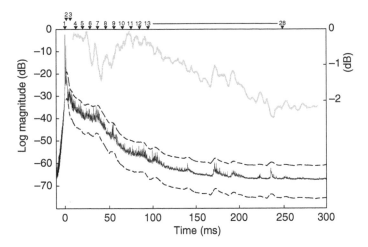

FIGURE 5.12 Four-day statistics of the medium impulse response measured on hydrophone R1. The black line is the log envelope, average response calculated from the median of the squared envelope of all measurements. The dashed lines are the 10th and 90th percentiles (smoothed). The gray line is the difference between the night and day averages. The arrows indicate, for each group, the mean travel time of the corresponding eigenrays predicted by the model.

is the first arrivals already seen in Figure 5.10. The remainder of the response is due to multiple reflection and scattering at the surface and bottom boundaries, which produce a long reverberation tail. The total duration of the response, defined by the time taken to return to the background noise level, is ≈300 ms.

Different groups of paths were identified from the average response and classified according to the number of boundary reflections:

- The first group (1) comprises the direct (D) and bottom-reflected (BR) waves, separated by 60 μs only (Figure 5.7C). Note that in the envelope representation of Figure 5.10 and Figure 5.12 these two interfering paths are not resolved. Their time-of-flight is ≈34 ms. The BR wave had a low attenuation, similar to the D wave, and a phase reversal of ≈180° due to bottom interaction at very low grazing angle: $\theta = 3.5° \ll \theta_c$.

- The second group (2) consists of the surface-reflected (SR) wave, with a pressure-release 180° phase shift (Figure 5.7C), and the associated combinations of bottom reflections (SBR) taking place near the source and receiver. The SBR waves with ray grazing angles $13° < \theta < 20° < \theta_c$ were attenuated and phase-shifted (Figure 5.11A). The arrival times relative to D were in the range 1 to 2.2 ms. These paths are consistently resolved in the geotime stacked plot of Figure 5.10.

- The following groups (3 to ∞) are multiple surface and bottom reflections (MSBR). These paths were strongly attenuated due to repeated bottom interaction at steeper angles $\theta > 28°$ (Figure 5.12). Figure 5.11A shows that the near-critical reflections (groups 3 to 5) experienced an angle-dependent phase shift and that the supercritical reflections were much more attenuated than the subcritical ones, except near normal incidence. The relative arrival times of the groups 3 to 10, whose rays are displayed in Figure 5.4, were in the range 4.7 to 57.5 ms.

Note that the indicated values of grazing angle and arrival time are not exact as they were determined from a range-independent model that did not account for the small changes of bathymetry along the transect.

The critical angle effect explains the waveguide nature of sound propagation in shallow water. The energy propagating within subcritical angles is referred to as the normal-mode field (or discrete spectrum) because the near-perfect reflectivity permits the existence of a set of discrete standing waves analogous to those of a vibrating string. Each mode corresponds to a pair of paths, which interfere constructively.[*]

---

[*] A mode can be conceived to correspond to a pair of plane waves incident on the boundaries at an angle θ and propagating in a zigzag fashion by successive reflections. For a pressure-release surface and a rigid bottom, the grazing angle θ and the mode number m are related by $\sin(\theta) = (m - 0.5)(\lambda/d)$.[40] The higher modes correspond to steeper angles.

The low-frequency cutoff of the waveguide formed by the water column bounded by the free surface and hard bottom is given by[37]

$$f_{0m} = \frac{(m - 0.5)c_w}{2d\left[1 - (c_w / c_b)^2\right]^{1/2}} \tag{5.6}$$

where $d = 8.5$ m is the average water depth and m is the mode number. At the frequency of 2 kHz the channel supported 15 discrete propagating modes. Only two modes were excited at the lower signal frequency $f_1 = 200$ Hz. The supercritical angle energy is referred to as the nearfield (or continuous spectrum) and is rapidly lost into the bottom. In contrast to the previous experiment (see Appendix 5.A) the discrete spectrum could be exploited here because of the much shorter range of transmission.

5.4.3   Energy Time Distribution of Medium Response

To quantify the acoustical effects of photosynthesis (and other environmental processes) the medium impulse response measurements, $g(t)$, were described by their energy distribution in time.

The D, BR, and SR arrivals were described by their *peak pressure-squared*. Figure 5.13A shows their relative energy vs. geotime. For D and BR, the rapid fluctuation was mostly caused by thermal micro-structure, turbulence and water currents, and interference effects between the two propagation paths. For SR, the fluctuation was also due to sea surface motion.

Since the (M)SBR arrival peaks were not well resolved on a signal-by-signal basis, it was necessary to resort to *time-integral-pressure-squared* calculation in the time window of interest, i.e., corresponding to a specific group of paths. These paths were subject to amplitude and phase fluctuations, which intensified with grazing angle and frequency because of the sea surface acting less and less as a perfect mirror. This resulted in a partial decorrelation of the coded signal used to estimate the medium impulse response. Hence, the individual paths of each group were not truly resolved and appeared as blobs of energy in the medium response. The late arrivals involving many bottom bounces (and weak echoes from distant reflectors) were masked by reverberation and ambient noise.[*] These were apparent only in the average responses.

By considering $|g(t)|^2$ as a density in time, the *fractional energy* in the time interval δt at time t is

$$e = \int_{\delta t} |g(t)|^2 \, dt \tag{5.7}$$

where $|g(t)|^2$ is the energy per unit time at time t.[41] By taking, without loss of generality, the *total energy*

$$E = \int |g(t)|^2 \, dt \tag{5.8}$$

equal to unity, the *mean time* can be defined in the usual way any average is defined:

$$\mu_t = \langle t \rangle = \int t |g(t)|^2 \, dt \tag{5.9}$$

and the *mean duration* is defined as two times the standard deviation, σ_t, given by

$$\sigma_t^2 = \int (t - \mu_t)^2 |g(t)|^2 \, dt = \langle t^2 \rangle - \langle t \rangle^2 \tag{5.10}$$

where $\langle t^2 \rangle$ is defined similarly to $\langle t \rangle$.

These descriptive statistics provide a gross characterization of where and how the received energy is spread in time. In our application the two moments were used as a robust measure of the cumulative attenuation encountered by the bottom-interacted paths relative to the non-interacted ones. For constant

---

[*] To reduce the effect of additive noise, multiple signals can be averaged coherently but over a limited time due to tidal modulation of the path lengths (time-of-flights).

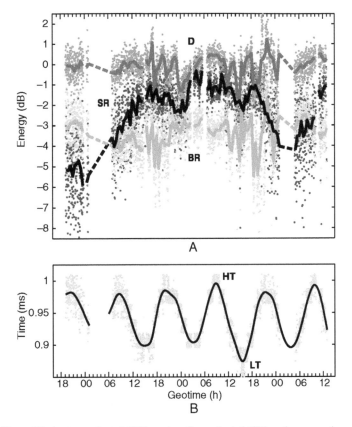

FIGURE 5.13 Direct (D), bottom-reflected (BR), and surface-reflected (SR) paths vs. geotime. (A) Normalized energy. Gray dots and lines: D path; light gray: BR path; black: SR path. The dots are the raw data and the lines connect the half-hour median averages. (B) Time-of-flight difference between the SR and D arrivals. Gray dots: raw data; solid line: smoothed data. LT and HT: low and high tide. R1 data.

energy of the D, BR, and SR paths, contained in the leading part, the mean duration increased with the energy of the MSBR paths and reverberation, contained in the tail (assuming the average power of ambient noise is constant). As shown later, excess attenuation due to the presence of gas bubbles near the bottom resulted in a mean-duration decrease.

5.4.4 Non-Photosynthesis-Related Effects

As the objective was to identify the effects of photosynthesis on sound transmission it was equally important to assess the effects of other environmental — physical and biological — factors.

5.4.4.1 Tide — The amplitude of Mediterranean tide, albeit small, represents a significant fraction of the very shallow water depth. The tide modulated the source–receiver and channel geometry, i.e., grazing angle at the boundaries and length of all SR paths.

The tide effect is most evident in the time-of-flight difference between the D and SR paths vs. geotime (Figure 5.13B). The amplitude, deduced from geometry and depth-average sound speed, is ≈40 cm. Sinusoidal patterns of extreme intensity in the received-signal spectra reveal tide-controlled interference effects between pairs of SBR arrivals (Figure 5.14). These relate to the classical Lloyd's mirror effect: interference between two continuous waves (CW) with one experiencing an additional 180° phase shift at the pressure-release surface.

The spectral shape including marked peaks and valleys, e.g., at 0.9 and 3 kHz, relate to the transmitting sensitivity response of the source (Figure 5.7B), and the overall decrease above 5 kHz is also due to the

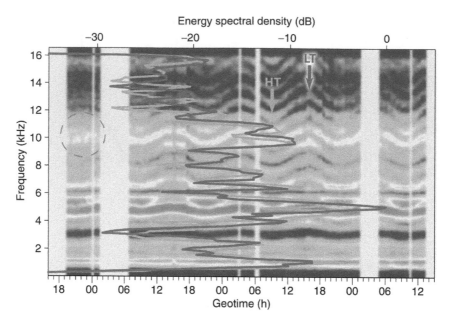

FIGURE 5.14 (**Color figure follows p. 332.**) Received signal spectrum vs. geotime. Normalized energy spectral density (left scale). The blue-to-red color map corresponds to the range –30 dB to 0 dB. The curves (top scale) are for high tide (light gray, 0900 hours) and low tide (dark gray, 1530 hours) on September 24, 1999; see Figure 5.13B. The spectra were computed from the raw acoustic data, i.e., not matched filtered or equalized, and median-averaged over half-hour periods. R1 data.

source directivity. At certain frequencies, absolute differences between low- and high-tide spectra were greater than 10 dB. There is a marked feature in the high-energy band centered about 5 kHz where a single peak at low tide was split into two peaks at high tide.

The received-signal energy in a narrow frequency band was not suitable to investigate the photosynthesis effect since it was too sensitive to the tidal effect. This required integration of the energy over a wide frequency band so that the time variation was independent of tidal modulation.

5.4.4.2   Sea Surface Motion — Frequency-dependent amplitude and phase fluctuations of the received acoustic signal were due to sea-surface roughness or wind speed. These resulted in a partial loss of correlation across the signal bandwidth, especially in the upper part of the transmitted spectrum.

On September 22, wind started to decrease followed by 2 days of calm sea and low wind speed conditions. Wind picked up again during the night on September 25.[*] The SR energy, shown in Figure 5.13A, followed exactly this pattern, being highly correlated with sea-surface roughness observations. SR energy was even lower than the BR during the rough-sea periods. In Figure 5.10, time spreading of the 1-ms delayed SR peak due to the generation of micro-multipaths is evident. The energies of the D and BR paths, which did not interact with the surface, showed no correlation with the amplitude of wind-generated waves. The time-of-flight difference (and interference) between these two arrivals was stable.

During the two calm days, the sea surface acted as a near perfect mirror for most of the transmitted frequencies. The acoustic variations were then determined principally by changes in the water-column and seagrass bed conditions. The rapid fluctuations of the MSBR energy due to sea surface motion were of smaller amplitude and timescales relative to the variations induced by photosynthesis. They were averaged out in the half-hour data block processing applied to analyze the main, longer-term, acoustic variations.

---

[*] Actually, these sea conditions determined the deployment and recovery of the acoustic instruments.

5.4.4.3 Water Temperature Profile — The speed of sound in the water column, which depends on temperature, plays the same role as the index of refraction plays in optics. The path of an acoustic ray is determined by its initial angle at the source and the sound speed structure. If sound speed is piecewise linear in depth and range independent, the paths follow either straight line segments or circular arcs.

Sun is the source of sea-surface heating and seafloor light irradiance. Solar heating modifies the water-sound speed profile (SSP) and solar radiation controls the seagrass-oxygen production. Since the two phenomena, physical and biological, occur contemporaneously, their respective acoustical effects are not separable on the basis of time delay or timescale differences, and thus deserve special attention.

Small, diurnal temperature (sound speed) variations involved the entire water column (Figure 5.8A). On September 24, their amplitudes at the sea surface and seafloor were, respectively, 0.6 and 0.2°C. During the 0200 to 0900 hours period, the column was perfectly isothermal. Then, a thin mixed layer and small thermocline developed during the day.[*]

The mild, gradual change of SSP (Figure 5.8A) had no noticeable contribution to the main, diurnal acoustic variations. These were attributed principally to gas production processes as demonstrated later. A first verification was to compare acoustic measurements at specific points of the temperature and dissolved-oxygen time series. For example, on September 24 the negative temperature gradient increased from zero to $\Delta T = 0.7°C$ ($\Delta c_w = 1.5$ m/s) in the period 1100 to 1830 hours (Figure 5.9A). During the same period, the oxygen concentration at the bottom increased by 0.4 mg/l and then returned to the same value, $C_O = 6$ mg/l (Figure 5.9B). The medium impulse responses observed at the beginning and end of the period had similar energy time distribution (e.g., the same mean duration $2\sigma_t \approx 18$ ms; Figure 5.18B).

Numerical propagation modeling with measured SSP inputs (Figure 5.8A) showed minor differences in the multipath character between the isospeed and mildly refracting conditions of the experiment.[**] Hence, points in time with different temperature but similar oxygen conditions had nearly identical acoustic responses. On the other hand, as shown later, substantial acoustic variations were observed during the isothermal periods at night, which indicated the influence of other environmental variables including the gas void fraction in the seagrass-matte layer.

5.5 Effects of Photosynthesis on Sound Propagation

The foregoing analysis has demonstrated that the variations of broadband acoustic energy, observed on a day scale, are essentially independent of the tidal cycle and changes in wind and subsurface temperature conditions. In this section, time series of dissolved oxygen concentration are interpreted together with the acoustic data to establish a causal relationship between photosynthesis and the diurnal acoustic variations.

5.5.1 Time Variation of Dissolved Oxygen

The water oxygen content was measured near the foliage to monitor the photosynthetic and respiratory activity (Figure 5.9B). The gas void fraction[***] in the seagrass layer, which influences acoustics, was not directly measured but was obviously related to the concentration of oxygen dissolved in the surrounding water.

During the two days of calm sea, September 23 and 24, production of air bubbles due to wave action was sufficiently small to detect the contribution of photosynthetic oxygen, in spite of the low-productivity season. The time evolution of the oxygen depth profile is correlated with the photosynthesis cycle (Figure 5.8B). During the rough-sea day, September 22, oxygen concentration near the surface was substantially higher due to wave action (not shown).

---

[*] Range dependence of the depth profile of temperature (sound speed) was negligible because of the short length and sheltered position of the S-R transect.
[**] The medium impulse responses were synthesized from the depth-dependent Green's function, which was evaluated at a number of discrete frequencies over the signal frequency range.
[***] The volume of gas in bubble form per volume of water.

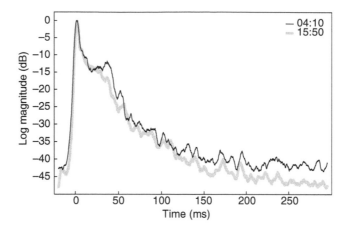

FIGURE 5.15 Comparison of a night and a day medium impulse responses. Black: 0410 hours; gray: 1550 hours. September 24, 1999. The responses were smoothed with a Savitsky–Golay FIR (polynomial) filter of degree 2 and frame size 10 ms. R1 data.

On September 24, the day- and depth-average oxygen concentration was $C_O \approx 6$ mg/l. After midday, oxygen increased above the average value up to a maximum of 6.5 mg/l at 1400 hours and then returned back to the average value during the remaining part of day (Figure 5.9B). The concentration peak at the surface was delayed by 2 h with respect to the bottom peak (not shown). During the night, there was a gradual decrease of bottom oxygen to a minimum of 5.5 mg/l at 0700 hours corresponding to the end of the respiratory phase.

5.5.2 Effect of Photosynthetic Bubbles on Multipaths

Multipaths refer to sound that is reflected coherently in the specular direction at the channel boundaries as shown in Figure 5.4. Their overall structure of arrival was resolved in the 4-day average response of Figure 5.12, where energy peaks are superimposed on the reverberation and ambient noise background. The peaks of the average response are slightly spread in time due to tidal modulation of the pathlengths and other effects of environmental variability.

Because the sea surface acted as a near-perfect mirror and the water column was nearly isospeed for most of the observation period, the observed time variations of the multipath character were mostly due to changes in the acoustic properties near the bottom. Information about photosynthesis was "accumulated" ("integrated") in the low-energy MSBR paths that interacted repeatedly with the hard bottom interface through the seagrass layer.

The most energetic D and SR paths were not influenced by the bottom conditions. The received-energy variations in Figure 5.13A show no evident cyclic behavior relatable to photosynthesis. At low grazing angles, most of the energy was reflected back into the water column and, as the angle increased, part of the energy was transmitted into the bottom (Figure 5.11A). The very low grazing angle BR path was much less sensitive to the seagrass and bottom interfaces than the higher-grazing-angle SBR and MSBR paths.

It should be emphasized that the higher-order paths were partially excited at the upper signal frequencies due to the directional response of the source. The EMF processing compensated for flat spectrum only on the source axis. Calibration data showed that the transmit voltage response at 60° off axis decreased by 1 to 13 dB in the frequency range 5 to 16 kHz. The frequency limit for full excitation of all intermediate angles $\theta < 70°$ was ≈ 3 kHz.

Figure 5.15 compares two snapshots of the medium impulse responses (smoothed log envelope) taken early morning before the plant respiratory phase and midafternoon during the photosynthesis phase. Differences in the multipath-reverberation character are noticeable, especially the disappearance in the afternoon of the main blobs of energy for $\tau < 100$ ms.

Figure 5.16 shows the overall geotime variability of the acoustic-impulse response. Each overlaid curve is the difference between a 0.5-h and the 3-night average responses (log envelope). The clustering

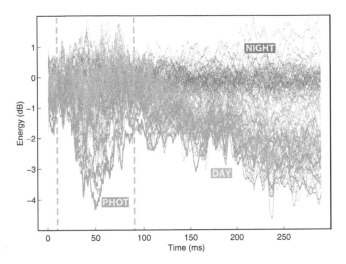

FIGURE 5.16 (Color figure follows p. 332.) Geotime variability of the medium acoustic-impulse response (log envelope) in the time-delay range 1 ms to 290 ms. Each line is the difference between a 0.5-hour and the 3-night median averages. Blue lines: night hours 0700–1930 hours, 4 days; orange lines: day hours, 4 days; green lines: most active photosynthesis hours 1300–1600 hours, September 24, 1999. Each average response was smoothed with a 10-ms polynomial filter. The vertical lines indicate the time window, which corresponds to intermediate grazing angles. R1 data.

of the responses reveals night and day regimes. Along the time axis, three regions can be isolated with distinct geotime variability. The regions correspond to the following groups of path and ranges of relative arrival time and grazing angle:

Subcritical:	2–3	$1.5\ \text{ms} < t < 10\ \text{ms}$	$\theta < 40°$
Intermediate:	4–12	$10\ \text{ms} < t < 90\ \text{ms}$	$40° < \theta < 70°$
Near-vertical:	13–∞	$t > 90\ \text{ms}$	$\theta > 70°$

(5.11)

as determined from the eigenray calculations. The leading part of the response, which includes only two to three bottom bounces, shows a lesser sensitivity to environmental variability and no diurnal trend. The tail, which is dominated by reverberation and contaminated by background noise, shows a marked day–night difference.

The middle part of the response, which includes predominantly reflected and scattered energy in the specular direction, shows a marked sensitivity to photosynthesis. The contour plot in Figure 5.17 shows the diurnal variations for that part of the response. At daylight, when gas void fraction increased in the leaf layer, the energy partly excited into the bottom and partly reradiated to the water column both decreased owing to scatter and absorption.

The omnidirectional scatter of the intervening bubble layer adds to the directional scatter of the rough hard bottom. The combined processes redistribute the incident energy in the water column, taking away, together with the absorption process, a larger portion of the energy transmitted in the specular direction. The resulting attenuation is expected to be strongly dependent on grazing angle and frequency.

In addition to the attenuation effect, the presence of bubbles in a near-bottom water layer caused ray refraction. The refraction index of that layer was frequency dependent, as the effect of bubbles on sound speed depends on the ratio of the excitation frequency to the bubble resonant frequency.[*] When bubble density increased near the bottom, rays were refracted and reflected at more nearly normal incidence.

It is reasonable to assume that most bubble sizes in the leaf water were much smaller that the bubble resonant size at the higher frequency of excitation $f_2 = 16$ kHz, i.e., a radius $a = 373\ \mu\text{m}$ at the depth $z = 8$ m. At frequencies well below the bubble resonances, the sound speed can be determined from the simple mixture theory. The low frequency, asymptotic value for a gas void fraction U is given by Wood's equation:[42]

---

[*] This frequency dependence does not exist in bubble-free water where sound speed depends only on temperature, salinity, and pressure.

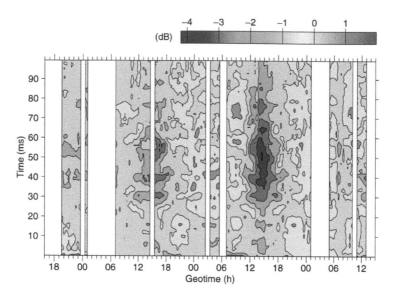

FIGURE 5.17 (Color figure follows p. 332.) Time distribution of multipath energy vs. geotime. The contour lines are from –4 dB to +1 dB in 1-dB steps. The levels are referenced to the three-night median-average. The displayed time interval comprises the multipath arrivals of groups 2 through 13. R1 data. See Figure 5.16.

$$c_B = \frac{\left(E_g E_w\right)^{1/2}}{\left[U\rho_g + (1-U)\rho_w\right]^{1/2}\left[UE_w + (1-U)E_g\right]^{1/2}} \tag{5.12}$$

where $E_g = \gamma p_A$ and $E_w = \rho_w c_w^2$ are the bulk moduli of elasticity of gas and water, $\gamma = 1.4$ is the ratio of specific heats of gas (air or oxygen), $\rho_g = 1.43$ g l^{-1} is the gas density (oxygen), $\rho_w = 1030$ kg m^{-3} is the water density and $p_A = p_{A0} + \rho_w g z = 1.81 \cdot 10^5$ Pa is the ambient pressure at a depth $z = 8$ m. From Snell's law, the modified grazing angle on the basalt interface as a function of the void fraction is given by

$$\theta_B(U) = \cos^{-1}\left[(c_B(U)/c_w)\cos(\theta_w)\right] \tag{5.13}$$

which is shown in Figure 5.11B for plausible values of the void fraction.

The greater sensitivity of the paths with intermediate grazing angles was due to the reflection loss vs. angle curve for a basalt half-space (Figure 5.11A) combined with the refraction effect in the seagrass layer (Figure 5.11B). For an impedance contrast increase due to sound speed (and density) smaller than bubble-free water, say, for $U = 5 \cdot 10^{-5}$:

1. The near-horizontal paths at the seagrass interface are refracted but remain in a low-loss region of the curve.
2. The near-critical paths move to a higher-loss region.
3. Most intermediate paths remain in the high-loss region.
4. The higher ones move to a lower-loss region.
5. The near-vertical paths remain in a medium-loss region.

The net result of (a) the loss redistribution among the paths, (b) their associated number of bottom bounces, and (c) the attenuation effect explains the shape of the diurnal variations in the middle part of the medium response envelope (gray line in Figure 5.12 and Figure 5.16).

The plot in Figure 5.11A shows the complexity of the reflection at the composite bottom. The loss is contoured as a function of angle and frequency. The prediction includes the refraction effect only, i.e., not the attenuation in the seagrass layer due to scattering and absorption. At the lower frequencies the loss is similar to the curve of basalt half-space. This is because here the seagrass layer is thin

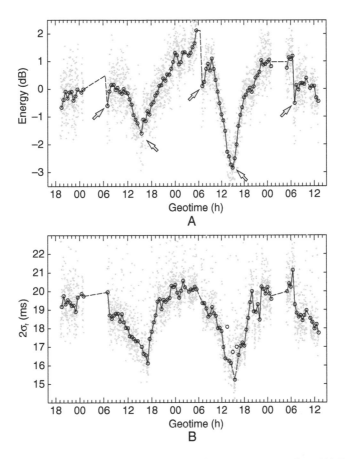

FIGURE 5.18 Descriptive statistics of the medium acoustic-impulse response vs. geotime. (A) Normalized fractional energy in the time window 35 to 45 ms, corresponding to the multipath group 7. (B) Mean duration. The gray dots are the raw data. The circles connected by solid lines are half-hour median averages. The dashed lines are interpreted missing points. The arrows are explained in the text.

compared to the acoustic wavelength ($H = 0.60$ m, $\lambda = 6.3$ m at $f_1 = 200$ Hz) and therefore acoustically transparent. The apparent critical angle is seen to decrease with frequency (from 40° to 35° at 1 kHz). The angle-dependent resonance pattern is evident, with quarter and half-wavelength layer effects regularly interspersed (on a linear frequency scale).

Figure 5.18A shows the geotime variation of the energy received in one group of multipaths: group 7 with a mean grazing angle of 61°. A remarkable feature is the similarity of shape between the acoustic and dissolved oxygen time series. The group-7 energy, reversed in the vertical, is overlaid to the plot of Figure 5.9B for direct comparison (dashed line). The well-defined minima of energy, $E = -3$ dB, and mean duration, $2\sigma_t = 15$ ms (Figure 5.18B) that occurred at 1500 hours, correspond to the maximum concentration of oxygen near the foliage. Also the maximum of energy coincides with the minimum of bottom oxygen at 0600 hours. The difference of energy between 0600 and 1500 hours, $E = -5$ dB, represents an excess attenuation of ≈0.7 dB per bottom bounce. A local minimum of energy was consistently observed at 0700 hours for the three days (gray arrows in Figure 5.18A) in correspondence to the oxygen minimum at the end of the respiratory phase. The excess attenuation was attributed to the concomitant flows of gas to the rhizomes and roots as discussed in the next section. Similar but smaller and smoother energy variations were observed for time windows that included a larger number of arrivals with intermediate grazing angles (not shown).

Figure 5.19 shows the relationships between the acoustic and environmental measurements: group-7 multipath energy vs. temperature in the upper part of the water column and dissolved oxygen content near the foliage. In Figure 5.19A, the great scatter of the data pairs indicates no obvious relationship

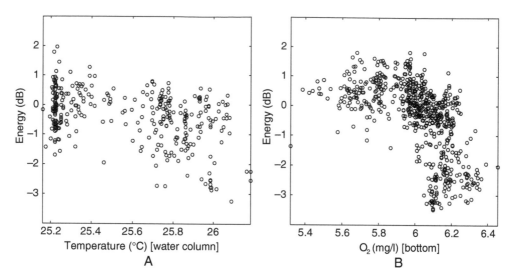

FIGURE 5.19 Acoustic vs. environmental data. The acoustic data are the energy propagated along intermediate grazing-angle paths (group 7) shown in Figure 5.18A. (A) Energy vs. temperature in the upper part of the water column ($z \leq 1$ m). (B) Energy vs. dissolved oxygen content near the foliage ($z > 8$ m). All data points available in both data sets at the same time (minute) are displayed.

between the small temperature (sound speed) gradient and the acoustic energy variations. In Figure 5.19B, there is a strong (nonlinear) relationship between the oxygen concentration and the received-energy variations. The energy decreased sharply when the water oxygen content rose above the nominal value of $C_O = 6$ mg/l.

In conclusion, the overall attenuation of the MSBR arrivals during the daylight hours was due to a near-bottom excess attenuation controlled by photosynthesis.

5.5.3 Effect on Reverberation

Reverberation refers to sound that is scattered away from the specular direction both in and out of the vertical plane containing the source and receivers. Inhomogeneities within the water body and at the channel boundaries form discontinuities in the physical properties of the medium and thereby intercept and reradiate a portion of the acoustic energy incident on them. The sum total of the contribution from all the scatterers, observed at a receiver distant from the source, is the bistatic reverberation.

As the specularly reflected multipaths, the reverberation was sensitive to photosynthesis. Figure 5.15 and Figure 5.16 show marked differences between night and day in the reverberation character, including the rate of decay. As mentioned earlier, the angle- and frequency-dependent bottom scattering strength was modified by the seagrass bubble layer. Although the reverberation component was not extracted from the multipath component, the overall decay time was grossly quantified by the mean duration of the medium impulse response, which decreased during the daylight hours (Figure 5.18B).

5.5.4 Effect on Ambient Noise

One of the principal features of the ambient noise was its marked time variability. Three major noise sources were identified: biologics, the wind or waves, and ships and other human-made activities at moderately close ranges. The diurnal variations in the apparent level of ambient noise were conjectured to be partly due to photosynthesis.

5.5.4.1 *Spectral Characteristics* — Figure 5.20 shows the diurnal variations of ambient-noise level and spectrum observed during the experiment. Abundance and diversity of animals in seagrass beds are known to increase at night due to immigration of reef fish and movement of diurnal planktivores

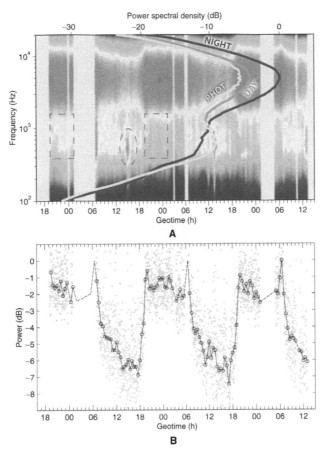

FIGURE 5.20 (Color figure follows p. 332.) Ambient noise vs. geotime. (A) Normalized energy spectral density. The color map corresponds to the range −30 dB (blue) to 0 dB (red). The spectra were median averaged over half-hour periods. The overlaid curves are time-mean averages. Dark gray: night hours 1930–0700 hours, 4 days; gray: day hours, 4 days; light gray: mid-afternoon hours 1400–1600 hours, September 24, 1999. The raw spectra were computed from reverberation-free data sections of 1-s duration taken before each probe signal was received. Dashed boxes: see text. (B) Normalized total power. The gray dots are the raw data. The circles connected by solid lines are half-hour median averages. The dashed lines are interpreted missing points. The data were affected by scuba diving noise during the periods 1440–1520 hours on September 23 and 1310–1340 hours on September 24 (dashed ellipsis).

from the water column to sheltering sites beneath the foliage. This behavior was remarkably well observed during the present experiment from diurnal variations of both ambient noise and volume scattering features.

The time variation of noise level resembles a square function with characteristic constants of exponential-like rise and decay times. When darkness approached, the level increased abruptly due to sounds produced by fish migrating into the seagrass bed to feed (Figure 5.20B). During the night, local biological sounds dominated the natural physical sounds and shipping noise resulting in a 4 to 6 dB level increase. At sunrise, the apparent level of ambient noise first decreased rapidly and then more slowly to a minimum before sunset. The rapid decay was due to the diminishing number and intensity of biological sound sources while the slow decay was attributed to the photosynthesis-driven sound attenuation characteristics of the plant bubble layer.

The biological origin of sound level increase at night is confirmed by a strengthening of the spectral band centered about 5 kHz (curves in Figure 5.20A). This modified the standard, spectral shape of deep-sea ambient noise,[40] which is also applicable to shallow water in the absence of biological and human-made noise. Comparison of night and day averages shows a transition frequency of 1 kHz about which the power spectral densities varied in an opposite way. Below that frequency, the densities were larger

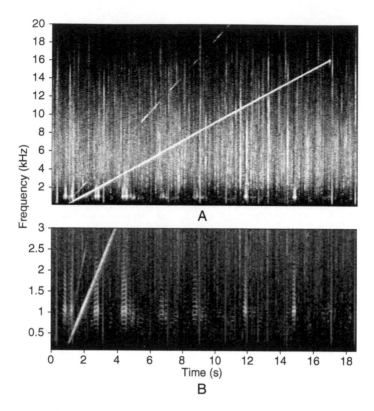

FIGURE 5.21 Spectrogram of ambient noise showing the transmitted chirp signal, biological transients, and breaking-wave events. (A) Normalized energy spectral density gray-coded in the range –60 dB to 0 dB (white). (B) Zoom of the top panel in the low frequency range showing details of fish stridulatory sounds.

during the day due to local shipping and diving activities supporting the experiment. In particular, the spectral prominences about 500 Hz on September 23 and 24 were due to the presence of divers in the vicinity of the transducers. It should be mentioned that, below the 1-kHz transition frequency, the very shallow water location appeared to be quieter than deep water because of the absence of deep-going favorable transmission paths.

In Figure 5.20B, the peaks at 1930 hours and sharper peaks at 0700 hours were likely caused by fish whose produced sound level is known to increase when they feed, usually at dusk and dawn, as observed here. The peaks at 1930 hours are also consistent with the shrimp diurnal cycle known to have a maximum level at sunset.

The contribution of sea-state noise, or "wind noise,"[43] is evident in the time history of ambient noise spectrum. Spectral broadening toward the lower frequencies was well correlated with the increased wind speed on September 22, as seen from the comparison of the 1900 to 0100 hours period of that day with the same periods of the quieter days (dashed boxes in Figure 5.20A).

5.5.4.2 *Time-Frequency Characteristics* — Figure 5.21 shows a spectrogram of ambient noise taken during the period a probe signal was received.

The linear frequency sweep (and second harmonic distortion) of the transmitted signal is well resolved above the background noise. Listening to the recordings revealed a variety of sounds produced by crustaceans and fish. All recordings were dominated by the characteristic noise of snapping shrimps: an uninterrupted crackle, resembling the sound of "frying fat" and known to range from 1 to 50 kHz. Over this continuous background noise, marked events with distinctive time-frequency features indicated the presence of different species of fish. Their sounds ranged in frequency from 50 Hz to 8 kHz. The lower part of the spectrogram, expanded in Figure 5.21B, shows a sequence of stridulatory sounds produced

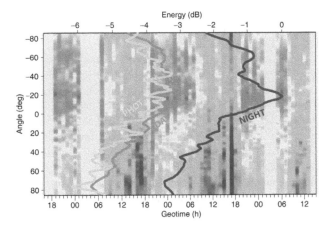

FIGURE 5.22 (Color figure follows p. 332.) Directional characteristics of ambient noise vs. geotime. Normalized correlation function (log envelope) between hydrophone signals R1 and R2. Negative and positive angles correspond to downgoing and upgoing energy, respectively. Color map: −10 dB (blue) to 0 dB (red). The overlaid lines are geotime mean-averages. dark gray: night hours, 1930–0700 hours, 4 days; gray: day hours, 4 days; light gray: midafternoon hours, 1400–1600 hours, September 24, 1999. The raw functions were computed from the same data sections as in Figure 5.20 and their envelopes were median-averaged over 1-h periods.

by an individual. They consist of frequency sweeps with a lower frequency of ≈500 Hz and rich harmonic content, lasting a fraction of second. Other transient events were fish colliding with the hydrophones and waves breaking on the nearby shore.

5.5.4.3 Directional Characteristics — The background noise was a combination of locally generated sounds and sounds that traveled over larger distances. The origins of the main sources were roughly determined from the temporal correlation between signals received on the pair of vertical hydrophones. Figure 5.22 shows the vertical directivity of ambient noise vs. geotime. The bearing corresponding to the largest correlation peak coincides with the measured average bottom slope of 15°. Here, the oxygen-bubble effect is noticed from the comparison of the night, day, and midafternoon averages. There was a stronger attenuation of ≈1 dB, at daytime, of the distant sounds originating near the shore relative to the closer ones, near broadside and toward endfire. The distant sounds propagating in the thin downslope waveguide were attenuated by repeated interaction with the *Posidonia* bubble patches covering of the entire inlet.

5.5.4.4 Other Observations — There are other interesting observations to be made from the comparison of biological noise and transmitted signal variability. Seagrass scattering and absorption were shown to be more effective at daytime during the active phases of photosynthesis, whereas fish swimbladder and macro zooplankton are biological scatterers that are more abundant during the night, especially at dusk and dawn. This form of volume scattering was characterized by its intermittency. The presence of fish schools along the transmission paths caused substantial acoustic effects. In Figure 5.13A, the rapid fluctuations of received energy are larger for the D and BR paths, which traverse, almost horizontally, a water layer near the foliage where fish tend to concentrate. The received energy was stabler for the SR path, which did not involve that scattering layer. Hence fish movement explains the greater scatter of the acoustic transmission data in the evening and early morning.

5.5.5 Gaseous Interchange of the Leaf Blade

The change of acoustic propagation features described in the previous section were explained by the variation of gas void fraction within the seagrass layer. The mechanisms described were adapted from a general account on gaseous movement in seagrasses, which does not deal specifically with *P. oceanica* species.[44]

From average leaf blade dimensions[21] and shoot densities,[23,45] the leaf-volume fraction is of the order $O(10^{-4})$ and the volume fraction of porous region beneath the cuticle is $O(10^{-5})$ (Figure 5.2A). The fraction of lacunar air spaces is much smaller.

As photosynthetic tissue is concentrated in the epidermidis, the epidermal cells experience a rapid build-up of oxygen at dark-to-light transition. The resulting diffusion gradient causes oxygen to diffuse toward the ambient medium and the lacunar system, in proportions that depend on the diffusion resistance of the outward and inward pathways, size of the leaf, and physical characteristics of seawater and lacunar gases. Under *in vivo* pressurized conditions, e.g., $p_A = 1.9 \cdot 10^5$ Pa at the 8.5-m average water depth of the present experiment, the unstirred layer, cuticle, and cell wall provide large resistance to diffusion.

Initially, at low irradiance, oxygen accumulates in the porous wall beneath the cuticle and in the small air spaces of the mesophyll (see Figure 5.2). This causes an increase of oxygen concentration and pressure within the leaf blade. Even under rapid stirring conditions, pressures greater than $2 \cdot 10^4$ Pa have been measured in comparable seagrasses. A continuous stream of bubbles produced from any cut surface of leaves, rhyzomes, or roots is observed by divers. The initial pressurization causes a transient mass flow of lacunar gas to the rhizomes and roots.

Then, with increasing irradiance, oxygen starts to diffuse into seawater and form bubbles, which adhere to the leaf blade. Previous acoustic measurements indicated that the phenomenon occurs in a matter of minutes.[11] Although free large gas bubbles in water tend to collapse by gas diffusion forced by surface tension or rise rapidly to the surface by buoyant action, the continuous oxygen supply and wall-adhesion effect maintain a large void fraction of oxygen bubbles near the sea bottom in addition to oxygen in the porous wall and gas in the lacunar system. At light, bubbles of visible size on the leaf blades and rising to the surface are commonly observed by divers.[*]

During steady-state photosynthesis, an equilibrium is established between oxygen production rate and processes of bubble formation and dissolution in seawater that mostly depend on the degree of stirring. The photosynthetic oxygen progressively enriches the water column (see Figure 5.9B) and, in spring and summer seasons, contributes to oxygen supersaturation in the surface layer.[24]

The sequence of events described is consistent with the acoustic observations establishing a causal relationship between the photosynthesis daily cycle and the main acoustic variations, i.e., the primary peaks and valleys in Figure 5.18A and B.

Two other factors influence the acoustic properties of the plant and matte layers: Vented gas of zooplankton and fish may contribute to the void fraction in the plant layer. Transient mass flows of gas to the rhizomes and roots at dark-to-light transition and the reverse at light-to-dark change the void fraction in the matte layer. The latter explains the secondary valleys in Figure 5.18A.

5.6 Conclusion

The experimental results presented in this chapter demonstrate that products of photosynthesis by *Posidonia* seagrass affect the transmission of low-frequency sound (<16 kHz) over the prairie.

The acoustic-channel propagation characteristics — multipath excess attenuation, reverberation decay time, and ambient noise level — were analyzed as a function of time of day. The main diurnal variations were explained by time-dependent scattering and absorption in the seagrass and matte layers. During daylight hours, bubbles of photosynthetic oxygen were formed on the leaf blade and, at dark-to-light transition, gas was released within the matte. The gaseous movements in the seagrass bed modified the interaction of acoustic energy with the rock substratum.

In two experiments conducted under totally different conditions the acoustic transmissions were highly sensitive to the photosynthesis cycle. This indicates that the inverse problem of determining oxygen and gas void fractions can be solved. Parameters such as surface density and photosynthetic efficiency of the *Posidonia* plants can be derived from the variations of inverted void fractions. This requires calibration

---

[*] Because of the omnipresence of particles, sometimes called "snow," it is difficult to positively identify bubbles of radius less than 40 μm by simple photography.

TABLE 5.1

Environmental, Geometrical and Acoustical Parameters of the Experiments at Scoglio Africa Island, May 1995, and Ustica Island, September 1999.

Location Date		Sc. Africa **May 1995**	Ustica **September 1999**
Vegetation characteristics			
Type		*P. oceanica*	*P. oceanica*
Epiphytic colonization		Low	High
Leaf density		High	Low
Average height	cm	75	60
Prairie dimensions	km	8.5 × 5	≈0.2 × 0.05
Matte sediments		Medium to coarse sands	Scarce
Thickness	cm	≈20	<5
Substratum		Sand	Volcanic basalt
Oxygen productivity			
Max. concentration	mg/l	9.5	6.5
Day-night difference	mg/l	2.5	1
Supersaturation		Yes	No
Measurement geometry			
Range	R (m)	1541	53
Average water depth	d (m)	25	8.5
Source depth	d_S (m)	21	6.45
Receiver depths	d_R (m)	4, 10, 16, 22	6.30, 7.70
Acoustic transmissions			
Frequency range	Π (kHz)	0.1–1.6	0.2–16
Pulse duration	Δt (s)	3	15.8
Time-bandwidth product	$\Delta t \Delta f$	$4.5 \cdot 10^3$	$2.5 \cdot 10^5$

Note: The oxygen concentrations were measured just above the plants. The day–night differences were calculated from the respective minimum and maximum of respiratory and photosynthesis phases.

of the acoustic measurements with *in situ* oceanographic data and comprehensive modeling of the seagrass scattering and absorption mechanisms in the audio-frequency band.

The proposed method opens interesting possibilities for ecological monitoring and surveillance, e.g., the non-intrusive, *in situ* study of the metabolism of certain submersed aquatic plants in response to environmental factors and stresses and the assessment of the global state of health of seagrass beds.

Acknowledgments

This research was partly supported by Prof. R. Catalano, Department of Geology and Geodesy, University of Palermo, and the Saclant Undersea Research Centre, La Spezia, Italy. I gratefully acknowledge Dr. M. Agate, the scientific divers C. Lo Iacono and M. Longo, the fisherman community of Ustica island, R. Ialuna, Centro Oceanologico Mediterraneo, and Prof. A. Stefanon, University Cà Foscari of Venezia, who contributed to the success of the data collection effort.

Appendix 5.A   Comparison with Earlier Experiments

For comparison purposes, the experimental conditions of the two sites investigated by the author are summarized in Table 5.1.

The site at Scoglio Africa is quite a different kind of environment than that considered in this chapter. The *Posidonia* plants and matte are well developed due to the gently sloping topography and sandy nature of the bottom; the prairie is particularly dense and extensive. At the site of Ustica the steep topography and volcanic rock do not favor the establishment of the plants.

Seasons of the measurements were also different. In September, fully developed epiphytes modify the plant metabolism resulting in lower oxygen production and no subsurface, supersaturation conditions.

In the SCOGLIO AFRICA 95 experiment, the range was 20 times larger and the upper frequency of sound transmission was 10 times smaller than in USTICA 99. There, propagation was entirely determined by bottom reflections at low grazing angles (discrete mode spectrum) while, here, higher grazing angles (continuous spectrum) were also important. For Scoglio, a theory was proposed based on the concept of trapping a portion of the sound in the low-speed waveguide formed by the bubble layer.[12] A similar effect was reported in Reference 46 for the propagation of ambient noise in an ocean-surface bubble layer. For Ustica, waveguiding in the secondary waveguide was not the dominant effect because of the much shorter range. Here the multiple bottom-interacted paths near and above critical angle were mostly effected by the bubble layer.

References

1. A. Caiti, J.-P. Hermand, S. Jesus, and M.B. Porter, Eds., *Experimental Acoustic Inversion Methods for Exploration of the Shallow Water Environment*, Kluwer Academic, Dordrecht, June 2000.
2. O. Diachok, A. Caiti, P. Gerstoft, and H. Schmidt, Eds., *Full Field Inversion Methods in Ocean and Seismo-Acoustics*, Vol. 12 of *Modern Approaches in Geophysics*, SACLANT Undersea Research Centre, Kluwer Academic, Dordrecht, June 1995.
3. Special issue on inversion techniques and the variability of sound propagation in shallow water, *IEEE J. Oceanic Eng.*, 21(4), 1996.
4. J.-P. Hermand, Broad-band geoacoustic inversion in shallow water from waveguide impulse response measurements on a single hydrophone: theory and experimental results, *IEEE J. Oceanic Eng.*, 24, 41–66, 1999.
5. P. Colantoni, Mapping of the *Posidonia oceanica* meadows, in F. Cinelli, E. Fresi, C. Lorenzi, and A. Mucedola, Eds., *La Posidonia oceanica*, Rivista Marittima, Rome, Dec. 1995, 116–122.
6. L. Chessa, E. Fresi, and C. Lorenzi, The state of health of a *Posidonia oceanica* prairie: study method, in F. Cinelli, E. Fresi, C. Lorenzi, and A. Mucedola, Eds., *La Posidonia oceanica*, Rivista Marittima, Rome, Dec. 1995, 78–83.
7. A. Peirano, N. Stoppelli, and C. Bianchi, Monitoring and study techniques of sea grasses in Liguria, in F. Cinelli, E. Fresi, C. Lorenzi, and A. Mucedola, Eds., *La Posidonia oceanica*, Rivista Marittima, Rome, Dec. 1995, 88–91.
8. E. McCarthy, Acoustic characterization of submerged aquatic vegetation, in *High Frequency Acoustics in Shallow Water*, N. Pace, E. Pouliquen, O. Bergem, and A. Lyon, Eds., SACLANTCEN Conference Proceedings, Lerici, Italy, 1997, 363–369.
9. A. Siccardi and R. Bozzano, A test at sea for measuring acoustic backscatter from marine vegetation, in A. Caiti, J.-P. Hermand, S. Jesus, and M.B. Porter, Eds., *Experimental Acoustic Inversion Methods for Exploration of the Shallow Water Environment*, Kluwer Academic, Dordrecht, June 2000, 145–159.
10. J.-P. Hermand, F. Spina, P. Nardini, and E. Baglioni, Yellow Shark broadband inversion experiments 1994–1995: environmental data compilation, CD-ROM YS-1, Saclant Undersea Research Centre, La Spezia, Italy, Dec. 1996.
11. J.-P. Hermand, P. Nascetti, and F. Cinelli, Inversion of acoustic waveguide propagation features to measure oxygen synthesis by *Posidonia oceanica*, in *Proceedings of the Oceans '98 IEEE/OES Conference*, C.O. Committee, Ed., Vol. II, IEEE Oceanic Engineering Society, IEEE, Piscataway, NJ, Sept. 1998, 919–926.

12. J.-P. Hermand, P. Nascetti, and F. Cinelli, Inverse acoustical determination of photosynthetic oxygen productivity of *Posidonia* seagrass, in A. Caiti, J.-P. Hermand, S. Jesus, and M.B. Porter, Eds., *Experimental Acoustic Inversion Methods for Exploration of the Shallow Water Environment*, Kluwer Academic, Dordrecht, June 2000, 125–144.

13. C.S. Clay and H. Medwin, *Acoustical Oceanography: Principles and Applications*. Wiley, New York, 1977.

14. H. Medwin and C.S. Clay, *Fundamentals of Acoustical Oceanography*. Academic Press, New York, 1998.

15. T. O'Hern, L. d'Agostino, and A. Acosta, Comparison of holographic and Coulter counter measurements of cavitation nuclei in the ocean, *Trans. ASME J. Fluids Eng.*, 110(2), 200–207, 1988.

16. H. Medwin, In-situ acoustic measurements of bubble populations in coastal waters, *J. Geophys. Res.*, 75, 599–611, 1970.

17. H. Medwin, In-situ measurements of microbubbles at sea, *J. Geophys. Res.*, 82, 971–976, 1977.

18. F. Fox, S. Curley, and G. Larson, Phase velocity and absorption measurements in water containing air bubbles, *J. Acoust. Soc. Am.*, 27(3), 534–539, 1955.

19. V.D. Grosso, New equation for the speed of sound in natural waters (with comparisons to other equations), *J. Acoust. Soc. Am.*, 56(4), 1084–1091, 1974.

20. A. Weissler and V. Del Grosso, The velocity of sound in sea water, *J. Acoust. Soc. Am.*, 23(2), 219–223, 1951.

21. P. Mariani Colombo, N. Rascio, and F. Cinelli, *Posidonia oceanica* (L.) Delile: a structural study of the photosynthetic apparatus, *Mar. Ecol.*, 4(2), 133–145, 1983.

22. M. Libes, Productivity-irradiance relationship of *Posidonia oceanica* and its epiphytes, *Aquat. Bot.*, 26, 285–306, 1986.

23. F. Cinelli, Mappatura delle praterie di *Posidonia oceanica* (L.) Delile intorno alle isole minore dell'Arcipelago Toscano, tech. rep., Ministero della Marina Mercantile, Ispettorato Centrale per la Difesa del Mare, 1992.

24. P. Nascetti, Rilievi stagionali e circadiani della produzione di ossigeno di praterie di *Posidonia oceanica* in supporto ad uno studio di inversione acustica, tesi di laurea in scienze biologiche, Facoltà di Scienze Fisiche, Matematiche e Naturali, Genova, 1998.

25. P. Colantoni, Sediments of the *Posidonia oceanica* meadows, in F. Cinelli, E. Fresi, C. Lorenzi, and A. Mucedola, Eds., *La Posidonia Oceanica*, (Roma), Rivista Marittima, Dec. 1995, 52–55.

26. T. Gardner and G. Sills, An examination of the parameters that govern the acoustic behavior of sea bed sediments containing gas bubbles, *J. Acoust. Soc. Am.*, 110(4), 1878–1889, 2001.

27. F. Wezel, D. Savelli, M. Bellagamba, M. Tramontana, and R. Bartole, Plio-quaternary depositional style of sedimentary basins along insular Tyrrhenian margins, in *Sedimentary Basins of Mediterranean Margins*, F. Wezel, Ed., Tecnoprint, Bologna, 1981, 239–269.

28. N. Calanchi, P. Colantoni, G. Gabbianelli, P. Rossi, and G. Serri, Phisiography of Anchise seamount and of submarine part of Ustica island (South Tyrrhenian): petrochemistry of dredged volcanic rocks and geochemical characteristics of the mantle sources, *Mineral. Petrogr. Acta*, 28, 215–241, 1984.

29. J.V. Vleck and D. Middleton, A theoretical comparison of the visual, aural and meter reception of pulsed signals in the presence of noise, *J. Appl. Phys.*, 17, 940–971, 1946.

30. C. Helstrom, *Statistical Theory of Signal Detection*. Pergamon Press, New York, 1960.

31. A. Whalen, *Detection of Signals in Noise*. Academic Press, New York, 1971.

32. H. Van Trees, *Detection, Estimation and Modulation Theory*, Vol. III. Wiley, New York, 1971.

33. A. Oppenheim and R. Schafer, *Digital Signal Processing*. Prentice-Hall, Englewood Cliffs, NJ, 1975.

34. A. Oppenheim and R. Schafer, *Discrete-Time Signal Processing*. Prentice-Hall, Englewood Cliffs, NJ, 1989.

35. J.-P. Hermand and L. Alberotanza, A novel acoustic tomography experiment in the lagoon of Venice, in *La Ricerca Scientifica per Venezia. Il Progetto Sistema Lagunare Veneziano. Modellistica del Sistema Lagunare. Studio di Impatto Ambientale*, Vol. II, Atti dell' Istituto Veneto di Science, Lettere ed Arti, Padova, 2000, 199–222.

36. J.-P. Hermand and W.I. Roderick, Acoustic model-based matched filter processing for fading time-dispersive ocean channels: theory and experiment, *IEEE J. Oceanic Eng.*, 18, 447–465, 1993.

37. F.B. Jensen, W.A. Kuperman, M.B. Porter, and H. Schmidt, *Computational Ocean Acoustics*. American Institute of Physics, Modern Acoustics and Signal Processing, New York, 1994.

38. E. Hamilton, Geoacoustic modeling of the seafloor, *J. Acoust. Soc. Amer.*, 68(5), 1313–1340, 1980.

39. H. Schmidt, OASES, technical document, Department of Ocean Engineering, Massachusetts Institute of Technology, Cambridge, MA, 1999.

40. R.J. Urick, *Principles of Underwater Sound*. McGraw-Hill, New York, 1983.

41. L. Cohen, *Time-Frequency Analysis*. Prentice-Hall, Englewood Cliffs, NJ, 1995.
42. A. Wood, *A Textbook of Sound*. Macmillan, New York, 1955.
43. G. Wenz, Acoustic ambient noise in the ocean: spectra and sources, *J. Acoust. Soc. Am.*, 34(12), 1936–1956, 1962.
44. A. Larkum, A. McComb, and S. Shepherd, *Biology of Seagrasses*. Elsevier, New York, 1989.
45. E. Balestri, L. Piazzi, S. Acunto, and F. Cinelli, Flowering and fruiting beds in Tuscany (Italy), in F. Cinelli, E. Fresi, C. Lorenzi, and A. Mucedola, Eds., *La Posidonia Oceanica*, Rivista Marittima, Rome, Dec. 1995, 28–30.
46. D. Farmer and S. Vagle, Waveguide propagation of ambient sound in the ocean-surface bubble layer, *J. Acoust. Soc. Am.*, 86(5), 1897–1908, 1989.
47. F. Cinelli, E. Fresi, C. Lorenzi, and A. Mucedola, Eds., *La Posidonia Oceanica*, Rivista Marittima, Rome, Dec. 1995.
48. M. Agate, G. Lo Cicero, M. Lucido, D. Zuccarello, A. Borruso, A. d'Argenio, and R. Catalano, Geological marine map of the Ustica island, tech. rep., Centro Interdipartimentale di Ricerche sull' Interazione Tecnologia-Ambiente, Gruppo di Geologia Marina, Dipartimento di Geologia e Geodesia, Università degli Studi di Palermo, Palermo, July 2001.

6

Multiscale in Situ Measurements of Intertidal Benthic Production and Respiration

Dominique Davoult, Aline Migné, and Nicolas Spilmont

CONTENTS

6.1 Introduction 97
6.2 Materials and Methods 98
6.3 Results 99
    6.3.1 *In Situ* Measurements and Estimation of Daily Potential Primary Production 99
    6.3.2 Seasonal Variations of Primary Production and Respiration 100
    6.3.3 Temporal Resolution of Measurements and Microscale Adaptation of
           Microphytobenthos to Variations in Irradiance 102
    6.3.4 Mesoscale Variations within the Gradient of Exposure 102
    6.3.5 Microscale Variability 103
6.4 Conclusion and Perspectives 104
Acknowledgments 105
References 106

6.1 Introduction

Although phytoplankton are considered to be responsible for the major part of marine primary production, both macrophytobenthos and microphytobenthos can play an important role in coastal ecosystems.[1,2] Microphytobenthos can provide as much as two thirds of total primary production in some estuaries[3] and macroalgae as much as three quarters of total primary production in some bays.[4] Such estimations need direct (*in situ* monitoring of changes in CO_2 or O_2) or indirect (modeling or laboratory incubations under artificial irradiance conditions) measurements of gross primary production and the knowledge of both temporal and spatial variability of the studied system.

Temporal variations can act at mesoscale (interannual and seasonal variations), small scale (e.g., meteorological or spring tides/neap tides variations) and microscale (e.g., tidal cycle or irradiance variations) and so can spatial variations (salinity gradient, granulometric heterogeneity such as variations in silt content, water retention, duration of emersion, etc.).

Our purpose was to use an *in situ* method for measuring net primary production and respiration, as opposed to experimental indirect methods[5-7] which are unable to estimate short time and space variations of these processes. The aim of this study was to accurately evaluate both trends and variability in primary production and respiration of sandy and muddy intertidal estuarine or marine flats in the eastern English Channel. Differences were expected along a gradient of exposure (characterized by the amount of mud within the sediment), particularly between exposed sandy beaches and sheltered estuarine mud flats. A closed-chamber method was used, based on CO_2 concentration measurements performed with an infrared gas analyzer.

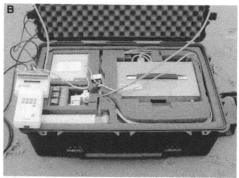

FIGURE 6.1 (A) The whole system during a light incubation in the Seine estuary. (B) Portable container with circuit of CO_2 analysis (a pump, a drying column, a flowmeter, an infrared gas analyzer) and data logger.

In this study, spatial variability was considered from the elementary surface of measurement of the closed chamber (0.126 m^2) to the comparison between several intertidal systems (an exposed sandy beach typical of the conditions in the eastern English Channel and two estuarine flats) and temporal variability was considered from the frequency of data logging (every 30 s for CO_2 concentration) to the annual scale for carbon budget estimation.

All measurements were made under emersed conditions as it was assumed that no primary production occurred during immersion because of strong light limitation due to particulate suspended matter, particularly in estuaries[8,9] (up to 120 mg l^{-1} in the Somme estuary[10]).

6.2 Materials and Methods

The closed chamber used for CO_2 flux measurement at the air–sediment interface was constructed using a dome (16.75 l, 40 cm in diameter) of transparent (or opaque) Perspex fitted on a crown wheel of stainless steel, which is pushed into the substrate to a depth of 10 cm, enclosing a volume of 24.92 l and a surface area of 0.126 m^2. A pump (Brailsford and Co., TD-2SA) maintained an airflow of about 2 l min^{-1} through the closed circuit. Variations of CO_2 concentration were measured with an infrared gas analyzer (LiCor Li-6251) and PAR (photosynthetically active radiations, 400 to 700 nm) inside the chamber was quantified with a quantum sensor (LiCor Li-192SA). Analyzer data (internal temperature, CO_2 concentration) as well as environmental data (PAR, temperature in the enclosure) were stored on a data logger (LiCor Li-1400). The logging frequency was 30 s for analyzer data and 1 min for environmental data. The whole system of analysis was placed in a portable container (Figure 6.1). More details on calibration and conditions of measurements are given in Migné et al.[11] Experiments were carried out in ambient light and darkness to estimate community net primary production and respiration, respectively. Gross community production was calculated from net production corrected with respiration.

One experiment consisted of a series of incubations under different conditions of irradiance during the day, from dawn to zenith or from zenith to twilight. One incubation occurred during 10 to 20 min according to the site and the season, depending on the rate of variation of CO_2 concentration within the chamber and on the duration of a stable (linear) signal to make consistent calculations of CO_2 fluxes. Respiration (dark incubation) was measured from one to seven times per day to deal with variability of respiration during emersion.

Measurements were performed during the year 2000–2001 on four stations located in three different sites along the French coast of the English Channel (Figure 6.2), chosen within a gradient from exposed to sheltered conditions: an exposed sandy beach at Wimereux (50°45.905′ N, 1°36.397′ E), a muddy-sand sediment (about 2% mud in the sediment) at Le Crotoy (Somme estuary, 50°13.554′ N, 1°36.449′ E), a muddy-sand sediment (about 15% mud in the sediment, 49°26.841′ N, 00°14.622′ E) and a sandy-mud sediment (about 50% mud in the sediment, 49°26.882′ N, 00°14.592′ E) near Le Havre (Seine estuary).

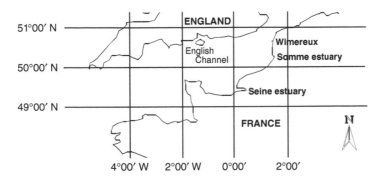

FIGURE 6.2 Location of the different sites of measurement.

All sites were located in the zone of retention, between mean high water of neap tides and mean tide level and were then subject to flooding (3 h per tidal cycle) twice a day. Measurements were made seasonally in the Seine estuary, monthly at Le Crotoy and Wimereux, with a higher frequency in this latter site during spring.

Annual carbon production due to gross primary production was estimated at Le Crotoy, using four production-irradiance curves (February, April, July, and October). An estimation was realized every day by taking into account a daily theoretical irradiance curve (see below in results), the combination of the times of sunrise, sunset and the duration of emersion, with a time step of 1 min.

6.3 Results

6.3.1 *In Situ* Measurements and Estimation of Daily Potential Primary Production

During each incubation, linear variations of CO_2 concentration (Figure 6.3) occurred after a period of stabilization (about 1 to 10 min), which was generally longer in dark incubations than in light ones. The slope of CO_2 concentration vs. time was calculated (least-square regression) from the linear part of each recording: it changed as a function of irradiance and a series of incubations within 1 day allowed us to fit a production-irradiance (P-I) curve (Figure 6.4), according to the following equation:[12]

$$P = P_{max}[1 - \exp(-I/I_k)] \tag{6.1}$$

where P = gross primary production (mgC m^{-2} h^{-1}); P_{max} = maximal gross primary production under saturating irradiance; I = irradiance (μmol m^{-2} s^{-1}); I_k = onset of saturating irradiance (μmol m^{-2} s^{-1}), determined as the point of inflection on the P-I curve.

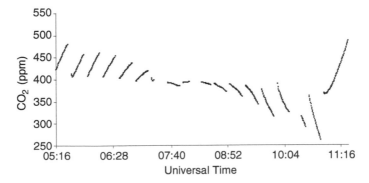

FIGURE 6.3 Example of variations of CO_2 concentration during successive light incubations under different irradiances from dawn to zenith; the last incubation is a dark one (Seine estuary, sandy mud, August 2001).

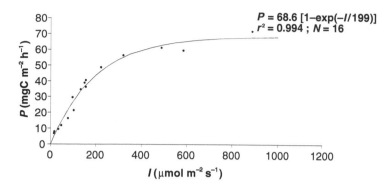

FIGURE 6.4 Example of gross production–irradiance curve fitted with results of field incubations presented in Figure 6.3.

A daily potential gross primary production can then be estimated taking into account both the theoretical daily irradiance curve and the duration of emersion on the site. A theoretical daily irradiance curve can be calculated as a function of time:[13]

$$I_t = I_{max} \cdot \sin(\pi t / DL) \tag{6.2}$$

where I_t = irradiance at time t (μmol m^{-2} s^{-1}); I_{max} = maximal irradiance (μmol m^{-2} s^{-1}); t = time (h); DL = day length (h).

Two examples were given for February and July at Le Crotoy. In both cases, we chose two close dates with the highest difference in daily carbon production within the month.

The P-I curve used in February was: $P = 6.69[1 - \exp(-I/102)]$ ($r^2 = 0.959$, $n = 9$). Calculations were made on February 6 and 16. The two daily irradiance curves were slightly different (from 7:16 to 16:55 U.T. on February 6, from 6:59 to 17:12 U.T. on February 16) but the period and the duration of emersion before the night were very different from one day to another (from 10:34 to 16:55 U.T. on February 6, from 6:59 to 15:56 U.T. on February 16), which led to a potential production 41.2% higher on February 16 (57.6 mgC m^{-2}) than on February 6 (40.8 mgC m^{-2}). This result indicated that not only the duration but also the timing of flooding (in the morning or at noon, for example) could be a major factor controlling daily gross primary production and so should be taken into account for budget estimation at longer timescales.

The P-I curve used in July was: $P = 97.71[1 - \exp(-I/310)]$ ($r^2 = 0.989$, $n = 12$). Calculations were made on July 10 and 15. The two daily irradiance curves were very close (from 4:00 to 19:52 U.T. on July 10, from 4:05 to 19:48 U.T. on July 15) but the period and duration of emersion before the night were different from one day to another (from 4:00 to 12:23 U.T. and from 15:23 to 19:52 U.T. on July 10, from 6:47 to 16:19 U.T. and from 19:19 to 19:48 U.T. on July 15), which led to a potential production 18.5% higher on July 10 (1111.2 mgC m^{-2}) than on July 15 (938.1 mgC m^{-2}). The lower relative difference observed in July is mainly due to the longer duration of the day but the absolute difference is higher in July (173.1 mgC m^{-2}) than in February (16.8 mgC m^{-2}) and so summer variations may play a significant role in the estimation of the annual carbon production.

We also calculated gross primary production, on the one hand using a theoretical irradiance curve, on the other hand using the actual irradiance data (Figure 6.5) during 3 h on February 16 and during 5 h on July 10. On February 16, calculations led to 19.9 mgC m^{-2} with the theoretical curve and 16.4 mgC m^{-2} with the actual data, that is 17.6% lower. On July 10, calculations led to 472.8 mgC m^{-2} with the theoretical curve and 336.6 mgC m^{-2} with the actual data, that is, 28.8% lower. The absolute difference (per hour) is of course higher in summer when production is higher than in winter, but the relative difference is also higher in summer. That may be explained by the higher value of I_k in July (310 μmol m^{-2} s^{-1} instead of 102 μmol m^{-2} s^{-1} in February): actual data of irradiance stayed above I_k in February even under a cloudy sky whereas irradiance recorded on July 10 stayed a long time below I_k and led to a low primary production under unsaturating irradiance.

6.3.2 Seasonal Variations of Primary Production and Respiration

No seasonal response seemed clearly to occur for gross primary production in the exposed sandy beach at Wimereux, certainly because of short-term instability of sediment, which did not allow microphytobenthic

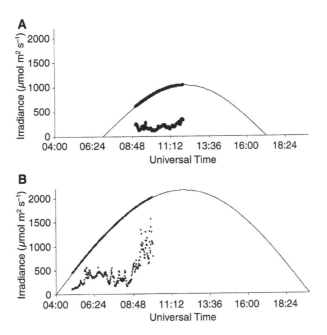

FIGURE 6.5 Theoretical irradiance curve and actual data recorded on (A) February 16th and (B) July 10th in Le Crotoy (Somme estuary).

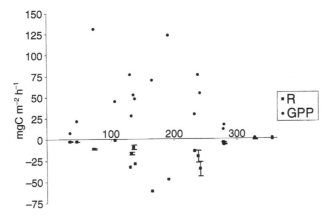

FIGURE 6.6 Seasonal variations of respiration and maximal gross primary production in Le Crotoy (Somme estuary) as a function of days.

resilience (chlorophyll *a*, or Chl*a*, concentration was quite variable and always less than 5 mgChl*a* m^{-2}) and so continuous production. Gross primary production remained low and highly variable all along the year, always less than 16 mgC m^{-2} h^{-1} (generally less than 5 mgC m^{-2} h^{-1}) as did respiration (always less than 7 mgC m^{-2} h^{-1}).

In the sheltered conditions of estuaries, both gross primary production and respiration remained higher and a seasonal trend in production and respiration could be observed. In Le Crotoy, for example, the community respiration showed a strong seasonal trend, varying from 0.5 mgC m^{-2} h^{-1} in December to 61.5 mgC m^{-2} h^{-1} in June (Figure 6.6). Variability of respiration remained low during the emersion, except in August when it varied with temperature (see standard deviations in Figure 6.6). A seasonal response was also observed for maximal gross primary production (Figure 6.6), with a minimum value in winter (1.1 mgC m^{-2} h^{-1} in November), a maximum value in spring (130.5 mgC m^{-2} h^{-1} in March), a decrease in spring to 27.2 mgC m^{-2} h^{-1} in May, and a relative maximum value at the beginning of summer (122.8 mgC m^{-2} h^{-1} in July). This seasonal pattern looks almost identical to that of respiration, with the

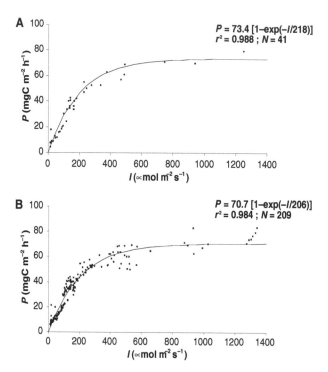

FIGURE 6.7 Gross production-irradiance curves derived from the curve of the Figure 6.4 and recalculated with periods of 5 min and 1 min of recordings, respectively.

notable exception of the spring maximum value. This very high production was due to a very high productivity and not to a very high Chl$a$ concentration (36.2 ± 9.2 mgChl$a$ m^{-2} in March vs. 228.5 ± 61.4 mgChl$a$ m^{-2} in July), which could indicate small phytoplanktonic cell deposits on the sediment surface following a water column spring bloom: actually, small cells showed higher productivity[14] and large blooms due to *Phaeocystis* sp. (Prymnesiophyceae) occurred in the eastern English Channel during spring every year.[15]

This strong seasonal response allowed us to estimate an annual potential gross primary production at Le Crotoy, using both seasonal P-I curves and day to day changes in irradiance and tidal variations, with the aim to take into account observed short time variability. This production was estimated to 140.3 gC m^{-2} y^{-1}.

6.3.3 Temporal Resolution of Measurements and Microscale Adaptation of Microphytobenthos to Variations in Irradiance

Previous calculations (P-I curves) have been made using trends (slopes of CO_2 concentration vs. time) during recordings from 10 to 20 min. It could be assumed that response of microphytobenthos to steady variations of irradiance might be recorded with a high frequency, as has already been shown under subtidal conditions.[16] Then, as logging frequency was 30 s for CO_2 concentration and 1 min for irradiance, CO_2 production was successively calculated during each period of 5 min of recordings, then during each period of 1 min, and new P-I curves were established according to Equation 6.1. Photosynthetic parameters of equations were very close as well as determination coefficients (e.g., in Figure 6.7). It clearly showed a progressive adaptation of microphytobenthos to continuous increasing or decreasing irradiance. On the contrary, when fast and irregular variations of irradiance occurred (cloudy conditions, for example), it was impossible to calculate a P-I curve in some of the experiments, whatever the timescale, because variations of CO_2 production did not always follow irradiance variations.

6.3.4 Mesoscale Variations within the Gradient of Exposure

Beyond seasonal variations, which can be seen, particularly for respiration, a significant variability occurred within the gradient of exposure (Figure 6.8), both for respiration and primary production. In

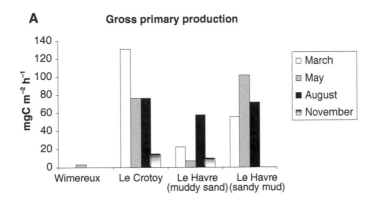

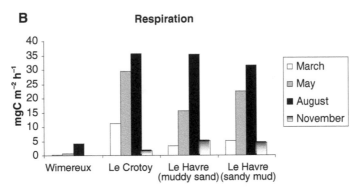

FIGURE 6.8 Comparison of (A) maximal gross primary production and (B) respiration measured on the four sites in March, May, August, and November.

the exposed conditions of the sandy marine beach of Wimereux, respiration and gross primary production remained very much lower than those of other locations all year. Gross primary production (Figure 6.8A) showed a greater variability, with the maximum value in March in Le Crotoy, in May in Wimereux and in the sandy-mud sediment in the Seine estuary, and in August in the muddy-sand sediment in the Seine estuary. It was higher in the sandy mud than in the muddy sand, except in winter conditions.

Neither gross production nor respiration therefore seemed to follow evenly the gradient of exposure, but they clearly indicated a higher benthic metabolism, both autotrophic and heterotrophic, in more or less muddy estuarine sediments.

6.3.5 Microscale Variability

A microscale variability experiment was carried out on three adjacent (a few centimeters from each other) areas (each 0.126 m²) in Le Crotoy in October. Three series of incubations were conducted, each series consisting of successive 15-min measurements of net production, under saturating irradiance, from 2 h before zenith to 2 h after. Irradiance during this period was 795 ($\pm$ 189) µmol m^{-2} s^{-1} ($\pm$ standard deviation). A single dark incubation was then performed at each area with the aim of comparing respirations, and to estimate gross primary production.

Results (Figure 6.9) did not show significant differences for gross primary production ($p > 0.05$, Kruskal–Wallis test) between areas, and variability within each area remained low (gross primary productions $\pm$ standard deviation: $P_1 = 42.9 \pm 2.9$ mgC m^{-2} h^{-1}, $P_2 = 45.8 \pm 0.4$ mgC m^{-2} h^{-1}, $P_3 = 41.2 \pm 0.8$ mgC m^{-2} h^{-1}, respectively). Respiration also varied little between locations (Figure 6.9). At this spatial scale, the intertidal system appeared relatively homogeneous. Surficial sediment seemed homogeneous and microscale patchiness of microphytobenthos, such as shown by Blanchard,[17] is integrated within the area of measurement: three measurements of Chl*a* concentration (1.6 cm of diameter, 1 cm deep) were realized within each of the three experimental areas; values varied from 55.5 to

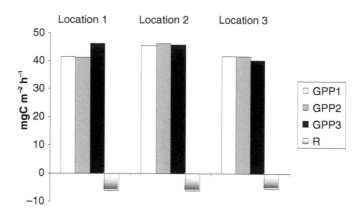

FIGURE 6.9 Comparison of respiration and gross primary production measured on three adjacent elementary areas in Le Crotoy (Somme estuary). GPP1: gross primary production during incubation 1, GPP2: gross primary production during incubation 2, GPP3: gross primary production during incubation 3, R: respiration.

85.2 mgChl$a$ m^{-2}, but there was no significant difference between areas ($p > 0.05$, Kruskal–Wallis test), with variability within each area higher than variability between areas (Chl$a$ concentration $\pm$ standard deviation: $C_1 = 79.9 \pm 7.5$ mg m^{-2}, $C_2 = 71.2 \pm 14.0$ mg m^{-2}, $C_3 = 63.6 \pm 11.8$ mg m^{-2}, respectively).

6.4 Conclusion and Perspectives

Coastal ecosystems are well known for their high physical, chemical, and biological variability,[18] mainly due to multiscale physical forcings (seasonal, tidal, mesoscale weather, and currents) and to the closeness of interfaces with other systems (continent, atmosphere, offshore ocean). Intertidal areas are subject to all these forcings and particularly to specific variations due to the alternance of emersion and immersion, which induces drastic thermal and irradiance changes. Consequently, short-term variations of intertidal primary production and respiration during a tidal cycle could be almost as dramatic as long-term variability such as seasonal or interannual processes.

In the present study, microalgal communities appear to be able to respond to gradual changes in light intensity over the course of a day following a relationship that can generally be described in a P-I curve. However, P-I curves cannot be easily constructed when unsteady changes in irradiance have occurred, under cloudy sky, for example. It could be due to several reasons such as a measurement artifact if changes in irradiance occur faster than irradiance data were collected, but this may be because the algal community cannot respond to a highly variable light environment (fast and short successive increases and decreases in light intensity).

The daily gross primary production within a given season, at least in winter, may vary more from one day to another due to the timing between the emersion and the sun course, rather than with the cloudiness of the sky. It might even vary as much from one day to another (in summer, for example) as between different seasons (between spring and autumn, for example), thus making the calculation of a monthly carbon budget difficult.

Carbon budget calculations are also made more difficult by poor knowledge of the role of seasonal development of microphytobenthic assemblages in the productivity of intertidal sediments and, on the other hand, the potential ability for deposited phytoplankton (such as during spring blooms) to carry on photosynthesis on intertidal sediments during emersion. In the present study, our calculations took into account day-to-day changes in irradiance and tidal variations and allowed us to integrate a part of short-time variability at Le Crotoy (Somme estuary). The estimation (140.3 gC m^{-2} y^{-1}) is rather high, in the upper part of estimates of primary production in intertidal sand beaches and mudflats.[18] However, small spatial variability is not yet well understood and measured enough to estimate the annual production of the whole system (see below).

Moreover, it appeared that the response of microphytobenthos to short-term variations in irradiance was not regular and depended on the variation patterns. Production regularly increased from dawn up to saturation or decreased toward twilight if irradiance increase or decrease was regular, but it did not seem to show fast adaptability when irradiance development was disturbed by cloudy periods before saturation.

Seasonal variations of respiration clearly occurred in the four sites, depending both on the abundance of heterotrophs (microbiota are assumed to be the major group responsible of heterotrophic activity) and on the temperature. No seasonal patterns in gross primary production occurred in the exposed sandy beach at Wimereux, whereas seasonal effects were strong but unsteady from one site to another in estuaries (maxima in spring at Le Crotoy and in summer in the Seine estuary).

Both marine sandy beaches and estuarine tidal flats showed a characteristic spatial distribution of zoobenthic communities at small scale, strongly correlated with geomorphological features, granulometric properties, and water retention characteristics of the sediment.[20,21] As zoobenthic zonation resulted from these characteristics and general dynamics of the system, it can be assumed that it also influenced the general distribution of microphytobenthos and so its production.[18] However, Guarini et al.[22] reported large patches of microphytobenthos in Marennes-Oléron Bay (Atlantic coast, France) distributed independently from the geomorphological structure and dominant gradient (emersion) of the area.

At a smaller scale of few tens of square meters, small variations in the level of sediment or of water content could act on production and respiration[23,24] and should be estimated. In the present study, although microscale variability occurred in Chl*a* concentration, which indicated a typical patchiness of microalgal communities,[25] gross photosynthesis and respiration were homogeneous on a larger scale of several square meters (Figure 6.9).

At the mesoscale, spatial variations of primary production clearly followed general features of structure and dynamics between, on the one hand, exposed sandy beaches and, on the other hand, sheltered mudflats. However, variations in production and respiration could not be simply related to the gradient of exposure as expressed by the granulometric gradient (percent of mud in sediment). This may be true even for cases where the abundance of bacteria increased with smaller sediment particle size and higher carbon and nitrogen occurring in finer sediments[18] or with nutrients available in the sediment.[8] Decreasing exposure and increasing sediment stability favor biodiversity and production in sheltered shores,[18] as has been shown by composite measures of exposure, particle size, and slope of the beach (the beach stability index of McLachlan,[19] for example). However, diversity generally declines with salinity in sheltered bays or estuarine gradients.

In the present study, it could be assumed that the greater production and respiration measured in muddy estuarine sediments than in the sandy exposed beach of Wimereux is due to the occurrence of a threshold energy level, mainly controlled by wave action, for microalgal colonization of sediment. As the exposed sandy beach at Wimereux is generally above that level, microphytobenthos cannot become well established so production and respiration rates remain low. On the contrary, the energy level could generally stay below a critical threshold in more sheltered estuarine areas and so favor sediment colonization and stabilization by microphytobenthos, which could then be more abundant and productive.

The next interesting step would be to test the effect of different patterns of irradiance variability by laboratory simulation to gain a better understanding of the adaptability of microphytobenthos.

Environmental factors, such as temperature variations and meiobenthic/microbenthic biomass variations, acting on seasonal variations of primary production and respiration, also need to be precisely taken into account. Furthermore, it appears important to deal with spatial distribution and photosynthetic characteristics of microphytobenthos in relation to standard features of intertidal zonation and other small-scale heterogeneity of sand and mudflats.

Further experiments will be conducted using three similar systems simultaneously, with a view to dealing with microscale (several square meters) and small-scale (from few tens of square meters within a facies to variability between facies) variations in the four studied sites.

Acknowledgments

This study has been supported by a grant (Action Thématique Innovante 99N50/0345) from the Institut National des Sciences de l'Univers (INSU) and by the scientific programs Seine-Aval 2 and PNEC

(Programme National Environnement Côtier). The authors thank the anonymous reviewers and Pete Strutton for their very helpful comments on the manuscript and Mr. and Mrs. Migné for providing croquet equipment.

References

1. Charpy-Roubaud, C. and A. Sournia, The comparative estimation of phytoplanktonic, microphytobenthic and macrophytobenthic primary production in the oceans. *Mar. Microbial Food Webs*, 4, 31, 1990.
2. Mann, K.H., Seaweeds: their productivity and strategy for growth. The role of large marine algae in coastal productivity is far more important than has been suspected. *Science*, 182, 975, 1973.
3. Asmus, R., Field measurements on seasonal variation of the activity of primary producers on a sandy tidal flat in the northern Wadden Sea. *Neth. J. Sea Res.*, 16, 389, 1982.
4. Mann, K.H., Ecological energetics of the seaweed zone in a marine bay on the Atlantic coast of Canada. II. Productivity of the seaweeds. *Mar. Biol.*, 14, 199, 1972.
5. Blanchard, G. and J.-M. Guarini, Studying the role of mud temperature on the hourly variation of the photosynthetic capacity of microphytobenthos in intertidal areas. *C. R. Acad. Sci. Paris*, 319, 1153, 1996.
6. Mortimer, R.J.G., M.D. Krom, P.G. Watson, P.E. Frickers, J.T. Davey, and R.J. Clifton, Sediment-water exchange of nutrients in the intertidal zone of the Humber estuary, U.K. *Mar. Pollut. Bull.*, 37, 261, 1998.
7. Hondeveld, B.J.M., R.P.M. Bak, W. van Raaphorst, and F.C. Van Duyl, Impact of grazing by benthic eucaryotic organisms on the nitrogen sediment-water exchange in the North Sea. *J. Sea Res.*, 41, 255, 1999.
8. Barranguet, C., J. Kronkamp, and J. Peene, Factors controlling primary production and photosynthetic characteristics of intertidal microphytobenthos. *Mar. Ecol. Prog. Ser.*, 173, 117, 1998.
9. Underwood, G.J.C. and J. Kronkamp, Primary production by phytoplankton and microphytobenthos in estuaries. *Adv. Ecol. Res.*, 29, 92, 1999.
10. Loquet, N., H. Rybarczyk, and B. Elkaïm, Echanges de sels nutritifs entre la zone côtière et un système estuarien intertidal: la Baie de Somme (Manche, France). *Oceanol. Acta*, 23, 47, 2000.
11. Migné, A., D. Davoult, N. Spilmont, D. Menu, G. Boucher, J.-P. Gattuso, and H. Rybarczyk, A closed-chamber CO_2 flux method for estimating intertidal primary production and respiration under emersed conditions. *Mar. Biol.*, 140, 865, 2002.
12. Webb, W.L., M. Newton, and D. Starr, Carbon dioxide exchange of *Almus rubra*: a mathematical method. *Oecologia*, 17, 281, 1974.
13. Lizon, F., L. Seuront, and Y. Lagadeuc, Photoadaptation and primary production study in tidally mixed coastal waters using a Lagrangian model. *Mar. Ecol. Prog. Ser.*, 169, 43, 1998.
14. Kirk, J.T.O., *Light and Photosynthesis in Aquatic Ecosystems*, 2nd ed., Cambridge University Press, Cambridge, U.K., 1994, chap. 9.
15. Gentilhomme, V. and F. Lizon, Seasonal cycle of nitrogen and phytoplankton biomass in a well-mixed coastal system (eastern English Channel). *Hydrobiologia*, 361, 191, 1998.
16. Boucher, G., J. Clavier, C. Hily, and J.-P. Gattuso, Contribution of soft-bottoms to the community metabolism (primary production and calcification) of a barrier reef flat (Moorea, French Polynesia). *J. Exp. Mar. Biol. Ecol.*, 225, 269, 1998.
17. Blanchard, G., Overlapping microscale dispersion patterns of meiofauna and microphytobenthos. *Mar. Ecol. Prog. Ser.*, 68, 101, 1990.
18. Raffaelli, D. and S. Hawkins, *Intertidal Ecology*, 2nd ed., Kluwer Academic, Dordrecht, the Netherlands, 1999, chap. 6.
19. McLachlan, A., Sand beach ecology, swash features relevant to the macrofauna. *J. Coastal Res.*, 8, 398, 1992.
20. Salvat, B., Les conditions hydrodynamiques interstitielles des sédiments meubles intertidaux et la répartition verticale de la faune endogée. *C. R. Acad. Sci. Paris*, 259, 1576, 1964.
21 McLachlan, A., Dissipative beaches and macrofauna communities on exposed intertidal sands, *J. Coast. Res.*, 6, 57, 1990.
22. Guarini, J.-M., et al., Dynamics of spatial patterns of microphytobenthic biomass: inferences from a geostatistical analysis of two comprehensive surveys in Marennes-Oléron Bay (France). *Mar. Ecol. Prog. Ser.*, 166, 131, 1998.

23. Toulmond, A., La respiration chez les Annélides. *Oceanis*, 3(7), 308, 1976.
24. Dye, A.H., Tidal fluctuations in biological oxygen demand in exposed sand beaches. *Estuarine Coastal Mar. Sci.*, 11, 1, 1980.
25. Seuront, L. and N. Spilmont, Self-organized criticality in intertidal microphytobenthos patch patterns. *Physica A*, 313, 513, 2002.

7

Spatially Extensive, High Resolution Images of Rocky Shore Communities

David R. Blakeway, Carlos D. Robles, David A. Fuentes, and Hong-Lie Qiu

CONTENTS

7.1 Introduction 109
    7.1.1 The Intertidal Context 109
    7.1.2 Limitations of Traditional Sampling Approaches 110
    7.1.3 Spatially Extensive High Resolution Images (SEHRI) 111
7.2 The System 111
7.3 Demonstration of Concept 113
    7.3.1 Using the SEHRI in a GIS Database 113
    7.3.2 Analysis of Patch Scale 116
    7.3.3 Retrospective Analysis with the SEHRI 118
7.4 Discussion 119
    7.4.1 Expanding the Utility of SEHRI 119
    7.4.2 Limitations of the Technique and Recommendations for Optimizing Results 120
    7.4.3 Photographic Equipment, Settings, and Conditions 120
7.5 Conclusions 121
Acknowledgments 121
References 121

7.1 Introduction

7.1.1 The Intertidal Context

The vivid spatial patterns of communities on rocky shores have long served as inspiration for ecological thought. Vertical zonation of mussels and barnacles, to cite a prominent example, is a focal point of the keystone predator and refuge hypotheses.[1,2] Similarly, the marked patchiness of mussel beds resulting from physical disturbances has prompted theory about the maintenance of species diversity.[3,4] Although some earlier descriptive studies of zonation considered how zones vary over scales of tens to hundreds of meters,[5] only recently has scale itself become a central issue for studies of spatial patterns. Many recent field studies are concerned with discovering (1) at what spatial scale certain patterns are apparent or (2) how processes interact over different scales to produce the range of patterns observed. As examples of the first, several studies[6–8] independently investigated the extent to which distributions of mussels exhibit fractal properties. As an example of the second, Petraitis and others[9,10] proposed that whether one of two alternative intertidal species assemblages (alternative stable states) becomes established depends on the spatial scale of physical disturbances. Most recently, applications of spatially explicit population models to intertidal communities examine the scalar relationships of certain spatial patterns and the processes underlying them.[11,12] Therefore, an examination of methods facilitating comparisons of pattern across spatial scale seems especially timely.

7.1.2   Limitations of Traditional Sampling Approaches

Field studies often rely on the conventional sampling method of stratified random survey. The basic sample unit is usually a quadrat, sometimes a point sample, placed regularly or randomly within strata (subregions within the sampling area), and from which abundance estimates (e.g., percent cover, density) are made. The strata are laid out as subdivisions of environmental features, such as shore level, distance along shore, aspect of the substratum, and so forth. Spatial patterns are constructed by comparing the abundance estimates among subdivisions and strata. The comparisons are represented in tabular form or graphically as histograms. The significance of the difference among categories is then tested with ANOVA. Thus, in both representation and analysis, environmental variation is cast into categories, and the spatial patterns emerging from this method must to some extent be determined by the initial definitions of those categories.

The stratified random sampling method is a straightforward means for testing the association of the abundance estimates with the preselected environmental features. However, we also see limitations of this approach, when applied to understanding spatial patterns over a range of scales. Surveying sample grids in the field is time-consuming, and time constraints imposed by the tides or project budget can severely limit the size, and hence the spatial extent, of the total sample. This is especially true for grids of quadrats for which estimates are made in the field. Data acquisition in the field can be speeded up, however, by photographing the sample quadrats ("photoquadrats," for example;[13,14] see Buckland et al.[15] for a discussion of quadrat-free sampling). Neither the point sampling nor quadrat methods record information about the spatial configuration of individuals or the spatial arrangement of species at the level of the individual sample itself, unless additional measures are taken (e.g., nearest neighbor analysis[16]).

Regardless of the method of acquisition, subsequent analyses of spatial variation above the level of the individual quadrat are constrained by the layout of the stratified random array. Frequently, technical or logistical limitations lead to differences in the intensity of sampling, and hence differences in pattern resolution, across spatial scales. Investigators focusing on mechanisms at relatively small scales sometimes pay less attention to patterns at larger scales. For example, Robles et al.[17] provide a relatively high-resolution view (12 shore levels over 4 m) of the vertical zonation of mussels, but a low-resolution view (three sites over 5 km) of how the zonation changes at locations with different levels of mussel recruitment. Even when studies are designed with comparisons of scale in mind, the levels of scales are predetermined, and the amount of spatial information actually extracted from the system is small. For example, in one of the first studies of spatial variation in recruitment at different spatial scales, Caffey[18] confirmed differences in spatial variance of barnacle recruitment at three predetermined levels, but the sampling array could not reveal the spatial patterns of recruits or adults over the landscape. To be fair, this was not the aim of the study, but the logical next step for studies investigating mechanisms generating variance at different scales is to look for clues in the spatial patterns themselves. Conventional sampling designs for the intertidal zone often do not allow the investigator to readily compare spatial patterns across scales.

As the previous passage implies, stratified random quadrat arrays are often set up to test specific relationships. If the investigator has chosen categories that span sufficient variation in relevant (causally related) factors, then marked spatial patterns are likely to emerge, and the subsequent ANOVA will be significant. However, relationships unknown at the time of planning the sampling design may go unnoticed in the subsequent analysis, or, if seen, might require more field trips to yield a significant description. Thus, retrospective analyses can be problematic.

The lack of retrospective capability may be especially troublesome when the aim is not just to determine whether spatial patterns correspond to certain environmental features but, rather, when the goal is a more general characterization of the community. Assessments of long-term environmental change or characterizations of baseline conditions for environmental injury require a comprehensive spatial depiction of the community.[19] It is the very selectivity of stratified random quadrat methods — omitting nontargeted species, spatial configurations, or certain areas of the landscape that later become the focus of concern — that is the method's chief drawback in *post hoc* analyses. For example, injury from an oil spill may be restricted to a narrow range of shore levels because heaviest deposition occurs along the waterline at the time of the spill. If prior baseline surveys defined strata by shore levels ("striped" alongshore transects), estimates of change could be exaggerated or underestimated, depending on whether the level of deposition happened to coincide with a sampling stratum[20,21] (see Miller and

Ambrose[22] for an evaluation of various sample designs in the intertidal zone, including vertical vs. horizontal transects). Such problems reveal the larger difficulty that the assessment of impacts may be determined inadvertently by *a priori* definitions of conventional sampling designs.[23]

7.1.3 Spatially Extensive High Resolution Images (SEHRI)

The foregoing limitations all involve a loss of information about pattern to varying degrees at different scales. While any observation of nature "extracts" a subset of information from the whole, the conventional methods discard more information than seems warranted, given recent advances in technology. What we require are methods providing data that allow the investigator to shift perspective easily among spatial scales. Such a database would provide sufficiently high resolution at any arbitrarily chosen scale so that patterns unrecognizable at one scale can be reliably seen at another. The information stored must be comprehensive, rather than selective, to allow *post hoc* explorations of relationships unsuspected at the time of acquisition. To do this one must preserve spatial relationships of all species across the range of scales. The techniques of acquisition must be convenient and cheap enough to allow spatially extensive surveys.

With increasingly efficient and inexpensive means of image processing and analysis, spatially extensive images present themselves as an effective alternative to traditional quadrat or point sampling in the field. Recent studies employing kite blimps[24] or low aerial photography[25,26] provide a look at patterns of community dominants over scales of tens to thousands of square meters. For example, Guichard et al.[24] took overlapping color photographs using a 35mm camera aboard a 6-m-long helium-filled blimp tethered at 50 to 80 m altitude. The resulting photo-mosaic image covered well over a thousand square meters with a resolution of about 0.02 m.

We developed a less expensive and more easily deployed method of obtaining a finer-scale photo-mosaic that provides sufficiently high resolution (<1 cm) to identify and measure small benthic organisms. We call the product Spatially Extensive High Resolution Images (SEHRI). The method employs a handheld pole with a 35mm camera atop. The method has the benefits of quick acquisition, high resolution, and yet an arbitrarily large areal extent of the total image. The photo-mosaic is assembled using recently developed freeware for desktop PCs.

To illustrate its utility, we demonstrate three applications with a SEHRI of mussel and barnacle populations on a rocky shore on Santa Catalina Island, California. In the first application, we combine the SEHRI with data acquired by other means (topography, wave action, mussel settlement), as layers in a geographical information system (GIS) database. Comparison of patterns on different layers suggests the process underlying the adult mussel distributions. Second, we compare the relative scales of patchiness apparent in the distributions of mussels and gooseneck barnacles, using semivariograms. Third, we analyze historical changes in mussel aggregation by comparing the recent SEHRI with an earlier panoramic photograph taken with a handheld camera. The oblique panoramic image is spatially registered to the SEHRI, and the relative fragmentation of the mussel bed quantified.

7.2 The System

A trial of the method was undertaken at Bird Rock, an islet of volcanic breccia offshore from Santa Catalina Island in the Southern California Bight (33°N 118°W, Figure 7.1). The western end of Bird Rock is exposed to the prevailing swell, and has developed a narrow wave-cut intertidal platform. The platform is occupied predominantly by a bed of California mussels, *Mytilus californianus*, and gooseneck barnacles, *Pollicipes polymerus*. Previous research along the entire intertidal platform (Robles, unpublished data) had established a roughly regular array of 115 points, identified by numbered plastic tags bolted to the rock. A 20 × 10 m section of the platform, containing 29 of the tagged sites, was selected for the photographic survey in May 2001. Monofilament fishing line was extended over the area in parallel lines, each 2.5 m apart, as a guide for the photography. A series of overlapping photographs was taken along the lines using the pole-mounted camera. The camera-to-ground distance was 3.2 m,

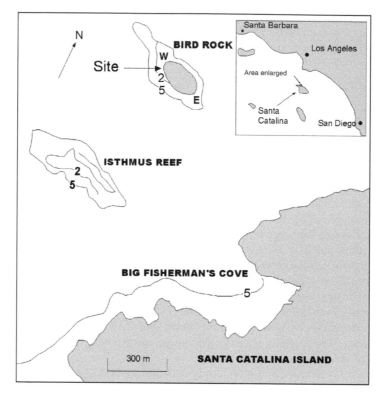

FIGURE 7.1 Location of the study site in Southern California.

and the field of view 3.3 × 2.1 m. A photograph was taken every 1.6 m along the monofilament lines; thus the photographs overlapped by approximately 50 cm on each border.

The photographic equipment comprised a 35mm Nikon SLR with a 35mm f2.5 lens, a motor drive, and a remote trigger cord (Figure 7.2). The camera and motor drive were mounted with a metal bracket on a Topcon® (Topcon America Corp., Pleasanton, CA) telescopic surveying pole. The bracket was bent so that the camera lens was oriented straight down when the pole was inclined 20° from vertical. The inclination of the pole kept the operator and equipment out of the image. A circular spirit level mounted on the pole at a 70° angle helped the operator keep the camera level when the pole was extended.

Standard 200ASA Kodak Elite Chrome transparency film was used for the 43 exposures covering the study area. Transparencies were scanned in RGB color at 2700 dpi, then the digital images were resampled to 1500 dpi and jpeg compressed in Adobe Photoshop® (Adobe Systems, Inc., San Jose, CA). The reduction in resolution to 1500 dpi was necessary to keep the images to a manageable size. We retained the 2700 dpi images for substitution back into the final mosaic once adequate computer power becomes available at a reasonable cost. The freeware Photoshop plug-in "Panorama Tools" was used to mosaic the 1500 dpi images. Panorama Tools assembles a mosaic based on user-defined control points in the images, and optimizes the fit of the images by compensating for differences in image scale, lens distortion, pitch, yaw, and roll. Both the mosaic assembly and image optimization functions can be carried out automatically, but we found it necessary to fine-tune the settings manually to obtain the best results. The manual positioning function also allowed us to register the image to the coordinate system obtained from the topographic survey. We did this by overlaying the mosaic on a background showing the x and y coordinates of the tagsites, and moving the individual images until the positions of the numbered tags corresponded as closely as possible to their surveyed coordinates. After these adjustments, the mean distance between corresponding control points in adjacent images was 12.2 cm (s.d. 8.1 cm, $n = 187$), and the mean error in tagsite position was 22.8 cm (s.d. 14.3 cm, $n = 25$).

The mosaic was output as a layered Photoshop file with each component image occupying a single layer (Figure 7.3). The layers were individually contrast stretched, sharpened with Photoshop's unsharp mask

FIGURE 7.2 Photographic equipment in operation in the intertidal zone.

function, and standardized for color balance, brightness, and contrast. After the image processing, the mosaic was flattened into a single layer and exported as a TIF file to the GIS program ArcInfo® (ESRI, Inc., Redlands, CA).

Three additional data layers were added to the ArcInfo database. These data — topography, wave action, and mussel settlement — had been previously recorded from the 29 tagsites in the mosaic. Topographic data was obtained by recording X, Y, and Z coordinates at each tagsite, and at additional points over the intertidal platform, with a Topcon TotalStation® (Topcon Corp.) survey system. As a measure of wave action, maximum bottom flow speeds were measured with spring dynamometers.[27] Each dynamometer recorded the maximum bottom flow speed at its tagsite during a 24-h (two high tides) period of deployment on February 19 and 20, 2000. Settlement data were recorded during a 4-day period in April 1999 using 10×10 cm fibrous pads attached at each tagsite (see King et al.[28] and Myers and Southgate[29] for variations on this technique). The settlement values represent the number of early post-metamorphic mussels found on each pad after the deployment period.

7.3 Demonstration of Concept

7.3.1 Using the SEHRI in a GIS Database

Few aspects of intertidal ecology can be comprehensively studied solely from an image, and we see SEHRI as useful complements to, rather than substitutes for, field observation, measurement, and experimentation. In this section we gain preliminary insights into factors influencing pattern formation in the landscape by comparing patterns of adult mussel distribution drawn from the SEHRI with GIS surfaces depicting topography, wave action, and mussel settlement rates. The surfaces were interpolated from the tagsite records by polynomial kriging.[30]

Comparison of the surfaces for topography and wave action (Figure 7.4A, B) suggests a relationship between the two. Prevailing swells, as occurred on the day of the dynamometer deployment, wash over the shore from the left to the right of the frame. As they do so, energy is dissipated, and bottom speeds decline toward higher shore levels. Minimal speeds occur high on the shore behind forward projecting contours. Maximum speeds occur at low shore levels where the topography appears to "funnel" the wave front. The mussel settlement rate appears quite variable in space (Figure 7.4C), but there is an evident trend for reduced settlement at higher shore levels.

FIGURE 7.3 (Color figure follows p. 332.) Photo-mosaic of the intertidal zone on the western end of Bird Rock, Santa Catalina Island, May 2001 (see Figure 7.1 for location). The lowest areas, occupied by the southern sea palm *Eisenia arborea* and the brown alga *Halidrys dioica*, occur along the left side of the image, and in a small pocket to the upper right. Immediately above this zone is a band of mixed turf algae, predominantly *Gigartina caniculata*, *Pterocladia capillacea*, and *Corallina officianalis*. Higher still is the mussel bed, which occupies most of the image, and gives way to bare rock in the upper intertidal. The large white rectangle delineates a 6 × 3 m subregion of the image used for an analysis of mussel and barnacle clumping. The inset is an enlarged portion of the image, showing details of individual mussel and barnacle distribution.

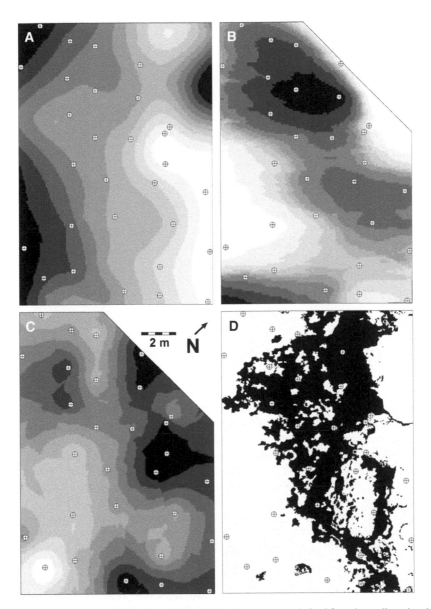

FIGURE 7.4 GIS surfaces for the Bird Rock study area. (A–C) Contour maps derived from data collected at the 29 tagged sites in the SEHRI. (D) The extent of the mussel bed at the same scale as the contour maps. Site locations are marked with crosses in each figure. (A) Topographic contours, ranging from 0.2 m (black) to 1.7 m (white) above mean lower low water. (B) Maximum wave speed contours, ranging from 3.5 m s^{-1} (black) to 7 m s^{-1} (white). The upper right corner is masked due to lack of data. (C) Mussel settlement contours, ranging from 0 (black) to 25 (white) mussels per 100 cm^2 of artificial substrate. The upper right corner is masked due to lack of data. (D) Configuration of the mussel and barnacle bed (in black) traced from Figure 7.3.

The most striking comparison, and one that appears to have the greatest biological significance, is that between patterns of juvenile settlement and adult distribution traced from the SEHRI (Figure 7.4C, D). Adults are absent from the portion of the area receiving the greatest input of juveniles (lower left of the frame of Figure 7.4C) and most abundant where mussel settlement is comparatively minimal. This suggests a spatially structured source of mortality acting on younger mussels. Earlier experimental caging studies[31] suggest that predation is the most important source of mortality, as predatory lobsters eliminated all uncaged juvenile mussels from the subarea of peak settlement (lower left corner of Figure 7.4C, site 6 in Robles[31]).

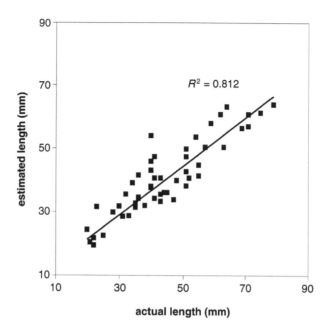

FIGURE 7.5 Correlation between mussel shell length estimated from the SEHRI and the actual length measured directly with vernier calipers.

Thus, comparisons across layers in the GIS database suggest processes of pattern formation. Greater spatial resolution and temporal replication of the points are required to be certain of the spatial patterns, and further experiments would be needed to confirm mechanisms.

The SEHRI has sufficiently high resolution to allow extraction of additional layers of GIS data. For example, spatial differences in the size frequency of benthic organisms can be quantified because even small individuals can be distinguished and measured. To test the accuracy of size measurements made from the SEHRI, we marked a sample of 50 mussels in the middle of the intertidal platform, and compared estimates of their shell length made from a scanned image (at 2700 dpi) with their actual length measured with vernier calipers in the field. As most of the mussels were oriented with their long axis vertical, we were unable to measure their lengths directly from the image. Instead, we measured width and inferred length using a polynomial equation derived from the dimensions of all 50 mussels. We obtained a good linear correlation ($R^2 = 0.812$) between the inferred and actual lengths, although the slope of the regression line was less than the ideal 1 (Figure 7.5). It should be noted that the equation we used to infer length from width will not be generally applicable at all sites, because the allometric relationship between width and length varies with environmental factors (e.g., Reference 32). Nevertheless, length-from-width measurements using a standardized equation should be at least good enough to allocate mussels into general size classes. The smallest mussels we measured were approximately 8 mm wide and 20 mm long. The resolving power of the SEHRI, tested by photographing a standard USAF 1951 test chart of black and white lines, is approximately 2 mm at 2700 dpi and 3 mm at 1500 dpi.

7.3.2   Analysis of Patch Scale

By magnifying specific selections of the SEHRI, successively closer views of the benthic community can be obtained. The first change of pattern one sees shifting from the broadest to intermediate scales is that clumps of mussels and barnacles emerge within the larger outline of the mussel bed.

To better understand the scales of patchiness of mussels and barnacles, we examined a 6 × 3 m subregion of the SEHRI, oriented perpendicular to the shore and covering most of the tidal range of the mussel bed (Figure 7.3). Within this area we counted the individual mussels and barnacles within

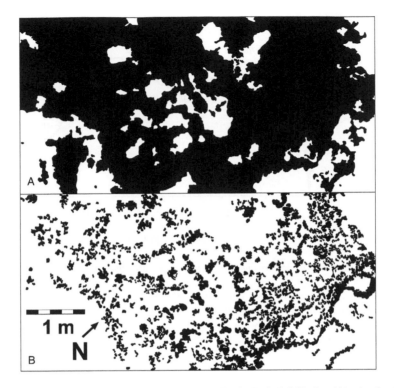

FIGURE 7.6 Cover of *M. californianus* (A) and *P. polymerus* (B), both shaded black, within the 6 × 3 m study area outlined in Figure 7.3.

contiguous 10 × 10 cm grid cells. The counts are likely to slightly underestimate the actual number of mussels, due to layering in the mussel bed. However, as most of the Bird Rock mussel bed is single layered, the potential for underestimation at this site is minimal. Counts of mussels in areas overlain by barnacles were estimated based on the density of surrounding mussels.

Mytilus californianus occupies 74% of the 18 m² area, of which 71% is a continuous bed with a few interior holes (Figure 7.6A). The *P. polymerus* population occupies 19.6% of the total area and is divided among 550 patches (Figure 7.6B). Almost all of the *P. polymerus* patches occur within the outlines of the *M. californianus* bed, and many of the patches are arranged in discontinuous "strings." These *P. polymerus* strings are generally aligned approximately along the topographic contours, and are best developed along the steep edge at the upper margin of the bed (lower right of Figure 7.6B).

Spatial patterns of mussel and barnacle distribution were described with semivariograms derived from the count data (Figure 7.7). Semivariograms quantify the autocorrelation among sample points as a function of distance (see Turner et al.[33] for a review of statistical procedures). The semivariograms in Figure 7.7 demonstrate some key differences in scaling between the two species. Both curves pass through the origin, indicating that there is no measurement error inherent in the sampling design (no "nugget effect"). The curve for *P. polymerus* rises abruptly to a definite sill, whereas that for *M. californianus* rises gradually and undergoes large amplitude fluctuations without establishing a sill. The shape of the *P. polymerus* curve suggests that the grain size of our analysis (10 × 10 cm grid cell size) is sufficiently fine and the extent of the sample area (the 6 × 3 m subregion) is sufficiently large to characterize this species. This reinforces a perception gained from the image itself: patches of the gooseneck barnacle appear as a repeated pattern over the area (Figure 7.6B). However, the 6 × 3 m area is not sufficient to produce a stable semivariogram for *M. californianus*. In Figure 7.6A, smaller patches appear to be juxtaposed with several large patches that are not completely encompassed by the bounds of the area. A much larger extent would be needed to characterize the patchiness of the mussels as a repeated pattern over the landscape.

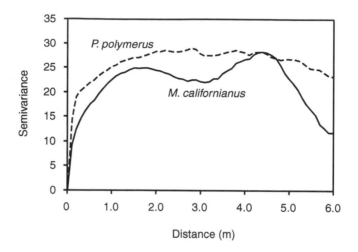

FIGURE 7.7 Semivariograms for *M. californianus* and *P. polymerus* within the 6 × 3 m study area.

The *P. polymerus* semivariogram does not indicate the apparent directionality in the distribution of this species, because it is calculated using all possible pairs of points at each distance. The technique of "directional binning" can potentially be used to distinguish directionality with semivariograms,[34] although we did not undertake these analyses.

7.3.3   Retrospective Analysis with the SEHRI

In addition to providing a baseline record for future monitoring of the benthic community, SEHRI may be used as a reference for retrospective analysis of community change. Although there are no earlier high-resolution aerial views of the Bird Rock platform, we have several earlier handheld 35mm photographs of the area, taken from the rocks above at an oblique angle of approximately 30°. Qualitatively, it is evident that the mussel bed has enlarged substantially since the first photographs were taken in 1990 and 1991. To compare the earlier configuration of the mussel bed to its contemporary configuration, we used the image processing package ENVI® (Research Systems, Inc., Boulder, CO) to warp an April 1991 photograph so that recognizable features in it were spatially registered as closely as possible to the corresponding features in the SEHRI. Mussel cover was determined within a 6 × 3 m subregion of the warped image, corresponding to the area used for the mussel and barnacle semivariograms (Figure 7.6). Total cover was 36%, comprising 73 patches (Figure 7.8). Although this value includes barnacles, it is

FIGURE 7.8 Cover of *M. californianus* and *P. polymerus* (in black) traced from the warped April 1991 photograph of the Bird Rock platform. The 6 × 3 m area corresponds to that shown in Figure 7.6.

approximately equivalent to mussel cover alone, because barnacles generally overlie mussels and therefore do not increase the projected surface area of the bed. A semivariogram could not be constructed from the warped image because its resolution was insufficient for counting. Instead, contagion analysis (a measure of clumping ranging from 0 to 1[35]) was employed to compare mussel distribution between 1991 and 2001. Not surprisingly, the two years show very different results: moderately dispersed 0.347 in 1991, and highly clumped 0.906 in 2001.

Some problems were evident in the output of the warping function (a nearest neighbor polynomial), including alignment artifacts and inaccurate matching of some registration points. These reduce the utility of the image for determining the exact positions, areas, and perimeters of mussel patches. Significant improvements in registration could likely be achieved using a more advanced registration method, such as the JUKE technique recently developed by Aschenwald et al.[36] for rectifying oblique photographs of mountainous terrain.

7.4 Discussion

7.4.1 Expanding the Utility of SEHRI

Progress in computing power and digital storage capacity makes recording and archiving SEHRI a practical supplement to, or for some purposes an alternative to, earlier methods of community characterization and monitoring. Recent software developments, such as the "Multi resolution Seamless Image Database®" (LizardTech, Inc., Santa Fe, NM) support the distribution and display of large raster images to researchers and managers worldwide via the Internet.

We suggest that SEHRI are particularly appropriate when there is a need to (1) archive detailed records of the community composition of a location, (2) analyze spatial pattern at different scales, or (3) make *post hoc* analyses of the community. All three needs are part of environmental impact assessment. Archiving the intertidal areas selected for reserves or recognized as subject to risk (i.e., near transport lanes) would support the reconstruction of baseline (pre-protection or pre-disturbance) conditions.[19] Understanding spatial patterns on arbitrary scales is crucial because both community composition and the agents of potential impacts are spatially patterned. Failure to understand the relationships among patterns of the community, the impact agent, and the sampling array employed to assess impact can seriously impair *post hoc* analyses.[23]

The analysis of spatial pattern across scale is also an integral part of the development and validation of Spatially Explicit Population Models (SEPMs). SEPMs originated in studies of terrestrial vegetation (reviews in References 38 through 40), but recent applications to intertidal communities (e.g., References 11, 12, and 41) suggest that this approach is well suited to understanding pattern formation in benthic communities. Model expressions formulate how processes at one scale may influence processes at another. The resulting output is a depiction of population and community patterns in idealized landscapes, and these patterns may be analyzed at several scales. To be sure, one can employ traditional quadrat sampling schemes to evaluate the output of SEPMs. For example, in a model of disturbance patterns within mussel beds, Wootton[12] compared predicted frequency distributions of gap sizes with observed distributions estimated from a stratified random quadrat array. However, SEHRI provide the opportunity to make direct comparisons of patterns at different scales. For example, Robles and Desharnais[11] present a cellular automaton model depicting mussel beds in an idealized landscape characterized by spatial variation in size-limited predation and prey productivity (settlement and growth). The output predicts community characteristics at the range of scales covered by SEHRI, including trends in mussel sizes and clustering at fine scales, sharp population boundaries at intermediate scales, and the shape of the entire mussel bed over the landscape.

In addition to evaluating model output, SEHRI may be potentially useful for model calibration. For example, time series images could be analyzed to determine the influence of local neighborhood conditions on community dynamics (e.g., recruitment, growth, predation, disturbance), and this information could be directly incorporated into spatial models.

7.4.2   Limitations of the Technique and Recommendations for Optimizing Results

Some spatial distortion is present in the Bird Rock SEHRI, evidenced by mismatches at the seams of the individual images. The worst mismatch between images occurs halfway up the right side of the SEHRI. This was due primarily to excessive pitch, yaw, and roll distortion introduced by poor camera orientation. The problem occurred in this area because the pole was awkwardly held over the small embayment that can be seen immediately above it in the SEHRI (Figure 7.3). Additionally, parallax distortions were exacerbated by the complex topography in this area.

Several steps could be taken to reduce spatial distortion in the photographs. A longer focal length lens would improve the perspective of the images, although it would also significantly increase the number of images required to cover a given area. A more practical way to minimize distortion may be to mount the camera on a free-swinging gymbal instead of the fixed bracket we used. This would ensure that the camera is always oriented vertically, thus eliminating pitch and yaw distortion.

Because the SEHRI is a horizontal two-dimensional projection of a three-dimensional surface, areas of sloping substrata are underrepresented. In cases where the distribution of organisms is slope dependent, this limitation could cause significant bias in abundance values derived from the SEHRI. However, suitable corrections can be applied to the image, if a detailed topography is available (see, for example, Reference 36). For the simple rectification of the panoramic photograph, we assumed that the area covered by both the oblique photograph and the normal SEHRI were flat planes, an assumption that holds reasonably well for approximately 80% of the platform. The photo-mosaic technique itself has the potential to support high-resolution topographic reconstructions through stereo analysis, which would allow correction of parallax biases. Accurate reconstruction requires that the overlap between adjacent component images is at least 60% (e.g., Reference 24).

An additional limitation is that some organisms may not be visible in the component images. Foster et al.[42] (see also References 13, 43, and 44) compared the accuracy and precision of photoquadrats relative to matched point sample quadrats recorded in the field. Given sufficiently high resolution, photoquadrats give accurate, precise measures of cover, and they, along with visual quadrat estimates, were less likely to miss rare species. However, the two-dimensional photographs cannot reveal understory layers, nor can they record minute or highly camouflaged species. For these, direct subsampling may be necessary. For example, in our first application we combined estimates of microscopic mussel settlement with the SEHRI.

Although obtaining the photographs is relatively fast, constructing an accurately registered mosaic is time-consuming. Two main aspects slow the process. First, opening, manipulating, and saving the large image files is memory intensive. This problem is likely to become less significant as computer power increases. Second, manual adjustment of the image positioning and image optimization functions is laborious. Manual adjustments are necessitated by the poor performance of the automatic assembly process. Satisfactory automatic assembly should be feasible if camera orientation errors (yaw, pitch, and roll) can be minimized, and variations in scale between images can be reduced. Orientation errors can potentially be reduced by modifying the photographic procedure, particularly by incorporating a gymbal camera mount. Variation in scale between images is difficult to eliminate, because it is an inherent aspect of using the pole camera technique in topographically complex areas. However, manually rescaling the images is relatively fast. Rapid image assembly should still be possible even if it remains necessary to rescale the images manually.

7.4.3   Photographic Equipment, Settings, and Conditions

We used a 35mm camera because its cost was relatively low, and its resolution sufficient for our purpose. A digital camera would be more convenient, not least because the quality and positioning of the images can be checked on location. At the time of this writing, the resolution of images from all but high-end professional models is currently inferior to that of images scanned from transparencies. However, advances in digital camera technology should soon overcome this limitation.

We preferred to photograph on days with mild overcast. These conditions produced diffuse light, which ensured a relatively narrow range of exposure in the images and preserved the greatest detail. We avoided days with direct sunlight, which increased the exposure range and caused loss of detail in highlights and shadows. The 200ASA Kodak Elite Chrome proved an adequate film choice, allowing shutter speeds faster than 1/30th of a second with the aperture set to f5.6 or f8. These aperture settings provided sufficient depth of field in most cases.

7.5 Conclusions

Photographic images of rocky intertidal environments contain an extraordinary amount of information about the composition and spatial arrangement of the benthic community. The recent development of powerful image stitching freeware, allied with continuing improvements in computer power and disk capacity, allow SEHRI to be assembled and stored on consumer desktop PCs. Spatial information can be readily extracted from SEHRI in a GIS framework. GIS also facilitate cross-data comparisons and the development of hypotheses regarding pattern generation in the benthic community. SEHRI should prove particularly useful for archiving detailed records of community composition, analyzing spatial pattern at different scales, and making *post hoc* analyses of community change.

Acknowledgments

We thank the staff of the Wrigley Institute of Marine Science, Santa Catalina Island. Field assistance was provided by Kathy Lazorov and Ricardo Lopez. The freeware image assembly program Panorama Tools was created by Helmut Dersch. Our research was supported by NSF award HRD9805529.

References

1. Paine, R.T., Intertidal community structure: experimental studies on the relationship between a dominant competitor and its principal predator, *Oecologia*, 15, 93–120, 1974.
2. Paine, R.T., Size limited predation: an observational and experimental approach with the *Mytilus–Pisaster* interaction, *Ecology*, 57, 858–873, 1976.
3. Dayton, P.K., Competition, disturbance, and community organization: the provision and subsequent utilization of space in a rocky intertidal community, *Ecol. Monogr.*, 41, 351–389, 1971.
4. Paine, R.T. and Levin, S.A., Intertidal landscapes: disturbance and the dynamics of pattern, *Ecol. Monogr.*, 51, 145–178, 1981.
5. Lewis, J.R., *The Ecology of Rocky Shores*, English University Press, London, 1964.
6. Commito, J.A. and Rusignuolo, B.R., Structural complexity in mussel beds: the fractal geometry of surface topography, *J. Exp. Mar. Biol. Ecol.*, 255(2), 133–152, 2000.
7. Schmid, P.E., Fractal properties of habitat and patch structure in benthic ecosystems, *Adv. Ecol. Res.*, 30, 339–401, 1999.
8. Snover, M.L. and Commito, J.A., The fractal geometry of *Mytilus edulis* L. spatial distribution in a soft-bottom system, *J. Exp. Mar. Biol. Ecol.*, 223(1), 53–64, 1998.
9. Petraitis, P.S. and Latham, R.E., The importance of scale in testing the origins of alternative community states, *Ecology*, 80(2), 429–442, 1999.
10. Petraitis, P.S. and Dudgeon, S.R., Experimental evidence for the origin of alternative communities on rocky intertidal shores, *Oikos*, 84(2), 239–245, 1999.
11. Robles, C.D. and Desharnais, R.A., History and current development of a paradigm of predation in rocky intertidal communities, *Ecology*, 83, 1521–1536, 2002.
12. Wootton, J.T., Local interactions predict large-scale patterns in empirically derived cellular automata, *Nature*, 413, 841–843, 2001.
13. Bohnsack, J.A., Photographic quantitative sampling of hard-bottom benthic communities, *Bull. Mar. Sci.*, 29, 242–252, 1979.

14. Whorff, J. and Griffing, L., A video recording and analysis system used to sample intertidal communities, *J. Exp. Mar. Biol. Ecol.*, 160, 1–12, 1992.

15. Buckland, S.T., Anderson, D. R., Burnham, K.P., and Laake, J.L., *Distance Sampling: Estimating the Abundance of Biological Populations*, Chapman & Hall, London, 1993.

16. Sokal, R.R. and Rohlf, F.J., *Biometry*, W.H. Freeman, New York, 1981.

17. Robles, C., Sherwood-Stephens, R., and Alvarado, M., Responses of a key intertidal predator to varying recruitment of its prey, *Ecology*, 76, 565–579, 1995.

18. Caffey, H.M., Spatial and temporal variation in settlement and recruitment of intertidal barnacles, *Ecol. Monogr.*, 55, 313–332, 1985.

19. Julius, B., Scaling compensatory restoration actions: guidance document for natural resource damage assessment under the Oil Pollution Act of 1990, Damage Assessment Center, National Oceanic and Atmospheric Administration, Silver Spring, MD, 1997.

20. Cubit, J.D. and Connor, J.L., Effects of the Bahia Las Minas oil spill on reef flat communities, in Long-Term Assessment of the Oil Spill at Bahia Las Minas, Panama, Synthesis Report, B.D. Keller and J.B.C. Jackson, Eds., U.S. Department of the Interior, Minerals and Management Service, Gulf of Mexico Regional Office, New Orleans, LA, 1993a, 131–242.

21. Cubit, J.D. and Connor, J.L., Effects of the 1986 Bahia Las Minas oil spill on reef flat communities, in *Proceedings of the 1993 International Oil Spill Conference*, Washington, D.C., 1993b, 329–334.

22. Miller, A.W. and Ambrose, R.F., Sampling patchy distributions: comparison of sampling designs for rocky intertidal habitats, *Mar. Ecol. Prog. Ser.*, 196, 1–14, 2000.

23. Peterson, C.H., McDonald, L.L., Green, R.H., and Erickson, W., Sampling design begets conclusions: the statistical basis for detection of injury to and recovery of shore-line communities after the "Exxon Valdez" oil spill, *Mar. Ecol. Prog. Ser.*, 210, 255–238, 2001.

24. Guichard, F., Bourget, E., and Agnard, J.-P., High-resolution remote sensing of intertidal ecosystems: a low-cost technique to link scale-dependent patterns and processes, *Limnol. Oceanogr.*, 45, 328–338, 2000.

25. Millat, G. and Herlyn, M., Documentation of intertidal mussel bed (*Mytilus edulis*) sites at the coast of Lower Saxony, *Senckenbergiana Maritima*, 29, 83–93, 1999.

26. Thome, D.M. and Thome, T.M., Radio-controlled model airplanes: inexpensive tools for low-level aerial photography, *Wildl. Soc. Bull.*, 28(2), 343–346, 2000.

27. Bell, E.C. and Denny, M.W., Quantifying "wave exposure": a simple device for recording maximum velocity and results of its use in several field studies, *J. Exp. Mar. Biol. Ecol.*, 181, 9–29, 1994.

28. King, P., McGrath, D., and Britton, W., Use of artificial substrates in monitoring mussel (*Mytilus edulis* L.) settlement on an exposed rocky shore in the west of Ireland, *J. Mar. Biol. Assoc. U.K.*, 70, 371–380, 1990.

29. Myers, A.A. and Southgate, T., Artificial substrates as a means of monitoring rocky shore cryptofauna, *J. Mar. Biol. Assoc. U.K.*, 60, 963–975, 1980.

30. Oliver, M.A. and Webster, R., Kriging: a method of interpolation for geographical information systems, *Int. J. Geogr. Inf. Syst.*, 4(3), 313–332, 1990.

31. Robles, C.D., Changing recruitment in constant species assemblages: implications for predation theory of intertidal communities, *Ecology*, 78, 1400–1414, 1997.

32. Kopp, J.C., Growth and the intertidal gradient in the sea mussel *Mytilus californianus* Conrad, 1837, *Veliger*, 22, 51–56, 1979.

33. Turner, S.J., O'Neill, R.V., Conley, W., Conley, M.R., and Humphries, H.C., Pattern and scale: statistics for landscape ecology, in *Quantitative Methods in Landscape Ecology*, M.G. Turner and R.H. Gardner, Eds., Springer, New York, 1991, 17–49.

34. Johnston, K., Ver Hoef, J.M., Krivoruchko, K., and Lucas, N., *Using ArcGIS Geostatistical Analyst*, Environmental Systems Research Institute, New York, 2001.

35. O'Neill, R.V., Krummel, J.R., Gardner, J.H., Sugihara, G., Jackson, B., DeAngelis, D.L., Milne, B.T., Turner, M.G., Zygmunt, B., Christensen, S.W., Dale, V.H., and Graham, R.L., Indices of landscape pattern, *Landscape Ecol.*, 1, 153–162, 1988.

36. Aschenwald, J., Leichter, K., Tasser, E., and Tappeiner, U., Spatio-temporal landscape analysis in mountainous terrain by means of small format photography: a methodological approach, *IEEE Trans. Geosci. Remote Sensing*, 39(4), 885–893, 2001.

37. Turner, M.G., Landscape ecology: the effect of pattern on process, *Annu. Rev. Ecol. Syst.*, 20, 171–198, 1989.

38. Hastings, A., Spatial heterogeneity and ecological models, *Ecology*, 71, 426–428, 1990.
39. Dunning, J.B., Stewart, D.J., Danielson, B.J., Noon, B.R., Root, T.L., Lamberson, R.H., and Stevens, E.E., Spatially explicit population models: current forms and future uses, *Ecol. Appl.*, 5, 3–11, 1995.
40. Tilman, D. and Kareiva, P., *Spatial Ecology: The Role of Space in Population Dynamics and Interspecific Interactions*, Princeton University Press, Princeton, NJ, 1997.
41. Burrows, T.M. and Hawkins, S.J., Modeling patch dynamics on rocky shores using determinate cellular automata, *Mar. Ecol. Prog. Ser.*, 167, 1–13, 1998.
42. Foster, M.S., Harrold, C., and Hardin, D.D., Point vs. photoquadrat estimates of the cover of sessile marine organisms, *J. Exp. Mar. Biol. Ecol.*, 146, 193–203, 1991.
43. Meese, R.J. and Tomich, P.A., Dots on the rocks: a comparison of per cent cover estimation methods, *J. Exp. Mar. Biol. Ecol.*, 165, 59–73, 1992.
44. Dethier, M.N., Graham, E.S., Cohen, S., and Tear, L.M., Visual versus random-point percent cover estimations: "objective" is not always better, *Mar. Ecol. Prog. Ser.*, 96, 93–100, 1993.

8

Food Web Dynamics in Stable Isotope Ecology: Time Integration of Different Trophic Levels

Catherine M. O'Reilly, Pieter Verburg, Robert E. Hecky, Pierre-Denis Plisnier, and Andrew S. Cohen

CONTENTS

8.1 Introduction 125
8.2 Methods 126
8.3 Results and Discussion...127
    8.3.1 Isotopic Structure of the Food Webs 127
    8.3.2 Temporal Fluctuations in the Nutrient Source 128
    8.3.3 Temporal Integration within the Food Web..129
8.4 Factors Sensitive to Time...129
    8.4.1 Direct Effects..130
    8.4.2 Indirect Effects 130
8.5 Conclusion..131
Acknowledgments..131
References...131

8.1 Introduction

Stable isotopes are becoming a standard analytical tool in food web ecology. Differences in carbon and nitrogen isotope ratios between consumers and their diet provide information on energy flows, nutrient sources, and trophic relationships. Typically, carbon provides information on the primary energy source (e.g., benthic vs. pelagic photosynthesis), while nitrogen allows discrimination among trophic levels. Relative enrichment with increasing trophic level often allows a better interpretation of dietary relationships than gut content analysis alone because stable isotopic ratios record material that is actually assimilated (Michener and Schell, 1994). A recent survey has shown an average enrichment of 0.05‰ ± 0.63 $\delta^{13}C$ and 3.49‰ ± 0.23 $\delta^{15}N$ in field studies (Vander Zanden and Rasmussen, 2001), and these values are similar to the frequently used average trophic fractionation values of 1‰ $\delta^{13}C$ and 3.4‰ $\delta^{15}N$ (Minagawa and Wada, 1984; Michener and Schell, 1994). The ease of stable isotope analyses makes them an appealing tool in ecology, but both sampling design and interpretation of the results should be undertaken carefully (Gannes et al., 1997; O'Reilly et al., 2002). An isotopic ratio of an organism represents its diet, but it should be remembered that this isotopic value is also time specific and is an average ratio related to tissue turnover rate and the life of the organism.

The pelagic food web in Lake Tanganyika, East Africa, provided an excellent example of how stable isotopic analyses of food web structure are not always straightforward. Lake Tanganyika is a deep (mean 570 m; max 1470 m), large (mean width 50 km; length 650 km) lake located a few degrees south of the equator (Figure 8.1). The lake is permanently stratified and is anoxic below 100 to 150 m. The pelagic food web is relatively simple, and trophic relationships have been previously established from

FIGURE 8.1 Location of Lake Tanganyika in East Africa. This study took place off shore from Kigoma, Tanzania.

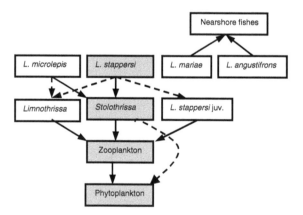

FIGURE 8.2 The pelagic food web of Lake Tanganyika. The heavy lines indicate major food preferences; the dashed lines indicate other prey relationships. Each of the *Lates* species includes *Stolothrissa tanganicae* and *Limnothrissa miodon* among their prey. The shaded boxes represent the dominant species of the pelagic zone and illustrate the linear food chain relationship. (Modified from Coulter, 1991.)

gut content analysis (Figure 8.2) (Coulter, 1991). The zooplankton are dominated by the copepod *Tropodiaptomus simplex*, which is a major dietary component of the two clupeid fish species, *Stolothrissa tanganicae* and *Limnothrissa miodon*. The upper trophic level is composed of two species that rely on nearshore fishes as a food source throughout their lives, *Lates angustifrons* and *L. mariae*. One species, *L. microlepis*, spend their larval and juvenile stages near shore and recruit to the pelagic as adults. Only one species, *L. stappersi*, has a fully pelagic life cycle. Thus, the dominant species of the pelagic food web form a linear food "chain" from phytoplankton to the copepod *T. simplex* to the zooplanktivorous *Stolothrissa* to the predatory *L. stappersi*. The isotopic structure of this food web should be a distinct sequential enrichment in the carbon and nitrogen isotopes with increasing trophic level.

8.2   Methods

Our study took place near Kigoma, Tanzania, during the dry season (May to September). Food web samples were collected two consecutive years, in 1999 (FW-A) and 2000 (FW-B). FW-B was collected by the second author over several sampling periods in August 2000. For this food web, particulate organic matter (POM) and zooplankton samples were taken at least 10 km offshore from Kigoma. POM was collected by filtering whole water samples ($n = 4$) on 0.45 μm quartz filters, which were then rinsed with 10% N HCl and distilled water. Zooplankton were collected using 100 m vertical tows ($n = 6$) with a 100 μm mesh and collected on 0.45 μm quartz filters. Fish specimens of *Stolothrissa* ($n = 7$; 71 to 85 mm), *Limnothrissa* ($n = 4$; 83 to 125 mm), *Lates stappersi* juveniles ($n = 7$; 69 to 136 mm), and *L. stappersi* ($n = 4$; 219 to 372 mm) were obtained from fishermen in the morning after they returned from fishing offshore from Kigoma. FW-A was collected by the first author, with all samples collected during the night of 1 August 1999, approximately 15 km offshore (4°50′ S, 29°29′ E). For FW-A, phytoplankton were collected using vertical tows with a 50 μm mesh net ($n = 6$). Samples were filtered through 100 μm mesh to remove zooplankton, then collected on 0.45 μm glass-fiber filters and rinsed with 10% N HCl and distilled water. Zooplankton were caught using 100 m vertical tows with a 100 μm mesh net ($n = 4$). They were placed in filtered lake water for 2 h to clear gut contents, then collected on 0.45 μm glass-fiber filters and rinsed with 0.01 N HCl and distilled water. Fish specimens of *Stolothrissa* ($n = 6$; 80 to 90 mm), *Limnothrissa* ($n = 5$; 107 to 120 mm), *Lates stappersi* juveniles ($n = 4$; 130 to 160 mm), and *L. stappersi* ($n = 4$; 262 to 322 mm) were obtained from local fishermen who were fishing adjacent to the site when the plankton samples were collected.

For all species except *Stolothrissa*, the stable isotopic analyses were done on a section of white muscle tissue from behind the dorsal fin. As *Stolothrissa* were too small to obtain a large muscle sample, the entire body was used after removing the head, tail, and viscera, with the spine additionally removed from the samples in FW-B. Fish samples were washed with distilled water, dried, and homogenized before analysis. All samples were dried at 50°C and stored wrapped in aluminum foil.

Samples were analyzed at the University of Waterloo Environmental Isotope Lab on an Isochrom Continuous Flow Stable Isotope Mass Spectrometer (Micromass) coupled to a Carla Erba Elemental Analyzer (CHNS-O EA1108). The isotope ratios are expressed in delta notation with respect to deviations from standard reference material (Pee Dee belemnite carbon and atmospheric nitrogen). Standard error is 0.2‰ for carbon and 0.3‰ for nitrogen. Statistical analyses were done using JMP IN (SAS Institute, Inc.).

8.3   Results and Discussion

8.3.1   Isotopic Structure of the Food Webs

The two isotopic studies produced radically different views of the pelagic food web. For one food web, the isotopic structure appeared as expected, with a gradual isotopic enrichment through the food chain from phytoplankton to the top predator (Figure 8.3 FW-B). For the other food web, the isotopic structure was not that of a linear food chain (Figure 8.3 FW-A). Although there was a general trend of carbon

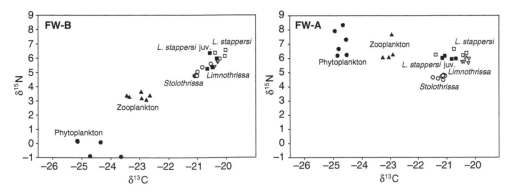

FIGURE 8.3 Isotopic structure of the food webs. FW-B shows the expected isotopic enrichment with trophic level increase. In contrast, FW-A shows a very different isotopic structure, with phytoplankton and zooplankton more enriched than would be expected.

enrichment in FW-A, there was a depletion of ^{15}N with trophic level increase. Between the two years the fish species had similar values, and the differences between the two food webs appeared in the lower trophic levels — the phytoplankton and zooplankton.

The isotopic analyses for the upper food web fish species were consistent with the known dietary preferences of these fish species. Both FW-A and FW-B had similar values. The primary prey item for adult *L. stappersi* is *Stolothrissa*, and this relationship was clearly seen in the isotope data, where *L. stappersi* was approximately 1‰ enriched in ^{13}C and 2‰ enriched in δ^{15}N relative to *Stolothrissa*. Juvenile *L. stappersi* were depleted in ^{15}N compared to the adults, which reflects the fact that their diet includes relatively more copepods (Mannini et al., 1999). The other clupeid species, *Limnothrissa*, includes smaller *Stolothrissa* in its diet, which is likely the reason for its isotopic enrichment relative to *Stolothrissa*.

Given the δ^{15}N values of the upper food web and the likely importance of atmospheric N fixation in Lake Tanganyika (Hecky, 1991), phytoplankton were expected to have a δ^{15}N value around 0‰, and zooplankton should be approximately 1‰ enriched in δ^{13}C and 3.4‰ enriched in ^{15}N relative to phytoplankton. The results from FW-B were consistent with these expectations. In contrast, FW-A had vastly different isotopic values for the lower food web that were also inconsistent with the ratios for its upper trophic levels. FW-A phytoplankton had a δ^{15}N of between 6 and 8‰, and there was no trophic level enrichment between phytoplankton and zooplankton. In addition, the lower food web had δ^{15}N values similar to or slightly enriched compared to the upper consumer levels in that food web. Why were zooplankton and phytoplankton in FW-A so enriched?

There are several possible reasons for a shift in isotopic ratios. In this case, the mechanism behind the enrichment in δ^{15}N must explain a shift of ~6‰ that occurred only in the lower trophic levels and is not apparent in the upper food web. A discrepancy of this magnitude cannot be explained easily by either a change in phytoplankton productivity rates or species composition and thus implies a change in nitrogen source.

8.3.2　Temporal Fluctuations in the Nutrient Source

Although the primary sources of new nitrogen are atmospheric deposition and biological nitrogen fixation, internal loading of deep-water nutrients is also an important nutrient source for the pelagic zone (Hecky, 1991). Lake Tanganyika is permanently stratified, but strong winds during the dry season cause seiche activity with a 28 to 36 day period, leading to episodic vertical metalimnion entrainment (Plisnier et al., 1999). Wind speeds vary diurnally, but daily mean speeds were significantly higher in the week preceding sampling (1.48 ± 0.09m s^{-1}) than in the week following sampling (0.81 ± 0.08 m s^{-1}), with four consecutive days where speeds were higher than the long-term seasonal mean (Johannes et al., 1999). Concurrently, nitrate concentration profiles showed upwelling of nitrate from 100 m deep around 28 July (Figure 8.4) (Johannes et al., 1999). These data provide strong evidence that an upwelling event occurred 4 to 6 days prior to sampling the food web.

This deep-water nitrogen is likely enriched in ^{15}N (François et al., 1996). The metalimnion of Lake Tanganyika has low oxygen levels and elevated nitrate concentrations (Hecky et al., 1991). As denitrification occurs in the suboxic section of the water column, the lighter isotope is selectively removed, and the remaining nitrate becomes increasingly enriched in ^{15}N. Field studies in temperate lakes have shown that fractionation during nitrogen assimilation by phytoplankton can be −4 to −5‰ if nitrogen is in excess (Fogel and Cifuentes, 1993), which implies that the nitrogen source for these phytoplankton would require a δ^{15}N of at least 10 to 14‰. Denitrification has fractionations in the range of 10 to 30‰ (Wada and Hattori, 1991), which would lead to an enriched nitrate pool in the suboxic metalimnion of Lake Tanganyika. For other lakes, investigators have measured deep-water nitrate values of 15.1‰ (Yoshioka et al., 1988) or have calculated values between 10 and 30‰ (Teranes and Bernasconi, 2000).

A short-term, episodic nutrient input from deeper water to the epilimnion would alter phytoplankton isotopic signals, eventually causing them to have δ^{15}N signatures in the range observed in this study. With high nitrate concentrations, discrimination against the heavier nitrogen isotope occurs and initially phytoplankton are depleted relative to dissolved nitrogen (Altabet and François, 1994). Following Rayleigh fractionation kinetics, the remaining nitrate becomes relatively enriched. As the algal bloom continues, however, demand for nitrogen remains high and discrimination against the heavier isotope

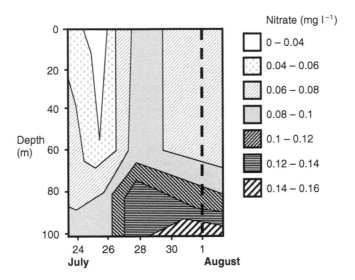

FIGURE 8.4 Nitrate concentration profiles. Nitrate concentrations from surface waters to the upper metalimnion off Kigoma Bay throughout the week prior to sampling. The dashed line indicates the sampling date. Upwelling of water from 100 m deep is indicated by the increase in nitrate concentrations around 28 July. (Data from Johannes et al., 1999.)

decreases as nitrogen concentrations decline. The pelagic zone of Lake Tanganyika is usually nitrogen limited (Hecky et al., 1991), implying that Rayleigh fractionation continues to near completion. Thus, phytoplankton eventually attain an isotope value similar to that of their nitrogen source.

8.3.3   Temporal Integration within the Food Web

While upwelling provides a mechanism for the enriched $\delta^{15}N$ values of the phytoplankton in FW-A, time-averaging explains the lack of trophic enrichment with respect to $\delta^{15}N$ for the consumer levels. The food web has different turnover times associated with each trophic level. Phytoplankton population growth rates are 1.2 day^{-1} or higher (Hecky, 1991), and thus their stable isotope signal represents carbon and nitrogen uptake and sources over the last few days. Preliminary evidence suggests that the time to full development for copepods in Lake Tanganyika is 31 to 45 days (Hyvonen, 1997); thus their isotopic ratios represent an average of phytoplankton consumed both pre- and post-upwelling. The clupeid *Stolothrissa* integrates diet over a period of several months to 1 year, and the predatory *L. stappersi* has a life span of several years (Coulter, 1991). The effect of this greater temporal integration was seen in the upper food web in Lake Tanganyika, where the short-term fluctuation in nutrient source was not apparent.

8.4   Factors Sensitive to Time

Recognizing that temporal integration occurs in an organism is important, particularly in the development of food web models or assignment of trophic level using isotopic signatures. Isotopic signals of primary producers are subject to greater variation than other trophic levels in this system because of constantly changing nutrient sources and concentrations. It is precisely because of this variation that food web modeling using isotopes is usually based on the isotope signal of the primary consumer, whose longer-term integration is assumed to reduce the short-term variability found in primary producers (i.e., Post et al., 2000). However, this work suggests that temporal variation may be significant at the primary consumer level and could affect assessment of relative trophic level.

There are several factors that are time sensitive and may affect the isotopic structure of a food web. Because a food web consists of organisms with a range of life spans, the upper trophic levels, with larger body sizes, will generally be integrating over longer time periods than lower trophic levels. A change in isotopic signature may be caused either directly, through changes in predators and/or their prey, or

TABLE 8.1

A Compilation of Factors That Are Time-Sensitive and That Have the Potential to Affect the Isotopic Signature of an Organism in a Way That Can Affect the Isotopic Structure of the Food Web

Directly	Change in Predator	Change in Prey
	Metabolic processes (starvation, age)	Immigration
	Migration	Diet quality
	Change in diet	
Indirectly	**Base of Food Web**	
	Change in productivity rate	
	Change in species composition	
	Change in nutrient or light availability	
	Change in nutrient source	
	Change in diet quality	

by indirectly affecting the nutrient base of the food web (Table 8.1). These indirect changes in the phytoplankton isotopic signature are then incorporated in the upper trophic levels at rates depending on tissue turnover.

8.4.1   Direct Effects

Several factors affect the isotopic signature of an organism directly. As a result of preferential conversion of ^{14}N in metabolic reactions, highly metabolized molecules should be enriched in ^{15}N (Minagawa and Wada, 1984). Consequently, starvation has been associated with enriched $\delta^{13}C$ and $\delta^{15}N$ values (Overman and Parrish, 2001; Oelbermann and Sheu, 2002). Although relatively few studies have been done on starvation, isotope ratios change on the order of 1.3‰ $\delta^{15}N$ and 1.5‰ $\delta^{13}C$ (Oelbermann and Scheu, 2002), which would be sufficient to confound trophic level placement. Possibly for metabolic reasons, age has been linked to increased $\delta^{15}N$ ratios for sole (Spies et al., 1989), cladocerans (Adams and Sterner, 2002), and wolf spiders (Oelbermann and Scheu, 2002) when dietary signatures were held constant. Age is also often associated with changes in dietary preferences, frequently leading to a trophic level shift from primary to higher-level consumer, which causes enrichment in both $\delta^{13}C$ and $\delta^{15}N$ (Vander Zanden et al., 1999). Migration of the predator for feeding purposes or seasonal migrations of prey might create a signature that does not reflect the predator's current environment (Hansson et al., 1997). For example, many fish species shift from benthic/littoral habitat to the pelagic at certain life stages, but may return to the littoral region for breeding purposes.

Other dietary aspects may affect an isotopic signature even though the prey items remain constant. Prey may immigrate from another locale characterized by different isotopic signatures (Hansson et al., 1997). Diet quality may also affect an organism's isotopic signature, and lower quality diets (relatively higher C:N ratios) induce nitrogen starvation, leading to enriched $\delta^{15}N$ values, up to or even greater than 3‰ (Adams and Sterner, 2000; Oelbermann and Scheu, 2002). Carbon isotope ratios may also become enriched with lower diet quality (Oelbermann and Scheu, 2002).

Recognizing that different tissue types (heart, muscles, etc.) have different turnover rates is also an important consideration (Pinnegar and Polunin, 1999). For upper trophic levels, white muscles tissue is usually the preferred tissue source because of the relatively slow turnover rate compared to other body tissues like the liver, kidney, etc. Fatty tissues are depleted in $\delta^{13}C$, and thus muscle tissue with a high lipid content may need to undergo lipid extraction to ensure that the isotopic signature reflects diet.

8.4.2   Indirect Effects

More indirect perturbations affect the nutrient base of the food web, and then are gradually integrated up the food web. One of the more obvious possible mechanisms is a shift in the isotopic composition of the nutrient source for the base of the food web, as illustrated in this study. The input of a new nitrogen source to an aquatic system is also often detectable, because nitrogen from sewage or soils has a much more enriched isotopic value than the natural variation of the system (Spies et al., 1989; Dover et al., 1992; Kendall, 1998; Tucker et al., 1999).

In addition, changes in light or relative nutrient availability or productivity rates can alter the signature of the overall phytoplankton pool (Farquhar et al., 1989; Fogel and Cifuentes, 1993; Laws et al., 1995). Nutrient availability can also have an effect on individual plankton species isotopic signatures, and lower light or nutrient availability reduces productivity rates, leading to a decrease in $\delta^{13}C$ ratios (Eek et al., 1999; Waser et al., 1999). Changes in relative nutrient availability often cause shifts in phytoplankton species composition, which can subsequently affect the isotopic signatures of the phytoplankton pool. For example, a shift toward nitrogen limitation may increase dominance by nitrogen-fixing species, whose $\delta^{15}N$ of around 0‰ may be markedly different from previously dominant plankton species that were using aqueous forms of nitrogen. These shifts are a common feature of temperate lakes during the summer, and time-averaging of this type has also been invoked in Lake Ontario, where the isotopic signature of late summer zooplankton may reflect phytoplankton consumed earlier in the summer (Leggett et al., 1999).

8.5 Conclusion

In summary, this study illustrates the importance of understanding the temporal resolution of different trophic levels and the effect of time-averaging in stable isotope ecology. In general, temporal integration increases from lower to upper trophic levels in a food web, as body size and life span increase. The interpretation of energy flow in food webs based on isotopic signatures could be incorrect without considering the possible effects of this time-averaging. The variability in time integration can also be exploited to learn more about ecosystem dynamics, through the use of either natural stable isotope abundance (Hansson et al., 1997) or tracers (Peterson, 1999).

Acknowledgments

This research was funded by NSF Grants ATM–9619458 (the Nyanza Project) and EAR–962277 to A.S. Cohen, a NSERC Grant to R.E. Hecky, and an Analysis of Biological Diversification Fellowship from the Department of Ecology and Evolutionary Biology, University of Arizona to C.M. O'Reilly. The authors wish to thank the Tanzania Fisheries Research Institute and the Nyanza Project.

References

Adams, T.S. and R.W. Sterner. 2000. The effect of dietary nitrogen content on trophic level ^{15}N enrichment. *Limnol. Oceanogr.,* 45: 601–607.

Altabet, M.A. and R. François. 1994. Sedimentary nitrogen isotopic ratio as a recorder for surface ocean nitrate utilization. *Global Biogeochem. Cycles,* 8: 103–116.

Coulter, G.W. 1991. Pelagic fish, in G.W. Coulter, Ed., *Lake Tanganyika and Its Life.* Oxford University Press, New York, 111–138.

Dover, C.L.V., J.F. Grassie, B. Fry, R.H. Garritt, and V.R. Starczak. 1992. Stable isotope evidence for entry of sewage-derived organic material into a deep-sea food web. *Nature,* 360: 153–156.

Eek, M.K., M.J. Whiticar, J.K.B. Bishop, and C.S. Wong. 1999. Influence of nutrients on carbon isotope fractionation by natural populations of Prymnesiophyte algae in NE Pacific. *Deep-Sea Res. II,* 46: 2863–2876.

Farquhar, G.D., J.R. Ehleringer, and K.T. Hubick. 1989. Carbon isotope discrimination and photosynthesis. *Annu. Rev. Plant Physiol. Plant Mol. Biol.,* 40: 503–537.

Fogel, M.L. and L.A. Cifuentes. 1993. Isotope fractionation during primary production, in M.H. Engel and S.A. Macko, Eds., *Organic Geochemistry*. Plenum Press, New York, 73–100.

François, R., C.H. Pilskaln, and M.A. Altabet. 1996. Seasonal variation in the nitrogen isotopic composition of sediment trap materials collected in Lake Malawi, in T.C. Johnson and E.O. Odada, Eds., *The Limnology, Climatology and Paleoclimatology of the East African Lakes*. Gordon & Breach, New York, 241–250.

Gannes, L.Z., D.M. O'Brien, and C. Martínez del Rio. 1997. Stable isotopes in animal ecology: assumptions, caveats, and a call for more laboratory experiments. *Ecology*, 78: 1271–1276.

Hansson, S., J.E. Hobbie, R. Elmgren, U. Larsson, B. Fry, and S. Johansson. 1997. The stable nitrogen isotope ratio as a marker of food-web interactions and fish migration. *Ecology*, 78: 2249–2257.

Hecky, R.E. 1991. The pelagic ecosystem, in G.W. Coulter, Ed., *Lake Tanganyika and Its Life*. Oxford University Press, New York, 90–110.

Hecky, R.E., R.H. Spigel, and G.W. Coulter. 1991. The nutrient regime, in G.W. Coulter, Ed., *Lake Tanganyika and Its Life*. Oxford University Press, New York, 76–89.

Hyvonen, K. 1997. Study on zooplankton development time at Lake Tanganyika, FAO/FINNIDA Research for the Management of the Fisheries of Lake Tanganyika. GCP/RAF/271/FIN-TD/78 (En), Bujumbura, Burundi.

Johannes, E., J. Nowak, M.G. Nzeyimana, and L. Wimba. 1999. An investigation of the short-term fluctuations in the water column and its relation to weather patterns: Kigoma Bay, Lake Tanganyika, in A. Cohen, Ed., *The Nyanza Project 1999 Annual Report*. Department of Geosciences, University of Arizona, Tucson.

Kendall, C. 1998. Tracing nitrogen sources and cycles in catchments, in C. Kendall and J.J. McDonnell, Eds., *Isotope Tracers in Catchment Hydrology*. Elsevier, New York, 519–576.

Laws, E.A., B.N. Popp, R.R. Bidigare, M.C. Kennicutt, and S.A. Macko. 1995. Dependence of phytoplankton carbon isotopic composition on growth rate and [CO2]aq: theoretical considerations and experimental results. *Geochim. Cosmochim. Acta*, 59: 1131–1138.

Leggett, M.F., M.R. Servos, R. Hesslein, O. Johannsson, E.S. Millard, and D.G. Dixon. 1999. Biogeochemical influences on the carbon isotope signatures of Lake Ontario biota. *Can. J. Fish. Aquat. Sci.*, 56: 2211–2218.

Mannini, P., I. Katonda, B. Kissaka, and P. Verberg. 1999. Feeding ecology of *Lates stappersi* in Lake Tanganyika. *Hydrobiologia*, 407: 131–139.

Michener, R.H. and D.M. Schell. 1994. Stable isotope ratios as tracers in marine aquatic food webs, in K. Lajtha and R. Michener, Eds., *Stable Isotopes in Ecology and Environmental Science*. Blackwell Scientific, Oxford, U.K., 138–157.

Minagawa, M. and E. Wada. 1984. Stepwise enrichment of ^{15}N along food chains: Further evidence and the relation between δ^{15}N and animal age. *Geochim. Cosmochim. Acta*, 48: 1135–1140.

Oelbermann, K. and S. Scheu. 2002. Stable isotope enrichment (δ^{15}N and δ^{13}C) in a generalist predator (*Pardosa lugubris*, Araneae: Lycosidae): effects of prey quality. *Oecologia*, 130: 337–344.

O'Reilly, C.M., R.E. Hecky, A.S. Cohen, and P.-D. Plisnier. 2002. Interpreting stable isotopes in food webs: recognizing the role of time-averaging at different trophic levels. *Limnol. Oceanogr.*, 47: 306–309.

Overman, N.C. and D.L. Parrish. 2001. Stable isotope composition of walleye: ^{15}N accumulation with age and area-specific differences in δ^{13}C. *Can. J. Fish. Aquat. Sci.*, 58: 1253–1260.

Peterson, B.J. 1999. Stable isotopes as tracers of organic matter input and transfer in benthic food webs: a review. *Acta Oecol.*, 20: 479–487.

Pinnegar, J.K. and N.V.C. Polunin. 1999. Differential fractionation of δ^{13}C and δ^{15}N among fish tissues: implications for the study of trophic interactions. *Functional Ecol.*, 13: 225–231.

Plisnier, P.-D., D. Chitamwebwa, L. Mwape, K. Tshibangu, V. Langenberg, and E. Coenen. 1999. Limnological annual cycle inferred from physical-chemical fluctuations at three stations of Lake Tanganyika. *Hydrobiologia*, 407: 45–58.

Post, D.M., M.L. Pace, and J.N.G. Hairston. 2000. Ecosystem size determines food-chain length in lakes. *Nature*, 405: 1047–1049.

Spies, R.B., H.P. Kruger, R. Ireland, and D.W. Dice. 1989. Stable isotope ratios and contaminant concentrations in a sewage-distorted food web. *Mar. Ecol. Prog. Ser.*, 59: 33–38.

Teranes, J.L. and S.M. Bernasconi. 2000. The record of nitrate utilization and productivity limitation provided by δ^{15}N values in lake organic matter — A study of sediment trap and core sediments from Baldeggersee, Switzerland. *Limnol. Oceanogr.*, 45: 801–813.

Tucker, J., N. Sheats, A.E. Giblin, C.S. Hopkinson, and J.P. Montoya. 1999. Using stable isotopes to trace sewage-derived material through Boston Harbor and Massachusetts Bay. *Mar. Environ. Res.,* 48: 353–375.

Vander Zanden, M.J., J.M. Casselman, and J.B. Rasmussen. 1999. Stable isotope evidence for the food web consequences of species invasions in lakes. *Nature,* 401: 464–467.

Vander Zanden, M.J. and J.B. Rasmussen. 2001. Variation in $\delta^{15}N$ and $\delta^{13}C$ trophic fractionation: implications for aquatic food web studies. *Limnol. Oceanogr.,* 46: 2061–2066.

Wada, E. and A. Hattori 1991. *Nitrogen in the Sea: Forms, Abundances, and Rate Processes.* CRC Press, Boca Raton, FL.

Waser, N.A., Z.M. Yu, K.D. Yin, B. Nielsen, P.J. Harrison, D.H. Turpin, and S.E. Calvert. 1999. Nitrogen isotopic fractionation during a simulated diatom spring bloom: importance of N-starvation in controlling fractionation. *Mar. Ecol. Prog. Ser.,* 179: 291–296.

Yoshioka, T., E. Wada, and Y. Saijo. 1988. Isotopic characterization of Lake Kizaki and Lake Suwa. *Jpn. J. Limnol.,* 49: 119–128.

9

Synchrotron-Based Infrared Imaging of Euglena gracilis Single Cells

Carol J. Hirschmugl, Maria Bunta, and Mario Giordano

CONTENTS

9.1 Introduction 135
9.2 Experimental Details 137
    9.2.1 Infrared Storage Ring Radiation 137
        9.2.1.1 IRSR Power 137
        9.2.1.2 IRSR Brightness 138
    9.2.2 Brightness Limited Experiment: Infrared Microspectroscopy 138
    9.2.3 Sample Preparation 138
    9.2.4 Spectral Absorption Bands 140
9.3 Data Analysis and Results 141
    9.3.1 Spectra 141
    9.3.2 IR Images 142
    9.3.3 Diffraction Limited Infrared Imaging 143
9.4 Discussion 144
9.5 Conclusion 145
Acknowledgments 145
References 146

9.1 Introduction

Giordano and co-workers[1] examined the major cellular constituents of the diatom *Chaetoceros muellerii* Lemmerman in response to nitrogen starvation. These authors pioneered the novel combination of infrared (IR) spectroscopy, with its chemical identification, with studies of algal physiology. Additionally, Giordano et al. compared these data with chemical analysis. However, due to the novelty of the methodology, they started with studying dried microalgal samples. Here we present IR microspectroscopy data for living *Euglena gracilis* single cells at high spatial resolution (5 μm) for cells in a moist environment.

IR vibrational spectroscopy is a well-established, nondestructive tool to examine and identify chemistry in biological samples. IR radiation is absorbed by molecules with functional groups vibrating at similar frequencies. Unique absorption signatures exist for different chemicals including, for example, sugars, lipids, and proteins. Fourier Transform Infrared (FTIR) microspectroscopy is a very efficient example of this methodology to investigate small samples (100 μm or less) that has become readily available with recent advances in optical and computer technology.

An IR microscope using a storage ring (synchrotron) radiation source produces high-quality, spatially resolved IR images. A recent study by Carr[2] showed that a diffraction-limited spatial resolution, which is at best equal to half the wavelength of light, can be obtained between 10 to 2.5 μm (1000 cm^{-1} to 4000 cm^{-1}).[2] Storage ring radiation is emitted from swift electrons (with velocities approximately equal

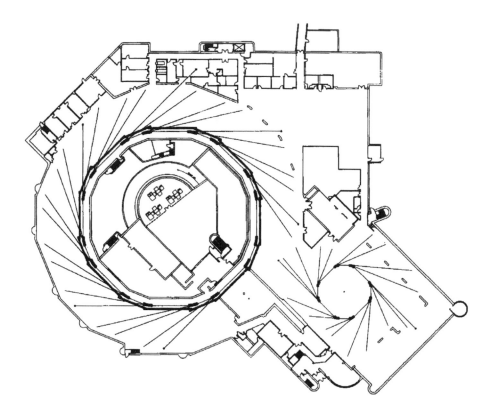

FIGURE 9.1 Overview of a synchrotron radiation facility, the National Synchrotron Light Source (NSLS) at Brookhaven National Laboratory. The figure shows the two storage rings of the NSLS, the 750 MeV vacuum ultraviolet ring on the top right, and the large 2.5 GeV x-ray ring on the bottom. Each ring is surrounded by a large number of beamlines that can support many experiments in parallel. (From Margaritondo, G., *Introduction to Synchrotron Radiation*, Oxford University Press, Oxford, 1988. With permission.)

to the speed of light) that are accelerated by a magnetic field in a high vacuum storage ring. A well-defined cone of broadband radiation tangential to the electron beam path (Figure 9.1) is generated. A source with these characteristics is crucial for challenging experiments with brightness-limited geometries, such as microspectroscopy at high spatial resolution.

Although most storage ring facilities are optimally designed to emit x-rays or ultraviolet (UV) radiation, all synchrotrons emit IR radiation. However, because IR wavelengths are much longer than x-rays and UV radiation, substantial modifications are typically required to extract IR radiation from storage rings. In most cases, IR beamlines have been constructed on storage rings optimized for UV radiation.

There are several sources of storage ring IR radiation available worldwide. IR microspectroscopy facilities have been established at storage rings in the U.S. (National Synchrotron Light Source, Brookhaven, NY; Lawrence Berkeley National Laboratory, Berkeley, CA; Aladdin, Stoughton, WI) and in Europe (Synchrotron Radiation Center, Daresbury, U.K.; LURE, Orsay, France). There are many other synchrotrons with IR capabilities that could accommodate IR microscopes located in the U.S. (NIST, Gathersburg, VA), Europe (MAX, Lund, Sweden; Bessy-2, Berlin, Germany), and Japan (UVSOR, Okasaki, Japan). In addition there are many facilities that are constructing or developing IR capabilities (CAMD, Baton Rouge, LA; Delta, Dortmund, Germany; ANKA, Karlsruhe, Germany; DAPHNE, Rome, Italy; SOLEIL, France; European Synchrotron Radiation Facility, Grenoble, France; Spring-8, Japan; SRRC, Taiwan; Hefei, China; Canadian Light Source, Saskatoon, Canada; Campinas, Brazil; Singapore). Most operating facilities accept experimental proposals semiannually, or more frequently. The proposals are typically peer-reviewed and allocated access to beamtime based on scientific merit.

A wide variety of materials has been examined using the powerful combination of the IR microscope and storage ring radiation source. For example, Jamin and co-workers[3] have produced a detailed study of the distribution of proteins and lipids within live single UN2.C3 cells derived from the mouse UN2 hybridoma B cells (20 by 20 μm). These authors obtained IR images of similar cells undergoing apoptosis and division. Examinations of the distribution of minerals and proteins, providing an understanding of the changes in bone composition, have been obtained by IR microspectroscopy studies of monkey bones after onset of osteoporosis and atherosclerosis.[4] Miller et al.[5] have also investigated the acid phosphate content and mineral crystallite perfection of human iliac bone around a human osteon, revealing dramatic changes in the bone composition within 30 μm from the center of an osteon. In another investigation using this powerful tool, the conversion of Cr(VI) to Cr(III) on a basaltic mineral surface was observed in the presence of mineral inhabiting microorganisms (endoliths).[6] In sum, an IR microspectroscope coupled with IR storage ring radiation has proved valuable in examining complex chemistry at relatively high spatial resolution.

Using an IR microspectroscope with a storage ring radiation source to examine algae represents a substantial advancement in the understanding of the ecophysiological role of these organisms and consequently of the aquatic ecosystems, in which algae are a very important player.[7] The responses of microalgae to stress induced by environmental changes are generally aimed at maintaining growth rates as close to optimum as possible.[8] To achieve this objective, cells must often substantially reorganize their components.[1,7,9–12] The study of the absolute and relative changes in the cellular pools of macromolecules is therefore essential for understanding the response of organisms to alterations in the environmental conditions. Unfortunately, most of the methods used for the assessment of the size of cellular pools of macromolecules and of their variations are invasive and introduce major perturbations of the system. For this reason FTIR spectroscopy represents a great hope for the future of ecophysiological and environmental studies, especially if, as the present work shows, it can be applied to live cells.

In this chapter, spatially resolved IR images of single cells of the microalga *E. gracilis* were measured using synchrotron radiation as a bright IR source. Infrared images of the distribution of lipids, proteins, and carbohydrates with a few micron spatial resolution for individual living cells were obtained for the first time. These results are consistent with previously understood biochemistry for *E. gracilis*. The two alga samples that are the focus of this paper were separated from an original batch culture 10 days apart. The overall trends observed for the two specimens are in qualitative agreement with the hypothesis that there was a significantly reduced concentration of nitrogen in the culture as a function of time.

9.2 Experimental Details

9.2.1 Infrared Storage Ring Radiation

Here we briefly introduce the characteristics of infrared storage ring radiation (IRSR) sources and compare them to a traditional lab-based globar (2000 K) source.

9.2.1.1 IRSR Power — The dipole radiation power emitted as a function of wavelength in the infrared region can be calculated using the following expression:[13]

$$P(\lambda) = 4.38 \times 10^{14} \times I \times \theta \times BW \times (\rho/\lambda)^{1/3} \text{ photons/s} \tag{9.1}$$

where I (amps) is the beam current, θ (rads) is the horizontal collection angle, BW is the bandwidth, λ(m) is the wavelength, and ρ(m) is the radius of the ring. (BW is either % bandwidth or $\Delta v/v$ bandwidth. For example, 0.1% BW means ±0.05 eV BW around 1 eV wavelength and is similar to that used for UV calculations. However, in the infrared it can be more insightful to use a bandwidth based on $\Delta v/v$, where 2 cm⁻¹ BW refers to a constant 2 cm⁻¹ window around each frequency). It is evident from Equation 9.1 that the emitted power is directly proportional to the beam current and the opening angle. In practice, the most powerful sources have been extracted from low-energy, small-radius storage rings, as these facilities can maintain higher beam current conditions and accommodate larger opening angles.

9.2.1.2   IRSR Brightness — The brightness, or brilliance, of a source is defined as the power per source area per angle into which it emits light, which requires an accurate representation of these characteristics of the source. Here we discuss the angle subtended by the source and the contributions to the horizontal and vertical source sizes, as functions of wavelength and storage ring extraction geometry, and give a general expression for the brightness of storage ring sources in the infrared.

The angle subtended by the source is given by the smaller of either the extraction angle or the characteristic natural opening angle, θ_{nat}, which is wavelength dependent: $\theta_{nat} = 1.66\lambda/\rho^{1/3}$.[13] Notice that the natural opening angle is proportional to wavelength to the third, which means that the angle is larger for longer wavelengths. In practice at smaller wavelengths, the natural opening angle is smaller than the extraction angle, and at larger wavelengths, the natural opening angle becomes greater than the extraction angle, which is determined by the physical limitations of the machine.

In general, there are three geometrical contributions[14] to the horizontal and vertical source sizes: (1) the intrinsic size of the electron beam itself s_i; (2) the projected (observed) size due to the large opening angle and hence extended source s_p; and (3) the diffraction-limited source size s_{diff}. The practical horizontal and vertical sizes are the sum of these contributions added in quadrature, thus for s_H (horizontal source size) $s_H = \sqrt{s_{Hdiff}^2 + s_{Hp}^2 + s_{Hi}^2}$. Frequently, both the horizontal and vertical sizes are dominated by that corresponding to the diffraction limit. The diffraction limit, s_{diff}, is given by the full width half maximum (FWHM) of λ/θ_{nat}. If the horizontal extraction angle is larger than the natural opening angle, then s_p, the projected source size, is an important contribution due to the curvature of the storage ring: the horizontal projected size is $\rho\theta_h^2/8$ and the vertical projected source size is approximately $\rho\theta_h\theta_{nat}/8$.

Recently, Murphy and Williams[15,16] have determined a universal expression for the brightness for all electron storage rings, assuming that the opening angle of the storage ring matches the natural radiation opening angles, and that the intrinsic source size is smaller than that due to diffraction:

$$B(\lambda) = 3.8 \times 10^{20} \, I \times BW/\lambda^2 \text{ photons / sec / mm}^2 \text{ / sr} \tag{9.2}$$

where I (amps) is the stored current, BW (%) is bandwidth, and λ (µm) is wavelength. The units for the brightness can be converted into a bandwidth of 2 cm^{-1} by multiplying by the correct factors, and into watts (by multiplying by $5.04 \times 10^{18} \times \lambda$ µm).

9.2.2   Brightness Limited Experiment: Infrared Microspectroscopy

An optimal optical design successfully focuses all of the available source photons onto the detector. Ideally, the experimental optical throughput (product of the area-angle acceptance of the experimental geometry) is equivalent to the source emittance (area-angle product of the source). This obeys Liouiville's theorem, which states that the brightness (power/emittance) of a source can, at best, be conserved through an optical path. Frequently, the limiting factor in designs for black-body sources is that the light emits in a 4π steradian solid angle, and is therefore difficult to collect completely with optical elements, and some of the flux from the source is unused. However, because the IRSR source has a small emittance angle and size, most or all of its flux can be captured by the optical system with a limited experimental throughput. The infrared microscope is one example of a brightness-limited experimental geometry, which is well matched to the emittance of a storage ring. In Figure 9.2 the signals for a 1 A, 90-mrad opening angle extraction port for a storage ring source and a 2000K black-body (lab-based) source that are collected by the optics for the IR microscope are compared to each other. Notably, the signals that arrive at the sample for the storage ring are between three and four orders of magnitude higher than for the black-body source.

9.2.3   Sample Preparation

All the experiments were performed on a batch culture of *Euglena gracilis*, UTEX 368 (UTEX, Austin, TX). The medium used (soilwater:PEA) was an adaptation of, e.g., Pringsheim's[17] biphasic soil-water medium using the basic formula of 1 teaspoon of dry garden soil, 1 garden pea, and 200 ml glass distilled

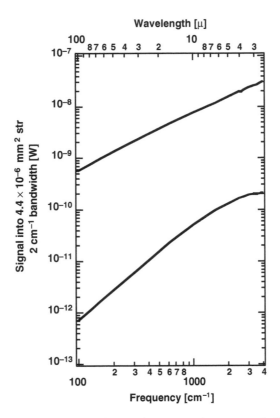

FIGURE 9.2 Graph comparing the infrared spectral output for a storage ring source and a black-body source.

water. This mixture was covered and steamed for 3 h on each of 2 consecutive days, then refrigerated (10°C) for 24 h or more and returned to room temperature before using. After the algae were delivered, they were maintained at 18°C ± 2°C, and exposed to approximately 150 μmol photons m^{-2} s^{-1} for 20 h a day. The soil medium was never exchanged or replenished after the cells were delivered from UTEX. The results from two algae are presented here and are representative of the results obtained from 25 individual cells. The two alga samples that are the focus of this chapter were separated from the original culture (200 ml) 10 days apart. The alga removed from the culture first (second) is referred to as alga #1 (alga #2) throughout this chapter. A pipette worth of algal suspension (~1 mm^3) was gathered from the sample, released on a gold-plated slide, and allowed to dry for 2 h at room temperature. While the algae were still moist, the slide was placed on the FTIR microscope stage in a nitrogen-purged environment to acquire spectra in an IR transparent atmosphere.

Spectra were collected on a FTIR spectrometer (Thermo Nicolet 560 bench equipped with a KBr beamsplitter and a liquid N$_2$ cooled MCT detector with a 0.25 by 0.25 mm detector element) coupled to an IR microscope (model Nicplan, Thermo Nicolet) located at a synchrotron facility and modified to accept IR radiation from the Aladdin storage ring. While this source provided more photons than a globar source, the power absorbed by the sample did not raise the temperature by more than 1°C.[18] The sample was placed on a computer-controlled *x–y* positioning stage (reproducibility better than 1 μm), which rastered the sample through the IR beam.

To obtain high-quality spectra, two performance criteria must be met: high throughput and an accurately defined sample area. Although it is simple to establish a small (microns) sampling area, setting apertures at the focal planes while assisted by visible radiation, the actual sampling area is larger for IR radiation because of diffraction effects. As the sample size decreases, it becomes even more important to define the sample area accurately.

The spectra were collected in the reflection-absorption geometry, using clean gold as a reference. A background spectrum was used to eliminate signals that were due to the spectrometer from the sample spectrum. Reflection spectra were divided by the background and transformed into infrared absorption spectra. After these manipulations, the amplitude of a vibrational band in the FTIR spectrum is given in units of absorbance. The results are presented in absorbance, where the incident radiation has been absorbed twice by the alga. The spectra were collected between 4000 and 700 cm^{-1} with a spectral resolution of 8 cm^{-1} and 512 scans co-added at each point. A Happ–Genzel apodization function was used and an automatic baseline function was applied to all the spectra in the map to correct sloping, curving, or otherwise undesirable baseline in the spectrum without specifying baseline points.

Spectra are taken with two apertures. Initially, one spectrum for the entire algae is collected with a 20 by 20 μm aperture. Then a series of spatially resolved spectra are collected with a 5 by 5 μm (and 6 by 6 μm) aperture for sample 1 (sample 2), rastering the algae in 5 μm (6 μm) steps through the focus of the beam. This collection of data is called an "area map." Finally, another spectrum is collected for the whole algae with a 20 by 20 μm aperture. The first and last spectrum (of the entire algae) were always similar, which indicates that the algae did not alter notably between the time of the first and last spectra collection for the spatially resolved map.

9.2.4 Spectral Absorption Bands

Band assignments are based on previous studies of whole cells, organelles, and macromolecules. Table 9.1 and Figure 9.3 summarize this information. The spectra of BSA, taken as an exemplary protein, are characterized by two intense bands at ~1650 cm^{-1} (amide I), due primarily to the amide carbonyl stretching vibrations, and ~1540 cm^{-1} (amide II), attributable to a combination of N–H bending and C–N vibrations in amide complexes. Minor bands are also observed at ~1450 and ~1400 cm^{-1}, which are attributable to the bending vibrations of CH_3 and CH_2 groups. Palmitic acid was used as a standard for lipids; its C=O mode of the side chain from ester carbonyl strongly absorbs at 1742 cm^{-1}, while bands between 3000 and 2800 cm^{-1} are due to C–H stretching vibrations, and bands at 1464 and 1438 cm^{-1} are assigned to the bending vibrations of CH_3- and CH_2-. The spectrum for starch displays intense bands at ~1024, ~1150, and ~1050 cm^{-1}, which are characteristic of C–O stretching vibrations from carbohydrates. With respect to glycogen,[19] however, starch has an additional strong absorption band at

TABLE 9.1

Band Assignments for FTIR Spectroscopy Used in This Study

Wavenumber Values / cm^{-1}	Assignment[a]	Ref.	Comments
30002800	ν C-H of saturated CH	Williams and Fleming, 1996	Primarily from lipids
~1740	ν C=O of ester functional groups primarily from lipids and fatty acids	Williams and Fleming, 1996; Zeroual et al., 1995	Primarily from phospholipids
~1650	ν C=O of amides associated with proteins	Nelson, 1991; Williams and Fleming, 1996	Usually called the amide I band; may also contain contributions from C=C stretches of olefinic and aromatic compounds
~1540	δ N-H of amides associated with proteins	Nelson, 1991; Williams and Fleming, 1996	Usually called the amide II band; may also contain contributions from C=N stretches
~1455	δ_{as} CH_3 and δ_{as} CH_2 of proteins	Zeroual et al., 1994	The positions of these assignments can vary in the literature
~1398	δ_s CH_3 and δ_s CH_2 of proteins, and ν_s C-O of COO^- groups	Nelson 1991, Zeroual et al., 1994	The positions of these assignments can vary in the literature
~1230	ν_{as} P=O of the phosphodiester backbone of nucleic acid (DNA and RNA)	Nelson, 1991; Zeroual et al., 1994	May also be due to the presence of phosphorylated proteins and polyphosphate storage products
~1200900	ν C-O-C of polysaccharides	Wong et al., 1991; Zeroual et al., 1994	Polysaccharide bands are a series of three bands in this frequency range
~800	Tetrapyrrole ring system bending	Katz, 1966	Primarily from chlorophyll

[a] ν_{as} = asymmetric stretch, ν_s = symmetric stretch, δ_{as} = asymmetric deformation (bend), δ_s = symmetric deformation (bend).

Source: Giordano, M. et al., *J. Phycol.*, 37, 271, 2001. With permission.

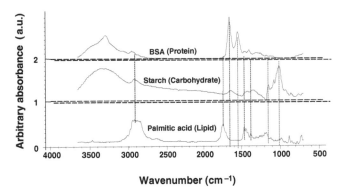

FIGURE 9.3 Infrared spectra for protein, lipid, and carbohydrate standards. (From Giordano, M. et al., *J. Phycol.*, 37, 271, 2001. With permission.)

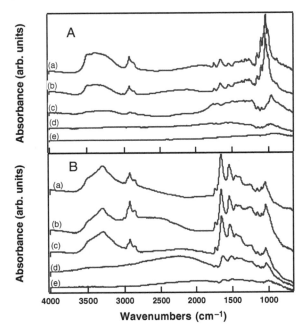

FIGURE 9.4 Five IR absorption spectra for sample 1 (A) and sample 2 (B) are shown. Spectrum a was taken with a 20 × 20 µm aperture and spectra b through e were taken with a 5 × 5 µm aperture. Spectra b through e represent IR absorption signatures observed at four different places on top of and around the alga. Notice, for example, the presence of the amide II band at 1545 cm⁻¹ and the phospholipid band at 1742 cm⁻¹, and that these intensities vary with position.

1040 cm⁻¹.[1] No paramylon standard is commercially available; however, it can be assumed that the overall appearance of the spectrum of such β-1,3 polymer of glucose does not appreciably differ from that of starch. Chlorophyll has one strong absorption band at 800 cm⁻¹, a slightly weaker band around 750 cm⁻¹, and a series of much weaker bands between 1000 and 4000 cm⁻¹.[20]

9.3 Data Analysis and Results

9.3.1 Spectra

In Figure 9.4 two series of infrared spectra are shown, one each for samples 1 (Figure 9.4A) and 2 (Figure 9.4B). In both cases, the top spectrum was taken with a 20 µm square aperture covering the entire algal cell, while the remaining spectra were taken with 6 µm (sample 1) and 5 µm (sample 2) square apertures at different positions over and next to the cells. Prominent absorption bands observed

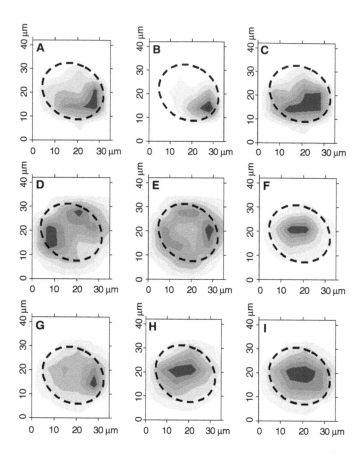

FIGURE 9.5 Nine IR images for sample 1 are shown. The IR images are derived from the peak strength as a function of position for the following peaks: (A) 1545 cm^{-1} (N–H bending and C–N stretching for protein group Amide II), (B) 870 cm^{-1} chlorophyll, (C) 1742 cm^{-1} (C–O stretch for phospholipids), (D) 973 cm^{-1} (unidentified), (E) 937 cm^{-1} (unidentified), (F) 1040 cm^{-1} (unassigned for paramylon), (G) 1147 cm^{-1} (CO stretching for sugar), (H) 1091 cm^{-1} (C–C stretching for sugar), and (I) 1014 cm^{-1} (C–O–H deformation for sugar). The backgrounds that were chosen for each IR image are between 1600 and 1500 for image A, between 2980 and 2872 for image B, between 1700 and 1780 for image C, and between 3750 and 680 cm^{-1} for images D through I.

in the spectra include bands assigned to proteins, carbohydrates, and lipids (see Table 9.1). The relative absorption strengths of the bands in spectrum a in Figure 9.4A and B are strikingly different. For example, the protein (1560) to carbohydrate (1040) ratio for sample 1 is approximately 1:0.2, whereas the same ratio for sample 2 is approximately 1:10. The relative strength of the lipid to protein signatures is also drastically different for these samples. In addition, the absorption bands in spectra b to e for both samples have notably varying strengths. This is indicative of the distribution of different constituents in and throughout the cell. IR images for individual absorption bands are obtained from similar spectra taken with the smaller apertures as the sample is rastered through the focus.

9.3.2   IR Images

The IR images of the algae are contour maps representing the relative strength of IR absorption features as a function of position. The peak or maximum (black) in a contour plot represents a strong absorption by the sample, and a dip or minimum (white) in the contour plot represents no absorption by the sample. They are generated from the spectra using the following method. A background line is defined by choosing two data points for each spectrum for the whole collection of spectra in a map. The strength for the contour plot at each pixel is the absorption strength of a given peak minus the selected baseline.

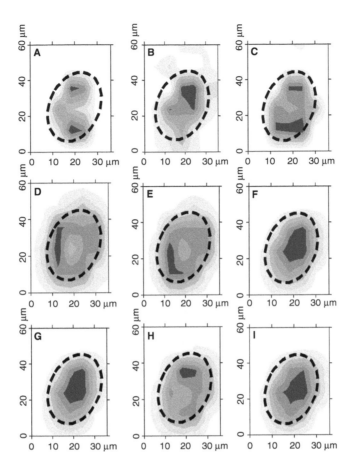

FIGURE 9.6 Nine IR images for sample 2 are shown. The IR images are derived from the peak strength as a function of position for the following peaks: (A) 1545 cm^{-1} (N–H bending and C–N stretching for protein group Amide II), (B) 870 cm^{-1} chlorophyll, (C) 1742 cm^{-1} (C–O stretch for phospholipids), (D) 975 cm^{-1} (unidentified), (E) 937 cm^{-1} (unidentified), (F) 1040 cm^{-1} (unassigned for paramylon), (G) 1152 cm^{-1} (CO stretching for sugar), (H) 1096 cm^{-1} (C–C stretching for sugar), and (I) 1023 cm^{-1} (C–O–H deformation for sugar). The backgrounds that were chosen for each IR image are between 1600 and 1500 for image A, between 2980 and 2872 for image B, between 1700 and 1780 for image C, and between 3750 and 680 cm^{-1} for images D through I.

A different set of points may be chosen for each contour map. The differences in spectral shapes can make it very difficult to choose a baseline. Appropriate baselines have been carefully chosen, and checked for accuracy by comparing the absorption strength in each individual spectrum with the absorption strength indicated in the contour maps. The end points of the baselines are indicated for each IR image in the figure captions. Nine IR images were generated for samples 1 and 2, respectively, in Figure 9.5 and Figure 9.6 for the following absorption signatures: (A) the amide II band, (B) a chlorophyll band, (C) two lipid bands, (D, E) two unidentified bands, (F) starch, and (G, H, I) sugar. These maps are similar to images obtained for the other individual cells that were examined. The superimposed black oval dotted line represents the position of the algae for each series.

9.3.3 Diffraction Limited Infrared Imaging

Diffraction limited IR images are produced from storage ring-based IR microspectroscopy[2] with a confocal microscope geometry and an aperture size that is equal to or smaller than half the wavelength of probing light. When the wavelength of light is comparable to an aperture it traverses, the light is diffracted — some portion of the intensity generates a pattern outside the original confined beam path.

The spatial resolution is determined by a convolution of the central burst of the Airy diffraction patterns for both apertures. For example, the infrared image for an absorption feature at 1000 cm^{-1} ($\lambda = 10$ μm) has a spatial resolution of 5 μm if the apertures are 5 μm or smaller. The entire spectral region (4000 to 800 cm^{-1}, or 2.5 to 11.5 μm) is collected with an aperture of 5 or 6 μm. All of the IR images in this chapter have a spatial resolution of 5 to 6 μm.

9.4 Discussion

The optical images of the *Euglena gracilis* in this experiment show a football shape that appears to be about a 1:2 ratio in width to length (12 by 25 μm). Scanning electron microscopy (SEM) images of uncoated, freeze-dried *E. gracilis* single cells that have been KMnO$_4$ fixed produce similar football-shape images.[21] Other SEM images of uncoated and gold-coated freeze-dried samples (that have not been fixed) have produced images that have a 1:8 ratio for the width to the length (5 by 40 μm), which is similar to observations of swimming algae.[17] Collins et al. did not discuss any reasons for the differences in these SEM images. It is possible that the cells under observation for the present work were experiencing "euglenoid" movements, typically observed in nonswimming cells and consisting of contraction and relaxation phases.[22] It is also possible that the cells encysted or formed palmelloid stages, due to the lack of water, which made them become spherical and surrounded by a gelatinous sheath.[23] The optical images obtained through the IR system, unfortunately, are not of sufficient quality to discern among these options. In particular, one cannot locate the orange-red eyespots, which could also be used to identify the orientation of the alga.[24]

The IR absorption spectra for the cells are similar to the spectra published by Giordano et al.,[1] allowing for interspecific differences, and exhibit absorption bands for proteins, lipids, and carbohydrates, which are all major constituents of algae. Spectrum a for sample 1 in Panel A of Figure 9.4 shows a high protein/carbohydrate ratio, whereas spectrum a for sample 2 in Panel B has a low protein/carbohydrate ratio. This qualitative finding is in very good agreement with the Giordano et al.[1] examination of protein/carbohydrate ratios as a function of exposure to nitrogen. This agreement is plausible because a high concentration of algae present in the culture vial at the beginning of the experiment would be expected to deplete the nitrogen availability over the experimental time frame of 10 days.

The IR images for the lipids and proteins for sample 1 (Panels A and C in Figure 9.5) show low-level absorption intensity distributed over the entire cell, and a higher, overlapping, absorption intensity located toward one end of the cell. The strongest band for chlorophyll is reported at 800 cm^{-1} by Katz et al.[20] There is a strong absorption band at 820 cm^{-1} observed in several spectra from the map for sample 1. The highest density for the image generated from this band corresponds to the highest density in the protein and lipid IR images. These contour maps are consistent with the known chemistry of *E. gracilis*.[25] In particular, *E. gracilis* is typically girdled by a pellicle (or periplast) located inside the lipid bilayer of the plasmalemma. The pellicle is composed mostly of protein with smaller concentrations of lipids and carbohydrates.[23] Cells of *Euglena* are not known to accumulate large amounts of lipids as storage products, even if occasional phospolipid granules are present in the cytosol; most of the lipids in the cells are therefore associated with membranes, a large part of which is represented by the endoplasmic reticulum and by the membrane systems of the chloroplasts (i.e., the three peripheral membranes and the thylakoids).[26,27] Proteins are also not expected to produce a strong signal in the cytosol. The highest proportion of protein is likely to be in the chloroplasts, and specifically in the pyrenoids, where the primary photosynthetic enzyme ribulose bisphosphate carboxylase (rubisco) can be very abundant.[28] Considering these facts, it is plausible that the concurrent maxima in the protein, chlorophyll, and lipid images correspond to chloroplasts. There appear to be one cluster of chloroplasts in sample 1 (Figure 9.5A to C), and several clusters of chloroplasts in sample 2 (Figure 9.6A to C).

The IR images for the absorption bands associated with carbohydrates (Figure 9.5F to I) can also be reconciled with *E. gracilis* biochemistry. The IR image for the absorption band at 1040 cm^{-1} (Figure 9.5F) shows a maximum that is confined to a small area, neighboring the maxima in the protein. In contrast, the IR images for sugars show less localized distributions with a weaker gradient. *Euglena gracilis* accumulates paramylon, a β-1-3 polymer of glucose. Paramylon production is often associated with pyrenoids, but free granules of this polysaccharide are also frequently observed in the cytosol. Generally speaking, therefore, paramylon is much more localized than soluble sugars. It is very likely that the maximum in Figure 9.5F is the site of a concentration of paramylon located next to the pyrenoid. In comparison to the distribution of the paramylon, the soluble sugars will be diffuse and found both in the chloroplasts, due to synthesis of glucose, and the cytosol, where sugar metabolism occurs. These expected intensity distributions are in agreement with the observations for sample 1. Further striking evidence can be found by comparing individual spectra from the map to spectra of standards shown in Figure 9.3. Spectrum b for sample 1 (Figure 9.4A) is the spectrum found at the maximum in the IR image for cell 1 (Figure 9.5F). Spectrum c for sample 1 (Figure 9.4A) is a spectrum found within the maximum in the IR image (Figure 9.5H), but not overlapping the maximum in Figure 9.5F. It is clear that the feature at 1040 cm^{-1} is very strong in spectrum b, and not so strong in spectrum c, and that the overall shape of the rest of these two spectra is very similar. These spectral fingerprints agree very well with the published data for starch,[1] which is very similar to paramylon, and glycogen,[19] respectively.

One striking qualitative difference between sample 1 and sample 2 is the extent to which the paramylon is confined (see Figures 9.5F and 9.6F). Sample 2 (N-starved cell) shows a larger distribution of paramylon than sample 1. This is a typical response of N-starved cells, in which C, which cannot be used for protein synthesis for the lack of N, is mostly allocated in storage polysaccharides. One can also compare the ratio of the areas of the maxima for the protein/paramylon in the two cells. Clearly, the protein/paramylon area ratio is much smaller for sample 2 (the N-starved cell) than for sample 1. This is similar to the ratio of the protein/carbohydrate peaks in the overall IR absorption spectra (spectra a in Figure 9.4A and B) of the samples that were discussed earlier.

Finally, the IR images for the absorption features observed at 973 and 937 cm^{-1} (Figures 9.5D and E and 9.6D and E) show a minima in the center of the algae with a maximum around the edge of the cell. The authors propose that these absorption bands could possibly originate from a gelatinous sheath: in fact, a thick mucilaginous matrix could coat the cells, if they have become encysted or are in a palmelloid phase; moreover, even vegetative *Euglena* cells have rows of "mucilage-producing bodies" under the pellicle extruding mucilage to the exterior through canals.[23] Recall that the diffraction-limited spatial resolution for this wavelength should be approximately 6 μm. It is clear that these absorption bands show some intensity distribution that is beyond 6 μm outside the edges of the cell. An IR spectrum of the main components of the growth medium was checked and found not to contain any IR absorption bands that overlap with these peaks. In addition, the spectrum associated with these bands shows no evidence of C–H stretches or CO stretches.

9.5 Conclusion

In summary, we have measured spatially resolved IR maps of single cells of *E. gracilis in situ*, using IR storage ring radiation as a powerful probe of the cell. IR spectra agree well with those observed in previous measurements of Giordano and co-workers.[1] There is qualitative agreement between the IR images and the biochemistry for *E. gracilis*. These initial measurements show the feasibility of studying chemistry *in situ* for living single cells, with the capability of monitoring the overall spectra and the changes in the chemistry within the cells. Furthermore, the changes in the IR spectra as a function of exposure to nutrients are in very good qualitative agreement with prior work by Giordano et al.[1] Importantly, this study demonstrates the potential to examine changes in the chemistry of living cells while modifying environmental stimuli.

Acknowledgments

The authors are indebted to Rudi Strickler, Andrey Skilirov, and Justin Holt for valuable discussions regarding this work. The authors are grateful for support from the Synchrotron Radiation Center (SRC) staff, in particular Bob Julian and Roger Hanson for invaluable assistance at the IR Beamline, and especially for access to beamtime around our limited schedules. Invaluable funding for the SRC is provided by the NSF under Grant DMR-0084402. Other funding for this project came from the University of Wisconsin–Milwaukee graduate school Advanced Analysis Facility summer fellowships for undergraduates (J.B.H.). The authors (C.J.H., M.B., and J.B.H.) are extremely grateful for the primary funding for this project that was provided by NSF Grant CHE-9984931.

References

1. Giordano, M., Kansiz, M., Heraud, P., Beardall, J., Wood, B., and McNaughton, D., Fourier transform infrared spectroscopy as a novel tool to investigate changes in intracellular macromolecular pools in the marine microalga *Chaetoceros muellerii* (Bacillariophyceae), *J. Phycol.*, 37, 271, 2001.

2. Carr, G.L., Resolution limits for infrared microspectroscopy explored with synchrotron radiation, *Review of Scientific Instruments*, 72, 1513, 2001.

3. Jamin, N., Dumas, P., Moncuit, J., Fridman, W.-H., Teillaud, J.-L., Carr, G.L., and Williams, G.P., Highly resolved chemical imaging of living cells by using synchrotron infrared microspectroscopy, *Proc. Natl. Acad. Sci.*, 95, 4837, 1998.

4. Miller, L.M., Hamerman, D., Chance, M.R., and Carlson, C.S., Analysis of bone protein and mineral composition in bone disease using synchrotron infrared microspectroscopy, *SPIE*, 3775, 104, 1999.

5. Miller, L.M., Vairavamurthy, V., Chance, M., Mendelsohn, R., Paschalis, E.P., Betts, F., and Boskey, A.L., *In situ* analysis of mineral content and crystallinity in bone using infrared microspectroscopy of the v4 PO_4^{3-} vibration, *Biochim. Biophys. Acta*, 1527, 11, 2001.

6. Holman, H.Y., Perry, D.L., Martin, M.C., Lamble, G.M., McKinney, W.R., and Hunter-Cevera, J.C., Real-time characterization of biogeochemical reduction of Cr(VI) on basalt surfaces by SR-FTIR imaging, *Geomicrobiol. J.*, 16, 307, 1999.

7. Falkowski, P.G., The role of phytoplankton photosynthesis in global biogeochemical cycles, *Photosynthesis Res.*, 39, 23, 1994.

8. Beardall, J. and Giordano, M., Ecological implications of algal CCMs and their regulation, *Funct. Plant Biol.*, 29(2–3), 225, 2002.

9. Giordano, M. and Bowes, G., Gas exchanges, metabolism, and morphology of *Dunaliella salina* in response to the CO2 concentration and nitrogen source used for growth, *Plant Physiol.*, 115, 1049, 1997.

10. Morris, I., Photosynthetic products, physiological state and phytoplankton growth, *Can. Bull. Fish. Aquat. Sci.*, 210, 83, 1981.

11. Turpin, D.H., Effects of inorganic N availability on algal photosynthesis and carbon metabolism, *J. Phycology*, 27, 14, 1991.

12. Giordano, M., Pezzoni, V., and Hell, R., Strategies for the allocation of resources under sulfur limitation in the green alga *Dunaliella salina*, *Plant Physiol.*, 124, 857, 2000.

13. Duncan, W.D. and Williams, G.P., Infra-red synchrotron radiation from electron storage rings, *Appl. Opt.*, 22, 2914, 1983.

14. Hirschmugl, C.J., Low Frequency Adsorbate Substrate Dynamics for CO/Cu Studied with Infrared Synchotron Radiation, Ph.D. thesis, Yale University, New Haven, CT, 1994.

15. Murphy, J., Personal communication, 1999.

16. Williams, G.P., Infrared synchrotron radiation, review of properties and perspectives, *SPIE*, 1775, 2, 1999.

17. Pringsheim, E.G., The biphasic or soil-water culture method for growing algae and flagellata, *J. Ecol.*, 33, 193, 1946.

18. Martin, M.C., Tsvetkova, N.M., Crowe, J.H., and McKinney, W.R., Negligible sample heating from synchrotron infrared beam, *Appl. Spectrosc.*, 55, 111, 2001.

19. Diem, M., Boydston-White, S., and Chiriboc, L., Infrared spectroscopy of cells and tissues: shining light onto a novel subject, *Appl. Spectrosc.*, 53, 148A–161A, 1999.

20. Katz, J.J., Dougherty, R.C., and Boucher, L.J., Infrared and nuclear magnetic resonance spectroscopy of chlorophyll, in *The Chlorophylls*, L.P. Vernon and G.R. Seely, Eds., Academic Press, New York, 1966, 367–377.

21. Collins, S.P., Pope, R.K., Scheetz, R.W., Ray, R.I., Wagner, P.A., and Little, B.J., Advantages of environmental scanning electron microscopy in studies of microorganisms, *Microsc. Res. Tech.*, 25, 398, 1993.

22. Mikolajczyk, E. and Kuznicki, L., Body contraction and ultrastructure of *Euglena, Acta Protozool.*, 20, 1, 1981.
23. Bold, H.C. and Wynne, M.J., *Introduction to the Algae*, 2nd ed., Prentice-Hall, Englewood Cliffs, NJ, 1985, 720.
24. Graham, L.E., Lee, W., and Wilcox, L.W., *Algae*, Prentice-Hall, Upper Saddle River, NJ, 2000, 166.
25. Arnott, H.J. and Walne, P.L., Observations in the fine structure of the pellicle pores of *Euglena granulata, Protoplasma*, 64, 330, 1967.
26. Gibbs, S.P., The chloroplast of *Euglena* may have evolved from symbiotic green alga, *Can. J. Bot.*, 56, 2883, 1978.
27. Gibbs, S.P., The chloroplast of some alga groups may have evolved from endosymbiotic eukaryotic algae, *Ann. N.Y. Acad. Sci.*, 361, 193, 1981.
28. Badger, M.R., Andrews, T.J., Whitney, S.M., Ludwig, M., Yellowlees, D.C., Leggat, W., and Price, G.D., The diversity and coevolution of *Rubisco*, plastids, pyrenoids and chloroplast-based CO_2 concentrating mechanisms in algae, *Can. J. Bot.*, 76, 1052, 1998.

10

Signaling during Mating in the Pelagic Copepod, Temora longicornis

Jeannette Yen, Anne C. Prusak, Michael Caun, Michael Doall, Jason Brown, and J. Rudi Strickler

CONTENTS

10.1 Introduction 149
10.2 Methods 150
    10.2.1 Trail Visualization 150
    10.2.2 Copepod Pheromones 151
10.3 Results and Discussion 151
    10.3.1 Scent Preferences 151
    10.3.2 Tracking Behavior 154
    10.3.3 Quantitative Analyses of Trail Structure and Odorant Levels 155
10.4 Conclusion 157
Acknowledgments 158
References 158

10.1 Introduction

In 1973, Katona depicted the mate-finding response of copepods as occurring when the male copepod detects the edge of the diffusing cloud of pheromone emanating from the female at a distance of 4 mm. Here, the process of diffusion is needed to *transport* the signal molecules to the sensors of the male copepod to alert him of the presence and location of a female copepod. However, using the equation for the characteristic diffusion time (Dusenbery, 1992; $t = r^2/4D$, where the diffusivity coefficient $D = 10^{-5}$ cm²/s for small chemical molecules), it would take approximately 45 min for the pheromone to diffuse this distance. It is unlikely the female copepod would remain in the same three-dimensional (3D) position in the ocean for that length of time. Instead, in 1998, we (Doall et al., 1998; Weissburg et al., 1998; Yen et al., 1998) reported that male copepods detect discrete odor trails left in the wake of the swimming female copepod (Figure 10.1). Within this low Reynolds number regime, viscosity limits the rate of diffusion of the odor trail to molecular processes. Diffusion does not transport the pheromone to the male and, instead, acts to *restrict* odor dispersion. The scent persists as a coherent trail, with little dilution of signal strength, and hence remains detectable for a period that gives enough time for the male to encounter it. In the case of the copepod *Temora longicornis*, this aquatic microcrustacean could find trails that were less than 10.3 s old. Trails were followed for distances as long as 13.8 cm (~100 body lengths), greatly extending the encounter volume of the copepod (Gerritsen and Strickler, 1977). Past studies showed that copepods could detect signals only within a few body lengths (see Haury and Yamazaki, 1995, for a review). Our findings show the perceptive distance can be 10 to 100 times greater.

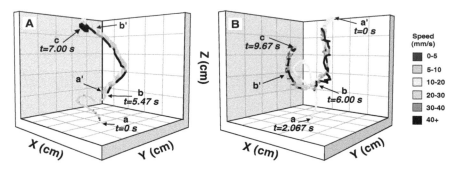

FIGURE 10.1 (Color figure follows p. 332.) (A) Mate-tracking by the copepod, *T. longicornis* (1.2 mm prosome length). The male copepod (thin trail) finds the trail of the female copepod (thick trail) when the trail is 5.47 s old and the female is 3.42 cm distant from him (nearly 30 body lengths). Upon encounter, the male spins to relocate the trail, then accelerates to catch up with the female. The male copepod follows the path of the female precisely in 3D space. When within 1 to 2 mm of the female, the male pauses quietly, then pounces swiftly to capture his mate for the transfer of gametes packaged in a flasklike spermatophore. During copulation, the mating pairs spin and remain together for a few seconds or more. (B) Mate-tracking by the copepod, *T. longicornis*: backtracking. The male copepod (thin trail) finds the trail of the female copepod (thick trail) because of a strong cross-plume odor gradient. The female is 1.30 cm distant from him. Upon encounter, the trail is 2.3 s old and the male follows the trail in the wrong direction, away from the female, because of a weak along-plume gradient. When initially following the trail, the male smoothly and closely follows the trajectory of the female. After 1.27 s when he is 24.4 mm from his mate and the trail is 6.7 s old, the male turns around and backtracks. When backtracking, he follows the female path erratically, casting back and forth over the trail. After reaching his intersection point, the male copepod resumes smooth close-following of the undisturbed female path. (From Doall, M. et al., *Philos. Trans. R. Soc. Lond.*, 353, 681, 1998. With permission.)

Documenting copepods following 3D aquatic trails provoked the question of the species specificity of chemical trails. A capability to discriminate the scent of conspecifics enables mate preference, important in maintaining species integrity (Palumbi, 1994; Orr and Smith, 1998; Higgie et al., 2000). For such widely dispersing species as aquatic marine organisms, an understanding of mate recognition, and the mechanisms for mate finding and mate choice, may be key steps in the cascade that enforces reproductive isolation in pelagic environments. To evaluate odor preference in copepods, we developed a bioassay, based on the behavioral response of trail following. The bioassay relies on trail visualization (a Schlieren optical path: Strickler and Hwang, 1998; Strickler, 1998) so that we can *see* the trail and *see* which trail the copepod follows. With this technique, we saw male copepods follow scent trails containing the odor of their conspecific female copepods. We also saw how the male disturbed the odor trail, making it less likely that another competing male would find and follow the trail.

10.2 Methods

10.2.1 Trail Visualization

To visualize the trail, we mixed high-molecular-weight dextran with filtered seawater to change the refractive index of the trail and used Schlieren optics to see the trail (Strickler and Hwang, 1998; Strickler, 1998). For this optical path, an infrared laser light, focused through a pinhole spatial filter, was expanded and diverted with a large (8-in.) spherical mirror to pass through and illuminate an experimental vessel filled with filtered seawater. The collimated light passed through the vessel, was collected on another large (8-in.) spherical mirror, and focused onto a small pointlike matched filter. The matched filter prevented the light from reaching the image plane and the CCD image looked black. When a phase object was introduced into the experimental vessel, it diffracted the light rays so the rays passed by the matched filter and reached the image plane. Phase objects, like the translucent copepod or the trail, were visualized as bright silhouettes against a black background of the image captured by the infrared-sensitive CCD video camera.

A 1-mm-wide trail was created by dispensing fluid from a fine pipette tip (Eppendorf) to flow down into the 4-l observation vessel filled with filtered seawater (28 ppt). The pipette was gravity fed, via thin

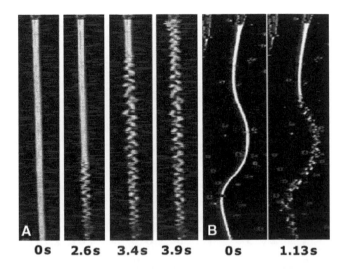

0s   2.6s   3.4s   3.9s   0s   1.13s

FIGURE 10.2 (Color figure follows p. 332.) (A) Visualized trail and upwardly directed trail following. A 1-mm-wide trail was created by dripping fluid from a fine pipette tip (Eppendorf) to flow into the 4-l observation vessel filled with filtered seawater (28 ppt). The pipette was gravity-fed, via thin tubing, from a beaker filled with a mixture of seawater and dextran (MW of 500,000 at 0.5 g/100 ml). The difference in the refractive index of the dextran–seawater mixture sinking down through the seawater could be detected, using a Schlieren optical path. The trail on the left is the undisturbed trail. The disturbed chevron-dotted trails on the right show how the trail structure changes at various time intervals after the male copepod *T. longicornis* follows the trail. Copepod indicated at the upper left on last trail, colored orange (B) Following a curvy trail: When following the scent in the trail, *T. longicornis* also can stay on the track of curved trails, making the same turns as taken by the trail. The undisturbed curved trail on left was followed for 1.13 s by the male copepod, colored orange in panel on right. Note how trail structure changes in wake of swimming male.

tubing, from a beaker filled with a mixture of seawater and dextran (MW of 500,000 at 0.5 g/100 ml). The difference in the refractive index of the dextran–seawater mixture sinking down through the seawater could be detected, using the Schlieren optical path. When this mixture passed through the plain seawater, a fine trail could be observed as a bright vertical line on the video image (Figure 10.2).

10.2.2   Copepod Pheromones

To test the attractiveness of waterborne odorants, different scents were obtained by putting different copepod types (stage, sex, species) in the dextran-labeled water and using this conditioned water to create the doubly labeled trails. For these scents, the same number of copepods (20) would be added to the same volume of dextran-labeled seawater (20 ml) in the beaker, to introduce comparable odorant levels to the conditioned water. Copepods that had spent the day in this dextran–seawater mixture had no detrimental effects as they lived for days after being transferred back to plain seawater with phytoplanktonic food (*Rhodomonas* sp.). The rate of inflow of the doubly labeled water, less than 3 mm/s and more often 1 mm/s, was adjusted by varying the height of the beaker of conditioned water that was gravity-fed into the tank. This small beaker was filled from a stock of unscented dextran-labeled seawater. The observation vessel held 20 to 50 male copepods to test their interest in the scented trails. Female copepods do not mate-track.

10.3   Results and Discussion

10.3.1   Scent Preferences

When offered the choice of trails with dextran-only vs. trails with dextran and female copepod scent, the male copepods followed the trail with the scent of their conspecific 80% of the time and the unscented

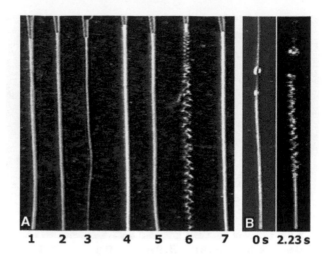

FIGURE 10.3 (Color figure follows p. 332.) (A) Scent preferences of *T. longicornis*. Male copepods of *T. longicornis* were offered seven choices of scent trails (from left to right: Female, Male, Acartia, Dextran, Male, Female, Acartia). The male copepod showed an 80% preference to follow trails scented with the odor of conspecific females (choice 6). Trails were followed to the source. The source of the odor was a stock of dextran-mixed filtered seawater within which 20 females were swimming. The conditioned water flowed through fine pipettes down through the observation vessel, filled with filtered seawater and 20 to 50 male copepods. Schlieren optics visualized the flow patterns. (B) Downwardly directed trail-following. The trail on the left is the undisturbed trail. The disturbed trail on the right was the same trail that was followed down to the source by a male copepod, appearing in the lower right, colored orange. Trails were created by allowing scented seawater to flow out of fine pipettes at the bottom of the observation vessel. The vessel was filled with a mixture of dextran (20 to 40 g/4 l of 500,000 MW dextran) and seawater to create the difference in refractive index necessary to visualize the trail using Schlieren optics. The plain seawater would float up to create the trail. Flow speed of 4.43 mm/s can be calculated by the upward movement of the bright bubble.

trail less than 10% of the time (Table 10.1A). The dextran in the trail did not preclude the male's interest in the trail with the female odors. To determine when, during the reproductive maturation period of the female, the scent was produced at sufficient concentrations, we offered the choice of dextran trails with the scent of conspecific females that had oocytes within their oviducts vs. females without oocytes. Chow-Fraser and Maly (1988) found for the freshwater copepod, *Diaptomus*, more males attempted to mate with gravid than with nongravid females. Here, we found no preference for *T. longicornis* based on scent-tracking (Table 10.1A). Barthélémy et al. (1998) note that a congener, *T. stylifera*, requires a separate fertilization event prior to each clutch of eggs, which may explain why both gravid and nongravid females were attractive to their male mates. Other copepods are able to store sperm for days to weeks; these copepods, which do not need to find mates for every clutch, may be more attractive as virgins or nongravid females (Kelly et al., 1998).

When given the choice to follow trails scented with the odor of conspecific females (combined gravid and nongravid), conspecific males, copepods of another species (*Acartia hudsonica*), or dextran-only trails, varying degrees of preference were discovered. Preference, evaluated as the likelihood to follow an encountered trail, could be ranked from most preferred to least interesting: conspecific females, conspecific males, the other species, the unscented trail (Figure 10.3, Table 10.1B). As an additional test of preference, we also measured the tracking speeds of the male along the trail (Table 10.1B). Males would track conspecific female-scented trails at 15 mm/s, which was faster than their tracking speed along trails scented with conspecific male odor. For the rare event of tracking the dextran-only unscented trail, the tracking speed (5 mm/s) was just above the flow speeds of the trail (1 to 3 mm/s). For the trails with higher flow speeds, there was a slightly greater probability that the male would show disinterest (nick, cross, or escape from the trail; Table 10.1). Flow speeds now are kept below 1 mm/s to reduce the likelihood that the hydromechanical flow disturbance would elicit an escape (Fields and Yen, 1997). The relative tracking speed on preferred trails (copepod swimming speed plus trail flow speed) was slightly slower than natural trail-following speeds (Doall et al., 1998).

TABLE 10.1

Scent Choice by Male Copepods of the Species *Temora longicornis*

Source of Scent	Preference = # follows/# encounter [%]	% Escapes	Tracking Speed	Flow Speed
A. Response to Trails with the Scent of Females of Different Reproductive State and of Males				
Gravid female Tl	43/54 [80%]	7.41%	13.9	2.77
Nongravid female Tl	21/30 [70%]	6.67%	19.2	3.07
Male Tl	4/20 [20%]	10.0%	17.3	2.54
Dextran-only	0/4 [0%]	23.8%	—	2.95
B. Response to Trails with Scent from Different Sexes and Species of Copepod				
Female Tl	31/39 [80%]	0	15.8	1.29
Male Tl	10/19 [52%]	0	10.1	1.28
Acartia tonsa	1/8 [12%]	0	13.6	1.0
Dextran-only	1/11 [9%]	0	5.7	1.32

Note: Random encounters by freely swimming male copepods of the species *T. longicornis* [*Tl*] with scent trails would evoke a response to the trail. Vertical scent trails were created from dextran-labeled seawater flowing down through fine pipette tips into the observation vessel filled with filtered seawater (28 ppt). Male copepods, given a choice of trails scented with the odors of different types of copepods, would follow the trail traveling at an accelerated swimming speed or show disinterest (escape from, nick, or just cross the trail). Preference for a trail type is presented here as (# trails followed/# trails of that type encountered). Disinterest is presented as (# escapes/# encounters). Swimming speeds while trail following were compared to flow speeds of the trail (mm/s). Experiments in A represent responses to the scent of conspecifics. Experiments in B include responses to scents from a copepod of another genus.

Even though dextran-only trails were rarely followed, the copepod does not avoid dextran because scented dextran trails were followed. In most cases, the male copepod would encounter the scented trail and follow the trail up to the source. On occasion, the male would follow the trail down and away from the source and frequently would turn around. To determine if gravity influenced tracking direction of the male copepod, we designed the reverse situation. Here, seawater with and without the scent of other copepods was used to form trails in a tank filled with dextran-labeled water (20 to 40 g dextran/4 l). The trails began now at the bottom of the tank, as the dextranless water would float up to the surface of the tank filled with dextran-labeled water. When the male encountered the trail, he would follow it 91.7% (11 trails followed/12 female trails encountered) of the time *to* the source, at the bottom of the tank (Figure 10.3).

From these analyses of trail following up or down to the source and the speed at which the trails were followed, we conclude that the behavior of mate-tracking by *T. longicornis* is chemically mediated and tracking direction is not determined by gravity. We found that most trails were followed to the source, indicating that the copepods are either able to detect the odor gradient or can detect the directional flow in the sheared slow flow of the manufactured trails. We also conclude that these male copepods can discern differences in scents between sexes and between species.

However, these male *T. longicornis* copepods do not follow female trails exclusively, questioning the specificity of the pheromone. Male copepods would not only follow female trails, but also male trails. Here, we would like to describe an event showing the adaptive value of following one's own track. When studying mate-tracking, we would start with male-only swarms of copepods (Doall et al., 1998). In these swarms, there was the infrequent event when a male would follow the trail of another male but upon contacting the male, he would release him immediately. More interestingly, another event in swarms of mixed sex showed a male copepod that followed the trail of the female, but somehow got derailed and wandered off the trail. After a couple seconds, he turned around, retraced his "steps," found the female trail, and followed it to her to form the spinning mating pair. Hence, an ability to self-track allowed this copepod to find his mate successfully.

The specificity of the diffusible pheromone is further questioned by our observations of male *T. longicornis* following the trails scented with the odor of another species, *Acartia tonsa*. This behavior

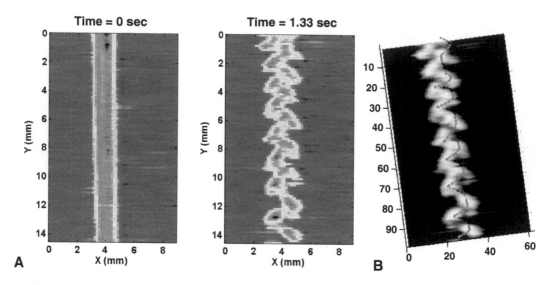

FIGURE 10.4 (Color figure follows p. 332.) Changes in scent trail shape caused by tracking behavior of male *T. longicornis*. (A) Trail structure changes from a smooth undisturbed vertical trail (left) to a disturbed trail (same portion of the trail shown on right) after the passage of the male copepod of *T. longicornis*. (B) Close-up of the disturbed trail. The red dotted line defines the hypothesized helical trail left by spiraling movements of the male while tracking the trail. The spiral is longer, thinner, and has a greater surface area than the smooth trail. The action of molecular diffusion can dilute the pheromone more quickly. Quantitative measurements of trail geometry (for Table 10.3 calculations) include: length of smooth and helical trail; radius of smooth and helical trail, slope of spiral structure of helix. For each 180° segment of the helix, the radius and slope were repetitively calculated to reconstruct the helix so it cuts through the brightest spots of the image. Scale: 6.8 pixels/mm.

suggests another possible scenario: as *Temora* is an omnivorous copepod, could it be responding to this scent as a possible means to track its prey? Prey tracking recently has been observed for catfish detecting the hydrodynamic wakes of goldfish prey (Pohlmann et al., 2001).

Remote detection via the diffusible pheromone is one means of mate recognition. Other possible moments in mate recognition involve contact pheromones (Snell and Carmona, 1994) or a key-and-lock fit of the complex coupling plate (Fleminger, 1975; Blades, 1977). However, there are copepod species that do not have coupling devices on their spermatophore that mirror the genital structure of the conspecific female and only have simple means to cement the base to the female genital area. It also is not known how many copepods secrete contact pheromones, as this has been confirmed for only one species of benthic harpacticoid copepod (Ting et al., 2000), which exhibits precopulatory mate-guarding over extended periods of time. It is likely that decisions at every step in the mating process contribute to final species recognition.

10.3.2 Tracking Behavior

When the male copepod of *T. longicornis* detects a trail, he follows it in either direction, going to the scent source — the female copepod — or away from her. In one example where the male correctly tracked to the female, the position of the male copepod was 1.02 ± 0.34 SD (*n* = 45) mm from the central axis of the 3D trajectory taken by the female seconds earlier, indicating spatially accurate tracking (Figure 10.1A). In another example (Figure 10.1B), the male found and followed the trail in the wrong way, away from the source of the scent: the female copepod. After 1.27 s, he reoriented and returned along the trail back to the female to capture her. Further analyses (Table 10.2) of this behavior of backtracking showed that when the male first follows the trail, his positions precisely overlap the positions in the 3D trajectory taken by the female seconds earlier. The distance from the central axis of the female trail is close to 1 mm. When he turns around, we find that the 3D trajectory of the male copepod no longer matches the path of the female. Instead, the male backtracks the disturbed trail very erratically,

TABLE 10.2

Kinematics of Backtracking in a Mate-Seeking Copepod, *Temora longicornis*

Mate-Tracking of *T. longicornis*	AWAY from Female (*n* = 39)	Back TO Female on Disturbed Trail (*n* = 34)	TO Female on Undisturbed Trail (*n* = 23)	Normal Swimming (*n* = 40)
Velocity (mm/s)	33.9 ± 9.7	50.4 ± 10.6	33.5 ± 12.4	14.4 ± 2.78
Course angle (*x–z*)	46.4 ± 41.5	73.3 ± 50.7	15.3 ± 15.6	
Course angle (*y–z*)	42.3 ± 37.0	71.3 ± 43.1	26.2 ± 18.5	
Track distance (mm)	1.16 ± 0.49	2.09 ± 1.14	1.15 ± 0.66	

Note: The velocity, course angle (in both *x–y* and *x–z* directions), and track distance (closest distance between 3D trajectory of the male copepod and central axis of female trajectory) are given for the event when the male copepod crossed the trail of the female copepod, but went the wrong way: away from the source of the scent, the female. While tracking and retracing his steps, course angles and track distance nearly double in value, indicating the erratic casting behavior exhibited by the male as he returns along the disturbed trail.

casting back and forth over the original location of the female trail at distances twice as far from the central axis of the path. When backtracking, he casts at an average course angle of more than 70° in either the *x–z* or *y–z* direction, which is greater than the average course angle of 46° taken when following the trail initially. A tightening of his course angles to 15° begins after passing his initial intersection with the trail. Here, he again precisely follows the trail, remaining at distances of close to 1 mm from the axis of the path. Weissburg et al. (1998) found that this pattern of counterturns enabled the copepod to stay near or within the central axis of the odor field.

The male copepod's casting behavior suggested that the trail was disturbed by his own swimming activity, making the trail more difficult to follow. Casting behavior also has been noted when the female copepod swims slowly or hovers (Figure 3b in Doall et al., 1998), producing a trail so thick, the male wanders back and forth between the edges. When the female copepod hops and creates a diffuse cloud, male casting behavior also is observed (Figure 5a in Doall et al., 1998) and the male can lose the trail, suggesting that the female hop diluted her scent to levels below the threshold sensitivity of his chemical receptors.

These observations led us to hypothesize that the trail structure has changed in the wake of the tracking by the male copepod. To test this hypothesis, we relied on our method of small-scale flow visualization to document changes in the structure of the trail, evoking the behavioral response of trail following by placing the scent of the female in the trail we created. By visualizing the trail, we saw how the tracking male disturbed the scented trail. Close analyses of the trail (Figure 10.4) show dramatic changes in trail structure in the wake of the tracking male copepod. The undisturbed trail is a smooth vertical line and we presume from the two-dimensional (2D) image that the trail is a 3D cylinder. When the male copepod intersects the trail, he spins and turns to relocate the trail. He then accelerates to three times higher than his normal swimming speed as he follows the trail. Detailed analysis of the structure of the disturbed trail reveals a herringbone 2D pattern. We can imagine that the male copepod swims around the trail, deforming it into a 3D helix. Tsuda and Miller (1998) saw a larger male copepod of the species *Calanus marshallae* that would tilt back and forth as he followed the scent trail of his female. The tilting appears to allow the copepod to insert one antennule into the odor trail and one antennule out of it, thus assessing location by bilateral comparison. As these were 2D observations of the large copepod, we make the assumption that this tilting occurs in 3D, suggesting that the copepod spiraled around the trail. Spiraling while swimming fast has been observed during the escape response of the large *Euchaeta norvegica* (Yen et al., 2002). Spiraling by the mate-searching copepod *Oithona davisae* has been suggested to help the male locate a pheromone source more accurately and to promote diffusion of the pheromone to prevent other males from pursuing the source (Uchima and Murano, 1988).

10.3.3 Quantitative Analyses of Trail Structure and Odorant Levels

To define the characteristics of the disturbed trail, we assume that the brightness of the Schlieren image represents the location of the chemical trail, where areas with similar brightness represent similar concentrations (Gries et al., 1999). Here, we assume the structure of the deformed trail to resemble a

TABLE 10.3

Geometric Changes in Trail Structure Dilutes Chemical Signal

Scent Trails	Smooth Trail (A)	Disturbed Trail (A)	Natural Trail (B)	Disturbed Trail (B)
Initial diameter[a]	0.14 cm ± 5.6% SE ($n = 82$)	0.083 cm ± 17.3% SE ($n = 15$)	0.05 cm	59%
Initial length	1.2 cm	2.9 cm	0.10 cm	240%[b]
Initial V	1.09×10^{-2} cm³	1.6×10^{-2} cm³	2.0×10^{-4} cm³	1.6×10^{-4} cm³
Initial SA	5.4×10^{-1} cm²	7.5×10^{-1} cm²	1.6×10^{-2} cm²	2.2×10^{-2} cm²
Initial C	100%	100%	100%	100%
Final radius	0.09 cm	0.06 cm	0.05 cm	0.04 cm
Final length	1.25 cm	2.91 cm	0.14 cm	0.277 cm
Final V	0.0318 cm³	0.0363 cm³	8.9×10^{-4} cm³	10.4×10^{-4} cm³
Final SA	0.707 cm²	1.15 cm²	4.0×10^{-2} cm²	6.1×10^{-2} cm²
Final C	59%	43%	22%	15.2%

Volume (V) = $\pi r^2 h$; surface area (SA) = $2\pi rh$.

Possible initial odorant concentration (C) = 10^{-5} M (Poulet and Ouellet, 1982).

r = radius of cylinder; h = length of cylinder (see Figure 10.4 for illustration).

Length increment = characteristic diffusion length = SQRT[$4Dt$]; t = time.

D = diffusion coefficient = 10^{-5} cm²/s for small chemical molecules (Jackson, 1980).

Note: The initial volume (V), surface area (SA), and pheromone concentration (C) of the smooth and disturbed trail mimic (A) and natural trail (B) were compared to the final volume (V), surface area, and concentration after 10 s of molecular diffusion (radius increases by 0.2 cm). The geometric measurements of the undisturbed natural trail are from Table 2 in Yen et al. (1998).

[a] Using MATLAB image analytical techniques, the edges were determined as that corresponding to a threshold luminance value of 40%. As the entire trail was only 10 pixels wide, different thresholds will yield different values.

[b] As noted in Table 10.1, when backtracking natural trails, the male casts at nearly twice the distance from the central axis of the trajectory of the female copepod. Similarly, after tracking, we found here that the trail expanded 2.37 times wider.

3D helix, where the relocation of odorant due to copepod movements producing an outward curve of the helix equals the amount relocated on an inward curve of the 3D structure. Using geometric and image processing techniques, we estimated the length, volume, and surface area of the scent trail before and after the copepod swam through it (Table 10.3). The undisturbed trail was assumed to be a cylinder. A section of the image of this cylinder was isolated; length and diameter measurements were taken and used to calculate volume and surface area. The disturbed trail was assumed to have a helical character. A section of the image of the disturbed trail corresponding to the section of the image of the undisturbed trail was isolated. This image section was thought of as a 2D projection of the helical trail onto the CCD imager: the bright spots of the image were assumed to indicate points where the helix crosses the plane of the image projection. Taking these bright spots as data points, an "adaptive helix" was fit to the image in 180° segments as follows. Two successive bright spots were located and their positions used to determine parameters for a helical line segment passing from the first to the second bright spot (Figure 10.4). Next the third bright spot was located and its position was used in conjunction with the second bright spot's position to generate the parameters for another helical line segment to link the second and third bright spots. Proceeding in a similar fashion produced a 3D helical line joining the bright spots of the scent trail. A consequence of the construction of this helical line was that each 180° segment was itself naturally divided into many smaller subsegments and this characteristic was exploited to calculate length, volume, and surface area. The length was found by summing the lengths of each individual subsegment. For the remaining measurements, each subsegment was taken to be the axis of a short cylinder, the diameter of which was found by measuring several such "diameters" in the image and averaging. The volume and surface area of the whole helical scent trail then were found by summing the volumes and surface areas of the individual cylinders thus defined.

We compared the structure of the trail before and after tracking (Table 10.3A) and found that the disturbed trail was 60% thinner, 2.4 times longer, with a surface area 40% greater than the original trail. Although the helix extends over the same linear length occupied by the trail, it expands to a larger helix width and therefore may change the probability of encounter. Once encountered, the next male would have to manage to stay on the track of a thinner trail and also follow a much longer trail. Copepods are able to follow curvy natural (Doall et al., 1998) and manufactured (Figure 10.2B) scent trails of the female copepod so if they found the helix, they could spend extra time following every helical loop or take longer casts suggesting exploration of a more diffuse odorant cloud.

Comparing the volume of the smooth trail to that of the helical trail, we found the volume of the trail decreased by only 17%. The similarity in trail volume suggests that, at these Reynolds numbers, turbulent eddy diffusion is limited by viscous forces. The Reynolds number for the male copepod, while swimming along natural trails, can reach up to values of 60 (maximum tracking velocity = 50 mm/s in contrast to the males' normal swimming velocity of 10 mm/s), quite different from the initial Reynolds number for the trail of 7 (swimming speed of female copepod = 6 mm/s). Here, the Reynolds number of the manufactured trail is close to 1 and that in the wake of the male is between 15 and 20. At these Reynolds numbers, diffusion acts by molecular processes only to slowly disperse the odor. The spiraling copepod reshapes the odor trail but there appears to be little dilution of trail contents.

To determine if the odorant levels would differ after diffusion from a smooth vs. helical trail, we calculated how much larger the trail would become after 10 s of molecular diffusion (radius increases by 0.2 cm after 10 s). It is possible that adjacent loops of the helix would diffuse and fuse to reform the smooth trail. To estimate the time needed for this change to occur, we measured the average separation distance between loops of the helix as 1.92 mm. The loops would need to diffuse 0.96 mm to meet the next loop. This would take more than 3 min. Because trails older than 10 s are rarely followed, it is not necessary to consider the coalescing of odors between loops of the helix. Instead, we considered the change in concentration if the helix were a straight cylinder to compare to the change in concentration of the undisturbed smooth trail. We found that, after 10 s, the final odorant concentration was 59% of the original level (= 100%) in the smooth trail and 43% of the initial concentration in the disturbed trail, with a 60% increase in surface area over the original trail. For natural mating trails, the oldest trail followed by the male *T. longicornis* was 10.3 s old (Doall et al., 1998). This suggests that trails of this age stimulate the copepod sensors just at their threshold. Trails any older are undetectable. Similarly, the backtracking copepod turned around when the trail age was 6.7 s old, close to the time limit beyond which diffusion reduced the concentration levels below the detection threshold. Considering the geometry of a natural trail (from Table 2 in Yen et al., 1998), we calculated that the odorant would decline to 22% of the original odorant levels excreted into the natural trail after 10 s. We conclude that 22% is the threshold for detection by copepod chemoreceptors. If we now disturb the natural trail in a geometrically similar fashion as was determined for the post-tracking manufactured scent trail, the odorant level in post-tracking helical natural trail would decline to 15.2% of the original levels after 10 s (Table 10.3B). This is less than the 22% threshold level. Therefore, in addition to becoming a longer and more convoluted trail, the odor became less distinct where odorant concentrations may drop below the threshold sensitivity of the male's receptors. These structural changes, along with the changes in odorant levels in the disturbed trail plus possible chemical degradation of pheromones over time, have the potential to confound the tracking ability of another chemoreceptive planker. Hence, by disturbing the trail due to his scent-tracking behavior, the male may lower the possibility that another competing mate or threatening predator will find his female.

10.4 Conclusion

To summarize, the study of mate-tracking, using this new behavioral bioassay along with the observation of 3D copepod trajectories in a 4-l container, indicates that *T. longicornis* is able to detect and follow scent trails. As he follows these pheromonal trails to his mate, the male copepod disturbs the signal and effectively lowers the possibility that another mate or chemoreceptive predator will find his female. However, for *T. longicornis*, mate recognition by remote detection of a diffusible pheromone is not certain. *Temora longicornis* appears capable of following different trail types. To understand this seeming lack of specificity, we consider the biology of sparse populations. Plankton, although common, form sparse populations because they are widespread at low densities; they seldom encounter each other and so are effectively rare (Gerritsen, 1980). In sparse populations, the probability that mates encounter each other is reduced, resulting in reduced birth rate and reduced population growth rates. If the population density is low enough, mating encounters are so rare that few individuals reproduce to sustain the population, leading to possible extinction. Because extinction is such a strong selective force, adaptations will be favored that increase the probability of mating encounters. Here it is adaptive to follow any trail because the probability of encountering *anything* is low for these small (1 to 10 mm), slowly swimming (1 mm/s), widely dispersed (1/m³), aquatic microcrustaceans with limited ranges for their sensory field (1 to 100 body lengths). As copepods are considered some of the most numerous multicellular organisms on earth (Humes, 1994), this study of their mating strategies has elucidated their reliance on unusually precise mechanisms for survival and dominance.

Acknowledgments

This work was supported by National Science Foundation Grants OCE-9314934 and OCE-9402910 to J.Y. and J.R.S., with continued support to J.Y. from IBN-9723960. All thank the editors: Laurent Seuront and Peter Strutton.

References

Barthélémy, R.-M., C. Cuoc, D. DeFaye, M. Brunet, and J. Mazza. 1998. Female genital structures in several families of Centropagoidae (Copepoda: Calanoida). *Philos. Trans. R. Soc. Lond. B,* 353: 721–736.

Blades, P.I. 1977. Mating behavior of *Centropages typicus* (Copepoda: Calanoida). *Mar. Biol.* (Berlin), 40: 47–64.

Chow-Fraser, P, and E.J. Maly. 1988. Aspects of mating, reproduction, and co-occurrence in three freshwater calanoid copepods. *Freshw. Biol.,* 19: 95–108.

Doall, M.H., S.P. Colin, J.R. Strickler, and J. Yen. 1998. Locating a mate in 3D: the case of *Temora longicornis. Philos. Trans. R. Soc. Lond.,* 353: 681–689.

Dusenbery, D.B. 1992. *Sensory Ecology.* New York: W.H. Freeman.

Fields, D.M. and J. Yen. 1997. The escape behavior of marine copepods in response to a quantifiable fluid mechanical disturbance. *J. Plankton Res.,* 19: 1289–1304.

Fleminger, A. 1975. Taxonomy, distribution, and polymorphism in the *Labidocera jollae* group with remarks on evolution within the group (Copepoda: Calanoida). *Proc. U.S. Natl. Mus.,* 120: 1–61.

Gerritsen, J. 1980. Sex and parthenogenesis in sparse populations. *Am. Nat.,* 115: 718–742.

Gerritsen, J. and J.R. Strickler. 1977. Encounter probabilities and community structure in zooplankton: a mathematical model. *J. Fish. Res. Board Can.,* 34: 73–82.

Gries, T., K. Johnk, D. Fields, and J.R. Strickler. 1999. Size and structure of "footprints" produced by *Daphnia:* impact of animal size and density gradients. *J. Plankton Res.,* 21: 509–523.

Haury, L.R. and H. Yamazaki. 1995. The dichotomy of scales in the perception and aggregation behavior of zooplankton. *J. Plankton Res.,* 17: 191–197.

Higgie, M., S. Chenoweth, and M.W. Blows. 2000. Natural selection and the reinforcement of mate recognition. *Science,* 290: 519–521.

Humes, A.G. 1994. How many copepods? *Hydrobiologia,* 292/293: 1–7.

Jackson, G.A. 1980. Phytoplankton growth and zooplankton grazing in oligotrophic oceans. *Nature,* 284: 439–441.

Katona, S.K. 1973. Evidence for sex pheromones in planktonic copepods. *Limnol. Oceanogr.,* 18: 574–583.

Kelly, L.S., T.W. Snell, and D.J. Lonsdale. 1998. Chemical communication during mating of the harpacticoid *Tigriopus japonicus. Philos. Trans. R. Soc. Lond.,* 353: 737–744.

Orr, M.R. and T.B. Smith. 1998. Ecology and speciation. *TREE,* 13: 502–506.

Palumbi, S.R. 1994. Genetic divergence, reproductive isolation and marine speciation. *Annu. Rev. Ecol. Syst.,* 24: 547–572.

Pohlmann, K., F.W. Grasso, and T. Breithaupt. 2001. Tracking wakes: the nocturnal predatory strategy of piscivorous catfish. *PNAS*, 98: 7371–7374.

Poulet, S.A. and G. Ouellet. 1982. The role of amino acids in the chemosensory swarming and feeding of marine copepods. *J. Plankton Res.*, 4: 341–361.

Prusak, A., M. Caun, M.H. Doall, J.R. Strickler, and J. Yen. 2001. Happy trails: a behavioral bioassay for copepod mating pheromones. *Am. Soc. Limnol. Oceanogr.* Abstract, February 2001.

Snell, T.W. and M.J. Carmona. 1994. Surface glycoproteins in copepods: potential signals for mate recognition. *Hydrobiologia*, 292/293: 255–264.

Strickler, J.R. 1998. Observing free-swimming copepods mating. *Philos. Trans. R. Soc. Lond. B*, 353: 671–680.

Strickler, J.R. and J.-S. Hwang. 1998. Matched spatial filters in long working distance microscopy of phase objects, in *Focus in Multidimensional Microscopy*, P.C. Cheng, P.P. Hwang, J.I. Wu, G. Wang, and H. Kim, Eds., River Edge, NJ: World Scientific.

Ting, J.H., L.S. Kelly, and T.W. Snell. 2000. Identification of sex, age and species-specific proteins on the surface of the harpacticoid copepod *Tigriopus japonicus*. *Mar. Biol.*, 137: 31–37.

Tsuda, A. and C.B. Miller. 1998. Mate-finding behaviour in *Calanus marshallae* Frost. *Philos. Trans. R. Soc. Lond. B*, 353: 713–720.

Uchima, M. and M. Murano. 1988. Mating behavior of the marine copepod *Oithona davisae*. *Mar. Biol.*, 99: 39–45.

Weissburg, M.J., M.H. Doall, and J. Yen. 1998. Following the invisible trail: mechanisms of chemosensory mate tracking by the copepod *Temora*. *Philos. Trans. R. Soc. Lond.*, 353: 701–712.

Yen, J., M.J. Weissburg, and M.H. Doall. 1998. The fluid physics of signal perception by a mate-tracking copepod. *Philos. Trans. R. Soc. Lond.*, 353: 787–804.

Yen, J., H. Browman, J.F. St.-Pierre, and M. Belanger. 2002. Role reversal: the gladiatorial match between the carnivorous copepod *Euchaeta norvegica* and Atlantic cod. Larval Fish Conf. abstract, Norway, July.

11

Experimental Validation of an Individual-Based Model for Zooplankton Swarming

Neil S. Banas, Dong-Ping Wang, and Jeannette Yen

CONTENTS

11.1 Introduction...161
11.2 Theory...163
    11.2.1 Differentiating between Swarming and Diffusion.......................................163
    11.2.2 Diffusion in an Aggregative Force Field..164
    11.2.3 Further Model Predictions..165
    11.2.4 Swarming in Two and Three Dimensions..166
    11.2.5 The Acceleration Field 166
11.3 Experiment..167
11.4 Analysis...168
    11.4.1 Constructing a Statistical Ensemble...168
    11.4.2 A Procedure for Testing Model Consistency..169
11.5 Results...170
    11.5.1 Velocity Distributions 170
    11.5.2 Velocity Autocorrelations and Fit Parameters...171
    11.5.3 Acceleration Fields 173
11.6 Discussion...174
    11.6.1 Model Consistency 174
    11.6.2 Model Interpretation 175
        11.6.2.1 Damping...175
        11.6.2.2 Excitation...177
        11.6.2.3 Concentrative Force 177
        11.6.2.4 Physical–Behavioral Balances 177
11.7 Conclusion 178
Acknowledgments...178
References...178

11.1 Introduction

The ecology of marine planktonic assemblages depends, in essential, intricate ways, on the behavior of individual zooplankters. Swarming behavior is among the most crucial, and also least charted, of the territories that span population and organismal biology in this way. On large scales in the ocean, and possibly in some small-scale environments like frontal zones, aggregation into patches is probably a physics-driven, passive process. At the same time, active swarming behavior — that is, a type of motion that resists dispersion without orienting or distributing animals in an organized way — is well known, and controls small-scale zooplankton distribution in the ocean to an unknown extent. Passive and active aggregations are essential for setting encounter rates between predators and prey, between grazers and patchily distributed food sources (Lasker, 1975; Davis et al., 1991) and between conspecifics in search of mates (Brandl and Fernando, 1971; Hebert et al., 1980; Gendron, 1992).

Swarming differs from schooling (which is known but rare among the zooplankton; Hamner et al., 1983) in that it is stochastic, unarrayed, and largely uncoordinated between individuals. At the same time swarming differs from truly random motion — that is, Brownian motion or diffusion — in that a swarm does not spread out over time and disperse as does a cloud of molecules or drifting particles. Behaviors that maintain swarms against diffusion may be either social and density dependent or nonsocial and density independent; we consider each category briefly in turn.

A variety of models for social swarm maintenance have been proposed, centered on mechanisms in which, for example, animals seek a target density (Grünbaum, 1994) or correlate their motion with their neighbors' motion (Yamazaki, 1993). The possibilities for such mechanisms in a given species are strongly constrained by sensory ability. Perhaps the most fundamental constraint is that the majority of zooplankton lack image-forming eyes (Eloffson, 1966) and, accordingly, do not appear to orient to their conspecifics visually. Long-distance interactions along scent trails have been observed in copepod swarms (Katona, 1973; Weissburg et al., 1998), and similar interactions are possible along trails in the shear or pressure fields (Fields and Yen, 1997), as has been observed in schools of Antarctic krill (Hamner et al., 1983). Nevertheless, most intraspecies communication in zooplankton appears to be local and intermittent, as opposed, say, to the long-range and constant visual coordination that gives fish schools their character. Leising and Yen (1997), for example, found five copepod species to be insensitive to the proximity of their conspecifics except at nearest-neighbor distances of a few body lengths. Swarms of this density are rarely found in nature (Alldredge et al., 1984). It is important to note that even intermittent social interactions may play a large role in swarm dynamics. Leising and Yen argue that the number density of the laboratory swarms they observe is controlled by avoidance reactions to chance close-range encounters in the swarm center.

A number of nonsocial aggregative behaviors have also been observed *in situ*. Phototaxis maintains swarms of the cyclopoid copepod *Dioithona oculata* in shafts of light between mangrove roots (Ambler et al., 1991), and similar responses to light gradients are known in several other species (Hamner and Carlton, 1979; Hebert, 1980; Ueda et al., 1983). Attraction to food odors, and increased turning in food patches, which aids foraging and increases forager density, have also been observed in a number of species (Williamson, 1981; Poulet and Ouellet, 1982; Tiselius, 1992; Bundy et al., 1993). Some zooplankton in fact may respond directly to water temperature and salinity (Wishner et al., 1988; Gallager et al., 1996).

This enormous range of social and individual behaviors does not consolidate readily into a simple, general account of zooplankton swarming. Still, a unifying thread runs through them: whatever the driving behavioral mechanism, the tendency that counters dispersion in a swarm is not just a collective, statistical property, but rather must be observable in each swarmer's individual motion as a hidden regularity. This is an experimentally powerful notion, for it suggests that we may be able to apprehend the dynamics of a large, observationally unwieldly aggregation by studying the behavior of a few typical individuals. Indeed, in a dynamic, rather than statistical (one could also say ethological, rather than ecological) approach to animal swarming, the larger aggregation may often be close to irrelevant. Writes Okubo (1986):

> There is an interesting observation by Bassler that a swarm could be reduced to a single individual of mosquito, *Culex pipiens* which yet continued to show the characteristic behavior of swarming dance [Clements, 1963]. Also Goldsmith et al. [1980]... noted that a single and a few midges did show movements characteristic of swarming by a large number of animals.

Provocatively, many zooplankton may behave like mosquitoes and midges in this respect, and form swarms in which interaction between individuals plays at most a secondary role. Yen and Bundock (1997), for example, found no social interaction within phototactic laboratory swarms of the harpactacoid copepod *Coullana canadensis*.

When interaction does occur, as noted above, it is generally chemical or rheotactic, rather than visual, and thus occurs slowly, along spatially torturous paths. These paths are very difficult to observe or map, and in a large aggregation, especially outside the laboratory, would not easily be differentiated from a diffuse, continuous sensory cue. Indeed, if such interactions are fundamental to the dynamics of a swarm (as opposed to simply being facilitated by the swarm, as mating encounters might be), then their importance is not particular but cumulative, statistical, parameterizable. While the details of social communication are crucially relevant to swarming biology on one level, for analyzing balances between dispersion and counterdispersion — for understanding the kinematics of an individual swarmer — it seems more apt to average over many encounters, and to model social effects as a net dispersive or concentrative tendency in each individual.

In this modeling approach, then, whether a swarm is socially or nonsocially driven, we regard it as an interaction not between animals, but between each animal and its local stimulus field. This approach has the advantage of generality. While one can parameterize a social, density-dependent response — a series of avoidance reactions, for example, or motion through a network of pheromonal trails — as an individual response to a steady cue, it is hard to imagine modeling, say, phototaxis by reversing the analogy.

Attempts to produce a general quantitative description of zooplankton swarming have been frustrated by a lack of marine observations. Okubo and Anderson (1984) present a simple and general individual-based model of steady-state swarming (see also Okubo, 1980, 1986), but write that for purposes of model verification, "no such data really exist in the marine field." They proceed, with reservations, by examining data from "aeroplankton," midges swarming in a forest clearing. Since the time of their publication, optical and acoustic technologies for observing and recording *in situ* zooplankton distribution and behavior have become available (Alldredge et al., 1984; Schultze et al., 1992; Smith et al., 1992; Davis et al., 1992). These and laboratory methods have occasionally been applied to the problem of aggregation dynamics. McGehee and Jaffe (1996), for example, examine the relationship between path curvature and swimming speed in a zooplankton patch observed through acoustic imaging; Buskey et al. (1995) used three-dimensional imaging in the laboratory to analyze the phototactic formation of a swarm in *Dioithona oculata*.

Still, no dynamic account has been given of the swarming motion of individual animals followed for sustained periods. In the present study we use optical methods developed by Strickler (1998) to evaluate Okubo and Anderson's model — the "aggregating random walk" (Yamazaki, 1993) — in the case of two species of zooplankton, the calanoid copepod *Temora longicornis* and the cladoceran *Daphnia magna*, swarming phototactically in the laboratory.

11.2 Theory

In this section we derive quantitative predictions of the aggregating random-walk model, which we can directly compare with data, beginning from kinematic first principles, following Okubo and Anderson (1984).

11.2.1 Differentiating between Swarming and Diffusion

Assume for simplicity that the swarming motion is one dimensional, in the x direction — we generalize to two and three dimensions later — and assume that the swarm has reached a steady state and is isotropic. If all the swarmers are responding to the attractive stimulus in the same way, then their paths will be centered on the same point: call this the origin. Then the spatial variance $\overline{x^2}$ of an individual's path also measures the size of the swarm.

A standard result in the theory of diffusion (Okubo, 1980) is that under these assumptions, for large t, $\overline{x^2}$ increases as

$$\overline{x^2} \to 2Dt \tag{11.1}$$

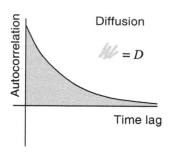
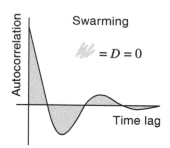

FIGURE 11.1 Schematic representation of the distinction in velocity autocorrelations between swarming and diffusion. The shaded area in each case is equal to the diffusion coefficient D.

where D is the diffusivity. D can be written as

$$D \equiv \overline{u^2} \int_0^\infty R(\tau) d\tau \tag{11.2}$$

where $u \equiv dx/dt$. R is the Lagrangian velocity autocorrelation coefficient, a function of time lag τ:

$$R(\tau) \equiv \frac{1}{\overline{u^2}} \overline{u(t)u(t+\tau)} \tag{11.3}$$

Eventually any individual's velocity decorrelates from its earlier values, since no animal, swarming, diffusing, or otherwise, moves in the same pattern forever, and so $R(\tau) \to 0$ for large τ. The shape of $R(\tau)$ as it tends to this asymptotic limit, however, is variable. In the case of simple diffusion, $R(\tau)$ decays exponentially. Its integral is a positive number, so that D, by Equation 11.2, is a positive number, and x^2, by Equation 11.1, increases linearly. In the case of swarming, in which trajectories do not spread over time, however, x^2 remains constant; D must then be zero; and $R(\tau)$ oscillates about the axis in such a way that its area converges likewise to zero (Figure 11.1).

Thus the shape of $R(\tau)$ — specifically, the presence or absence of an axis crossing — is the key to distinguishing kinematically between swarming animals and diffusing ones. $R(\tau)$ can be calculated directly from position and velocity data by Equation 11.3.

11.2.2 Diffusion in an Aggregative Force Field

We can recast the problem in dynamic Newtonian terms, again following Okubo and Anderson, as a balance between diffusion and a deterministic concentrative force. In this model the concentrative force, an inward acceleration that grows with distance from an attractor at the swarm center, keeps the swarmers from dispersing, while diffusive motion keeps the swarmers from collapsing onto the central attractor.

The resulting equation of motion can be written:

$$du/dt = A - ku - \omega^2 x \tag{11.4}$$

The first two terms on the right-hand side constitute a standard "random-flight" model of diffusion: k is a frictional coefficient in a Stokes model of drag, and $A(t)$ is a random acceleration, some sort of white noise, with a delta-function autocorrelation and finite power $B \equiv \overline{Au}$. Note that we can interpret this random excitation and frictional damping either as true external, turbulent forces acting upon a passive particle, or simply as accelerations that express behavior patterns.

The third term on the right-hand side, the concentrative force, can be interpreted either as a harmonic (linear) restoring force, such as the force that gravity exerts on a pendulum, or as a local approximation to a more complicated, anharmonic restoring force. Okubo and Anderson suggest anharmonicity as a mechanism for maintaining a uniform, rather than Gaussian, density distribution through the swarm center. Our experiment neither confirms nor refutes this idea, and so in the interest of empiricism we have retained only the harmonic concentrative term. We view our dynamic model (Equation 11.4) as a first-order approximation to a behavior that may well involve a number of unmodeled, higher-order effects, such as advection of momentum, in both the diffusive and concentrative motions. Higher-order models may be necessary to parameterize the effects of foraging, mate-finding, escapes, and other behaviors simultaneous with swarming.

As shown by Okubo (1986), Equation 11.4 does indeed yield a velocity autocorrelation of the form required by kinematic considerations, as discussed above and illustrated in Figure 11.1:

$$R(\tau) = e^{-k\tau/2}\left(\cos\omega_1 t - \frac{k}{2\omega_1}\sin\omega_1 t\right) \tag{11.5}$$

where

$$\omega_1^2 = \omega^2 - k^2/4 \tag{11.6}$$

The integral of $R(\tau)$ is identically zero. Note that as the attractive forcing weakens — that is, as $\omega \to 0$ — Equations 11.4 and 11.5 both approach the results for simple diffusion:

$$du/dt = A - ku \tag{11.7}$$

$$R(\tau) = e^{-k\tau} \tag{11.8}$$

In practice, we necessarily calculate the autocorrelation $R(\tau)$ on discrete time series, either from experiment or from numerical simulation, and such a discrete autocorrelation is not fully equivalent to Equation 11.5, derived from continuous dynamics. Yamazaki and Okubo (1995) show that the discrete autocorrelation does not integrate to zero, i.e., that discretization introduces an apparent (but artificial) net diffusion rate. This mathematical inconsistency is potentially a limit on the precision with which we can determine k and ω from observations.

11.2.3 Further Model Predictions

In this model, a swarm is characterized primarily by position variance $\overline{x^2}$ (a measure of the size of the swarm), velocity variance $\overline{u^2}$ (a measure of the kinetic energy of its members) and by k, ω, and B (measures of the strength of the damping, attractive, and excitational forces). We can derive from Equation 11.4 some useful and experimentally verifiable relationships between these parameters.

Multiplying Equation 11.4 by u and taking the time average, we obtain

$$\overline{u\frac{du}{dt}} = -\overline{ku^2} - \omega^2\overline{xu} + \overline{Au} \tag{11.9}$$

$\overline{Au}$ is the power of the random forcing, or B. Since we assume the swarm is in a steady state, both $u(du/dt)$ and $x(dx/dt)$ are zero. Thus, Equation 11.9 becomes

$$\overline{u^2} = \frac{B}{k} \tag{11.10}$$

Note that this relationship is independent of ω and therefore true of purely diffusive motion as well. Thus the attractive force has no bearing on the kinetic energy of a swarm. Nor does it influence the speed distribution of the swarmers: an individual subject to Equation 11.3 follows the Maxwellian speed distribution for free particles (like gas molecules):

$$p(u) = \frac{1}{\sqrt{2\pi\overline{u^2}}}\exp\left(-\frac{u^2}{\overline{u^2}}\right) \tag{11.11}$$

where $p(u)$ is the fraction of a statistical ensemble found between velocities u and $u + du$ (Okubo and Anderson, 1984).

The strength of the attractive force does, however, have a very direct bearing on the steady-state swarm size x^2. For large t, x^2 approaches the value

$$\overline{x^2} = \frac{\overline{u^2}}{\omega^2} \tag{11.12}$$

(Uhlenbeck and Ornstein, 1930). Thus swarm size is inversely proportional to the strength of the attractive forcing. When ω is large — that is, when the attractive tendency increases rapidly with distance from the swarm center — the swarm is concentrated tightly. As ω approaches zero — the case of pure diffusion — the swarm's limiting size increases toward infinity.

11.2.4   Swarming in Two and Three Dimensions

These results can easily be generalized to more than one dimension. Assume, for example, that the same forces that act in the x direction act in the y. Because our model involves no coupling of motion along different axes, the results so far derived still hold, and $\overline{y^2}$ and $\overline{v^2}$ (where $v \equiv dy/dt$) are related to k and ω by expressions analogous to Equation 11.10 and 11.12. We can then write expressions for the two-dimensional parameters of interest:

$$\overline{V^2} \equiv \overline{u^2} + \overline{v^2} = \frac{2B}{k} \tag{11.13}$$

$$\overline{r^2} \equiv \overline{x^2} + \overline{y^2} = \frac{\overline{V^2}}{\omega^2} \tag{11.14}$$

The generalization to three dimensions is analogous.

Velocity distributions for two- and three-dimensional swarms are simply the two- and three-dimensional Maxwellian distributions. In two dimensions,

$$p(V) = \frac{2}{\overline{V^2}} V \exp\left(-\frac{V^2}{\overline{V^2}}\right) \tag{11.15}$$

Note that animals swarming in a three-dimensional ocean are not necessarily engaged in three-dimensional swarming. The number of dimensions in which the swarming model applies depends on the symmetries of the attractor. If the attractor is planar (for example, a front or thin layer containing high food concentrations: Yoder et al., 1994; Hanson and Donaghay, 1998), then the swarming forces do indeed act in only one dimension, the vertical, and motion in the two horizontal dimensions, in which the attractor is isotropic, is likely to be diffusive. If the attractor is one dimensional (for example, the light shaft in our experiment, or a light shaft through mangrove roots: Ambler et al., 1991), then we might expect swarming motion in two dimensions and diffusive motion in the third. Only if the attractor is a point or spherically symmetrical (for example, a diffusing chemical signal in an isotropic environment) does the swarming model apply to three dimensions.

11.2.5   The Acceleration Field

Finally, returning to the one-dimensional case, we can derive predictions about the mean and variance of the spatially varying acceleration field.

Taking the mean of Equation 11.4, holding x constant, we obtain

$$\overline{a(x)} = -\omega^2 x \tag{11.16}$$

where $a \equiv du/dt$. Thus, the mean acceleration field consists of the center-directed concentrative force alone.

Fluctuations about this mean consist of the random-flight accelerations $A(t) - ku$. Okubo (1986) derives an expression for the expected variance of these fluctuations, $\overline{a^2}$. The quantity we actually calculate from data, however, is not acceleration but a finite-difference approximation to acceleration:

$$\frac{\Delta u}{\Delta t} = \frac{u(t_0 + \Delta t) - u(t_0)}{\Delta t} \tag{11.17}$$

where Δt is the sampling period, and the theoretical variance of $\Delta u/\Delta t$ is not the same as that of the acceleration itself. (Note that the distinction does not affect the predicted mean acceleration field, or the predictions concerning the velocity field given above.) Without loss of generality, set $t_0 = 0$. Then the finite-difference acceleration variance can be written

$$\begin{aligned}
\frac{\overline{\Delta u^2}}{\Delta t^2} &= \frac{1}{\Delta t^2} \overline{\left[u(t_0 + \Delta t) - u(t_0)\right]\left[u(t_0 + \Delta t) - u(t_0)\right]} \\
&= \frac{1}{\Delta t^2}\left(2\overline{u^2} + 2\overline{u(t_0)u(t_0 + \Delta t)}\right)
\end{aligned} \tag{11.18}$$

since the velocity variance is assumed constant in time. The second term can be evaluated using the definition of the autocorrelation (Equation 11.3) and the diffusive result (Equation 11.8), so that

$$\frac{\overline{\Delta u^2}}{\Delta t^2} = \frac{2\overline{u^2}}{\Delta t^2}\left(1 - e^{-k\Delta t}\right) \tag{11.19}$$

predicts the variance about the mean acceleration $\overline{a(x)}$ for a time series sampled with a time step Δt.

Just as the acceleration field has a finite variance about the mean, so does the autocorrelation curve $R(\tau)$. Only the mean autocorrelation was derived above. Because of the complexity of the problem, however, in the data analysis below we estimate the theoretical variance of the autocorrelation numerically, through direct integration of the equation of motion (Equation 11.4), rather than deriving an expression for it analytically.

11.3 Experiment

Data were collected for homogeneous groups of two species, the freshwater cladoceran *Daphnia magna* (Hussussian et al., 1993) and the calanoid copepod *Temora longicornus*. In each experiment the animals were placed in a small ($10 \times 10 \times 15$ cm) Plexiglas tank, a light down the center axis of the tank turned on, and the trajectories of the animals swarming to the light recorded on videotape and then digitized.

The geometry of the tank-and-light system is shown in Figure 11.2. Description of the optical methods employed can be found in Strickler and Hwang (1998) and Strickler (1998); digitization methods are

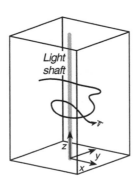

FIGURE 11.2 Geometry of the tank and light system.

described in Doall et al. (1998). The light source in the *Daphnia* experiment was a 3-mW argon laser (wavelength 488 and 514.5 nm), which produced a collimated beam 4 mm wide throughout the water column. The *Temora* experiment used a 2-mm-diameter fiber-optic light guide, projecting light from a halogen lamp at the top of the tank. This produced a narrow cone-shaped beam with more attenuation through the water column, although the beam reached clearly to the bottom of the tank.

All three axes of motion were captured for the *Temora* experiment, using a pair of orthogonal projections. One camera recorded motion in the *x–z* plane, and a second, motion in the *y–z* plane, so that by matching *z* motions one could recover full three-dimensional trajectories. The *Daphnia* experiment was recorded with an earlier generation of the optical system and thus contains only a single *x–z* projection. Because of the radial symmetry of the light source, and the fact that our model predicts no coupling of motion between orthogonal axes, the lack of a third dimension in the *Daphnia* data does not hinder the statistical analysis.

Phototaxis in marine and freshwater macrozooplankton is well known, primarily in the context of light-gradient-driven diel migration (Russell, 1934; Stearns, 1975; Bollens et al., 1994). The behavioral significance of the strong positive phototaxis that zooplankters often show in the laboratory, however, is poorly understood. Some animals placed in the tank showed no attraction to the light, and the response of other animals was intermittent. Some of the *Temora* simply remained at the bottom of the tank, while others collected at the surface, where the fiber-optic illumination was intensified. Flux between these subpopulations and the swarm in the mid-water column was continual. Trajectories for analysis were taken from the animals that remained in the swarm for longer than one sample period.

These swarms tended to be small, generally consisting of fewer than ten animals, so that mean nearest-neighbor distances were many body lengths. This low density is consistent with our treatment of swarming behavior not as a density-dependent, social interaction, but rather as an individual response to a sensory cue. The simultaneity or nonsimultaneity of the swarmers' motions does not enter into the analysis, because each animal is in effect swarming alone.

11.4 Analysis

11.4.1 Constructing a Statistical Ensemble

Trajectories were sampled at 1-s intervals. This time step was chosen both to resolve the fundamental timescales of motion (the damping time $1/k$ and the attractive force period $2\pi/\omega$, which were estimated by an initial calculation of the velocity autocorrelation) and simultaneously to filter out higher-frequency behaviors such as avoidance reactions and trail following (Weissburg et al., 1998). Long-time step sampling is an unrefined method for low-pass-filtering a trajectory record; but a test performed on *Temora* data sampled at 0.2 s suggested that the choice of this method over tapered low-pass filtering of a higher-resolution data set had a negligible effect on the analysis. Longer-time step sampling makes the digitization of trajectories (which was done by hand here rather than by computer program to ensure accuracy) far less labor intensive.

Recorded trajectories were divided into uniform 20-s segments for the analysis. This length of time was chosen to balance two considerations. The segments had to be long enough to resolve the inherent dynamic timescales mentioned above, and at the same time as short as possible — i.e., as close to the autocorrelation timescale as possible — to ensure uniformity of behavior within each record. They thus become independent samples in the statistical sense. Within each 20-s segment, velocities and accelerations were calculated using the finite-difference equations

$$u(t) \sim \frac{x(t) - x(t - \Delta t)}{\Delta t} \tag{11.20}$$

$$a(t) = \frac{u(t + \Delta t) - u(t)}{\Delta t} = \frac{x(t + \Delta t) - 2x(t) + x(t - \Delta t)}{\Delta t^2} \tag{11.21}$$

Even among the animals that remained suspended in the water column, not all were engaged in swarming behavior. A few simply drifted through the swarm without, in a kinematic sense, being part of it, like waiters carrying trays across a lively ballroom. The shape of the velocity autocorrelation, as illustrated in Figure 11.1, quantitatively distinguishes swarmers from drifters, i.e., slow diffusers. As our goal is not to test a hypothesis about the phototactic behavior of *Daphnia* and *Temora*, but rather to describe the swarming that a light gradient *can* evoke in these animals, we winnowed the trajectory set to remove those animals who were not participating. (Because this winnowing eliminated only a small number of candidate trajectories, as described below, it is fair to assume that the swarming dynamics thus described account for the bulk of the animals' response to this stimulus.) We eliminated from both the *Daphnia* and *Temora* data sets trajectory samples whose velocity autocorrelation did not cross the time axis, as well as samples whose mean position lay more than two standard deviations from the *z* axis, the swarm center. These criteria eliminated 4 of 24 *Temora* trajectories and 2 of 60 *Daphnia* trajectories. Some diffusers were also eliminated by eye before digitization of trajectories began. Lateral projections of all remaining *Daphnia* and *Temora* trajectories are shown superimposed in Figure 11.3A and Figure 11.4A, B.

The final step in creating a uniform statistical ensemble of trajectories is to verify that the process captured is kinematically stationary. Figure 11.3B and Figure 11.4C show the means and standard deviations of position and velocity for both data sets. Note that there are no strong outliers, and that the variances of the ensembles are on the same order as the variances of individual trajectories within them.

11.4.2 A Procedure for Testing Model Consistency

We proceed as follows for both the *Daphnia* and *Temora* experiments.

1. Verify that the *velocity distribution* is Maxwellian. This confirms (in the absence of an independent measure of the excitation variance B) the *kinetic energy balance* of Equation 11.10.

2. Calculate the *velocity autocorrelation* $R(\tau)$, and verify that its form fits the kinematic requirement illustrated in Figure 11.1.

   Fit $R(\tau)$ to find ω and k. This is done by minimizing, by inspection, the square-error function:

   $$J(\omega,k) = \sum_{\tau} \left(R_{observed}(\tau) - R_{theoretical}(\tau) \right)^2 \qquad (11.22)$$

   With these parameters, directly integrate Equation 11.4 to create a simulated ensemble of velocity autocorrelations, and thus a numerical prediction of the autocorrelation *variance* (just as Equation 11.5 gives an analytical prediction for the autocorrelation *mean*).

3. With ω from step 2, confirm the *relation between swarm size and kinetic energy* predicted by Equation 11.12.

4. Compare the *mean and variance of the spatial acceleration field* with those predicted by Equations 11.16 and 11.23.

5. Examine *motion in the z (vertical, along-laser) direction,* which we have not to this point discussed. In this dimension the attractor is more-or-less isotropic and we might expect diffusive, not swarming, dynamics. The animals' vertical motion can be compared with theory under the assumption that the diffusion-driven parameters $\overline{u^2}$, B, and k will be the same in the *z* direction as they are in the *x* and *y* (cross-laser, swarming) directions. Here we follow an abbreviated version of the outline above.

   Look for a Maxwellian *velocity distribution,* as in step 1.

   Calculate the *velocity autocorrelation,* compare its ensemble *mean* with Equation 11.8, and compare its ensemble *variance* with numerical results, as in step 2.

   Calculate the *mean and variance of the acceleration field* as in step 4. We now expect the mean to be zero (by Equation 11.16, with $\omega = 0$) and the variance to be as predicted by Equation 11.19.

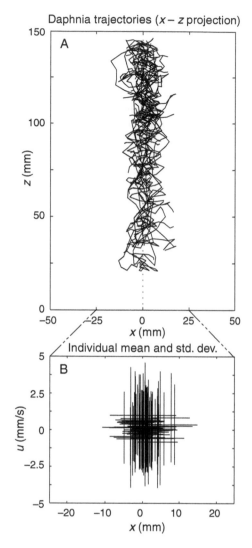

FIGURE 11.3 (A) Superimposition of all 58 *Daphnia* trajectories included in the analysis, shown in *x–z* projection. Plot boundaries indicate the edges of the tank, and the dotted line shows the position of the light shaft. (B) Position (horizontal axis) and velocity (vertical axis) statistics for each trajectory sample. The crosshairs, each representing a single animal trajectory, are centered on the mean horizontal position and velocity, and indicate standard deviations by their extent.

11.5 Results

Results for the *Daphnia* experiment are shown in Figure 11.5 and results for *Temora* in Figure 11.6. Note that the calculations for the horizontal dimensions in the *Temora* data are two dimensional, as both horizontal components of motion were captured and analyzed, and one dimensional for the *Daphnia* data. The parameters k, ω, and B, derived from the horizontal analysis, along with rms velocity and position (which represent u^2 and x^2), are summarized for both species in Table 11.1. Also given are rms positions (i.e., swarm sizes) predicted by Equation 11.12 and the error between these predictions and the values observed.

11.5.1 Velocity Distributions

Figure 11.5A, B and Figure 11.6A, B show observed velocity distributions for the horizontal (swarming) and vertical (diffusive) dimensions, along with theoretical curves based on Equations 11.11 and 11.15.

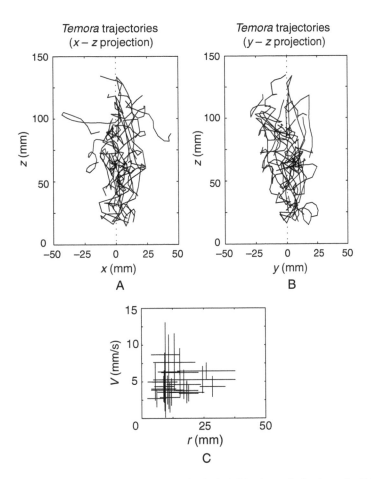

FIGURE 11.4 (Superimposition of the 20 *Temora* trajectories included in the analysis, shown in (A) *x–z* and (B) *y–z* projection. Plot boundaries indicate the edges of the tank, and the dotted line shows the position of the light shaft. (C) Position (horizontal axis) and velocity (vertical axis) statistics for each trajectory sample. The crosshairs, each representing a single animal trajectory, are centered on the mean horizontal position and velocity, and indicate standard deviations by their extent.

All velocity fields are close to stationary $(\overline{u}^2 << \overline{u^2})$. Velocity distributions for both species lie close to the predicted Maxwellian curves, with $r^2 = 0.65$ and 0.98 for the horizontal and vertical axes of the *Daphnia* motion and $r^2 = 0.91$ and 0.86 for the horizontal and vertical axes for *Temora*. The horizontal *Daphnia* distribution is more platykurtic than predicted, and the horizontal *Temora* distribution more sharply peaked, but these deviations could be either statistical artifacts or true dynamic effects related to these species' swimming styles, and are second-order effects in either case. The first-order match with theory supports our assumption of a Stokes form for drag and a stochastic, spatially symmetrical excitation.

Note also that, with the exception of a possible upward drift along the laser axis in the *Temora* data (Figure 11.6B), the velocity fields are very close to symmetrical, and that velocity variances are similar, for each animal, in the horizontal and vertical dimensions. In fact, velocity variances are indistinguishable for the two horizontal dimensions in the *Temora* data ($\overline{u^2} = 16.7$ mm²/s², $\overline{v^2} = 16.6$ mm²/s²). These patterns support the spatial symmetries we have assumed and our supposition of dynamic consistency between the along-laser and cross-laser diffusion processes. Note that an upward drift in the *Temora* experiment would be consistent with a weak phototactic response to the attenuation of the light shaft through the water column.

11.5.2 Velocity Autocorrelations and Fit Parameters

Figure 11.5C, D show observed and theoretical velocity autocorrelations for the two dimensions of *Daphnia* motion. The horizontal theoretical mean is a fit to the data shown, while the vertical theoretical

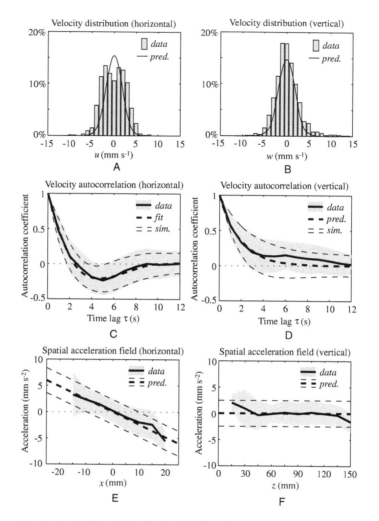

FIGURE 11.5 Results for *Daphnia*. Velocity distributions (A, B), velocity autocorrelations (C, D), and acceleration fields (E, F) for the horizontal (A, C, E) and vertical (B, D, F) axes of motion. In all panels, thick lines indicate means and shaded areas and thin dotted lines indicate standard deviations. Full explanations of plotted quantities are given in the text.

mean is not a fit but a prediction based on Equation 11.8 with k from the horizontal analysis. Theoretical standard deviations are calculated from numerical simulation. Figure 11.6C, D show autocorrelations for *Temora* in the same format, except that here the observed horizontal velocity autocorrelation (Figure 11.6C) is the average of the x and y autocorrelations, a quantity that is invariant under rotation of the horizontal axes.

Horizontal autocorrelations for both species are fit very closely by theoretical curves (*Daphnia*, $r^2 = 0.995$; *Temora*, $r^2 = 0.90$). This match is a strong validation of the kinematic theory underlying our analysis. Variance of individuals around the ensemble mean for each species is close to the level that simulation predicts (*Daphnia*, $r^2 = 0.95$; *Temora*, $r^2 = 0.71$). Note that in contrast to the underdamped, highly orbital trajectories of members of the midge swarm analyzed by Okubo and Anderson (1984), both the *Daphnia* and *Temora* swarms appear to be near critical damping, with individuals' velocities decorrelating almost entirely in less than one period of the harmonic attractive force.

Values of k derived from the horizontal swarming motion suggest decaying autocorrelation curves for the vertical motion that correspond well with observations (Figure 11.5D and Figure 11.6D), confirming our supposition that motion along the laser axis is primarily random and diffusive, and that the same damping force acts on horizontal and vertical swimming. The *Daphnia* vertical autocorrelation, however, has a longer tail than predicted, suggesting that the animals tend to drift, actively or passively, along the

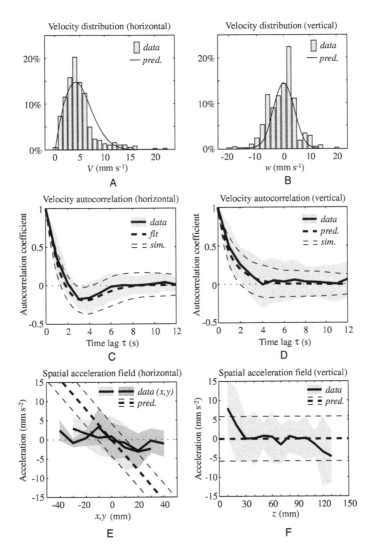

FIGURE 11.6 Results for *Temora*. Velocity distributions (A, B), velocity autocorrelations (C, D), and acceleration fields (E, F) for the horizontal (A, C, E) and vertical (B, D, F) axes of motion. In all panels, thick lines indicate means and shaded areas and thin dotted lines indicate standard deviations. Full explanations of plotted quantities are given in the text.

laser axis, in combination with their diffusive behavior. The *Temora* vertical autocorrelation may or may not indicate the same tendency.

Fitting the horizontal autocorrelations for k and ω lets us test the consistency of the theoretical relationship (Equation 11.12) between swarm size, swimming speed, and the strength of the concentrative force. Table 11.1 gives observed and predicted values of the swarm size for both species. They agree — i.e., the relationship between parameters is consistent with theory — to 4% for *Daphnia* and 45% for *Temora*.

11.5.3 Acceleration Fields

Figure 11.5E, F and Figure 11.6E, F show the observed acceleration fields with theoretical means and standard deviations from Equations 11.16 and 11.19. The x and y acceleration fields for *Temora* (Figure 11.6E) are superimposed.

The horizontal *Daphnia* acceleration field (Figure 11.5E) agrees well with theory in both its mean slope (which is not significantly different from the predicted value of zero) and its mean standard

TABLE 11.1

Dynamic and Kinematic Parameters along the Axes of Swarming Motion for the *Daphnia* and *Temora* Experiments

		Source	*Daphnia*	*Temora*
Damping coefficient	k	Fit to velocity autocorrelation	0.54 s^{-1}	0.80 s^{-1}
Concentration coefficient	w	Fit to velocity autocorrelation	0.49 s^{-1}	0.63 s^{-1}
Power of excitation	B	Equations 11.10, 11.13	3.6 mm^2 s^{-3}	13.3 mm^2 s^{-3}
Kinetic energy	$\sqrt{\overline{u^2}}$	Observed	2.6 mm s^{-1}	
(rms velocity)	$\sqrt{\overline{u^2} + \overline{v^2}}$			5.8 mm s^{-1}
Swarm radius	$\sqrt{\overline{x^2}}$	Observed	5.5 mm	
(rms position)	$\sqrt{\overline{x^2} + \overline{y^2}}$			16.6 mm
Predicted swarm radius	$\sqrt{\overline{x^2}}_{pred}$	Equation 11.12	5.3 mm	
	$\sqrt{\overline{x^2} + \overline{y^2}}_{pred}$	Equation 11.14		9.1 mm
Error in swarm size (observed–predicted)			4%	45%

deviation (error = 11%). The vertical *Daphnia* acceleration field (Figure 11.5F) shows similar agreement (error in mean standard deviation = 8.9%), although superimposed on the diffusive field are an upward mean acceleration near the bottom of the tank and a downward mean near the top. These perturbations suggest that the animals feel the top and bottom boundaries of the domain, perhaps hydrodynamically, and turn away from them. The vertical *Temora* acceleration field suggests the same behavior.

The variances of the *Temora* acceleration fields are similar to those predicted by our random-flight model (error in mean standard deviation 2.2% for the *x* direction, 30% for *y*, 11% for *z*), but the horizontal mean field is much smaller than predicted. In the absence of other results this might be taken to mean the absence of a concentrative acceleration and thus primarily diffusive behavior, but the horizontal velocity autocorrelation (Figure 11.6C) suggests a swarming balance very strongly. Instead, we attribute the poor definition of the mean horizontal *Temora* acceleration field to noise in the data, i.e., a combination of sampling error and real, high-speed behaviors not accounted for by our model. There are several reasons to suspect this type of error. The *Temora*, unlike the *Daphnia*, make their turns quickly, generally in less than the sampling time step of 1 s. Their trajectories wobble and jitter at high ($\sim$10 Hz) frequencies, in ways that may be related to the trail-following that Weissburg et al. (1998) observed in the same series of *Temora* experiments. Furthermore, in general, successive derivatives of a data set (calculation of the acceleration requires two) amplifies noise at the high-frequency end of a power spectrum. One would expect a moderate level of noise in the acceleration field to wash out the small mean field but have only a secondary effect on the field's variance, as is observed.

11.6 Discussion

11.6.1 Model Consistency

Our goal here has been to provide an internally consistent model description of a steady-state zooplankton swarm, by a method that could be applied generally to the problem of assessing physical and biological controls on zooplankton aggregation. The analysis above verifies the suitability of the model presented here to the artificially stimulated swarms we have described. It also suggests the level of quantitative agreement we can expect for real organisms, which do not move like idealized particles and inevitably are engaged in other behaviors at the same time they are swarming: errors up to a factor of two, with closer correspondences in most measures. The median unexplained variance $(1 - r^2)$ for all theoretical fits for which it was calculable was 11%.

For many purposes, high-precision agreement with data may not actually make a descriptive model any more *useful* than one with moderate-precision agreement. This is especially true for a problem like zooplankton aggregation, in which the parameters of ultimate interest — number densities and encounter rates — vary over several orders of magnitude. Nevertheless, higher-frequency sampling or more involved frequency-domain filtering might reduce errors in laboratory results, by suppressing sampling error and isolating swarming motion from other behaviors more precisely. It appears, however, that in the present study — perhaps for models that entail strongly stochastic behavior in general — larger sample sizes would not. Variations in individual behavior around the ensemble mean are substantial — this is most visible in the velocity autocorrelations (Figure 11.5C and Figure 11.6C) — but this variation is not significantly greater than what the model dynamics predict, suggesting that in a larger ensemble the mean would be no more precisely defined.

Figure 11.7 shows individual velocity autocorrelations for 36 *Daphnia* trajectories, selected at random out of the full set of 58 for illustrative purposes, and for comparison 36 autocorrelations numerically that were simulated using the momentum Equation 11.4 and fit values of ω and k. Only a few of the observed individual trajectories actually match the ensemble average (Figure 11.5C) in form; those that do suggest orbital periods (i.e., values of ω) and decorrelation times (i.e., values of k) that vary by an order of magnitude. Just as many records appear to decorrelate within a few seconds, suggesting no easily describable kinematic pattern thereafter. What is significant here is that all these forms for the autocorrelation appear among the simulated trajectories as well. Because of the inherent stochastic element in this motion, a single dynamic balance with a single set of parameters can produce what appears to be a wide variety of individual behaviors. Note also the common element in these variegated trajectories: all satisfy the kinematic requirement for swarming, illustrated in Figure 11.1, that their autocorrelations integrate to a value near zero.

11.6.2 Model Interpretation

A number of physical or biological interpretations are possible for each of the terms in our model Equation 11.4, and thus we have so far left the meaning of these terms very general. Clearly, the strict Newtonian interpretation — external forces incident upon a passive particle — is insufficient in our experiment, although this interpretation might sometimes be appropriate for animals in a more energetic and complex environment than our placid laboratory tank: plankton who live up to their Greek name and simply drift. In general, however, each of the force terms in Equation 11.4 can be given either a physical or a behavioral interpretation:

11.6.2.1   Damping — The damping $-ku$ may represent viscous drag, here assumed to obey Stokes' law to keep the model linear, although Okubo and Anderson (1984) note that this may not be the right form of drag for large and fast-moving zooplankton. Alternatively, it could represent a decorrelative behavior, akin to the tendency of many zooplankton to turn more often in a food patch (Williamson, 1981; Buskey, 1984; Price, 1989; McGehee and Jaffe, 1996).

An estimate of hydrodynamic drag on the animals in our experiment suggests the behavioral interpretation. Assume that the damping $-ku$ is a linearization of a more realistic and more widely applicable quadratic drag law, so that

$$-ku \sim \frac{1}{2}\frac{C_d \rho_w A u^2}{V \rho_z} \tag{11.23}$$

(Haury and Weihs, 1976) and

$$k \sim \frac{1}{2}C_d \frac{\rho_w}{\rho_z} \frac{A}{V} u \tag{11.24}$$

where u is the velocity, V the volume, A the frontal cross section, ρ_z the density, and C_d the drag coefficient of a zooplankter moving through water of density ρ_w. We can assume that the ratio of water to animal

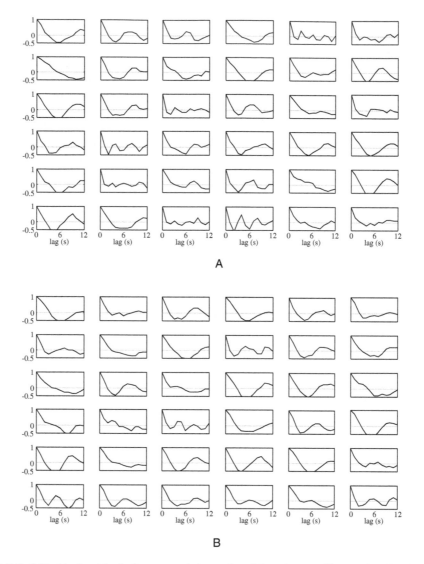

FIGURE 11.7 Individual horizontal velocity autocorrelations, selected from the ensemble at random, for (A) *Daphnia* and (B) numerically simulated *Daphnia*.

density is O(1), and estimate the ratio of volume to frontal area as body length l. Drag coefficients are a function of Reynolds number:

$$\mathrm{Re} \equiv \frac{ul}{v} \qquad (11.25)$$

where v is the dynamic viscosity of seawater, $\sim 10^{-6}$ m^2 s^{-1}, and we have assumed that body length and diameter are similar. For the animals in our experiment (*Daphnia*: $u \sim 3$ mm s^{-1}, $l \sim 3$ mm; *Temora*: $u \sim 6$ mm s^{-1}, $l \sim 1$ mm; Yen et al., 1998), Re is between 1 and 10, values for which observations of zooplankton suggest $C_d \sim 10$ or higher (Haury and Weihs, 1976). Thus, for both *Daphnia* and *Temora*, a low estimate for the effective Stokes coefficient is $k \sim 10$ s^{-1}.

This is fully an order of magnitude larger than the k values associated with swarming in our experiment (Table 11.1). In other words, hydrodynamic drag appears to operate here at a timescale $1/k$ much shorter than the timescale of swarming motion, indeed much shorter than our sampling interval of 1 s. While drag may well be important in the dynamics of avoidance reactions between the swarmers, trail-following, rapid changes of direction, and the like, it does not appear to have explicit effects in the regime of motion we have been considering. Thus the damping of forward motion that limits the energy and spatial extent of the *Daphnia* and *Temora* swarms appears to be not passive and fluid mechanical, but rather active and behavioral, a tendency consistent with the classical "area-restricted search" model of Tinbergen et al. (1967).

11.6.2.2 *Excitation* —

The excitation $A(t)$ clearly is behavioral in our experiment, since the background flow field is negligible, but one could also model turbulent dispersion through this term. Indeed, the kinematic approach of our model does not differentiate between dispersion by behavior and dispersion by the environment, and thus it lets one express observations of excitational behavior in a form directly comparable with standard turbulence formulations. In the absence of an aggregative tendency (that is, with $\omega = 0$), Equations 11.2 and 11.8 predict that a swarm will disperse with a diffusivity

$$D = \overline{u^2} \int R(\tau)d\tau \sim \frac{\overline{u^2}}{k} \qquad (11.26)$$

For our *Daphnia* and *Temora* swarms, for example, this (behavioral) diffusivity is 10^{-6} to 10^{-5} m^2 s^{-1}, similar to background diapycnal diffusivities in the open ocean (Gregg et al., 1999). This comparison is simply illustrative and is not meant to measure the ability of these animals to aggregate against turbulence, as they do not appear to be swimming at capacity here: *Temora*, for example, have been observed to lunge at speeds more than ten times the rms velocity of the swarmers we have described (Doall et al., 1998). Yamazaki and Squires (1996) find that most zooplankton swim at speeds greater than or comparable to the velocity fluctuations associated with upper-ocean turbulence.

11.6.2.3 *Concentrative Force* —

The concentrative force $-\omega^2 x$ represents phototaxis in our experiments and, indeed, is likely to represent a behavioral response like phototaxis or chemotaxis in most physical environments. There are important exceptions, however: fronts from river plumes (e.g., Mackas and Louttit, 1988), Langmuir circulations (Stavn, 1971), and internal waves (Shanks, 1985) have all been found to cause passive aggregation as well. These physical mechanisms, which generally consist of advection by a convergent flow field, may not be represented well by the harmonic restoring force that our model employs.

11.6.2.4 *Physical–Behavioral Balances* —

In summary, these possibilities suggest three principal balances of physical and biological processes to which the "aggregating random-walk" model could be applied:

1. Aggregative behavior, which maintains high animal concentrations against turbulent dispersion (as, for example, Tiselius et al. [1994] observed in a thin layer at a sharp pycnocline)
2. Dispersive behavior, which limits animal concentrations in a region of convergent flow (perhaps similar to the maintenance of nearest-neighbor distance by avoidance reaction observed by Leising and Yen, 1997)
3. A pure balance of aggregative and dispersive behaviors in a quiescent environment

Note that while so far we have discussed the excitational and concentrative forces as separate processes, we might also interpret them as mathematically paired components of a single swarming behavior: a behavior better described as a *spatial gradient in excitation*. (Spatial variations in eddy diffusivity are thought to cause anisotropic tracer distributions in the ocean [Armi and Haidvogel, 1982], and Mullen [1989] models fish dispersal and abundance using a variable diffusivity proportional to saturation of local carrying capacity.) This interpretation allows the possibility of a conceptually simple link between the gradient in stimulus (say, light intensity or chemical concentration) and a gradient in the behavioral response.

11.7 Conclusion

The members of phototactic swarms of *Daphnia* and *Temora* in the laboratory are found to follow, to first order, the dynamics of randomly diffusing particles subject to a linear restoring force. This behavioral model is found to fit the data to within ~10 to 20% overall. Larger errors, attributed to sampling method and unmodeled behaviors, occur in some measures of the *Temora* acceleration field and spatial distribution, but much closer matches (errors of 0.5 to 10%) are found in several fundamental kinematic measures: most notably, the form of the velocity autocorrelation along the axes of swarming motion.

In these experiments the concentrative force and diffusive excitation are individual behavioral responses to a stable environmental stimulus. In other circumstances, the same model dynamics could represent a variety of balances of physical accelerations like turbulence, drag, and flow convergence; individual responses to environmental cues; and interaction between swarm members.

The environmental and intraspecies cues that initiate individual swarming behaviors are varied and not well understood. A natural continuation of the laboratory experiments described above would be to investigate the effects of variations in the swarm stimulus (weaker and stronger light gradients, chemical and hydrodynamic cues) and the swarm demographics (species, life stage, physiological state, presence of predators, density, and so on.) One could also analyze *in situ* observations of zooplankton swarms in similar terms, and through this model draw inferences, in both directions, between individual behavior on the small scale and the dynamics of patches on the large scale.

Without further elaboration the model we have been considering makes no predictions about the outcome of such a series of observations. The benchmarks this model proposes, however — the form of the velocity autocorrelation and acceleration field; the parameters k, ω, and B, which index the strength of the forces which must balance to produce a stable swarm — are likely to clarify the role of behavior in maintaining zooplankton aggregations, as well as help identify the sensory pathways through which swarming zooplankton organize their environment. Thus we propose the zooplankton-swarming model discussed above not as a hypothesis concerning these problems, but as a mathematical language in which to pose the specific behavioral questions that experiment and ethology can answer.

Acknowledgments

This work was supported by National Science Foundation OCE-9314934 and OCE-9723960. N. Banas was partially supported by the Ward and Dorothy Melville Summer Research Fellowship at the Marine Sciences Research Center at the State University of New York at Stony Brook. Many thanks to J.R. Strickler, who generously provided the *Daphnia* data.

References

Alldredge, A., Robison, B., Fleminger, A., Torres, J., King, J., and Hamner, W.M. 1984. Direct sampling and *in situ* observation of a persistent copepod aggregation in the mesopelagic zone of the Santa Barbara Basin. *Mar. Biol.*, 80: 75–81.

Ambler, J.W., Ferrari, F.D., and Fornshell, J.A. 1991. Population structure and swarm formation of the cyclopoid copepod *Dioithona oculata* near mangrove cays. *J. Plankton Res.*, 13: 1257–1272.

Armi, L. and Haidvogel, D.B. 1982. Effects of variable and anisotropic diffusivities in a steady-state diffusion model. *J. Phys. Oceanogr.*, 12: 785–794.

Bollens, S.M., Frost, B.W., and Cordell, J.R. 1994. Chemical, mechanical and visual cues in the vertical migration behavior of the marine planktonic copepod *Acartia hudsonica*. *J. Plankton Res.*, 16: 555–564.

Brandl, Z. and Fernando, C.H. 1971. Microaggregation of the cladoceran *Ceridaphnia affinis* Lilljeborg with a possible reason for microaggregation of zooplankton. *Can. J. Zool.*, 49: 775.

Bundy, M.H., Gross, T.F., Coughlin, D.J., and Strickler, J.R. 1993. Quantifying copepod searching efficiency using swimming pattern and perceptive ability. *Bull. Mar. Sci.*, 53: 15–28.

Buskey, E.J. 1984. Swimming pattern as an indicator of the roles of copepod sensory systems in the recognition of food. *Mar. Biol.,* 79: 165–175.

Buskey, E.J., Peterson, J.O., and Ambler, J.W. 1995. The role of photoreception in the swarming behavior of the copepod *Dioithona oculata. Mar. Freshw. Behav. Physiol.,* 26: 273–285.

Clements, A.N. 1963. *The Physiology of Mosquitoes.* Pergamon Press, Oxford.

Davis, C.S., Flierl, G.R., Wiebe, P.H., and Franks, P.J.S. 1991. Micropatchiness, turbulence, and recruitment in plankton. *J. Mar. Res.,* 49: 109–151.

Davis, C.S., Gallager, S.M., and Solow, A.R. 1992. Microaggregations of oceanic plankton observed by towed video microscopy. *Science,* 257: 230–232.

Doall, M.H., Colin, S.P., Strickler, J.R., and Yen, J. 1998. Locating a mate in 3D: the case of *Temora longicornis. Philos. Trans. R. Soc. London B,* 353: 681–689.

Eloffson, R. 1966. The nauplius eye and frontal organs of the non-malacostraca (Crustacea). *Sarsia,* 25: 1–128.

Fields, D.M. and Yen, J. 1997. The escape behavior of marine copepods in response to a quantifiable fluid mechanical disturbance. *J. Plankton Res.,* 19: 1289–1304.

Gallager, S.M., Davis, C.S., Epstein, A.W., Solow, A., and Beardsley, R.C. 1996. High-resolution observations of plankton spatial distributions correlated with hydrography in the Great South Channel, Georges Bank. *Deep-Sea Res.,* 43: 1627–1663.

Gendron, D. 1992. Population structure of daytime surface swarms of *Nyctiphanes simplex* (Crustacea: Euphausiacea) in the Gulf of California, Mexico. *Mar. Ecol. Prog. Ser.,* 87: 1–6.

Goldsmith, A., Chiang, H.C., and Okubo, A. 1980. Turning motion of individual midges, *Anarete pritchardi,* in swarms. *Ann. Entomol. Soc. Am.,* 73: 526–528.

Gregg, M.C., Winkel, D.W., MacKinnon, J.A., and Lien, R.-C. 1999. Mixing over shelves and slopes. *'Aha Huliko'a Proceedings, Hawaiian Winter Workshop,* P. Muller and D. Henderson, Eds., University of Hawaii, 35–42.

Grünbaum, D. 1994. Translating stochastic density-dependent individual behavior with sensory constraints to an Eulerian model of animal swarming. *J. Math. Biol.,* 33: 139–161.

Hamner, W.M. and Carlton, J.H., 1979. Copepod swarms: attributes and role in coral reef ecosystems. *Limnol. Oceanogr.,* 24: 1–14.

Hamner, W.M., Hamner, P.P., Strand, S.W., and Gilmer, R.W. 1983. Behavior of Antarctic krill *Euphausia superba:* chemoreception, feeding, schooling, and molting. *Science,* 220: 433–435.

Hanson, A.K. and Donaghay, P.L. 1998. Micro- to fine-scale chemical gradients and layers in stratified coastal waters. *Oceanography,* 11: 10–17.

Haury, L. and Weihs, D. 1976. Energetically efficient swimming behavior of negatively buoyant zooplankton. *Limnol. Oceanogr.,* 21: 797–803.

Hebert, P.D.N., Good, A.G., and Mort, M.A. 1980. Induced swarming in the predatory copepod *Heterocope septentrionalis. Limnol. Oceanogr.,* 25: 747–750.

Hussussian, G., Yen, J., and Strickler, J.R. 1993. Digitized data from 1993 experiments on swarming in *Daphnia magna.* Special report 47, Center for Great Lake Studies, Milwaukee, WI, 48 pp.

Katona, S.K. 1973. Evidence for sex pheromones in planktonic copepods. *Limnol. Oceanogr.,* 18: 574–583.

Lasker, R. 1975. Field criteria for survival of anchovy larvae: the relation between inshore chlorophyll maximum layers and successful first feeding. *Fish. Bull.,* 73: 453–462.

Leising, A.W. and Yen, J. 1997. Spacing mechanisms within light-induced copepod swarms. *Mar. Ecol. Prog. Ser.,* 155: 127–135.

Mackas, D.L. and Louttit, G.C. 1988. Aggregation of the copepod *Neocalanus plumchrus* at the margin of the Fraser River plume in the Strait of Georgia. *Bull. Mar. Sci.,* 43: 810–824.

McGehee, D. and Jaffe, J. 1996. Three-dimensional swimming behavior of individual zooplankters: observations using the acoustical imaging system FishTV. ICES *J. Mar. Sci.,* 53: 363–369.

Mullen, A.J. 1989. Aggregation of fish through variable diffusivity. *Fish. Bull.,* 8: 353–362.

Okubo, A. 1980. *Diffusion and Ecological Problems: Mathematical Models.* Springer-Verlag, Berlin, 254 pp.

Okubo, A. 1986. Dynamical aspects of animal grouping: swarms, schools, flocks and herds. *Adv. Biophys.,* 22: 1–94.

Okubo, A. and Anderson, J.J. 1984. Mathematical models for zooplankton swarms: their formation and maintenance. *Eos,* 65: 731–732.

Poulet, S.A. and Ouellet, G. 1982. The role of amino acids in the chemosensory swarming and feeding of marine copepods. *J. Plankton Res.,* 4: 341–361.

Price, H.J. 1989. Swimming behavior of krill in response to algal patches: a mesocosm study. *Limnol. Oceanogr.,* 34: 649–659.

Russell, F.S. 1934. The vertical distribution of marine macroplankton. XII. Some observations on the vertical distribution of *Calanus finmarchicus* in relation to light intensity. *J. Mar. Biol. Assoc. U.K.,* 19: 569–584.

Schultze, P.C., Strickler, J.R., Bergström, B.I., Donaghay, P.C., Gallager, S., Haney, J.F., Hargreaves, B.R., Kils, U., Paffenhöfer, G.-A., Richman, S., Vanderplo, E.G., Welsch, W., Wethey, D., and Yen, J. 1992. Video systems for *in situ* studies of zooplankton. *Arch. Hydrobiol. Beih. Erg. Limnol.,* 36: 1–21.

Shanks, A.L. 1985. Behavioral basis of internal-wave-induced shoreward transport of megalopae of the crab *Pachygraspus crassipes. Mar. Ecol. Prog. Ser.,* 24: 289–95.

Smith, S.L., Pieper, R.E., Moore, M.V., Rudstam, L.G., Greene, C.H., Zamon, J.E., Flagg, C.N., and Williamson, C.E. 1992. Acoustic techniques for the *in situ* observation of zooplankton. *Arch. Hydrobiol. Beih.,* 36: 23–43.

Stavn, R.H. 1971. The horizontal-vertical distribution hypothesis: Langmuir circulations and *Daphnia* distributions. *Limnol. Oceanogr.,* 16: 453–466.

Stearns, S.C. 1975. Light responses of *Daphnia pulex. Limnol. Oceanogr.,* 20: 564–570.

Strickler, J.R. 1998. Observing free-swimming copepods mating. *Philos. Trans. R. Soc. London B,* 353: 671–680.

Strickler, J.R. and Hwang, J.-S. 1998. Matched spatial filters in long working distance microscopy of phase objects. In *Focus on Multidimensional Microscopy,* P.C. Cheng, P.P. Hwang, J.L. Wu, G. Wang, and H. Kim, Eds., World Scientific, River Edge, NJ.

Tinbergen, N., Impekoven, M., and Franck, D. 1967. An experiment on spacing-out as a defense against predation. *Behaviour,* 28: 207–321.

Tiselius, P. 1992. Behavior of *Acartia tonsa* in patchy food environments. *Limnol. Oceanogr.,* 37: 1640–1651.

Tiselius, P., Nielsen, G., and Nielsen, T.G. 1994. Microscale patchiness of plankton within a sharp pycnocline. *J. Plankton Res.,* 16: 543–554.

Ueda, H., Kuwahara, A., Tanaka, M., and Azeta, M. 1983. Underwater observations on copepod swarms in temperate and subtropical waters. *Mar. Ecol. Prog. Ser.,* 11: 165–171.

Uhlenbeck, G.E. and Ornstein, L.S. 1930. On the theory of the Brownian motion. *Phys. Rev.,* 36: 823–841.

Weissburg, M.J., Doall, M.H., and Yen, J. 1998. Following the invisible trail: mechanisms of chemosensory mate tracking by the copepod *Temora longicornus. Philos. Trans. R. Soc. London B,* 353: 701–712.

Williamson, C. 1981. Foraging behavior of a freshwater copepod: frequency changes in looping behavior at high and low prey densities. *Oecologia,* 50: 332–336.

Wishner, K., Durbin, E., Durbin, A., Macaulay, M., Winn, H., and Kenney, R. 1988. Copepod patches and right whales in the Great South Channel off New England. *Bull. Mar. Sci.,* 43: 825–844.

Yamazaki, H. 1993. Lagrangian study of planktonic organisms: perspectives. *Bull. Mar. Sci.,* 53: 265–278.

Yamazaki, H. and Okubo, A. 1995. A simulation of grouping: an aggregating random walk. *Ecol. Modelling,* 79: 159–165.

Yamazaki, H. and Squires, K.D. 1996. Comparison of oceanic turbulence and copepod swimming. *Mar. Ecol. Prog. Ser.,* 144: 299–301.

Yen, J. and Bundock, E.A. 1997. Aggregative behavior in zooplankton: phototactic swarming in four developmental stages of *Coullana canadensis* (Copepoda, Harpacticoida). In *Animal Groups in Three Dimensions,* J. Parrish and W. Hamner, Eds., Cambridge University Press, New York, 143–162.

Yen, J., Weissburg, M.J., and Doall, M.H. 1998. The fluid physics of signal perception by mate-tracking copepods. *Philos. Trans. R. Soc. London B,* 353: 787–804.

Yoder, J.A., Ackleson, S.G., Barber, R.T., Flament, P., and Balch, W.M. 1994. A line in the sea. *Nature,* 371: 689–692.

Section II

Analysis

12

On Skipjack Tuna Dynamics: Similarity at Several Scales

Aldo P. Solari, Jose Juan Castro, and Carlos Bas

CONTENTS

12.1 Introduction 183
12.2 Data 185
12.3 Methods 185
12.4 Results 186
12.5 Discussion 192
    12.5.1 The Proposed Equations 193
    12.5.2 The Phase Spaces 193
    12.5.3 Variable Carrying Capacity (Ceilings, K_i) 193
    12.5.4 Minimum Populations (Floors, P_i) 194
    12.5.5 Multiple (Stable) Equilibria 194
    12.5.6 Compensatory and Depensatory Dynamics 194
    12.5.7 Extinction of the Commercial Fishery 195
    12.5.8 Migration through a Fractal Marine System 195
Acknowledgments 196
References 197

12.1 Introduction

The tuna resources in the Eastern Central Atlantic have been the object of both intensive fishery for over 30 years and numerous studies conducted under the coordination of the International Commission for the Conservation of Atlantic Tunas (ICCAT) (Fonteneau and Marcille, 1993). The skipjack tuna (*Katsuwonus pelamis*, henceforth referred as "skipjack") supports an important commercial fishery across the Eastern Atlantic from the Gulf of Guinea to the southwestern Irish coast (ICCAT, 1986). Tag recovery studies have indicated that skipjack migration routes lie from the southeast toward the north and northwest Atlantic (Ovchinnikov et al., 1988) and catches on the highly migratory tuna stocks are due to multigear fisheries (Fonteneau, 1991). Also, both the spatiotemporal distribution and abundance of skipjack tuna have been related to causes such as environmental requirements and feeding (Ramos et al., 1991), upper ocean dynamics (Ramos and Sangrá, 1992), hydro climatic factors (Pagavino and Gaertner, 1994), prey abundance (Roger, 1994a, b; Roger and Marchal, 1994), thermal habitat (Boehlert and Mundy, 1994), schooling behavior (Bayliff, 1988; Hilborn, 1991), mesoscale frontal ocean and upwelling dynamics (Fiedler and Bernard, 1987; Ramos et al., 1991), as well as several other aspects beyond the scope of the present chapter. Skipjack tuna appears to be able to adapt the feeding strategy to environmental conditions preying upon what it encounters (Roger, 1994a, b) and the 18°C isotherm and 3 ml oxygen per liter isoline are considered lower limiting factors (Piton and Roy, 1983). The exploitation rate on most tuna stocks has been constantly increasing and assessments have been inefficient in estimating the real maximum sustainable yield of those stocks (Fonteneau, 1997). In the Atlantic, tuna catches were suggested to be both underestimated and misreported (Wise, 1985) and, despite the high level of fishing effort, recruitment overfishing has never been suggested for skipjack (Fonteneau, 1987).

Skipjack data are, generally, studied in light of the early population models (henceforth referred as "classical approaches/models") by Beverton and Holt (1957) and Ricker (1954). These models are still being used to provide quantitative advice to fishery managers (Gulland, 1989) and proposed extinction curves where recruitment either reaches an asymptotic maximum (Beverton–Holt) or becomes low at high spawning stock sizes (Ricker). These classical approaches became widely accepted functions to describe the stock-recruitment relationship and they introduced a general, nonlinear framework into fish population dynamics. However, they had a limited capacity both to include key factors of specific situations and to link internal (population) and external (environmental) dynamics to each other. This lack of specificity has been discussed in the literature by Clark (1976), De Angelis (1988), Fogarty (1993), and other authors. Also, classical models assumed that populations under exploitation were naturally limited in a way that will permit them to respond in a compensatory way to fishing (Beverton and Holt, 1957; Ricker, 1975; Cushing, 1977; Rothschild, 1986). The classical models excluded dynamic features critical to understand the mechanics behind the data (linking and transition mechanisms between steady states, system behavior, extinction of the commercial fishery, environmental interactions, among other factors). However, two more advanced frameworks were proposed during the 1970s and 1980s. On the one hand, Paulik (1973) described an overall spawner-recruit model, which was formed from the concatenation of survivorship functions. This approach could exhibit multiple (stable) equilibria and complex dynamics and was the result of a multiplicative process where the initial egg production could be modified by nonlinear functions specific to each life stage and cohort population size: the main shortcoming was the interdependency between the functions due to the multiplicative nature of the model. On the other hand, Shepherd (1982) unified the dome-shaped and asymptotic approaches proposed by Ricker (1954) and Beverton and Holt (1957), respectively, but could incorporate neither multiple stable equilibria nor depensatory dynamics. In our view, there was an urgent need to develop a flexible framework that would allow us both to ask better questions and to understand causal mechanisms to dynamic patterns behind the data.

In previous papers by Solari et al. (1997), Bas et al. (1999), and Castro et al. (1999), we proposed recruitment both to the population (influx of juveniles to the adult population), area (migration of cohorts/individuals into fishery areas), and fishery (dynamics of the fishery) as a system or summation of nonlinear functions (multiple steady states) with dynamic features ranging from chaos (when external conditions are extremely benign), through a range of relatively stable, converging cycles (as external stress increases), to a quasi-standstill state with no clear oscillations (when the minimum viable population is being approached). The system was suggested to have the capacity to evolve persistently and return within a wide range of equilibrium states allowing for multiple carrying capacities as well as density-dependent (compensation and depensation due population numbers), density-independent (compensation and depensation due environmental fluctuations and fisheries), and inverse-density-dependent (per capita reproductive success and recruitment declines at low population levels) coupled mechanics. Our new, dynamic framework was justifiable on an *ad hoc* basis because of the flexibility it afforded. Also, it offered some conceptual advantages over classical approaches as it allowed for (1) multiple equilibria, which could be independent from each other and operate at the same time (no mathematical interdependence between the functions due to the additive nature of the approach); (2) either higher or lower equilibria could be incorporated into the system; (3) transitions between equilibria due to density-dependent and density-independent oscillations could be linked; and, among other features, (4) several maxima and minima and depensatory dynamics could be described in the same relationship allowing for simultaneous equilibrium states at different spatiotemporal scales and substocks. This model allowed us both to approach the dynamics of the skipjack from a new perspective and investigate whether there was any dynamic similarity, at several spatial scales, in the captures of this migratory stock. The aims of the present study were to (1) analyze three independent skipjack fishery landing series representing catches from three different spatial scales; (2) determine whether there may be any similarity between the series; and (3) discuss new concepts to study the evolution of both recruitment-to-the-area and the dynamics and future approaches to skipjack populations in the Eastern Central Atlantic.

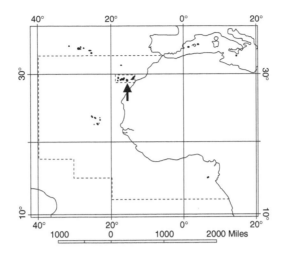

FIGURE 12.1 Three spatial scales of skipjack tuna sampling. The CECAF (Committee for Eastern Central Atlantic Fisheries) Division 34 (larger area indicated by the dashed line; from Gibraltar to the Congo River, Lat. 36°00′N–6°04′36″S, Long. 12°19′48″E–5°36′W); the Canary Islands archipelago (minor area indicated by the dashed line, Lat. 29°40′N–27°10′N, Long. 13°W–18°20′W), and the Port of Mogan (local waters off the southern shore, island of Gran Canaria, Lat. 27°55′N–Long. 15°47′W, indicated by the arrow). (Map modified from FAO, 2001.)

12.2 Data

The skipjack fishery series analyzed herein (annual catches in metric tonnes, Tn) were the following:

1. Landings due to a local bait fishery at the Port of Mogan (Lat. 27°55′N, Long. 15°47′W, henceforth referred as the "Mogan series"), island of Gran Canaria (Canary Islands, Spain), years 1980–1996 according to Hernández-García et al. (1998)
2. Overall pooled landings due to local bait fisheries for the whole of the Canary Islands area (Eastern Central Atlantic, Lat. 29°40′N–27°10′N, Long. 13°W–18°20′W, henceforth referred as "Canarian series"), years 1975–1993 according to Ariz et al. (1995)
3. Pooled landings due to multigear (bait, long-line, and purse-seine) both oceanic and coastal fisheries within the CECAF (Committee for Eastern Central Atlantic Fisheries) Division 34 (from Gibraltar to the Congo River, Lat. 36°00′N–6°04′36″S, Long. 12°19′48″E–5°36′W; henceforth referred to as the "CECAF series"), years 1972–1996 according to FISHSTAT/FAO (1999).

Figure 12.1 shows these spatial scales (map modified after FAO, 2001): a point (waters off the Port of Mogan), a minor area (waters within the Canary Islands archipelago), and a relatively large ocean area (the CECAF area 34). Figure 12.2 shows the skipjack series from each location.

12.3 Methods

We standardized the series to the same scale (Z values with mean = 0) to facilitate both the analyses and visual comparison. We used both t-tests and autocorrelations to determine the homogeneity between the series and indications of autosimilarity. Also, the Welch method (after Oppenheim and Schafer, 1975) was used to estimate the spectral density. The phase spaces (stock-in-area against recruits-to-the-area) were obtained by plotting data values from a certain year (N_t) against values the year after (N_{t+1}). Cross-correlations were used to determine the degree of correspondence between the series. Furthermore, data values were fitted both by linear regressions through the origin (to determine the "replacement line" or recruitment needed to replace the stock-at-spatial-location) and sixth-order polynomials (to describe the

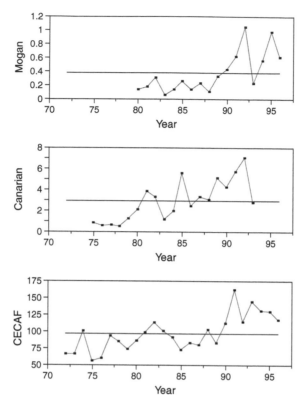

FIGURE 12.2 Skipjack tuna series (Tn*10^3) from a local bait fishery at the Port of Mogan (island of Gran Canaria, Canary Islands, years 1980–1996), after Hernández-García et al. (1998); overall pooled landings due to local bait fisheries for the whole of the Canary Islands area (years 1975–1993), after Ariz et al. (1995); and pooled landings due to multigear (bait, long-line, and purse-seine) both oceanic and coastal fisheries within the CECAF Division 34 (years 1972–1996), after FISHSTAT/FAO (1999). The catches represent sampling series from three significantly different spatial scales. The straight lines indicate the mean of the series.

dynamic features of the systems). To set the final, schematic example, we simulated sinusoidal waves with an arbitrary noise to represent the proposed system.

12.4 Results

We start this section with a general figure showing the difference between the classical approaches based on a unique equilibrium and our new, nonlinear framework. This preliminary step is helpful for understanding the rest of the chapter and it shows some of the guidelines to apply our multiple steady-state model to the plane N_t against N_{t+1} attempting to understand the dynamic features we propose.

On the one hand, the Shepherd (1982) functional form is given by

$$R = \frac{\alpha \cdot S}{1 + \left(\dfrac{S}{K}\right)^{\delta}} \tag{12.1}$$

where R is recruitment, S is the spawning stock abundance, and K the threshold abundance above which density-dependent effects dominate (i.e., the carrying capacity). The parameters α and δ are referred to as the slope at the origin and degree of compensation involved, respectively. This approach could unify, within a single framework, both the classical dome-shaped (for $\delta > 1$) and asymptotic (for $\delta = 1$) functional forms proposed by Ricker (1954) and Beverton and Holt (1957), respectively. An example for each of the models is shown in Figure 12.3.

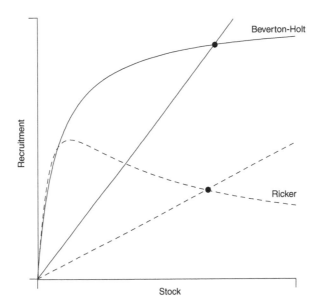

FIGURE 12.3 Two examples of the Shepherd (1982) function, which unified, within a single framework, the classical dome-shaped and asymptotic functions proposed by Ricker (1954, dashed curve) and Beverton–Holt (1957, continuous curve). The stock-recruitment relationship is assumed to (1) be governed by a single stable equilibrium (dot) shown by the intersection of the function with a simple regression through the origin (i.e., the replacement line); (2) respond solely in a compensatory way to fishing, and (3) become either asymptotic (Beverton–Holt) or be limited by a single carrying capacity (i.e., threshold abundance above which density-dependent effects dominate; Ricker).

On the other hand, recruitment (R) in our framework is defined (Equation 12.2) as the summation of nonlinear functions of spawning stock, S, given by

$$R \cong \sum_{i=1}^{m} \frac{a_i \cdot (S)}{(S - b_i)^2 + c_i} \tag{12.2}$$

where the entries $i = 1 \dots m$ represent the number of equilibrium states in the stock-recruitment (SR) system, with m the highest equilibrium where the SR relationship reaches the ceiling or maximum allowable carrying capacity. Equilibrium states are controlled by the coefficients a_i (slope of the curve at the origin), with b_i and c_i the density-dependent mortality entries. For example, a_i fulfills a similar function to the natural rate of increase in the logistic equation. These coefficients define each equilibrium state and their values may be fixed. Also, values of b_i define the ranges of spawning stock for which equilibrium states may arise. A graphical representation of a dynamic system with m equilibrium states is shown in Figure 12.4. This new approach can be applied to the plane N_t, N_{t+1} to describe dynamics in both migration and catches (recruitment to the area) in skipjack and link local and mesoscalar trends from different spatial scales, as well. An m number of oscillatory phenomena ranging from limit cycles to chaos and inverse density dependence are allowed in this system, which may be approximated either by least squares using Equation 12.2 or by polynomial regressions incorporating three constants for each equilibrium state. The objective of the sixth-order fittings we use is to describe in a familiar way to all readers the multiple steady-state cases we approach. Further aspects of this new model are well detailed in Solari et al. (1997).

We have assumed a positive relationship between the number of juveniles being recruited to the population and those entering the area of the fishery; this implies that the number of recruits-to-the-area in a migratory stock may increase as recruitment increases in the remote nursery areas. Also, we have regarded the analyzed series as statistically independent as a result of the different sources and relatively large geographic range of the areas concerned; paired t-tests ($p < 0.001$ in all of the cases) showed the series may represent three significantly different levels of recruitment. Furthermore, in spite of the limited degrees of freedom in all of the series, the autocorrelation values were 0.42 ($p = 0.06$), 0.52 ($p = 0.01$,

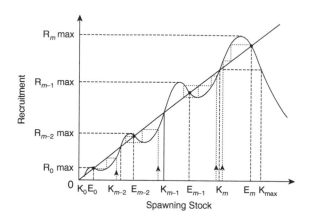

FIGURE 12.4 Graphical representation of a dynamic system with m equilibrium states (Equation 12.2) proposed by Solari et al. (1997). This new approach can be applied to the plane N_t vs. N_{t+1} to describe dynamics in both migration and catches (recruitment to the area) in skipjack tuna (*K. pelamis*) at three spatial scales (CECAF Area 34, Canary Islands area, and Port of Mogan, Central East Atlantic). K_m, K_{m-1}, K_{m-2} represent both the minimum viable populations for the equilibrium states m, m^{-1}, and m^{-2} and the carrying capacities for their immediate lower equilibria, respectively. E_m, E_{m-1}, E_{m-2}, E_0 represent equilibria around which the SR relationship may turn in density-dependent compensation and depensation phases. K_{max} is the maximum allowable carrying capacity and any values of stock surpassing this ceiling will induce a shift toward lower equilibria. K_0 is the floor or minimum viable population below which the SR relationship may tend to zero (extinction of commercial fishery). System persistence and local stability are shown in all three cases of stability analyses (dotted lines) while $K_0 < S < K_{max}$ and $R(K_0) < R < R_{max}$. An m number of oscillatory phenomena ranging from limit cycles to chaos and inverse density-dependence are allowed in this system. Each equilibrium state may represent stock and recruitment to the area in a certain spatial location and link local and mesocalar dynamics.

lag = 1) and 0.56 (p = 0.003, lag = 1) for the Mogan, Canarian, and CECAF series, respectively. These results may suggest that there is a certain autosimilarity or "memory" in the series implying that the skipjack stock a certain year may depend on the abundance in preceding years. Moreover, the spectral density of the series detected maxima around the periods of 3 to 4 years in all of three series (Figure 12.5). Only 14 years were common to all series and, consequently, the spectral analyses should be interpreted independently for each time series. Likewise, the cross-correlations showed a certain degree of correspondence between the series; we tested for several time lags and the highest obtained values were 0.84 (Mogan-Canarias series, lag = 0), 0.56 (Canarian-CECAF series, lag = 1), and 0.68 (Mogan-CECAF series, lag = 1). The lag 1 between the CECAF and Canarian and Mogan series may be a consequence of the reduction of the spatial scale.

Figure 12.6 through Figure 12.8 show the phase spaces for the Mogan, Canarian, and CECAF series, respectively. The linear regression through the origin represents both the recruitment needed to replace the stock-at-spatial-location and overall equilibrium values. Furthermore, the polynomial regressions describe the dynamic features, which may be common to all of the three cases: (1) a relatively high equilibrium state (indicated by "A") with high levels of both captures and recruitment where maxima and minima diverge and (2) a relatively low equilibrium state (indicated by "B") with lower levels of both captures and recruitment where maxima and minima converge. A summary of results from the linear and polynomial regressions is shown on Table 12.1. Also, there may be an indication of a third equilibrium state both in the cases shown in Figure 12.7 (for years 1975 through 1978) and Figure 12.8 (between A and B). Figure 12.6 through Figure 12.8 show the principal results in this chapter. The plane N_t, N_{t+1} fitted by a third-degree polynomial and a simple regression may allow us to easily understand the dynamics behind the data and both describe and link them through our multiple steady-state approach. Other nonlinear models may be used both to fit the data and to obtain several equilibrium states. However, our approach can be used as an *ad hoc* model because it allows great flexibility and may link and explain most population dynamic phenomena (compensation, depensation, density dependence, density independence, inverse density depensation, and dynamic system behavior) in a relatively simple framework taking into consideration different spatial scales and substocks. No references describing such a dynamic similarity at several spatial scales were found in the literature

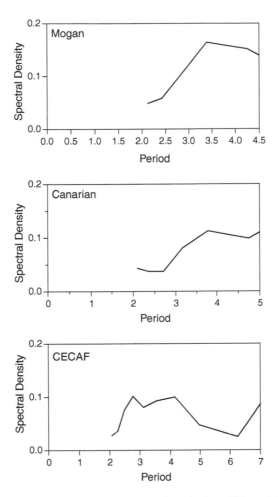

FIGURE 12.5 The spectral densities of three skipjack tuna series from the Port of Mogan (island of Gran Canaria, Canary Islands, years 1980–1996, Mogan series, after Hernández-García et al., 1998), the whole Canary Islands area (years 1975–1993, Canarian series, after Ariz et al., 1995), and the CECAF Division 34 (years 1972–1996, CECAF series, after FISHSTAT/FAO, 1999).

on skipjack tuna and we were able to detect this dynamic autosimilarity while interpreting the data in light of our model.

To illustrate the dynamic features we have proposed, we show, in Figure 12.9, a schematic example on arbitrary data (sinusoidal waves plus noise) of a theoretical in-area stock and recruitment system where the following features are described:

1. The linear regression represents an overall replacement line: while the system evolves above the linear fit, compensation operates and numbers grow, whereas depensation operates under the replacement line implying that numbers decrease.

2. A high and a low equilibrium state represented by A and B, respectively. These orbits of stability (indicated by the dashed ellipses) are caused by oscillations due to density-dependent compensatory and depensatory phases (indicated by the arrows on the ellipses). Also, the polynomial regression describes the two-steady-state system, the evolution of equilibria and the shift between equilibria as both the floor of equilibrium A and the carrying capacity of equilibrium B are approached (indicated by the dot).

3. Also, density-independent transitions may occur due to changes in the environment and fishing mortality: as the lower equilibrium state approaches the ceiling or particular carrying capacity

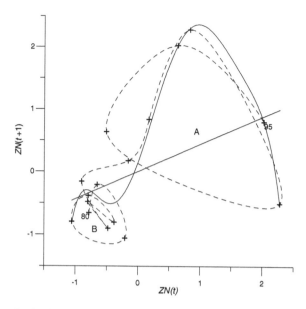

FIGURE 12.6 Phase space for the skipjack fishery landing series at the Port of Mogan (island of Gran Canaria, Canary Islands). The linear regression through the origin represents both the recruitment needed to replace the stock-at-spatial-location and overall equilibrium values. The sixth-degree polynomial regression describes the evolution of the high and low steady-states indicated by "A" and "B," respectively. Z indicates standardized values and N_t and N_{t+1} the generation of the values; 80 and 95 indicate the start and end year of the plotted values.

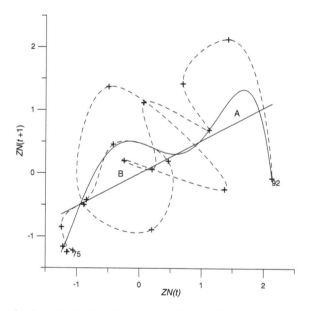

FIGURE 12.7 Phase space for the skipjack fishery landing series from the Canary Islands area (Eastern Central Atlantic). The linear regression through the origin represents both the recruitment needed to replace the stock-at-spatial-location and overall equilibrium values. The sixth-degree polynomial regression describes the evolution of the high and low steady states indicated by "A" and "B," respectively. Z indicates standardized values and N_t and N_{t+1} the generation of the values; 75 and 92 indicate the start and end year of the plotted values.

(indicated by the dot), the system may shift toward a higher equilibrium through a density-independent compensatory phase. Moreover, while density-dependent depensation operates in the higher equilibrium state and the floor of A is approached, density independent depensation may shift the system toward the lower equilibrium state.

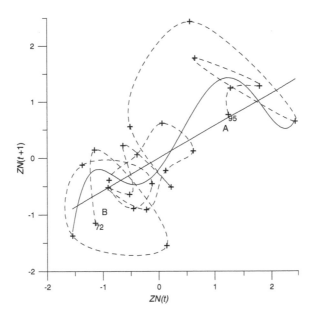

FIGURE 12.8 Phase space for the skipjack fishery landing series from the CECAF Division 34 (Eastern Central Atlantic). The linear regression through the origin represents both the recruitment needed to replace the stock-at-spatial-location and overall equilibrium values. The sixth-degree polynomial regression describes the evolution of the high and low steady states indicated by "A" and "B," respectively. Z indicates standardized values and N_t and N_{t+1} the generation of the values; 72 and 95 indicate the start and end year of the plotted values.

TABLE 12.1

Summary of Results from Data Fits upon the Steady-State Systems Proposed for the Skipjack Tuna Series (standardized values) from Three Spatial Scales (Port of Mogan, Canary Islands area, and CECAF Division 34)

DF	Linear Trend			Polynomial (6th order)		
	R	*F*	*p<*	*R*	*F*	*p<*
15	0.44	3.35	0.09	0.92	7.77	0.01
17	0.54	6.55	0.05	0.77	2.63	0.08
23	0.59	11.62	0.01	0.70	2.66	0.06

DF = degrees of freedom, R = regression value, F = F value, p = probability.

4. Maxima and minima converge as the system evolves toward the lower equilibrium and diverge as it shifts toward the higher steady state. Density dependence may operate similarly in both equilibrium states but at different levels of numbers. On the one hand, oscillations may be larger in high equilibrium states as the system evolves toward the maximum carrying capacity of the system (a critical value or overall ceiling above which the trajectory enters depensation, K_{max}); on the other hand, oscillations may become lower as the system evolves toward a minimum viable population (a critical value around which oscillations become small or non-existent implying the extinction of the commercial fishery).

The proposed two-steady-state system may be described either by a multiplicative equation such as

$$R_a \cong f_1(S_A) \cdot f_2(S_B) \tag{12.3}$$

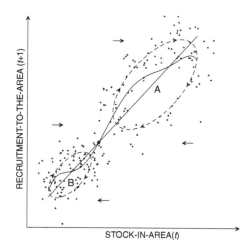

FIGURE 12.9 Theoretical in-area stock and recruitment system proposed for three spatial scales in the Eastern Central Atlantic (Port of Mogan, Canary Islands area and CECAF Division 34). The linear regression represents the replacement line and the polynomial fit describes the dynamic evolution of the system. A and B are the high and low equilibrium states, respectively. The dot represents the transition point between the steady states, being the floor of A and carrying capacity of B, respectively. Density-independent compensation and depensation are represented by the arrows → and ←, respectively. Orbits of stability are indicated by the dashed ellipses on which arrows represent density-dependent compensation and depensation. Data values are arbitrary and were generated by sinusoidal waves plus noise.

or an additive model such as

$$R_a \cong f_1(S_A) + f_2(S_B)$$ (12.4)

where R_a represents recruitment, $f_{1,2}$ are nonlinear, three parameter functions, and $S_{A,B}$ are the stock-in-area for the equilibrium states A and B, respectively.

12.5 Discussion

Although skipjack harvesting is subject to some form of international regulation, control is minimal and we have regarded it as an open-access fishery: exploitation may be carried out in international waters on the migratory stock without any effective policy enforcement. In the present approach, recruitment-to-the-area is a key concept, which describes a migrating population or stock entering the area of operation of the fishery. This concept is used, in part, because of the lack of data on juveniles being recruited to the adult population and migrating from the nursery areas in the Gulf of Guinea. Also, we lack fishing effort data for all of the series. Within the Canary Islands area, however, fishing effort may be assumed relatively stable during the time span of the Mogan and Canarian series. Also, the largest skipjack individuals enter the area of the fishery twice a year as they return from the Azores to the Gulf of Guinea passing through the Canary Islands area a second time. Furthermore, it may be argued that both the classical models and second-order polynomial regressions could be more appropriate to fit the data, meeting mainstream modeling criteria. However, such classical approaches are inappropriate to describe the linked dynamics of multiple steady-state systems (density-dependent and density-independent compensation and depensation, orbits of stability, multiple carrying capacities, and autosimilarity at several scales). We need both conceptual frameworks and statistical techniques that will allow us to understand the dynamics behind the data and link changes in skipjack stocks to fluctuations in the environment. In our view, the system consists of multiple steady states, distinct regimes, qualitatively similar, which should be dynamically linked both to each other and to the environment. A minimum of three constants are needed to describe each equilibrium state and allow the linkage between the steady states. Furthermore, while the classical models and second-order fittings may assume that residuals are either random or solely caused by noise, we have assumed that residuals may be a combination of both signal and noise. Although dynamic structures may be an artifact of smoothing techniques (i.e., we may obtain cyclic-like patterns after several steps of smoothing upon random data), the temporal evolution and the structures observed in the phase spaces upon the standardized data showed that a multi-steady-state system may be more appropriate to understand the mechanics behind the system. Also, the theoretical criteria we put forward may be useful to explain the processes governing both recruitment and stock dynamics in skipjack tuna in the Atlantic and to develop new approaches for the preventive management of the migratory stock.

12.5.1 The Proposed Equations

On the one hand, a multiplicative approach formed from a concatenation of functions could exhibit multiple (stable) equilibria and complex dynamics: stock and recruitment may be described by the result of a multiplicative process where the initial number of recruits could be modified by nonlinear functions specific to each substock at a certain spatial location. Such an approach could be an extension of the model proposed by Paulik (1973) for an overall spawner-recruit system. However, the drawback of the multiplicative approach is the interdependence between the functions which would imply the collapse of the system once any "near-zero" recruitment occurs or whenever "outliers" or extreme fluctuations appear to control any one of the steady states, any one year. On the other hand, an additive approach such as an extension to the model proposed by Solari et al. (1997), where the stock-recruitment system is described as a summation of nonlinear functions, is extremely flexible and may show complex dynamics for each of the steady states. Although equilibria are linked, they are mathematically independent and the model may describe a wide range of dynamic situations (chaos, cycles, quasi-cycles, standstills). Both of these approaches may be approximated by sixth-order polynomial regressions for a two-steady-state system. Nevertheless, the determination of the generic model functions are beyond the scope of this chapter and several alternatives may be viable either as extensions of aforementioned models or new *ad hoc* functions that may describe the data within a dynamic framework.

12.5.2 The Phase Spaces

These relationships may be assumed to describe the dynamics of the exploited skipjack population irrespective of the fishing effort, assuming the following criteria: fishing mortality may either (1) become asymptotic as the fishery approaches the so-called zero net value (i.e., economic overfishing resulting in benefits reduced to zero followed by a stabilizing reduction in the fishing effort, as suggested by Clark, 1976) or (2) cycle due to a backward bending (depensatory) yield against effort relationship because of either biological overfishing (as described by Pitcher and Parrish, 1993) or reductions in recruitment as a result of either environmental perturbations or the combined effect between the environment and fishing mortality during depensatory trends (as described by Solari et al., 1997). Both of these assumptions imply that fishing mortality may reach a ceiling that may either be asymptotic during periods of relative environmental stability or follow depensatory trends toward lower equilibrium states as the external environment becomes less benign.

12.5.3 Variable Carrying Capacity (Ceilings, K_i)

This is a central concept in our criteria. While carrying capacity is considered as a single value in the classical approaches, we assume (1) it may be multiple and a threshold between equilibrium states; (2) it will link different equilibria and each particular steady state will show a particular ceiling or value of carrying capacity, and (3) it may be quantitatively different at different spatial scales while remaining similar, qualitatively. In our view, it may be more realistic to consider a population parameter, such as K_i, as variable. On the one hand, the skipjack population migrates through relatively large geographic ranges where it will encounter a continuous transition scheme with a multiplicity of external perturbations; on the other hand, density-independent inputs will determine different levels of numbers recruited implying a particular K_i for each orbit of stability.

12.5.4   Minimum Populations (Floors, P_i)

As numbers decrease, due to either fishing mortality or external perturbations, or to the combined effects from both of these factors, each steady state may, gradually, shift toward a critical value or unstable equilibrium under which stock and recruitment will "jump" onto a lower, relatively stable equilibrium state. Also, the per capita reproductive success may decline at lower population levels implying that reduced numbers of individuals are recruited to the area of the fishery. Floors may be approached through either density-dependent or density-independent depensation; both of these depensations combined may generate rapid shifts toward lower equilibria. Furthermore, the proposed system may contain an overall minimum viable population under which (1) no oscillations in stock and recruitment will be detected and (2) the commercial fishery may cease.

12.5.5   Multiple (Stable) Equilibria

There is no evidence in the field data to assume the dynamics of the skipjack system are governed by a single attractor and a global carrying capacity and that residuals could solely be a consequence of either random processes or noise. The observed structures and temporal evolution in the data may rather suggest that the skipjack system is governed by at least two (or multiple) attractors (and repellors), which are dynamically linked by multiple carrying capacities and minimum populations through which stock and recruitment may, persistently, evolve and return between a wide range of equilibrium states allowing for stable, periodic, and chaotic dynamics. The trajectories of these steady states may turn in orbits of stability determined both by density-dependent compensatory and depensatory phases. The equilibrium states may be linked through floors (or minimum threshold values) and ceilings (or maximum threshold values) that appear during transitions determined by the combined effect from fishing mortality and environmental fluctuations; these critical values may be regarded as the minimum population for the higher equilibrium state and the carrying capacity for the lower equilibrium. As recruitment reaches K_i, the system will "jump" onto the higher equilibrium, whereas it will enter the lower equilibrium as P_i is approached. Also, we may observe that equilibrium states (1) converge as they tend either to zero or to an overall minimum viable population (K_0) and (2) diverge as they tend to the overall ceiling of the system or maximum carrying capacity (K_{max}). Multiple, linked equilibrium states both within and between relevant spatial scales may describe the dynamics of migratory skipjack substocks. Also, classical approaches may describe different unlinked regimes but will not explain the dynamics behind the data. The idea of a dynamic continuum is appealing to describe the phase-space and temporal evolution of a persistent system. Rothschild (1986) suggested that populations reduced by fishing or anthropogenic substances that compensate for reductions in vital rates may easily transit among stable, periodic, and chaotic population dynamics. Garcia (1998) and Sharp et al. (1983) suggested that the Hokkaido sardine series were characterized by loops and proposed an oscillating system consisting of two strange attractors, linked by some transitional shifts, operating at two different levels of spawners and recruits. Furthermore, Berg and Getz (1988) suggested that stock and recruitment, in a sardine-like population, moved along a path or attractor in some higher-dimension coordinate system; Conan (1994) observed that lobster and snow crab landings in Atlantic Canada may follow two orbits of stability or cycles; Powers (1989) suggested chaotic behavior for a two-species system of fish, and Tyutyunov et al. (1993) demonstrated cycles of different periods and chaos in population dynamics of perch from ten lakes. Moreover, Caddy (1998) pointed out several other cases, in semienclosed areas, where stock and recruitment dynamics could be linked to oscillatory phenomena: (1) an apparent 9- to 18-year periodicity for the Bay of Fundy scallop stocks (Caddy, 1979); (2) a 12-year fishing-effort-independent periodicity in the landings of both hake and red mullet at the island of Mallorca in the Mediterranean Sea (Astudillo and Caddy, 1986); and (3) a 12- to 13-year oscillatory pattern in the catches of the Adriatic sardine.

12.5.6   Compensatory and Depensatory Dynamics

In each steady state, oscillations may be due to both density-dependent compensation (numbers increase) and depensation (numbers decrease); stock and recruitment will be affected by short- and medium-term external perturbations, oceanographic diffusion–advection processes, migration between schools, availability of food items, fishing mortality, and catch-effort oscillations, as well as several other both internal (population) and external (environmental, fishing related) factors that will determine the temporal evolution of an equilibrium state. Also, transitions between equilibria may be caused either by density-independent compensation (as the environment becomes more benign and recruitment increases) or density-independent depensation (as external stress increases). While fishing mortality may be incremented during density-independent compensation, reductions may not be enough while density-independent depensation is operating. Also, the combined effects of fishing mortality during both density-dependent and -independent depensation may imply the rapid shift toward both the floor of an equilibrium and, hence, a lower steady state.

12.5.7 Extinction of the Commercial Fishery

Fonteneau (1987) observed that recruitment overfishing has never been suggested for skipjack in the Atlantic. The steady states observed in the skipjack system appear to be persistent within relatively wide ranges of stability. Also, the stability of equilibrium states may be further enhanced by catch and effort oscillations as economical overfishing is being approached. However, several mechanisms could generate both recruitment overfishing and the (temporal) extinction of the fishery. On the one hand, environmental medium-term disturbances in the nursery grounds may cause the skipjack system to shift toward low equilibria with decreasing amplitude between maxima and minima; on the other hand, both types of depensation combined with a relatively constant fishing mortality may imply that the skipjack system evolves toward an overall minimum viable population with no oscillations.

12.5.8 Migration through a Fractal Marine System

While the studied time series showed quantitatively different qualities such as dynamic similarity observed in the two-equilibrium system, memory and periodicity may be similar features to all of the cases. These results may open up an interesting field of work in the research on exploited skipjack populations in the Atlantic. The correspondence between the series and similarity in the phase spaces may suggest that stock and recruitment relationships may be caused by deterministic mechanisms with similar dynamics at several spatial scales. This may imply that we could (1) estimate complex processes governing recruitment and migration in skipjack; (2) link population processes between different spatial scales relevant to the dynamics of the migratory stock; (3) forecast recruitment in wider areas from local series at certain spatial locations, and (4) estimate future recruitment in minor spatial locations from overall CECAF series taking into consideration the detected time lags. To discuss these ideas further, we simulated a self-similar system through the iteration of the function $f(x) = x^2 + m$. We allowed the function the random choice between two possible inverses (+1 or –1) and let the iteration run until we obtained an arbitrary number of data points ($N = 19851$). The data were standardized and the system is shown in Figure 12.10. All variables (R, IM), the initial value of the parameter (m), and the number of iterations were arbitrary. To construct a dynamically referenced description of the data, we fitted the output to a linear regression, a sixth-degree polynomial, a cubic spline (to show more detailed local dynamics), and 50 and 95% bivariate ellipses (also, confidence intervals), as well. Furthermore, we sampled the simulated series both randomly and sequentially to 10, 5, 1, and 0.1% of the total number of points (to resemble different spatial scales or sampling windows) and, in all cases, we obtained similar tendencies: as in the skipjack fishery, the simulated system showed results with similar dynamic patterns at different sampling windows. Although the tendencies remain similar at several scales, we may obtain different levels of numbers depending on the quadrant in which we carry out the sampling. As the skipjack stock migrates through the ocean, it will be affected by a multiplicity of external perturbations of dynamic nature. There is an increasing body of evidence suggesting that the upper ocean layer through which the skipjack is recruited and migrates may be both of fractal nature and affected by multifractal processes: the spatial distribution of foam and white caps (Kerman and Szeto, 1994), wind-wave breaking (Raizer et al., 1994) and breaking of waves (Kerman and Bernier, 1994), distribution of sea surface temperature (Fu, 1994, 1995), isotherm lengths and patterns of the sea surface temperature in mesoscale

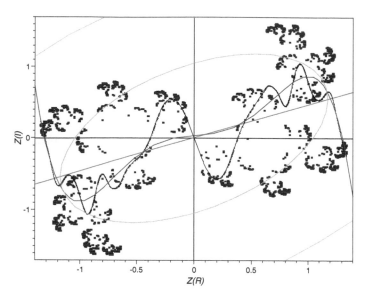

FIGURE 12.10 A theoretical self-similar system resembling a stock-in-area [$Z(R)$] recruitment-to-the-area [$Z(I)$] relationship. Data values ($N = 19851$) are dynamically referenced through linear regression, a sixth-degree polynomial, a cubic spline (to show more detailed local dynamics) and 50% (inner) and 95% (outer) bivariate ellipses (also, confidence intervals). The random and sequential sampling of 10, 5, 1, and 0.1% of the total number of points (resembling captures from different spatial scales) showed similar tendencies: as in the skipjack fishery, the simulated system shows similar dynamic patterns at different sampling windows. Different levels of numbers are obtained depending on the quadrant sampled. Iteration and data, parameter values, and functions are arbitrary.

turbulence (Bunimovich et al., 1993), and fractal behavior of the temperature isolines and properties of frontal regions (Marullo et al., 1993) may be examples of factors that determine recruitment patterns and spatial distributions in fishery areas. Also, such fractal structures in the ocean may explain the dynamic similarities we have proposed for the skipjack system at the three spatial scales. If a skipjack stock diffuses-advects through a fractal ocean with similar properties at several scales, we might expect the equilibrium states we observe in the skipjack system to show a certain degree of autosimilarity, as well. Assuming the theoretical criteria proposed in our framework, we could develop a method based on remotely sensed data with which we could estimate the dynamics of one or several substocks of skipjack from a few spatial windows. Block et al. (2001) reported electronic, satellite-tracked tag recovery data on both vertical-diagonal and transatlantic (Gulf of Mexico–Mediterranean Sea) migration of bluefin tuna. The incorporation of data that show depth boundaries in tuna migration combined with sea surface temperature, recruitment-to-the-area and catches may allow for the determination of the dynamic three-dimensional system (latitude, longitude, depth) through which tuna migrates: a multi-steady-state framework as proposed in this chapter may incorporate all of the variables to describe this hyperspace. Such an approach could be critical both for the conservation of tuna and the preventive control of the fishery. Also, it may become the ground for the development of a formal theory for both system behavior and migrations in skipjack and other tunas. Classical approaches assume a single spatial scale, an equilibrium state, and a sole value of carrying capacity. It is critical to realize that a dynamic framework will enable us to understand both the temporal evolution of the skipjack system and the causal mechanisms behind the data.

Acknowledgments

John Caddy and Serge Garcia (FAO/Fishery Resources Division, Rome, Italy) and Sami Souissi and Laurent Seuront (Ecosystem Complexity Research Group, Université des Sciences et Technologies de Lille, France) are acknowledged for their encouragement and comments. This study was funded, in part, by the Ministry of Fisheries and Agriculture of the Canarian Government (Consejería de Agricultura y Pesca del Gobierno Autónomo de Canarias) through the Grant 540/1998 to the Las Palmas University Foundation (Fundación Universitaria de Las Palmas).

References

Ariz, J., R. Delgado de Molina, J.C. Santana, and A. Delgado de Molina. 1995. Datos estadísticos de la pesquería de túnidos de las islas Canarias durante el período 1975–93. Colección de documentos científicos, ICCAT, 44(2).

Astudillo, A. and J. Caddy. 1986. Periodicidad de los desembarcos de merluza (*Merluccius merluccius*) y salmonete (*Mullus* sp.) en la isla de Mallorca, in *Int. Symp. Long Term Changes Mar. Fish Pop.*, V.T. Wyatt and M.G. Larrañeta, Eds., Vigo, Spain.

Atkinson, C.A. 1987. A nonlinear programming approach to the analyses of perturbed marine ecosystems under model parameter uncertainty. *Ecol. Modelling* 35:1–28.

Bas, C., A.P. Solari, and J.M. Martín. 1999. Considerations over a new recruitment model for exploited fish populations. *R. Acad. Sci. Barcelona* 58(5):157–183.

Bayliff, W.H. 1988. Integrity of schools of skipjack tuna, *Katsuwonus pelamis*, in the eastern Pacific Ocean, as determined from tagging data. *Fish. Bull.* 86:631–643.

Berg, M. and W. Getz. 1988. Stability of discrete age-structured and aggregated delay-difference population models. *J. Math. Biol.* 26:551–581.

Beverton, R.J. and S.J. Holt. 1957. On the dynamics of exploited fish populations. Ministry of Agriculture, Fisheries and Food, London. Fisheries Investigation Series 2(19).

Block, B.A., H. Dewar, S.B. Blackwell, T.D. Williams, E.D. Prince, C.J. Farwell, A. Boustany, S.L. Teo, A. Seitz, A. Wall, and D. Fudge. 2001. Migratory movements, depth preferences, and thermal biology of Atlantic bluefin tuna. *Science* 293:1267.

Boehlert, G.W. and B.C. Mundy. 1994. Vertical and onshore–offshore distributional patterns of tuna larvae in relation to physical habitat features. *Mar. Ecol. Prog. Ser.* 107:1–13.

Bunimovich, L.A., A.G. Ostrovskii, and S. Umatani. 1993. Observations of the fractal properties of the Japan sea surface temperature patterns. *Int. J. Remote Sens.* 11:2185–2201.

Caddy, J. 1979. Long-term trends and evidence for production cycles in the Bay of Fundy scallop fishery. *Rapp. P.-V. Réun. Cons. Int. Explor. Mer.* 175:97–108.

Caddy, J. 1998. Personal communication. Letter from the Chief, Marine Resource Service, Fishery Resources Division, Food and Agriculture Organization (FAO), Rome, Italy.

Castro, J.J., A.P. Solari, J.M. Martín González, and C. Bas. 1999. Recruitment to the fishery of the skipjack tuna *Katsuwonus pelamis* in Canary Islands: application of a new conceptual model. Fisheries Resource Group, University of Las Palmas. Report for the project funded by the Ministry of Fisheries of the Canarian Government (Consejería de Pesca del Gobierno Autónomo de Canarias) through the Las Palmas University Foundation (Grant 540, December, 1998). In Spanish.

Clark, C. 1976. *Mathematical Bioeconomics: The Optimal Management of Renewable Resources*, John Wiley & Sons, New York.

Conan, G. 1994. Can simple linear correlation satisfactorily detect environmental or interspecific effects on fisheries landings in a chaotic oceanic universe? ICES-Council Meeting, P/8.

Conrad, M. 1986. What is the use of chaos? in *Chaos: Nonlinear Science, Theory and Applications*, A.V. Holden, Ed., Manchester University Press, U.K., 1–14.

Cushing, D.H. 1977. The problems of stock and recruitment, in *Fish Population Dynamics,* J.A. Gulland, Ed., John Wiley & Sons, New York, 116–133,

De Angelis, D.L. 1988. Strategies and difficulties of applying models to aquatic populations food webs. *Ecol. Modelling* 43:57–73.

FAO. 2001. Map of the CECAF Division 34. Food and Agriculture Organization, Fisheries Circular 835 Rev. 1. Summary Information on the Role of International Fishery Bodies with Regard to the Conservation and Management of Living Resources of the High Seas, prepared by M.J. Savini in 1991. FAO-Fisheries Home Page at http://www.fao.org/fi/body/body.asp.

Fiedler, P.C. and H.J. Bernard. 1987. Tuna aggregation and feeding near fronts observed in satellite imagery. *Cont. Shelf Res.* 7:871–881.

FISHSTAT/FAO. 1999. FISH STAT Plus version 2.0. FAO Fisheries Department. Fishery Information, Data and Statistics Unit (FIDI). Database of Nominal catches and landings reported to regional commissions. http://www.fao.org/WAICENT/FAOINFO/FISHERY/struct/fidif.htm.

Fogarty, M. 1993. Recruitment in randomly varying environments. *ICES J. Mar. Sci.* 50:247–260.

Fonteneau, A. 1987. Competition between tuna fisheries — critical review based on Atlantic examples. Collective Volume of Working Documents, Expert Consultation on Stock Assessment of Tunas in the Indian Ocean, Colombo, Sri Lanka, 4–8 December, 1986. FAO/UNDP Indo-Pacific Tuna Development and Management Programme, 195–213.

Fonteneau, A. 1991. Modelización, gestión y ordenación de las pesquerías atuneras del Atlántico centro-este. Modelling and management of the tuna fisheries in the central-eastern Atlantic. International Committee for the Conservation of Atlantic Tunas, ICCAT, 37:344–387.

Fonteneau, A. 1997. A critical review of tuna stocks and fisheries trends world-wide, and why most tuna stocks are not yet overexploited, in *Developing and Sustaining World Fisheries Resources. The State of Science and Management*, D.A. Hancock, D.C. Smith, A. Grant, and J.P. Beumer, Eds., Commonwealth Scientific and Industrial Research Organ (CSIRO), Collingwood, Australia, 39–48.

Fonteneau, A. and J. Marcille. 1993. Resources, fishing and biology of the tropical tunas of the eastern Central Atlantic. Fish. Tech. Pap., (FAO), 292, 354 pp.

Fu, Y. 1994. Relationship between sea surface temperature and typhoon analysed by fractal dimension. *Trans. Oceanol. Limnol.* 3:10–17.

Fu, Y. 1995. Fractal analysis and forecast of monthly average sea surface temperature. *Mar. Forecasts* 1:49–54.

Garcia, S. 1998. Personal communication. Letter from the Director of the Fishery Resources Division, Fisheries Department, Food and Agriculture Organization (FAO), Rome, Italy.

Gulland, G.A. 1989. Fish populations and their management. *J. Fish Biol.* 35(Suppl. A:1–9). C.E. Hollingworth and A.R. Margetts, Eds., Fisheries Society of the British Isles, Sawson, Cambridge, U.K.

Hernández-García, V., J.L. Hernández-López, and J.J. Castro. 1998. The octopus (*Octopus vulgaris*) in the small-scale trap fishery off the Canary Islands (Central-East Atlantic). *Fish. Res.* 35:183–189.

Hilborn, R. 1991. Modelling the stability of fish schools: exchange of individual fish between schools of skipjack tuna *(Katsuwonus pelamis)*. Can. J. Fish. Aquat. Sci. 48:1081–1091.

ICCAT. 1986. *Proceedings of the International Commission for the Conservation of Atlantic Tunas (ICCAT) Conference on the International Skipjack Year Program*. P. Symon, P. Miyake, and G. Sakagawa, Eds., 388 pp.

Kerman, B. and L. Bernier. 1994. Multifractal representation of breaking waves on the ocean surface. *J. Geophys. Res. Oceans* C8:16179–16196.

Kerman, B. and K. Szeto. 1994. Fractal properties of white caps. *Atmos. Ocean* 32:531–551.

Marullo, S., A. Provenzale, R. Santoleri, and B. Villone. 1993. Fractal fronts in the Mediterranean Sea. *Ann. Geophys. Atmos. Hydrospheres Space Sci.* 2(3):111–118.

Oppenheim, A.V. and R.W. Schafer. 1975. *Digital Signal Processing*, Prentice-Hall, Englewood Cliffs, NJ, 556.

Ovchinnikov, V.V., V.Z. Gaikov, Y.P. Fedoseev, and V.G. Shcheglov. 1988. Main results of realization of the Soviet program of tuna tagging in the Atlantic Ocean. Collective Volume of Working Documents, FAO/UNDP Indo-Pacific Tuna Development and Management Programme, Colombo, Sri Lanka, 224–226.

Pagavino, M. and D. Gaertner. 1994. Variación espacio-temporal de las capturas de atunes aleta amarilla y listado realizadas por la flota venezolana de superficie en el Mar Caribe, entre 1988 y 1992 [Spatio-temporal variations of the yellowfin and skipjack catches by the Venezuelan surface fleet in the Caribbean Sea, from 1988 to 1992]. Meeting of the ICCAT Working Group to Evaluate Atlantic Yellowfin Tuna, Tenerife, Canary Islands, Spain, 3–9 June 1993. ICCAT 42(2):314–318.

Paulik, G.J. 1973. Studies of the possible form of the stock-recruitment curve. *Rapp. P.-V. Reun. Cons. Int. Explor. Mer.* 164:302–315.

Pitcher, T. and J. Parrish. 1993. In *Behaviour of Teleost Fishes*, Fish and Fisheries Series 7, Chapman & Hall, New York, 372–374.

Piton, B. and C. Roy. 1983. Annee internationale listao: Donnees d'environnement pour la periode juin, juillet et aout 1981 dans le golfe de Guinee [International "listao" (skipjack) year: environmental data for the period June, July and August 1981 in the Gulf of Guinea]. *Collect. Vol. Sci. Pap.* 18(1):205:253.

Powers, J.E. 1989. Multispecies models and chaotic dynamics. ICES-MSM, P/21.

Raizer, V.Y., V.M. Novikov, and T.Y. Bocharova. 1994. The geometrical and fractal properties of visible radiances associated with breaking waves in the ocean. *Ann. Geophys. Atmos. Hydrospheres Space Sci.* 12:1229–1233.

Ramos, A. and P. Sangrá. 1992. Características oceanograficas en el area de Canarias: Relación con la pesquería de listado *(Katsuwonus pelamis)* [Oceanographic characteristics of the area of the Canary Islands: relationship with the skipjack fishery *(Katsuwonus pelamis)*]. Meeting of the ICCAT Standing Committee on Research and Statistics, Madrid, Nov. 1991. ICCAT 39(1):289–296.

Ramos, A., I. Ramírez, and J. Pajuelo. 1991. Aspectos biológicos del *Katsuwonus pelamis* en aguas del Archipiélago Canario: Reproducción [Biological aspects of *Katsuwonus pelamis* in waters of the Canary Islands: Reproduction]. Meeting of the ICCAT Standing Committee on Research and Statistics, Madrid, Nov. 1990. ICCAT 35(1):14–21.

Ramos, A., P. Sangrá, A. Hernandez-Guerra, and M. Cantón. 1991. Large and small scale relationship between skipjack tuna *(Katsuwonus pelamis)* and oceanographic features observed from satellite imagery in the Canary Islands area. Meeting of the International Council for the Exploration of the Sea, La Rochelle, France, 26 Sept.–4 Oct. 1991. ICES C. M. Pap. L:78.

Ricker, W.E. 1954. Stock and recruitment. *J. Fish. Res. Board Can.* 11:559–623.

Ricker, W.E. 1975. Computation and interpretation of biological statistics of fish populations. *Bull. Fish. Res. Board Can.* 191:382 pp.

Roger, C. 1994a. Relationships among yellowfin and skipjack tuna, their prey-fish and plankton in the tropical western Indian Ocean. *Fish. Oceanogr.* 3(2):133–141.

Roger, C. 1994b. On feeding conditions for surface tunas (yellowfin, *Thunnus albacares* and skipjack, *Katsuwonus pelamis*) in the western Indian Ocean. *Proceedings of the 5th Expert Consultation on Indian Ocean Tunas*, Seychelles, 4–8 Oct. 1993. FAO/UNDP Indo-Pacific Tuna Development and Management Programme, IPTP 1994, 8:131–135.

Roger, C. and E. Marchal. 1994. Mise en evidence de conditions favorisant l'abondance des albacores, *Thunnus albacares*, et des listaos, *Katsuwonus pelamis*, dans l'Atlantique equatorial est [Conditions favoring the abundance of yellowfin *(Thunnus albacares)* and skipjack *(Katsuwonus pelamis)* in the eastern equatorial Atlantic]. Meeting of the ICCAT Working Group to Evaluate Atlantic Yellowfin Tuna, Tenerife, Canary Islands, Spain, 3–9 June 1993. ICCAT 42(2):237–248.

Rothschild, B.J. 1986. *Dynamics of Marine Fish Populations*, Harvard University Press, Cambridge, MA.

Sharp, G., J. Csirke, and S. Garcia. 1983. Modelling fisheries: what was the question? *FAO Fish. Rep.* 291(3):1177–1224.

Shepherd, J.G. 1982. A versatile new stock-recruitment relationship for fisheries, and the construction of sustainable yield curves. *J. Cons. Int. Explor. Mer* 40:67–75.

Solari, A.P., J.M. Martín-González, and C. Bas. 1997. Stock and recruitment in Baltic cod *(Gadus morhua)*: a new, non-linear approach. *ICES J. Mar. Sci.* 54:427–443.

Tyutyunov, Y., R. Arditi, B. Buettiker, Y. Dombrovsky, and E. Staub. 1993. Modelling fluctuations and optimal harvesting in perch populations. *Ecol. Modelling* 69:19–42.

Wise, J. 1985. Probable underestimates and misreporting of Atlantic small tuna catches, with suggestions for improvement. Meeting of the ICCAT Standing Committee on Research and Statistics, Palma de Mallorca, Spain, Nov. 1985. *Collect. Vol. Sci. Pap.*, 25:324–332.

13

The Temporal Scaling of Environmental Variability in Rivers and Lakes

Hélène Cyr, Peter J. Dillon, and Julie E. Parker

CONTENTS

13.1 Introduction..201
13.2 Methods...203
13.3 Results...205
13.4 Discussion...208
    13.4.1 Environmental Variability Is Higher in Rivers Than Lakes 208
    13.4.2 The Scaling of Environmental Variability in Rivers and Small Lakes 209
    13.4.3 Changes in the Scaling of Environmental Variability at Different Timescales...........210
    13.4.4 Ecological Implications 210
13.5 Summary 210
Acknowledgments..211
References...211

13.1 Introduction

Environmental variability, which we define as the variability in any environmental factor that affects the distribution, growth, and survival of organisms in nature (e.g., temperature, water, light, nutrients, or food), occurs at many temporal and spatial scales. Temperature at a given location, for example, varies on very short timescales (minutes to hours) depending among other things on cloud cover and shading; it also varies diurnally with colder night temperatures; it varies between days according to local weather conditions; it varies seasonally; it varies between years, decades, centuries, and millennia. Temperature also varies over short distances depending on microtopography and at larger spatial scales. The spatial and temporal sources of variability in temperature and other geophysical variables are being studied intensively in climatology (e.g., Koscielny-Bunde et al., 1998; Weber and Talkner, 2001), oceanography (e.g., Powell, 1989; Zang and Wunsch, 2001), and in river hydrology and geomorphology (e.g., Rodriguez-Iturbe and Rinaldo, 1997; Hubert, 2001). Striking similarities in the scaling of variability are emerging across a wide range of fields (e.g., Halley, 1996; Havlin et al., 1999).

Several approaches are being used to describe variability at different scales in time series (Weber and Talkner, 2001) and these methodological differences (e.g., method of analysis, type of data, data averaging technique) complicate comparisons among studies and among types of ecosystems (e.g., Weber, 1993; Scheuring and Zeöld, 2001). In this study, we focus on the temporal aspect of environmental variability in lakes and rivers using spectral analysis, a classical approach used to compare the amount of variability at different scales (e.g., Platt and Denman, 1975; Halley, 1996).

Spectral analysis is used to partition the variability in a time series into different frequencies (or timescales). This allows a comparison of the magnitude of variability at different timescales, e.g., variability in common events (low frequency or short timescale) compared to rare events (high frequency or long timescale). Interestingly, the variability (or noise) in a wide range of environmental time series has been shown to follow the so-called $1/f$-noise relationships (i.e., relationships of the form $S(f) \propto f^\gamma$, where $S(f)$ is a spectral density function, f is frequency, and where $-\gamma$, the scaling exponent, indicates the "color" of the noise; e.g., Mandelbrot and Wallis, 1969; Steele, 1985; Pelletier and Turcotte, 1997). A scaling exponent $(-\gamma)$ of 0 suggests an equal amount of variability at all frequencies (i.e., called white noise, by analogy to white light). Scaling exponents of -1, -2, and ≤-2 represent pink, brown (or red), and black noise, respectively. Steeper negative slopes (i.e., redder spectra) suggest that progressively more variability is found in rare (low frequencies) events relative to more common (high frequencies) events. Different scaling exponents also suggest different rates of increase in temporal variability over time (e.g., exponential increase for pink noise with $-\gamma = -1$; see Figs. 1 and 3 in Halley, 1996).

The temporal scaling of environmental variability in natural systems is still poorly understood, but it is becoming increasingly clear that environmental variability often does not scale as white noise (Mandelbrot and Wallis, 1969; Steele, 1985; Cuddington and Yodzis, 1999). Atmospheric warming, global changes in rainfall patterns, eutrophication, and so on, all increase the amount of variability of important environmental parameters through time, and produce reddened noise. But even without major human disturbances, environmental variability is expected to scale differently in different types of ecosystems. Steele (1985) first suggested this possibility after comparing the scaling of temperature variability on land and in the oceans. He showed that air temperature variability in England was approximately constant (i.e., white noise) over ecological timescales (1 to ~100 years), while sea level, a surrogate for whole ocean temperature, became more variable the longer the timescale (i.e., reddened noise). His conclusions for terrestrial systems were recently supported by extensive comparisons of long-term air temperature records around the world (Pelletier, 1997; Vasseur and Yodzis, in press). Steele's comparison was also extended by Cyr and Cyr (in press), who showed that the scaling of temperature variability becomes systematically more reddened from land to rivers to lakes of increasing sizes and to the oceans. There is clear empirical evidence of a systematic shift in the scaling of ambient temperature among different types of natural ecosystems.

Because ambient temperature affects everything an organism, especially a poikilothermic organism, does (e.g., locomotion, feeding, respiration, growth, reproduction; Cossins and Bowler, 1987), it clearly has the potential to affect population dynamics, albeit in complex ways (see discussion), through changes in birth rates and survival (Laakso et al., 2001). However, many other factors also regulate the growth, reproduction, and survival of organisms in nature. How do they scale compared to temperature? Pelletier and Turcotte (1997) compared 49 long time series of annual rainfall and 636 time series of monthly river discharge, and found them to scale very similarly to air temperature. These results suggest that two of the major factors regulating terrestrial organisms (temperature and water) scale similarly, and that similar scaling applies to rivers and, perhaps, to small lakes that are strongly affected by runoffs from the land.

In aquatic ecosystems, different environmental variables are also linked and likely to scale similarly. For example, in rivers the flux of nutrients and of sediments depends on rainfall patterns, river discharge, and air temperature (which affects decomposition, water runoff, and leaching from the land), as well as on the geomorphology and structure of the watershed (e.g., soil types, gradient, proportion of agricultural, urban, and forested areas; Howarth et al., 1996; Russell et al., 1998; Arheimer and Lidén, 2000; Ekholm et al., 2000; McKee et al., 2000). In lakes, the seasonal thermal stratification regulates the mixing and availability of nutrients within the lake, while stream/river inflows and runoff from the landscape regulate the inputs of chemicals and particles to the lake (e.g., Yan et al., 1996; Reynolds and Davies, 2001). Therefore, the scaling of variability in temperature, rainfall, and river discharge is likely to be a good indicator of the scaling of variability in other important factors that shape an organism's environment.

In this study, we compare the temporal scaling of variability in water temperature and river flow to that of four other important environmental factors: two nutrients — total nitrogen (TN) and total phosphorus (TP) — that are well known to limit primary production, dissolved organic carbon (DOC), which affects light penetration and algal productivity and which provides food for microbial communities, and suspended solids, which limit light penetration and algal productivity and which affect the feeding efficiency of invertebrates and fish. The goal of this study is to test whether potentially important

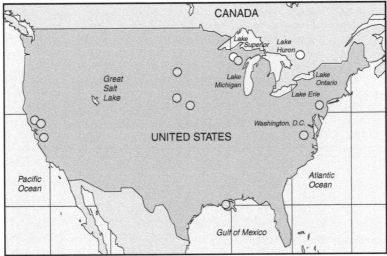

FIGURE 13.1 Map showing the approximate location of each sampling site. Gray circles are rivers; white circles are lake regions. Site characteristics are listed in Table 13.1.

environmental factors show reddened patterns of variability (i.e., with higher variability at longer timescales) similar to those reported for water temperature and river flow. We are not testing how different environmental variables relate to each other within particular types of ecosystems.

13.2 Methods

Our analyses are based on long-term continuous data sets from diverse rivers and small lakes. The requirement for continuous year-round data severely limits the number of sites that can be compared. Most aquatic data are only available for part of the year (e.g., period of open water) and long-term data sets (≥8 consecutive years) are rare (also see Duarte et al., 1992; Dickson, 1995). We compiled long-term (8 to 18 consecutive years) monthly data for 12 rivers in five hydrologic regions of the U.S. (National Stream Water Quality Monitoring Networks, U.S. Geological Survey Digital Data Series DDS-37, 1996) and for 11 lakes from three long-term ecological research centers in North America (MOEE Dorset Environmental Science Center, ON, Canada; North Temperate Lakes Long Term Ecological Research site, Center for Limnology, University of Wisconsin–Madison, WI, U.S.A., http://www.limnology.wisc.edu) and in the U.K. (CEH Windermere Laboratory; Figure 13.1, Table 13.1).

The National Stream Water Quality Monitoring Networks of the U.S. Geological Survey provides physical, chemical, and biological data at numerous stations on all major river basins in the U.S. (NASQAN, National Stream Quality Accounting Networks) and on 63 smaller and less impacted watersheds (HBN, Hydrologic Benchmark Network). From these, we found 12 rivers where total nutrients had been sampled at least monthly for a minimum of 8 consecutive years, and found only 1 river where chlorophyll concentration was measured monthly for almost 15 years. Drastic reductions

TABLE 13.1

Site Characteristics and Sampling Periods Used in the Analysis

Lake	Country	Lat	Long	Lake Area (ha)	Max. Depth (m)	Mean Depth (m)	Median TN Conc. (mg l⁻¹)	Median TP Conc. (µg l⁻¹)	Strat	Years
Heney	Canada	45.08	79.06	21	5.8	3.3	0.26	6	P	1980–92
Red Chalk, Main	Canada	45.11	78.56	44	38.0	16.7	0.25	4	D	1976–92
Harp	Canada	45.23	79.07	71	37.5	13.3	0.30	6	D	1976–93
Dickie	Canada	45.09	79.05	94	12.0	5.0	0.33	9	D	1977–92
Crystal Bog	USA	46.00	89.60	1	2.5	1.7	0.49	14	P	1986–00
Crystal Lake	USA	46.00	89.62	37	20.4	10.4	0.16	5	D	1986–00
Sparkling	USA	46.00	89.70	64	20.0	10.9	0.22	6	D	1986–00
Allequash	USA	46.00	89.62	168	8.0	2.9	0.29	14	D	1986–00
Big Muskellunge	USA	46.00	89.62	396	21.3	7.5	0.35	7	D	1986–00
Trout	USA	46.00	89.67	1607	35.7	14.6	0.20	6	D	1986–00
Derwent Water	UK	54.57	3.15	535	22.0	5.5	NA	10	P	1991–00

River	State (USA)	Lat	Long	Drainage Area (km²)	Median TN Conc. (mg l⁻¹)	Median TP Conc. (µg l⁻¹)	Years	USGS Site Code
Popple	WI	45.75	88.45	224	NA	30	1971–82	R4-04063700
Maumee	OH	41.50	83.70	10,187	5.00	220	1973–81	R4-04193500
Cuyahoga	OH	41.38	81.62	1,138	3.80	390	1974–81	R4-04208000
Missouri	IA	42.48	96.40	512,671	0.54	60	1974–81	R10-06486000
Republican	KS	39.35	97.12	39,497	2.00	320	1973–81	R10-06856600
Kansas	KS	38.98	94.95	96,168	NA	320	1973–81	R10-06892350
Susquehanna	PA	40.25	76.88	38,785	1.40	70	1973–81	R2-01570500
James	VA	37.67	78.08	10,070	0.64	90	1974–81	R2-02035000
Russian	CA	38.50	122.92	2,153	0.95	160	1974–81	R18-11467000
Sacramento	CA	38.45	121.50	37,823	0.52	120	1973–81	R18-11447650
San Joaquin	CA	37.67	121.25	21,784	NA	240	1973–81	R18-11303500
Atchafalaya	LA	29.70	91.20	NA	NA	140	1980–95	R8-07381600

Note: Lake stratification (Strat) is D for dimictic and P for polymictic. Lakes are ordered by region and lake surface area. Rivers are ordered by latitude and drainage basin (first part of the Site Code). NA, not available.

in sampling occurred in the early 1980s due to program cutbacks and to the increasing cost of samples, so our analyses focus on river data from the 1970s. We found sufficient data from only one station on a relatively pristine watershed (Popple River, WI). All other stations are located on large watersheds that have experienced impact by humans in multiple ways. To test whether the position of a station along a river affects our results, we compared the scaling of variability in water temperature and flow (nutrient data were not available) at six stations along the Missouri River (Toston, MT, USGS site R10-06054500; Landusky, MT, site R10-06115200; Culbertson, MT, site R10-06185500; Sioux City, IA, site R10-06486000; St. Joseph, MO, site R10-06818000; Hermann, MO, site R10-06934500). The Missouri River runs for ~4000 km and has seven mainstream reservoirs along its course. We included stations upstream, between, and downstream from these dams, but did not include any station located right at a dam.

Relatively pristine lakes were selected for this analysis to avoid clear human-induced trends in the environmental variables of interest. The lakes are from three regions. The Ontario lakes are located on Precambrian Shield with a very thin layer of glacial till. The Wisconsin lakes are located on a thin layer of sedimentary rock overlaid by infertile glacial deposits and are strongly influenced by groundwater inputs. Derwent Water, U.K. is located in an infertile mountainous catchment. These lakes are all unproductive (oligotrophic to mesotrophic status; median TP < 15 µg/l, median chlorophyll concentration < 10 µg/l).

Because most data in lakes and rivers are still collected manually (i.e., water samples are collected and analyzed on site or sent for analysis), sampling frequency varies among sites and among variables, and is not perfectly regular. Water temperature and flow (calculated from water level measurements and calibrated monthly) are recorded automatically in rivers, but other variables (e.g., nutrients, algal biomass, lake temperature) are measured much less frequently. To standardize our analyses across sites and variables, we used the data collected on the closest sampling date from a regular date each month (e.g., 15th of the month), regardless of sampling frequency. This procedure allows us to obtain comparable monthly values for all variables.

All environmental variables considered in this study showed seasonal variation at some sites. This seasonal component of variability is well understood, and is removed from the analysis to allow detection of other scales of variability in the data (e.g., Steele, 1985; Pelletier, 1997; Talkner and Weber, 2000). We also followed this approach and removed the seasonal variability in each time series by subtracting from each observation the deviation from the overall average measured for that month over the whole sampling period (e.g., deviation from the mean value of all January samples; Weber and Talkner, 2001). This study does not deal with trends in variability caused by clear anthropogenic impacts (e.g., changes in nutrient concentrations during eutrophication or nutrient abatement programs). Linear trends in environmental data were only found for TP concentrations in Dickie, Harp and Red Chalk lakes and were removed. Our analysis focuses on the nonseasonal stationary components of environmental variability.

The temporal scaling of environmental variability was measured using spectral analysis on the seasonally detrended time series (Jenkins and Watts, 1968; Chatfield, 1989). We smoothed all spectra using a Parzen window with a truncation point (M) of approximately $2 \times n^{0.5}$ (n is the number of observations), as suggested by Chatfield (1989). Truncation points between 45 and 59 were used depending on the length of each time series. We then fitted a straight line to each $\log_{10}$-transformed power spectrum with simple linear regression analysis. The slope of the $\log_{10}$-transformed power spectrum is equivalent to the exponent $(-\gamma)$ in the $1/f$-noise relationship (Halley 1996).

We compared slopes of power spectra among environmental variables using Kruskal–Wallis tests with multiple comparisons, and we compared slopes between lakes of different regions using Mann–Whitney tests (Conover, 1999). All analyses were done using Statistica for Windows, version 5.

13.3 Results

The power spectra for water temperature overlapped greatly between the rivers (open symbols, dashed lines) and small lakes (closed symbols, solid line) in our data set (Figure 13.2A). At comparable latitudes, water temperature variability was generally higher in rivers than in lakes. In contrast, the smallest (drainage area < 2500 km²) and warmest (minimum temperature > 5°C; Russian, Sacramento, San Joaquin, Lower Atchafalaya) rivers showed lower temperature variability, and were more comparable to the northern lakes (dashed lines with open symbols overlapping with solid lines with closed symbols, Figure 13.2A). In lakes, the long-term temperature variability was slightly higher in small polymictic basins (Crystal Bog and Heney; solid triangles in Figure 13.2A) than in stratified basins.

In contrast, the spectra for nutrient concentration (TN, TP; Figure 13.2B, C) clearly showed higher variability in rivers, especially for TP concentrations (Figure 13.2C). The most pristine reference river (Popple; open inverted triangles in Figure 13.2C) shows an intermediate level of variability in TP concentrations, between the highly eutrophic rivers and the oligo-mesotrophic lakes. Unfortunately, TN was rarely measured in Popple River and could not be included in this analysis.

The seven indicators of environmental variability we compared showed slightly reddened noise at the subannual timescale, with median scaling exponents (or slopes of power spectra) between –0.57 and –0.15 (Figure 13.3). The power spectra were clearly nonlinear (Figure 13.2), but our analysis is heavily weighted on subannual timescales (see Section 13.2). The implications of nonlinearities in the spectra are discussed below.

All environmental variables scaled similarly in rivers, but not in lakes (Figure 13.3). In rivers, temperature, nutrient concentrations (TN, TP), suspended solids (SS), and flow scaled similarly, with median exponents between –0.36 to –0.15 (no significant difference among variables, Kruskal–Wallis test, $H = 4.2$, $P = 0.4$, $n = 46$; shaded bars in Figure 13.3). There were a few exceptions with much steeper slopes than the average (shown by the long whiskers on the box plots in Figure 13.3). The

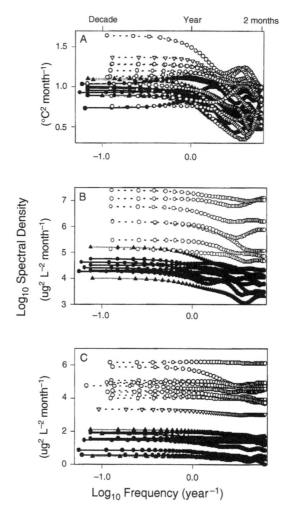

FIGURE 13.2 Power spectra for (A) water temperature, (B) TN concentration, and (C) TP concentration in rivers (open symbols and dashed lines) and lakes (closed symbols and solid lines). Open circles are rivers from the NASQAN USGS data set, open triangles are for the only relatively pristine river (Popple) from the HBN USGS data set, closed circles are for dimictic lakes, and closed triangles are for polymictic lakes. See Table 13.1 for site characteristics.

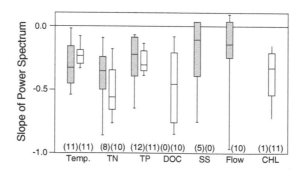

FIGURE 13.3 Box plots of the slopes of power spectra for seven environmental variables in rivers (gray bars) and lakes (open bars). Each box shows the 50th percentile (middle line), the 25th and 75th percentiles (full extent of box), and the 10th and 90th percentile (whiskers). The number of sites used for each variable is indicated in parentheses below the bars. Temp. is water temperature, TN is total nitrogen concentration, TP is total phosphorus concentration, DOC is dissolved organic carbon concentration, SS is suspended solid concentration, CHL is chlorophyll concentration (an indicator of algal biomass). Power spectra with slopes of zero represent white noise, slopes of -1 represent pink noise.

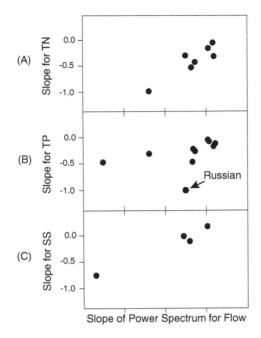

FIGURE 13.4 Relationships between the slopes of power spectra for (A) TN concentration, (B) TP concentration, and (C) SS concentration in rivers compared to the slopes of power spectra for river flow.

exceptions were found in the agricultural and highly regulated portion of the Missouri River (Sioux City, IA) for TN variability (slope = –1.01), in the intermittently flowing Russian River for TP variability (slope = –1.03), and in the San Joaquin River for SS (suspended solids) variability (slope = –0.76). The slopes describing the scaling of TN, TP, and SS were very closely related to the slopes for river flow (Figure 13.4). These correlations between scaling exponents, of course, do not imply causal relationships.

We tested whether the position of sampling stations along a river affected our estimates of scaling exponents by comparing scaling exponents for water temperature and for flow at six stations along the ~4000 km of the Missouri River. The scaling exponents for water temperature were very consistent among stations (–0.54 to –0.40) but more variable for flow (–0.77 to –0.29). There was no systematic trend in these scaling exponents as one moves downstream along the Missouri River (Figure 13.5).

In lakes, the slopes of power spectra also ranged from 0 to –1, but differed significantly among environmental variables (Kruskal–Wallis test, $H = 10.4$, $P = 0.03$, $n = 53$). The scaling exponents for water temperature and TP variability were very similar among lakes (i.e., narrow range of power spectrum slopes, open bars in Figure 13.3) and clearly within the range of scaling exponents found for these variables in rivers (Figure 13.3). Interestingly, the slopes of power spectra for TN were significantly steeper than those for water temperature and for TP (Kruskal–Wallis test with multiple comparisons, $P = 0.003$ and $P = 0.03$, respectively; n values shown in Figure 13.3), and the slopes for DOC were significantly steeper than those for water temperature (Kruskal–Wallis, $P = 0.02$). Moreover, the power spectrum slopes for TN were at least as variable among lakes as among rivers, despite the very small range of lake characteristics selected for this study (i.e., relatively unimpacted oligo-mesotrophic lakes from similar geological areas; Table 13.1). The slopes of power spectra for TN were lower in the Ontario lakes (range: –0.81 to –0.59) than in the Wisconsin lakes (range: –0.66 to –0.18; Mann–Whitney test, $H = 4.5$, $P = 0.03$, $n = 10$), suggesting a regional difference in the scaling of TN variability. TN was not measured in Derwent Water, but the scaling exponent for dissolved TN in this lake was well within the range of scaling exponents measured in the Ontario lakes (slope of power spectrum = –0.73). In contrast, the slopes of power spectra for DOC concentrations ranged from 0 to –1 in both Ontario and Wisconsin lakes (ranges: –0.94 to –0.12 and –0.77 to –0.05, respectively; Mann–Whitney test, $H = 0.7$, $P = 0.4$, $n = 10$), suggesting among lake rather than regional differences in the scaling of DOC variability. The lakes showed less consistency than the rivers in the scaling of different environmental variables.

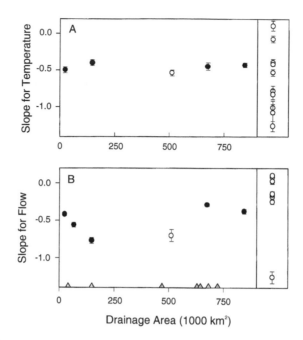

FIGURE 13.5 Slopes of power spectra for (A) water temperature and (B) flow at five to six stations along the Missouri River. The sampling sites are ordered in terms of their drainage area, and the approximate position of seven major dams along the Missouri River are shown with gray triangles in B. Slopes from Sioux City, IA, are used to represent the Missouri River in Figure 13.3, and are shown as open circles. The slopes of power spectra measured in other rivers are shown for comparison (open circles in right panel).

The variability in algal biomass (CHL, measured as chlorophyll concentration) scaled similarly to other environmental factors in lakes (Figure 13.3). We found a narrow range of power spectrum slopes for CHL in the six Wisconsin lakes (–0.35 to –0.12), but greater differences among the four Ontario lakes (–0.83 to –0.22). The slopes of power spectra were not statistically different between the two regions (Mann–Whitney test, $H = 3.7$, $P = 0.06$, $n = 10$), but these results should be tested further given the low P-value. The scaling of CHL in Derwent Water was most similar to the Ontario lakes (slope = –0.58).

13.4 Discussion

13.4.1 Environmental Variability Is Higher in Rivers Than Lakes

The magnitude of variability in water temperature and in nutrient concentration was generally higher in rivers than in lakes (Figure 13.2). Water temperature in surface-fed rivers is closely related to the air temperature over the drainage basin and, not surprisingly, is more variable than lake surface temperature (Linacre, 1969; Straškraba, 1980). Nutrient concentrations were also clearly more variable, at all timescales, in rivers than in lakes (Figure 13.2B, C). This is partly due to differences in mean nutrient concentrations between the rivers and lakes included in our data set (Table 13.1; Chételat and Pick, 2001). Cattaneo and Prairie (1995) also reported that for a given mean concentration, TP was two to three times more variable from week to week in rivers than in lakes. Our results extend their finding to a wider range of temporal scales, and suggest a similar difference in the variability in TN concentrations between rivers and lakes.

Several studies have proposed sampling guidelines to account for the intra-annual variability in physical, chemical, and biological variables (e.g., France and Peters, 1992; Cattaneo and Prairie, 1995). Our results suggest that long-term variability (inter-annual and longer timescales, up to the order of a decade) is at least as important as intra-annual variability. This emphasizes the importance of long-term data sets.

13.4.2 The Scaling of Environmental Variability in Rivers and Small Lakes

The scaling exponents for water temperature and river flow (Figure 13.3) are well within the range reported in other studies. Cyr and Cyr (submitted) compared the scaling of temperature variability in different types of ecosystems around the world, and reported exponents between 0 and –0.5 in rivers and between –0.2 to –0.6 in lakes. Pelletier and Turcotte (1997) reported a steeper exponent (–0.50) for the average spectrum of 636 monthly river discharge series. This discrepancy can easily be explained by their use of monthly mean discharge instead of individual flow measurements. Much steeper exponents (≤2) have been reported for these variables (see review in Cuddington and Yodzis, 1999), but these analyses are based on measurements collected at different timescales (e.g., every second for several hours, annual means) and are not strictly comparable (see discussion below about the shape of environmental power spectra).

The five environmental variables we compared in rivers showed a surprisingly small range of scaling exponents (median slopes from –0.36 to –0.15; Figure 13.3), especially given the wide diversity of rivers included in our analysis (Table 13.1, Figure 13.1). The scaling exponents for TN, TP, and SS were closely related to the exponents for river flow (Figure 13.4). This is not surprising as river discharge is one of the most important parameters that affects the concentration of nutrients and of SS in rivers (e.g., Correll et al., 1999; Arheimer and Lidén, 2000; McKee et al., 2000). Other important factors include the potential for nutrient release and erosion from the catchment (land-use pattern, air temperature; e.g., Howarth et al., 1996; Ekholm et al., 2000; Chapman et al., 2001) and the transformation and immobilization of nutrients and particulates in the watershed (e.g., denitrification, sedimentation; Russell et al., 1998; McKee et al., 2000). The similarities we observed in the scaling of variability among physical and chemical variables in rivers (Figure 13.3 and Figure 13.4) do not imply direct causation in individual watersheds, but are generally consistent with the important linkage between the variables.

In contrast to rivers, the scaling of variability in lakes differed among environmental variables (Figure 13.3). The scaling of variability in surface temperature and TP concentrations was very consistent among the three lake regions. TP measurements are notoriously variable in fresh water (e.g., Håkanson and Peters, 1995; Cattaneo and Prairie, 1995) and may generally show whiter noise patterns compared to other chemical variables. The small range of scaling exponents for TP may also reflect the narrow range of lakes included in this analysis (oligo-mesotrophic lakes) and the tight biological control of TP in these phosphorus-limited lakes.

The scaling exponents for TN and DOC were on average steeper than the exponents for temperature and TP, and showed a broader range of values among lakes (Figure 13.3). TN variability was clearly whiter in Wisconsin lakes than in Ontario lakes and, perhaps, in Derwent Water. This is likely due to large groundwater inputs into the Wisconsin lakes, which stabilize their chemical composition over long periods of time (Webster et al., 2000). The scaling exponents for DOC also differed greatly among lakes, but there were no clear differences between the Wisconsin and the Ontario lakes. DOC concentrations in lakes depend on inputs from the watershed, but are also greatly influenced by photolytic, chemical, and microbial degradation (Yan et al., 1996; Dillon and Molot, 1997; Molot and Dillon, 1997). Our results suggest that the scaling of DOC variability varies among lakes within a region, and is largely determined by within-lake processes.

The scaling exponents for algal biomass (CHL) ranged widely among lakes (–0.83 to –0.12) and did not differ significantly between regions. George et al. (2000) also showed very little spatial coherence in seasonal algal biomass and zooplankton density among lakes of the English Lake District. Rusak et al. (2002) compared the variability in annual zooplankton biomass between lakes from the Dorset Environmental Science Centre in Ontario, the North Temperate Lake LTER site in Wisconsin, and the Experimental Lakes Area in northwestern Ontario, and found more variability in annual zooplankton biomass among lakes than among regions. Soranno et al. (1999) and Webster et al. (2000) also report low synchrony in annual chlorophyll *a* concentration among lakes in six different regions. These studies suggest that the phytoplankton and zooplankton show little spatial synchrony among lakes. Their temporal variability also scales differently among lakes.

The variability in physical, chemical, and biological characteristics of rivers and small lakes is slightly reddened, mostly at subannual timescales. In rivers, the scaling exponents for temperature and water flow overlap greatly with those of other potentially important environmental variables (TN, TP, SS). Similar results are found on land, where there are striking similarities in the average power spectra for air temperature (slope = –0.43), rainfall (slope = –0.52), and tree ring chronologies (slope = –0.54; Pelletier and Turcotte, 1997) at the 10 to 100 years timescale. In contrast, different environmental variables scale differently in lakes, with more negative scaling exponents, on average, for TN and DOC than for temperature and TP. The implications of such differences in the scaling of environmental variables on population dynamics and on communities remain to be explored.

13.4.3   Changes in the Scaling of Environmental Variability at Different Timescales

All power spectra showed a plateau in spectral density at interannual and longer timescales (Figure 13.2). Although some of our river time series are short (8 consecutive years of monthly data), the plateaus we observe are unlikely to be simply artifacts of the analysis. Such plateaus are expected on theoretical grounds for water temperature (Frankignoul and Hasselmann, 1977; Steele and Henderson, 1994), and have been observed in other rivers and lakes with longer time series (up to 75 consecutive years of monthly temperature data; Cyr and Cyr, submitted). These plateaus suggest that environmental variability changes color at different temporal scales, a result that is consistent with observations from other studies in the oceans (e.g., Wunsch, 1981) and on land (e.g., Pelletier, 1997; Weber and Talkner, 2001; Inchausti and Halley, 2002). Models should be based on scaling exponents of environmental variability ($-\gamma$) that are measured over timescales relevant to the processes of interest.

13.4.4   Ecological Implications

Ecologists are actively debating the importance of the temporal scaling of environmental variability on population dynamics (e.g., Ripa et al., 1998; Petchey, 2000; Laakso et al., 2001), probabilities of population extinction (e.g., Cuddington and Yodzis, 1999; Morales, 1999; Ripa and Lundberg, 2000), dispersal and metapopulation dynamics (Travis, 2001), and on the structure of communities (Litchman, 1998; Litchman and Klausmeier, 2001). Traditionally, theoretical models have assumed equal magnitudes of environmental variability at all timescales (i.e., white noise). More recently, several studies have shown that including a different structure of environmental variability (e.g., red, brown, or black noise) in different parts of these models (e.g., on growth rate or carrying capacity) can alter the outcomes in significant ways (e.g., Morales, 1999; Ripa and Lundberg, 2000; Heino et al., 2000).

The dynamics of natural populations is clearly reddened (Ariño and Pimm, 1995; Inchausti and Halley, 2002) and several studies have suggested that reddened environmental variability could lead to such population dynamics (e.g., Kaitala et al., 1997). The impact of environmental variability on biological processes is complex and likely difficult to disentangle (Steele and Henderson, 1994; Laakso et al., 2001). A few experimental studies have shown that the effect of environmental variability is most important around the nonlinear portions of the biological response curves (i.e., when environmental conditions fluctuate between limiting and saturating or inhibiting levels; Litchman, 2000; Petchey, 2000). Laakso et al. (2001) reached similar conclusions, but also found that biological responses to slightly reddened environmental noise are robust to differences in the timing of the environmental signal and in the shapes of the biological response curves. This suggests that the impacts of environmental noise on organisms may be easiest to understand in rivers and small lakes, where we find slightly reddened environmental variability. We found, however, that the scaling of environmental variability was not constant at all timescales, and the theoretical implications of these changes remain to be explored. A better understanding of the structure of temporal (and spatial) variability in nature is essential to develop more realistic ecological models.

13.5   Summary

Recent models and experiments show that the temporal scaling (or noise color) of environmental variability can affect the dynamics and persistence of populations and the structure of communities.

These studies explore the effect of different noise colors, without much evidence regarding the actual scaling of environmental variability in nature. It is now well established that temperature variability is relatively constant over ecological timescales (1 to 100 years) on land, but increases with longer timescales in aquatic systems, with progressively larger increases from rivers to lakes to the Great Lakes to the oceans. Ambient temperature, however, is only one factor that regulates the growth, reproduction, and survival of organisms in nature. In this study, we use spectral analysis to compare the temporal scaling of variability in seven ecologically important environmental factors — temperature, nutrients (TN, TP), DOC, SS, river flow, algal biomass — in 12 rivers and 11 lakes. All variables showed reddened noise, mostly at the subannual timescales. In rivers, the scaling of variability in flow was a good indicator of the scaling of variability in nutrients (TN, TP) and SS. In lakes, TP showed relatively more short-term variability than TN (i.e., whiter signal). The scaling of variability in TN differed significantly between the groundwater-fed lakes in Wisconsin compared to the surface-fed lakes in Ontario or the U.K., whereas the scaling of variability in DOC varied among lakes, not regions. The scaling of environmental variability is more consistent among variables in rivers than in lakes. The ecological implications of these results remain to be explored.

Acknowledgments

This work would not have been possible without the extensive long-term data collected by the U.S. Geological Survey (http://water.usgs.gov) and the North Temperate Lakes Long-Term Ecological Research program supported by the National Science Foundation and coordinated by the Center for Limnology, University of Wisconsin–Madison, WI, U.S.A. (http://www.limnology.wisc.edu). We thank P. Inchausti and an anonymous reviewer for their useful comments. This research was funded by a Natural Sciences and Engineering Research Council of Canada operating grant to H.C.

References

Arheimer, B. and R. Lidén. 2000. Nitrogen and phosphorus concentrations from agricultural catchments — influence of spatial and temporal variables. *J. Hydrol.* 227:140–159.

Ariño, A. and S.L. Pimm. 1995. On the nature of population extremes. *Evol. Ecol.* 9:429–443.

Cattaneo, A. and Y.T. Prairie. 1995. Temporal variability in the chemical characteristics along the Riviere de l'Achigan: How many samples are necessary to describe stream chemistry? *Can. J. Fish. Aquat. Sci.* 52:828–835.

Chapman, P.J., A.C. Edwards, and M.S. Cresser. 2001. The nitrogen composition of streams in upland Scotland: some regional and seasonal differences. *Sci. Total Environ.* 265:65–83.

Chatfield, C. 1989. *The Analysis of Time Series.* Chapman & Hall, London.

Chételat, J. and F.R. Pick. 2001. Temporal variability of water chemistry in flowing waters of the northeastern United States: does river size matter? *J. North Am. Benthol. Soc.* 20:331–346.

Conover, W.J. 1999. *Practical Nonparametric Statistics.* 3rd ed. John Wiley & Sons, New York.

Correll, D.L., T.E. Jordan, and D.E. Weller. 1999. Effects of precipitation and air temperature on phosphorus fluxes from Rhode River watersheds. *J. Environ. Qual.* 28:144–154.

Cossins, A.R. and K. Bowler. 1987. *Temperature Biology of Animals.* Chapman & Hall, London.

Cuddington, K.M. and P. Yodzis. 1999. Black noise and population persistence. *Proc. R. Soc. Lond. B* 266:969–973.

Cyr, H. and I. Cyr. Temporal scaling of temperature variability from land to oceans. *Evol. Ecol. Res.,* in press.

Dickson, R.R. 1995. The natural history of time series, in Powell, T.M. and J.H. Steele, Eds., *Ecological Time Series.* Chapman & Hall, New York, 70–98.

Dillon, P.J. and L.A. Molot. 1997. Effect of landscape form on export of dissolved organic carbon, iron, and phosphorus from forested stream catchments. *Water Resour. Res.* 33:2591–2600.

Duarte, C.M., J. Cebrían, and N. Marba. 1992. Uncertainty of detecting sea change. *Nature* 356:190.

Ekholm, P. et al. 2000. Relationship between catchment characteristics and nutrient concentrations in an agricultural river system. *Water Res.* 34:3709–3716.

France, R.L. and R.H. Peters. 1992. Temporal variance function for total phosphorus concentration. *Can. J. Fish. Aquat. Sci.* 49:975–977.

Frankignoul, C. and K. Hasselmann. 1977. Stochastic climate models, Part II: Application to sea-surface temperature anomalies and thermocline variability. *Tellus* 29:289–305.

George, D.G., J.F. Talling, and E. Rigg. 2000. Factors influencing the temporal coherence of five lakes in the English Lake District. *Freshw. Biol.* 43:449–461.

Håkanson, L. and R.H. Peters. 1995. *Predictive Limnology — Methods For Predictive Modelling.* SPB Academic Publishers, Amsterdam.

Halley, J.M. 1996. Ecology, evolution and 1/f-noise. *TREE* 11:33–37.

Havlin, S. et al. 1999. Scaling in nature: from DNA through heartbeats to weather. *Physica A* 273: 46–69.

Heino, M., J. Ripa, and V. Kaitala. 2000. Extinction risk under coloured environmental noise. *Ecography* 23:177–184.

Howarth, R.W. et al. 1996. Regional nitrogen budgets and riverine N and P fluxes for the drainages to the North Atlantic Ocean: natural and human influences. *Biogeochemistry* 35:75–139.

Hubert, P. 2001. Multifractals as a tool to overcome scale problems in hydrology. *Hydrol. Sci. J.* 46:897–905.

Inchausti, P. and J. Halley. 2001. Investigating long-term ecological variability using the Global Population Dynamics Database. *Science* 293:655–657.

Inchausti, P. and J. Halley. 2002. The long-term temporal variability and spectral color of animal populations. *Evol. Ecol. Res.* 4:1033–1048.

Jenkins, G.M. and D.G. Watts. 1968. *Spectral Analysis and Its Applications.* Holden-Day, San Francisco.

Kaitala, V., J. Ylikarjula, E. Ranta, and P. Lundberg. 1997. Population dynamics and the colour of environmental noise. *Proc. R. Soc. Lond. B* 264:943–948.

Koscielny-Bunde, E., H.E. Roman, A. Bunde, S. Havlin, and H.J. Schellnhuber. 1998. Long-range power-law correlations in local daily temperature fluctuations. *Philos. Mag. B* 77:1331–1340.

Laakso, J., V. Kaitala, and E. Ranta. 2001. How does environmental variation translate into biological processes? *Oikos* 92:119–122.

Linacre, E.T. 1969. Empirical relationships involving the global radiation intensity and ambient temperature at various latitudes and altitudes. *Arch. Meteorol. Geophs. Bioklimatol. Ser. B* 17:1–20.

Litchman, E. 1998. Population and community responses of phytoplankton to fluctuating light. *Oecologia* 117:247–257.

Litchman, E. and C.A. Klausmeier. 2001. Competition of phytoplankton under fluctuating light. *Am. Nat.* 157:170–187.

Mandelbrot, B.B. and J.R. Wallis. 1969. Some long-run properties of geophysical records. *Water Resour. Res.* 5:321–340.

McKee, L., B. Eyre, and S. Hossain. 2000. Intra- and interannual export of nitrogen and phosphorus in the subtropical Richmond River catchment, Australia. *Hydrol. Proc.* 14:1787–1809.

Molot, L.A. and P.J. Dillon. 1997. Photolytic regulation of dissolved organic carbon in northern lakes. *Global Biogeochem. Cycles* 11:357–365.

Morales, J.M. 1999. Viability in a pink environment: why "white noise" models can be dangerous. *Ecol. Lett.* 2:228–232.

Pelletier, J.D. 1997. Analysis and modeling of the natural variability of climate. *J. Climate* 10:1331–1342.

Pelletier, J.D. and D.L. Turcotte. 1997. Long-range persistence in climatological and hydrological time series: analysis, modeling and application to drought hazard assessment. *J. Hydrol.* 203:198–208.

Petchey, O.L. 2000. Environmental colour affects aspects of single-species population dynamics. *Proc. R. Soc. Lond. B* 267:747–754.

Platt, T. and K.L. Denman. 1975. Spectral analysis in ecology. *Annu. Rev. Ecol. Syst.* 6:189–210.

Powell, T.M. 1989. Physical and biological scales of variability in lakes, estuaries, and the coastal ocean, in Roughgarden, J., R.M. May, and S.A. Levin, Eds., *Perspectives in Ecological Theory.* Princeton University Press, Princeton, NJ, 157–180.

Reynolds, C.S. and P.S. Davies. 2001. Sources and bioavailability of phosphorus fractions in freshwaters: a British perspective. *Biol. Rev.* 76:27–64.

Ripa, J. and P. Lundberg. 2000. The route to extinction in variable environments. *Oikos* 90:89–96.

Ripa, J., P. Lundberg, and V. Kaitala. 1998. A general theory of environmental noise in ecological food webs. *Am. Nat.* 151:256–263.

Rodriguez-Iturbe, I. and A. Rinaldo. 1997. *Fractal River Basins: Chance and Self-Organization*. Cambridge University Press, Cambridge.

Rusak, J.A. et al. 2002. Temporal, spatial, and taxonomic patterns of crustacean zooplankton variability in unmanipulated north-temperate lakes. *Limnol. Oceanogr.* 47:613–625.

Russell, M.A., D.E. Walling, B.W. Webb, and R. Bearne. 1998. The composition of nutrient fluxes from contrasting UK river basins. *Hydrol. Proc.* 12:1461–1482.

Scheuring, I. and O.E. Zeöld. 2001. Data estimation and the colour of time series. *J. Theor. Biol.* 213:427–434.

Soranno, P.A. et al. 1999. Spatial variation among lakes within landscapes: ecological organization along lake chains. *Ecosystems* 2:395–410.

Steele, J.H. 1985. A comparison of terrestrial and marine ecological systems. *Nature* 313:355–358.

Steele, J.H. and E.W. Henderson. 1994. Coupling between physical and biological scales. *Philos. Trans. R. Soc. Lond. B* 343:5–9.

Straškraba, M. 1980. The effects of physical variables on freshwater production: analyses based on models, in E.D. Le Cren and K. H. Lowe-McConnell, Eds., *The Functioning of Freshwater Ecosystems*. Cambridge University Press, Cambridge, 13–84.

Talkner, P. and R.O. Weber. 2000. Power spectrum and detrended fluctuation analysis: application to daily temperatures. *Phys. Rev. E* 62:150–160.

Travis, J.M.J. 2001. The color of noise and the evolution of dispersal. *Ecol. Res.* 16:157–163.

Vasseur, D.A. and P. Yodzis. The color of environmental noise. *Ecology*, in press.

Weber, R.O. 1993. Influence of different daily mean formulas on monthly and annual averages of temperature. *Theor. Appl. Climatol.* 47:205–213.

Weber, R.O. and P. Talkner. 2001. Spectra and correlations of climate data from days to decades. *J. Geophys. Res.* 106:20131–20144.

Webster, K.E. et al. 2000. Structuring features of lake districts: landscape controls on lake chemical responses to drought. *Freshwater Biol.* 43:499–515.

Wunsch, C. 1981. Low-frequency variability of the sea, in B.A. Warren and C. Wunsch, Eds., *Evolution of Physical Oceanography*. MIT Press, Cambridge, MA, 342–376.

Yan, N.D., W. Keller, N.M. Scully, D.R.S. Lean, and P.J. Dillon. 1996. Increased UV-B penetration in a lake owing to drought-induced acidification. *Nature* 381:141–143.

Zang, X. and C. Wunsch. 2001. Spectral description of low-frequency oceanic variability. *J. Phys. Oceanogr.* 31:3073–3095.

14

Biogeochemical Variability at the Sea Surface: How It Is Linked to Process Response Times

Amala Mahadevan and Janet W. Campbell

CONTENTS

14.1 Introduction 215
14.2 Tracer Distributions and Transport 215
14.3 Quantifying Variability 218
     14.3.1 Analysis of Satellite Data 220
     14.3.2 Analysis of Model Fields 220
14.4 Modeling and Results 222
     14.4.1 How Patchiness Relates to Response Time τ 222
     14.4.2 Dependence of the Vertical Distribution of a Tracer on τ 224
     14.4.3 Resolution Requirements for Tracers with Different τ 225
14.5 Discussion 225
14.6 Conclusions 226
Acknowledgments 226
References 226

14.1 Introduction

The sea surface distributions of many biogeochemical tracers are highly correlated in space and time on meso (10 to 100 km) and smaller scales, but their scales of variability differ. Some tracers like sea surface chlorophyll (Chl) are patchier or finer scaled than others, such as sea surface temperature (SST) (Figure 14.1). This work aims to characterize the spatial distributions of various tracers in terms of a variance-based measure of their patchiness. Using a scaling argument and a numerical model, we relate the patchiness of a tracer distribution to the characteristic response time τ of the tracer to processes that alter its concentration in the upper ocean. This enables us to relate the distributions of different tracers in the upper ocean and provide an estimate for the relative size of the grid spacing needed to observe or model different tracers. We also suggest a scaling relationship to infer the mean rate of upwelling in a region from the mean vertical concentration profile of a tracer and *a priori* knowledge of the characteristic response time τ of the tracer.

14.2 Tracer Distributions and Transport

Many biogeochemical tracers in the ocean, like dissolved inorganic carbon (DIC), oxygen, nitrate, phosphate, dissolved organic nitrogen and carbon (DON and DOC), as well as physical variables like temperature and salinity, change rapidly with depth in the upper 500 or so meters of the ocean (Figure 14.2). Tracer

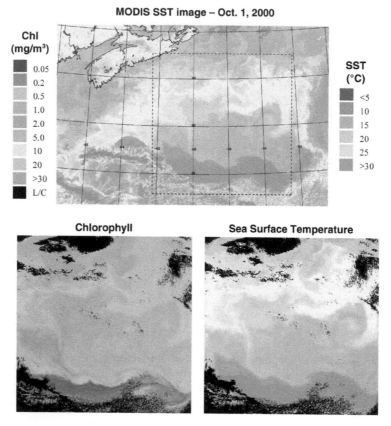

FIGURE 14.1 (Color figure follows p. 332.) Simultaneous satellite images of Chl (lower left) and SST (upper center and lower right) in the western Atlantic Ocean south of Nova Scotia acquired by the Moderate-resolution Imaging Spectro-radiometer (MODIS) on October 1, 2000. The region shown in the two lower panels is indicated by a dotted outline in the upper panel and is approximately 512 × 512 km². The resolution of the data is approximately 1 × 1 km². The color scale is logarithmic for Chl, but linear for SST. Black areas in the two lower panels are clouds.

concentration gradients in the thermocline are orders of magnitude larger in the vertical than in the horizontal. This is largely because these tracers are modified in the upper ocean by various processes (such as phytoplankton production, air–sea gas exchange, heat fluxes, and evaporation–precipitation), but rates of vertical exchange that convey these changes to the interior are extremely small. The mean vertical tracer concentration profiles result from a balance between the rate at which the tracer is modified by nonconservative processes that act primarily in the upper ocean, and the rate at which it is transported or mixed vertically. A small rate of vertical exchange (as compared to rate of modification) results in a steep profile, while a higher rate results in a more homogeneous vertical distribution.

Vertical transport across the thermocline is vital to the exchange of properties between the atmospher-ically forced surface ocean and its interior. Within the thermocline, rates of small-scale diapycnal mixing are far too weak to account for the observed rates of exchange of properties. Vertical transport is thought to occur on meso- and smaller-scales largely via the advection associated with the strain and divergence of the horizontal flow field. Sloping isopycnal surfaces, which may be termed fronts, act as pathways for vertical motion (Figure 14.3). Thus, vertical velocities are pronounced at fronts (Pollard and Regier, 1992; Voorhis and Bruce, 1982), or along the edges of meanders and eddies (Levy et al., 2001). Upwelling velocities can be of the order of tens of meters per day, but away from boundaries and topography, it is generally episodic and occurs on scales smaller than the internal Rossby radius of deformation that are associated with the sub-mesoscale dynamics (1 to 10 km in the horizontal). Convective and shear-induced mixing facilitates communication between the upper thermocline, lower mixed layer, and the surface ocean, exposing to the surface substances transported isopycnally up to the mixed-layer base.

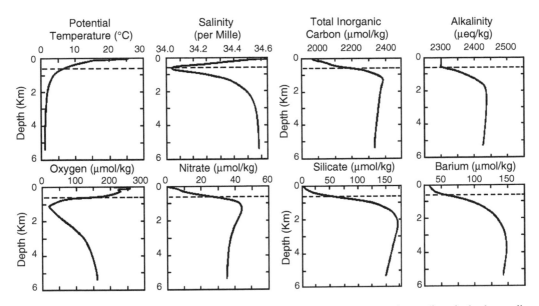

FIGURE 14.2 Typical vertical profiles from the Pacific Ocean showing strong concentration gradients in the thermocline. (Reproduced from GEOSECS data displayed on the Web.)

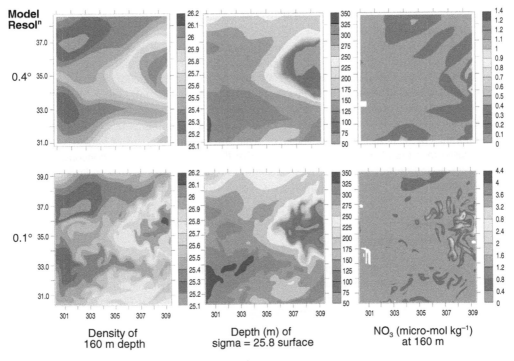

FIGURE 14.3 (Color figure follows p. 332.) Model potential density and nitrate fields from a $10° \times 10°$ region representative of the oligotrophic subtropics near Bermuda depict that nitrate is fluxed into the euphotic zone where isopycnals outcrop. The fields are shown at two different model resolutions to demonstrate the importance of resolving the small scales (less than the internal Rossby radius) in capturing vertical advective transport. The dimensions of the domain are in degrees latitude and longitude.

Because the tracer concentration gradients are typically very large in the vertical, the upwelling of water from the thermocline to the mixed layer introduces a different concentration at the surface. Therefore, in many regions of the pelagic ocean, variations in the surface distribution of tracers are induced largely by upwelling. While upwelling introduces tracer anomalies in the surface, upper ocean processes typically tend to annihilate the anomalies and restore the concentration to its background surface value. In the oligotrophic ocean, for example, phytoplankton nutrients such as nitrate and phosphate are depleted in the surface ocean. Frontal upwelling brings nutrients to the surface layer at certain places, simultaneously introducing an anomalous signature in temperature, DIC, O_2, etc. The increased nutrient supply results in new production of phytoplankton, reflected in increased Chl values. As the upwelled nutrients are consumed, the surface is restored to its oligotrophic state. Biological production also consumes DIC and liberates O_2, tending to annihilate the DIC and O_2 signals introduced by upwelling. In addition, air–sea gas exchange tends to equilibrate the surface concentrations of dissolved O_2 and CO_2 with the atmosphere. The timescale of equilibration depends on the wind speed and surface conditions (affecting the so-called piston velocity) and the mixed-layer depth. Although gas exchange equilibrates the mixed layer O_2 with the atmosphere typically in a month's time, DIC takes almost a year to attain equilibrium because of the reactions buffering CO_2 in seawater. The remineralization of sinking organic matter at depth maintains the large-scale vertical gradients, but it is mainly upwelling and the upper ocean processes that affect the surface distribution characteristics of the various biogeochemical substances. Surface anomalies are consequently advected at the surface and bear the signature of stirring by surface eddies and currents (Figure 14.1).

As many biogeochemical tracers possess a strong vertical concentration gradient, are affected by the same physical dynamics and often even the same upper-ocean nonconservative processes, it is no surprise that their sea surface distributions are highly correlated, particularly on spatial scales of 10 to 100 km and temporal scales of weeks. It thus seems plausible that by relating various tracers and processes, one might be able to diagnose the distribution of a biogeochemical tracer such as DIC or O_2 from remotely sensed variables like SST and Chl. This work takes a first step in this direction. We begin by analyzing various tracer distributions and finding a diagnostic by which to differentiate them. We then explain the observed differences in their distributions.

A significant and quantifiable difference among various tracers is their spatial heterogeneity. Variables like Chl are finer scaled or patchier than others, such as SST. In model simulations, we find that the variability of O_2 is finer scaled than that of DIC. We believe that such differences arise primarily because of the varied response of the tracers to processes that alter their concentration at the surface. In the pelagic ocean, upwelling occurs on fine scales (~1 km). In the meso- and submeso-scale range (1 to 100 km), tracers that are modified or restored rapidly at the surface tend to be finer scaled or patchier than those that equilibrate slowly. In this work, we analyze tracer distributions in terms of variance and suggest a variance-based measure of patchiness. We characterize the tracer's response to nonconservative upper ocean processes in terms of a characteristic (e-folding) response time τ, and show how the patchiness of the distribution of the tracer in the upper ocean varies with τ (Mahadevan and Campbell, 2002).

14.3 Quantifying Variability

We use a variance-based empirical approach to characterize the spatial heterogeneity of a tracer at the sea surface and demonstrate the technique using data from remote sensing and model simulations. In contrast to methods such as spectral analysis (Denman and Platt, 1975; Gower et al., 1980), semi-variogram analysis (Yoder et al., 1987, 1993; Glover et al., 2000; Deschamps et al., 1981) and auto-correlation analysis (Campbell and Esaias, 1985), this method does not invoke the assumption of isotropy in the two-dimensional spatial distributions. We calculate the variance associated with a characteristic length scale L, as the average variance contained within regions of area L^2. Beginning with the largest scale L_1, which is the size of the whole domain, we first compute the total variance within this domain, V_1. Next, we partition the domain into four quadrants, and calculate the average variance V_2, associated with length scale $L_1/2$. We continue to partition each subdomain into quadrants with successively smaller dimensions: $L_1/4$, $L_1/8$, ... and compute the average variance at each scale. The average variance within

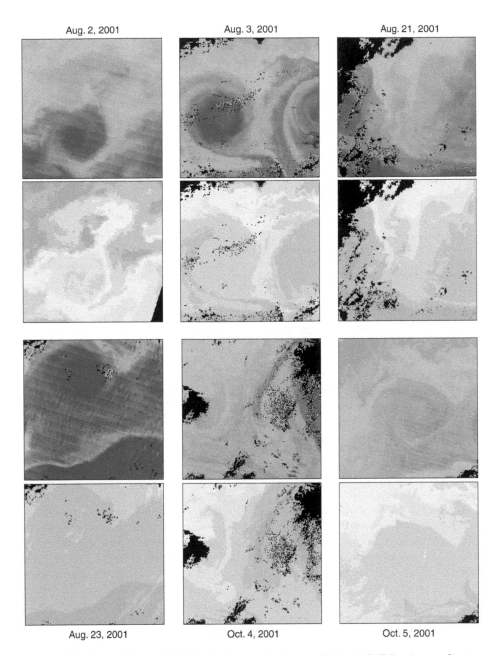

FIGURE 14.4 (Color figure follows p. 332.) Simultaneous satellite images of chlorophyll (Chl) and sea-surface temperature (SST) for domains of size 256×256 km^2 and resolution 1×1 km^2 in the western Atlantic continental shelf and slope region acquired by MODIS on six dates between August 2 and October 5, 2001. In each case, the Chl image is above the SST image. Color scales are the same as those shown in Figure 14.1.

the length scale L, $V(L)$, can be normalized by the total variance within the domain V_1, to facilitate the comparison of different variables like SST and Chl.

This simple empirical technique relates to the more familiar method involving the variance (or power) spectrum. A variance spectrum *partitions* the total variance into components at wavenumbers $\nu = 1/L$, and is normally expressed as a power law function of the wavenumber. Our method expresses the average variance within areas of size L^2, and thus all variance at length scales *less than or equal to L*, or wavenumbers greater than or equal to ν. Since the variance spectrum is the incremental change in variance V for each wavenumber, it would be analogous to the derivative of V with respect to ν.

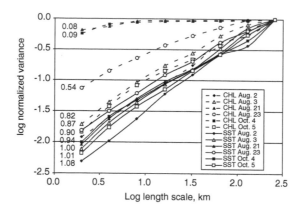

FIGURE 14.5 Variance-scale (*V–L*) relationships in SST (solid lines) and Chl (dashed lines) for the MODIS satellite images shown in Figure 14.4. The slopes *p* are indicated to the left of each curve. The difference in the Chl and SST distributions can be diagnosed from the slopes of these lines. The shallower slope of the Chl (as compared to SST) curves indicates that Chl is patchier than SST.

14.3.1 Analysis of Satellite Data

We have applied this method to satellite-derived distributions of Chl and SST from the Moderate-resolution Imaging Spectroradiometer (MODIS) (Esaias et al., 1998), and to Chl from the Sea-Viewing Wide Field-of-view Sensor (SeaWiFS) (McClain et al., 1998) matched with SST from the Advanced Very High Resolution Radiometer (AVHRR) (McClain et al., 1985). The satellite data are approximately at 1×1 km^2 spatial resolution and are analyzed for areas approximately 256×256 km^2 in size. In the regions analyzed, upwelling that brings colder waters to the surface also brings nutrients that fuel biological production. Hence, high Chl is generally associated with lower temperatures. A sampling of the several SST and Chl views analyzed is shown in Figure 14.4. For both SST and Chl, the plot of *V(L)* vs. *L* on log–log axes (Figure 14.5) tends to be more or less linear, suggesting a power law relationship of the form:

$$V \sim L^p \tag{14.1}$$

where $p > 0$. Here we have normalized the variance by the total variance V_1, since we are more concerned with the exponent *p*, which reflects the degree of fine-scale structure or patchiness in the field.

If most of the variance in a tracer field exists at small scales, then the exponent *p* will be relatively small. That is, *V(L)* increases with *L* much more rapidly at small *L* than at large *L*, as compared to the case when the spatial variability is at large scales. In each of the cases analyzed, the slope of the variance curve for Chl is clearly smaller than that for SST, demonstrating that Chl is finer scaled or patchier than SST over this range of scales (2 to 256 km). When *p* is constant over a range of scales, we infer that there is one dominant process operating over that range of scales. While the actual variance in a variable may vary by as much as two orders of magnitude from one time or place to another, the exponent *p* varies much less. The exponent *p* does not usually exceed 1, indicating that the increase in *V* with *L* slows with increasing *L*, i.e., the curvature of the *V(L)* curve in linear space is negative.

The relationship between these *V–L* curves and the variance spectrum is seen by taking the derivative of Equation 14.1 with respect to the wavenumber ν, where ν = 1/L, as follows:

$$dV/d\nu \sim d(\nu^{-p})/d\nu \sim -p\nu^{-p-1} \tag{14.2}$$

The *p* exponents for Chl and SST in Figure 14.5 thus translate into variance spectrum exponents between −1.1 and −2.1, which are similar to values reported by others.

14.3.2 Analysis of Model Fields

We next apply this method to analyze model-generated surface fields of SST, new production, DIC, dissolved oxygen, and two idealized tracers that resemble H$_2$O$_2$ and DOC (Figure 14.6). The model is

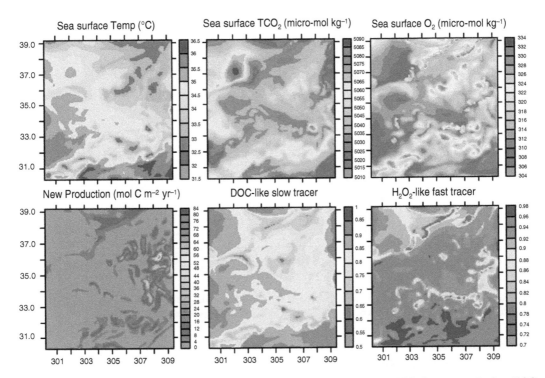

FIGURE 14.6 (Color figure follows p. 332.) Sea surface distribution of temperature, DIC, O₂, new production, DOC, and H₂O₂ from a limited region model of a $10° \times 10°$ region in the subtropical Atlantic near Bermuda (Mahadevan and Archer, 2000). The model was driven at the open boundaries by dynamic fields from a global circulation model (Semtner and Chervin, 1992). The picture depicts a typical snapshot view of the ocean surface in the autumn, modeled at $0.1°$ resolution. The different tracers exhibit different scales of variability.

a fully three-dimensional nonhydrostatic model that is deployed at $0.1°$ resolution within a $10° \times 10°$ region of the subtropical gyre near Bermuda (Mahadevan and Archer, 2000). It is forced at the open boundaries with time-dependent flow fields from the coarser resolution global circulation model of Semtner and Chervin (1992). The biogeochemical tracers in the model are initialized to resemble data from the Bermuda Atlantic Time Series site, and salinity and temperature are restored to monthly climatological values from Levitus (1982). The upwelling of nitrate fuels new production, which occurs within the euphotic layer with the uptake of the upwelled nitrate and DIC and the release of oxygen. The characteristic (e-folding) timescale for biological new production is taken to be 3 days. Oxygen and DIC are also altered by air–sea gas exchange that occurs with a constant piston velocity of 3 m per day. But CO_2 buffering slows the air–sea equilibration of DIC by a factor of 10 (the Revelle factor) as compared to O_2. Remineralization of the organic matter at depth restores the dissolved nitrate and carbon in the ocean, but the surface distributions of these tracers are clearly affected by the net biological production near the surface, rather than the remineralization at depth. The two idealized tracers that resemble H_2O_2 and DOC are initialized with an exponential profile that ranges from 1 at the surface to 0 at depth over an e-folding depth of 100 m. They are linearly restored to this steady-state profile, but one of these tracers (H_2O_2) is fast acting and has an e-folding time of 3 days, while the other (DOC) is restored with an e-folding time of 60 days. From viewing the surface model fields (Figure 14.6), it is evident that the distributions of new production, oxygen, and the fast-acting H_2O_2-like tracer are finer scaled than those of SST, DIC, and the slower-acting DOC.

Analysis of the surface fields with the variance method (Figure 14.7) confirms that the H_2O_2-like tracer, new production, and oxygen are patchier (have smaller p) than the DOC-like tracer, SST, and DIC. The patchier tracers are those that equilibrate more quickly at the surface, or respond more rapidly to processes that alter them. We therefore seek to characterize tracer patchiness in terms of the characteristic response time of the tracer to processes that alter it.

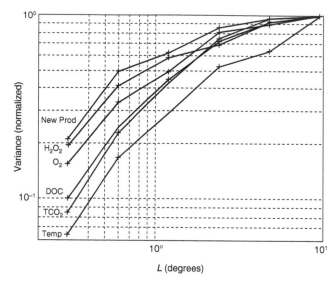

FIGURE 14.7 Log–log plot of the average variance V vs. L for the model fields shown in Figure 14.6. The spatial heterogeneity of each tracer is characterized by its slope p; smaller p corresponds to greater patchiness.

14.4 Modeling and Results

14.4.1 How Patchiness Relates to Response Time τ

A simplified model for the distribution of a tracer, whose concentration is denoted by c, is given by

$$\frac{\partial c}{\partial t} + u_H \bullet \nabla_H c + w \partial_z c = S \tag{14.3}$$

where u_H and w denote the horizontal and vertical velocity of the fluid, ∇_H is the horizontal gradient operator, ∂z is the vertical gradient operator, and S is a source or sink term to account for the modification of the tracer by nonconservative processes. S can often be written in terms of a characteristic timescale τ and concentration c as $S = -\tau^{-1}c$. In the case of a nutrient-like tracer depleted by biological uptake within the euphotic layer, τ is a characteristic e-folding timescale for biological production. Similarly, for air–sea gas flux modeled in terms of a wind speed–dependent piston velocity k and air–sea concentration gradient Δc, the characteristic time scale is $\tau = (\Delta z/k)(\Delta c/c)$, where Δz is the mixed layer depth. Heat flux and evaporation–precipitation are often accounted for by restoring the temperature and salinity fields at the surface in a model to observed values with a characteristic restoration timescale τ. Thus for any process, one can identify a characteristic response time τ. Our aim is to characterize the distribution of the tracer as a function of τ.

We do this within the framework of a model. The above model (Equation 14.3) is used with $S = -\tau^{-1}c$ in the upper 95 m to represent the biological consumption of nutrient in the euphotic layer of the oligotrophic ocean. The nutrient-like tracer concentration c is initialized to resemble nitrate; it is abundant at depth, has a strong concentration gradient in the thermocline, and is depleted in the euphotic layer (taken to be the upper 95 m in the model), but is horizontally uniform at initialization. The tracer model is coupled to a three-dimensional dynamic model configured to simulate an upper ocean front. The model domain is a periodic channel 258 km in zonal extent and 285 km in meridional extent with solid north–south boundaries and periodic east–west boundaries. The grid spacing is approximately 4 km and 4.45 km in the zonal and meridional directions, respectively. In the vertical, we maintain the uppermost layer as 60 m, while the layers below vary in thickness from 16 m in the thermocline to 100 m at depth. The model domain is initialized with fresher water in the southern half of the channel to create an across-channel density front that is accompanied by an east–west geostrophic jet. As the model simulation progresses, the jet and front meander and set up a more complex flow that we assume to be representative of the ocean. Nutrient is upwelled into the euphotic zone (upper 95 m), where it is consumed at a rate $S = -\tau^{-1}c$ due to biological production. We use six different values of τ, 2.5, 5, 10, 20, 40, and 80 days, in our numerical experiments to examine the relation between τ and the surface distribution of the tracer. The tracers are identically initialized and subject to the same dynamics; they vary only in their response times τ.

If we assume a balance between the terms that represent the upwelling of nutrient into the euphotic layer and its uptake in Equation 14.1, we can write

$$\frac{W}{h}\left(c' - c_\infty\right) \sim -\frac{c'}{\tau} \tag{14.4}$$

where c' and c_∞ are the concentration of the tracer anomaly at the surface and the concentration at a depth h below the mixed layer. Both c' and c_∞ are normalized by the mean surface concentration. Thus $(c' - c_\infty)/h$ represents the vertical concentration gradient in the thermocline, and W is the characteristic vertical velocity. Rearranging terms in Equation 14.4 gives

$$\left(1 - \frac{c_\infty}{c'}\right) \sim -\frac{1}{\overline{\tau}}$$

where $\overline{\tau} = \tau W / h$ is the ratio of the tracer response time to the vertical advection timescale. (Its inverse $h / (\tau W)$ is known as the Damköhler number.) Taking the logarithm of this relation gives

$$\log c' \sim \log c_\infty + \log \overline{\tau}. \tag{14.5}$$

The variance $V \sim c'^2$ and patchiness $p = \log V/\log L$. Therefore, by assuming that p is constant over the range of length scales considered, we infer that

$$p \sim \log c_\infty + \log \overline{\tau}. \tag{14.6}$$

This suggests that when several tracers are subject to the same flow (W/h is the same for the tracers), we can relate their spatial heterogeneity (measured by p) to their response time τ. The index p not only increases with τ, but is more sensitive to τ when τ is relatively small.

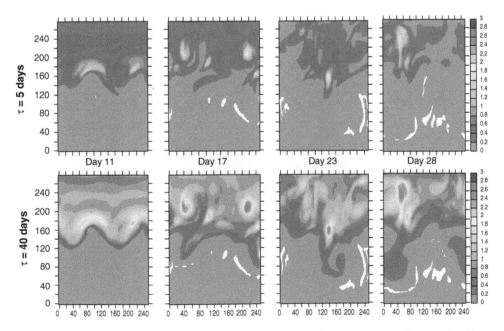

FIGURE 14.8 (Color figure follows p. 332.) Concurrent views of the surface concentration of the nutrient-like tracers with different response times τ, averaged over the upper 95 m in the model. The tracer with smaller τ is patchier. The domain dimensions are in kilometers.

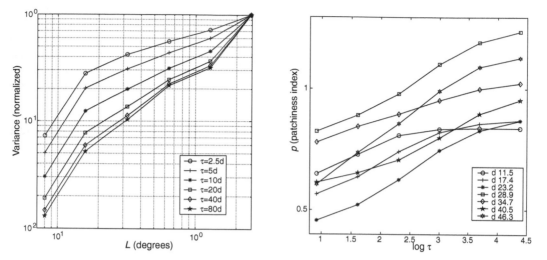

FIGURE 14.9 (A) Log–log plots of the average variance *V* vs. *L* generated from concurrent snapshots of the tracers with different τ. The analysis demonstrates that tracers with small τ have shallower *V*–*L* curves or smaller values of *p*. (B) The slopes *p* of the *V*–*L* analysis plotted vs. log τ, at different times (days) in the simulation to verify the scaling relation (Equation 14.6).

Model Results: We now use results from the model described above to verify the scaling relation (Equation 14.6). In Figure 14.8 we show snapshots of the average tracer concentration in the upper 95 m of the ocean for the nutrient-like tracers that differ in their characteristic response times τ. The tracer with smallest τ is the patchiest; because the tracer at the surface is rapidly obliterated by biological consumption, only recently upwelled tracer is visible. As τ is increased, the tracer patches at the surface persist for longer and expand as they continue to be fed by upwelling from the subsurface while also being advected by the horizontal flow field. The variance analysis (curves of *V(L)* in log–log space) for tracers with different τ in Figure 14.9A show a systematic increase in *p* (decrease in patchiness) with response time τ. We estimate the slope *p* of the curves in Figure 14.9A using their central portions and plot this patchiness measure *p* vs. log τ at different instances during the simulation (Figure 14.9B). The plots are linear and support the scaling relation 14.6. The deviation from linearity is most pronounced at early stages in the simulation for tracers with large τ that have not attained their full surface concentration.

14.4.2 Dependence of the Vertical Distribution of a Tracer on τ

The time-averaged concentration profiles of tracers, like those described here, are set largely by the balance between the rate of upwelling and their rate of alteration in the upper ocean. By averaging Equation 14.3 in the horizontal and assuming that over long times, the vertical concentration profiles of these tracers are in steady state, we find that the average concentration in the surface layer $\bar{c}$ varies as

$$\left(D/h^2\right)\left(\bar{c} - c_\infty\right) \sim -\bar{c}/\tau. \tag{14.7}$$

Here *D* represents an effective diffusivity that accounts for the averaged vertical transport, *h* represents a scale height over which the transport occurs, and h^2/D is the *mean* upwelling timescale (different from the *local* upwelling timescale *h*/*W*). Rearranging Equation 14.7 gives

$$\bar{c}/c_\infty = \left(1 + h^2/(D\tau)\right)^{-1} \tag{14.8}$$

a relation linking the mean vertical concentration profile characteristics of the tracer to the ratio of its characteristic response time and mean upwelling timescale. We can now use the results of the model to verify Equation 14.8. Because the variable parameter in the model is τ, we rewrite Equation 14.8 as $\tau^{-1} = (c_\infty/\bar{c} - 1)D/h^2$ and plot $c_\infty D/\bar{c}h^2$ vs. τ^{-1} at different times in the model simulation (Figure 14.10). D/h^2 varies with time, but at each time in the simulation $c_\infty/\bar{c}$ scales with τ^{-1}. The plots collapse into

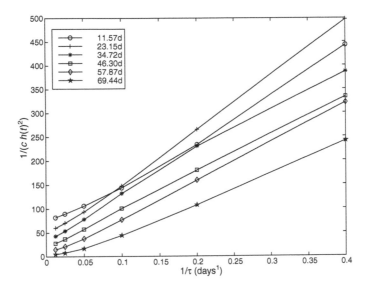

FIGURE 14.10 The relation between the mean sea surface concentration of a tracer $\bar{c}$ and τ is tested by plotting $c_\infty D/\bar{c}h^2$ vs. τ^{-1} at different times in the simulation. $\bar{c}$ is taken to be the mean over the upper 95 m in the model. D and c_∞ are taken to be constant, and h is assumed to increase linearly with time as this makes the plots collapse into a narrow region.

a narrow region if we take h to increase linearly with time as the simulation progresses. The relation (Equation 14.8) implies that in regions where the time-averaged horizontal concentration gradients of a tracer are non-zero, we can use their vertical distributions to infer the mean rate of upwelling or effective diffusivity D given a characteristic e-folding timescale τ with which the tracer is altered in the upper ocean.

14.4.3 Resolution Requirements for Tracers with Different τ

Tracers with short response timescales τ are finer scaled and patchier than those with large τ. Thus, they need to be sampled or modeled at higher resolution if the intention is to capture their variability. If, for example, we wish to capture a fixed proportion (say, 80%) of the variance in a region by sampling or modeling, an appropriate choice for the (nondimensional) horizontal grid size Δ is given by substituting $V = 0.8$ and $L = \Delta$ in Equation 14.1. Taking into consideration the negative sign generated by log(0.8), this implies that, $\log \Delta \sim -1/p$, which when combined with Equation 14.6 gives

$$\Delta \sim \exp(-1 / (\log \bar{\tau} + \log c_\infty)). \tag{14.9}$$

Thus, two tracers that differ in their response times by a factor of 10 (and do not differ in their normalized depth concentrations c_∞), differ in their model resolution requirement by a factor of 4. If biological production were negligible and air–sea exchange were the dominant process modifying DIC and O_2 at the sea surface, and if the normalized depth concentrations are assumed to be nearly the same, a factor of 10 difference in their equilibration times (1 month for O_2 vs. 10 months for DIC) would imply that O_2 requires four times higher resolution for observing or modeling than DIC. Similarly, Chl, which responds on a timescale of approximately 3 days, requires four times more resolution than SST, which responds to surface heat fluxes on a timescale of approximately 30 days. This explains why biological variables that are typically altered on short time scales are particularly difficult to observe and model in the ocean as compared to physical variables like temperature and salinity.

14.5 Discussion

The modification of tracers by nonconservative upper ocean processes can generally be parameterized in terms of a characteristic timescale τ and the tracer concentration c. Thus, even though the model used to derive the above results is simple, it can be generalized to various tracers and processes. When multiple processes modify a tracer, the effective characteristic timescale τ_{eff}, is the geometric mean of the individual process time scales τ. Since patchiness of a tracer scales inversely as the index p, and $p \sim \log \tau$, tracer distributions are more sensitive to changes in τ when τ is small, but relatively insensitive when τ is large. In an ecosystem carbon cycle model, the variability in the DIC field is more sensitive to the choice of the biological time constants than to the piston velocity that determines the timescale of air–sea exchange, because the timescales of biological production are typically much smaller than those for air–sea exchange.

The relationships described here in terms of spatial distributions can be extended to the temporal context using intermittency as the analogue of patchiness. Further work needs to be done in verifying these ideas with oceanographic data. Since high spatial resolution *in situ* data are difficult to obtain, it may be more feasible to examine these ideas in the temporal context with time-series data.

The idea that patchiness decreases (p increases) with increasing response time contradicts the concept that patchiness increases by the generation of fine-scale filaments formed by stirring a two-dimensional field (Abraham, 1998). The theory on two-dimensional stirring neglects the effect of upwelling, but this work considers the effect of continued upwelling on the surface fields. Although these results might be somewhat modified by including the effect of horizontal stirring, we believe that the surface signal of tracers that possess a steep concentration gradient across the thermocline is largely dominated by upwelling. Thus, these results quantify, to first order, the characteristics of the sea surface distribution of a tracer in response to ocean dynamics and nonconservative processes.

14.6 Conclusions

The analysis of satellite SST and Chl data over a range of length scales (2 to 256 km) shows that $V(L)$, the average variance contained in regions of area L^2, scales as $V(L) \sim L^p$. We propose the exponent p as a measure of patchiness; smaller p implies a larger percentage of the variance exists at small scales, and corresponds to greater patchiness. Satellite data shows that $p(\text{Chl}) < p(\text{SST})$, i.e., surface Chl is patchier than SST at meso- and submeso-scales. Likewise, other tracer distributions in the surface ocean differ in their spatial heterogeneity. The tracers considered have steep vertical concentration gradients, and surface concentration anomalies are introduced by upwelling. Differences in the spatial heterogeneity (and p) result from differences in the characteristic response times τ with which the tracers are modified by nonconservative processes in the upper ocean. Using a scaling argument and numerical model, we show that $p \sim \log \tau$. Tracers with relatively small τ have fine-scaled surface distributions, require higher resolution for modeling and sampling and have lower mean concentrations at the surface. For tracers differing in τ, the grid spacing required scales as $\Delta \sim \exp(-1 / \log \tau)$, and the ratio of the mean surface to deep concentrations scales as $c / c_\infty \sim 1 / (1 + \tau^{-1})$.

Acknowledgments

We thank T. Moore for help with the analysis of satellite data. The SeaWiFS and MODIS data were obtained from the NASA Goddard DAAC. This work was sponsored by ONR (N00014-00-C-0079) and NASA (NAS5-96063 and NAG5-11258).

References

Abraham, E., The generation of plankton patchiness by turbulent stirring, *Nature*, 391, 577–580, 1998.
Campbell, J.W. and Esaias, W.E., Spatial patterns in temperature and chlorophyll on Nantucket Schoals from airborne remote sensing data, May 7–9, 1981, *J. Mar. Res.*, 1985, 43, 139–161.
Denman, K.L. and Platt, T., Spectral analysis in ecology, *Annu. Rev. Ecol. Syst.*, 6, 189–210, 1975.

Deschamps, P.Y., Frouin, R., and Wald, L., Satellite determination of the mesoscale variability of the sea surface temperature, *J. Phys. Oceanogr.*, 11, 864–870, 1981.

Esaias, W.E. et al., 1998, An overview of MODIS capabilities for ocean science observations, *IEEE Trans. Geosci. Remote Sensing*, S6(4), 1250–1265, 1998.

Glover, D.M., Doney, S.C., Mariano, A.J., Evans, R.H., and McCue, S.J., Mesoscale variability in time series data: satellite-based estimates of the U.S. JGOFS Bermuda Atlantic Time-series Study (BATS) site, *J. Geophys. Res.*, 107(C8), 7-1 to 7-21, 2000.

Gower, J.F.R., Denman, K.L., and Holyer, R.J., Phytoplankton patchiness indicates the fluctuation spectrum of mesoscale oceanic structure, *Nature*, 288, 157–159, 1980.

Levitus, S., Climatological Atlas of the World Ocean, NOAA Professional Paper 13, 1982, 173 pp.

Levy, M., Klein, P., and Treguier, A.-M., Impacts of sub-mesoscale physics on production and subduction of phytoplankton in an oligotrophic regime, *J. Mar. Res.*, 59, 535–565, 2001.

Mahadevan, A. and Archer, D., Modeling the impact of fronts and mesoscale circulation on the nutrient supply and biogeochemistry of the upper ocean, *J. Geophs. Res.*, 105, 1209–1225, 2000.

Mahadevan, A. and Campbell, J.W., Biogeochemical patchiness at the sea surface, *Geophys. Res. Lett.*, 29(19), 1926, doi: 10.1029/2001GLO14116, 2002.

McClain, E.P., Pichel, W.G., and Walton, C.C., Comparative performance of AVHRR-based multichannel sea surface temperatures, *Geophys. Res.*, 90(C), 11587–11601, 1985.

McClain, C.R., Cleave, M.L., Feldman, G.C., Gregg, W.W., Hooker, S.B., and Kurling, N., Science quality SeaWiFS data for global biosphere research, *Sea Technol.*, 10–16, Sept. 1998.

Pollard, R.T. and Regier, L.A., Vorticity and vertical circulation at an ocean front, *J. Phys. Oceanogr.*, 22, 609–625, 1992.

Semtner, A.J., Jr. and Chervin, R.M., Ocean general circulation from a global eddy-resolving model, *J. Geophs. Res.*, 97(C4), 5493–5550, 1992.

Voorhis, A.D. and Bruce, J.G., Small-scale surface stirring and frontogenesis in the subtropical convergence zone of the western North Atlantic, *J. Mar. Res.*, 40, S801–S821, 1982.

Yoder, J.A., McClain, C.R., Blanton, J.O., and Oey, L.-Y., Spatial scales in CZCS-chlorophyll imagery of the southeastern U.S. continental shelf, *Limnol. Oceanogr.*, 32, 929–941, 1987.

Yoder, J.A., Aiken, J., Swift, R.N., Hoge, F.E., and Stegmann, P.M., Spatial variability in the near-surface chlorophyll *a* fluorescence measured by the Airborne Oceanographic Lidar (AOL), *Deep Sea Res. II*, 40, 37–53, 1993.

15

Challenges in the Analysis and Simulation of Benthic Community Patterns

Mark P. Johnson

CONTENTS

15.1 Empirical and Theoretical Treatments of Spatial Scale in Benthic Ecology 229
    15.1.1 Rarity of Spatially Explicit Models for Benthic Systems 230
15.2 Robust Predictions from Spatial Modeling 231
15.3 Comparing Markov Matrix and Cellular Automata Approaches to Analyzing Benthic Data 231
    15.3.1 Nonspatial (Point) Transition Matrix Models 233
    15.3.2 Spatial Transition Matrix Models 234
    15.3.3 Comparison of Empirically Defined Alternative Models 235
15.4 Extending the Spatial CA Framework 236
15.5 Conclusions 239
Acknowledgment 240
References 240

15.1 Empirical and Theoretical Treatments of Spatial Scale in Benthic Ecology

The composition and dynamics of benthic communities reflect the interplay of factors that operate at a range of scales. Variability at almost every scale of observation is likely to affect benthic species. For example, hydrodynamic gradients exist from the centimeter scale of the benthic boundary layer to ocean basin scale circulation patterns. The settlement of benthic species from planktonic life history stages will reflect both the large-scale and small-scale influences on propagule supply. Many benthic species have limited mobility as adults, so individuals may only interact with other individuals within a relatively short distance. However, population dynamic processes such as mortality may also be composed of elements at quite different scales. For example, mortality of barnacles can be caused by both the crowding effects of neighbors and mobile predators such as whelks or crabs.

Given that there seems no basis for assuming any "correct" scale of observation (Levin, 1992), the empirical response has been to characterize variability at a number of scales. This form of pattern identification can be considered a prerequisite for subsequent studies on process (Underwood et al., 2000). Many studies of spatial scale have used nested analysis of variance (ANOVA, e.g., Jenkins et al., 2001; Lindegarth et al., 1995; Morrisey et al., 1992). This may reflect the familiarity of the ANOVA approach from experimental hypothesis testing in benthic ecology. Indeed, many experimental manipulations also include explicit considerations of scale and scaling effects (Thrush et al., 1997; Fernandes et al., 1999). Spatial autocorrelation and fractal analyses have also been used to characterize spatial pattern in benthic systems (Rossi et al., 1992; Underwood and Chapman, 1996; Johnson et al., 1997;

Maynou et al., 1998; Snover and Commito, 1998; Kostylev and Erlandson, 2001). Although patterns may have been considered at different times, many studies of spatial scale have taken a snapshot view: spatial pattern at one point in time has not been mapped onto the spatial pattern at other times. The snapshot approach restricts further investigation into issues of turnover and dynamics. However, with the growing interest (Koenig, 1999) in defining the spatial scales over which population dynamics are linked (synchronous), there are likely to be more spatiotemporal studies of benthic populations and communities in the future. These studies of population synchrony are important as they define the spatial structure of populations (Koenig, 1999; Johnson, 2001). The degree of synchrony between local populations has implications for conservation biology. If local populations are asynchronous, then a local extinction event may be reversed by individuals supplied from another, healthy, population. Large-scale loss of a species is more likely where local populations are synchronous and no rescue effects occur (Harrison and Quinn, 1989; Earn et al., 2000).

The inclusion of explicit treatments of spatial scale in much of the empirical research on benthic systems has not been paralleled by extensive theoretical work on the same systems. Influential models do exist for space-limited benthic systems (Roughgarden et al., 1985; Bence and Nisbet, 1989; Possingham et al., 1994) and patch dynamics on rocky shores (Paine and Levin, 1981). However, these models do not have an explicit treatment for space: the variables in the models are not differentiated on the basis of their relative locations (although see Possingham and Roughgarden, 1990, and Alexander and Roughgarden, 1996, for extensions of the framework to include spatial population structure along a coastline). Simulations of benthic communities with a one-dimensional representation of spatial location have been used to look at the development of intertidal zonation along environmental gradients (Wilson and Nisbet, 1997; Johnson et al., 1998a). Spatially explicit models of benthic systems on two dimensional lattices have generally shown interactions between processes at different scales. Local interactions can lead to a large-scale pattern (Burrows and Hawkins, 1998; Wooton, 2001a) and the predictions of spatially explicit and nonspatial models differ (Pascual and Levin, 1999; Johnson, 2000).

15.1.1   Rarity of Spatially Explicit Models for Benthic Systems

Despite the observation that "space matters" and an explosion of interest in spatial ecology (Tilman and Kareiva, 1997), there are a number of reasons why spatially explicit models of benthic communities may be uncommon. The lack of system-specific models partly reflects the manner in which spatial theory has developed. Spatially explicit models tend to be caricature or generic models that attempt to capture the essential features of the system (Keeling, 1999). This approach improves the conceptual understanding of systems and allows numerical experiments that would be difficult or destructive in a real system (Keeling, 1999). The use of generic models improves communication between theoreticians as there can be clarity about techniques and general conclusions without debate on the individual nature of biological interactions in particular systems.

The rarity of system-specific models can also be explained by considering the problems associated with an imaginary spatially explicit model for a benthic community. The model is as realistic as possible, with a number of interacting species influenced by stochastic variation in processes such as recruitment. Simulation output resembles the patterns seen in the real system. However, formal testing of the model would involve collecting large amounts of detailed spatial data from the field (independently from that used to derive the model). As the model contains stochastic processes, a large number of repeated simulations are needed to define the potential behavior of the system. Given the range of potential outputs that the model may produce, it is difficult to envisage how a limited number of spatial data sets could be used to falsify the model. Both the collection of data and repeated simulations are time-consuming. More importantly, we are not likely to be interested in the detailed spatial arrangement of species in the benthic community. Only a subset of model predictions (such as the mean abundance of a species) is likely to be both of interest to applied research and testable. Hence the time required to develop a model for a specific system may not be justified in the end results.

15.2   Robust Predictions from Spatial Modeling

There is a tension between the observation that spatial effects can be important and the difficulties involved in testing detailed spatially explicit simulations. However, if our understanding of benthic community patterns is to be addressed, a way of resolving this tension is needed. There have been two approaches to this problem, which can be loosely classified as theory based and data based.

Theory-based approaches to spatially explicit modeling are extremely diverse and include reaction-diffusion and partial differential equations. It is difficult, however, to construct a mathematically tractable model that is also applicable to particular ecological systems such as different benthic communities (Tilman and Kareiva, 1997). In a recent development, researchers have used "pair approximation" techniques to provide analytically tractable models (Levin and Pacala, 1997; Rand, 1999; see Snyder and Nisbet, 2000, for a critique and alternative approach). The idea behind pair approximation is that the equation for a nonspatial process can be extended to a spatial system by using functions that approximate the average neighborhood structure in a spatial model. Hence, in contrast to a model where the same equation is repeated at a large number of locations, the small-scale spatial detail is included in a limited number of equations. The pair approximation approach therefore facilitates investigation of model behavior more efficiently than would be the case in a simulation. As yet these models still tend to be generic, and may thus ignore important features of benthic systems. For example, a common assumption is that dispersal is a local process (Levin and Pacala, 1997; although see Pascual and Levin, 1999). This contrasts to the characterization of many benthic populations as open (Roughgarden et al., 1985; Caley et al., 1996): new recruits may be supplied by sources at some distance from the local population.

In contrast to the development of generic descriptions in the theory-based approach, the data-based approach involves case studies of specific systems. Ideally, a number of alternative models with different treatments of space will be tested against field observations. This approach has the advantage that movement to more complex models is justified only where there are improvements in predictive ability. The benefits of building a sequence of models are further outlined in Hilborn and Mangel (1997).

As it seems impractical to develop a large number of spatially explicit models for different benthic systems, the challenge in the analysis and simulation of benthic populations is to combine the theory-based and data-based approaches to produce a set of methodological approaches that can be used to investigate and contrast different systems. Although this viewpoint is not novel, there remain few examples of synergy between theoretical and empirical approaches for benthic systems. A notable exception is the work of Wootton (2001b) on intertidal mussel beds. The approach taken in the first section below mirrors that of Wootton (2001b) in that multispecies Markov models are used as the basis for comparing spatial and nonspatial models. A slightly different approach is taken in the second section, where a more complex model is used to suggest methods for distinguishing between alternative hypotheses using field data. The analyses presented use a broad interpretation of "benthic" that includes the rocky intertidal. Rocky shores are generally considered more tractable than sandy or muddy systems. For example, it is far easier to fix locations and organisms are generally not subsurface in rocky systems. On a conceptual level, however, there is nothing to prevent the application of spatial models to sandy or muddy systems (although the scales of processes such as adult mobility are likely to differ with increasing mobility of the sediment).

15.3   Comparing Markov Matrix and Cellular Automata Approaches to Analyzing Benthic Data

What approaches are there available to move beyond generic models and statistical pattern identification in the analysis of benthic systems? A first task is to recap on the potential shortcomings of theory-based and wholly empirical approaches. The generic nature of certain theoretical approaches has been detailed above. Potential limitations of statistical pattern analysis (e.g., spatial autocorrelations, nested ANOVA) are restrictions on generalization from results and a lack of sensitivity tests of conclusions. Assessment of a pattern, once identified, can be limited to rhetorical arguments about the interaction of processes at different scales. Follow-up experiments can be difficult to design as the alternative, spatially explicit, hypotheses are not always intuitive. By examining the consequences of different assumptions, models can extend experimental results to create appropriate hypotheses. Existing techniques for incorporating field data into a modeling framework include Markov transition matrix models and cellular automata. Other techniques exist, probably dependent on the ingenuity of the investigator. Markov models and cellular automata, however, have several advantages. They are well known and relatively simple to apply. Hence, different investigators can use them and compare results in a common format. Markov models have the potential for sensitivity testing; they also form an appropriate nonspatial null model for comparison with data and spatially explicit alternative models. Cellular automata can be used to investigate neighborhood effects and can be used to identify scaling properties of systems (e.g., power law relationships in patch geometry; Pascual et al., 2002). Here I emphasize cellular automata as they can be "twinned" with experimental procedures at the same scale: within shore patches and external forcing at the grid scale (cf. spatial replication between shores). Other techniques exist for spatial modeling, for example, applications of geographical information systems (GIS) at the landscape scale. However, GIS applications are probably closer to statistical pattern analysis in that the scope for sensitivity tests and experimental investigation of predictions is limited.

Cellular automata (CA) and Markov transition matrix approaches have underlying similarities and yet they are generally used in completely different ways. Both approaches use a discrete description of time and state. Temporal dynamics in both frameworks are usually first order: state at time $t + 1$ is dependent on state at time t. Such transitions may be entirely deterministic or occur with a specified probability. Where the two approaches differ is that CA includes a discrete representation of space, typically visualized as a grid of square or hexagonal cells. The cells in the neighborhood of an individual location on the CA grid influence the transition between states at that location from one time step to the next. Applications of CA usually stress simplicity at the expense of biological realism (Molofsky, 1994; Rand and Wilson, 1995) but cite speed of computation and heuristic value (Phipps, 1992; Ermentrout and Edelstein-Keshet, 1993). In comparison, Markov transition matrix models are frequently derived directly from field data and are used to examine characteristic processes in the observed communities (Horn, 1975; Usher, 1979; Callaway and Davis, 1993; Tanner et al., 1994).

In theory, it is straightforward to reconcile the issues of spatial dependence and empiricism that transition matrices and CA, respectively, ignore. By constructing a CA using observed local transition probabilities, it is possible to compare models containing local interactions with nonspatial models. A problem with this approach is the data requirement needed to parameterize even a simple CA. For example, a cell in a system with four states and eight neighbors would have 4^8 (65,536) possible neighborhood configurations. It would be practically impossible to empirically define a transition probability associated with each neighborhood configuration. However, given information about the important interactions in a system, effort can be concentrated on defining a limited number of transitions.

An example of a relatively well studied system is the mosaic of macroalgal (mostly *Fucus* spp.) patches on smooth moderately exposed rocky shores in the northeast Atlantic (Hawkins et al., 1992). Spatial structure and patch dynamics in this system are thought to be driven by limpet grazing (Hartnoll and Hawkins, 1985; Johnson et al., 1997). Spatial autocorrelation studies have suggested an algal patch length scale of approximately 1 m in this system. Time series from a quadrat of similar dimensions to the patch scale show multiannual variations in algal cover, with limpet densities tending to lag these fluctuations. The conceptual model developed for this system is based on the interaction between limpet grazing pressure and the recruitment of algae. Limpets are aggregated in clumps on the shore and the uneven spatial distribution of grazing pressure leads to the formation of new patches of algae in areas where there are few limpets. The spatial mosaic of algal patches formed by uneven grazing pressure is dynamic (Figure 15.1). Adult limpets relocate to established patches of algae. This generates changes in grazing pressure and allows new patches of algae to be generated elsewhere on the shore. Older patches of algae do not regenerate, possibly because of the increased local density of limpets associated with them. Hence the shore is patchy, but the locations of patches change, creating the multiannual fluctuations seen at the patch scale.

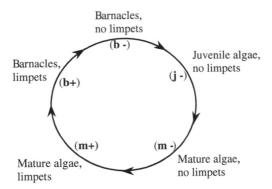

FIGURE 15.1 Idealized cycle at the patch scale on moderately exposed shores in the northeast Atlantic. Spatial variation in limpet grazing pressure allows recruitment of juvenile algae to the shore. Patches eventually decay. The aggregation of limpets in aging patches of algae changes the spatial pattern of grazing pressure, allowing new patches to be formed elsewhere on the shore.

The proposed mechanism for the patch mosaics on moderately exposed rocky shores in the northeast Atlantic implies that the effort in deriving spatial transition rules can be concentrated on defining how they are affected by the local limpet density. By constructing a traditional nonspatial transition matrix model it is possible to test if the system dynamics are at least a first-order Markov process. Empirically derived CA rules with and without a local limpet presence can be tested to investigate whether limpets do actually affect local state transitions. Spatial and nonspatial Markov processes can be compared to test whether local interactions alter the projected dynamics of the system.

15.3.1 Nonspatial (Point) Transition Matrix Models

Transition matrix models are defined by marking out fixed sites, defining states, and recording the transitions between states in a defined time period. In work carried out in the Isle of Man (methods described more fully in Johnson et al., 1997) the fixed sites were 0.01 m^2 square "cells" in permanently marked 5 × 5 m quadrats (2500 cells per quadrat) and the time step was 1 year. If a cell contained algae, a distinction was made between "mature" and "juvenile" cells. A juvenile cell was one where algal frond lengths did not exceed 0.1 m and reproductive structures were absent. Barnacle cover outside algal patches was variable. If a cell contained no barnacles at all it was classed as bare rock. Coralline red algae were generally associated with small rock pools. If the areal cover of coralline red algae exceeded that of barnacles, a cell was classified as "coralline red." The presence or absence of adult limpets was recorded (shell diameter >15 mm) for each of the five basic classification states (barnacle, juvenile, mature, coralline red, and rock).

Transition matrices take the form:

$$\mathbf{A} = \begin{pmatrix} p_{11} & p_{12} & p_{13} & \cdots & p_{1n} \\ p_{21} & p_{22} & p_{23} & \cdots & p_{2n} \\ p_{31} & p_{32} & p_{33} & \cdots & p_{3n} \\ \vdots & \vdots & \vdots & \vdots & \vdots \\ p_{n1} & p_{n2} & p_{n3} & \cdots & p_{nn} \end{pmatrix} \qquad (15.1)$$

where p_{jk} is the probability of transition from state k to state j with each time step. Transition probabilities are derived from a frequency table of state k to j changes. The frequency of each change from one state to another is divided by the column total to give the probability of each transition. Transitions are tested for interdependence (with the null hypothesis being that transitions are independent, i.e., random, and therefore the process is not Markovian) using a likelihood ratio test, with $-2 \ln \lambda$ compared to χ^2 with $(m-1)^2$ degrees of freedom (Usher, 1979):

$$-2\ln\lambda = 2\sum_{j=1}^{m}\sum_{k=1}^{m} n_{jk}\ln\left(\frac{p_{jk}}{p_j}\right) \qquad (15.2)$$

where

$$p_j = \sum_{k=1}^{m}\frac{n_{jk}}{\displaystyle\sum_{j=1}^{m}\sum_{k=1}^{m} n_{jk}} \qquad (15.3)$$

n_{jk} = number of transitions from state k to j in the original data matrix
p_{jk} = probability of transition from state k to j
p_j = sum of transition probabilities to state j
m = order of the transition matrix (number of rows)

The sum effect of all transitions over a time step is found by the multiplication:

$$\mathbf{Ax}_{(t)} = \mathbf{x}_{(t+1)} \qquad (15.4)$$

where $\mathbf{x}_{(t)}$ is a column vector containing frequencies of separate cell state at time t. With transition matrices, repeated multiplication by $\mathbf{A}$ generally causes the community composition to asymptotically approach a stable state distribution defined by the right eigenvector of $\mathbf{A}$ (Tanner et al., 1994).

The temporal scales of processes can be investigated from metrics derived from transition matrices. For example, the rate of convergence to a stable stage structure is governed by the damping ratio, ρ (Tanner et al., 1994; Caswell, 2001):

$$\rho = \lambda_1 / |\lambda_2| \qquad (15.5)$$

where λ_j is an eigenvalue of the transition matrix. As matrix columns sum to one, the first eigenvalue is always one. A convergence timescale is given by t_x, the time taken for the contribution of the first eigenvalue to be x times as great as the contribution from the second eigenvalue (Caswell, 2001):

$$t_x = \ln(x) / \ln(\rho) \qquad (15.6)$$

15.3.2 Spatial Transition Matrix Models

Maps of adjacent 0.01 m² cells allow spatial transition rules to be defined. The effect of limpets on the transitions occurring in their neighborhood can be tested by deriving two separate transition matrices: one for transitions when limpets were present in at least one of the neighboring eight cells and one matrix for cell transitions occurring in the absence of limpets in surrounding cells. The significance of differences between "local limpets" and "no local limpets" transition matrices can be examined using (Usher, 1979; Tanner et al., 1994):

$$-2\ln\lambda = 2\sum_{j=1}^{m}\sum_{k=1}^{m}\sum_{L=1}^{L} n_{jk}(L)\ln\left(\frac{p_{jk}(L)}{p_{jk}}\right) \qquad (15.7)$$

where
L = number of transition matrices associated with limpet grazing effects (= 2)
$n_{jk}(L)$ = number of k to j transitions recorded for matrix L
$p_{jk}(L)$ = transition probability from k to j in matrix L
p_{jk} = transition probability from k to j if L matrices are pooled

The likelihood ratio is compared to χ^2 with $m(m-1)(L-1)$ degrees of freedom and a null hypothesis that there is no difference between matrices dependent on the presence or absence of limpets in the eight cell neighborhood.

TABLE 15.1

Matrix of Transition Probabilities for Quadrat *a* Surveyed in the Isle of Man

	b+	b–	j+	j–	m+	m–	cr+	cr–
b+	0.024	0.036	0.000	0.015	0.028	0.010	0.100	0.039
b–	0.040	0.105	0.021	0.024	0.028	0.030	0.100	0.078
j+	0.079	0.042	0.021	0.039	0.056	0.035	0.050	0.024
j–	0.333	0.224	0.128	0.119	0.139	0.055	0.050	0.083
m+	0.095	0.072	0.149	0.124	0.250	0.199	0.000	0.044
m–	0.397	0.468	0.670	0.671	0.500	0.662	0.200	0.150
cr+	0.000	0.004	0.000	0.003	0.000	0.005	0.050	0.044
cr–	0.032	0.048	0.011	0.006	0.000	0.005	0.450	0.539

Note: Cells are classified as barnacle occupied (b), juvenile *Fucus* (j), mature *Fucus* (m), and coralline red algae (cr). Bare rock was not recorded in cells at this site. + or – modifiers indicate the presence or absence of limpets in the cells.

It is not possible to iterate the spatial model using matrix multiplication as the choice of transition probability is dependent on local conditions. Spatial transition matrices were therefore investigated using CA simulations. These simulations were based on 50×50 square cell grids with periodic boundary conditions (cells on one edge of the grid are considered to be neighbors to cells on the opposite edge of the grid). As the CA rules are derived empirically from counts of 0.01 m² cells, spatial simulations represent an area of 25 m². Cell state transitions at each time step were based on probabilities drawn from a matrix chosen according to the neighborhood state ("local limpets" or "no local limpets"). Simulations were stochastic as random numbers were used to generate cell state transitions based on the probabilities in the appropriate matrix (the spatial model was what is sometimes referred to as a "probabilistic CA").

15.3.3 Comparison of Empirically Defined Alternative Models

Point and spatial transition matrices were derived for three separate 25 m² quadrats at different sites in the Isle of Man (hereafter referred to as sites *a*, *b*, and *c*). At each site, likelihood ratio tests supported the application of Markov matrices to the observed transitions (Equation 15.2, $p < 0.001$ in all cases). Hence the matrices contain information about a nonrandom process of transitions at each site.

An example point transition matrix is shown in Table 15.1. The pattern of transitions reflects parts of the patch cycle proposed by Hartnoll and Hawkins (1985). For example, the majority of barnacle-classified cells became occupied by algae. Most cells classed as juvenile algae were recorded as mature algae in the following year. The predicted dynamics rapidly approached equilibrium, with convergence time scales (t_{10}) of 4.17, 1.51, and 1.23 years for sites *a*, *b*, and *c*, respectively. This implies a high degree of resilience at two of the sites with recovery to the equilibrium state within 2 years of a perturbation. It is not clear what features make site *a* recover more slowly than the other sites. One possibility currently under investigation is that variation in dynamics reflects differences in surface topography.

The spatial transition matrices for the "no local limpets" and "local limpet" cases were significantly different at all three sites (Equation 15.7, $p < 0.05$). This supports the hypothesis (Hartnoll and Hawkins, 1985) that the spatial pattern of limpet grazing affects interactions on the shore. There was some variation between sites, but the transition frequencies reflected the influences of limpets on transitions to algal cover. For example, at the site with the largest difference between matrices, 63% of all transitions were to algal occupied states in the "no local limpets" matrix compared to 53% in the "local limpets" case.

As has been shown elsewhere (Wootton, 2001b), predictions of the matrix models fit the observed state frequencies reasonably well (explaining between 45 and 97% of the variation in frequencies; Figure 15.2). G tests show that the fit of the models is closer than would be expected for randomly generated frequencies, but that there were still departures between model predictions and observed frequencies (Table 15.2). The discrepancy between predicted and observed frequencies was generally not reduced by using a spatial rather than a point model. In addition, the predictions of spatial and point models were not significantly different for site *c*. Despite the detection of spatial effects associated with limpets, the increase in model complexity from point to spatial models was not justified by a better fit to the data.

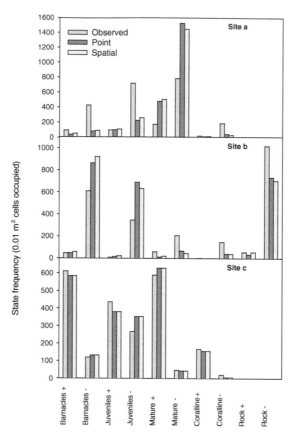

FIGURE 15.2 Comparison of observed and predicted cell state frequencies in 25 m² sampling quadrats. Observed frequencies are the average of separate annual samples. Predicted frequencies are from point or spatial transition matrix models.

The spatial model may still have some heuristic value if it generates a dynamic pattern of states in simulations. Techniques for investigating spatiotemporal pattern include calculating correlations between sites at different distances from each other (Koenig, 1999). An alternative approach used in scaling investigations of spatial models (De Roos et al., 1991; Rand, 1994; Rand and Wilson, 1995) is to compare the dynamics of cell frequencies in "windows" of different sizes on the simulation grid. For any probabilistic CA, cell state frequencies will fluctuate with time. The standard deviation of a time series taken from a window of $L \times L$ grid cells will decrease with increasing L (tending to zero at very large window sizes). For a stochastic process, the reduction in standard deviation with window size will generally be proportional to $1/L$ (Keeling, 1999). However, if a model contains coherent patch structures, there will be deviations from the $1/L$ line predicted for a stochastic process. If the patches are long-lived structures with respect to the time series, then standard deviations taken from windows smaller or equal to the patch scale will be less variable than expected. The expected scaling behavior is seen in time series drawn from a probabilistic version of the point model (transitions occur to populations of $L \times L$ cells with probabilities drawn from the nonspatial matrix for site a; Figure 15.3). The relationship between standard deviation of time series and window size was the same in probabilistic point and spatial models (ANCOVA, p no difference between slopes > 0.5). Hence there is no evidence that patch structures are formed at any scale in the spatial model.

15.4 Extending the Spatial CA Framework

The derivation of a spatial matrix model demonstrated that the local density of limpets affected the transitions between states on the shore. However, the empirically derived CA failed to generate spatial

TABLE 15.2

Comparisons between the Observed Frequencies of Different States, the Predictions of Point and Spatial Models, and Community Frequencies Generated Randomly

	Observed	Point Model	Spatial Model
Site a			
Point model	2005.41		
Spatial model	1883.29	22.17	
Random	**3533.15**	**6120.27**	**5907.50**
Site b			
Point model	753.58		
Spatial model	784.86	39.36	
Random	**6350.90**	**5576.49**	**4895.45**
Site c			
Point model	42.01		
Spatial model	56.41	4.77	
Random	**3166.67**	**3826.65**	**3766.11**

Note: G tests (Sokal and Rohlf, 1995) are used as measures of goodness of fit (Wootton, 2001b). Scores for the random model communities are averages of 250 independently generated tests. Lower G test values imply a better match between the frequencies being compared. Numbers in bold indicate significant differences between the frequencies being compared.

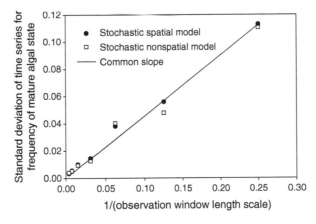

FIGURE 15.3 Standard deviation of mature algal frequencies in time series collected at different spatial scales. Observation window length scales range from 4 to 256 cells. The common slope is a statistically significant regression passing through the origin.

pattern or improve model predictions of state frequencies when compared to a nonspatial model. Wootton (2001a) in a study of intertidal mussel beds also found that empirically derived CA simulations with local interactions (but without locally propagated disturbances) did not produce patterning. The absence of spatial pattern in the CA models may reflect that spatial structures are sensitive to the stochastic nature of transitions between states (Rohani et al., 1997). The spatial transition rules for the *Fucus* mosaic and mussel bed were defined from field data. This implies that it is not possible to scale up from observations at small scales to patterns at large scales. There are, however, two reasons this conclusion may be premature. It may be that the CA framework is too crude a method to characterize the local interactions in the intertidal. The CA models also did not include "historical" effects, despite the

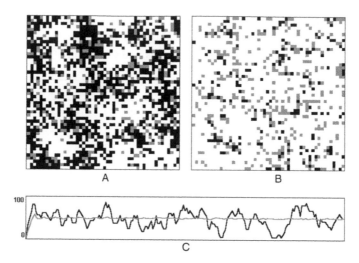

FIGURE 15.4 Screen grab of simulation output from the artificial ecology of the limpet–*Fucus* mosaic. The spatial plots show (A) *Fucus* distribution (white–empty, gray–juvenile, black–mature) and (B) limpet occupancy (white–empty, gray–one limpet, black–more than one limpet). The time series (500 time steps) of algal cover (C) shows records from the patch scale (black line) and the grid scale (gray line).

observation that history can intensify local interactions in probabilistic CA, leading to pattern formation (Hendry and McGlade, 1995).

In the context of Markov transition matrices, historical effects are modifications to the transition probabilities based on the state of the system at lags exceeding one time step (models include second and higher order processes, Tanner et al., 1996). Hence the age of particular states can affect their transition probabilities. For example, not all mussel beds are equivalent. Waves are more likely to remove old, multilayered beds (Wootton, 2001a). In the *Fucus* mosaic, patches of algae persist for 5 years before they break down (Southward, 1956). Tanner et al. (1996) demonstrated that historical effects could be detected in coral communities, although these effects did not affect overall community composition in comparison to first-order models.

Incorporating a more sophisticated representation of local grazing interactions and historical effects into CA simulations requires a framework variously known as mobile cellular automata, lattice gas model, or artificial ecology (Ermentrout and Edelstein-Keshet, 1993; Keeling, 1999). Time, space, and state are still discrete, but the artificial ecology formulation allows simulated organisms to move around the grid. This is a more flexible method of representing aggregations of mobile organisms than a conventional CA.

An artificial ecology for the *Fucus* patch mosaic can be based on the spatial effect of individual limpets on the probability that new patches of algae will be formed. This relationship can be defined from maps of limpet and algal location. The maps previously used for transition matrices have a minimum spatial scale below the average distance that limpets forage from their semipermanent home scar (0.4 m; Hartnoll and Wright, 1977). Hence the grazing effects should extend over several 0.01 m² cells. Stepwise logistic regression using increasing distances from the target cell was used to define the strength and the range of limpet effects on the probability of a cell containing juvenile algae (Johnson et al., 1997). This information was then used to simulate the *Fucus* mosaic in a 50 × 50 cell grid, equivalent to the scale of the maps made in the field. Each time step the distribution of limpets defined the probability of juvenile *Fucus* establishing in any unoccupied cell on the grid. As on the shore, limpets potentially relocated to new home scars each year, creating a dynamic pattern of grazing pressure. Simulations of this artificial ecology created realistic mosaic patterns (Figure 15.4). A fuller description of the model, including investigation of the roles of limpet movement and habitat preferences is given in Johnson et al. (1998b).

An advantage of the empirically defined rules for the artificial ecology is that the scales in the simulations are clearly defined. This facilitates more demanding confrontations with data than is possible with more generic spatial models. For example, the spatial autocorrelation produced in simulations gives

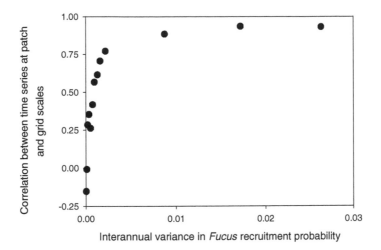

FIGURE 15.5 Correlations between *Fucus* abundance at patch and grid spatial scales with increasing levels of variability in grid scale *Fucus* recruitment probability. Time series were 500 time steps long with the first 50 time steps excluded to remove transient behavior.

a patch length scale of approximately 0.4 m, compared to a patch scale of 0.8 m for the same location in the field. The better definition of patches in the field may reflect heterogeneity in limpet grazing efficiency or algal recruitment probability associated with small-scale topographic features. These features could be investigated further by looking at local deviations (residuals) from the fitted regression of recruitment probability to grazer density (see Sokal et al., 1998, for a related approach to defining structures with local spatial autocorrelation).

Reflection on scale in the artificial ecology draws attention to the lack of scaling in the original time series. Although records of fluctuation in *Fucus* cover and limpet abundance exist for a period of over 20 years, the spatial scale of observations is limited to a 2 m² permanent quadrat. From a fixed scale of observation, it is not clear whether the small-scale process of limpet grazing really drives the fluctuations in algal cover or whether the fluctuations reflect larger-scale processes such as interannual variability in recruitment success across the entire shore (Gunnill, 1980; Lively et al., 1993). These alternatives can be tested by looking at the correlation between small and large scales. Where local grazing processes are important, the patches cycle independently of *Fucus* abundance at the large scale (Figure 15.2C). In contrast, if the recruitment of *Fucus* is unpredictable at large scales, the dynamics at the patch scale and the large scale become correlated (Figure 15.5). This observation suggested a novel way of using a photographic time series of the entire shore in the Isle of Man to examine the influence of small-scale processes on *Fucus* abundance (Johnson et al., 1998b). A consistent ranking of the photographs was produced after presenting them in random order to seven different observers. The correlation between this ranking and the abundance of *Fucus* in records from the 2 m² quadrat was low (0.237, $p > 0.5$). Hence field observations suggest that local processes are important in the temporal dynamics of the *Fucus* mosaic on rocky shores in the northeast Atlantic.

15.5 Conclusions

Research on the limpet–*Fucus* mosaic and mussel beds (Wootton, 2001b) suggests that it is possible to combine empirical and theoretical approaches directly to improve the understanding of processes in benthic communities. Empirical description of model rules facilitates model testing, while the models themselves can be used to derive new ways of testing field data. The tools applied here have different strengths and weaknesses, but they can generally be applied to analysis of the same data set. Contrasts between the predictions of the different methods may provide a fuller understanding of any community than application of a single approach.

Markov transition matrices appear to produce reasonable first approximations of community composition. This may reflect the relatively open nature of many benthic communities. The transition rate to a certain state (say, mussel occupied) may not be affected by the number of sites already occupied by mussels as the larvae come from elsewhere (the population is open). Under these circumstances, the frequency-invariant nature of transition probabilities may not be an issue. Further research on Markov models is needed to characterize the features that would result in inaccurate projections of community composition. Algorithms are needed for parsimonious selection of community states in the matrices as well as investigations of spatial grain (the optimal size of the "cells" in models). The sensitivity of communities to particular species transitions is an interesting area. Tanner et al. (1994) present a sensitivity analysis that may be technically invalid: perturbations to transition probabilities cannot be examined independently of one another due to the constraint on column totals in the transition matrix to sum to 1. Wootton (2001b) suggests an alternative method of sensitivity analysis when looking at the loss of species from a community. The temporal scaling of community dynamics provided by the convergence timescales may be a useful way of classifying community resilience to perturbations. It would be interesting to test this approach using data from the time series that exist from experimental perturbations of rocky shore communities (e.g., Dye, 1998).

The spatial transition matrix models appear to offer fewer insights on community pattern. Despite the demonstration of a spatial component to transition probabilities based on the presence or absence of limpets in adjoining cells, there were no improvements in predictive power in comparison to nonspatial models. In addition the CA approach did not generate spatial pattern. However, spatial transition matrix models are relatively easy to derive as an alternative to point models. The two matrices derived are a subset of a very large number of possible spatial transition rules. Even if the rules do not generate pattern, they can be used as part of a number of methods of investigating structuring processes in communities (e.g., Law et al., 1997), although some techniques may be restricted (Freckleton and Watkinson, 2000) to species with limited dispersal of propagules. One area where simpler spatial transition matrix models may be appropriate is in communities where space-occupying individuals or colonies grow out horizontally so that effects on neighbors are strong. Encrusting communities of groups such as bryozoans (Barnes and Dick, 2000) could be an example of this.

The most flexible approach to modeling communities is to use an artificial ecology. There are dangers of producing a sophisticated "realistic" model that is intuitively satisfying yet fails to provide insights on the dynamics and spatial scales of real communities. A potential check on this is to embed empirically defined rules and scales within the model. Hence it should be clear where to look for any scaling behavior or patterns derived in the model. A potential restriction on wider application of these techniques is that benthic ecologists have tended not to collect data repeatedly on regularly spaced grids. However, repeated data collection at the same sites can be a powerful technique for identifying pattern and process (Bouma et al., 2001). It has been suggested that regularly spaced samples can complement the more common ANOVA-based hierarchical techniques (Underwood and Chapman, 1996). If use of both survey approaches becomes more common, this will increase the opportunities to investigate scaling in empirically based models of benthic communities.

Acknowledgment

Steve Hawkins and Mike Burrows provided fruitful discussion on aspects of this work.

References

Alexander, S.E. and Roughgarden, J., Larval transport and population dynamics of intertidal barnacles: a coupled benthic/oceanic model, *Ecol. Monogr.*, 66, 259, 1996.

Barnes, D.K.A. and Dick, M.H., Overgrowth competition in encrusting bryozoan assemblages of the intertidal and infralittoral zones of Alaska, *Mar. Biol.*, 136, 813, 2000.

Bence, J.R. and Nisbet, R.M., Space-limited recruitment in open systems: the importance of time delays, *Ecology*, 70, 1434, 1989.

Bouma, H., de Vries, P.P., Duiker, J.M.C., Herman, P.M.J., and Wolff, W.J., Migration of the bivalve *Macoma balthica* on a highly dynamic tidal flat in the Westerschelde estuary, The Netherlands, *Mar. Ecol. Prog. Ser.*, 224, 157, 2001.

Burrows, M.T. and Hawkins, S.J., Modelling patch dynamics on rocky shores using deterministic cellular automata, *Mar. Ecol. Prog. Ser.*, 167, 1, 1998.

Caley, M.J., Carr, M.H., Hixon, M.A., Hughes, T.P., Jones, G.P., and Menge, B.A., Recruitment and the local dynamics of open marine populations, *Annu. Rev. Ecol. Syst.*, 27, 477, 1996.

Callaway, R.M. and Davis, F.W., Vegetation dynamics, fire and the physical environment in coastal central California, *Ecology*, 74, 1567, 1993.

Caswell, H., *Matrix Population Models*, 2nd ed., Sinauer Associates, Sunderland, MA, 2001, chap. 4.

De Roos, A.M., McCauley, E., and Wilson, W.G.,. Mobility versus density-limited predator-prey dynamics on different spatial scales, *Proc. R. Soc. Lond. B*, 246, 117, 1991.

Dye, A.H., Dynamics of rocky intertidal communities: analyses of long time series from South African shores, *Estuarine Coastal Shelf Sci.*, 46, 287, 1998.

Earn, D.J.D., Levin, S.A., and Rohani, P., Coherence and conservation, *Science*, 290, 1360, 2000.

Ermentrout, G.B. and Edelstein-Keshet, L., Cellular automata approaches to biological modelling, *J. Theor. Biol.*, 160, 97, 1993.

Fernandes, T.F., Huxham, M., and Piper, S.R., Predator caging experiments: a test of the importance of scale, *J. Exp. Mar. Biol. Ecol.*, 241, 137, 1999.

Freckleton, R.P. and Watkinson, A.R., On detecting and measuring competition in spatially structured plant communities, *Ecol. Lett.*, 3, 423, 2000.

Gunnill, F.C., Recruitment and standing stocks in populations of one green alga and five brown algae in the intertidal zone near La Jolla, California during 1973–1977, *Mar. Ecol. Prog. Ser.*, 3, 231, 1980.

Harrison, S. and Quinn, J.F., Correlated environments and the persistence of metapopulations, *Oikos*, 56, 293, 1989.

Hartnoll, R.G. and Hawkins, S.J., Patchiness and fluctuations on moderately exposed rocky shores, *Ophelia*, 24, 53, 1985.

Hartnoll, R.G. and Wright, J.R., Foraging movements and homing in the limpet *Patella vulgata* L., *Anim. Behav.*, 25, 806, 1977.

Hawkins, S.J., Hartnoll, R.G., Kain(Jones), J.M., and Norton, T.A., Plant-animal interactions on hard substrata in the north-east Atlantic, in *Plant-Animal Interactions in the Marine Benthos*, John, D.M., Hawkins, S.J., and Price, J.H., Eds., Systematics Association Special Vol. 46, Clarendon Press, Oxford, U.K., 1992, 1–32.

Hendry, R.J. and McGlade, J.M., The role of memory in ecological systems, *Proc. R. Soc. Lond. B*, 259, 153, 1995.

Hilborn, R. and Mangel, M., *The Ecological Detective*, Princeton University Press, Princeton, NJ, 1997, chap. 1.

Horn, H.S., Forest succession, *Sci. Am.*, 232, 90, 1975.

Jenkins, S.R., Aberg, P., Cervin, G., Coleman, R.A., Delany, J., Hawkins, S.J., Hyder, K., Myers, A.A., Paula, J., Power, A.M., Range, P., and Hartnoll, R.G., Population dynamics of the intertidal barnacle *Semibalanus balanoides* at three European locations: spatial scales of variability, *Mar. Ecol. Prog. Ser.*, 217, 207, 2001.

Johnson, M., A re-evaluation of density dependent population cycles in open systems, *Am. Nat.*, 155, 36, 2000.

Johnson, M.P., Metapopulation dynamics of *Tigriopus brevicornis* (Harpacticoida) in intertidal rock pools. *Mar. Ecol. Prog. Ser.*, 211, 215, 2001.

Johnson, M.P., Burrows, M.T., Hartnoll, R.G., and Hawkins, S.J., Spatial structure on moderately exposed rocky shores: patch scales and the interactions between limpets and algae, *Mar. Ecol. Prog. Ser.*, 160, 209, 1997.

Johnson, M.P., Hawkins, S.J., Hartnoll, R.G., and Norton, T.A., The establishment of fucoid zonation on algal dominated rocky shores: hypotheses derived from a simulation model, *Functional Ecol.*, 12, 259, 1998a.

Johnson, M.P., Burrows, M.T., and Hawkins, S.J., Individual based simulations of the direct and indirect effects of limpets on a rocky shore *Fucus* mosaic, *Mar. Ecol. Prog. Ser.*, 169, 179, 1998b.

Keeling, M., Spatial models of interacting populations, in *Advanced Ecological Theory*, McGlade, J., Ed., Blackwell Science, Oxford, U.K., 1999, chap. 4.

Koenig, W.D., Spatial autocorrelation of ecological phenomena, *Trends Ecol. Evol.*, 14, 22, 1999.

Kostylev, V. and Erlandson, J., A fractal approach for detecting spatial hierarchy and structure on mussel beds, *J. Mar. Biol.*, 139, 497, 2001.

Law, R., Herben, T., and Dieckmann, U., Non-manipulative estimates of competition coefficients in a montane grassland community, *J. Ecol.*, 85, 505, 1997.

Levin, S.A., The problem of pattern and scale in ecology, *Ecology*, 73, 1943, 1992.

Levin, S.A. and Pacala, S.W., Theories of simplification and scaling of spatially distributed processes, in *Spatial Ecology*, Tilman, D. and Kareiva, P., Eds., Princeton University Press, Princeton, NJ, 1997, chap. 12.

Lindegarth, M., Andre, C., and Jonsson, P.R., Analysis of the spatial variability in abundance and age structure of two infaunal bivalves, *Cerastoderma edule* and *C. lamarcki*, using hierarchical sampling programmes, *Mar. Ecol. Prog. Ser.*, 116, 85, 1995.

Lively, C.M., Raimondi, P.T., and Delph, L.F., Intertidal community structure: space-time interactions in the Northern Gulf of California, *Ecology*, 74, 162, 1993.

Maynou, F.X., Sarda, F., and Conan, G.Y., Assessment of the spatial structure and biomass evaluation of *Nephrops norvegicas* (L.) populations in the northwestern Mediterranean by geostatistics, *ICES J. Mar. Sci.*, 55, 102, 1998.

Molofsky, J., Population dynamics and pattern formation in theoretical populations, *Ecology*, 75, 30, 1994.

Morrisey, D.J., Howitt, L., Underwood, A.J., and Stark, J.S., Spatial variation in soft sediment benthos, *Mar. Ecol. Prog. Ser.*, 81, 197, 1992.

Paine, R.T. and Levin, S.A., Intertidal landscapes: disturbance and the dynamics of pattern, *Ecol. Monogr.*, 51, 145, 1981.

Pascual, M. and Levin, S.A., Spatial scaling in a benthic population model with density-dependent disturbance, *Theor. Popul. Biol.*, 56, 106, 1999.

Pascual, M., Roy, M., Guichard, F., and Flierl, G., Cluster size distributions: signatures of self-organization in spatial ecologies, *Philos. Trans. R. Soc. Lond. B*, 357, 657, 2002.

Phipps, M.J., From local to global: the lesson of cellular automata, in *Individual-Based Models and Approaches in Ecology: Populations, Communities and Ecosystems*, DeAngelis, D.L. and Gross, L.J., Eds., Chapman & Hall, London, 1992, 165–187.

Possingham, H.P. and Roughgarden, J., Spatial population dynamics of a marine organism with a complex life cycle, *Ecology*, 71, 973, 1990.

Possingham, H., Tuljapurkar, S.D., Roughgarden, J., and Wilks, M., Population cycling in space limited organisms subject to density dependent predation, *Am. Nat.*, 143, 563, 1994.

Rand, D.A., Measuring and characterizing spatial patterns, dynamics and chaos in spatially extended dynamical systems and ecologies, *Philos. Trans. R. Soc. Lond. A*, 348, 498, 1994.

Rand, D., Correlation equations and pair approximations for spatial ecologies, in *Advanced Ecological Theory*, Mcglade, J., Ed., Blackwell Science, Oxford, U.K., 1999, chap. 5.

Rand, D.A. and Wilson, H.B., Using spatio-temporal chaos and intermediate-scale determinism to quantify spatially extended ecosystems, *Proc. R. Soc. Lond. B*, 259, 111, 1995.

Rohani, P., Lewis, T.J., Grünbaum, D., and Ruxton, G.D., Spatial self-organization in ecology: pretty patterns or robust reality? *Trends. Ecol. Evol.*, 12, 70, 1997.

Rossi, R.E., Mulla, D.J., Journel, A.G., and Franz, E.H., Geostatistical tools for modeling and interpreting ecological spatial dependence, *Ecol. Monogr.*, 62, 277, 1992.

Roughgarden, J., Iwasa, Y., and Baxter, C., Demographic theory for an open marine population with space-limited recruitment, *Ecology*, 66, 54, 1985.

Snover, M.L. and Commito, J.A., The fractal geometry of *Mytilus edulis* L. spatial distribution in a soft-bottom system, *J. Exp. Mar. Biol. Ecol.*, 223, 53, 1998.

Snyder, R.E. and Nisbet, R.M., Spatial structure and fluctuations in the contact process and related models, *Bull. Math. Biol.*, 62, 959, 2000.

Sokal, R.R. and Rohlf, F.J., *Biometry*, 3rd ed., W.H. Freeman, New York, 1996, chap. 17.

Sokal, R.R., Oden, N.L., and Thomson, B.A., Local spatial autocorrelation in biological variables, *Biol. J. Linn. Soc.*, 65, 41, 1998.

Southward, A.J., The population balance between limpets and seaweeds on wave beaten rocky shores, *Annu. Rep. Mar. Biol. Stat. Port. Erin.*, 68, 20, 1956.

Tanner, J.E., Hughes, T.P., and Connell, J.H., Species coexistence, keystone species and succession: a sensitivity analysis, *Ecology*, 75, 2204, 1994.

Tanner, J.E., Hughes, T.P., and Connell, J.H., The role of history in community dynamics: a modelling approach, *Ecology*, 77, 108, 1996.

Thrush, S.F., Cummings, V.J., Dayton, P.K., Ford, R., Grant, J., Hewitt, J.E., Hines, A.H., Lawrie, S.M., Pridmore, R.D., Legendre, P., McArdle, B.H., Schneider, D.C., Turner, S.J., Whitlatch, R.B., and Wilkinson, M.R., Matching the outcome of small-scale density manipulation experiments with larger scale patterns: an example of bivalve adult/juvenile interactions, *J. Exp. Mar. Biol. Ecol.*, 216, 153, 1997.

Tilman, D. and Kareiva, P., *Spatial Ecology*, Princeton University Press, Princeton, NJ, 1997.

Underwood, A.J. and Chapman, M.G., Scales of spatial patterns of distribution of intertidal invertebrates, *Oecologia*, 107, 212, 1996.

Underwood, A.J., Chapman, M.G., and Connell, S.D., Observations in ecology: you can't make progress on processes without understanding the patterns, *J. Exp. Mar. Biol. Ecol.*, 250, 97, 2000.

Usher, M.B., Markovian approaches to ecological succession, *J. Anim. Ecol.*, 48, 413, 1979.

Wilson, W.G. and Nisbet, R.M., Cooperation and competition along smooth environmental gradients, *Ecology*, 78, 2004, 1997.

Wootton, J.T., Local interactions predict large-scale pattern in empirically derived cellular automata, *Nature*, 413, 841, 2001a.

Wootton, J.T., Prediction in complex communities: analysis of empirically defined Markov models, *Ecology*, 82, 580, 2001b.

16

Fractal Dimension Estimation in Studies of Epiphytal and Epilithic Communities: Strengths and Weaknesses

John Davenport

CONTENTS

16.1 Introduction..245
16.2 Fractal Analysis and Biology 248
16.3 Fractal Dimensions in Ecology 249
16.4 How Is D Estimated?..251
16.5 Areal Fractal Dimensions of Intertidal Rocky Substrata — An Investigation........252
16.6 Value of Fractal Dimension Estimation to Marine Ecological Study 253
16.7 Limitations of Fractal Analysis 254
Acknowledgments..255
References..255

16.1 Introduction

Newton rules biology (*but Euclid doesn't!*)

(with apologies to Pennycuick)

It is many years since Mandelbrot[1] published his *The Fractal Geometry of Nature*. However, the significance of this seminal work has still to reach many biologists and ecologists, so some basic principles need to be rehearsed before consideration of the use of fractal analysis in aquatic ecology. Fractal geometry extends beyond the familiar Euclidean geometry of lines and curves, and has its roots in the 19th century (see Lesmoir-Gordon et al.[2] for a recent popular account), but remained the province of mathematicians until Mandelbrot's intervention. He relied heavily on an obscure publication by Richardson,[3] who had noted that published values for the length of geographical borders between countries differed between sources. Richardson found that such structures were usually measured from maps by the use of dividers — and that the total length resulting from such measurements varied depending on the scale of the map and the length of the step at which the dividers were set, as long as the borders were based on natural features, rather than being perfectly Euclidian political boundaries. The shorter the step, the longer the total length measured. He found that plotting log step length against log total length resulted in a straight line (Figure 16.1), provided that the dividers were not set too close together or too far apart. Mandelbrot[4] published coastline information (Figure 16.2) derived from Richardson's studies, and noted that the length of such structures tended toward the infinite, because of the phenomenon of "self-similarity" (Figure 16.3). For coastlines, for example,

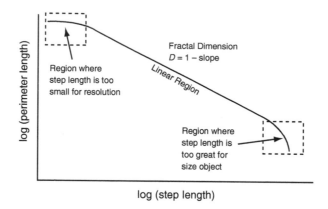

FIGURE 16.1 Richardson plot. (After Richardson[3] and Mandelbrot.[4])

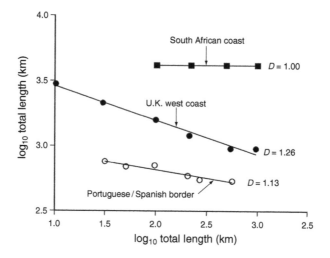

FIGURE 16.2 Richardson plots of geographical boundaries. (Redrawn and calculated from Richardson[3] and Mandelbrot.[4])

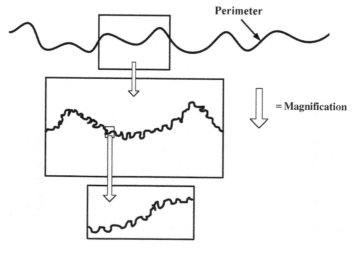

FIGURE 16.3 Diagram to illustrate phenomenon of self-similarity (as applied, for example, to coastlines by Mandelbrot[1,4]).

the complexity evident in charts will repeatedly become evident if a section of that coastline is studied in finer and finer detail until the outlines of individual grains of silt and sand are being traced, or beyond that until bacterial cells and protein molecules are evident. The upshot of this is that, with finer and finer measurement, the coastline length does not converge to some fixed "true" value, but keeps increasing, essentially forever. Coastlines are "fractal" (shapes that are detailed at all scales), a term coined by Mandelbrot.

These considerations apply to many natural objects and to areas and volumes as well as lines. A coastline does not have a length, nor does a human lung have an area or a volume; instead, they have "fractal extents."[5] Statements such as "the Nile has a length of 6670 km" or "human lungs have the surface area of a tennis court," although widely believed, are fundamentally erroneous — in the latter case not least because *both* lungs and tennis courts are fractal objects!

Fractal lines derived from natural objects or mathematicians' ingenuity differ from Euclidean lines in that they cannot be differentiated or integrated; they are not susceptible to calculus. However, values can be derived from them that are of utility. The commonest information is that of "fractal dimension" D. Figure 16.1 shows that fractal dimensions may be calculated from Richardson plots as $D = 1 -$ slope. There are many other methods of calculating fractal dimensions (see Russ[6] for review). Richardson plots are often the easiest to deal with intuitively in biological/ecological situations. A Euclidean curve or straight line will not vary in total length with step size (provided that step size is not too large to follow curves), so a Richardson plot will be a horizontal line and the slope value will be zero (so $D = 1$). An infinitely complex and self-similar line will have a slope of -1 ($-45°$ to the horizontal) so that $D = 2$. Values of 2 are only achieved by space-filling and completely self-similar mathematicians' fractal lines, but the perimeters of natural objects have D values somewhere between 1 and 2. For example, the U.K. coastline has a fractal dimension of about 1.26, while a typical cloud outline has a D of 1.35.[2] One of the more complex natural objects reported so far is the multiply branched, fine filamentous seaweed *Desmarestia menziesii* which has a D of 1.51 to 1.83 at step lengths of 1 to 8 cm (Figure 16.4).[7] For areas, a Euclidean area (flat or smoothly curved) will have $D = 2$; a completely self-similar complex area will have $D = 3$.

Another common method of estimation of D in ecology is by use of the boundary-grid method.[1,8,9] In this case square-section grids are laid over images of objects and the numbers of squares entered (N) by the profile of the object counted. This is repeated with grids of different sizes (square side length n) and is a process well suited to processing of digital images. Fractal dimension is calculated from

$$N = kn^{-D}$$

where k is a constant. D is easily estimated as the negative slope of a log–log plot of N upon n.

It must be stressed that, while fractal dimensions are measures of a certain sort of complexity, complex objects need not be fractal at all. A seaweed holdfast, for example, is in ordinary terminology a complex structure, but certainly the large holdfast of the basket kelp *Macrocystis pyrifera* is composed of a meshwork of tubular elements that are themselves virtually Euclidean (Davenport, unpublished data) and there is no hint of self-similarity — the essence of fractal objects — unless the holdfast is filled with silt and stones that provide that sort of complexity. Euclidean measures of complexity, such as circularity (circularity = $P2/4\pi A$, where P = perimeter and A = area; e.g., Park et al.[10]), surface roughness, average profile amplitude, and indices such as the potential settling site (PSS) index used by Hills et al.[11] in their barnacle settlement studies, are not rendered obsolete by fractal analysis.

Early work on plants generally made the assumption that a given plant had a representative single fractal dimension. However, it is evident that, if a wide range of scales are considered, this is far from true for seaweeds (Table 16.1, Figure 16.4 and Figure 16.5), with surfaces tending to become Euclidean at small scale. Marine macroalgae are therefore said to exhibit "mixed fractal" characteristics.[7] Penny-cuick,[5] when considering islands, noted that one with a rugged coastline could have a smooth vertical profile, so that a coastline D would differ from an elevation D. Biological objects can show similar disparities. Plants, both aquatic and terrestrial, are particularly prone to this as selection favors flat surfaces for the gathering of sunlight. So, while branching and leaf/frond serration may yield high D in some directions, the leaves, leaflets, and fronds may be almost completely Euclidean flat surfaces

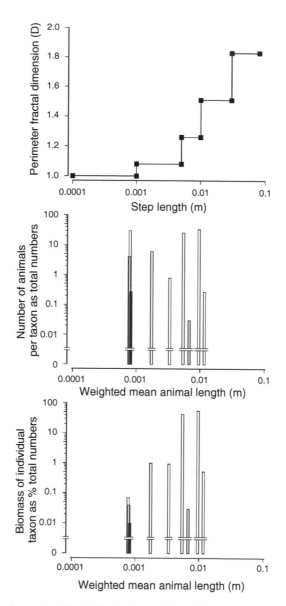

FIGURE 16.4 Fractal dimensions and epiphytal faunal characteristics of *D. menziesii*. (From Davenport, J. et al., *Mar. Ecol. Prog. Ser.,* 136, 245, 1996. With permission.)

(compare Table 16.1 and Table 16.2). Objects that are fractal in two dimensions, but not in a third, are described as "anisotropic."

16.2   Fractal Analysis and Biology

An exhaustive review of the use of fractals in biology is far beyond the scope of the present chapter. However, fractal analysis is now widespread (see Nonnenmacher et al.[12] for review) and highly varied, as may be illustrated by a few examples. In anatomy and paleontology it has been used to compare skull suture anatomy between mammals (e.g., Long and Long[13]), or to compare and characterize vascular networks (e.g., Herman et al.[14]), and in medicine it has been used to analyze the rhythmicity of eye movements in schizophrenic and normal patients.[15]

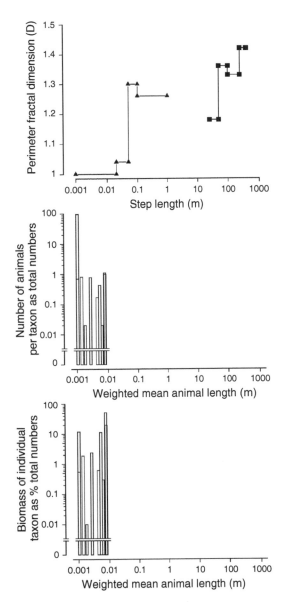

FIGURE 16.5 Fractal dimensions and epiphytal faunal characteristics of *Macrocystis pyrifera*. (From Davenport, J. et al., *Mar. Ecol. Prog. Ser.*, 136, 245, 1996. With permission.)

16.3 Fractal Dimensions in Ecology

Early applications included estimates of coral reef fractal dimension[16,17] and the derivation of a positive relationship between bald eagle nesting frequency and increasing coastline complexity.[18] The major ecological applications of fractal geometry initially centered on the links between plant (both terrestrial and aquatic) fractal geometry and associated faunal community structure (e.g., Morse et al.,[8] Lawton,[19] Shorrocks et al.,[20] Gunnarsson,[21] Gee and Warwick,[22,23] Davenport et al.,[8,24] Hooper[25]). In general terms, such studies have shown an association between high fractal dimensions of vegetation and greater diversity of animal community,[22] and/or greater relative abundance of smaller animals.[8,20–23] The utility of such studies is discussed in more detail later.

TABLE 16.1

Fractal Dimensions (*D*) of Perimeters of Images of Four Macroalgae from Sub-Antarctic South Georgia Measured over Various Scales

Macroalgae	Step Length Range (m)	Mean Perimeter *D*	SD
Macrocystis pyrifera	250–400	1.42	0.14
(kelp bed outlines)	100–250	1.33	0.05
	50–100	1.36	0.02
	25–50	1.18	0.03
Macrocystis pyrifera	0.1–1.0	1.26	0.04
(individual plants)	0.05–0.1	1.30	0.03
	0.02–0.05	1.04	0.00
	0.001–0.02	1.00	0.00
Desmarestia menziesii	0.03–0.08	1.83	0.10
(individual plants)	0.01–0.03	1.51	0.01
	0.005–0.01	1.26	0.01
	0.001–0.005	1.08	0.00
	0.0001–0.001	1.00	0.00
Schizoseris condensata	0.01–0.05	1.56	0.07
(individual plants)	0.005–0.01	1.34	0.02
	0.001–0.005	1.31	0.00
	0.0002–0.001	1.05	0.00
	0.00005–0.0002	1.04	0.00
Palmaria georgica	0.05–0.1	1.37	0.02
(individual plants)	0.01–0.05	1.41	0.02
	0.0025–0.01	1.17	0.01
	0.001–0.0025	1.13	0.01
	0.0001–0.001	1.00	0.00

Source: Davenport, J. et al., *Mar. Ecol. Prog. Ser.,* 136, 245, 1996. With permission.

TABLE 16.2

Cross-Frond Fractal Dimensions (*D*) of Three Macroalgae Measured over Various Scales

Macroalgae	Step Length Range (m)	Mean Cross-Frond *D*	SD
Macrocystis pyrifera	0.02–0.05	1.00	0.01
	0.001–0.02	1.00	0.00
	0.00025–0.001	1.03	0.01
	0.0001–0.00025	1.00	0.00
	0.00005–0.0001	1.04	0.00
	0.000001–0.00005	1.00	0.00
Schizoseris condensata	0.00001–0.01	1.00	0.00
	0.000001–0.00001	1.10	0.00
Palmaria georgica	0.00001–0.1	1.00	0.00
	0.000001–0.0001	1.01	0.00

Source: Davenport, J. et al., *Mar. Ecol. Prog. Ser.,* 136, 245, 1996. With permission.

More recently, aquatic ecologists have shifted to fractal analysis of a wider range of sorts of habitat complexity. At small scale a particularly elegant study was conducted by Hills et al.[11] who investigated settlement behavior in the barnacle *Semibalanus balanoides*. They used replicated epoxy surfaces that simulated solid substrata of varied complexity. They were able to demonstrate that cyprids of the barnacle selected sites on the basis of Euclidean measures of surface complexity and were oblivious to fractal detail. The presence of fouling animals on soft substrata and intertidal rock has also attracted attention; a particularly interesting paper is that of Kostylev et al.,[26] who compared the distributions of various morphs of the snail *Littorina saxatilis* on mussel and barnacle patches, demonstrating greater abundance associated with higher *D*, but also showing that snail size increased with *D* (against expectation), because higher *D* values were found for mussel patches — where interstices were large enough to act as refuges.

Fractal analysis is also a mainstay of landscape ecology (Milne[27,28]) allowing the examination of spatial and temporal complexity to discover how ecological phenomena change steadily, but predictably, at multiple scales. Another aspect of fractal use that has an impact on wide areas of ecology is that of the study of movement by animals. Provided that information is available (e.g., by videophotography, radio-tracking, or remote sensing by satellite), it is possible to reconstruct paths of animals employed during foraging or migration. These paths can then be subject to fractal analysis. This has been done at many scales, from the foraging of marine ciliates[29] in relation to food patch availability, to the movements of polar bears in relation to the fractal dimensions of sea ice.[30] Landscape ecology can use this approach in the study of foraging herbivores, while it has resonance in marine ecology with investigation of foraging and trail following by intertidal gastropods (e.g., Erlandson and Kostylev[31]).

16.4   How Is *D* Estimated?

Measurement of true surface fractal dimension of objects is difficult and measuring techniques currently rely heavily on assessment of boundary complexity of two-dimensional images extracted from three-dimensional objects.[6] Thus measured *D* is a good estimate of its overall complexity if an object is isotropic, i.e., similarly complex in three dimensions as in two, but not if it is anisotropic, i.e., its complexity in the third dimension is different from that in the other two. To illustrate the process of estimation, a particularly complex example is given here for the basket kelp of the Southern Hemisphere, *Macrocystis pyrifera*. *Macrocystis* is reputedly the largest alga in the world and occurs in extensive beds that can be kilometers in extent. The process used to determine its fractal dimension over a wide range of scales was as follows.[7] Three whole plants were collected. Each, in turn, was laid out with minimal overlapping of blades on flat ground and photographed from a platform 6 m high, using a 10 m tape to provide a scale. A sequence of eight 35 mm color transparencies was taken (50 mm lens) to yield a montage of the whole plant. Next, three photographs of randomly chosen parts of the plants were taken against a 1 m measure with an 80 to 200 mm lens. Finally, with a macro lens, three randomly chosen parts of weed were photographed with a 50 mm macro lens so that a full frame occupied 0.1 m. Randomly chosen blades of each plant (complete with pneumatocyst and piece of stipe) were preserved in 2% seawater-formalin and returned to the laboratory where images were obtained by photocopying, macrophotography, and microscopy. Vertical aerial photographs (taken by 152 mm lens from a height of about 3000 m) yielded images of whole *Macrocystis* beds that were also susceptible to magnification. Sections of fronds were cut with a sharp blade and mounted on either glass slides (for microscopic investigation) or aluminum stubs (for analysis by scanning electron microscopy) so that cross-frond *D* could be estimated.

Two-dimensional images for estimate of perimeter *D* were obtained from each plant (or part of plant) by combinations of direct photocopying of plant material (using both enlarging and shrinking as appropriate), microscope/*camera lucida* drawings of plant pieces or projected 35 mm slides in the case of whole/part *Macrocystis* plants or whole kelp beds. Precise magnifications were chosen pragmatically. Perimeter *D* for each plant image at each magnification was measured by the "walking dividers" method and construction of a Richardson plot (it could equally have been determined by the boundary-grid technique). The dividers were walked with alternation of the swing (i.e., clockwise then anticlockwise rotation) to avoid bias. A total of five replicate perimeter measurements, started from different randomly selected points were made from each magnification, and for a given image, at least five points in the straight line region of the Richardson plot for each of the three plants were regressed to calculate *D*. This allowed assessment of fractal dimension over six orders of magnitude of step length (Figure 16.5). Cross-frond *D* was measured over a smaller range of steps (Table 16.2).

Sometimes considerable ingenuity has been needed to collect images. Particularly noteworthy are the recent studies of Commito and Rusignuolo.[32] Snover and Commito[33] had already shown that the outlines of soft-bottom mussel beds were fractal (confirmed more recently for intertidal rock mussel patches by Kostylev and Erlandson.[34] Commito and Rusignuolo wanted to look at mussel bed surface topography. To do this they encased portions of mussel beds in plaster of Paris, then sawed the resultant casts to yield surface profiles that were coated with graphite, then scanned into a computer for analysis based on the boundary-grid method. Work at smaller scales has created additional problems. Kostylev et al.[26] used contour gauges with 1 mm pins to record profiles of rocky shores that possessed mussel and barnacle cover, while Hills et al.[11] who studied settlement of barnacle cypris larvae used a laser profilometer device to record the profiles of their manufactured epoxy settlement panels.

16.5 Areal Fractal Dimensions of Intertidal Rocky Substrata — An Investigation

An exploratory investigation was conducted by the author in September 2001 on the southern coast of County Cork, Ireland. This was designed to establish first whether it was feasible to make area-based estimations of fractal dimension directly without collecting images, and second to determine whether visibly different complexities of rock surface showed significant differences in fractal dimension. This was a necessary preliminary to any investigation of the effects of fractal dimension on epilithic faunas and floras. A rigid 1 m² aluminum quadrat of the design shown in Figure 16.6 was used, together with a family of fine-pointed metal dividers set to the following step lengths: 200, 150, 120, 100, 80, 50, and 20 mm.

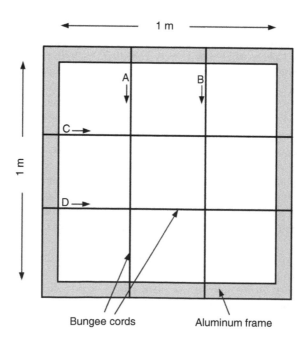

FIGURE 16.6 Quadrat design for rock surface fractal estimation.

TABLE 16.3

Raw Transect Data Collected from a Rough Intertidal Rock Surface
on the Upper Shore at Garretstown, Co. Cork, Ireland

Divider Step Length (mm)	A Length (mm)	B Length (mm)	C Length (mm)	D Length (mm)
200	1040	1020	1030	1090
150	1030	1040	1022	1092
100	1130	1135	1165	1070
80	1050	1043	1049	1078
50	1090	1048	1110	1120
20	1048	1020	1102	1122

Note: A 1 m^2 quadrat was used. A through D refer to bungee cord transects (see Figure 16.6).

Three rock surfaces were investigated, all on the upper shore where visible macrofauna were limited to the mobile gastropod *Littorina saxatilis*. This was done to avoid the complication of biogenic hard material (barnacles, mussels, saddle oysters, serpulid worms, etc.). The first was a smooth, gently undulating surface with no littorinids, and the second a smooth surface with a few large cracks that contained *L. saxatilis*. The last was a very rough fractured shale surface with many facets, crevices, and cracks and plentiful specimens of *L. saxatilis*. All three surfaces featured hard, nonporous rock. In each case the quadrat was placed in random orientation on the surface and weighted with lead weights at the corners to prevent movement. Dividers (of each step length) were walked along each of the four quadrat bungee cords in turn and the number of steps counted until another step was impossible; at this point the distance from the end of the last step to the inner edge of the quadrat was measured with a ruler and added to the cumulative step distance recorded. Data collection was complete when seven step values had been obtained for each of cords A to D (28 in total). This was a time-consuming process, particularly for the rough surface. Measurement from all three quadrat placements occupied two people for about 3 h. For mathematical analysis, total lengths for different step lengths established for cords A and C were multiplied to give seven estimates of total rock area enclosed by the quadrat. This process was repeated for cords B and D. The mean of the two sets of area estimations was then Richardson-plotted against the square of the divider step length (on a double log$_{10}$ basis), and a regression equation was obtained. From the slope of the resultant line, the areal fractal dimension was established. The results were as follows: smooth surface $D = 2.02$ (SD 0.014), smooth surface with cracks $D = 2.08$ (SD 0.089), rough shale surface $D = 2.01$ (SD 0.063). Effectively, all three surfaces were near Euclidean ($D = 2$) and certainly indistinguishable from each other. Individual bungee cord transect profile D did not exceed 1.12 even on roughest part of the shale surface. Transect lengths on the rough surface were greater than 1 m (reflecting the roughness), but the transect length was little affected by step length (Table 16.3). This finding reinforces the fact that fractal dimension values reflect a certain sort of complexity (that incorporates self-similarity) and are not a measure of complexity per se. The exercise showed that it was possible to measure fractal dimension of rock surfaces, and to do so on an areal basis. However, the basic message from the exercise is that rock surfaces are unlikely to be fractal to an extent where comparative exercises are worthwhile — unless enriched by the presence of barnacle cover or mussel patches (cf. Kostylev et al.[26]). Just as vegetational studies show that there are scales at which plant structures are fractally complex, and scales at which they are not, it seems probable that intertidal rock surfaces are Euclidean intervals in a scale sequence ranging from fractally complex sediments to complex coastlines. This perhaps stems from smoothing and polishing by wave action in combination with sediment load.

16.6 Value of Fractal Dimension Estimation to Marine Ecological Study

Complexity of habitat structure has profound effects on the nature of the ecology of marine habitats. Increased complexity alters flow rates over and through the habitat: it provides increased possibilities of attachment, shade, and hiding places. Quantitative measurement of structural complexity greatly enhances the possibilities of attributing patterns of epiphytal or epilithic assemblage composition to features of that complexity (e.g., Davenport et al.,[7] Kostylev et al.[26]). In the future it is possible to envisage that increasing accuracy of complexity estimation (including the subset of fractal complexity), combined with better measurement of complexity of use (e.g., by foragers and their predators), will generate important new information.

An interesting question arising from work on epiphytal and epilithic faunas concerns whether epiphytal or epilithic animals inhabit Euclidean or fractal domains.[7] By this is meant: Is their size such that are they likely to perceive the environment as a simple or complex one? Comparison of Figure 16.4 and Figure 16.5 illustrates this question. *Desmarestia* is a complex red seaweed with numerous small leaflets a few millimeters across. *Desmarestia* plants are normally no more than 0.5 m high and do not form beds. From Figure 16.4 it may be seen that many small animals (predominantly harpacticoid copepods) are of a size (<1 mm) where their immediate environment is likely to be uniformly flat and Euclidean. However, the bulk of the epiphytal biomass is composed of animals around 1 cm in body length, which means that their bodies are of comparable dimension to *Desmarestia* structure where D is 1.1 to 1.6. For such animals the seaweed surrounding them will undoubtedly be perceived as complex. The situation for *Macrocystis* is very different. This huge alga is fractally complex ($D > 1.3$) at scales from 10 cm to 1 km, as fish, seals, and human SCUBA divers undoubtedly find as they swim through its complex meshwork of blades, stipes, and pneumatophores. However, all of the epiphytal faunal assemblage is composed of animals <1 cm in length — for them the habitat must appear to be simple and Euclidean.

Hills et al.[11] provide a penetrating analysis of the relationship between surface texture and settlement in barnacles. It has long been known that barnacle cyprids perceive surface texture and use textural characteristics to make "decisions" about settlement (e.g., Crisp and Barnes,[35] Le Tourneux and Bourget,[36] Hills and Thomason[37]). Hills et al.[11] were able to demonstrate that the cyprids perceived Euclidean textural forms close to their body size in dimension rather than responding to fractal clues.

16.7 Limitations of Fractal Analysis

The analysis of rock surfaces presented here suggests that such surfaces are not susceptible to useful fractal analysis at least in the centimeter range. This finding is similar to the results of a series of investigations on coral reefs in the early 1980s. Bradbury and Reichelt[16] initially (and erroneously) calculated that coral reef structure was fractally complex with high contour D values (step range 10 to 1000 cm); this was correlated by them with the known complexity of the associated ecosystems. Mark[17] pointed out their computational error and Bradbury et al.[38] reinterpreted and extended the data, finding low contour D values of 1.05 to 1.15 indicating that coral reefs are actually very smooth and near Euclidean, essentially putting a stop to further fractal analysis in that environment. However, it is evident that rocks enriched with biogenic material are a fruitful source of further study (cf. Kostylev et al.[26]), although care must be taken not to underestimate the fractal complexity of material such as mature mussel patches that contain many voids and overhangs.[32]

There are limitations in vegetational studies, as well. At present, a major problem lies in reliable comparisons between plants that are complex in three dimensions with those that are complex only in two. Davenport et al.[24] compared the epiphytal assemblages of a range of lower-shore algae, demonstrating that coralline turf with a high D had a far higher level of biomass and species diversity than neighboring green and brown algae. The epiphytic fauna of the turf was also much less disturbed by emersion. However, the green and brown algae have a far more two-dimensional structure and tend to collapse during emersion. The real differences in fractal complexity between the types of algae were undoubtedly underestimated. Many studies of intertidal algae have so far ignored the changes in form associated with the emersion–immersion cycle. Ideally, three-dimensional images need to be collected, perhaps by stereophotography,[25] but neither hardware to collect information nor software to analyze it subsequently are readily available at present.

A fundamental limitation of the great majority of fractal investigations conducted so far was identified by Hills et al.[11] They point out that almost all such work has demonstrated correlation rather than causality; only experimental work can reveal the latter. Unfortunately, experimental manipulation of environmental fractal dimension is generally not feasible, save at the relatively small scales employed by Hills et al. so this basic deficiency is a weakness of fractal investigations that needs recognition by all who use them.

Acknowledgments

Much of the author's work with fractals stemmed from studies in South Georgia and Australia. These were supported by grants from the Royal Society and the TranAntarctic Association. The author thanks David Walton and Bill Block of the British Antarctic Survey for facilitating his visit to Husvik, South Georgia, and Alan Butler for his hospitality at the University of Adelaide. In Ireland thanks are due to Bob MacNamara and Julia Davenport for their help in the investigation of rock surface fractal dimension.

References

1. Mandelbrot, B., *The Fractal Geometry of Nature*, W. H. Freeman, New York, 1977.
2. Lesmoir-Gordon, N., Rood, W., and Rodney, R., *Introducing Fractal Geometry*, Icon Books, Cambridge, U.K., 2000.
3. Richardson, L.F., The problem of contiguity: an appendix of statistics of deadly quarrels. *Gen. Syst. Ybk.*, 6, 139, 1961.
4. Mandelbrot, B.B., How long is the coast of Britain? Statistical self-similarity and fractional dimension. *Science*, 156, 636, 1967.
5. Pennycuick, C.J., *Newton Rules Biology*, Oxford University Press, Oxford, 1992.
6. Russ, J.C., *Fractal Surfaces*, Plenum Press, New York, 1994.
7. Davenport, J., Pugh, P.J.A., and McKechnie, J., Mixed fractals and anisotropy in subantarctic marine macroalgae from South Georgia: implications for epifaunal biomass and abundance. *Mar. Ecol. Prog. Ser.*, 136, 245, 1996.
8. Morse, D.R., Lawton, J.H., Dodson, M.M., and Williamson, M.H., Fractal dimension of vegetation and the distribution of arthropod body lengths. *Nature*, 314, 731, 1985.
9. Sugihara, G. and May, R.M., Applications of fractals in ecology. *Trends Ecol. Evol.*, 5, 79, 1990.
10. Park, K., Mao, F.W., and Park, H., Morphological characterization of surface-induced platelet activation. *Biomaterials*, 11, 24, 1990.
11. Hills, J.M., Thomason, J.C., and Muhl, J., Settlement of barnacle larvae is governed by Euclidean and not fractal characteristics. *Funct. Ecol.*, 13, 868, 1999.
12. Nonnenmacher, T.F., Losa, G.A., and Weibel, E.R., *Fractals in Biology and Medicine*, Birkhäuser, Cambridge, 1994.
13. Long, C.A. and Long, J.E., Fractal dimensions of cranial structures and wave-forms. *Acta Anat.*, 145, 201, 1992.
14. Herman, P., Kocsis, L., and Eke, A., Fractal branching pattern in the pial vasculature in the cat. *J. Cereb. Blood Flow Metab.*, 21, 741, 2001.
15. Yokoyama, H., Niwa, S., Itoh, K., and Mazuka, R., Fractal property of eye movements in schizophrenia. *Biol. Cyber.*, 75, 137, 1996.
16. Bradbury, R.H. and Reichelt, R.E., Fractal dimension of a coral reef at ecological scales. *Mar. Ecol. Prog. Ser.*, 10, 169, 1983.
17. Mark, D.M., Fractal dimension of a coral reef at ecological scales: a discussion. *Mar. Ecol. Prog. Ser.*, 14, 293, 1984.
18. Pennycuick, C.J. and Kline, N.C., Units of measurement for fractal extent, applied to the coastal distribution of bald eagle nests in the Aleutian Islands, Alaska. *Oecologia* (Berlin), 68, 254, 1986.

19. Lawton, J.H., Surface availability and insect community structure: the effects of architecture and fractal dimension of plants, in *Insects and the Plant Surface*, Juniper, B.E. and Southwood, T.R.E., Eds., Edward Arnold, London, 1986, 317–331.

20. Shorrocks, B., Marsters, J., Ward, I., and Evennett, P.J., The fractal dimension of lichens and the distribution of arthropod body lengths. *Funct. Ecol.*, 5, 457, 1991.

21. Gunnarsson, B., Fractal dimension of plants and body size distribution in spiders. *Funct. Ecol.*, 6, 636, 1992.

22. Gee, J.M. and Warwick, R.M., Metazoan community structure in relation to the fractal dimensions of marine macroalgae. *Mar. Ecol. Prog. Ser.*, 103, 141, 1994.

23. Gee, J.M. and Warwick, R.M., Body-size distribution in a marine metazoan community and the fractal dimensions of macroalgae. *J. Exp. Mar. Biol. Ecol.*, 178, 247, 1994.

24. Davenport, J., Butler, A., and Cheshire, A., Epifaunal composition and fractal dimensions of marine plants in relation to emersion. *J. Mar. Biol. Assoc. U.K.*, 79, 351, 1999.

25. Hooper, G., Effects of Algal Structure on Associated Motile Epifaunal Communities. Ph.D. thesis, University of London, 2001.

26. Kostylev, V., Erlandson, J., and Johannesson, K., Microdistribution of the polymorphic snail *Littorina saxatilis* (Olivi) in a patchy rocky shore habitat. *Ophelia*, 47, 1, 1997.

27. Milne, B.T., Lessons from applying fractal models to landscape pattern, in *Quantitative Methods in Landscape Ecology*, Turner, M.G. and Gardner, R.H., Eds., Springer-Verlag, New York, 1991, 199–235.

28. Milne, B.T., Application of fractal geometry in wildlife biology, in *Wildlife and Landscape Ecology: Effects of Pattern and Scale*, Bissonette, J.A., Ed., Springer-Verlag, New York, 1997, 32–68.

29. Jonsson, P.R. and Johansson, M., Swimming behaviour, patch exploitation and dispersal capacity of a marine benthic ciliate in flume flow. *J. Mar. Exp. Mar. Biol. Ecol.*, 215, 135, 1997.

30. Ferguson, S.H., Taylor, M.K., Born, E.W., and Messier, F., Fractals, sea-ice landscape and spatial patterns of polar bears. *J. Biogeogr.*, 25, 1081, 1998.

31. Erlandson, J. and Kostylev, V., Trail following, speed and fractal dimension of movement in a marine prosobranch, *Littorina littorea*, during a mating and a nonmating season. *Mar. Biol.*, 122, 87, 1995.

32. Commito, J.A. and Rusignuolo, B.R., Structural complexity in mussel beds: the fractal geometry of surface topography. *J. Mar. Exp. Mar. Biol. Ecol.*, 255, 133, 2000.

33. Snover, M.L. and Commito, J.A., The fractal geometry of *Mytilus edulis* L. spatial distribution in a soft-bottom system. *J. Mar. Exp. Mar. Biol. Ecol.*, 223, 53, 1998.

34. Kostylev, V. and Erlandson, J., A fractal approach for detecting spatial hierarchy and structure on mussel beds. *Mar. Biol.*, 139, 497, 2001.

35. Crisp, D.J. and Barnes, H., The orientation and distribution of barnacles at settlement with particular reference to surface contour. *J. Anim. Ecol.*, 23, 142, 1954.

36. Le Tourneux, F. and Bourget, E., Importance of physical and biological settlement cues used at different spatial scales by the larvae of *Semibalanus balanoides. Mar. Biol.*, 97, 57, 1988.

37. Hills, J.M. and Thomason, J.C., A multi-scale analysis of settlement density and pattern dynamics of the barnacle, *Semibalanus balanoides. Mar. Ecol. Prog. Ser.*, 138, 103, 1996.

38. Bradbury, R.H., Reichelt, R.E., and Green, D.G., Fractals in ecology: methods and interpretation. *Mar. Ecol. Prog. Ser.*, 14, 295, 1984.

17

Rank-Size Analysis and Vertical Phytoplankton Distribution Patterns

James G. Mitchell

CONTENTS

17.1 Introduction..257
     17.1.1 Data Set Size in Biological Oceanography..257
     17.1.2 Initial Analysis..258
     17.1.3 Rank-Size Method...258
     17.1.4 Recent Rank-Size Use 259
     17.1.5 Rank-Size Interpretation...259
17.2 Low Resolution and Times Series Fluorescence Profiles.............................263
     17.2.1 An Idealized Rank-Size Slope 268
17.3 Conclusion 275
Acknowledgments...277
References..277

17.1 Introduction

17.1.1 Data Set Size in Biological Oceanography

One of the marks of modern ecology, particularly for phytoplankton in aquatic environments, is the increasing size of data sets. This began with fluorometry data in the early 1970s (Platt, 1972) and satellite data, such as the coastal zone color scanner, in the late 1970s (Feldman et al., 1984). The trend toward large data sets has been continued by programs such as JGOFS, with the HOTS and BATS data sets as specific examples. What distinguishes these programs is the large data sets and the large number of scientists and technicians working on the data. They may contribute in the form of designing hardware, physical or electronic smoothing devices, initial filtering, interpretation or provision in some archival format (e.g., SeaWifs and BATS). This spreads the data inundation and prevents an individual from drowning in data preparation, reduction, analysis, and interpretation.

As electronics have become cheaper and personal computers more powerful, individual scientists increasingly have access to instruments that produce large data sets. Additionally, the Web has allowed individual scientists easy access to archival data sets. The number of data points referred to in the term "large data sets" is ever increasing and varies markedly from discipline to discipline. This chapter is concerned primarily with data sets that have at least 10,000 data points, particularly those that are a time or spatial series. There are many programs, such as SPSS and Matlab, which facilitate the rapid use of methods, including fast Fourier transformation, to provide power spectra. The implicit assumption of a Gaussian distribution is rarely tested and unless one is extremely fortunate, the data set usually requires some massaging (e.g., despiking and detrending) before the analysis can be performed. In addition, exactly

0-8493-1344-9/04/$0.00+$1.50
© 2004 by CRC Press LLC

what happens in these processes and which options to choose (e.g., Tukey–Hamming vs. Bartlett's method) require careful consideration if it is to be done properly. Some data sets that are inherently nonstationary or samples that are of variable spatial or temporal intervals are not suitable for some of these powerful methods, and require further processing or interpretation, or altogether different software programs.

17.1.2   Initial Analysis

The first question for the individual scientists confronted with multiple large data sets is which, if any, of those data sets contain meaningful data and are worthwhile spending the time on further analysis. Approaching data sets in this way is luxurious because it implies that there is sufficient data to discard flawed data sets, incomplete or failed experiments. Such luxury may be an unanticipated bonus to electronic data collection and allow improved data quality by moving away from the concept, often produced under pressure, that all data must be used and are useful, no matter how much time, interpolation, extrapolation, transformation, and *post hoc* assuming must be done. If the experimental design is such that discarding any data set would bias the results or be considered selective data use, then what is still needed is an understanding of the quality of the data.

Although large data sets conjure visions of satellite acquisition, measuring the distribution and interaction of pico- to microplankton over distances relevant to the interaction of individual plankton can potentially produce large data sets. The large number of plankton in the ocean makes applying individual interactions to ocean-scale processes difficult to grasp. This perspective has eluded many. Hutchinson's (1961) proposal that a paradox existed because there was high plankton diversity in a homogeneous environment completely fails to grasp the potential for heterogeneity created by the environment and by individual plankton behavior. The fundamental nature of behavior as a potential generator of heterogeneity in unicellular plankton has only recently been realized (Young et al., 2001; Long and Azam, 2001).

Measuring the interaction of individual microplankton is still beyond present technical capabilities. Progress toward this goal is marked by the increasing resolution to which microplankton distributions can be measured (Duarte and Vaqué, 1992; Seymour et al., 2000; Long and Azam, 2001). The limiting feature of all of these microscale distribution measurements is that they contain too few samples to allow good analysis or the formation of a reliable picture of what an average microscale distribution looks like. High-resolution fluorometry is used to describe chlorophyll distributions for entire water columns at the centimeter scale. The first result of such an undertaking is the production of tens of thousands of data points. The focus of this chapter is to show what the initial data processing of these large data sets tells us about microscale phytoplankton distributions.

17.1.3   Rank-Size Method

The methodological goals of this chapter are to compare spectral analysis, rank-size analysis, and qualitative descriptions to see what insight each of these can produce in their own right. There is an extensive literature on spectral analysis (reviewed by Chatfield, 1989). Here, spectral analysis is employed as a basis for comparison of rank-size analysis. The latter is the superset of what is often referred to as Zipf's law. This type of analysis first came to prominence in the literature when Pareto (1895, in Bak, 1999) and Zipf (1949) used it for examining economics and human population dynamics. The simplicity and ease appealed in the pre-computer, pre-electronic calculator era. Since that time, the appeal has continued and the process has seen widespread usage in areas such as human demographics and physics.

Rank-size analysis often produces so-called power laws. Zipf's law can be described as

$$s \propto r^{-b}$$

where s is the size of a value, r is the rank of that value, and b is the slope. This power law is classically described as Zipf when b is approximately 1. Similarly, Pareto's law, when the number of values larger than a given value is an inverse function of that value, is a cumulative version of Zipf. Specifically,

$$s(X > x) \propto x^{-b}$$

TABLE 17.1

The Slopes for Systems Studied Using Rank-Size Analysis

System	Slope	Ref.
All U.S. businesses	−1.059	Axtell, 2001
Words	−0.74 to −0.9	Cizrók et al., 1995
Words	−1.33	Perline, 1996
DNA	−0.206 to −0.537	Mantegna et al., 1995
Percolation lattice	−0.550 to −0.576	Watanabe, 1996
Open ion channels	−1.24	Mercik et al., 1999
Closed ion channels	−4.16	Mercik et al., 1999
Gene expression	−0.54 to −0.84	Ramsden and Vohradský, 1998

where X is the cumulative value larger than a given x. In the most general form, X is binned into exponentially wider bins as a probability density function. Zipf and Pareto have been described as separate power laws (Faloutsos et al., 1999). However, their form is identical by inspection and the two approaches are functionally interchangeable (Adamic, 2000). The significance of this interchangeability is that the probability density functions of Pareto's law are more simply expressed as a simple Zipf-style ranking of size. For this reason, the analysis in this chapter is confined to the simpler method.

17.1.4 Recent Rank-Size Use

Ranking by size has found considerable popularity in physical sciences, economics, and molecular biology. The power law exponents for various systems are shown in Table 17.1. This is far from an exhaustive list, but makes the point that many papers claiming Zipf distributions produce slopes that deviate significantly from −1. Classically, a Zipf distribution has a slope of −1. Zipf is applied to such a wide range of slopes that it is often synonymous with "power law" as Table 17.1 makes clear. For the sake of clarity and precision, the term "power law" replaces the historical terms of Zipf and Pareto in the rest of the text.

In Table 17.1, Axtell (2001) is notable for being close to −1 and for pointing out the importance of using large data sets for the analysis. The validity of Axtell's (2001) analysis and those others is not in doubt, but it is reasonable to ask the extent to which these analyses in economics and other fields can be applied to plankton. At a fundamental level there is considerable homology. Axtell (2001) was concerned with the distribution of growing businesses and pointed out that random growth processes of individuals converge on a power law distribution. In city formation, random pairwise interactions are enough to generate power law distributions (Marsili and Zhang, 1998). The key processes are random growth and interactions among individuals. At the conceptual level this is directly applicable to ocean plankton, because they grow and interact as individuals. The interactions among plankton include nutrient competition, reproduction, and grazing. Pairwise interactions for reproduction can cause aggregations or patchiness similar to that modeled for cities (Marsili and Zhang, 1998; Young et al., 2001).

17.1.5 Rank-Size Interpretation

The appeals of ranking values by size in a data set are the simplicity and the regularity of the behavior. Figure 17.1 shows the process. Figure 17.1A is a scattergram of 10,000 random points. Figure 17.1B shows the same points ranked in order with the highest value assigned the first rank, the second highest the second rank, and so on. With linear axes, random numbers give a straight line. This is representative of all values occurring with equal frequency; that is, no value is more likely to be more common than any other value. Sometimes this is formalized by binning similar values or creating a probability density distribution (Cizrók et al., 1995), but is only necessary when sample size is small or sizes are discrete values. On a log–log plot the straight line becomes a right square with a rounded corner (Figure 17.1C). Similar lines are produced for random walks (Figure 17.2). Figure 17.2 shows the tendency of a random

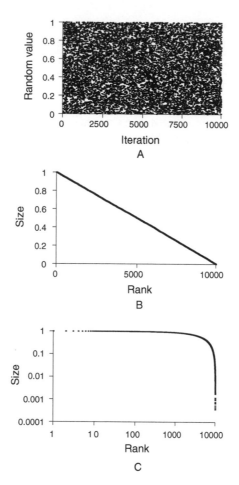

FIGURE 17.1 The effect of rank size on random numbers. (A) The distribution of 10,000 random numbers with the random value at each iteration plotted against that iteration. (B) The ranking of the 10,000 numbers. (C) The same ranking as B, but on a log–log plot.

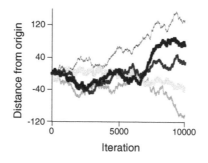

FIGURE 17.2 Five of 20 random walks of 10,000 iterations each. The heavy black line corresponds to the heavy black line in Figure 17.3.

walk to explore a narrow value range before moving off to explore another value range. Large changes relative to individual random walk steps tend to be gradual, that is, contain many points. The steepest multiple point change is shown on the heavy black line in Figure 17.2. The effect of this jump on the rank-size analysis is shown in Figure 17.3.

When interpreting rank-size graphs, random walks are useful for emphasizing how random variation produces local deviations around a straight line. That variation, however, does not produce a ranking

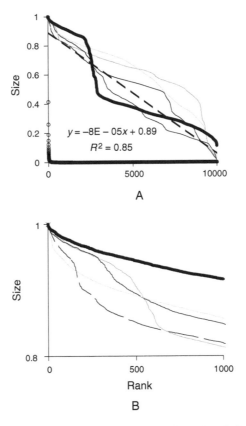

FIGURE 17.3 Rank sizing for five random walks. (A) The dashed line is the best fit for the random walks compared to the power law represented by the circles along the axes, which is included for comparison. (B) The top ranked 1000 points. The purpose of choosing the top 10% of points in real data is to avoid noise contamination that often occurs at the bottom ranks and to avoid chronic undersampling of the rarest events (Seuront et al., 1997, 1999). Choosing the top 10% with a random walk is heuristic as they are self-similar.

that resembles a power function. Although this is intuitively obvious, making it visually obvious allows for the addition of detailed comment. Figure 17.3 shows five rank-size graphs from among 20 random walks. The extremes in variation for the 20 were chosen, with the heavy black line indicating the most extreme jump. The step function of the heavy black line indicates the random walk visiting one particular range of values and then changing to another range quickly. The other, gentler, deviations from the dashed line represent the random walk visiting all size categories more or less equally. The self-similar nature of the random walk is reflected in Figure 17.3B, where the expansion of scale shows a step similar to that in Figure 17.3A. Such deviations will be important for discerning noise in phytoplankton distributions later in the chapter.

On a log–log plot, a power function is a straight line. If varying amounts of noise are added to the power function, then the noise causes a rightward departure from the straight line at a rank proportional to the amount of noise added (Figure 17.4A). Measuring the point of departure for a variety of noise levels recovers the original power function as would be expected (Figure 17.4B). The potential value of this is that it indicates that the curve shape can potentially provide quick information about the extent to which noise contaminates or contributes to the measured signal. This is shown for an extreme case in Figure 17.5, where the noise makes a relatively small contribution to the signal. The unranked data show influence of noise only at the higher rank numbers.

The departure of lines to the right of the power, as in Figure 17.1C and Figure 17.4A, is explained by the addition of noise. Alternatively, a downward departure from the power function to the left of the line is commonly seen in real data sets and indicates a fundamental change in the power law itself, rather than screening by noise. Figure 17.6 shows functions that fall below the power function line.

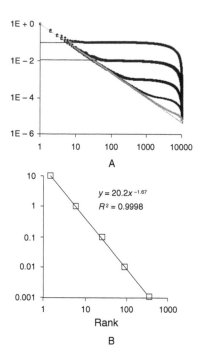

A

B

FIGURE 17.4 The addition of noise to a power law on log–log plots. (A) The thin diagonal line is a power law ($x^{-1.333}$). Each thick black line emerging to the right represents the power law with added noise of 10, 1, 0.1, 0.01, and 0.001% (from top to bottom) of the maximum value of 1. The two thin horizontal lines intercepting the *y*-axis are lines describing sets of random numbers. The salient feature is that points with different amounts of noise leave the power law line at different ranks. (B) The rank at which the noisy line first leaves the power law line. The implication is that, for a given sample size, the rank of departure is an indication of the amount of noise in the system.

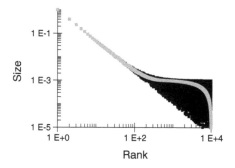

FIGURE 17.5 A comparison of a noisy power function and that same function ranked.

The set of points that fall just below the power function line were generated by adding a small random fluctuation (i.e., low frequency noise) to the power function; this effectively reduced the few high points. The bottom line in Figure 17.6 was generated by subtracting a competing power function of similar magnitude to the original. The characteristic rounded corner of random data appears in the lower right portion of the line. Figure 17.6 shows what happens to a rank-size graph when power functions are unstable or set against one another. From this it is possible to hypothesize the ecological implications of such modifications. For phytoplankton distributions, it is easy to imagine mixing or changing nutrient concentrations altering distribution and intensity to the extent that the exponent for a set of data varies due to natural processes. Simpler still, perhaps, is the reasoning that if phytoplankton characters such as growth or distribution follow a power law, then mortality processes such as grazing and lysis may well follow a similar but competing power law. Thus, if such power-function behavior can be shown in phytoplankton, the removal of the first rank large-size values could be interpreted as

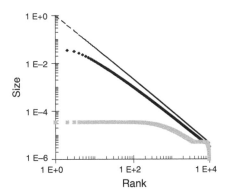

FIGURE 17.6 Competing power laws on a log–log plot. A rank-size graph for three power law functions. The straight line is a power law with an exponent of –1.333. The slightly flattened series of black diamonds is the same power law with a random number between zero and one subtracted from the original exponent. The horizontal gray squares show the same power law, but with a competing power law subtracted from the original.

an indication of a predator–prey cycle, a power law predator–prey cycle. This would require a great deal of checking and experimenting, but would be a useful model in that it could work spatially as well as temporally, adding another approach to studying phytoplankton dynamics. This assumes that effects from mixing, infection, and nutrient starvation can be teased out from those of grazing. Alternatively, perhaps a more unified approach would provide a better predictor of overall phytoplankton dynamics.

17.2 Low Resolution and Times Series Fluorescence Profiles

As pointed out in Figure 17.4, the use of rank size can quickly reveal noise or show the level of noise. Figure 17.7 shows a CTD profile from station *S* on 14 January 1999 (BATS GF124C1F). The obvious feature of the profile is the fluorescence maximum at a pressure of 83 dB. The majority of the water column is dominated by what appears to be background noise. The slight decrease from 0.06 to below 0.05 arbitrary fluorescence units is consistent with consumption, breakdown, and transformation of phytoplankton pigments with increasing depth. Beyond this there is no apparent pattern. Expanding a haphazardly chosen section of the profile (Figure 17.7B) shows no apparent pattern beyond random fluctuations. Peaks are single point maxima, or maxima with one or two transition points between a maximum and background level.

For the time being, the large fluorescence maximum is ignored and the focus will be on just the deep fluorescence background. Rank-size distribution of the profile section marked by the right brace in Figure 17.7 produces a curve on a log–log plot in which a power law (straight line) is a poor fit (Figure 17.8A). The reason for this is that the rolloff above 1000 points due to oversampling pulls the slope down. For small data sets this may not be a problem. However, for large data sets that significantly oversample the effect can be severe as illustrated in Figure 17.8A. Confining the data analyzed to the top 600 points shows an improved power law fit (Figure 17.8B). Continuing the data analyzed to the top 100 points further improves the fit (Figure 17.8C). Below 100 points the r^2 value of the power fit continues to improve. The message here is that the power law fit improves for this data set as the lower, baseline values are increasingly excluded. This repeated best-fit process is similar to what was seen when different amounts of noise were added to a power function (Figure 17.4). In short, the process of rank-sizing groups the noise due to oversampling at the high ranks. Subsequent exclusion of these higher ranks can substantially improve the fit and, finally, may be more indicative of the real structural signal.

The points in Figure 17.8B and C show terrace-like structures (parallel, but offset, horizontal lines). The reasons for this are that the data are near the measurement resolution of the instrument or the data processing caused binning of the measurements. Given the extensive data processing, instrument settings, and the deployment method of the CTD, all these limitations are likely (Knap et al., 1994). Two advantages of rank-size distributions are that they readily reveal this binning and that they effectively

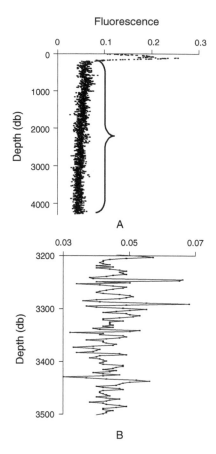

FIGURE 17.7 A fluorescence profile from the Bermuda Atlantic Time Series (BATS). (A) The entire upper 400 m of the profile. The right brace indicates the portion of the profile rank sized in the next figure. (B) An expanded section of the profile, showing the level of structure/noise among individual points. Fluorescence data have arbitrary units.

maximize the utility of resolution-limited data by grouping the visits to each bin to provide what is essentially a time spent at each bin. This time per bin causes local deviations from the best-fit line (Figure 17.8C), but overall does not greatly influence the best fit.

Ranking an entire chlorophyll profile, such as that in Figure 17.7, initially appears unproductive because what is recovered is a monotonic, neater version of the original vertical profile (Figure 17.9A). However, closer examination shows a step in the distribution. The line at the high end, or rare-event end, of the graph is clearly a rearrangement of the chlorophyll maximum (Figure 17.9B). The lower-level line is the deep fluorescence background, while the transition between the two levels is composed of the rises on each side of the chlorophyll maximum. The log–log plot also shows that the fit of the power law to the data is poor, specifically because of the transition region. The chlorophyll maximum is usually taken as the structure in the profile, or at least the most structured part of the profile. Using the Zipfian maxim of more structure with larger exponent, this common assumption can be tested, at least qualitatively. Figure17.9C shows separate power law fits to the first 50 points and the points between 100 and 1000. The exponent of the second step is significantly greater than the first. This implies that there is more structure in the deep fluorescence background than there is in the chlorophyll maximum. Both exponents are low compared to Zipf's −1 exponent, but different from zero. So, within the chlorophyll maximum there is some structure. The common assumption still holds, however, in the sense that the transition region of Figure 17.9C has an exponent (not fit for clarity purposes) of −1.26 ($r^2 = 0.98$). This is the source of structure and shows that the structure is due to an elevated fluorescence signal. The greater interest would be to have a −1 exponent for the whole maximum, as that would indicate structure throughout the peak.

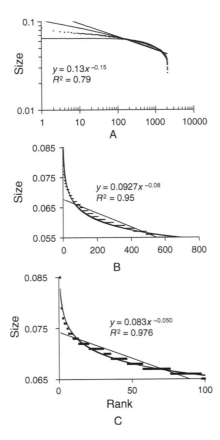

FIGURE 17.8 Rank sizing of the BATS fluorescence profile in Figure 17.7 for the region below the chlorophyll *a* maximum. (A) Ranking for the entire profile marked by a right brace in Figure 17.7. Graphs are rank-sized for subsets of the 600 (B) and 100 (C) highest values. Note that B and C have linear axes. The *y*-axis (size) signifies the rearrangement of the spatial pattern and is derived from fluorescence data with arbitrary units.

Figure 17.8 and Figure 17.9 are an analysis of a single BATS profile. Integrating many profiles from the same location provides a synopsis of that site. For that more integrated, time series approach, Figure 17.10 uses all the HOTS chlorophyll *a* profiles from 1988 through 2000. There is a clear chlorophyll maximum at approximately 100 m over these 13 years. The points are distributed over depth in bands that reflect preferred sampling depths. The preferred sampling depths result in depth-specific time series over the 13 years. This allows examination of how rank size performs over long periods. More complex algorithms have been used for analysis of a single-depth phytoplankton time series or a single transect, but these require extensive computer coding and are sometimes difficult to parameterize and interpret (Sugihara and May, 1990; Strutton et al., 1997a,b).

Here, the purpose is to test whether any coherent signal can be obtained from examining the depth-specific time series using rank size, or whether the result is a linear distribution indicating noise. The entire data set is rank-sized as two graphs in Figure 17.11. The first 500 points follow a power law with a shallow slope of –0.16. The remaining 2500 points fit a logarithmic distribution with a slope of about –0.09.

Figure 17.12A shows rank size of the 25 m time series for chlorophyll *a* for the HOTS data. A power law was the best fit and gave a slope of –0.32. Figure 17.12B is a combination of the chlorophyll *a* averaged over the 13 years and the exponent for the power law fit of rank-sized data for that depth. The error bars on the chlorophyll *a* are 95% confidence intervals with the data taken from 1.5 m on either side of a given depth. Zeros were not included and missing data values were ignored. Because rank sizing makes no assumptions about spatial or temporal relationships, the usual requirement for time series analysis, such as in fast Fourier transformations where data points are distributed regularly in time or space, is unnecessary here. Furthermore, there is no requirement for stationarity.

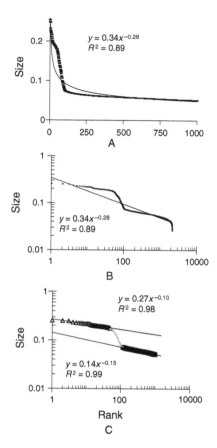

FIGURE 17.9 Rank sizing of the entire fluorescence profile in Figure 17.7. (A) Plotted with linear axes. (B) The same data plotted with logarithmic axes. Notice the stair-step structure. (C) The data in three sections, the highest ranks with a slope of –0.10, the transition region with a slope of –1.26, and the data of ranks from 125 to 1000 that have a slope of –0.15. The noise falloff at greater than about 1000 data points has been excluded here. The y-axis (size) signifies the rearrangement of the spatial pattern and is derived from fluorescence data with arbitrary units.

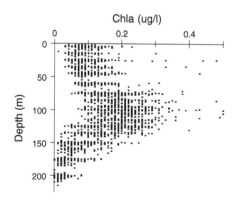

FIGURE 17.10 Chlorophyll *a* profiles of the Hawaiian Ocean Time Series (HOTS) from October 1988 through December 2000. There are 3700 measurements in the data set. The vertical white lines separating the data points are indicative of the measurement binning that occurs when the chlorophyll *a* concentration is near the resolution limit of the method. The horizontal groupings of the data points, such as those seen most clearly at 150, 175, and 200 m represent preferred sampling depths. These groupings constitute a depth-specific time series.

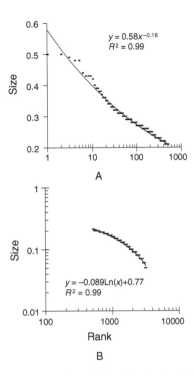

FIGURE 17.11 (A) Rank size of the 500 highest chlorophyll *a* values for the HOTS profiles, those above approximately 0.2 µg chl*a*/l. Note the rolling off of the highest ranks below the best-fit power law line. (B) The lowest 3000 points, generally those below 0.2 µg chl*a*/l. The *y*-axis (size) signifies the rearrangement of the spatial pattern and is derived from fluorescence data with arbitrary units.

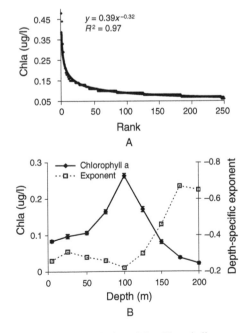

FIGURE 17.12 Depth-specific rank sizing. (A) Rank size of the chlorophyll *a* measurements at 25 m. (B) The solid diamonds on the solid line indicate mean chlorophyll *a* concentration. The error bars are 95% confidence intervals. The hollow squares on the dashed line are the exponent of depth-specific rank sizing. The exponent value can be read on the right *y*-axis. There is a distinct increase in the exponent value below the chlorophyll *a* maximum.

The squares on a dashed line in Figure 17.12B are the exponents for each depth, and are from the scale on the right axis. The significant feature is that the exponent is relatively constant for the upper 100 m, at about –0.26, but rises rapidly between 100 and 175 m to plateau at about –0.66. How are we to interpret this change? There are numerous papers in other fields suggesting that structure increases as the exponent moves away from zero (Marsili and Zhang, 1998; Bak, 1999; Axtell, 2001). A slope of –1 is taken to be aggregation and has been observed repeatedly in cities (Marsili and Zhang, 1998; Malacarne et al., 2002). Using a similar line of reasoning for the exponents in the HOTS profiles indicates that aggregation increases below the depth of the chlorophyll *a* maximum at 100 m. This is biologically reasonable in that below the chlorophyll maximum phytoplankton might be expected to be increasingly found in fecal pellets and marine snow (Riser et al., 2002). This shows some promise for using the rank-size exponent for assessing aggregation, but in this case required roughly 13 years to obtain enough data points for the analysis. The HOTS profiles demonstrate the potential of the rank-sized data. The section below on high-resolution fluorometry demonstrates how sufficient data can be generated in short periods to look at comparatively transient phenomena such as single aggregation events.

17.2.1   An Idealized Rank-Size Slope

If the chlorophyll maxima in Figure 17.7 and Figure 17.12 do not have a rank-size slope of –1, it is reasonable to ask what a slope of –1 would look like. Figure 17.13 shows a simulated peak that very closely matches the –1 slope when it is ranked by size. The relative fluorescence in Figure 17.13 has been arbitrarily set between 0 and 1. There are two important points to be noted from Figure 17.13. The first is that the peak shape is leptokurtic or very spiky. This is consistent with the usual interpretation of the –1 slope as an indicator of an aggregated nonrandom distribution. This true hallmark of the Zipf distribution is found in many human creations, ranging from the conceptual example of word frequency use to the physical example of how city sizes are distributed (Perline, 1996; Marsili and Zhang, 1998). The interpretation here is that a fluorescence peak approximating –1 would represent a form of aggregation; some would say organized aggregation. Thus, finding a fluorescence peak of this sort could be interpreted as an indication of local organization. It is important to note that the use of the term *aggregation* here does not mean a discrete particle. A discrete particle would show up as a single point at 1 on the schematic presented, with the remainder of the points at some background level near zero. The –1 aggregation is an intense region with a cloud of decreasing fluorescence around it. If these were to exist, they might be indicative of phytoplankton swarming around a nutrient source, dispersing from an exhausted particle, dispersing from an ungrazed area or perhaps a disintegrating fecal pellet.

The second important feature of the –1 slope is that high resolution is needed to detect it. At the low simulated relative fluorescence levels of Figure 17.13 this means that there is a great deal of redundant data that must be collected to view the few peak points. This, too, is consistent with the use and interpretation of rank-size analysis. Generally, rank-size analysis is used to process data where observations provide important information about the distribution of the data or the strength of a particular process. This is evident through the graphical emphasis on the few high points. Oceanographically, rare points of concentrated phytoplankton might be key points, the intensity and frequency of which strongly influence population or community dynamics. Here, we somewhat arbitrarily designate "rare" as occurring on the order of 1% or less of the time. A clear consequence of this is that large data sets are needed to perform statistics on rare events. Alternatively, the data sets could be tested relative to each other and their background levels.

To achieve sufficiently high resolution for examining the implicit hypothesis that small-scale hot spots are as important for phytoplankton as Azam (1998) proposes that they are for bacteria, it is necessary to use high-resolution profiling fluorometers. Standard CTD profiles and the associated profiling fluorometers fail on three levels. First, the tether to the ship introduces uncertainty and variation in the fall rate. Second, the cages that protect these devices may premix or otherwise contaminate or distort the signal, erasing or spreading the rare event, or rounding or smearing a –1 event. Third, excitation and detector design may average small intense fluorescence sources over large volumes. These are design limitations rather than theoretical barriers. Hence, the recent development of higher-resolution fluorometers than are commonly used is not surprising.

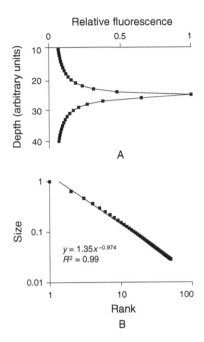

FIGURE 17.13 A simulated chlorophyll *a* peak with a rank-size slope near –1. (A) The simulated peak. (B) A rank sizing of the entire data set. The absence of baseline noise means there is no falloff, as seen in Figure 17.1B and in Figure 17.8A.

Wolk et al. (2001, 2002) describe a free-fall device that measures *in vivo* fluorescence with a 2 mm sampling interval. This device uses blue LEDs at the bottom of a 2 m hydrodynamically shaped cylinder. The LEDs face outward to put the sampling volume, approximately 1 ml, outside of the boundary layer of the falling device and its protrusions. A sample of the result is shown in Figure 17.14, taken from the Seto Inland Sea of Japan. The profile consists of 23,000 sequential points taken between 5.38 and 53.64 m. Shallower data were removed because a few meters are required for the fall speed to stabilize and for initial lateral precession to damp out. Using the standard CTD or bottle sampling method, the chlorophyll profile would probably look uniform. Indeed, here the standard deviation is 11% of the mean, suggesting overall uniformity. The profile, however, clearly shows spikes and other structures that are unevenly distributed across the profile.

Fitting a power law to all ranked data is often misleading. Noise associated with the largest and smallest ranked values at each end of the data set biases the fit. Similarly, breaks also bias or distort the fit. Figure 17.15 illustrates this and is a rank sizing of the profile, showing a power law r^2 value of only 0.90 (cf. an $r^2 > 0.97$ for most other profiles). However, there are three inflections in the ordered data. If these are used as boundaries and the material between them fitted to power laws and straight lines, the results are much tighter (Figure 17.16). Power law slopes of –0.133 and –0.106 are bounded by straight lines. These are the best fits for each of the four subsections. The curve at the lower end of the distribution is consistent with noise associated with a background or baseline level. The curves at the high end of the distribution could be noise from the paucity of high values and/or the presence of competing power laws. These two limitations may be interconnected in the sense that a factor such as copepod grazing may be responsible for the paucity of high values. This makes them difficult or impossible to disentangle without further investigation.

Dividing the total distribution into four sections based on best fit and inflections in the line is most meaningful if the different sections are qualitatively distinct, so that what is being detected is structurally and perhaps biologically meaningful. Figure 17.17 shows profile subsections associated with high-, medium-, and low-level fluorescence values. There are many notable features of these subsections. Figure 17.17A is a 0.7 m section that consists of approximately 350 points. The graph shows two high peaks, each composed of many points, indicating that these peaks are not single point maxima and that they are laterally extensive for about 10 cm. The two peaks are marginally asymmetrical, with the

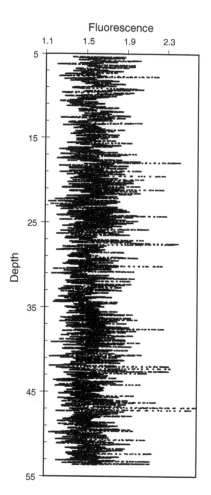

FIGURE 17.14 Turbomap fluorescence profile of 23,000 points. Fluorescence data have arbitrary units.

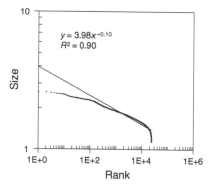

FIGURE 17.15 Rank sizing of the fluorescence measurements in Figure 17.14. The *y*-axis (size) signifies the rearrangement of the spatial pattern and is derived from fluorescence data with arbitrary units.

intensity difference between measurements greater on the shallow side of the peak than on the deep side of the peak. Given that the fall rate is constant, increased separation corresponds to increased gradient steepness. Perhaps more important than the shape of the peaks is the presence of secondary peaks. Specifically, in this case, it should be noted that the secondary peaks of each main peak are more similar to each other in number, position, and size than would be expected for random peaks. The presence of

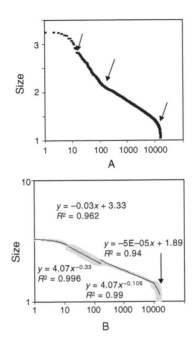

FIGURE 17.16 Rank sizing of the fluorescence measurements in Figure 17.14. A uses a linear *y*-axis and arrows to emphasize separation of data and changes in slope. B breaks the line into four sections. The lowest- and highest-ranked sections show linear best fits. The two middle sections show power law best fits. The *y*-axis (size) signifies the rearrangement of the spatial pattern and is derived from fluorescence data with arbitrary units.

such extensive and regular variation does violence to the concept of oceanographic consistency, but is nonetheless present and at a scale that is relevant to plankton. The generation of such regularity may be due to the interaction of turbulence with patches of phytoplankton. Certainly, the importance of such stirring effects has been pointed out previously (Abraham, 1998). Phytoplankton patches, by definition, rise above background concentrations and the expectation is that turbulence eventually mixes them to homogeneity. However, in the intervening period such a lack of homogeneity may have ecological significance to phytoplankton worth investigating. Furthermore, biological processes may aid in creating or maintaining inhomogeneities (Kessler, 1986).

Comparing Figure 17.17A and B shows the differences in magnitude and structure between points around large and medium peaks. The peaks in Figure 17.17B are approximately equal in size and spacing, and their range is less than twice the standard deviation of all data in the figure. In Figure 17.17C, distinct peaks are few and about equal to the standard deviation for the entire profile. Figure 17.17C is only a 0.2 m section because the low, even baseline sections are rare compared to the extent of intermediate and high sections of peaks and plateaus. Mechanistically, this may indicate that processes that disperse or reduce phytoplankton are rare or do not occur at background concentrations. Grazing, for example, may be most likely to occur at peaks where phytoplankton concentrations are high. Rank sizing of the data (Figure 17.18) from the profile in Figure 17.17 shows best fits as a power law for Figure 17.18A but as linear relationships for Figure 17.18B and C. The linear fits are consistent with less structure and variation at the medium and low levels of fluorescence.

Over 1.5 m sections, larger structures can be seen. The dominant feature in Figure 17.19A is a group of four large peaks. Similarly, the dominant features in Figure 17.19B are four peaks and their shoulder peaks. To the right of these four peaks is a set of four even smaller peaks and their shoulder peaks. This self-similarity at multiple scales is the hallmark of a scale-free system and an indication that the data are amenable to initial rank-size analysis as well as fractal and multifractal analysis (Seuront et al., 1997, 1999). The rank sizing of these data subsets (Figure 17.20) shows a greater slope associated with the structures in Figure 17.20A than in Figure 17.20B. In Figure 17.20A, however, the power law behavior does not expand toward the higher values so the fit is worse and the points near the origin fall under the

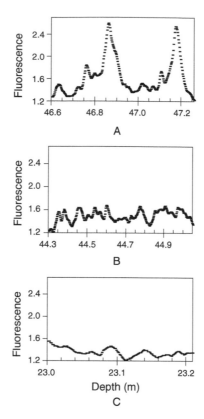

FIGURE 17.17 Turbomap profiles. (A) Fluorescence profile across 0.7 m. (B) Fluorescence profile across 0.7 m. (C) Fluorescence profile across 0.2 m. Each dash is a measurement. There were 500 measurements per meter. Fluorescence data have arbitrary units.

line, again possibly indicating that competing processes, such as grazing, are removing high values. Testing such a conjecture experimentally would require a thorough understanding of the distributions and dynamics of phytoplankton at microscales. The profiles in Figure 17.14, Figure 17.17, and Figure 17.19 have a 2 mm sampling interval and there is no reason to believe that this is the smallest interval needed to see all of the structure present. On the contrary, the following work indicates that higher resolution is possible and suggests that it may be necessary to understand the smallest-scale distributions and dynamics. Because these smallest scales are those at which individual plankton interact, such high-resolution measurements may be crucial to advancing our understanding of plankton dynamics.

The salient feature of Figure 17.21 is the presence of peaks at multiple scales essentially all the way down to the measuring resolution. The profiles were taken at the Port River Estuary in Adelaide, South Australia. Figure 17.21A shows a 0.5 m section of approximately 3000 points, with measurements at intervals of 0.170 mm. Measurements were made using a Fluoromap fluorometer from Alec Electronics (Kobe, Japan). The excitation source was a blue-LED laser and the sensed volume was 0.5 μl, with a high intensity core of 0.1 μl. Spectral analysis of entire data sets of 4,000 to 30,000 points consistently showed a flattened noise signal at period 3. To minimize the influence of the noise, a five-point symmetrical running mean was applied to the data before it was plotted in Figure 17.21. Figures 17.21B and C are 10 and 2 cm subsections of Figure 17.21A. Each graph shows a particular feature. Figure 17.21A has a wandering baseline on which are set a few large spikes. Over the 10 cm in Figure 17.21B, spikes are shown as multipoint peaks. Figure 17.21B also illustrates well-developed peaks at multiple scales. Points in Figure 17.21C are clearly highly autocorrelated, with some indication of a signal down to less than a dozen points. These peaks are all about 2 mm wide. This width probably results from a combination of applying a running mean and the thin, cylindrical shape of the sensed volume (approximately $0.2 \times 0.25 \times 12$ mm). An apparent resolution smaller than the largest dimension is achieved

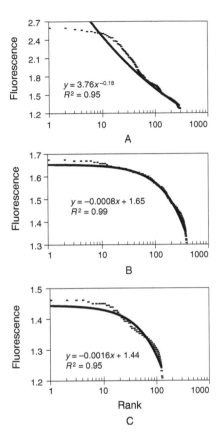

FIGURE 17.18 Rank-size graphs and best-line fits for the three fluorescence profiles in Figure 17.17. The *y*-axis (size) signifies the rearrangement of the spatial pattern and is derived from fluorescence data with arbitrary units.

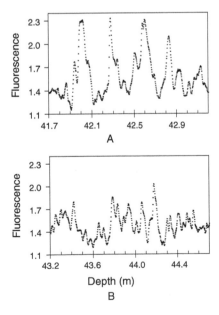

FIGURE 17.19 Turbomap profiles over 1.5 m intervals showing four grouped, large peaks (A), four grouped, medium peaks (B), and four grouped, small peaks (B, to the right of the medium peaks). Note the relatively even spacing and consistent height among peaks within a group. Fluorescence data have arbitrary units.

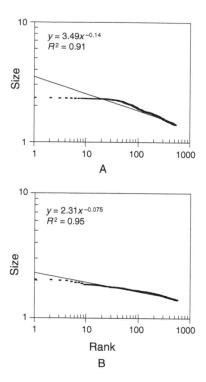

FIGURE 17.20 Rank-size graphs and best-line fits for the two fluorescence profiles shown in Figure 17.19. The *y*-axis (size) signifies the rearrangement of the spatial pattern and is derived from fluorescence data with arbitrary units.

because the shape of the excitation sensor head forces the flow through the beam at an oblique angle to the long axis.

The immediate objection to the peaks in Figure 17.21 is that the largest peak in Figure 17.21A is only about a 20% increase over the baseline signal, while the largest peak in Figure 17.21C is only about 10% of the baseline signal. Can such small relative changes have ecological relevance? It is too soon to definitively answer that question, but the possibility that these changes are ecologically relevant cannot be precluded *a priori*. The fluorometer measures the presence of fluorescent compounds, presumably primarily chlorophyll *a* in the ocean. While the peak changes are small compared to the total fluorescent signal, it may be that they represent a much larger species-specific fraction. For example, the Port River, Adelaide has a high cyanobacteria concentration with intermittent eukaryotic algal blooms. The high baseline may represent the cyanobacterial populations with the peaks representing eukaryotic algal distributions. Fluoromap does detect rare large peaks (Figure 17.22). These are not discussed further here, as the potential causes of these peaks, such as the presence of particles, aggregations, and phytoplankton swarms, cannot be determined at this time. Figure 17.23 is a rank sizing of the three graphs in Figure 17.21. The fits are remarkably tight, particularly in Figure 17.23B, C where the number of points is relatively small. The primary deviation comes from the falloff at the high fluorescence values, suggesting the possibility of a competing power law.

In determining what the profiles mean, it is worth keeping in mind that the measured volume in Figure 17.23A for all measurements over the whole profile totals somewhere between 300 to 3000 µl. This is a small volume compared to traditional phytoplankton counting and fluorescence measuring methods. However, the use of such small volumes to estimate phytoplankton distributions does have a precedent in the widespread use of flow cytometry, which routinely uses less than 1 ml for biomass estimates (Waters and Mitchell, 2002). Furthermore, no effort was made to take this profile at a location in the water column where phytoplankton biomass might be high. Instead, the function of this profile is to act as a test for rank sizing. Indeed, the fluoromap is a prototype free-fall fluorometer. A full-scale description of the fluoromap results are in preparation.

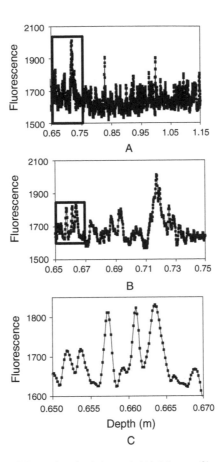

FIGURE 17.21 Fluoromap profiles of decreasing depth interval. (A) 0.5 m profile section. (B) A 0.1 m profile section. (C) A 0.02 m (2 cm) profile section. The black boxes along the left axes in A and B show the area enlarged in the graph below it. There were 5000 measurements per meter. Fluorescence data have arbitrary units.

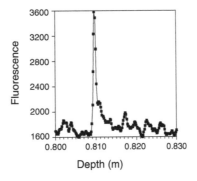

FIGURE 17.22 A fluoromap profile section extending over 0.03 m or 3 cm. The dash line is a five point running mean. Fluorescence data have arbitrary units.

17.3 Conclusion

There are two primary lessons from applying rank sizing to plankton profiles. The first and perhaps most useful, albeit incidental, is that it helps drive one to strive for large data sets of high resolution. To that end, it takes us beyond the habitual sampling and analysis protocols and points to a path of looking at

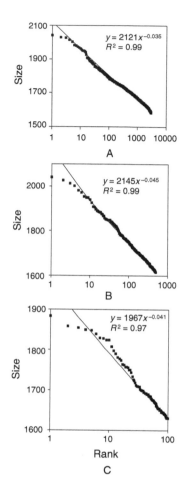

FIGURE 17.23 Rank sizing of the fluoromap profiles shown in Figure 17.21. The *y*-axis (size) signifies the rearrangement of the spatial pattern and is derived from fluorescence data with arbitrary units.

scales and mechanisms relevant to plankton rather than convenient to scientists and ships. The second lesson is the analytical advantages of the rank-size method. There is no need for regular sampling intervals. The data need not be stationary. There is no requirement that the data conform to a particular distribution. Overall, there is an intuitive character to the method compared to fast Fourier transformations and nearest neighbor algorithms. Despite the method's simplicity, the shape of the rank size appears to provide information on the signal-to-noise ratio, the presence of processes such as grazing (through competing power law rolloff), and aggregation (through a change in the power law exponent with depth).

For phytoplankton ecology, rank-size analysis may help speed the analysis of very large data sets necessary for characterizing microscale distributions. Historically the concern at the microscale has been to find the largest possible phytoplankton gradients (Mitchell and Fuhrman, 1989; Yoder et al., 1994; Seymour et al., 2000; Waters and Mitchell, 2002). These large gradients constitute a small portion of the marine phytoplankton and are usually transient. The average phytoplankton more often experiences or crosses small gradients or makes up weak patches of the sorts described here. The significance of the smallest, but most frequent gradients has been overlooked. Continually present differences in selection pressure smaller than 0.01% per generation have been shown to be enough to drive competition and evolution in microbial communities (Lenski et al., 1991). As such, the key to understanding plankton distributions and dynamics is likely to be the most commonplace rather than the most dramatic of gradients and interactions. Rank sizing is a first step in detecting structure and encouraging a higher resolution examination of those common phytoplankton distributions.

Acknowledgments

Brenda Kranz and Laurent Seuront provided continued support, useful comments, and essential editing on the manuscript. Thanks are also due to the BATS and HOTS programs for making their data freely available on the Internet. H. Yamazaki and Alec Electronics provided access to data and the prototype fluoromap fluorometer. This work was supported financially and infrastructurally by the Japan Society for the Promotion of Science, the Australian Research Council, and Flinders University of South Australia.

References

Abraham, E.R., The generation of plankton patchiness by turbulent stirring. *Nature*, 391, 577, 1998.

Adamic, L.A., Zipf, Power-laws, and Pareto — a ranking tutorial, available at http://www.hpl.hp.com/shl/papers/ranking/, 2000.

Axtell, R.L., Zipf distribution of U.S. firm sizes. *Science*, 293, 1818, 2001.

Azam, F., Microbial control of oceanic carbon flux: The plot thickens. *Science*, 280, 694, 1998.

Bak, P., *How Nature Works: The Science of Self-Organised Criticality.* Springer-Verlag, New York, 1999.

Chatfield, C., *The Analysis of Time Series: An Introduction.* 4th ed., Chapman & Hall, London, 1989.

Cizrók, A., Mantegna, R.N., Havlin, S., and Stanley, E., Correlations in binary sequences and a generalized Zipf analysis. *Phys. Rev. E*, 52, 446, 1995.

Duarte, C.M. and Vaqué, D., Scale dependence of bacterioplankton patchiness. *Mar. Ecol. Prog. Ser.*, 84, 95, 1992.

Faloutsos, M., Faloutsos, P., and Faloutsos, C., On power-law relationships of the Internet topology. *Comput. Commun. Rev.* (ACM SIGCOMM), 29, 251, 1999.

Feldman, G.C., Clark, D., and Halpern, D., Satellite color observations of the phytoplankton distribution in the eastern Equatorial Pacific during the1982–83 El Niño. *Science*, 226, 1069, 1984.

Hutchinson, G.E., The paradox of the plankton. *Am. Nat.*, 95, 137, 1961.

Kessler, J., Individual and collective dynamics of swimming cells. *J. Fluid Mech.*, 173, 191, 1986.

Knap, A. et al., Eds., Protocols for the Joint Global Ocean Flux Study (JGOFS) Core Measurements. JGOFS Report 19 as IOC Manual 29, UNESCO, Cat. 99739, 1994.

Lenski, R., Rose, M.R., Simpson, S.C., and Tadler, S.C., Long-term experimental evolution in Escherichia coli. I. Adaptation and divergence during 2,000 generations. *Am. Nat.*, 138, 1315, 1991.

Long, R.A. and Azam, F., Microscale patchiness of bacterioplankton assemblage richness in seawater. *Aquat. Microb. Ecol.*, 26, 103, 2001.

Malacarne, L.C., Mendes, R.S., and Lenzi, E.K., q-Exponential distribution in urban agglomeration. *Phys. Rev. E*, 65, 017106, 2002.

Mantegna, R.N., Buldyrev, S.V., Goldberger, A.L, and Havlin, S., Systematic analysis of coding and noncoding DNA sequences using methods of statistical linguistics. *Phys. Rev. E*, 52, 2939, 1995.

Marsili, M. and Zhang, Y.-C., Interacting individuals leading to Zipf's law. *Phys. Rev. Lett.*, 80, 2741, 1998.

Mercik, S., Weron, K., and Siwy, Z., Statistical analysis of ionic current fluctuations in membrane channels. *Phys. Rev. E*, 60, 7343, 1999.

Mitchell, J.G. and Fuhrman, J.A., Centimeter scale vertical heterogeneity in bacteria and chlorophyll *a*. *Mar. Ecol. Prog. Ser.*, 54, 141, 1989.

Perline, R., Zipf's law, the central limit theorem, and the random division of the unit interval. *Phys. Rev. E*, 54, 220, 1996.

Platt, T., Local phytoplankton abundance and turbulence. *Deep-Sea Res.*, 19, 183, 1972.

Ramsden, J.J. and Vohradský, J., Zipf-like behavior in prokaryotic protein expression. *Phys. Rev. E*, 58, 7777, 1998.

Riser, C.W., Wassmann, P., Olli, K., Pasternak, A., and Arashkevich, E., Seasonal variation in production, retention and export of zooplankton faecal pellets in the marginal ice zone and central Barents Sea. *J. Mar. Syst.*, 38, 175, 2002.

Seuront, L. and Lagadeuc, Y., Characterisation of space-time variability in stratified and mixed coastal waters (Baie des Chaleurs, Quebec, Canada): application of fractal theory. *Mar. Ecol. Prog. Ser.*, 159, 81, 1997.

Seuront, L., Schmitt, F., Lagadeuc, Y., Shcertzer, D., and Lovejoy, S., Multifractal analysis as a tool to characterize multiscale inhomogeneous patterns. Example of phytoplankton distribution in turbulent coastal waters. *J. Plankton Res.*, 21, 877, 1999.

Seymour, J., Mitchell, J.G., Pearson, L.P., and Waters, R.L., Heterogeneity in bacterioplankton abundance from 4.5 millimeter resolution sampling. *Aquat. Microb. Ecol.*, 22, 143, 2000.

Strutton, P.G., Mitchell, J.G., and Parslow, J.S., Using non-linear analysis to compare the spatial structure of chlorophyll with passive tracers. *J. Plankton Res.*, 19, 1553, 1997a.

Strutton, P.G., Mitchell, J.G, Parslow, J.S., and Greene, R.M., Phytoplankton patchiness: quantifying the biological contribution via fast repetition rate fluorometry. *J. Plankton Res.*, 19, 1265, 1997b.

Sugihara, G. and May, R.M., Nonlinear forecasting as a way of distinguishing chaos from measurement error in time series. *Nature*, 34, 734, 1990.

Watanabe, M.S., Zipf's law in percolation. *Phys. Rev. E*, 53, 4187, 1996.

Waters, R.L. and Mitchell, J.G., The centimetre-scale spatial structure of estuarine *in vivo* fluorescence profiles. *Mar. Ecol. Prog. Ser.*, 237, 51, 2002.

Wolk, F., Seuront, L., and Yamazaki, H., Spatial resolution of a new micro-optical probe for chlorophyll and turbidity. *J. Tokyo Univ. Fish.*, 87, 13, 2001.

Wolk, F., Yamazaki, H., Seuront, L., and Lueck, R.G., A new free-fall profiler for measuring biophysical microstructure. *J. Atmos. Oceanic Technol.*, 19, 780, 2002.

Yoder, J.A. et al. A line in the sea. *Nature*, 371, 689, 1994.

Young, W.R., Roberts, A.J., and Stuhne, G., Reproductive pair correlations and the clustering of organisms. *Nature*, 412, 328, 2001.

Zipf, G.K., *Human Behavior and the Principle of Least Effort*. Addison-Wesley, Cambridge, 1949.

18

An Introduction to Wavelets

Igor M. Dremin, Oleg V. Ivanov, and Vladimir A. Nechitailo

CONTENTS

18.1   Introduction..279
18.2   Wavelets for Beginners...281
18.3   Basic Notions and Haar Wavelets 284
18.4   Multiresolution Analysis and Daubechies Wavelets 286
18.5   Fast Wavelet Transform...288
18.6   The Fourier and Wavelet Transforms...290
18.7   Technicalities 291
18.8   Scaling..293
18.9   Applications 293
18.10 Conclusions..296
References..296

18.1 Introduction

Wavelets have become a necessary mathematical tool in many investigations. They are used in those cases when the result of the analysis of a particular *signal** should contain not only the list of its typical frequencies (scales) but also knowledge of the definite local coordinates where these properties are important. Thus, analysis and processing of different classes of nonstationary (in time) or inhomogeneous (in space) signals is the main field of applications of wavelet analysis.

The practical applications of wavelet analysis are extremely wide ranging from aviation, engines to medicine, image compression, etc. Some of them are described in what follows. In particular, it can be used for analysis of any signal in aquatic ecology or oceanology as well; however, it has not been yet widely applied in this field. For example, the dolphin or bat sonar signals can be fully deciphered only with the help of wavelets because of the fast-changing set of frequencies typical for them. The fractal structure of any biological object can be also studied. The aim of this chapter is to attract attention of the specialists working in biology, oceanology, and aquatic ecology to this new perspective mathematical method, in particular, to the scaling analysis provided by wavelets.

The wavelet basis is formed by using dilations and translations of a particular function defined on a finite interval. Its finiteness is crucial for the locality property of wavelet analysis. Commonly used wavelets generate a complete orthonormal system of functions with a finite support constructed in such a way. That is why by changing the scale (dilations) they can distinguish the local characteristics of a signal at various scales, and by translations they cover the whole region in which it is studied. Because

---

* The notion of a signal is used here for any ordered set of numerically recorded information about some processes, objects, functions, etc. The signal can be a function of some coordinates, be it the time, the space, or any other (in general, n-dimensional) scale.

of the completeness of the system, they also allow for inverse transformation to be properly done. In analysis of nonstationary signals, the locality property of wavelets gives a substantial advantage over Fourier transform, which provides us with knowledge only of the global frequencies (scales) of the object under investigation because the system of the basic functions used (sine, cosine, or imaginary exponential functions) is defined over an infinite interval.

The literature devoted to wavelets is highly voluminous, and one can easily obtain a lot of references by sending the corresponding request to Internet Web sites. Mathematical problems are treated in many monographs in detail (e.g., see References 1 through 5). Introductory courses on wavelets can be found in several books;[6–9] review papers adapted for physicists and practical users have been published in *Physics-Uspekhi Journal.*[10,11]

It has been proved that any function can be written as a superposition of wavelets, and there exists a numerically stable algorithm to compute the coefficients for such an expansion. Moreover, these coefficients completely characterize the function, and it is possible to reconstruct it in a numerically stable way by knowing these coefficients. Because of their unique properties, wavelets were used in functional analysis in mathematics, in studies of (multi)fractal properties, singularities, and local oscillations of functions, for solving some differential equations, for investigation of inhomogeneous processes involving widely different scales of interacting perturbations, for pattern recognition, for image and sound compression, for digital geometry processing, for solving many problems of physics, biology, medicine, technique, etc. (see recently published books[12–15]). This list is by no means exhaustive.

The programs exploiting the wavelet transform are widely used now not only for scientific research but for commercial projects as well. Some have been even described in books (e.g., see Reference 16). At the same time, the direct transition from pure mathematics to computer programming and applications is nontrivial and asks often for the individual approach to the problem under investigation and for a specific choice of wavelet used. Our main objective here is to describe in a suitable way the bridge that relates mathematical wavelet constructions to practical signal processing. Specifically, practical applications considered by Grossman and Morlet[17,18] have led to fast progress of the wavelet theory related to the work of Meyer, Daubechies, and others.

The main bulk of papers dealing with practical applications of wavelet analysis uses so-called discrete wavelets, which are our main concern here. The discrete wavelets look strange to those accustomed to analytical calculations because they cannot be represented by analytical expressions (except for the simplest one) or by solutions of some differential equations, and instead are given numerically as solutions of definite functional equations containing rescaling and translations. Moreover, in practical calculations their direct form is not even required, and only the numerical values of the coefficients of the functional equation are used. Thus, the wavelet basis is defined by the iterative algorithm of the dilation and translation of a single function. This leads to a very important procedure called multiresolution analysis, which gives rise to the multiscale local analysis of the signal and fast numerical algorithms. Each scale contains an independent non-overlapping set of information about the signal in the form of wavelet coefficients, which are determined from an iterative procedure called the fast wavelet transform. In combination, they provide its complete *analysis* and simplify the *diagnosis* of the underlying processes.

After such an analysis has been done, one can *compress* (if necessary) the resulting data by omitting some inessential part of the encoded information. This is done with the help of the so-called quantization procedure, which commonly allocates different weights to various wavelet coefficients obtained. In particular, it helps erase some statistical fluctuations and, therefore, increases the role of the dynamic features of a signal. This can, however, falsify the diagnostic if the compression is done inappropriately. Usually, accurate compression gives rise to a substantial reduction of the required computer *storage* memory and *transmission* facilities and, consequently, to a lower expenditure. The number of vanishing moments of wavelets is important at this stage. Unfortunately, the compression introduces unavoidable systematic errors. The mistakes one has made will consist of multiples of the deleted wavelet coefficients, and, therefore, the regularity properties of a signal play an essential role. *Reconstruction* after such compression schemes is then no longer perfect. These two objectives are clearly antagonistic. Nevertheless, when one tries to reconstruct the initial signal, the inverse transformation (*synthesis*) happens to be rather stable and reproduces its most important characteristics if proper methods are applied. The regularity properties of wavelets used also become crucial at the reconstruction stage. The distortions of the reconstucted signal due to quantization can be kept small, although significant compression ratios are attained. Since the part of the signal that is not reconstructed is often called noise, in essence, what we are doing is denoising the signals. At this stage, the superiority of the discrete wavelets becomes especially clear.

Thus, the objectives of signal processing consist of accurate analysis with help of the transform, effective coding, fast transmission, and, finally, careful reconstruction (at the transmission destination point) of the initial signal. Sometimes the first stage of signal analysis and diagnosis is enough for the problem to be solved and the anticipated goals to be achieved.

One should, however, stress that, even though this method is very powerful, the goals of wavelet analysis are rather modest. This helps us describe and reveal some features, otherwise hidden in a signal, but it does not pretend to explain the underlying dynamics and physical origin although it may give some crucial hints to it. Wavelets present a new stage in optimization of this description providing, in many cases, the best known representation of a signal. With the help of wavelets, we merely see things a little more clearly. To understand the dynamics, standard approaches introduce models assumed to be driving the mechanisms generating the observations. To define the optimality of the algorithms of the wavelet transform, some (still debatable) energy and entropy criteria have been developed. They are internal to the algorithm itself. However, the choice of the best algorithm is also tied to the objective goal of its practical use, i.e., to some external criteria. That is why in practical applications one should submit the performance of a "theoretically optimal algorithm" to the judgments of experts and users to estimate its advantage over the previously developed ones.

Despite very active research and impressive results, the versatility of wavelet analysis implies that these studies are presumably not yet in their final form. We try to describe the situation in its *status nascendi*.

18.2 Wavelets for Beginners

Each signal can be characterized by its averaged (over some intervals) values (trend) and by its variations around this trend. Let us call these variations fluctuations independently of their nature, be they of dynamic, stochastic, psychological, physiological, or any other origin. When processing a signal, we are interested in its fluctuations at various scales because from these we can learn about their origin. The goal of wavelet analysis is to provide tools for such processing.

Actually, physicists dealing with experimental histograms analyze their data at different scales when averaging over different size intervals. This is a particular example of a simplified wavelet analysis treated in this section. To be more definite, let us consider the situation when an experimentalist measures some function $f(x)$ within the interval $0 \leq x \leq 1$, and the best resolution obtained with the measuring device is limited by 1/16th of the whole interval. Thus, the result consists of 16 numbers representing the mean values of $f(x)$ in each of these bins and can be plotted as a 16-bin histogram shown in the upper part of Figure 18.1. It can be represented by the following formula:

$$f(x) = \sum_{k=0}^{15} s_{4,k} \varphi_{4,k}(x) \tag{18.1}$$

where $s_{4,k} = f(k/16)/4$, and $\varphi_{4,k}$ is defined as a step-like block of the unit norm (i.e., of height 4 and width 1/16) different from zero only within the kth bin. For an arbitrary j, we impose the condition $\int dx \, | \varphi_{j,k} |^2 = 1$, where the integral is taken over the intervals of the lengths $\Delta x_j = 1/2^j$ and, therefore, $\varphi_{j,k}$ have the following form $\varphi_{j,k} = 2^{j/2} \varphi(2^j x - k)$ with φ denoting a step-like function of the unit height over such an interval. The label 4 is related to the total number of such intervals in our example. At the next coarser level the average over the two neighboring bins is taken as is depicted in the histogram just below the initial one in Figure 18.1. Up to the normalization factor, we will denote it as $s_{3,k}$ and the difference between the two levels shown to the right of this histogram as $d_{3,k}$. To be more explicit, let us write down the normalized sums and differences for an arbitrary level j as

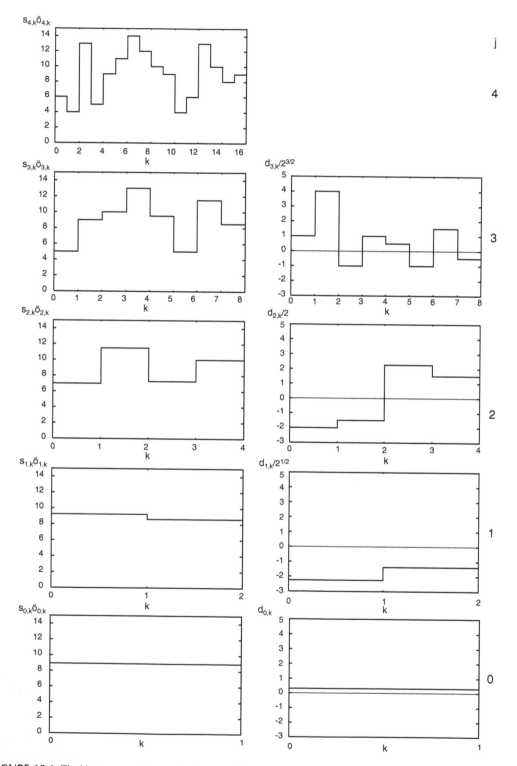

FIGURE 18.1 The histogram and its wavelet decomposition. The initial histogram is shown in the upper part of the figure. It corresponds to the level $j = 4$ with 16 bins (Equation 18.1). The intervals are labeled on the abscissa axis at their left-hand sides. The next level $j = 3$ is shown below. The mean values over two neighboring intervals of the previous level are shown at the left-hand side. They correspond to eight terms in the first sum in Equation 18.4. At the right-hand side, the wavelet coefficients $d_{3,k}$ are shown. Other graphs for the levels $j = 2,1,0$ are obtained in a similar way.

$$s_{j-1,k} = \frac{1}{\sqrt{2}}[s_{j,2k} + s_{j,2k+1}]$$

$$d_{j-1,k} = \frac{1}{\sqrt{2}}[s_{j,2k} - s_{j,2k+1}]$$

$$\text{(18.2)}$$

or for the backward transform (synthesis)

$$s_{j,2k} = \frac{1}{\sqrt{2}}[s_{j-1,k} + d_{j-1,k}]$$

$$s_{j,2k+1} = \frac{1}{\sqrt{2}}[s_{j-1,k} - d_{j-1,k}]$$

$$\text{(18.3)}$$

Since, for the dyadic partition considered, this difference has opposite signs in the neighboring bins of the previous fine level, we introduce the function ψ which is 1 and -1, correspondingly, in these bins and the normalized functions $\psi_{j,k} = 2^{j/2}\psi(2^j x - k)$. This allows us to represent the same function $f(x)$ as

$$f(x) = \sum_{k=0}^{7} s_{3,k}\varphi_{3,k}(x) + \sum_{k=0}^{7} d_{3,k}\psi_{3,k}(x)$$

$$\text{(18.4)}$$

We proceed farther in the same manner to the sparser levels 2, 1, and 0 with averaging done over the interval lengths $1/4$, $1/2$, and 1, correspondingly. This is shown in the subsequent drawings in Figure 18.1. The transition from s-values of a fine level to d-values at the coarser level is shown by arrows. The sparsest level with the mean value of f over the whole interval denoted as $s_{0,0}$ provides

$$f(x) = s_{0,0}\varphi_{0,0}(x) + d_{0,0}(x)\psi_{0,0}(x) + \sum_{k=0}^{1} d_{1,k}\psi_{1,k}(x)$$

$$+ \sum_{k=0}^{3} d_{2,k}\psi_{2,k}(x) + \sum_{k=0}^{7} d_{3,k}\psi_{3,k}(x)$$

$$\text{(18.5)}$$

The functions $\varphi_{0,0}(x)$ and $\psi_{0,0}(x)$ are shown in Figure 18.2. The functions $\varphi_{j,k}(x)$ and $\psi_{j,k}(x)$ are normalized by the conservation of the norm, dilated and translated versions of them. In the next section we give explicit formulae for them in a particular case of Haar scaling functions and wavelets. In practical signal processing, these functions (and more sophisticated versions of them) are often called low- and high-pass filters, correspondingly, because they filter the large- and small-scale components of a signal. The subsequent terms in Equation 18.5 show the fluctuations (differences $d_{j,k}$) at finer and finer levels with larger j. In all the cases (Equations 18.1 through 18.5) one needs exactly 16 coefficients to represent the function. In general, there are 2^j coefficients $s_{j,k}$ and $2^{j_n} - 2^j$ coefficients $d_{j,k}$, where j_n denotes the finest resolution level (in the above example, $j_n = 4$).

All the above representations of the function $f(x)$ (Equations 18.1 through 18.5) are mathematically equivalent. However, the last, representing the wavelet analyzed function directly, reveals the fluctuation structure of the signal at different scales j and various locations k present in a set of coefficients $d_{j,k}$ whereas the original form (Equation 18.1) hides the fluctuation patterns in the background of a general

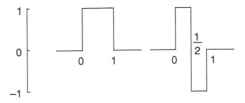

FIGURE 18.2 Haar scaling function $\varphi(x) \equiv \varphi_{0,0}(x)$ and "mother" wavelet $\psi(x) \equiv \psi_{0,0}(x)$.

trend. The final form (Equation 18.5) contains the overall average of the signal depicted by $s_{0,0}$ and all its fluctuations with their scales and positions well labeled by 15 normalized coefficients $d_{j,k}$ while the initial histogram shows only the normalized average values $s_{j,k}$ in the 16 bins studied. Moreover, in practical applications the latter wavelet representation is preferred because for rather smooth functions, strongly varying only at some discrete values of their arguments, many of the high-resolution d-coefficients in relations similar to Equation 18.5 are close to zero (compared to the "informative" d-coefficients) and can be discarded. Bands of zeros (or close to zero values) indicate those regions where the function is fairly smooth.

At first sight, this simplified example looks somewhat trivial. However, for more complicated functions and more data points with some elaborate forms of wavelets it leads to a detailed analysis of a signal and to possible strong compression with subsequent good-quality restoration. This example also provides an illustration of the very important feature of the whole approach with successively coarser and coarser approximations to *f*, which is called multiresolution analysis and discussed in more detail below.

18.3   Basic Notions and Haar Wavelets

To analyze any signal, one should, first of all, choose the corresponding basis, i.e., the set of functions to be considered as "functional coordinates." In most cases we deal with signals represented by the square integrable functions defined on the real axis. For nonstationary signals, e.g., the location of that moment when the frequency characteristics have abruptly been changed is crucial. Therefore, the basis should have a compact support; i.e., it should be defined on the finite region. The wavelets have this property. Nevertheless, with them it is possible to span the whole space by translation of the dilated versions of a definite function. That is why every signal can be decomposed in the wavelet series (or integral). Each frequency component is studied with a resolution matched to its scale.

Let us try to construct functions satisfying the above criteria. An educated guess would be to relate the function $\varphi(x)$ to its dilated and translated version. The simplest linear relation with $2M$ coefficients is

$$\varphi(x) = \sqrt{2} \sum_{k=0}^{2M-1} h_k \varphi(2x - k) \qquad (18.6)$$

with the dyadic dilation 2 and integer translation k. At first sight, the chosen normalization of the coefficients h_k with the "extracted" factor $\sqrt{;2}$ looks somewhat arbitrary. Actually, it is defined *a posteriori* by the traditional form of fast algorithms for their calculation (see Equations $s_{j+1,k} = \sum_m h_m s_{j,2k+m}$ 18.34 and

$d_{j+1,k} = \sum_m g_m s_{j,2k+m}$ 18.35 below) and normalization of functions $\varphi_{j,k}(x)$, $\psi_{j,k}(x)$. It is used in all the books

cited above. However, sometimes (see Reference 2, chapter 7) it is replaced by $c_k = \sqrt{2}h_k$.

For discrete values of the dilation and translation parameters one obtains discrete wavelets. The value of the dilation factor determines the size of cells in the lattice chosen. The integer M defines the number of coefficients and the length of the wavelet support. They are interrelated because from the definition of h_k for orthonormal bases:

$$h_k = \sqrt{2} \int dx \varphi(x) \overline{\varphi}(2x - k) \qquad (18.7)$$

it follows that only finitely many h_k are nonzero if φ has a finite support. The normalization condition is chosen as

$$\int_{-\infty}^{\infty} dx \varphi(x) = 1 \qquad (18.8)$$

The function $\varphi(x)$ obtained from the solution of this equation is called a scaling function.* If the scaling function is known, one can form a "mother wavelet" (or a basic wavelet) $\psi(x)$ according to

$$\psi(x) = \sqrt{2} \sum_{k=0}^{2M-1} g_k \varphi(2x - k) \tag{18.9}$$

where

$$g_k = (-1)^k h_{2M-k-1} \tag{18.10}$$

The simplest example would be for $M = 1$ with two nonzero coefficients h_k equal to $1/\sqrt{2}$, i.e., the equation leading to the Haar scaling function $\varphi_H(x)$:

$$\varphi_H(x) = \varphi_H(2x) + \varphi_H(2x + 1) \tag{18.11}$$

We easily solve this functional equation:

$$\varphi_H(x) = \theta(x) + \theta(1 - x) \tag{18.12}$$

where $\theta(x)$ is the Heaviside step-function equal to 1 at positive arguments and 0 at negative ones. The additional boundary condition is $\varphi_H(0) = 1$, $\varphi_H(1) = 0$. This condition is important for the simplicity of the whole procedure of computing the wavelet coefficients when two neighboring intervals are considered.

The "mother wavelet" is

$$\psi_H(x) = \theta(x)\theta(1 - 2x) - \theta(2x - 1)\theta(1 - x) \tag{18.13}$$

with boundary values defined as $\psi_H(0) = 1$, $\psi_H(1/2) = -1$, $\psi_H(1) = 0$. This is the Haar wavelet[19] known since 1910 and used in the functional analysis. This example has been considered in the previous section for the histogram decomposition. Both the scaling function $\psi_H(x)$ and the "mother wavelet" $\psi_H(x)$ are shown in Figure 18.2. This is the first of a family of compactly supported orthonormal wavelets $_M\psi : \psi_H =_1 \psi$. It possesses the locality property since its support $2M - 1 = 1$ is compact.

The dilated and translated versions of the scaling function φ and the "mother wavelet" ψ

$$\varphi_{j,k} = 2^{j/2} \varphi(2^j x - k) \tag{18.14}$$

$$\psi_{j,k} = 2^{j/2} \psi(2^j x - k) \tag{18.15}$$

form the orthonormal basis as can be (easily for Haar wavelets) checked.** The choice of 2^j with the integer valued j as a scaling factor leads to the unique and self-consistent procedure of computing the wavelet coefficients. In principle, there exists an algorithm of derivation of the compact support wavelets with an arbitrary rational number in place of 2. However, only for this factor, it has been shown that there exists the explicit algorithm with the regularity of a wavelet increasing linearly with its support. For example, for the factor 3 the regularity index only grows logarithmically. The factor 2 is probably distinguished here as well as in music where octaves play a crucial role.

The Haar wavelet oscillates so that

$$\int_{-\infty}^{\infty} dx \psi(x) = 0 \tag{18.16}$$

This condition is common for all the wavelets. It is called the oscillation or cancellation condition. From it, the origin of the name wavelet becomes clear. One can describe a "wavelet" as a function that oscillates within some interval like a wave but is then localized by damping outside this interval. This is a necessary condition for wavelets to form an unconditional (stable) basis. We conclude that for special choices of coefficients h_k one obtains the specific forms of "mother" wavelets, which give rise to orthonormal bases.

---

* It is also often called a "father wavelet," but we do not use this term.
** We return to the general case and therefore omit the index H because the same formula will be used for other wavelets.

One may decompose any function f of $L^2(R)$ at any resolution level j_n in a series

$$f = \sum_k s_{j_n,k} \varphi_{j_n,k} + \sum_{j \geq j_n,k} d_{j,k} \psi_{j,k} \tag{18.17}$$

At the finest resolution level $j_n = j_{max}$ only s-coefficients are left, and one obtains the scaling-function representation:

$$f(x) = \sum_k s_{j_{max},k} \varphi_{j_{max},k} \tag{18.18}$$

In the case of the Haar wavelets it corresponds to the initial experimental histogram with the finest resolution. Because we are interested in its analysis at varying resolutions, this form is used as an initial input only. The final representation of the same data (Equation 18.17) shows all the fluctuations in the signal. The wavelet coefficients $s_{j,k}$ and $d_{j,k}$ can be calculated as

$$s_{j,k} = \int dx f(x) \varphi_{j,k}(x) \tag{18.19}$$

$$d_{j,k} = \int dx f(x) \psi_{j,k}(x) \tag{18.20}$$

However, in practice their values are determined from the fast wavelet transform described below.

In reference to the particular case of the Haar wavelet, considered above, these coefficients are often referred as sums (s) and differences (d), thus related to mean values and fluctuations. Only the second term in Equation 18.17 is often considered, and the result is often called wavelet expansion. For the histogram interpretation, neglect of first sum would imply that one is not interested in average values but only in the histogram shape determined by fluctuations at different scales. Any function can be approximated to a precision $2^{j/2}$ (i.e., to an arbitrary high precision at $j \to -\infty$) by a finite linear combination of Haar wavelets.

18.4 Multiresolution Analysis and Daubechies Wavelets

Although Haar wavelets provide a good tutorial example of an orthonormal basis, they suffer from several deficiencies. One is the bad analytic behavior with the abrupt change at the interval bounds, i.e., their bad regularity properties. By this we mean that all finite rank moments of the Haar wavelet are different from zero — only its zeroth moment, i.e., the integral (Equation 18.16) of the function itself is zero. This shows that this wavelet is not orthogonal to any polynomial apart from a trivial constant. The Haar wavelet does not have good time-frequency localization. Its Fourier transform decays like $|\omega|^{-1}$ for $\omega \to \infty$.

The goal is to find a general class of those functions that would satisfy the requirements of locality, regularity, and oscillatory behavior. Note that in some particular cases the orthonormality property sometimes can be relaxed. They should be simple enough in the sense that they are of being sufficiently explicit and regular to be completely determined by their samples on the lattice defined by the factors 2^j.

The general approach that respects these properties is known as the multiresolution approximation. A rigorous mathematical definition is given in the above-cited monographs. One can define the notion of wavelets so that the functions $2^{j/2}\psi(2^j x - k)$ are the wavelets (generated by the "mother" ψ), possessing the regularity, the localization, and the oscillation properties. By varying j we can resolve signal properties at different scales, while k shows the location of the analyzed region.

We show how the program of the multiresolution analysis works in practice when applied to the problem of finding out the coefficients of any filter h_k and g_k. They can be directly obtained from the definition and properties of the discrete wavelets. These coefficients are defined by relations 18.6 and 18.9

$$\varphi(x) = \sqrt{2} \sum_k h_k \varphi(2x - k)$$

$$\psi(x) = \sqrt{2} \sum_k g_k \varphi(2x - k)$$

(18.21)

where $\sum_k |h_k|^2 < \infty$. The orthogonality of the scaling functions defined by the relation

$$\int dx \varphi(x) \varphi(x - m) = 0$$

(18.22)

leads to the following equation for the coefficients:

$$\sum_k h_k h_{k+2m} = \delta_{0m}$$

(18.23)

The orthogonality of wavelets to the scaling functions

$$\int dx \psi(x) \varphi(x - m) = 0$$

(18.24)

gives the equation

$$\sum_k h_k g_{k+2m} = 0$$

(18.25)

having a solution of the form:

$$g_k = (-1)^k h_{2M-1-k}$$

(18.26)

Thus, the coefficients g_k for wavelets are directly defined by the scaling function coefficients h_k.

Another condition of the orthogonality of wavelets to all polynomials up to the power $(M - 1)$, defining its regularity and oscillatory behavior

$$\int dx x^n \psi(x) = 0, \qquad n = 0, ..., (M - 1)$$

(18.27)

provides the relation

$$\sum_k k^n g_k = 0$$

(18.28)

giving rise to

$$\sum_k (-1)^k k^n h_k = 0$$

(18.29)

when the formula 18.26 is taken into account.

The normalization condition

$$\int dx \varphi(x) = 1$$

(18.30)

can be rewritten as another equation for h_k:

$$\sum_k h_k = \sqrt{2}$$

(18.31)

Let us write Equations 18.23, 18.29, and 18.31 for $M = 2$ explicitly:

$$h_0 h_2 + h_1 h_3 = 0$$

$$h_0 - h_1 + h_2 - h_3 = 0$$

$$-h_1 + 2h_2 - 3h_3 = 0$$

$$h_0 + h_1 + h_2 + h_3 = \sqrt{2}$$

The solution of this system is

$$h_3 = \frac{1}{4\sqrt{2}}(1 \pm \sqrt{3})$$

$$h_2 = \frac{1}{2\sqrt{2}} + h_3$$

$$h_1 = \frac{1}{\sqrt{2}} - h_3 \tag{18.32}$$

$$h_0 = \frac{1}{2\sqrt{2}} - h_3$$

which, in the case of the minus sign for h_3, corresponds to the well-known filter:

$$h_0 = \frac{1}{4\sqrt{2}}(1 + \sqrt{3})$$

$$h_1 = \frac{1}{4\sqrt{2}}(3 + \sqrt{3})$$

$$h_2 = \frac{1}{4\sqrt{2}}(3 - \sqrt{3}) \tag{18.33}$$

$$h_3 = \frac{1}{4\sqrt{2}}(1 - \sqrt{3})$$

These coefficients define the simplest D^4 (or $_2\psi$) wavelet from the famous family of orthonormal Daubechies wavelets with finite support. It is shown in the upper part of Figure 18.3 by the dotted line with the corresponding scaling function shown by the solid line. Some other higher-rank wavelets are also shown there. It is clear from this figure (especially, for D^4) that wavelets are smoother at some points than at others.

For the filters of higher order in M, i.e., for higher-rank Daubechies wavelets, the coefficients can be obtained in an analogous manner. The wavelet support is equal to $2M - 1$. It is wider than for the Haar wavelets. However, the regularity properties are better. The higher-order wavelets are smoother compared to D^4 as seen in Figure 18.3.

18.5 Fast Wavelet Transform

Fast wavelet transform allows us to proceed with all the computation within a short time interval because it uses the simple iterative procedure. Therefore, it is crucial for all work with wavelets.

The coefficients $s_{j,k}$ and $d_{j,k}$ carry information about the content of the signal at various scales and can be calculated directly using Equations 18.19 and 18.20. However, this algorithm is inconvenient for numerical computations because it requires many (N^2) operations, where N denotes a number of the sampled values of the function. We describe a faster algorithm. This is clear from Figure 18.4, and the

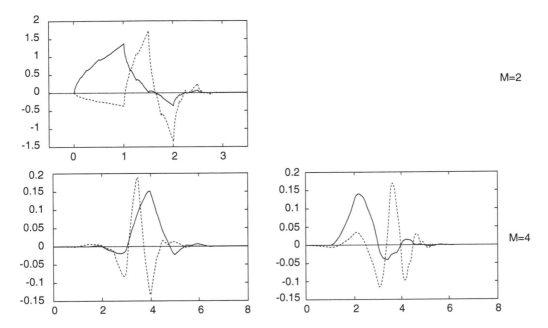

FIGURE 18.3 Daubechies scaling functions (solid lines) and wavelets (dotted lines) for $M = 2,4$.

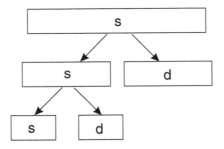

FIGURE 18.4 Fast wavelet transform algorithm.

fast algorithm formulas are presented below. In practical calculations, the coefficients h_k are used without referring to the shapes of wavelets.

In real situations with digitized signals, we have to deal with finite sets of points. Thus, there always exists the finest level of resolution where each interval contains only a single number. Correspondingly, the sums over k will reach finite limits. It is convenient to reverse the level indexation assuming that the label of this fine scale is $j = 0$. It is then easy to compute the wavelet coefficients for sparser resolutions $j \geq 1$.

Multiresolution analysis naturally leads to a hierarchical and fast scheme for the computation of the wavelet coefficients of a given function. The functional Equations 18.6 and 18.9 and the formulas for the wavelet coefficients Equations 18.19 and 18.20 give rise, in the case of Haar wavelets, to the relations (Equation 18.2), or for the backward transform (synthesis) to Equation 18.3.

In general, we can obtain the iterative formulas of the fast wavelet transform

$$s_{j+1,k} = \sum_m h_m s_{j,2k+m} \tag{18.34}$$

$$d_{j+1,k} = \sum_m g_m s_{j,2k+m} \tag{18.35}$$

where

$$s_{0,k} = \int dx f(x)\varphi(x-k) \qquad (18.36)$$

These equations yield fast algorithms (the so-called pyramid algorithms) for computing the wavelet coefficients, asking now just for $O(N)$ operations to be done. Starting from $s_{0,k}$ we compute all other coefficients provided the coefficients h_m, g_m are known. The explicit shape of the wavelet is not used in this case any more.

The remaining problem lies in the initial data. If an explicit expression for $f(x)$ is available, the coefficients $s_{0,k}$ may be evaluated directly according to Equation 18.36. But this is not so in the situation when only discrete values are available. In the simplest approach they are chosen as $s_{0,k} = f(k)$.

18.6   The Fourier and Wavelet Transforms

As has been stressed already, the wavelet transform is superior to the Fourier transform, first of all, due to the locality property of wavelets. The Fourier transform uses sine, cosine, or imaginary exponential functions as the main basis. It is spread over the entire real axis whereas the wavelet basis is localized. An attempt to overcome these difficulties and improve time localization while still using the same basis functions is made by the so-called windowed Fourier transform. The signal $f(t)$ is considered within some time interval (window) only. However, all windows have the same width.

In contrast, the wavelets ψ automatically provide the time (or spatial location) resolution window adapted to the problem studied, i.e., to its essential frequencies (scales). Namely, let t_0, δ and ω_0, δ_ω be the centers and the effective widths of the wavelet basic function $\psi(t)$ and its Fourier transform. Then for the wavelet family $\psi_{j,k}(t)$ and, correspondingly, for wavelet coefficients, the center and the width of the window along the t-axis are given by $2^j(t_0 + k)$ and $2^j\delta$. Along the ω-axis they are equal to $2^{-j}\omega_0$ and $2^{-j}\delta_\omega$. Thus the ratios of widths to the center position along each axis do not depend on the scale. This means that the wavelet window resolves both the location and the frequency in fixed proportions to their central values. For the high-frequency component of the signal it leads to a quite large frequency extension of the window whereas the time location interval is squeezed so that the Heisenberg uncertainty relation is not violated. That is why wavelet windows can be called Heisenberg windows. Correspondingly, the low-frequency signals do not require small time intervals and admit a wide window extension along the time axis. Thus, wavelets well localize the low-frequency "details" on the frequency axis and the high-frequency ones on the time axis. This ability of wavelets to find a perfect compromise between time localization and frequency localization by automatically choosing the widths of the windows along the time and frequency axes well adjusted to the location of their centers is crucial for their success in signal analysis. The wavelet transform cuts the signal (functions, operators, etc.) into different frequency components, and then studies each component with a resolution matched to its scale providing a good tool for time-frequency (position-scale) localization. That is why wavelets can zoom in on singularities or transients (an extreme version of very short-lived high-frequency features) in signals, whereas the windowed Fourier functions cannot. In terms of traditional signal analysis, the filters associated with the windowed Fourier transform are constant bandwidth filters whereas the wavelets may be seen as constant relative bandwidth filters whose widths in both variables linearly depend on their positions.

In Figure 18.5 we show the difference between these two approaches. The figure demonstrates the constant shape of the windowed Fourier transform region and the varying shape (with a constant area) of the wavelet transform region. The density of localization centers is homogeneous for the windowed Fourier transform whereas it changes for the wavelet transform so that at low frequencies the centers are far apart in time and become much denser for high frequencies.

The wavelet coefficients are negligible in the regions where the function is smooth. That is why wavelet series with plenty of nonzero coefficients represent really pathological functions, whereas "normal" functions have "sparse" or "lacunary" wavelet series and are easy to compress. On the other hand, the Fourier series of the usual functions have a lot of nonzero coefficients, whereas "lacunary" Fourier series represent pathological functions.

Thus these two types of analysis can be considered as complementary rather than overlapping.

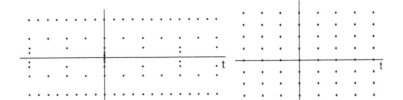

FIGURE 18.5 The lattices of time-frequency localization for the wavelet transform (left) and windowed Fourier transform (right).

18.7 Technicalities

One can already start the signal analysis with the above procedures. However, there are several technical problems that should be mentioned. At some length they are described in the cited monographs and review papers.

- The number of possible wavelets at our disposal is much larger than the above examples show. Let us mention coiflets, splines, frames, wavelet packets, etc. Usually, one chooses for the analysis the particular basis that yields the minimum entropy.

- Multiresolution analysis can be performed in more than one dimension. In two dimensions, dilations of the resulting orthonormal wavelet basis control both variables simultaneously, and the two wavelets are given by the following expression:

$$2^j \psi(2^j x - k, 2^j y - l), \qquad j, k, l \in Z \qquad (18.37)$$

where ψ is no longer a single function; on the contrary, it consists of three elementary wavelets. To obtain an orthonormal basis of W_0 one has to use in this case three families $\varphi(x - k)\psi(y - l)$, $\psi(x - k)\varphi(y - l)$, and $\psi(x - k)\psi(y - l)$. Then the two-dimensional wavelets are $2^j\varphi(2^j x - k)\psi(2^j y - l)$, $2^j\psi(2^j x - k)\varphi(2^j y - l)$, $2^j\psi(2^j x - k)\psi(2^j y - l)$. In the two-dimensional plane, the analysis is done along the horizontal, vertical, and diagonal strips with the same resolution in accordance with these three wavelets. Figure 18.6 shows how this construction appears. The schematic representation of this procedure in the left-hand side of the figure demonstrates how the corresponding wavelet coefficients are distributed for different resolution levels $j = 1$ (and 2). In the figure, a set of geometrical objects is decomposed into two layers. We clearly see how vertical, horizontal, and diagonal edges are emphasized in the corresponding regions. We should also notice that the horizontal strip is resolved into two strips at a definite resolution level.

- The study of many operators acting on a space of functions or distributions becomes simple when suitable wavelets are used because these operators can be approximately diagonalized with respect to this basis. Orthonormal wavelet bases provide a unique example of a basis with nontrivial diagonal, or almost-diagonal, operators. The operator action on the wavelet series representing some function does not have uncontrollable sequences; i.e., wavelet decompositions are robust. We can describe precisely what happens to the initial series under the operator action and how it is transformed. In a certain sense, wavelets are stable under the operations of integration and differentiation. That is why wavelets, used as a basis set, allow us to solve differential equations characterized by widely different length scales found in many areas of physics and chemistry. Moreover, wavelets reappear as eigenfunctions of certain operators. The so-called nonstandard matrix multiplication is a useful procedure for dealing with operators.

- Analysis of any signal includes finding the regions of its regular and singular behavior. One of the main features of wavelet analysis is its capacity of performing a very precise local analysis of the regularity properties of functions. This allows us to investigate, characterize, and easily distinguish some specific local behaviors such as approximate self-similarities and very strong oscillatory features such as those of the indefinitely oscillating functions, which

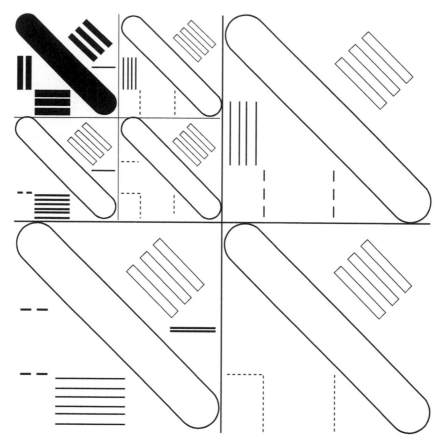

FIGURE 18.6 Example of the wavelet analysis of a two-dimensional plot. One sees that either horizontal or vertical details of the plot are more clearly resolved in the corresponding coefficients. Also, the small or large size (correlation length) details are better resolved depending on the level chosen.

are closely related to the so-called "chirps" of the form $x^\alpha \sin(1/x^\beta)$ (reminding us of bird, bat, or dolphin sonar signals with very sharp oscillations that accelerate at some point x_0). Chirps are well known to everyone dealing with modern radar and sonar technology. Two-microlocal analysis is used[20] to reveal the pointwise behavior of any function from the properties of its wavelet coefficients. The two-microlocal space $C^{s,s'}(x_0)$ of the real-valued n-dimensional functions f (distributions) is defined by the following simple decay condition on their wavelet coefficients $d_{j,k}$

$$\left| d_{j,k}(x_0) \right| \le C 2^{-\left(\frac{n}{2}+s\right)j} + \left(1 + \left|2^j x_0 - k\right|\right)^{-s'} \tag{18.38}$$

where s and s' are two real numbers. This is a very important extension of the Hölder conditions. The two-microlocal condition is a local counterpart of the usual uniform condition. It expresses the singular behavior of the function itself at the point x_0 in terms of the k-dependence of its wavelet coefficients at the same point.

- In signal analysis, real-life applications produce only sequences of numbers due to the discretization of continuous time signals. This procedure is called the sampling of analog signals. The behavior of wavelet coefficients across scales provides a good way to describe the regularity of functions whose samples coincide with the observations at a given resolution. Moreover, to save computing time, we can use, not a complete set of wavelet coefficients $d_{j,k}$, but only a part of them omitting small coefficients not exceeding some threshold value (ε). This standard estimation method is called estimation by coefficient thresholding.

18.8 Scaling

By scaling, one usually implies the self-similarity of the analyzed object or event. In turbulence, e.g., this is revealed as "whorls inside whorls inside whorls," which leads to the powerlike behavior of some distributions.

More generally, we say that some signals (objects) possess self-similar (fractal) properties, which means that by changing the scale we observe at a new scale features similar to those previously noticed at other scales. Since wavelet analysis consists of studying the signal at various scales by calculating the scalar product of the analyzing wavelet and the signal explored, it is well suited to revealing the fractal peculiarities. In terms of wavelet coefficients it implies that their higher moments behave in a powerlike manner with the scale changing. Specifically, let us consider the sum Z_q of the qth moments of the coefficients of the wavelet transform at various scales j:

$$Z_q(j) = \sum_k |d_{j,k}|^q \tag{18.39}$$

where the sum is over the maxima of $|d_{j,k}|$. Then it was shown[21,22] that for a fractal signal this sum should behave as

$$Z_q(j) \propto 2^{j[\tau(q)+\frac{q}{2}]} \tag{18.40}$$

i.e.,

$$\log Z_q(j) \propto j[\tau(q) + \frac{q}{2}] \tag{18.41}$$

Thus, the necessary condition for a signal to possess fractal properties is the linear dependence of $\log Z_q(j)$ on the level number j. If this requirement is fulfilled the dependence of τ on q shows whether the signal is monofractal or multifractal. Monofractal signals are characterized by a single dimension and, therefore, by a linear dependence of τ on q, whereas multifractal ones are described by a set of such dimensions, i.e., by nonlinear functions $\tau(q)$. Monofractal signals are homogeneous, in the sense that they have the same scaling properties throughout the entire signal. Multifractal signals, on the other hand, can be decomposed into many subsets characterized by different local dimensions, quantified by a weight function. The wavelet transform, if done with wavelets possessing the appropriate number of vanishing moments, removes the lowest polynomial trends that could cause traditional box-counting techniques to fail in quantifying the local scaling of the signal. The function $\tau(q)$ can be considered a scale-independent measure of the fractal signal. It can be further related to the Renyi dimensions, Hurst and Hölder exponents. The range of validity of the multifractal formalism for functions can be elucidated with the help of two-microlocal methods generalized to the higher moments of wavelet coefficients. Thus, wavelet analysis goes very far beyond the limits of the traditional analysis, which uses the language of correlation functions (see, e.g., Reference 23) in approaching much deeper correlation levels.

The scaling properties of the biological objects (e.g., the fractal structure of tree leaves, etc.) can be studied in the same manner.

18.9 Applications

Wavelets have become widely used in mathematics and in physics for purely theoretical studies in functional calculus, renormalization in gauge theories, conformal field theories, nonlinear chaoticity, and in more practical fields like quasicrystals, meteorology, acoustics, seismology, nonlinear dynamics of accelerators, turbulence, the structure of surfaces, cosmic ray jets, solar wind, galactic structure, cosmological density fluctuations, dark matter, gravitational waves, etc. In fact, they have also become very

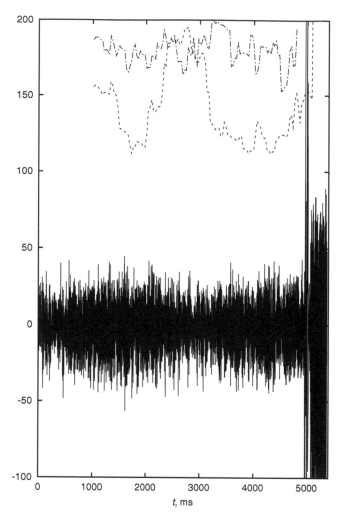

FIGURE 18.7 Signal of the pressure sensor (the solid line) and the dispersion of its wavelet coefficients (the dashed line). The time variation of the pressure in the engine compressor (the irregular solid line) has been wavelet-analyzed. The dispersion of the wavelet coefficients (the dashed line) shows the maximum and the remarkable drop prior to the drastic increase of the pressure providing the precursor of this malfunction. The shuffled set of the data does not show such an effect for the dispersion of the wavelet coefficients (the upper curve) pointing to its dynamic origin.

popular in many other fields of applied science, especially for pattern recognition, image compression, data classification, etc. (see, e.g., Reference 11 and our Web site http://awavelet.lpi.ru). Surely, they can be used for analysis of any digitized signals (events, objects) in oceanology, aquatic ecology, etc., as well. Here we describe just two examples of wavelet application to analysis of one- and two-dimensional objects.

It is tempting to begin with analysis of signals depending on a single variable. The example is provided by the time variation of the pressure in an aircraft compressor shown in Figure 18.7. The aim of the analysis of this signal is motivated by the desire to find the precursors of a very dangerous effect (stall+surge) in engines leading to their destruction. It happened that the dispersion of the wavelet coefficients shown by the dashed line in Figure 18.7 drops before the dangerous high pressure appears in the compressor of the engine. This can serve as a precursor of this effect. No such a drop is seen in the upper dash-dotted line, which shows the similar dispersion for the random signal obtained from the initial signal by shuffling its values at a different time. These curves show the internal correlations at the different scales existing in the primary signal possess a very complicated structure. Moreover, the described approach to scaling phenomena was applied in this case. It showed violation of scaling properties at the scale that appeared to be most dangerous. Just at this scale the function (Equation 18.41) showed the strong scaling violation. Let us mention that a similar procedure has been quite successful in analysis of other engines and of heartbeat intervals and diagnosis of a disease. The same scaling analysis can be used in oceanology, aquatic ecology, etc. However, it is still awaiting application.

Two-dimensional wavelet analysis can be also used in all these fields for recognition of object shapes. It has been applied, e.g., for pattern recognition of fingerprints (this has helped save a good deal of computer memory, in particular) and of erythrocytes and their classification. Here, we briefly describe how it was successfully used for both analysis and synthesis of various images.

Let us recall that any image in a computer must be digitized and saved as a bitmap or, in other words, as a matrix, each element of which describes the color of the point in the original image. The number of elements of the matrix (image points) depends on the resolution chosen in the digital procedure. A bitmap is used for the subsequent reproduction of the image on a screen, a printer, etc. However, it is not desirable to store it in such a form because this would require a huge computer capacity. That is why, at present, numerous coding algorithms (compression) for a bitmap have been developed, whose effectiveness depends on the image characteristics. All these algorithms belong to the two categories — they either code with a loss of information or without any loss of it (in that case the original bitmap can be completely recovered by the decoding procedure). In more general applications, the latter algorithm is often called as the data archivation.

As an example, we consider a compression algorithm with the loss of information. "The loss of information" means in this case that the restored image is not completely identical to the original image, but this difference is practically indistinguishable by a human eye.

At present, most computer-stored images (in particular, those used in Internet) with continuous tone are coded with the help of the algorithm JPEG. The main stages of this algorithm are as follows. One splits the image into matrices with 8×8 dots. For each matrix, a discrete cosine-transform is performed. The obtained frequency matrices are subjected to a "quantization" procedure when the most crucial elements for the visual recognition are chosen according to a special "weight" table compiled beforehand by specialists after their collective decision was achieved. This is the only stage where the "loss" of information occurs. Then the transformed matrix with the chosen frequencies (scales) is compacted and coded by the so-called entropy method (also called arithmetic or Huffmann method).

The applied algorithm differs from that described by use of the wavelets instead of the windowed cosine-transform and by the transform of the whole image instead of the 8×8 matrix only. Figure 18.8 demonstrates the original image and the two final images restored after similar compression according to JPEG and wavelet algorithms. It is easily seen that the quality of the wavelet image is noticeably higher than for JPEG for practically the same size of the coded files. The requirement of the same quality for both algorithms leads to file sizes 1.5 to 2 times smaller for the wavelet algorithm, which could be crucial for the transmission of the image, especially if the transmission line capacity is limited.

Many other examples can be found in the cited literature and in Web sites.

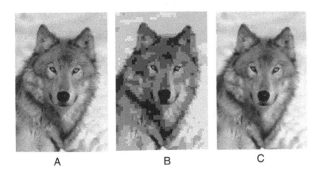

A   B   C

FIGURE 18.8 (A) Original photo (the file size is 461,760 bytes). (B) Photo reconstructed after compression according to the JPEG-algorithm (the file size is 3511 bytes). (C) Photo reconstructed after compression according to the wavelet algorithm (the file size is 3519 bytes). The better quality of the wavelet transform is clearly seen when comparing the original image (A) and the two images restored after similar compression by the windowed Fourier transform (B) and the wavelet transform (C).

18.10 Conclusions

The beauty of the mathematical construction of the wavelet transformation and its utility in practical applications attract researchers from both pure and applied science. Moreover, the commercial outcome of this research has become quite important. We have outlined a minor part of the activity in this field. However, we hope that the general trends in the development of this subject became comprehended and appreciated and this method will be fruitful for applications in aquatic ecology and oceanology.

References

1. Meyer, Y., *Wavelets and Operators,* Cambridge: Cambridge University Press, 1992.
2. Daubechies, I., *Ten Lectures on Wavelets,* Philadelphia: SIAM, 1991.
3. Meyer, Y. and Coifman, R., *Wavelets, Calderon-Zygmund and Multilinear Operators,* Cambridge: Cambridge University Press, 1997.
4. Meyer, Y., *Wavelets: Algorithms and Applications,* Philadelphia: SIAM, 1993.
5. Meyer, Y. and Rogues, S., Eds., *Progress in Wavelet Analysis and Applications*, Gif-sur-Yvette: Editions Frontieres, 1993.
6. Chui, C.K., *An Introduction to Wavelets*, San Diego: Academic Press, 1992.
7. Hernandez, E. and Weiss, G., *A First Course on Wavelets*, Boca Raton, FL: CRC Press, 1997.
8. Kaiser, G., *A Friendly Guide to Wavelets*, Boston: Birkhauser, 1994.
9. *Wavelets: An Elementary Treatment of Theory and Applications,* T. Koornwinder, Ed., Singapore: World Scientific, 1993.
10. Astafyeva, N.M., *Physics-Uspekhi,* 39, 1085, 1996.
11. Dremin, I.M., Ivanov, O.V., and Nechitailo, V.A., *Physics-Uspekhi,* 44, 447, 2001.
12. Van den Berg, J.C., Ed., *Wavelets in Physics,* Cambridge: Cambridge University Press, 1998.
13. Mallat, S., *A Wavelet Tour of Signal Processing*, New York: Academic Press, 1998.
14. Erlebacher, G., Hussaini, M.Y., and Jameson, L.M., *Wavelets Theory and Applications*, Oxford: Oxford University Press, 1996.
15. Aldroubi, A. and Unser, M., Eds., *Wavelets in Medicine and Biology,* Boca Raton, FL: CRC Press, 1994.
16. Carmona, R., Hwang, W.-L., and Torresani, B., *Practical Time-Frequency Analysis*, San Diego: Academic Press, 1998.
17. Grossman, A. and Morlet, J., In *Mathematica+Physics, Lectures on Recent Results,* Vol. 1, L. Streit, Ed., Singapore: World Scientific, 1985.
18. Morlet, J., Arens, G., Fourgeau, E., and Giard, D., *Geophysics,* 47, 203, 222, 1982.
19. Haar, A., *Math. Ann.,* 69, 331, 1910.
20. Jaffard, S. and Meyer, Y., *Mem. Am. Math. Soc.,* 123(587), 1996.
21. Muzy, J.F., Bacry, E., and Arneodo, A., *Phys. Rev. Lett.,* 67, 3515, 1991; *Int. J. Bifurc. Chaos,* 4, 245, 1994.
22. Arneodo, A., d'Aubenton-Carafa, Y., and Thermes, C., *Physica* (Amsterdam), 96D, 291, 1996.
23. De Wolf, E., Dremin, I.M., and Kittel, W., *Phys. Rep.,* 270, 1, 1996.

19

Fractal Characterization of Local Hydrographic and Biological Scales of Patchiness on Georges Bank

Karen E. Fisher, Peter H. Wiebe, and Bruce D. Malamud

CONTENTS

19.1 Introduction 297
19.2 Fractal Character of Underway Oceanographic Data 300
    19.2.1 U.S. GLOBEC Broadscale Surveys on Georges Bank 300
19.3 Wavelet-Based Variance Spectra 301
    19.3.1 Wavelet Transform 301
    19.3.2 Implementation 303
    19.3.3 Wavelet Analysis 305
    19.3.4 2D and 3D Representations of the Wavelet Transform 306
    19.3.5 Wavelet-Based Spectra 308
19.4 Seasonal Power Law Behavior 309
19.5 Hydrographic Region Power Law Behavior 310
    19.5.1 Well-Mixed Crest of the Bank 310
    19.5.2 Crossing Fronts of the Bank 311
    19.5.3 Stratified Flanks of the Bank 313
19.6 Fractal Interpolation 313
    19.6.1 Fractal Interpolation Equations 314
    19.6.2 Fractally Interpolated 2D Fluorescence Fields 315
    19.6.3 Implications for Future Work 317
19.7 Conclusions 317
Acknowledgments 318
References 318

19.1 Introduction

Plankton patchiness, defined by Cushing as the pattern of spatial extents of density-delimited clusters,[1] is intermittent in time and space. Depending on the scale of observation, the degree of plankton intermittency at any given location varies. In a dynamic ocean shelf environment, such as Georges Bank, Northwest Atlantic, the variability ranges from steady, gradual changes with scale to abrupt transitions. Temporal scales shorter than tidal cycles, longer than seasons, and at every scale in between influence environmental variability, and ultimately spatial variability in plankton distributions. Spatial variability in hydrography further enhances differences between regions. Transects from the GLOBEC Northwest Atlantic Broadscale survey of June 1999 illustrate the character of one-dimensional (1D) fluorescence data records reflecting rapid changes in hydrography on the southern flank of Georges Bank (Figure 19.1).

FIGURE 19.1 **(Color figure follows p. 332.)** Portion of the surface fluorescence record from GLOBEC Broadscale Survey in June 1999 (AL9906). Transects move onto Georges Bank, from deeper (500 m), stratified water into shallower (40 m), well-mixed water, going into the page, crossing the tidal mixing front. The center across-isobath transect onto the bank is 56 nautical miles long, going from Station 8 (started at 2345 on June 16 1999) to Station 11 (ended at 1402 on June 17 1999) of the survey; the distance along isobath from Station 7 to 8 is 28 nautical miles. All station locations are shown in Figure 19.3, and data are archived at http://globec.whoi.edu. The range in raw voltage linearly scales from 0 to 5.

Intermittency in the distribution of plankton in space and time has substantial impact on the flux of materials within and out of the mixed layer. Increasing accuracy in models of plankton dynamics requires that model input fields reflect the heterogeneous distributions observed in oceanographic data. Inaccuracy in plankton mapping based on survey data has a negative effect on both the accuracy of biomass estimates, and the characterization of system variability. For the GLOBEC surveys, the quantitative ramifications of incomplete reconstruction of patch structure have been investigated, and found to account for 40% of the error.[2] The spatial and temporal overlap of plankton at various trophic levels ultimately constrains the flux from one trophic level to the next. Therefore, methods allowing the characterization of the degree to which trophic levels exhibit contemporaneous patchiness has direct application to improving large-scale models of biologically mediated fluxes. Such methods are not possible without first realizing reliable means of constructing model input fields that preserve the variability of the observations on which they are based.

A data series can be examined either in the spatial domain or in the frequency domain. Both Fourier and wavelet transforms examine the intensity of variability at different frequencies. Of the two, wavelets have the advantage of examining this intensity locally, which allows distributions to be characterized across regions with significant variability. Wavelets are robust when examining nonstationary data series. Broadly speaking nonstationary systems where wavelets are superior candidates for conducting analysis include those with systematic changes in the mean (trends), systematic changes in variance, and/or periodic components.[3,4] Wavelets are discrete filters passed over a series of data. Wavelets are localized in both sample space (or time) and frequency, and provide flexibility because they can be constructed specifically to further optimize resolution in either sample space or frequency. Wavelets provide an attractive alternative to Fourier transforms for producing succinct characterizations of spatial

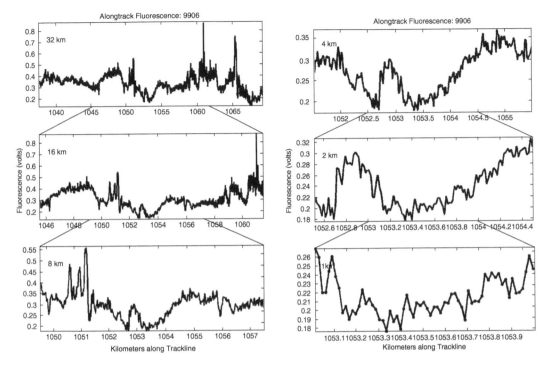

FIGURE 19.2 Increasing resolution of section from June 1999 fluorescence record. Length of section ranges from 32 km (top left) between Stations 24 and 25, to 1 km (bottom right). For the 1 km section, individual points are plotted to highlight relations of adjacent concentrations. Sections are centered on the transect heading nearly north toward Station 25 (see station map shown in Figure 19.3).

distributions.[3–5] A variety of wavelets can be constructed, each with different trade-offs regarding resolution in frequency and sample space.[6,7] Torrence and Campo[5] provide a concise, informative introduction to these trade-offs, with application to El-Niño Southern Oscillation time series. Once spatial distributions of the observed data are characterized suitably, the question becomes how to realistically characterize unobserved regions, based on the available information.

Fractal data series appear the same at different scales in time or space,[8] as shown in Figure 19.2. The wavelet transform can provide a local fractal measure of such data. This property can then be used to create fractal simulations of observed distributions. Fractal simulations permit input of spatially located frequency characterizations, based on fractal attributes of the data at each scale observed, to inform the model. In addition to providing a visually realistic view of observed distribution, fractal simulations allow quantitative bounds to be determined for biomass estimates and illustrate the plausible range in distributions that could result in a given set of observations.

In this chapter, the fractal character of biological and physical distributions is determined around Georges Bank, Northwest Atlantic using the wavelet transform on 1D fluorescence, salinity, and temperature observations during GLOBEC Broadscale survey cruises in January, March, and June of 1998 and 1999. The parameter β, the slope of the wavelet-based power spectrum, is used throughout to characterize fractal behavior. Spectral slopes, β, are analyzed for seasonality, based on the entire cruise track, and local spatial variability, based on transects in various hydrographic regimes. The existence of critical horizontal spatial scales is documented, manifest in the wavelet determined power spectra as breaks in slope. The 1D β values observed along a 100-km transect on the southern flank of Georges Bank in June 1999 are then applied to the task of synthesizing two-dimensional (2D) fields of fluorescence to assess variability at scales smaller than the station separation of the survey. Fractal interpolation is used, constrained by the determined β values, to reconstruct fluorescence fields with the observed patchiness.

19.2　Fractal Character of Underway Oceanographic Data

Fractals have self-similar structure, such that a small subset has similar properties to the whole, as was described for coastlines by Mandelbrot.[9] Fractals can also be based on self-affine structure, which is simply self-similar structure that varies on different scales with changes in direction.[4] In self-similar fractals, the *x*- and *y*-axes scale the same way, whereas in self-affine fractals the *x*- and *y*-axes scale differently and have distributions that are generally not isotropic.

Data series obtained from natural systems are often self-affine. Fractal assessments of natural distributions have been applied to numerous systems, including plankton dynamics,[10,11] landscape weathering,[8] atmospheric events,[12] turbulent fluids,[7] subgrid-scale turbulence,[13] stream flows,[14] and patchily distributed populations.[15,16] Fractals exist whenever a power law distribution is found or, in other words, any time that plotting the variance vs. the scale of a data set forms a straight line on a log–log plot.[17]

A general impression of whether a data series is fractal, exhibiting similar characteristics at a variety of scales, can be gained by looking at increasingly smaller segments of the series. Illustrated in Figure 19.2 is fluorescence data from June 1999 centering on a section across the Northeast Channel separating Georges Bank from the Scotian Shelf. These points suggest that the relationship of values at larger scales (here up to 32 km length of transect) and smaller scales (here down to 17 m distance between samples) are quite different. Persistence, or memory, is the property of the data set that measures this relationship. Persistence exists when adjacent values of a data series are on average closer to each other than for a random series (e.g., a white noise).[4]

At the 32 km scale, the data values in Figure 19.2 show a general correlation of each value with surrounding values. This illustrates a strong persistence at the large scales, where big values tend to follow big, and little values tend to follow little — like values cluster. At the 1 km scale, the data values in Figure 19.2 show a more random pattern. As further explained in Malamud,[4] a self-affine fractal exhibits the same persistence at all scales (long-range persistence). These fluorescence data are classic examples of self-affine, fractal structure with finite bounds; the fractal structure will be seen to be change at scales around 1 km, exhibiting a different level of persistence at scales larger than this value. The spatial structure of the spectral energy distribution pattern for given parameters around Georges Bank can be used to gain insight into the organization of the system over a wide range of scales.

19.2.1　U.S. GLOBEC Broadscale Surveys on Georges Bank

The U.S. GLOBEC Broadscale Survey in the Northwest Atlantic included 41 stations with bottom depths ranging from less than 40 m on top of Georges Bank to over 500 m off the flanks, as shown in Figure 19.3. The survey was designed to sample sites in both seasonally stratified water and in the central well-mixed region. The well-mixed region is delineated by a tidal mixing front often found near the 60 m isobath. The gradient across the front intensifies seasonally as stratification progresses in the deeper waters along the flanks of the bank. In January, March, and June 1998 and 1999, underway data were obtained over the majority of the trackline by the Greene Bomber tow body shown at bottom right in Figure 19.3. These six cruises towed the Greene Bomber at 3 m depth, providing calibrated fluorescence, salinity, and temperature measurements.

The underway Environmental Sensing System (ESS) surveys captured the seasonal transition from winter to summer states on Georges Bank. Each survey took 10 to 12 days to complete. The underway survey was designed to complement the intensive station work with continuous estimates of spatial distributions of plankton, and to provide a link between the physical and biological distributions. The ESS data were obtained at 4 s intervals, averaging 15–17 m apart when towed while steaming 6 to 7 kts along transects.[18] Complete records of the fluorescence data gathered from these six cruises are shown in Figure 19.4. Data series like these have the potential to resolve aspects of the controls and ramifications of plankton patchiness over a variety of spatial scales and in regions with rapidly varying hydrographic properties. The method of fractal interpolation allows stochastic development of modeled 2D fields consistent with observed 1D tracklines that incorporate rare events (extreme peaks), and subsequently results in probability density functions with more robust tails.[21,22]

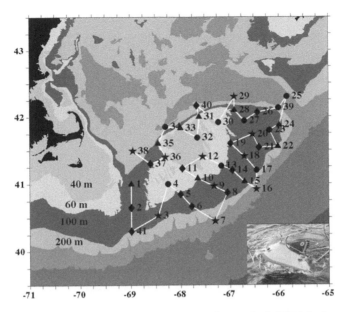

FIGURE 19.3 Stations of the U.S. GLOBEC Broadscale surveys on Georges Bank, NW Atlantic, and the Greene Bomber tow body (bottom right). Surveys were run monthly from January to June between 1995 and 1999, towing along the trackline at 3 m depth, off the port or starboard quarter. Temperature, conductivity, and fluorescence sensors are attached to the stainless steel framework 28 cm forward of the nose, outside of the tow body; main section houses two downward-looking acoustic transducers at 120 and 420 kHz.

19.3 Wavelet-Based Variance Spectra

Malamud and Turcotte[4] review the methods available for analysis of the frequency domain (spectral analysis), and subsequent fractal characterization of natural data series such as those shown in Figure 19.1 and Figure 19.4. They conclude that wavelet-based spectra are a preferable approach because wavelets are localized, and can be directly applied to data from anisotropic fields. Wavelet-based spectra can also be obtained from nonstationary fields where, broadly speaking, the mean or the variance changes with the length of the interval considered.

19.3.1 Wavelet Transform

The wavelet transform is a filter g passed over a data series $f(x)$. The generalized form of the 1D wavelet transform is

$$W(x, a_n) = \frac{1}{\sqrt{a_n}} \int_{-\infty}^{\infty} g\left(\frac{x' - x}{a_n}\right) f(x') dx' \tag{19.1}$$

where f is the function of the spatial position x, a_n is the scale being analyzed, and g is the filter. A variety of filters have been developed and tested, with different advantages in discriminating features of the data. The only requirement for designing a wavelet filter is that it must be continuous and integrate to zero:

$$\int_{-\infty}^{\infty} g(x') dx' = 0 \tag{19.2}$$

The Mexican Hat filter, illustrated in Figure 19.5, has been used to determine features of naturally occurring distributions such as river flow and planetary topography.[4,14] The general shape of the Mexican

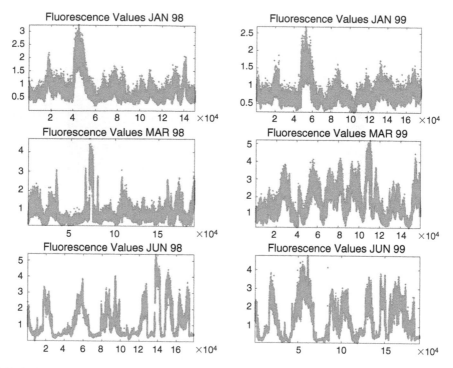

FIGURE 19.4 Fluorescence observed on GLOBEC Broadscale Survey trackline (Figure 19.3) around Georges Bank, NW Atlantic in 1998 (left) and 1999 (right) in: January (top), March (middle), and June (bottom). Fluorescence values are plotted in raw voltage ranging from 0 to 5 for each observation (x-axis is observation number). Higher values tend to be on top of the bank in the well-mixed region, and the number of peaks in the overall record roughly corresponds to the number of times the survey transect crosses the tidal mixing front into the well-mixed area. Evident features include variability in overall values (highest in March 1999 and lowest in January 1999), changes in variance within a survey, and shifts in total variance among seasons.

Hat filter is not unlike the character of frontal features encountered in the along-track data obtained on the GLOBEC Broadscale cruises. The Mexican Hat filter has the form:

$$g(x') = \left(\frac{1}{2\pi}\right)^{\frac{1}{2}}(1 - x'^2)e^{\frac{-x'^2}{2}} \tag{19.3}$$

Substituting Equation 19.3 into Equation 19.1 gives the following form:[4]

$$W(x, a_n) = \left(\frac{1}{2a_n\pi}\right)^{\frac{1}{2}} \int_{-\infty}^{\infty}\left[1 - \left(\frac{x'-x}{a_n}\right)^2\right]e^{-(x'-x)^2/2a^2} f(x')dx \tag{19.4}$$

The amplitude of the wavelet transform $W(x, a_n)$ at each scale a_n reflects the relation between adjacent areas at that scale; it measures the intensity of each contributing frequency. Figure 19.5 illustrates two Mexican Hat wavelets, at two different spatial resolutions, and two portions of the trackline signal taken from Figure 19.1. Suppose the taller wavelet is of scale $a_n = 1$. The 17 values in this wavelet are convolved with the data series. Convolution at $a_n = 1$ first multiplies the first 17 points in the data series with the value of each point in the wavelet filter, and then sums the products. The convolution produces a single value of $W(x, a_n)$ that is assigned the same spatial position as that of the data point located at the midpoint of the filter. The filter is then shifted to the right by one data point, and the process is repeated. This is continued until the end of the data series, resulting in $W(x, a_n)$ values beginning at $x = 8$, and ending at $x = n - 8$ (i.e., a vector shorter than the data vector by the width of the filter at each resolution).

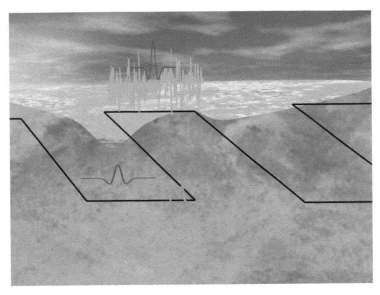

FIGURE 19.5 (Color figure follows p. 332.) Portions of the fluorescence record from two distinct regions along the survey track line (textures derived from satellite ocean color observations of Georges Bank obtained by SeaWiFS). Mexican Hat wavelets (in blue) at two resolutions shown during convolution with two sections taken from the fluorescence data (in green) along the survey track in Figure 19.1. The taller wavelet is shown with data from the well-mixed region that appears noisy in character; this wavelet convolves fewer points at each step, and therefore resolves smaller spatial scales. The wider wavelet is shown with data from the stratified region that appears persistent in character; it convolves more points at each step and therefore resolves larger spatial scales. Wavelets are normalized to have the same total energy at all scales. Note that practical limitations on scales that can be analyzed are imposed by the shape of the analyzing wavelet, the shape of the survey track, and the data sample spacing.

For a white noise, similar in character to the well-mixed data of Figure 19.5 with values equally uncorrelated at all scales, the wavelet magnitudes will not change with scale. Conversely, for a Brownian motion, similar in character to the stratified data of Figure 19.5, the energy associated with larger spatial scales will be higher as the wavelet encounters persistent peaks and valleys on a scale similar to the wavelet scale.

Following the analysis of Malamud and Turcotte,[4] self-affine series can be characterized succinctly because such series have a power law dependence of the power spectral density function on frequency f:

$$S(f) \propto f^{-\beta} \tag{19.5}$$

The exponent β here is the negative of the slope of the power spectrum of the series, and $S(f)$ is the sum of the square of the Fourier coefficients. The relation in Equation 19.5 defines a self-affine fractal in the same way a self-similar fractal is defined.[4] The variance V_n of the wavelet transform similarly has a power law dependence on scale a_n (inverse frequency):

$$V_n \propto a_n{}^{\beta} \tag{19.6}$$

Calculating the variance V_n of the wavelet transform at each scale a_n and plotting it as a function of scale examined produces a power law exponent β. A power law with exponent β (e.g., $y = x^2$ has $\beta = 2$), when plotted in log–log space (with spatial scale on the x-axis, not frequency), will produce a straight line with slope β (e.g., $\log y = 2 \log x$). The exponent β produced in wavelet variance analysis has the same magnitude, but opposite sign, from Fourier power-spectral analysis performed after detrending and windowing.

19.3.2 Implementation

As an example, three sections of data from the southern flank of Georges Bank were selected from distinct hydrographic regimes: well-mixed, tidal mixing front, and well-stratified waters. A comparison

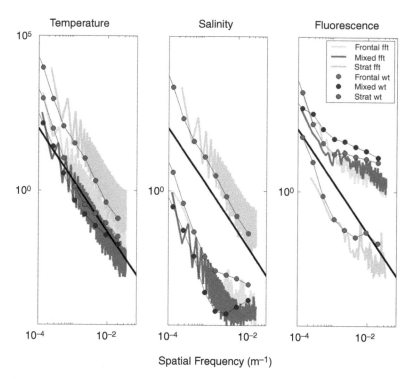

FIGURE 19.6 Results of conventional FFT and wavelet transforms of sections of data taken from well-mixed, frontal, and stratified regions on the southern flank of Georges Bank for temperature (left), salinity (middle), and fluorescence (right). Black reference line has slope −5/3. Note that total power in both transforms must theoretically be equal, but the nonstationary signals must be modified to perform the FFT; slopes β are similar between transforms, but concentrations of power (peaks) differ due to filter shape. Arbitrary units on the *y*-axis are the same for all three variables, so relative power spectral density is preserved.

of conventional fast Fourier transform (FFT) and wavelet-based transforms is shown in Figure 19.6 for the three hydrographic regimes, and spectra of both transforms are plotted relative to frequency. Plotting the wavelet variance in this manner provides a powerful characterization strategy for underway oceanographic measurements. The parameter β is empirically related to fractal dimension D by the relation $D = (5-\beta)/2$ for β between 1 and 3.[4]

Many of the features found in these spectra were found to exist on Georges Bank in earlier work.[27,28] In particular, the clear flattening of fluorescence spectra at scales less than 1 km confirms the earlier result, and reiterates the implication that fine-scale (less than 1 km) patchiness is not entirely controlled by turbulent energy dissipation. By contrast, the larger scales of the fluorescence spectra have slopes considerably closer to the $k^{-5/3}$ power law often observed in the temperature spectra, and evident here. Especially prominent is the complete mirror of the total power arrangement in the three regions between the physical variables, temperature, and salinity, and the biological variable fluorescence. For temperature and salinity, the total power is highest in the stratified area, intermediate in the frontal region, and lowest in the well-mixed area. For fluorescence, the situation is exactly reversed, and in addition, the total power in the wavelet transform appears to trace the top of the Fourier transform envelope, as opposed to tracing the bottom for the physical variables. However, in all cases, the frontal and well-mixed areas cluster together relative to the stratified area.

With wavelet-derived variance spectra, local patchiness can be quantitatively determined for underway observations of temperature, salinity, and fluorescence on commensurate scales, at each observation point. Scales of physical and biological processes contributing to population structures can then be compared directly over scales of meters to hundreds of kilometers. Recall that the slopes β of wavelet variance spectra measure persistence, or the relationship between adjacent values in a data series.[4] Flatter slopes, with β values near 0, are uncorrelated; a β = 0 is a white noise. Fluorescence spectra for the well-mixed and frontal areas are flatter at smaller scales (Figure 19.6). Steeper slopes, with β values greater than 1 have persistence; a $\beta = 2$ is a Brownian motion; steeper slopes have increasingly higher persistence. Fluorescence spectra for the stratified area have steeper slopes, approaching the steepness of the $k^{-5/3}$ reference line (i.e., $\beta = 1.667$). Persistence, in turn, relates to the fractal dimension of a data series. Steeper β are smoother series, with fractal dimensions that approach 1 (a line), while flatter β are rougher series, with fractal dimensions that approach 2 (a space-filling line). This can be visualized by looking at the relative steepness of the transforms (β values) for the stratified (steep spectra and smooth data) and well-mixed (flat spectra and rough data) areas in Figure 19.6, and comparing them to the stratified and well-mixed clips in Figure 19.5, or the data itself, shown in Figure 19.1.

The main goal here is to achieve characterization of patches along the trackline for a spatial reconstruction. The smallest resolution of the Mexican Hat requires 17 points (basis –8 to 8) to ensure the filter values integrate to near zero. The *resonance* scale at minimum dilation $a_n = 1$ is about equal to four times the data spacing for the Mexican Hat (roughly the zero crossing of the mainlobe). Hence, a dilation factor of $a_n = 1$ in the Mexican Hat–derived data series convolves 17 data points and picks up features with a minimum spatial scale of 50 to 60 m, while a dilation factor of $a_n = 128$ convolves 2176 data points and picks up features with a minimum spatial scale of about 3.8 km. The analyses that follow of fluorescence, salinity, and temperature data focus on spatial scales between 50 m and 30 km.

The numerical wavelet convolution is flexible. Instead of convolving a single wavelet sequentially with each series of points in the data series, $f(x)$, wavelets of a given dilation can be linked end to end to form a wavelet train long enough to convolve the entire data series, reducing the number of iterations in the calculation to the number of points convolved at each resolution. Since the wavelet train is localized, the value of the convolution with each individual wavelet in the train is assigned to the center data point of that particular convolution. The power found at a large front influences the magnitude of the transform only within the neighborhood defined at that particular scale by the size of the wavelet filter. Dilating the wavelet scale by powers of two (doubling the number of points convolved each time) between calculations makes for efficient processing of large data series, often yielding enough resolution to determine whether distributions follow power law behavior. Analyzing a continuous range of scales leads to increased redundancy at large scales, but lessens the impact of aperiodic shifts in the series on the overall spectra.

19.3.3 Wavelet Analysis

Wavelet analysis is well localized in space (or time), as the finite wavelet is sequentially convolved with each section of the data series, in contrast to the infinite sine waves used in traditional Fourier analysis. The number of points convolved increases as larger scales are assessed. However, to keep the total power within the filter constant, the wavelet filter becomes flatter when stretched to longer scales, and taller when compressed to shorter scales, as illustrated at two resolutions in Figure 19.5. A basic feature of wavelets, like any other filter, is a tendency to *resonate* most strongly with like-shaped features in the analyzed data. Highest values of the transform are obtained where data are encountered with both a scale and a shape similar to the spatial (or temporal) scope of the filter. Dilations (changes in scale) of the filter are arbitrarily chosen to resolve features of interest without undue overlap in analyses. The minimum resolution is based numerically on the smallest resolution meeting the *admissibility* requirement that the filter mean be zero.[7]

Different wavelets optimize resolution of different features, subject to the trade-off that a wavelet well localized in sample space (or time) is less well localized in frequency, and vice versa (Table 19.1).[5,18] Real wavelets, like the Mexican Hat, are optimized for location of features in sample space. The real transform used here for determining β values is a convolution of the data with the negative of the second derivative of the Gaussian (DoG-2) distribution, or Mexican Hat wavelet. The Mexican Hat wavelet has an intuitive relation to the original sample space. By contrast, complex wavelets, such as the Morlet wavelet, have better localization in frequency space. The Morlet wavelet transform is a convolution of the data with a Gaussian modulated plane wave. Because of its smaller *footprint* in Fourier space, the Morlet wavelet gives more resolution at smaller spatial (or temporal) frequencies than the Mexican Hat, but blurs the location in sample space (or time). Both these wavelets produce non-orthogonal, discrete

TABLE 19.1

Comparison of Morlet and Mexican Hat Wavelets

Wavelet	Equation	Wavelength	Stronger Resolution
Morlet (nondimensional frequency: $\omega_0 = 6$)	$\pi^{-\frac{1}{4}}e^{i\omega_0 x}e^{-\frac{x^2}{2}}$	$\dfrac{4a\pi}{\omega_0 + \sqrt{2+\omega_o^2}}$	Frequency
Mexican Hat (second derivative of Gaussian: $m = 2$)	$\left(\dfrac{1}{2\pi}\right)^{\frac{1}{2}}(1-x'^2)e^{\frac{-x'^2}{2}}$	$\dfrac{2a\pi}{\sqrt{m+\dfrac{1}{2}}}$	Space

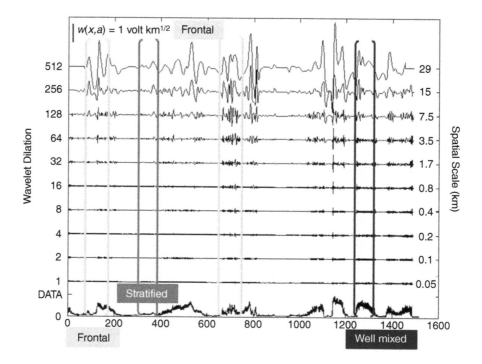

FIGURE 19.7 Example of Mexican Hat wavelet transform $W(x, a_n)$ illustrated in 2D for fluorescence data in June 1998. The data series is at the bottom, with a scale in volts; above are the ten horizontal series resulting from the convolution of wavelets at that scale with the data (Equation 19.4). The magnitude of the wavelet transform for each scale a_n is indicated by the scale bar; each dilation of a_n (left axis) is also related to its maximum resonance spatial scale (right axis). Well-mixed, frontal, and stratified areas within the transform that qualitatively exemplify the major hydrographic structures of Georges Bank are marked. Note that holding the page with the top angling away gives roughly the same view as the 3D representations below.

wavelet transforms. Because an aperiodic shift can produce a different wavelet spectrum in orthogonal analysis, the continuous approach that can be applied in non-orthogonal analysis is appropriate for data series analysis (see Torrence and Campo[5] for discussion of orthogonal and non-orthogonal wavelet choice). The Mexican Hat and Morlet wavelets, and their associated Fourier wavelengths, are shown in Table 19.1.

19.3.4   2D and 3D Representations of the Wavelet Transform

The wavelet transform calculated using scales that ascend by powers of two yields a spatial map of the wavelet magnitudes at each resolution showing the contributions to variance locally at each scale. As an example, the Mexican Hat transform for June 1998 fluorescence data is shown in Figure 19.7. The original data are convolved with ten different scales $a_n = 1, 2, 4, 8, ..., 512$, of the Mexican Hat wavelet

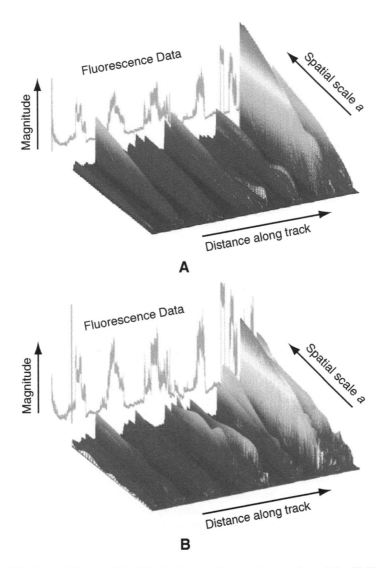

FIGURE 19.8 (Color figure follows p. 332.) 3D visualization showing absolute values of the 1D Mexican Hat wavelet (A) and Morlet wavelet (B) transforms $W(x, a_n)$ for fluorescence in June 1998, at spatial scales up to 4 km. Heading into the page, scales range from $a_n = 2$ to $a_n = 64$, and the convolutions are calculated at even integer values of a_n. The 3D mesh colors high values resulting from the convolution red, and low values blue, shown in the 2D transform for the intermediate powers of two. Transforms are over the same 1500 km of cruise track (heading left to right) shown in Figure 19.7. The fluorescence data are shown in the background.

using Equation 19.4. The wavelet variance of $W(x, a_n)$ at small scales (small a_n) in the fluorescence transform in Figure 19.7 generally has similar amplitudes as scale increases, while at larger scales (large a_n) it has increasing amplitudes as scale increases. This indicates that larger-scale wavelets found features similar to themselves in the fluorescence data.

When calculated continuously, the absolute value of the wavelet convolution $W(x, a_n)$ can be visualized in three dimensions, showing the existence of features such as narrow fronts, small-scale variability that does not extend to the larger scales, and other localized tendencies of the transform (Figure 19.8). For comparison, both the Mexican Hat and Morlet wavelet transforms are shown, illustrating the differences in their detection of peaks in power. The superior frequency resolution of the Morlet wavelet leads to more peaks at any given location at different spatial scales (inverse frequency). This wavelet has advantages for investigation of issues such as dominant patch (or grain) size. With so many data points

TABLE 19.2

Relations of β and Other Fractal Measures

Fractal Measure	Anti-persistent Shear	White Noise Random	1/f Noise	$k^{-5/3}$ Inertial Subrange 1D Turbulence	Brownian Walk	Persistent 2D Turbulence
β	−1	0	1	1.667	2	3
H	NA	NA	0	0.334	0.5	1
D	(2) ←——————— (2)			1.667	3/2	(1)
Shape	Area ←——————————————————— Squiggle				Mountains	Line

along the x-axis, it is not possible to see the accompanying loss in resolution of sample space of the Morlet wavelet relative to the Mexican Hat.

Particularly notable in both transforms, when viewed in this format, is the range in magnitude of the wavelet transform $W(x, a_n)$. The dramatically higher transform values occur late in the cruise on the northern flank of Georges Bank (where x is large). In addition, structure in the small spatial scale values is more striking in the 3D Mexican Hat wavelet representation calculated at each integer value of a_n than in the 2D visualization at powers of two of a_n. The discrete filter allows spatial and temporal resolution to be maintained in the course of analysis. Changing spatiotemporal resolution can be viewed as *binning* energy; the discrete filter does not leak power from one frequency to another. Locally, large variations, such as fronts, do not have to be partitioned separately from other data, as their influence is concentrated near their location. As the goal here is to determine power law characterization well localized in space, the remainder of the chapter focuses on results only of the Mexican Hat.

19.3.5 Wavelet-Based Spectra

Past studies of spectral attributes of plankton, based on power-spectral analysis, and its interpretation, are well summarized in papers by Platt[19] and Powell and Okubo.[20] While the implications of the pioneering spectral work were intriguing, the limitations on the methods available to produce localized spectra led to tedious processing requirements, and fairly involved preprocessing of data using *a priori* information. Since the variance calculated from the wavelet transform $W(x, a_n)$ value is localized, continuing this line of investigation is now considerably less daunting. Squaring the absolute values of the wavelet magnitude yields the localized power spectrum (i.e., the square of Figure 19.8). Wavelet-based spectra suggest a powerful tool for continuing the investigation of inferring controlling processes from the resulting plankton patterns set out by the careful work based on Fourier spectra over the last three decades.

In Table 19.2, fractal measure β is the slope of a wavelet power spectrum (plotted vs. spatial scale), or the negative of the slope of an FFT power spectrum (plotted vs. frequency). For persistent series with values of β between 1 and 3, empirical relations have been derived. The Hausdorff measure $H = (\beta - 1)/2$ is defined for series in this range. $D = (5 - \beta)/2$ is the fractal dimension that can be obtained. Fractional dimensions D of increasingly space-filling non-Euclidean series values of D fall between Euclidean values. The relationship of each measure to β is defined only in the heavily boxed squares; the arrows indicate that at the limits the relationship to dimension becomes unresolvable using the empirically derived relations as the signals become space filling. As a measure of relative persistence in the signal, the utility of β remains undiminished throughout the range of −1 to 3 and beyond.[4]

The relations of fractal measures β and D are shown in Table 19.2 for the range of persistence values often encountered in physical or biological data series. As summarized in Table 19.2, a spectrum of pure white noise has $\beta = 0$, showing the same power at all dilations; a spectrum of Brownian motion has $\beta = 2$, showing increasing power in the variance at large scales. Spectra of energy distribution (or temperature) are expected to have $\beta = 5/3$ ($\beta = 1.667$) in the inertial subrange, $\beta = 7/3$ ($\beta = 2.333$) in sheared flows, and $\beta = 3$ in 2D turbulence. The slope of $\beta = 1$ has been theoretically shown to result under certain conditions, from reactions similar to spatially varying growth (or loss) rates of a population.[19,20]

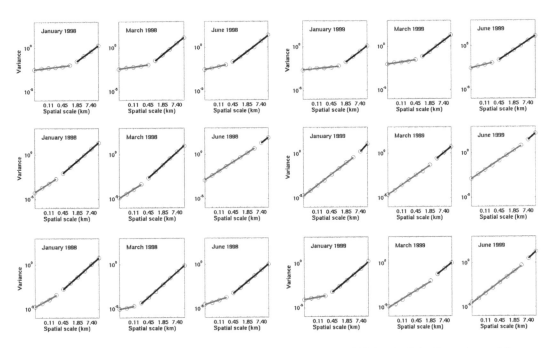

FIGURE 19.9 Wavelet variance spectra for fluorescence (top), temperature (middle), and salinity (bottom) around Georges Bank, NW Atlantic. Best-fit lines are shown for the dual fit maximizing r^2. Slopes are given in Table 19.3.

19.4 Seasonal Power Law Behavior

The bankwide wavelet variance spectra, based on the Mexican Hat wavelet transform, from six surveys are shown in Figure 19.9. Given the tendency of wavelet determined spectra from the GLOBEC data to qualitatively exhibit two or more power law distributions, three *best slopes* were fit to each spectrum. The three slopes were determined by linear regression: the best fit line over all ten dilations, and the two best fit lines that maximized the combined r^2 fit. The minimum resolvable scale of the slope break is at 220 m (first three dilations), the maximum at 7.4 km. The algorithm fits two lines at all possible scale breaks, and determines slopes for the combination yielding the maximum r^2. The β values that describe the best fit for smaller scales (up to the break) will be designated β_1, and the β values that describe the best fit for larger scales will be designated β_2.

The seasonal spectra for fluorescence exhibit a strikingly consistent behavior in both 1998 and 1999 (Figure 19.9). The wavelet-based spectra show the fluorescence spectra tend to have different behavior at different spatial scales. Single best fit lines have r^2 values from 0.83 to 0.962, while fitting two slopes to each spectra increases the fit to an r^2 range of 0.90 to 0.99. Smaller scales tend to be between white noise and $1/f$ noise in character. These smaller scale β_1 values range from 0.2 to 0.7 at spatial scales less than about a kilometer. Fluorescence spectra tend to steepen at scales greater than a kilometer, and larger scale β_2 values range from 1.8 to 2.2. A distinct seasonal trend also appears (Table 19.3). The scale of the breaks between steep large scale and flatter short scales decreases seasonally the same way in both years, from 1.8 km in January to 0.9 km in March, to 0.45 km in June. The slopes of β_1 are remarkably similar between years for each month observed. The values of β_1 increase in both years from about 0.3 in January, to about 0.4 in March, and finally reach 0.6 in June.

By contrast, temperature spectra show much smaller changes in slope, although they do increase in total variance in June of both years (Figure 19.9). Temperature spectra for Georges Bank can be reasonably generalized using a single best-fit over scales ranging from 50 m to 30 km (Table 19.3). Over the six cruises, the overall spectra have slopes ranging from $\beta = 1.7$ to $\beta = 2.2$, in the range of slopes expected for subinertial range turbulence spectra $\beta = 5/3$ (1.667) and shear flow turbulence spectra $\beta = 7/3$ (2.333).

TABLE 19.3

Slopes for Seasonal Spectra on Georges Bank

	January 1998	March 1998	June 1998	January 1999	March 1999	June 1999
Fluorescence						
β_{all}	0.955	1.330	1.468	0.958	1.215	1.400
β_1	0.318	0.376	0.593	0.269	0.397	0.662
β_2	1.984	2.208	2.028	2.038	2.052	1.838
Temperature						
β_{all}	2.094	2.228	1.731	2.106	1.977	1.857
β_1	1.661	1.697	1.682	2.037	1.817	1.775
β_2	2.283	2.427	1.973	2.797	2.204	2.356
Salinity						
β_{all}	2.092	1.934	1.739	1.624	1.859	2.110
β_1	1.484	0.530	0.872	0.460	1.631	1.999
β_2	2.354	2.305	2.207	2.240	2.146	2.531

Note: Slopes of the best-fit lines are listed for each of the following categories: β_{all} is the slope over all scales; β_1 is the slope at small scales (light lines in Figure 19.9); and β_2 is the slope at large scales (dark lines in Figure 19.9).

Although a single line fits all six temperature spectra with r^2 values between 0.995 and 0.999, fitting two lines to the spectra results in data values which tend to be flatter at small spatial scales, and steeper at large ones. Salinity spectra vary considerably, but do not appear to have a dominant seasonal component within that variation (Figure 19.9). These spectra reflect influences of water from other regions such as the Scotian Shelf (relatively fresh), or the Gulf Stream (relatively salty) that arrive on an irregular basis as well as seasonal changes in circulation.

19.5 Hydrographic Region Power Law Behavior

The vertical conductivity, temperature, depth (CTD) profiles shown in Figure 19.10 illustrate the rapid changes in hydrographic regime of cross-isobath transects of the Broadscale Survey. Spectra from well-mixed (Station 12), frontal (Station 14), and stratified (Station 16) horizontal transects between these stations show equally distinct character. The well-mixed, frontal, and stratified regions used to separate bank waters for the FFT-wavelet spectral comparison in Figure 19.6 show clear differences in the behavior of their characteristic spectral slopes, both among these areas, and among variables within them.

In the fluorescence spectra, the spatial scale at which the spectra steepen changes significantly between the stratified and well-mixed regions. Among regions, differences also exist in how fast the magnitudes increase (i.e., how steep the spectra will become), and in the magnitudes themselves (i.e., how much power will be in the variance spectrum at each scale). With practice, spectral slopes can almost be estimated by eye within the marked regions of Figure 19.7, and in other characteristic areas. Because the wavelet variance is localized, it is possible to pull any desired subset along the trackline, and analyze the variance therein. Examples of subsets for three sections with quite distinct hydrography indicate the importance of localized analysis in dynamic hydrographic regions.

19.5.1 Well-Mixed Crest of the Bank

In the shallow 40 to 50 m depths common on the bank crest, mixing dominates the physical hydrography, and temperature and fluorescence profiles typically look similar to Station 12 in Figure 19.10. The tem-

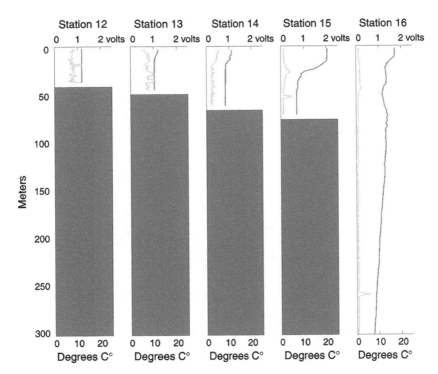

FIGURE 19.10 CTD profile of fluorescence (light line, on left) and temperature (dark line, on right) from Station 12 in 40 m of water on top of the bank, to Station 16 in over 500 m of water off the southern flank. Top axis is fluorescence scale (raw volts), bottom axis is temperature scale (°C); *y*-axis is depth (meters).

perature and salinity vary little, and their plots are highly coherent lines with very small trends for tens of nautical miles. However, the fluorescence response of the phytoplankton along the same nearly isothermal, isohaline trackline tends to look much less coherent. Fluorescence spectra on the crest have a single spectral slope between 50 m and 6.4 km in scale for which $\beta = 0.75$ (Figure 19.11). With the higher contribution of variance from small scales, total spectral power in the fluorescence signal turns out to be higher than the spectral power in either salinity or temperature for both the well-mixed and frontal regions.

Unlike fluorescence, both temperature and salinity show a clear break in their spectra for the well-mixed area (Figure 19.11). Moreover, all three spectra are unique; no two spectra look similar for this section. The spectral slopes of temperature are considerably steeper at large scales, $\beta_2 = 2.7$, and less steep, $\beta_1 = 1$, at scales smaller than 400 m. Salinity has a similar large scale slope, with $\beta_2 = 2.9$, but has a clear negative trend of $\beta_1 = -0.7$, indicating more power at smaller scales, possible evidence of an inverse enstrophy cascade (from small to large scales) at scales between 50 and 200 m. Such a cascade suggests an input of energy at small scales that then flows to larger scales, the reverse of the more commonly observed model where energy cascades from larger to smaller scales. The wavelet transform picked up several instances of this phenomenon during the course of the Georges Bank broadscales considered here, and although it is a bit outside the scope of this chapter, it is another indication of the value of localized analysis of spectral properties.

19.5.2 Crossing Fronts of the Bank

Moving from the well-mixed zone on top of the bank into water 60 to 80 m deep, there is a reasonably persistent tidal mixing front, which is well established on this June cruise. The section between Stations 13 and 14 just crosses this front, and features indicative of frontal structure (i.e., enhanced surface fluorescence) reflect this in Figure 19.10. Comparison of horizontal records shows that the fluorescence usually responds with a peak toward the crest from the point where the temperature record shows a minimum. The best-fit aggregate slopes for this area are shown in Figure 19.12. Unlike the previous

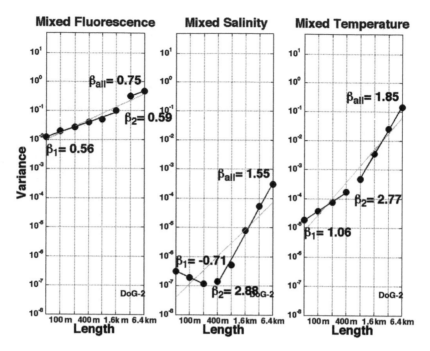

FIGURE 19.11 Best-fit slopes for mixed area spectra, integrated along the southern flank section. Black dots indicate the wavelet variance calculated at each discrete scale; light lines indicate the single best-fit slope; dark lines indicate the best-fit β_1 and β_2 lines.

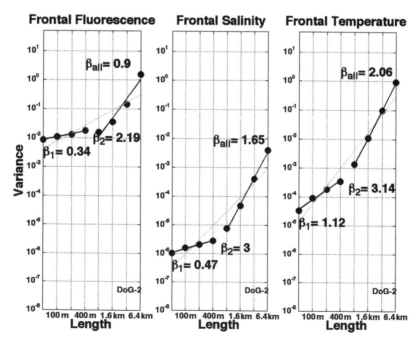

FIGURE 19.12 Best-fit slopes for frontal area spectra, integrated along the southern flank section. Black dots indicate the wavelet variance calculated at each discrete scale; light lines indicate the single best-fit slope; dark lines indicate the best-fit β_1 and β_2 lines.

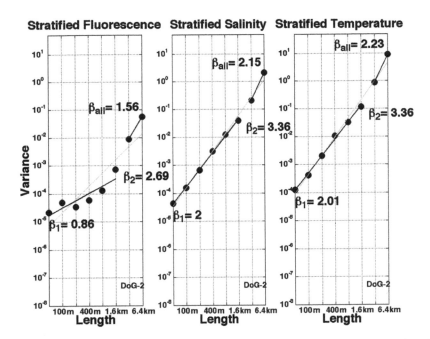

FIGURE 19.13 Best-fit slopes for stratified area spectra, integrated along the southern flank section. Black dots indicate the wavelet variance calculated at each discrete scale; light lines indicate the single best-fit slope; dark lines indicate the best-fit β_1 and β_2 lines.

section in the mixed area, all three spectra now show similar trends. The scale of the break in all three spectra is at 400 m. Temperature has the steepest large scale slope, at $\beta_2 = 3.14$, followed closely by salinity, at $\beta_2 = 3.0$, and then fluorescence, at $\beta_2 = 2.19$. Small-scale slopes of fluorescence and salinity are both less than $\beta_1 = 0.5$, while of temperature the slope is close to $\beta_1 = 1$.

19.5.3 Stratified Flanks of the Bank

In water from 100 to more than 500 m depth, stratification and currents can create rapid gradients in the horizontal variation of physical hydrography. The temperature and salinity records are marked by small fronts, and collocated changes in the small-scale variability at scales of tens of meters. The fluorescence response, indicating phytoplankton reactions to these same fronts is not as obvious. In the fluorescence spectrum integrated along the section shown in Figure 19.13 there is a peak in power at 100 m, and no particularly convincing power law behavior at small scales. Both the overall and best-fit large-scale slopes of fluorescence ($\beta = 1.6$ and $\beta = 2.7$) are somewhat less steep than those of salinity and temperature (both close to $\beta_2 = 2.3$ and $\beta_2 = 3.3$). Large-scale β_2 values for salinity and temperature, at scales larger than 1.6 km, are in the range of the theoretical spectra for 2D turbulence ($\beta = 3$); the water is strongly stratified, and the mixed layer is quite shallow, based on the CTD profile in Figure 19.10 for Station 16. At smaller scales, these steep slopes contrast sharply with the small scale slope for fluorescence, $\beta_1 = 0.86$.

19.6 Fractal Interpolation

In most oceanographic research, aside from satellite-based measurements, the goal of synoptic sampling is still far from our grasp. Despite the breadth of the U.S. GLOBEC Broadscale underway survey, the data described above undersample the system substantially. Station sampling of the GLOBEC Broadscale survey is nominally at a 40 km scale (Figure 19.3). McGillicuddy et al.[2] conclude that mapping error, defined as the degree to which the true variability is not represented by the sampling grid, produces errors of 40% relative to perfectly known fields. Model-based evaluation of errors associated with undersampling of the survey indicates that to estimate biomass to within better than 50% of the actual value, subgridscale parameterizations must be made.[2] Better performance depends on characterization of the patchiness at scales smaller than the station grid size, resolving the fine-scale variability of plankton distribution.

One method of addressing mapping error is fractal interpolation. The simulation technique of fractal interpolation makes use of the statistical self-similarity of fractal distributions. The algorithm presented by Voss[21] was used to illustrate Mandelbrot[8] and create examples of computer-generated landscapes. Strahle[22] applied fractal interpolation to turbulent combustion data, aiming to synthesize more accurate tails for probability distributions. Iterating realizations of a strongly nonlinear process represents a powerful tool for characterizing the influence of rare events. Since β is a fractal measure, it can be used in fractal interpolation to develop plausible predictions of distributions both at unobserved scales and in unobserved regions. Furthermore, there is no requirement of stationarity in fractal interpolation.

As many power law distributions are obtained from nonstationary fields, using the inherent statistics of the distributions themselves is a step toward realism in synthesized fields. This realism has both intuitive and quantitative utility. With a minimal amount of preprocessing, and therefore a minimal amount of presumption, the fractal dimension of a data series can be obtained from wavelet analysis, as shown above. Particularly in the case of biological populations, there may be more than one fractal dimension associated with a distribution over a finite range of scales. Because fractal interpolation is by nature an iterative function, these multiple dimensions can be easily included in the reconstructed field. The result is a field that preserves the intermittency of observations. Fractal interpolation realistically characterizes many data sets, and distributes the variance throughout the synthesized field in a way similar to the actual distribution of variance observed. Adjacent data points at distant positions in the field have the same relationship to one another as those actually observed. Multiple realizations of fractal fields can be constructed and used to assess the likelihood of rare events. In addition, fractal interpolation provides a visual tool for interpreting data.

19.6.1 Fractal Interpolation Equations

Fractal interpolation has been shown to be an effective way to extend 1D data series in turbulence studies, as developed by Strahle.[22,23] Applications have been made to 2D fields in a variety of areas: topography (terrain roughness), meteorology (rainfall), medicine (imaging and reconstruction), and turbulence (subgrid-scale fields).[8,12,13,24,25] However, these studies deal with signals having slopes β of 1 to 3, corresponding to the Hausdorff parameter H ranging from 0 to 1 (see Table 19.2). The relation of β to H within this range is defined as:

$$\beta = 2H + 1 \tag{19.7}$$

The midpoint displacement algorithm was shown by Voss[21] to faithfully produce noises throughout this range, simply by scaling the variance of a randomly sampled normal distribution at each successive midpoint, or other arbitrary distance r, by Equation 19.8:

$$r^{2Hn} \tag{19.8}$$

at each successive iteration n, hereafter referred to as the *Voss algorithm*.[21] For some scales of fluorescence and salinity, and for temperature in most regions, this works well. If spectra have both β_1 and β_2 values between 1 and 3, the Voss algorithm can used to synthesize fields similar to the data observations throughout all scales of the fractal iteration. However, the smaller scales of fluorescence in particular tend to be characterized by slopes β_1 ranging from 0 to 1, out of the range that can be directly synthesized using this characterization.

Slopes between 0 and 1 lack the degree of persistence required for the Voss algorithm. This range falls outside the range where there is sufficient persistence (where H is defined). Methods of approaching construction of series with breaks from slopes close to 2 to slopes between 0 and 1 are somewhat limited.

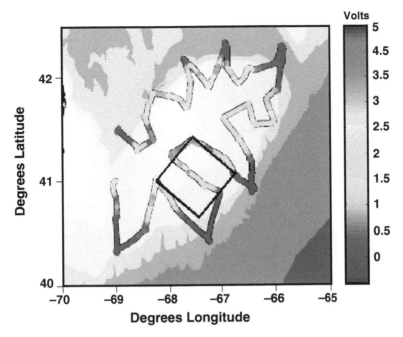

FIGURE 19.14 (**Color figure follows p. 332.**) Region of fractal interpolation for southern flank superimposed over the observations of raw fluorescence voltage observed at 3 m depth by tow body Greene Bomber on GLOBEC Broadscale cruise AL9906, June 1999.

The most commonly suggested approach is differentiating a series with β values between 2 and 3 (resulting in β values of 0 to 1), rescaling the variance appropriately, and then summing with a series with slope 1 to 3.[21] The scaling was found to be problematic to constrain using this method, and the break in the spectra that result are not as sharp as those observed in the spectra found here.

A numerical simulation of series that change from large scales with β_2 between 1 and 2 to small scales with β_1 between 0 and 1 can be achieved iteratively by *turning off* the Voss algorithm (freezing the n value) in Equation 19.8 at the iteration (n value) corresponding to the observed break, and then scaling the variance at each successively smaller ($r = \frac{1}{2}$ of previous scale) steps by the factor in Equation 19.9:

$$\left(\frac{1}{\beta}\right)^{\frac{1}{5}} \tag{19.9}$$

where β is the observed slope of the spectra at that scale. This factor was empirically determined, and results in increasing the size of the stochastic contribution at each iteration relative to the Voss algorithm, effectively reducing the persistence. This value converges to within 0.2 of the desired simulated β, but an analytical solution would be highly preferable. Freezing the n value of the Voss scale, but continuing to scale the variance by the same factor at each iteration keeps the total base-level variance in the remainder of the signal consistent with the total variance at the break, and creates in the synthesized series the same sharp bend observed in spectra determined from shiptrack data.

19.6.2 Fractally Interpolated 2D Fluorescence Fields

For the section on the southern flank of Georges Bank shown in Figure 19.14, observed fluorescence distributions have been interpolated into fractal fields for the final GLOBEC Broadscale cruise, in June 1999. This cruise had the most complete fluorescence coverage of the entire bank and did not have gaps in the transects used in the interpolation process. Each fractal interpolation incorporates the same data from three Broadscale Survey transects. The transect covering Broadscale Survey Stations 4 to

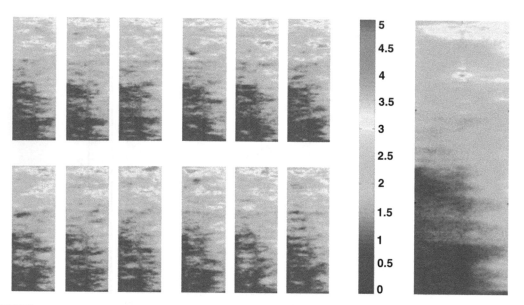

FIGURE 19.15 (Color figure follows p. 332.) Twelve iterations of the southern flank fractal fluorescence field for June 1999 Broadscale survey (left), and the mean field constructed from 15 iterations (right). Each iteration was based on a different random seed for that realization on synthetic June 1999 fields. The *x*-axis direction is along-isobath; the *y*-axis direction is across-isobath. Range for individual pixels in all fields is from 0 to 5 V, shown on the color bar. Note that all fields, the individual realizations and the mean, have the same data anchors from the survey as the right and left boundaries, and down the center. In a completely realistic field, these anchors would not visibly stand out.

7 functions as the leftmost boundary; the transect covering Broadscale Survey Stations 12 to 15 functions as the rightmost boundary. Both these transects move across isobaths from the top of the bank to the flanks. The midsection transect covers Stations 8 to 11, moving across isobaths onto the bank in time. The fields were constructed using the observed β_1 values along this center (target) trackline for the series from Station 8 to Station 11. For this initial evaluation, the β_1 values were held constant over each transect between stations. As a result of the short length of these series, local large-scale β_2 values were not well constrained, so the bankwide average value of $\beta_2 = 2$ was applied. This value was applied at each iteration of Equation 19.8, until the observed scale of the break was reached. After the break (if one occurs at that location), the local β_1 values were used to scale the variance for each iteration, using Equation 19.9 if the values were below 1.

Observations of correlation scales along and across isobaths on Georges Bank show that correlation lengths are strongly dependent on the gradient of bathymetry. Along-isobath correlation is higher by a half to one order of magnitude.[2] Because resolution along transects is 15 m for fluorescence, this initial implementation resolves pixels at 15 m resolution in the across bank direction, and 150 m along the bank. Future implementations with additional computational resources will proceed to higher along-isobath resolution, which should be constrained by data from other cruises on the bank that have resolved the variation in this direction (Broadscale cruises have limited data along isobaths).

These results are essentially experimental syntheses of potential fields capable of generating the observed values. They serve to illustrate the application of the statistical methods discussed above, and provide a basis for future work that will spatially index the data and constrain both the along- and across-isobath directions. The results of randomly seeded fractal interpolations, and the resulting mean field constructed from 15 such iterations for June 1999, are shown in Figure 19.15. Each individual realization represents a field capable of generating the observed three tracklines. These fields have realistic variance, based on the fractal characteristics of the observed data. For the mean field, total integrated fluorescence units (volts) are well within the range of the 15 individually iterated fields (4.29 ± 0.07 megavolts, or 1.6%). However, the variance of the mean field is more than 10% lower than the variance in any of the individually iterated fields. This results because the mean field is much smoother. In addition, numerous symmetrical artifacts occur around the tracklines when the average is taken.

19.6.3 Implications for Future Work

Steele and Henderson[26] note that without variability in plankton fields, there can be no viable commercial fishery. Losing variability in fields used as inputs in predictive models for phenomena such as larval fish survival has a variety of implications for the outcome of these models, which must then be compensated by some form of subgridscale parameterization. Future work will focus on activities to validate the realism of the simulated fields produced here. Additional information from station data (e.g., chlorophyll data) will be used to further refine the simulations. The evaluation will develop techniques to quantitatively compare contemporaneous ocean color measurements and stochastic simulations as ensembles, using simulated pigment fields at the same scale as SeaWiFS and MODIS observations (approximately 1 km). This evaluation of the synopticity of the simulated fields will guide the process of creating gridded fields of all of Georges Bank.

19.7 Conclusions

Spectral variance determined around Georges Bank from Mexican Hat wavelet convolutions results in characteristic seasonal power spectra for fluorescence distributions that are distinct from both temperature and salinity spectra. The power law distributions for each of the variables are tightly constrained. Characteristics of the spectra show that seasonality strongly influences the fluorescence spectra. The seasonal fluorescence spectra for both 1998 and 1999 are characterized by a progression of breaks in scale that decreases steadily in both years from 1.8 km in January, to 0.9 km in March, to 500 m in June, while overall slopes steepen. Salinity spectra have the largest range in slopes and change in character considerably throughout the six cruises without a clear seasonal response. By contrast, temperature spectra increase in total power in June of both years, but otherwise show less variability than salinity or fluorescence, although the slopes obtained for both small- and large-scale temperature spectra are consistent with expectations derived from theoretical spectra.

Local processes, such as frontal upwelling, mixing, and transport events, are associated with rapid spatial changes in spectral properties. Local power spectra for distinct hydrographic regions of Georges Bank were determined: the tidally well-mixed top; the frontal region near the 60 m isobath; and the stratified region on the flanks. The differing spatial structures produced in these regimes reflect both the magnitude and arrangement of energy distributions. The variation of spectral slopes β can be as great between stations as between seasons.

Fractal interpolation of transects on Georges Bank provides a way to apply observed variability to the task of characterizing variability between survey stations. The example above illustrates a method for using fractal characterization of 1D underway data to constrain 2D fields. The exercise in fractal synthesis also illustrates the loss of variability in fields generated using mean characteristics. The summing of 15 iterations creates a smooth field with symmetric artifacts. The iterated fields individually represent fields capable of generating the observed distribution along tracklines; the mean field does not. The importance of adequately representing the patchiness of plankton populations for ecological models such as the adjoint model on Georges Bank has been underscored by McGillicuddy et al.[2] Fractal interpolation represents a way both to preserve the variance from observed fields in larger dimension fields and to estimate probabilities associated with events that are spatially or temporally dependent on overlap. Reconstructing plankton fields utilizing the observed variance is particularly important for applications involving trophic interactions, because fluxes of material that are biologically mediated are strongly related to plankton spatial patterns. Unevenly distributed, patchy prey are subject to different pressures from predators than are uniform distributions. Moreover, theoretical studies have shown that uniformly distributed prey can be far more easily forced to extinction.

Localized wavelet analysis provides a powerful tool for assessing the fractal distribution of populations with various degrees of intermittency, across as wide a range of spatial and temporal scales as contained in any given data set. Fractal interpolation provides a way to incorporate observed intermittency into higher dimension fields. Currently, fractal interpolation techniques for slopes β below 1 are limited to empirical solutions; future work in this area would be of benefit for characterizing plankton populations whose spectral distributions often contain regions with β values between 0 and 1. Each iteration of the fractal simulation provides a visually realistic view of fluorescence distribution. The variation between simulations illustrates the plausible range in distributions that can result in a given set of observations. The mean of a number of iterations converges to the distribution of the probability density function. The method of fractal interpolation efficiently incorporates observed 1D patchiness into 2D fields.

Acknowledgments

This work was supported by NASA Headquarters under the Earth System Science Fellowship Grant NGT5-50218 and by NSF OCE9940880. This is Woods Hole Oceanographic Institution contribution 10781, and U.S. GLOBEC contribution number 355.

References

1. D.H. Cushing. Patchiness. *Rapp. Cons. Int. Explor. Mer*, 153:152–163, 1962.
2. D.J. McGillicuddy, Jr., D.R. Lynch, P.H. Wiebe, J. Runge, E.G. Durbin, W.C. Gentleman, and C.S. Davis. Evaluating the synopticity of the US GLOBEC Georges Bank broad-scale sampling pattern with observational system simulation experiments. *Deep Sea Res. II*, 48(1–3):1971–2001, 2001.
3. D.H. Chatfield, *Analysis of Time Series*, 5th ed. Chapman & Hall, London, 1996.
4. B.D. Malamud and D.L. Turcotte. Self affine time series: I. Generation and analysis. *Adv. Geophys.*, 40:1–90, 1999.
5. C. Torrence and G.P. Campo. A practical guide to wavelet analysis. *Bull. Am. Meteorol. Soc.*, 79(1):61–78, 1998.
6. I. Daubechies. *Ten Lectures on Wavelets*. Society for Industrial and Applied Mathematics, Philadelphia, 1992, 357 pp.
7. M. Farge. Wavelet transforms and their applications to turbulence. *Annu. Rev. Fluid Mech.*, 24:395–457, 1992.
8. B.B. Mandelbrot. *The Fractal Geometry of Nature*. W. H. Freeman, San Francisco, 1983.
9. B. Mandelbrot. How long is the coast of Britain? Statistical self-similarity and fractional dimension. *Science*, 156(3775):636–638, 1967.
10. L. Seuront and Y. Lagadeuc. Multiscale patchiness of the calanoid copepod *Temora longicornis* in a turbulent coastal sea. *J. Plankton Res.*, 23(10):1137–1145, 2001.
11. L. Seuront, F. Schmitt, Y. Lagadeuc, D. Schertzer, and S. Lovejoy. Universal multifractal analysis as a tool to characterize multiscale intermittent patterns: examples of phytoplankton distribution in turbulent coastal waters. *J. Plankton Res.*, 21(5):877–922, 1999.
12. R. Bindlish and A.P. Barros. Disaggregation of rainfall for one-way coupling of atmospheric and hydrological models in regions of complex terrain. *Global Planetary Change*, 25:111–132, 2000.
13. A. Scotti and C. Meneveau. A fractal model for large eddy simulation of turbulent flow. *Physica D*, 127:198–232, 1999.
14. L.C. Smith, D.L. Turcotte, and B.L. Isaacs. Stream flow characterization and feature detection using a discrete wavelet transform. *Hydrol. Proc.*, 12:233–249, 1998.
15. B.L. Li. Fractal geometry applications in description and analysis of patch patterns and patch dynamics. *Ecol. Modelling*, 132:33–50, 2000.
16. S.A. Levin and R.T. Paine. Disturbance, patch formation and community structure. *Sci. Mar.*, 65:171–179, 2001.
17. D.L. Turcotte. *Fractal and Chaos in Geology and Geophysics*. Cambridge University Press, New York, 1997.
18. R/V Albatross AL9906. June Cruise Report US GLOBEC Northwest Atlantic, available at http:globec.whoi.edu, 1999.
19. T. Platt. Spectral analysis of spatial structure in phytoplankton populations, in J.H. Steele, Ed., *Spatial Patterns in Plankton Communities*, Plenum Press, New York, 1978, 73–83.
20. T.M. Powell and A. Okubo. Turbulence, diffusion and patchiness in the sea, *Philos. Trans. R. Soc. Lond. B*, 343:11–18, 1994.

21. R. Voss. Random fractal forgeries, in R.A. Earnshaw, Ed., *Fundamental Algorithms for Computer Graphics*, Vol. F17 of NATO ASI Series, Springer-Verlag, Berlin, 1985, 805–835.

22. W.C. Strahle. Turbulent combustion data analysis using fractals. *AIAA J.*, 29(3):409–417, 1991.

23. M.A. Marvasti and W.C. Strahle. Fractal geometry analysis of turbulent data. *Signal Proc.*, 41:191–201, 1995.

24. A.I. Penn and M.H. Loew. Estimating fractal dimension with fractal interpolation function models. *IEEE Trans. Med. Imag.*, 16:930–937, 1997.

25. O.I. Craciunescu, S.K. Das, J.M. Poulson, and T.V. Samulski. Three-dimensional tumor perfusion reconstruction using fractal interpolation functions. *IEEE Trans. Biomed. Eng.*, 48(4):462–473, 2001.

26. J.H. Steele and E.W. Henderson. Simple model for plankton patchiness. *J. Plankton Res.*, 14(10):1397–1403, 1992.

27. D.G. Mountain and M.H. Taylor. Fluorescence structure in the region of the tidal mixing front on the southern flank of Georges Bank. *Deep-Sea Res.*, 43(7—8):1831–1853, 1996.

28. P.H. Wiebe, D.G. Mountain, T.K. Stanton, C.H. Greene, G. Lough, S. Kaartvedt, J. Dawson, and N. Copley. Acoustical study of the spatial distribution of plankton on Georges Bank and the relationship between volume backscattering strength and the taxonomic composition of the plankton. *Deep-Sea Res.*, 43(7–8):1971–2001, 1996.

20

Orientation of Sea Fans Perpendicular to the Flow

Thomas Osborn and Gary K. Ostrander

CONTENTS

20.1   Introduction 321
20.2   The Rayleigh Disc 322
20.3   Symmetry 324
20.4   Sea Fan Growth 325
20.5   Effects of the Bottom Boundary Layer 326
20.6   Orientation Relative to the Local Topography 326
20.7   Discussion and Conclusions 326
References 327

20.1 Introduction

The orientation of sea fans and sea pens (Figure 20.1) to the flow has been attributed to hydrodynamic phenomena without any detailed explanation of the reasons (e.g., Laborel and Vacelet, 1958; Theodor and Denizot, 1965). There are two possible mechanisms through which the natural torque on the fan can cause the population to align with the axis of the dominant flow.

1. The torque applied to the stalk by the flow is sufficient to reorient the fan perpendicular to the flow as it grows, or
2. The survivability of the fans with correct orientation (within certain angular bands) is dramatically better than the fans with improper orientation.

Wainwright and Dillon (1969) quantitatively verified the long-held hypothesis that sea fan orientation was not random and, in fact, sea fans preferentially orient so that their blades are perpendicular to the flow of the water. They found that the larger the sea fan the less the deviation of an individual fan from the mean orientation of all the fans examined. Among the largest sea fans, those equal to or greater than 50 cm in height, none varied by more than 20° from the mean orientation of all those examined ($n = 58$). In fact, the average variation was only slightly more than 8° from the mean (calculated from data in Wainwright and Dillon, 1969). The investigators also concluded that small fans (<20 cm) are oriented randomly and that there is an increase in the degree of preferred orientation with increasing size.

Wainwright and Dillon (1969) also resolved the issue of the mechanism utilized by the sea fans for achieving the preferred orientation as they increased their size. Through examination of transverse sections of the axial skeleton, it was determined that the age of the axial centers can be inferred from the number of axial cortex layers surrounding each center. Centers with multiple layers are much older. Comparative examination of the alignment of the youngest and oldest centers revealed age-dependent changes in the orientation of sea fans as they increased in size. Our own observations suggest that the

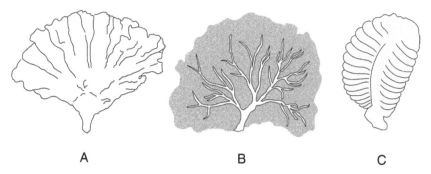

FIGURE 20.1 Schematic drawing of sea fan (A), *Udotea* algae (B), and sea pen (C).

tendency toward a random orientation relative to water flow among small sea fans (<20 cm) may relate to the many and varied turbulent flows generated by the highly variable reef structure close to the reef surface. As the fans grow they escape this microenvironment and align to the more consistent water flow above the roughness elements of the bottom flow.

The explanation for how hydrodynamic forces control the orientation of the planar and symmetric organisms like sea fans and sea pens can be found in relatively simple hydrodynamic theory for inviscid flow around a symmetric disc. The organisms are subjected to a torque, which rotates them into an orientation perpendicular to the flow and this orientation is a stable position even in a flow that oscillates back and forth. The model of a symmetric disc, free to rotate about a vertical central axis, in a horizontal flow is called a Rayleigh disc.

20.2 The Rayleigh Disc

Starting with a solid disc that is free to pivot about an axis perpendicular to the flow (as shown in Figure 20.2), one can ask what forces and torques are applied to this disc by the flow. In inviscid flow, there is no viscous force on the object and the effect of the flow is to create a pressure distribution across the fan. It is the net force, due to this pressure distribution, that exerts the torque about the axis. Thus, with an inviscid model and the associated potential flow, similar to many calculations of airfoils and airplane wings, the pressure distribution over the surface of a solid symmetric disc can be calculated. Integrating the pressure distribution over the surface produces the force applied to the disc by the flow. The torque is then the force times the moment arm. It can be a lengthy calculation and Faber (1995) uses a sophisticated mathematical technique and calculates the torque, M, directly:

$$M = \frac{4}{3}\rho a^3 U^2 \sin(2\Omega)$$

This result was found originally by Konig (1891). Rayleigh (1896) used it to measure the intensity of sound by placing a disc (hence, the name Rayleigh disc) in the mouth of a resonant cavity and determining the torque as a function of angle.

The torque is zero at 90° intervals and in those four positions the plate is in equilibrium. However, as can be seen in Figure 20.3 (where the torque is plotted as a function of angle) at both 0° and 180°, a small deviation from the equilibrium position leads to a torque away from the equilibrium position (this point was made by Wainwright and Dillon). At 90° and –90°, a small deviation leads to a torque favoring return to the equilibrium position. Hence, the equilibrium is stable at +90° and –90° while it is unstable at 0° and 180°. A plate, free to rotate, orients itself perpendicularly to the direction of the flow. Because the stable equilibria are with the plate perpendicular to the flow, in an oscillatory, but collinear flow, the plate remains stationary as the flow reverses direction. It is this characteristic that has made the Rayleigh disc so useful for measuring the intensity of sound. The plate did not rotate as the flow reversed and one only needed to measure the average torque, as a function of angle, to obtain the mean squared velocity.

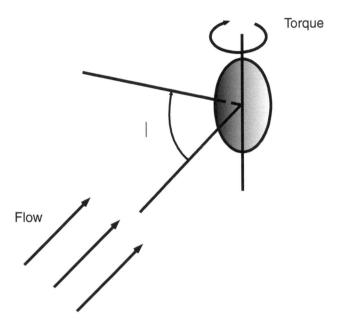

FIGURE 20.2 Schematic of Rayleigh disc. The disc is free to rotate about a vertical axis with the angle of rotation relative to the incoming flow indicated by Ω.

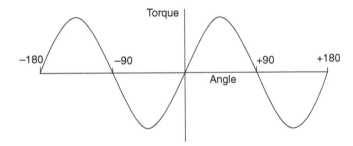

FIGURE 20.3 Torque vs. angle of orientation for a Rayleigh disc. The torque is in arbitrary units as a function of angle of rotation of the Rayleigh disc. Both torque and angle are measured as positive in the clockwise direction when viewed from above with the angle of zero orientation corresponding with the edge of the disc aligned with the flow.

Potential flow solutions cannot accommodate the effects of viscosity. The inviscid solution for the solid disc has no flow through the disc, no viscous boundary layer, and no separation of the flow. Potential theory predicts above ambient pressure on the backside of the fan. This pressure recovery on the backside of the fan does not occur in the real world as the flow behind the fan separates from the structure and leaves a turbulent wake behind the fan. The inviscid solution for the Rayleigh disc aligned perpendicularly to the flow has no pressure drop across the structure. In actuality, the flow around the fan "separates" from the edges of the fan and this effect leads to a pressure below the ambient on the back of the fan. This low pressure on the back, relative to the front, is the source of the drag exerted by the flow on the fan. The pressure gradient across the fan blades, induced by the flow and the flow separation, leads to flow through the fan, which is important in the feeding process.

Flow separation and the flow through the fan affect the pressure distribution and hence the magnitude of the torque. However, these details do not change the basic response of the fan to the flow: the sign of the torque as a function of angle and the zero crossings of the distribution. When in any of the four orientations at 90° intervals, the flow is symmetric and so the torque must be the same in either direction and, hence, zero. That observation, combined with (1) the requirement that the torque is a continuous function of angle, (2) that there are no further zero crossings, and (3) the equilibria at 0° and 180° are unstable to small perturbations, then the ±90° orientations must be stable equilibria points.

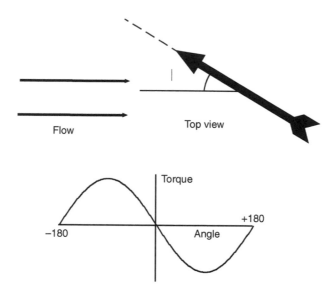

FIGURE 20.4 Schematic of weathervane and torque vs. angle. The weathervane is stable only when pointing into the flow.

For a symmetric fan, the net force due to the flow will be applied at some point on the upstream side of the axis of rotation leading to a torque that will rotate the fan toward the ±90° stability locations. It is symmetry that makes the torque zero at the ±90° as well as at 0° and 180°. Because (1) there are only four equilibrium points, (2) the 0° and 180° orientations are unstable equilibria, and (3) the torque is a continuous function of angle; the slope of the torque vs. orientation angle curve makes the ±90° locations stable equilibria. We are now brought to the question of the role of symmetry and its importance.

20.3 Symmetry

What would happen if the object under consideration were not laterally symmetric? When do we transition from something that behaves like a Rayleigh disc to something that acts like a weathervane? Consider the torque vs. orientation for a weathervane (Figure 20.4). Here the equilibrium at 0° is stable and the equilibria at ±180° are unstable and there are no intermediate equilibria. Thus, the weathervane is always oriented with the nose toward the wind (as per its the design criteria). The difference in response to reversing currents is dramatic.

As shown in Figure 20.5 trimming a bit off of one side of a symmetric fan at the equilibrium position allows that edge to rotate into the direction of the flow. Because taking off just a tiny bit of the blade cannot remove the equilibrium position altogether, it must just shift it a bit back toward 0°. Hence, the torque vs. orientation curve stays almost the same but the equilibria points (formerly at ±90°) are moved toward the origin. Trimming a bit more moves the stable equilibria points even closer to the origin. With continued trimming, the stable equilibria will eventually reach the origin, the number of equilibrium points will decrease from four to two, the slope of the curve will reverse at the origin, and the structure will behave like a weathervane. This obviously occurs at the point where the application point for the total force vector moves to the axis of rotation. Clearly a dramatic modification, by cutting off a significant fraction of one side of a sea fan would turn it into a weathervane. A similar effect would occur if the axis of rotation were moved to one side and the axis of symmetry were no longer the axis of rotation.

These considerations show that symmetry of the force distribution on the fan is valuable in making the equilibrium orientation the same for bidirectional flow. With slight asymmetry the fan feels a rotational torque some of the time when exposed to a bidirectional flow, unless the stalk is displaced to the side to compensate. Conversely, a perfectly symmetric fan can align perpendicularly to both flow directions and feel no torque about the vertical axis. A symmetric fan with a central stalk is the simplest design to avoid fluctuating torques in bidirectional flow.

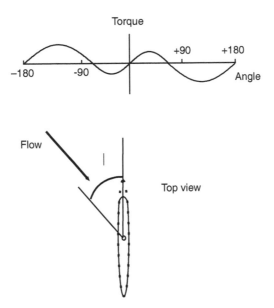

FIGURE 20.5 Schematic of asymmetric fan and torque vs. angle of orientation. The top view of the asymmetrical fan is indicated by the solid line compared to the original shape shown with the dots. The torque vs. angle plot shows the modification of the zero crossings due to the changed shape.

The oscillatory nature of a wave environment brings about current reversals every half cycle of the wave. Thus, the region of separated flow alternates between the sides of the fan, and, in fact, the flow about the fan will be very time dependent. However, the fan is not required to rotate 180° every time the current reverses the way a weathervane would. Thus, the fan feels no torque with flow from either direction, a very advantageous situation. Let's now consider how sea fans develop their symmetric shape.

20.4 Sea Fan Growth

In a manner similar to most members of the Phylum Cnidaria, sexual reproduction among colonial organisms such as sea fans involves the discharge of eggs and sperm into the water column and the formation of a planktonic larvae. The duration of the planktonic stage is species specific, and if a suitable substrate is available, the planula larva will settle, attach, and continue development. The resulting colonial organism (Figure 20.1A) can grow to considerable size via a process of continuous asexual reproduction.

The growth and development of individual colonial organisms such as the polyps of sea fans likely results from a combination of genetic factors and environmental interactions and is referred to as ontogeny. The ontogeny of the individual polyp is constrained by the physical and chemical processes as they govern the strength of biological materials and the rates of gas, nutrient, and waste product exchange (Lasker and Sánchez, 2002). Likewise, the ultimate size of the colony is limited by physical constraints of the environment and constituent materials. The constraints on the colony are not necessarily the same as those governing the individual organism, and as such considerable variation exists in sea fans, even among those of the same species in the same general area. Thus, the term astogeny can be applied to the process of colonial development (Pauchut et al., 1991).

Genes are the genetic factors that directly impact ontogeny and ultimately astogeny. Although all the intricacies are not well understood, it is known that the Hox gene cluster specifies positional information along the anterior–posterior axes of developing invertebrate (e.g., Duboule and Dolle, 1989; de Rosa et al., 1999) and vertebrate (e.g., Graham et al., 1989) embryos including Cnidaria (Shenk et al., 1993a, b). For example, it was recently demonstrated that differential expression of the Hox gene *Cnox-2* along the aboral–oral axis of colonial hydroids correlates with polyp polymorphism (Cartwright et al., 1999). While these and presumably other genes are clearly critical for polyp ontogeny, the mechanisms responsible for colony assembly are largely unknown. Although it is significant to note that among taxa with very different individuals, colonies can exhibit nearly identical organization (Sánchez, unpublished data). This suggests that either heretofore unknown genetic mechanisms are responsible for colonial astogeny or, perhaps, environmental factors have shaped the evolution of multiple species. Given that the environmental factors most frequently encountered by sea fans exhibit temporal variability on an irregular basis (e.g., current force, direct of water flow, wave action, nutrient availability), astogenic differences among multiple colonies in a given area may not be unexpected.

20.5 Effects of the Bottom Boundary Layer

The flow in the bottom boundary layer is turbulent, with the speed decreasing toward the bottom (Grant and Madsen, 1986). The boundary layer is due to the drag from the irregular shapes and structures that form the bottom. Protrusions and irregular shapes have flow separation behind them and shed "eddies" in their wakes. The near wake of a feature contains fluctuations with longitudinal scales on the order of five times the size of the causative feature. The irregular nature of the bottom in many reef areas makes for irregular flow in both speed and direction for heights that are comparable to the roughness scale of the bottom. Thus, as noted by Wainwright and Dillon (1969), sea fans that are short, or comparable in height, to the bottom roughness will be buffeted by the eddies that are shed from those structures and unlikely to align with the mean flow, which they cannot discern amid the fluctuating velocities of the wakes. In this region the bottom stress is generated by the form drag of the "roughness elements" and is showing up as separated flow and associated wakes.

The flexibility of some fans allows them to bend over in the downstream direction and thus streamline themselves by presenting a smaller cross-sectional area to the flow as well as protecting themselves by being in a region of lesser flow. Both these effects reduce the strain on the central stalk. However, our observations have shown that even the rigid fan species do not break at the central stalk, but rather the substrate fails and the fans and a piece of the reef are ripped from the bottom. The fans have developed a structure that is able to withstand the drag force generated by the flow most of the time. They are susceptible to hurricane-generated waves and other unusual events. They are unlikely to be damaged by the torque around their central axis, which orients them, as it is much less than the torque around their base trying to snap them off.

20.6 Orientation Relative to the Local Topography

When waves shoal and feel the bottom, their phase speed becomes dependent on the water depth, and they move more slowly the shallower the water becomes. This effect tends to align the crests of the waves with the depth contours of the bottom topography, making the orbital motion perpendicular to the bottom contours. Thus, sea fans at the outer edges of reefs are perpendicular to the wave motion and parallel to the bottom topography. The local topography in a reef can affect the motion in such a manner that the flow in the two halves of the wave cycle are not parallel. Channels in the reef and large mounds strongly modify the local flow. In such cases, the sea fans would presumably align with the direction of minimum net torque associated with the period of time in which they grew.

20.7 Discussion and Conclusions

Because of the torque created by the flow around them, sea fans and other laterally symmetric, planar structures tend to align perpendicularly to a steady mean flow or a linear oscillatory flow. This torque is due to the hydrodynamics of flow about planar objects and predates the evolution of the fan. Thus, the fans evolved their structure and feeding apparatus in the presence of this orientation, which is driven by physical processes. Although this orientation, at first glance, appears to maximize the net force applied to the structures, it does minimize the torque and may protect the structure from large forces. For, in fact, it is possible for a wing to generate a lift force (at angles of attack of 15° to 20°) that is larger than the drag on that wing when it is perpendicular to the flow. The marine abiotic environment has clearly influenced sea fan shape and orientation tendencies on and near coral reefs. This has led to the evolution of foraging strategies among the sea fans (e.g., Riedl and Forstner, 1968) that are "optimal" in response to the physical environment and other constraints. Water flow, perpendicular to sea fan orientation, is a physical factor crucial to the evolution of sea fan morphology. The significantly lower flow through the fan compared to the flow around the fan is essential. Within the rigid structure of the fan are tiny fragile feeding polyps that cannot function, or even exist, without the structural support and protection of the sea fan. The flow rate over these polyps and hydrodynamic forces experienced by them are limited by the form of the structure and enables these polyps to function in the high-energy environments where the sea fans are located.

Do other, similarly shaped aquatic organisms such as sea pens and marine algae (Figure 20.1) orient to the flow in a similar manner? The orientation of sea pens (Figure 20.1C) has also been attributed to hydrodynamic phenomena (e.g., Magnus, 1966). These are unique organisms compared to sea fans because they are not sessile and attached to hard substrates. Instead, they are found "rooted" in soft sand. As such, they have the ability to move and can exploit the ever-changing hydrodynamic variability found close to the substrate by rotating their entire bodies. In this arena, water velocity tends to be reduced because of impinging biotic and abiotic structures close to the substrate. The sea pen niche is significantly different from that of the semirigid sea fans as their soft, water-filled bodies would not withstand the strong flow encountered by sea fans. Sea pens may compensate for reduced feeding opportunities they encounter in the low-energy environment by their ability to change orientation into the water flow and in doing so increase the probability of encounter with small food items in the water column. Conversely, sea fans tend to fare best when found attached to structures that allow for their protrusion well into the water column and well above the ocean floor. While orienting into the water flow increases their foraging ability, it is unlikely this behavior evolved in response to limited food resources. Sea fans are often found at very high densities on or around reefs and in the immediate vicinity of other gorgonids or corals with similar dietary requirements.

Among sessile organisms, the *Udotea* algae present a striking resemblance to sea fans (Figure 20.1B). Our recent examination of the orientation of two species of marine algae (*U. cyathiformis* and *U. occidentalis*) shaped like the sea fans (Figure 20.1B) suggest the mechanisms modulating sea fan orientation may not be working in these organisms. We observed 134 individuals from these two species, located in various environments of differential wave energy around San Salvador Island, The Bahamas; the individuals failed to exhibit significant differences in orientation relative to current or predominant wave action (personal observations). This is not entirely surprising when the height of the algal species is considered. Maximal mean heights of *Udotea* spp. above the substrate were less than 8 cm and surrounding species (especially *Thalassia testudinum*, which was most abundant) extended up to 12 cm above the substrate and were highly mobile. Thus, it was apparent that they influenced and modulated the flow of water around the *Udotea*. Moreover, wave action in these eel grass beds creates ridges in the sand running parallel to shore that further modulate water flow around such low-lying species as *Udotea*. Thus, the factors modulating the morphology of these species remains to be investigated. They do not, however, appear to orient in any particular direction relative to water flow. Instead, their broad thallus may function to maximize surface area for photosynthesis and play a significant role in the recently reported phenomena of synchronous mass spawning by green algae on coral reefs (Clifton, 1997).

References

Cartwright, P., Bowsher, J., and Buss, L.W., Expression of a Hox gene, *Cnox-2*, and the division of labor in a colonial hydroid, *Proc. Natl. Acad. Sci. U.S.A.*, 96, 2183, 1999.

Clifton, K.E., Mass spawning by green algae on coral reefs, *Science*, 275, 116, 1997.

de Rosa, R., Grenier, J.K., Andreeva, T., Cook, C.E., Andoutte, A., Akam, M., Carroll, S.B., and Balavoine, G., Hox genes in brachiopods and priapulids and protostome evolution, *Nature*, 399, 772, 1999.

Duboule, D. and Dolle, P., The structural and functional organization of the murine HOX gene family resembles that of *Drosophila* homeotic genes, *EMBO J.*, 8, 1497, 1989.

Faber, T.E., *Fluid Dynamics for Physicists*, Cambridge University Press, Cambridge, 1995.

Graham, A., Papalopulu, N., and Drumlauf, R., The murine and *Drosophila* homeobox gene complexes have common features of organization and expression, *Cell*, 57, 367, 1989.

Grant, W.D., and Madsen, O.S., The Continental-Shelf bottom boundary layer, *Annu. Rev. Fluid Mech.*, 18, 265, 1986.

Konig, W., *Ann. Phys. Phys. Chem.*, 43, 43, 1891.

Laborel, J. and Vacelet, J., Etude des peuplements d'une grotte sous-marine du golfe de Marseille, *Bull. Inst. Oceanogr. Monaco*, 55, 1206, 1958.

Lasker, H. and Sánchez, J., Astogeny and allometry of modular colonial organisms, in R.N. Hughes, Ed., *Reproductive Biology of Invertebrates*, Vol. 11, *Progress in Asexual Reproduction*, John Wiley & Sons, Oxford, 2002.

Magnus, D.B.E., Zur Ökologie einer nachtaktiven Flachwasser-Seefeder (Octocorallia, Pennatularia) im Roten Meer, *Veröff. Inst. Meeresforsch. Bremerh.*, 2, 369, 1966.

Pauchut, J.F., Cuffey, R.F., and Anstey, R.L., The concepts of astogeny and ontogeny in stenolaemate bryozoans, and the illustration in colonies of *Tabulipora carbonaria* from the lower Permian of Kansas, *J. Paleontol.*, 65, 213, 1991.

Rayleigh, J.W.S., *The Theory of Sound*, Macmillan, London, 1896.

Riedl, R. and Forstner, H., Wasserbewegung im Mikrobereich des Benthos, *Sarsia*, 34, 163, 1968.

Shenk, M.A., Bode, H.R., and Steele, R.E., Expression of *Cnox-2*, a HOM/HOX homeobox gene in *Hydra*, is correlated with axial pattern formation, *Development*, 117, 657, 1993a.

Shenk, M.A., Gee, L., Steele R.E., and Bode, H.R., Expression of *Cnox-2*, a HOM/HOX gene, is suppressed during head formation in *Hydra*, *Dev. Biol.*, 160, 108, 1993b.

Theodor, J. and Denizot, M., Contribution a l'etude des Gorgones, I. A Propos de l'orientation d'organismes marins fixes vegetaux et animaux en fonction du courant, *Vie Milieu Ser. A Biol. Mar.*, 16, 237, 1965.

Wainwright, S.A. and Dillon, J.R., On the orientation of sea fans (genus *Gorgonia*), *Biol. Bull.*, 136, 130, 1969.

21

Why Are Large, Delicate, Gelatinous Organisms So Successful in the Ocean's Interior?

Thomas Osborn and Richard T. Barber

CONTENTS

21.1   Introduction...329
21.2   Analysis..329
21.3   Discussion...331
21.4   Conclusion 331
References...332

21.1 Introduction

A recent accomplishment of biological oceanography is the discovery that the mesopelagic ocean has an extremely abundant fauna of large, delicate, gelatinous organisms (Figure 21.1). What makes this body plan so successful in the ocean's interior? That it has been exploited by several phyla emphasizes its adaptive advantage. The classical large, gossamer, gelatinous phyla are the "jellies," cnidaria and ctenophores (Matsumoto and Robison, 1992; Robison et. al., 1998), but recent exploration of the mesopelagic realm (200 to 1000 m) by both manned and unmanned vehicles has found other taxa, such as tunicates (larvaceans), siphonophores, and mollusks (pteropods), that exploit this life form by producing large feeding structures (Hamner and Robison, 1992; Hopcroft and Robison, 1999). The usual size of some of these delicate organisms ranges from about 0.3 to 3.0 m, but sonic imaging reveals some mid-water siphonophores over 30 m in length (Robison, 1993, 1995). These magnificent animals are never represented intact in the mangled specimens harvested by nets. Our discussion of gelatinous predators below pertains to mid-water siphonophores and the other, smaller, gelatinous organisms.

21.2 Analysis

The large, gossamer design has one severe vulnerability: habitats with strong shearing motions destroy these delicate organisms. As a result, this design is absent in coastal waters, least prevalent in turbulent polar waters, and only occasionally abundant in pelagic surface waters. It is most prevalent in the lower-energy mesopelagic realm. Given that vulnerability to destruction by shearing motions is such a liability, what asset allows this design to be so successful?

If the physical character of a habitat permits an organism to survive, success then consists of eating well and not being eaten. Eating well leads to growth and reproduction; being eaten is a terminal event. We propose that the large, gossamer design, compared to more compact designs, both improves feeding and reduces predation.

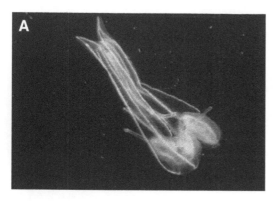
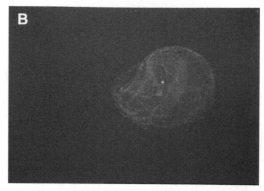

FIGURE 21.1 (Color figure follows p. 332.) Two large, delicate, gelatinous organisms representative of the gossamer design discussed in this report. (A) *Kiyohimea usagi*, a lobate ctenophore recently described by Matsumoto and Robison (1992). (Image copyright. 1996 MBARI.) (B) *Mesochordaeus erythrocephalus*, a mesopelagic larvacean recently described by Hopcroft and Robison (1999). (Image copyright, 1991 MBARI.)

Large, gelatinous organisms have very little food value for their size (or little organic biomass per unit volume), making them an unrewarding target for predators big enough to engulf them (Robison, 1999). Furthermore, the large, gelatinous forms are dangerous targets for small predators that might benefit from the rather limited meal provided by the large jellies, but small predators run the risk of becoming entangled in the gelatinous material, nematocysts, or blastocysts (Robison, 1999). Hence, by spreading out organic carbon content over a relatively large volume, gelatinous organisms can have the same biomass as a copepod or amphipod, but avoid the predation that nutritionally dense zooplankton undergo.

What is the advantage of the large, gelatinous life form for feeding? We propose that a larger size increases the capture rate for food, while keeping the body mass low restrains the gross metabolic demand. The rate of food accumulation increases monotonically with the relative motion between predator and prey, and this relative motion can arise from prey, predator, or the fluid environment (Saffman and Turner, 1956; Strickler, 1982; Rothschild and Osborn, 1988; Osborn, 1996). In the ocean, at feeding scales of millimeters to meters, relative motion between two points in the fluid is due to turbulence and internal waves and, on average, increases with separation:

$$\delta u \propto L \cdot \frac{\partial u}{\partial r}$$

where δu represents the relative motion, L the separation between the center of the predator and the prey and $\partial u/\partial r$ scales the velocity change with distance (due to internal waves and turbulence). Thus, even while allowing for all the details of shape and flexibility of oceanic organisms, one still expects the relative motion of predator and prey to scale with some factor of the distance between their centers of mass. It is intuitive that L scales with the size of the predator, l, and larger predators have larger relative motion with respect to their prey.

The volume from which the predator removes its food also increases with the size of the predator. For example, the cross section of the searched area is proportional to some length scale of the predator squared, and this length scale is almost certainly related to the size scale of the predator, l. This cross section might be the swept area for a swimming predator or, for an ambush predator, the surface area of the scanned region that is at the limit of its perception. With food capture proportional to the relative motion (length scale times local velocity shear) times the cross-sectional area (l^2, length scale squared), the feeding effectiveness increases as the volume, i.e., proportionally to l^3. However, to take advantage of this relationship and have a net gain, organisms have evolved to increase in size, but not in mass; then the gain in feeding is not offset by a proportional increase in metabolic demand. Thus, pelagic and mesopelagic gelatinous organisms have exploited the relationship of size, feeding, and respiration by increasing size much more than mass.

21.3 Discussion

There are many different genera and species of gelatinous organisms and, while the foregoing arguments hold for many of them, the details vary among the species. Many details are not understood, others are as yet unobserved and much can be learned about oceanic physics by studying the ecology and physiology of these creatures. Their structure, function, and life cycles contain information about the physics on scales from centimeters to meters, precisely the range where the transition from turbulence to internal waves occurs, i.e., where stratification becomes significant. Copepod and phytoplankton interactions are strongly affected by turbulence and feeding currents at scales where viscosity plays an important role. Fish live and feed in a higher Reynolds number environment using vision in their predation; for them viscosity is not a relevant parameter and motion of the water is relatively unimportant. The large, gelatinous organisms have size scales comparable to those of many fish, but they do not experience a high Reynolds number environment because they have little motion relative to the water. They live and feed at scales well above the viscous scale, but do not have the high-energy consumption body plan of fish and most other organisms that live above the viscous scale. We suggest that large, gelatinous organisms are nature's evolutionary response to the ocean's internal wave dynamics.

The evolutionary success of these large, delicate organisms leads to questions involving other physical–biological interactions. For example, what is the velocity distribution around these relatively freely floating animals? Or, what is the second-order velocity structure function following a particle that is moving with the water? Even if these organisms "follow the water," they average flow over their volume, and disturb it, because of their relatively large size. D'Asaro and Lien (2000) use a neutrally buoyant float to follow the water and study the internal wave and turbulent field from a quasi-Lagrangian perspective. That is the same physical perspective that these gelatinous organisms have of the mesopelagic ocean, and we know very little about how the ocean looks from this perspective.

How do these organisms maintain their shape in the mesopelagic realm? Because they exhibit characteristic, but changeable, shapes (Robison, 1999), it appears that they are not frequently stretched or deformed. They are robust enough to withstand the forces imposed by the ambient physical fields, that is, the straining motion of the internal waves on meter scales and turbulence on smaller scales. Can this apparent robustness be calibrated to provide information on the local physics? Conversely, does the local flow field ever rip these organisms apart? Horizontal transects of shear in the thermocline (Itsweire et al., 1989) near Monterey Canyon show 15-m-thick layers of high shear rotating at near inertial frequency. How deep into the ocean this shear propagates and how it changes with depth are unknown, but such events could have serious consequences for large, gelatinous structures.

What is the role of orientation in large, gelatinous organisms and how is it maintained? Mesopelagic larvaceans, such as *Bathochordaeus*, secrete a house that appears to serve as an umbrella, protecting them from the rain of large particles (Hamner and Robison, 1992). Do these organisms maintain their vertical orientation by passive ballasting or active behavior? Parallel questions arise concerning how these organisms maintain their characteristic depths in the mesopelagic water column with its subtle vertical gradients in physical properties. The obvious ability to maintain both orientation and depth suggests that these organisms are somewhat more complex than generally assumed.

The availability of observations by manned and unmanned submersibles has enabled us to begin to understand these magnificent organisms and how they have adapted to the physics of the ocean's interior, but there is much to be learned about this important component of the mesopelagic ocean. Further work will yield new insight on the ecology, biogeochemistry, and physics of this unique environment.

21.4 Conclusion

We suggest that the large, delicate, gelatinous design is successful in the ocean's interior because large size confers a feeding benefit that cannot be obtained with a more compact body design and because large size coupled with small mass reduces predation by providing an unrewarding meal for big predators and a danger to small predators.

References

D'Asaro, E.A. and R. Lien. 2000. Lagrangian measurements of waves and turbulence in stratified flows. *J. Phys. Oceanogr.,* 30: 641–655.

Hamner, W.M. and B.H. Robison. 1992. *In situ* observations of giant appendicularians in Monterey Bay. *Deep-Sea Res.,* 39: 1299–1313.

Hopcroft, R.R. and B.H. Robison. 1999. A new mesopelagic larvacean, *Mesochordaeus erythrocephalus,* sp. nov., from Monterey Bay, with a description of its filtering house. *J. Plankton Res.,* 21: 1923–1937.

Itsweire, E.C., T.R. Osborn, and T.P. Stanton. 1989. Horizontal distribution and characteristics of shear layers in the seasonal thermocline. *J. Phys. Oceanogr.,* 19: 301–320.

Matsumoto, G.I. and B.H. Robison. 1992. *Kiyohimea usagi,* a new species of lobate ctenophore from the Monterey Submarine Canyon. *Bull. Mar. Sci.,* 51: 19–29.

Osborn, T.R. 1996. The role of turbulent diffusion for copepods with feeding currents. *J. Plankton Res.,* 18(2): 185–195.

Robison, B.H. 1993. Midwater research methods with MBARI's ROV. *Mar. Technol. Soc. J.,* 28: 32–39.

Robison, B.H. 1995. Light in the ocean's midwaters. *Sci. Am.,* 273: 60–64.

Robison, B.H. 1999. Shape-change behavior by mesopelagic animals. *Mar. Freshw. Behav. Physiol.,* 32: 17–25.

Robison, B.H., K.R. Reisenbichler, R.E. Sherlock, J.M.B. Silguero, and F.P. Chavez. 1998. Seasonal abundance of the siphonophore, *Nanomia bijuga,* in Monterey Bay. *Deep-Sea Res.,* 45: 1741–1752.

Rothschild, B.J. and T.R. Osborn. 1988. The effect of turbulence on plankton contact rates. *J. Plankton Res.,* 10(3): 465–474.

Saffman, P.G. and J.S. Turner. 1956. On the collision of drops in turbulent clouds. *J. Fluid Mech.,* 1: 16–30.

Strickler, J.R. 1982. Calanoid copepods, feeding currents, and the role of gravity. *Science,* 218:158–160.

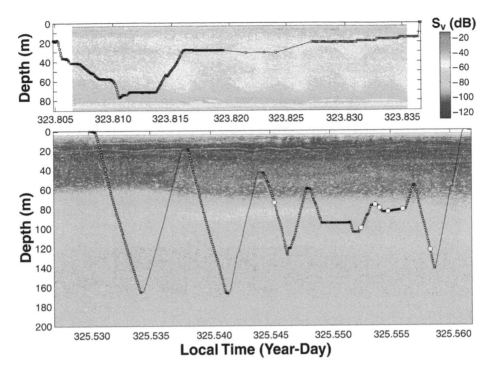

COLOR FIGURE 2.4 Tme-depth record of ZOOVIS deployments on November 19 in Hoeya Sound (top) and on November 21 farther up the Inlet (bottom). The trajectory of ZOOVIS is represented as a black line and locations where images were acquired are indicated by small white squares on the line. The trajectory is superimposed on the acoustic record (volume scattering strength) from the ship's 100 kHz echosounder. The yellow circles in the lower panel indicate locations where small euphausiids were imaged.

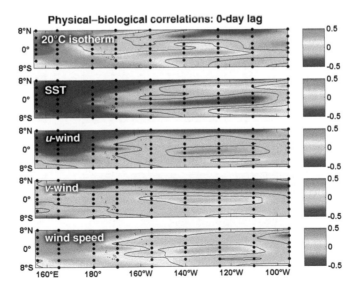

COLOR FIGURE 4.4 Maps of the equatorial Pacific showing the correlation between SeaWiFS daily chlorophyll and mooring-derived physical parameters for a lag of 0 days. The contours indicate the 95% significance level ($r = \pm 0.1$) for the correlation coefficient. The parameters represented are the depth of the 20°C isotherm ($Z_{20°C}$ or thermocline depth, m), sea surface temperature (SST, °C), the zonal or east–west wind vector (u-wind, m s^{-1}), the meridional or north–south wind vector (v-wind, m s^{-1}), and wind speed or scalar wind (m s^{-1}).

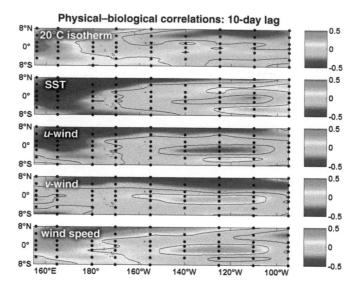

COLOR FIGURE 4.5 Maps of the equatorial Pacific showing the correlation between SeaWiFS daily chlorophyll and mooring-derived physical parameters for a lag of 10 days. The contours indicate the 95% significance level ($r = \pm 0.1$) for the correlation coefficient. The parameters represented are the depth of the 20°C isotherm ($Z_{20°C}$ or thermocline depth, m), sea surface temperature (SST, °C), the zonal or east–west wind vector (*u*-wind, m s^{-1}), the meridional or north–south wind vector (*v*-wind, m s^{-1}), and wind speed or scalar wind (m s^{-1}).

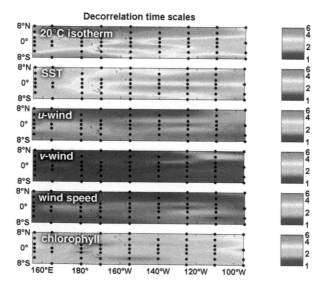

COLOR FIGURE 4.7 Maps of the equatorial Pacific showing the decorrelation scales of mooring-derived physical parameters and SeaWiFS daily chlorophyll. The parameters represented are the depth of the 20°C isotherm ($Z_{20°C}$ or thermocline depth, m), sea surface temperature (SST, °C), the zonal or east–west wind vector (*u*-wind, m s^{-1}), the meridional or north–south wind vector (*v*-wind, m s^{-1}), wind speed or scalar wind (m s^{-1}), and chlorophyll (mg m^{-3}).

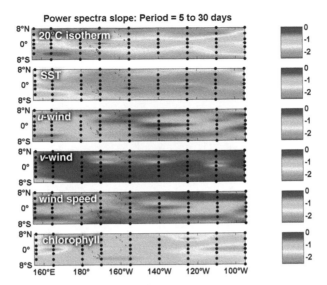

COLOR FIGURE 4.9 Maps of the equatorial Pacific showing the mean slope of the power spectra for each buoy location. The color scale has been adjusted so that a steeper negative slope is red, indicating more variability at longer timescales. The parameters represented are the depth of the 20°C isotherm ($Z_{20°C}$ or thermocline depth, m), sea surface temperature (SST, °C), the zonal or east–west wind vector (u-wind, m s^{-1}), the meridional or north–south wind vector (v-wind, m s^{-1}), wind speed or scalar wind (m s^{-1}), and chlorophyll (mg m^{-3}).

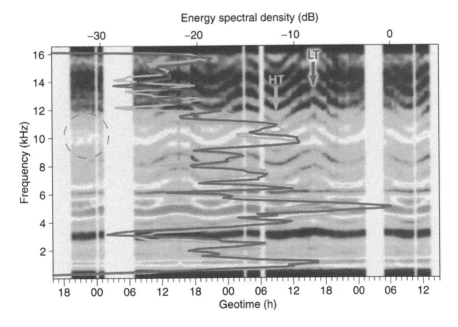

COLOR FIGURE 5.14 Received signal spectrum vs. geotime. Normalized energy spectral density (left scale). The blue-to-red color map corresponds to the range −30 dB to 0 dB. The curves (top scale) are for high tide (light gray, 0900 hours) and low tide (dark gray, 1530 hours) on September 24, 1999; see Figure 5.13B. The spectra were computed from the raw acoustic data, i.e., not matched filtered or equalized, and median-averaged over half-hour periods. R1 data.

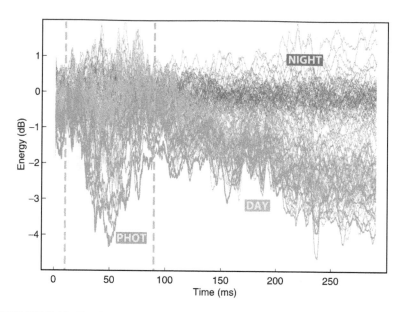

COLOR FIGURE 5.16 Geotime variability of the medium acoustic-impulse response (log envelope) in the time-delay range 1 ms to 290 ms. Each line is the difference between a 0.5-hour and the 3-night median averages. Blue lines: night hours 0700–1930 hours, 4 days; orange lines: day hours, 4 days; green lines: most active photosynthesis hours 1300–1600 hours, September 24, 1999. Each average response was smoothed with a 10-ms polynomial filter. The vertical lines indicate the time window, which corresponds to intermediate grazing angles. R1 data.

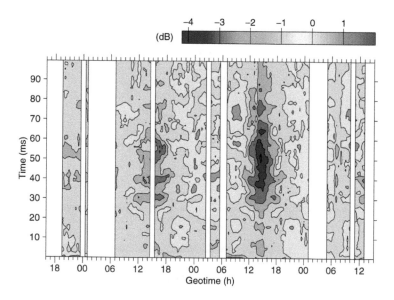

COLOR FIGURE 5.17 Time distribution of multipath energy vs. geotime. The contour lines are from −4 dB to +1 dB in 1-dB steps. The levels are referenced to the three-night median-average. The displayed time interval comprises the multipath arrivals of groups 2 through 13. R1 data. See Color Figure 5.16.

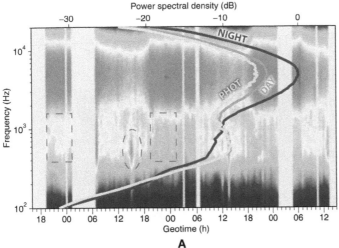

A

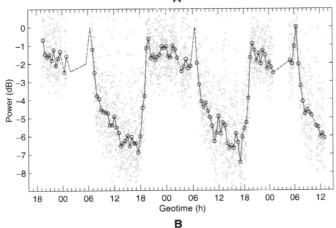

B

COLOR FIGURE 5.20
Ambient noise vs. geotime. (A) Normalized energy spectral density. The color map corresponds to the range −30 dB (blue) to 0 dB (red). The spectra were median averaged over half-hour periods. The overlaid curves are time-mean averages. Dark gray: night hours 1930–0700 hours, 4 days; gray: day hours, 4 days; light gray: mid-afternoon hours1400–1600 hours, September 24, 1999. The raw spectra were computed from reverberation-free data sections of 1-s duration taken before each probe signal was received. Dashed boxes: see text. (B) Normalized total power. The gray dots are the raw data. The circles connected by solid lines are half-hour median averages. The dashed lines are interpreted missing points. The data were affected by scuba diving noise during the periods 1440–1520 hours on September 23 and 1310–1340 hours on September 24 (dashed ellipsis).

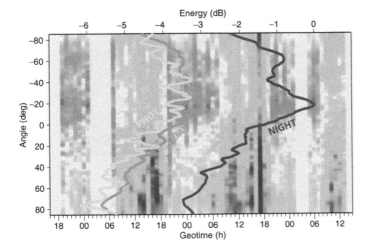

COLOR FIGURE 5.22
Directional characteristics of ambient noise vs. geotime. Normalized correlation function (log envelope) between hydrophone signals R1 and R2. Negative and positive angles correspond to downgoing and upgoing energy, respectively. Color map: −10 dB (blue) to 0 dB (red). The overlaid lines are geotime mean-averages. Dark gray: night hours, 1930–0700 hours, 4 days; gray: day hours, 4 days; light gray: midafternoon hours, 1400–1600 hours, September 24, 1999. The raw functions were computed from the same data sections as in Color Figure 5.20 and their envelopes were median-averaged over 1-h periods.

COLOR FIGURE 7.3 Photo-mosaic of the intertidal zone on the western end of Bird Rock, Santa Catalina Island, May 2001 (see Figure 7.1 for location). The lowest areas, occupied by the southern sea palm *Eisenia arborea* and the brown alga *Halidrys dioica*, occur along the left side of the image, and in a small pocket to the upper right. Immediately above this zone is a band of mixed turf algae, predominantly *Gigartina caniculata*, *Pterocladia capillacea*, and *Corallina officianalis*. Higher still is the mussel bed, which occupies most of the image, and gives way to bare rock in the upper intertidal. The large white rectangle delineates a 6 × 3 m subregion of the image used for an analysis of mussel and barnacle clumping. The inset is an enlarged portion of the image, showing details of individual mussel and barnacle distribution.

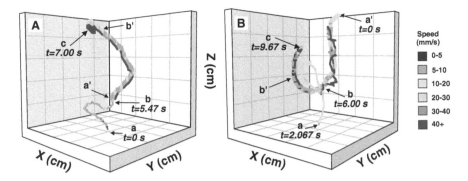

COLOR FIGURE 10.1 (A) Mate-tracking by the copepod, *T. longicornis* (1.2 mm prosome length). The male copepod (thin trail) finds the trail of the female copepod (thick trail) when the trail is 5.47 s old and the female is 3.42 cm distant from him (nearly 30 body lengths). Upon encounter, the male spins to relocate the trail, then accelerates to catch up with the female. The male copepod follows the path of the female precisely in 3D space. When within 1 to 2 mm of the female, the male pauses quietly, then pounces swiftly to capture his mate for the transfer of gametes packaged in a flasklike spermatophore. During copulation, the mating pairs spin and remain together for a few seconds or more. (B) Mate-tracking by the copepod, *T. longicornis*: backtracking. The male copepod (thin trail) finds the trail of the female copepod (thick trail) because of a strong cross-plume odor gradient. The female is 1.30 cm distant from him. Upon encounter, the trail is 2.3 s old and the male follows the trail in the wrong direction, away from the female, because of a weak along-plume gradient. When initially following the trail, the male smoothly and closely follows the trajectory of the female. After 1.27 s when he is 24.4 mm from his mate and the trail is 6.7 s old, the male turns around and backtracks. When backtracking, he follows the female path erratically, casting back and forth over the trail. After reaching his intersection point, the male copepod resumes smooth close-following of the undisturbed female path. (From Doall, M. et al., *Philos. Trans. R. Soc. Lond.*, 353, 681, 1998. With permission.)

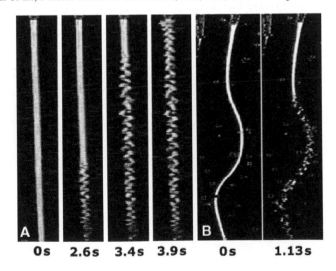

COLOR FIGURE 10.2 (A) Visualized trail and upwardly directed trail following. A 1-mm-wide trail was created by dripping fluid from a fine pipette tip (Eppendorf) to flow into the 4-l observation vessel filled with filtered seawater (28 ppt). The pipette was gravity-fed, via thin tubing, from a beaker filled with a mixture of seawater and dextran. The difference in the refractive index of the dextran–seawater mixture sinking down through the seawater could be detected, using a Schlieren optical path. The trail on the left is the undisturbed trail. The disturbed chevron-dotted trails on the right show how the trail structure changes at various time intervals after the male copepod *T. longicornis* follows the trail. Copepod indicated at the upper left on last trail, colored orange. (B) Following a curvy trail: When following the scent in the trail, *T. longicornis* also can stay on the track of curved trails, making the same turns as taken by the trail. The undisturbed curved trail on left was followed for 1.13 s by the male copepod, colored orange in panel on right. Note how trail structure changes in wake of swimming male.

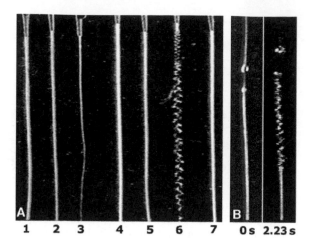

1 2 3 4 5 6 7 0 s 2.23 s

COLOR FIGURE 10.3 (A) Scent preferences of *T. longicornis*. Male copepods of *T. longicornis* were offered seven choices of scent trails (from left to right: Female, Male, Acartia, Dextran, Male, Female, Acartia). The male copepod showed an 80% preference to follow trails scented with the odor of con-specific females (choice 6). Trails were followed to the source. The source of the odor was a stock of dextran-mixed filtered seawater within which 20 females were swimming. The conditioned water flowed through fine pipettes down through the observation vessel, filled with filtered seawater and 20 to 50 male copepods. Schlieren optics visualized the flow patterns. (B) Downwardly directed trail-following. The trail on the left is the undisturbed trail. The disturbed trail on the right was the same trail that was followed down to the source by a male copepod, appearing in the lower right, colored orange. Trails were created by allowing scented seawater to flow out of fine pipettes at the bottom of the obser-vation vessel. The vessel was filled with a mixture of dextran and seawater to create the difference in refractive index necessary to visualize the trail using Schlieren optics. The plain seawater would float up to create the trail. Flow speed of 4.43 mm/s can be calculated by the upward movement of the bright bubble.

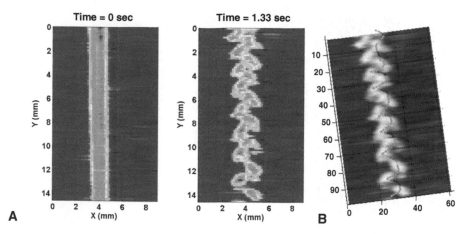

COLOR FIGURE 10.4 Changes in scent trail shape caused by tracking behavior of male *T. longicor-nis*. (A) Trail structure changes from a smooth undisturbed vertical trail (left) to a disturbed trail (same portion of the trail shown on right) after the passage of the male copepod of *T. longicornis*. (B) Close-up of the disturbed trail. The red dotted line defines the hypothesized helical trail left by spiraling move-ments of the male while tracking the trail. The spiral is longer, thinner, and has a greater surface area than the smooth trail. The action of molecular diffusion can dilute the pheromone more quickly. Qualitative measurements of trail geometry (for Table 10.3 calculations) include: length of smooth and helical trail, radius of smooth and helical trail, slope of spiral structure of helix. For each 180° segment of the helix, the radius and slope were repetitively calculated to reconstruct the helix so it cuts through the brightest spots of the image. Scale: 6.8 pixels/mm.

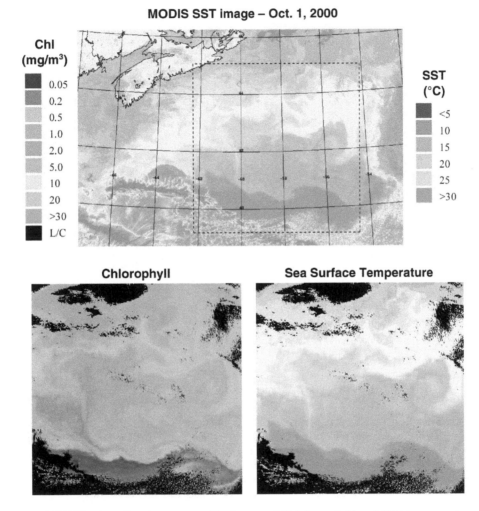

COLOR FIGURE 14.1 Simultaneous satellite images of Chl (lower left) and SST (upper center and lower right) in the western Atlantic Ocean south of Nova Scotia acquired by the Moderate-resolution Imaging Spectroradiometer (MODIS) on October 1, 2000. The region shown in the two lower panels is indicated by a dotted outline in the upper panel and is approximately 512×512 km^2. The resolution of the data is approximately 1×1 km^2. The color scale is logarithmic for Chl, but linear for SST. Black areas in the two lower panels are clouds.

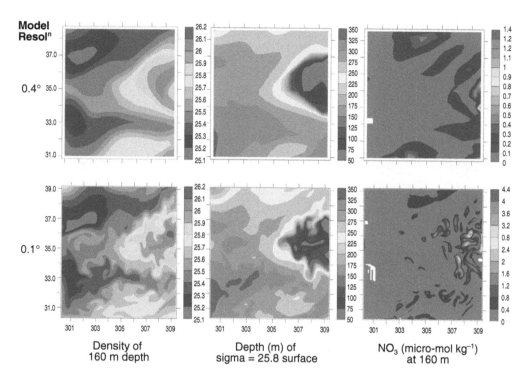

COLOR FIGURE 14.3 Model potential density and nitrate fields from a $10° \times 10°$ region representative of the oligotrophic subtropics near Bermuda depict that nitrate is fluxed into the euphotic zone where isopycnals outcrop. The fields are shown at two different model resolutions to demonstrate the importance of resolving the small scales (less than the internal Rossby radius) in capturing vertical advective transport. The dimensions of the domain are in degrees latitude and longitude.

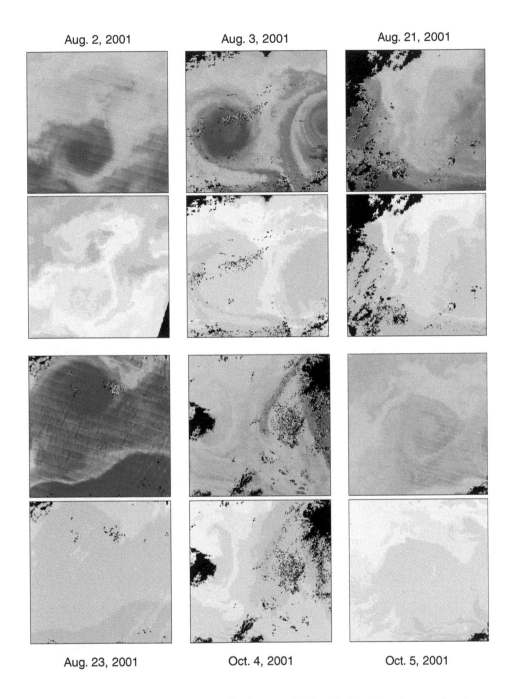

Aug. 2, 2001   Aug. 3, 2001   Aug. 21, 2001

Aug. 23, 2001   Oct. 4, 2001   Oct. 5, 2001

COLOR FIGURE 14.4 Simultaneous satellite images of chlorophyll (Chl) and sea-surface temperature (SST) for domains of size 256×256 km^2 and resolution 1×1 km^2 in the western Atlantic continental shelf and slope region acquired by MODIS on six dates between August 2 and October 5, 2001.
In each case, the Chl image is above the SST image. Color scales are the same as those shown in Color
Figure 14.1.

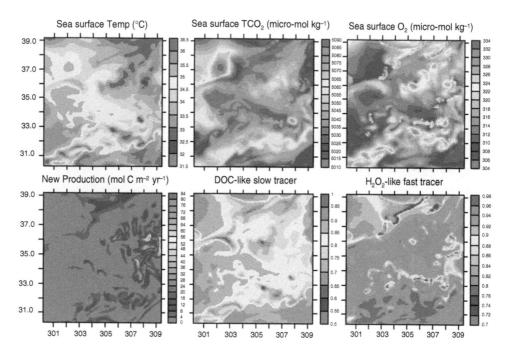

COLOR FIGURE 14.6 Sea surface distribution of temperature, DIC, O_2, new production, DOC, and H_2O_2 from a limited region model of a $10° \times 10°$ region in the subtropical Atlantic near Bermuda (Mahadevan and Archer, 2000). The model was driven at the open boundaries by dynamic fields from a global circulation model (Semtner and Chervin, 1992). The picture depicts a typical snapshot view of the ocean surface in the autumn, modeled at $0.1°$ resolution. The different tracers exhibit different scales of variability.

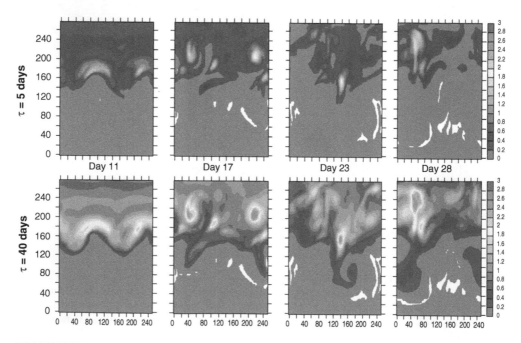

COLOR FIGURE 14.8 Concurrent views of the surface concentration of the nutrient-like tracers with different response times τ, averaged over the upper 95 m in the model. The tracer with smaller τ is patchier. The domain dimensions are in kilometers.

COLOR FIGURE 19.1 Illustrates spatial changes in character of a portion of the surface fluorescence record from GLOBEC Broadscale Survey in June 1995. Transects move onto Georges Bank, from deeper (500 m), stratified water into shallower (40 m), well-mixed water, going into the page, crossing the tidal mixing front. The center across-isobath transect onto the bank is 56 nautical miles long, going from Station 8 (started at 2345 on June 16 1999) to Station 11 (ended at 1402 on June 17 1999) of the survey; the distance along isobath from Station 7 to 8 is 28 nautical miles. All station locations are shown in Figure 19.3, and data are archived at http://globec.whoi.edu. The range in raw voltage linearly scales from 0 to 5.

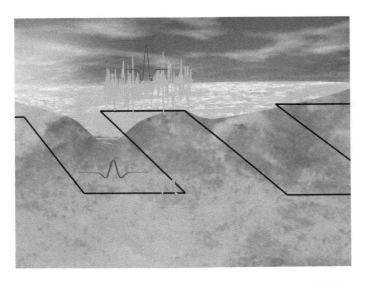

COLOR FIGURE 19.5 Portions of the fluorescence record from two distinct regions along the survey track line (textures derived from satellite ocean color observations of Georges Bank obtained by SeaWiFS). Mexican Hat wavelets (in blue) at two resolutions shown during convolution with two sections taken from the fluorescence data (in green) along the survey track in Color Figure 19.1. The taller wavelet is shown with data from the well-mixed region that appears noisy in character; this wavelet convolves fewer points at each step, and therefore resolves smaller spatial scales. The wider wavelet is shown with data from the stratified region that appears persistent in character; it convolves more points at each step and therefore resolves larger spatial scales. Wavelets are normalized to have the same total energy at all scales. Note that practical limitations on scales that can be analyzed are imposed by the shape of the analyzing wavelet, the shape of the survey track, and the data sample spacing.

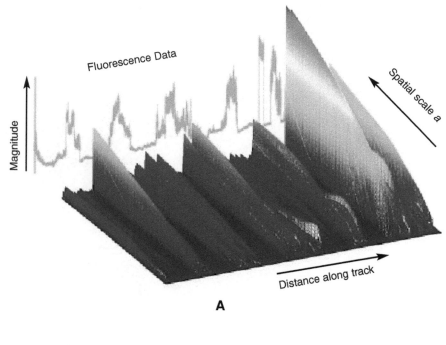

A

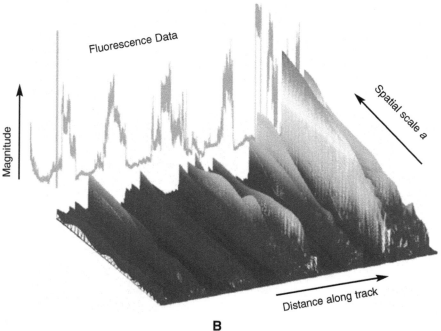

B

COLOR FIGURE 19.8 3D visualization showing absolute values of the 1D Mexican Hat wavelet (A) and Morlet wavelet (B) transforms $W(x, a_n)$ for fluorescence in June 1998, at spatial scales up to 4 km. Heading into the page, scales range from $a_n = 2$ to $a_n = 64$, and the convolutions are calculated at even integer values of a_n. The 3D mesh colors high values resulting from the convolution red, and low values blue, shown in the 2D transform for the intermediate powers of two. Transforms are over the same 1500 km of cruise track (heading left to right) shown in Figure 19.7. The fluorescence data are shown in the background.

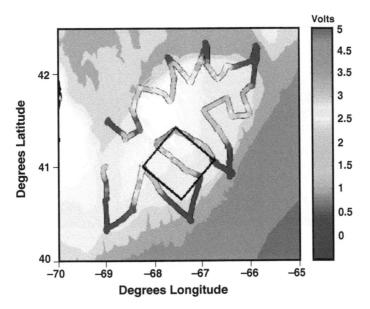

COLOR FIGURE 19.14 Region of fractal interpolation for southern flank superimposed over the observations of raw fluorescence voltage at 3 m depth by tow body Greene Bomber on GLOBEC Broadscale cruise AL9906, June 1999.

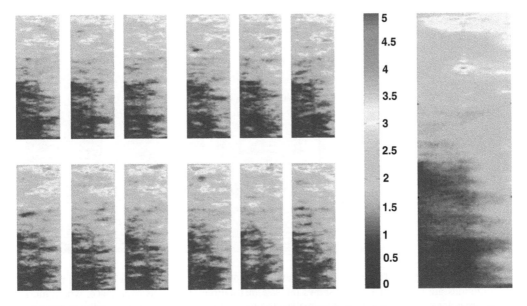

COLOR FIGURE 19.15 Twelve iterations of the southern flank fractal fluorescence field for June 1999 Broadscale survey (left), and the mean field constructed from 15 iterations (right). Each iteration was based on a different random seed for that realization of synthetic June 1999 fields. The x-axis direction is along-isobath; the y-axis direction is across-isobath. Range for individual pixels in all fields is from 0 to 5 V, shown on the colorbar. Note that all fields, the individual realizations and the mean, have the same data anchors from the survey as the right and left boundaries, and down the center. In a completely realistic field, these anchors would not visibly stand out.

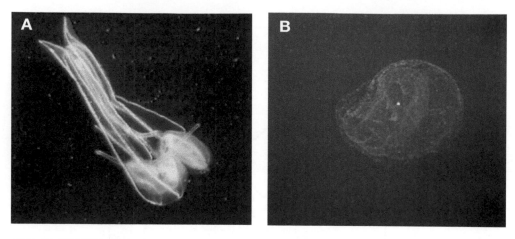

COLOR FIGURE 21.1 Two large, delicate, gelatinous organisms representative of the gossamer design discussed in this report. (A) *Kiyohimea usagi*, a lobate ctenophore recently described by Matsumoto and Robison (1992). (Image copyright, 1996 MBARI.) (B) *Mesochordaeus erythrocephalus*, a *mesopelagic larvacean* recently described by Hopcroft and Robison (1999). (Image copyright, 1991 MBARI.)

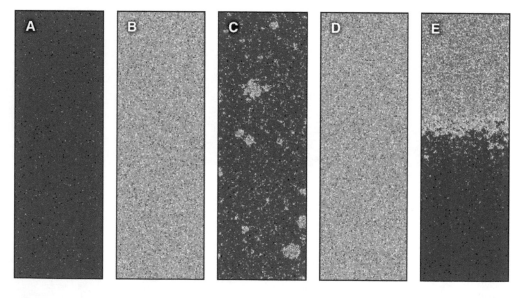

COLOR FIGURE 27.4 Two possible spatial effects are observed in the CA model when the mean field assumptions are relaxed. Each plot shows the size distribution of mussels over space, where mussel sizes are color-coded from blue for the smallest through red for the largest. The size distributions over space are shown for the mean field approximation ($h = \infty$) at (A) the lower and (B) the upper equilibria. (C) When the neighborhood size is small ($h = 1$) and the seastar attack rate is reduced ($\theta = 0.8$), "islands" of larger individuals develop at the lower equilibrium. (D) When a smooth linear gradient in the seastar attack rate is imposed from $\theta = 0$ at the upper boundary of the lattice to $\theta = 2$ at the lower boundary, and the mean field approximation ($h = \infty$) is used, the lattice equilibrates at the upper equilibrium. (E) When the same gradient in θ is combined with a small neighborhood effect ($h = 1$), the result is a sharp transition in prey cover similar to what is seen in many natural mussel beds. This predation gradient is what might be expected from differences in foraging times at different tidal heights.

22

Quantifying Zooplankton Swimming Behavior: The Question of Scale

Laurent Seuront, Matthew C. Brewer, and J. Rudi Strickler

CONTENTS

22.1 Introduction 333
22.2 Recording Swimming Paths 337
    22.2.1 Culture of *Daphnia* and Algae 337
    22.2.2 Recording Three-Dimensional Swimming Behavior 337
22.3 Characterizing Swimming Paths 338
    22.3.1 Swimming Path and Fractal Dimension 339
    22.3.2 Measuring Fractal Dimensions 339
        22.3.2.1 Compass Method 340
        22.3.2.2 Box-Counting Method 341
22.4 Testing the Robustness of Fractal Dimension Estimates 341
    22.4.1 On the Scale Dependence of Fractal Dimensions 341
    22.4.2 Toward an Objective Identification of Scaling Range 342
    22.4.3 Robustness of Fractal Dimension Estimating Algorithms 346
    22.4.4 Two-Dimensional vs. Three-Dimensional Fractal Dimension Estimates 348
22.5 Comparing Zooplankton Behavior with the Structure of Their Surrounding Environment 350
22.6 Conclusion 351
References 353

22.1 Introduction

Animals typically search for food, hosts, and sexual partners and avoid predators in complex, spatially and temporally structured environments. The resulting movements have implications for the optimization of food search patterns, energy investment, habitat selection, and territorial and social behaviors (Morse, 1980; Pyke, 1984; Stevens and Krebs, 1986; Bell, 1991; Turchin, 1998; Boinski and Garber, 2000). In zooplankton ecology, examples come from the wide spectrum of swimming behaviors related to the species (Tiselius and Jonsson, 1990), the age (Coughlin et al., 1992; van Duren and Videler, 1995; Fisher et al., 2000; Titelman, 2001), the prey density (Tiselius, 1992; Bundy et al., 1993; Dowling et al., 2000), the presence of a predator or a conspecific (van Duren and Videler, 1996; Tiselius et al., 1997; Titelman, 2001), the sex of individuals (van Duren and Videler, 1995; Brewer, 1998; Strickler, 1998), the information imparted into the surrounding water by a swimming animal (Yen and Strickler, 1996; Gries et al., 1999), including both chemical (Yen et al., 1998; Weissburg et al., 1998) and hydromechanical (Costello et al., 1990; Marrasé et al., 1990; Hwang and Strickler, 1994; Hwang et al., 1994; Brewer, 1998) stimuli. Moreover, considering that environmental complexity affects the movement patterns of animals (e.g., Wiens and Milne, 1989; Boinski and Garber, 2000) and the recent advances demonstrating the heterogeneous nature of physical and biological patterns and processes at scales relevant to individual organisms

(Mitchell and Fuhrman, 1989; Squires and Yamazaki, 1995; Cowles et al., 1998; Seuront et al., 1996a, b, 1999), there is a genuine need to establish a reference framework that will link pure behavioral observations, the qualitative and quantitative nature of environment complexity, and zooplankton trophodynamic hypotheses (Turchin, 1991; Keiyu et al., 1994; Seuront, 2001; Seuront et al., 2001; Schmitt and Seuront, 2001, 2002).

Despite the growing number of studies, analyses of the movement patterns of aquatic organisms are still far less common than of terrestrial organisms, primarily because of the difficulty in obtaining accurate records of the displacement of aquatic organisms, which unlike terrestrial organisms, move through a volume and therefore require systems capable of recording three-dimensional (3D) data. This is a nontrivial problem, and has resulted in many investigations of zooplankton behavior recording only two-dimensional (2D) swimming paths. Moreover, even using video systems capable of recording 3D data, there are still problems of scale resulting from the small size of planktonic organisms. Gathering 3D coordinates for zooplankton involves a trade-off between resolution and extent, typically presenting researchers with two alternatives in the collection of data: (1) high spatial resolution, but for short duration, or (2) longer time series, but at low spatial resolution. To our knowledge, the methods used in only four studies have permitted the collection of 3D swimming data at both high spatial resolution and for long periods (Coughlin et al., 1992; Bundy et al., 1993; Brewer, 1996; Schmitt and Seuront, 2001).

Even when the collection of 3D data is possible, behavioral ecologists face another, more fundamental, problem — the accurate quantitative description of animal paths. The difficulty in quantifying animal movement results from the fact that most of the quantitative metrics commonly applied in behavioral studies, e.g., path length, turning angle, turning rate, and net to gross displacement ratio (NGDR), are scale dependent. That is, the metrics will take on different values depending on the physical or temporal scale at which they are measured. This problem was recognized by Dodson et al. (1997), who reported that *Daphnia* swimming speeds were typically two to four times greater when measured at 30 Hz than when measured at 1 Hz. Although they acknowledged the scale dependence of their metrics, they did not propose an objective way to deal with this dilemma. The scale dependence inherent in most metrics results in there being no single scale at which swimming paths can be unambiguously described. Thus, there is no single scale at which swimming behaviors can be compared without leading to arbitrary, potentially spurious conclusions. Furthermore, because individual studies typically record behaviors at different temporal resolutions, (i.e., ranging from 0.01 to 50 Hz; Table 22.1), their results cannot be accurately compared. Despite the clear difficulties associated with scale-dependent metrics, as far as we know only six studies have analyzed plankton swimming behavior in a scale-independent framework (Coughlin et al., 1992; Bundy et al., 1993; Brewer, 1996; Jonsson and Johansson, 1997; Dowling et al., 2000; Schmitt and Seuront, 2001).

Mandelbrot (1977, 1983) introduced the term *fractal* to characterize spatial or temporal phenomena that are continuous but, because of their complexity, not differentiable. Unlike more familiar Euclidean constructs, every attempt to split a fractal into smaller pieces results in the resolution of more structure. As a consequence, in fractal constructs the detail is similar to the whole; i.e., there is no characteristic scale. Fractal objects and processes are therefore said to display "self-invariant" properties (e.g., Hastings and Sugihara, 1993), and can be further defined as being either "self-similar" or "self-affine." Self-similar objects are isotropic (the same in all three spatial dimensions) upon rescaling, whereas rescaling of self-affine objects is direction dependent (anisotropic). Thus, a trace of zooplankton motion in 3D space is self-similar, whereas a 2D trace, such as the plot of the x-coordinate of an organism's movement as a function of time, is self-affine (for more details see Schroeder, 1991). Regardless, fractal analysis presents a new way of addressing questions about structures and scales in ecological systems. In particular, self-invariant patterns and processes can be described by a (non-integer) fractal dimension, which can be viewed as a measure of complexity, or as an index of the scale dependence. The fractal dimension, D, characterizes a range of scales over which similar patterns and/or processes are operating across that range of scales. However, if there exists a critical scale beyond which a further increase results in a shift in the fractal dimension, or a loss of fractal structure, this may define a transition zone where the environmental properties or constraints acting upon a given system are probably changing rapidly, between two different hierarchical levels in which different patterns and/or processes are operating (Frontier, 1987; Seuront and Lagadeuc, 1997).

TABLE 22.1

Literature Survey of Zooplankton Behavioral Survey, Arranged in Chronological Order

Organism	View	Variable	Metrics [temporal scale]	Ref.
Daphnia	2D, side	Relative light intensity	Speed, position in the water column [–]	Ringelberg (1964)
Cyclops	2D, side	Light	Speed [0.1 Hz]	Strickler (1970)
Daphnia	2D, top	Polarized light	Speed, NGDR, IDT, turning rate [30 Hz]	Wilson and Greaves (1979)
Mesocyclops	2D, top	Prey patches	Speed, loops/min [0.016 Hz]	Williamson (1981)
Daphnia	3D	Angular light distribution	Speed, NGDR [–]	Buchanan et al. (1982)
Daphnia	2D, side	Food concentration	Speed [0.1 Hz]	Porter et al. (1982)
Acartia	2D, top	Bioluminescent dinoflagellates	Speed, NGDR, bursts [15 Hz]	Buskey et al. (1983)
Pseudocalanus	2D, top	Food concentration and odors	Speed, NGDR, bursts, pauses [15 Hz]	Buskey (1984)
Diaptomus	2D, top	Predators and competitors	Speed, NGDR, time between jumps [30 Hz]	Wong et al. (1986)
Daphnia	3D	Food concentration	Speed, turning rate, ground covered [30 Hz]	Young and Getty (1987)
Favella	2D, side	Food patches	Speed, NGDR, turning rate [15 Hz]	Buskey and Stoecker (1988)
Thysanoessa	3D	Algal patches	Speed, NGDR, bursts, % sinking [2 Hz]	Price (1989)
Six calanoids	2D, side	Light, food type	Speed, foraging mode [12.5 Hz]	Tiselius and Jonsson (1990)
Polyphemu	2D, top	Predator–prey interaction	Speed, turning rate, meander [1 Hz]	Young and Taylor (1990)
Bosmina	2D, top	Predator–prey interaction	Speed, turning rate, meander [1 Hz]	Young and Taylor (1990)
Daphnia	3D	Body size	Speed, displacement angle, NGDR, stroke velocity, sinking speed [30 Hz]	Dodson and Ramcharan (1991)
Diaptomus	3D	Predator	Speed, jump length, angle of motion [20 Hz]	Ramcharan and Sprules (1991)
Diaptomus	2D, top	Conspecific	Speed, NGDR [–]	Van Leeuwen and Maly (1991)
Acartia	2D, side	Turbulence	Speed, foraging activity and behavior [25 Hz]	Saiz and Alcaraz (1992)
Amphiprion	3D	Food concentration	Speed, NGDR, turning angles, **fractal dimension** [10–15 Hz]	Coughlin et al. (1992)
Acartia	2D, side	Food patches	Speed, vertical position, jump frequency, NGDR [0.1 Hz]	Tiselius (1992)
Centropages	3D	Food concentration	Speed, NGDR, Realized Encounter Volume, i.e., **fractal dimension** [30 Hz]	Bundy et al. (1993)
Various species	2D, side	Species	Speed, NGDR, rate of change in direction [15–30 Hz]	Buskey et al. (1993)
Diaptomus	2D, top	Gravid females	Speed, NGDR [–]	Maly et al. (1994)
Acartia	3D	Food, turbulence	Speed, behavioral observations [30 Hz]	Saiz (1994)
Brachionus	2D, top	Toxic stress	Speed, sinuosity, behavioral observations [25 Hz]	Charoy et al. (1995)
Temora	3D	Food concentration	Speed, NGDR, behavioral observations [50 Hz]	Van Duren and Videler (1995)
Dioithona	2D, side	Light, water flow	Speed, rate of change in directions [30 Hz]	Buskey et al. (1996)
Oithona	2D	Developmental stage	Speed, behavioral observations [30Hz]	Paffenhöfer et al. (1996)
Temora	2D, 3D	Predators, conspecific	Speed, NGDR, behavioral observations [50Hz]	Van Duren and Videler (1996)
Daphnia	3D	Food concentration, light, temperature	Speed, turning angle, turning rate, NGDR, **fractal dimension** [10Hz]	Brewer (1996)
Daphnia	2D, top	Food concentration	Speed [–]	Larsson and Kleiven (1996)

TABLE 22.1 (continued)

Literature Survey of Zooplankton Behavioral Survey, Arranged in Chronological Order

Organism	View	Variable	Metrics [temporal scale]	Ref.
Daphnia	3D	Light, food concentration, vessel size	Speed, turning angle [30 Hz]	Dodson et al. (1997)
Acartia	2D	Predators	Encounter rates [–]	Tiselius et al. (1997)
Euplotes	2D, top	Food patches	Speed, motility, **fractal dimension** [–]	Jonsson and Johansson (1997)
Protoperidinium	2D, top	Food type	Speed, rate of change of direction, behavioral observations [15 Hz]	Buskey (1997)
Centropages	2D	Turbulence, food concentration	% swimming, swimming behavior, jumps [25 Hz]	Caparroy et al. (1998)
Cyclops	3D	Conspecific	Speed, distance between male and female [60 Hz]	Strickler (1998)
Daphnia	3D	Predators	Speed, turning angle, behavioral observations [30 Hz]	O'Keefe et al. (1998)
Lates calcarifers	2D, top	Food concentration	Pause duration, distance traveled between pauses, travel duration, developmental stage, **fractal dimension** [25 Hz]	Dowling et al. (2000)
Pomacentrus	1D, side	Age	Speed [–]	Fisher et al. (2000)
Sphaeramia	1D, side	Age	Speed [–]	Fisher et al. (2000)
Amphiprion	1D, side	Age	Speed [–]	Fisher et al. (2000)
Acartia	2D, side	Predator	Speed, reaction distance, jumps [60 Hz]	Suchman (2000)
Acartia	3D, side	Age, predators	Speed, jump directionality, frequency, length, and speed [–]	Titelman (2001)
Temora	3D, side	Age, predators	Speed, jump directionality, frequency, length, and speed [–]	Titelman (2001)
Temora	3D	Female	**Multifractal parameters** [12.5Hz]	Schmitt and Seuront (2001, 2002)

Note: Values in parentheses are the temporal scale at which the listed metrics are calculated in each study.

Because of its scale-independent nature, in recent years fractal geometry has been used to investigate a surprisingly varied set of phenomena including electrochemical deposition (Mach et al., 1994), the structure of physiological systems such as bronchial trees (Shlesinger and West, 1991) and Hiss–Purkinje cardiac conduction (Goldberger et al., 1985), DNA sequences (Provata and Almirantis, 2000), the growth of bacterial colonies (Tang et al., 2001), taxonomic schemes (Burlando, 1990, 1993), and clusters of galaxies and stars (Wu et al., 1999; Pietronero and Labini, 2000). In addition, Frontier (1987), Sugihara and May (1990), and Seuront (1998) describe numerous possible ecological applications of fractals. Fractals have been used to describe habitat complexity (Bradbury and Reichelt, 1983; Bradbury et al., 1984; Gee and Warwick, 1994a, b), species diversity (Frontier, 1985, 1994), the shape of marine particles (Li et al., 1998), growth processes of benthic fauna (Abraham, 2001), the space–time distribution of phytoplankton biomass (Seuront and Lagadeuc, 1997, 1998), movements of marine invertebrates (Erlandson and Kostylev, 1995) or estuarine vertebrates (Dowling et al., 2000), movements of terrestrial invertebrates (Gautestad and Mysterud, 1993; Wiens et al., 1995), and the search paths of small (Cody, 1971; Pyke, 1981) and large (Siniff and Jenssen, 1969; Van Ballemberghe, 1983; Bascompte and Vilà, 1997) terrestrial vertebrates.

Yet even as the use of fractal dimensions has increased, their reliability for use in the quantification of animal paths has recently been questioned. Turchin (1996, 1998) argued on the basis of a simulated path generated by a correlated random walk that fractal dimensions are themselves scale dependent, varying continuously as a function of scale. Similarly, Tsonis and Elsner (1995) emphasized that scaling regions are subjectively estimated and are often the result of the generic property of the quantity to increase or decrease monotonically as the scale approaches zero, regardless of the geometry of the object (see, e.g., Davenport, Chapter 16, this volume). As a consequence, they proposed a standard procedure for dealing with fractal dimension estimates. Additionally, considering that many analyses of 3D behavior are carried out using 2D data (Table 22.1), we address here the question of whether fractal dimensions estimated from 2D trajectories can be reliably used to estimate the 3D structure of swimming paths.

This comes into question for two conceptually distinct reasons. First, following Mandelbrot (1983), the lower (D_f +1) and upper ($2D_f$) limits of the 3D fractal dimension can be estimated from a 2D of D_f characterizing the same pattern. However, the broadness of the resulting limits (e.g., if the limits are 2.4 to 2.8) and evidence suggesting that extrapolation to higher dimensions (e.g., from 2D to 3D) is invalid (Roy et al., 1987; Huang and Turcotte, 1989) restrict the reliability of this approach. Second, while swimming organisms move in essence in three dimensions, nothing *a priori* ensures the isotropy of their paths. Anisotropy in swimming can be generated by specific food patterns via patch exploitation strategies (Leising and Franks, 2000; Leising, 2001; Seuront et al., 2001), prey switching behavior (Kiørboe et al., 1996; Caparroy et al., 1998), species-specific behavior (Tiselius and Jonsson, 1990), or the effect of gravity (Strickler, 1982). Path isotropy is nevertheless an absolute requirement for extrapolating 2D to 3D behavior, and therefore must be carefully checked.

Given the increasing use of fractal dimensions in ecology in light of continuing questions regarding their utility, the aims of this chapter are to (1) demonstrate the unreliability of scale-dependent metrics and highlight the advantages of a scale-independent framework to characterize plankton swimming behavior, (2) address, in detail, the proposed limitations of fractal analysis, and (3) introduce an efficient statistical framework that will ensure the existence of a scaling range and the subsequent robustness of fractal dimension estimates. A set of high-resolution 3D trajectories of a common freshwater zooplankter, *Daphnia pulex*, are used throughout the chapter as ecological examples to illustrate the concepts presented.

22.2 Recording Swimming Paths

Gathering 3D coordinates remains the first (and major) limitation for *in vitro* zooplankton behavioral studies because it entails collection of 3D swimming data both at high spatial resolution and for extended periods of time. From the few studies available in the literature (see, e.g., Coughlin et al., 1992; Bundy et al., 1993; Schmitt and Seuront, 2001), it can nevertheless be seen that the methods used are conceptually similar to the one introduced below.

22.2.1 Culture of *Daphnia* and Algae

A clone of *D. pulex* was cultured in aged tap water under cool white fluorescent bulbs, in a 16/8 light–dark cycle. The cultures were maintained at the experimental temperature (20°C) and fed every day with a 1:1 mixture of the green algae *Ankistrodesmus* sp. and *Scenedesmus* sp. at a final concentration of about 5×10^5 cells ml^{-1}. Algae were grown in multiple 250 ml batch cultures under cool white fluorescent bulbs, in an 18/6 light–dark cycle, at 20°C, in Bold's Basal Medium.

22.2.2 Recording Three-Dimensional Swimming Behavior

All paths analyzed in this chapter are the movements of solitary *D. pulex* swimming in the 5-l (18 × 18 × 15.5 cm high) Plexiglas recording vessel of the CritterSpy™, a high-resolution 3D recording system. All recordings were made with animals swimming in an algal concentration of 5×10^4 cells ml^{-1}, which is an intermediate food concentration, well below *D. pulex*'s incipient limiting concentration (Lampert, 1987). The test chamber was illuminated with a diffused, fiber-optic light placed 0.5 m directly overhead that resulted in an illumination of about 12 μEm^{-2}s^{-1} in the vessel, approximately equal to full daylight. At least 1 h prior to experiments, adult, gravid females 2.1 ± 0.2 (mm) were transferred from their culturing vessels and acclimated to experimental light and food conditions in holding vessels. A single animal was then transferred from its holding vessel to the recording chamber with a large-bore pipette and allowed to acclimate for at least 10 min before recording began.

The CritterSpy uses a Schleiren optical system consisting of a collimated red laser beam (λ = 623 nm) which serves as the light source for two orthogonally mounted video cameras, two frame number generators, two 20-in. video monitors, and two VHS videocassette recorders; see Strickler (1985) and Bundy et al. (1993) for further details. This system simultaneously records orthogonal front (XZ) and side (YZ) views of the experimental chamber as dark field images. To run the system, two operators

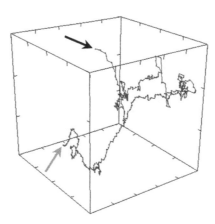

FIGURE 22.1 Example of a 3D pathway of *D. pulex*. Path is shown to scale in the CritterSpy's 5-l recording vessel.

view the camera images in real time. As the animal swam away from the center of the other camera's view (marked with crosshairs on the monitors), one operator used a trackball (X and Z dimensions) and the other a rotating cylinder (Y dimension) to bring the animal back into the center of both views. The actual recentering of the image was achieved via three computer-controlled linear positioning motors (one for each axis) that moved the entire optical system in response to the operators' input. A computer recorded the motor movements necessary to keep the animal centered in the two views as X, Y, and Z coordinates. Because the computer only recorded coordinates when the trackball or cylinder were moved, the coordinates were recorded at an uneven sampling rate (ranging from about 5 to 15 Hz). Paths were then interpolated to produce an even time interval (10 Hz) between successive position measurements. The 10 Hz rate is rapid enough that coordinates recorded at that temporal scale are the result of very small movements of the crosshairs corresponding to *Daphnia*'s characteristic hop-and-sink behavior.

Each individual *Daphnia* was recorded swimming for at least 30 min, after which the videotapes were reviewed and valid segments were identified for analysis. Valid segments consisted of video in which the animals were swimming freely, at least two body lengths away from any of the chamber's walls or the surface, and the animals were always within one half body length of the crosshairs in the center of the video monitors. To ensure that there would be a significant range of scales in each path, we only used paths that were at least 30 s in duration. After identifying valid sequences, the frame numbers imprinted on the video were used to isolate the corresponding time interval from the 3D coordinate data stored on the computer. These time series of coordinates formed the 3D trajectories used in our analysis (Figure 22.1).

22.3 Characterizing Swimming Paths

Movement paths may be characterized by a variety of measures (Figure 22.2; Table 22.1), including: path length (the total distance traveled, or gross displacement), move length (the distance traveled between consecutive points in time), move duration (time interval between successive pauses, as well as between successive spatial points), speed (move length divided by move duration), turning angle (the difference in direction between two successive moves), turning rate (turning angle divided by move duration), net displacement (the linear distance between starting and ending point, often used as a metric when making comparisons with diffusion or correlated random walk models; e.g., Kareiva and Shigesada, 1983; McCulloch and Cain, 1989; Turchin, 1991; Johnson et al., 1992a), NGDR (net to gross displacement ratio; Wilson and Greaves, 1979), and fractal dimension. For paths recorded at fixed time intervals, move duration is a constant. As discussed above, the values of all the metrics except fractal dimension (see below) are implicitly a function of their measurement scale (Figure 22.2). The scale dependence of these ratio metrics, i.e., the path length and the turning angle (Figure 22.3), implies that there is no single scale at which swimming paths can be unambiguously described.

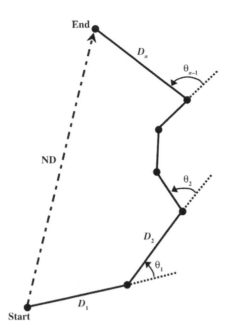

FIGURE 22.2 Illustration of the different metrics used in characterizing movement pathways. The mean displacement is the mean of distances D_i traveled per time interval, the net displacement ND is the straight-line distance between the initial and final locations, and the growth displacement is the sum of the distances D_i. The mean turning angle is the trigonometric mean of angles θ_i formed by changes in direction between steps.

22.3.1 Swimming Path and Fractal Dimension

Fractal analysis of animal behavior is based on the premise that the fractal dimension can serve as a scale-independent descriptor of the path an organism takes as it moves. The philosophy behind fractals is as follows: if an organism moves along a completely linear path, then the actual distance traveled, L, equals the displacement between the start and the finish, δ. The relationship between these two variables is linear. In other words, if we assume a power law relating L to δ, i.e., $L^D = \delta$, then the exponent $D = 1$. According to this power law, if the path deviates from linearity, that is, becomes tortuous, the exponent will then be greater than 1. In the extreme example of "curviness," i.e., for the case of Brownian motion, $D = 2$ (Mandelbrot, 1983). It appears that D provides a measure of the path "tortuosity," or "complexity," with the extreme cases delineated by linear and Brownian movement, respectively, and real-life cases expected to fall between these extremes of $D = 1$ and $D = 2$.

We reiterate Turchin et al.'s (1991) position on the importance of using individual organisms, and not steps within an individual path, as statistical replicates. Analyses performed on the above measures are termed second-order statistics because they collapse information from many observations of an individual into a single measure (Batschelet, 1981). In addition, we suggest that strict limits be placed on the minimum number of moves used to obtain a statistic for a given trajectory. Specifically, the minimum number of observed moves should set the number of observed moves from which statistics are gathered from any of the paths.

22.3.2 Measuring Fractal Dimensions

Practical approaches to measuring D using real data have not yet been standardized (e.g., Hastings and Sugihara, 1993). Some investigators plot the net squared displacement as a function of time (Johnson et al., 1992a), a practice that has a solid basis in random walk theory (see, e.g., Tarafdar et al., 2001). Others construct plots of the apparent path length vs. ruler length (With, 1994a). Here, as recommended by Fielding (1992) and Hastings and Sugihara (1993) to ensure the reliability of the fractal dimension estimates, we used two different, but conceptually similar, methods.

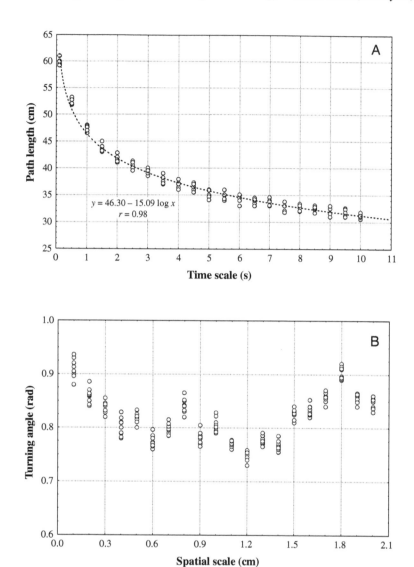

FIGURE 22.3 Scale dependence of path length (A) and turning angle (B) obtained from a 3D swimming path of *D. pulex.*

22.3.2.1 *Compass Method* —

Using this procedure, the fractal dimension D_c is estimated by measuring the length L of a path at various scale values δ. The procedure is analogous to moving a set of dividers (like a drawing compass) of fixed length δ along the path (Figure 22.4A). The estimated length of the path is the product of N (number of compass dividers required to "cover" the object) and the scale factor δ. The number of dividers necessary to cover the object then increases with decreasing measurement scale, giving rise to the power law relationship:

$$L(\delta) = k_1 \delta^m \tag{22.1}$$

where δ is the measurement scale, $L(\delta)$ is the measured length of the path, $L(\delta) = N\delta$, and k_1 is a constant. Practically, the fractal dimension D_c is estimated from the slope m of the log–log plot of $L(\delta)$ vs. δ for various values of δ, where

$$D_c = 1 - m \tag{22.2}$$

Hereafter, the fractal dimension D_c will be referred to as the "compass dimension."

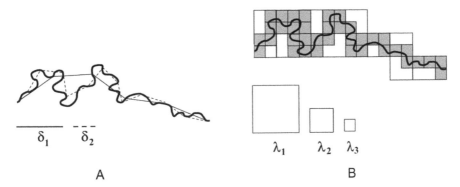

FIGURE 22.4 2D illustration of the compass (A) and the box-counting (B) methods used to describe swimming path complexity with fractal dimension. Two steps of the analyses are shown, using two different characteristic scales: the divider lengths δ_1 and δ_2, and the box sizes λ_1, λ_2, and λ_3.

22.3.2.2 Box-Counting Method — This procedure, like the compass method, can be used to measure the fractal dimension of a curve (Longley and Batty, 1989). In addition, it can be applied to overlapping curves (Peitgen et al., 1992) and structures lacking strict self-similar properties such as vegetation (Morse et al., 1985). Formally, the method finds the "λ cover" of the object, i.e., the number of pixels of length λ (or circles of radius λ) required to cover the object (Voss, 1988). A more practical alternative is to superimpose a regular grid of pixels of length λ on the object and count the number of "occupied" pixels (Figure 22.4B). This procedure is repeated using different values of λ. The volume occupied by a path is then estimated with a series of counting boxes spanning a range of volumes down to some small fraction of the entire volume (Figure 22.4B). The number of occupied boxes increases with decreasing box size, leading to the following power law relationship (Loehle, 1990):

$$N(\lambda) = k_2 \lambda^{-D_b} \tag{22.3}$$

where λ is the box size, $N(\lambda)$ is the number of boxes occupied by the path, k_2 is a constant, and D_b is the box-counting fractal dimension, referred to hereafter as the "box dimension." D_b is estimated from the slope of the linear trend of the log–log plot of $N(\lambda)$ vs. λ.

The fractal dimension of a data set is thus measured by making sure that the data have large scale-independent characteristics.

22.4 Testing the Robustness of Fractal Dimension Estimates

22.4.1 On the Scale Dependence of Fractal Dimensions

Generally, the key assumption regarding the fractal dimension is that it is a scale-independent parameter. Strictly speaking, this means that, in a particular environment, if we calculate D for the swimming behavior of an organism based on paths that are several centimeters long, we will arrive at the same value of D for paths measured at a scale of meters to hundreds of meters. This is central to one of the main issues faced by landscape ecologists; understanding how to meaningfully extrapolate ecological information across spatial scales (Gardner et al., 1989; Turner and Gardner, 1991). This scale-independence issue has been addressed in detail by Turchin (1996) who argued, based on a simulated path of 10,000 steps generated by a correlated random walk, that the fractal dimension, rather than being scale invariant, instead varies in a curvilinear fashion from $D = 1$ at very small spatial scales, to $D = 2$ at very large spatial scales (see also Davenport, Chapter 16, this volume; his Figure 16.1). However, we propose that these two features are simply artifacts of the algorithms used to estimate dimensions and can be explained accordingly. Because of the limited number of data points as the measurement scale approaches the resolution of the data, the path becomes more and more linear, and thus $D \rightarrow 1$. Alternatively, at larger scales most, if not all, of the available boxes have a high probability of including a portion of the path,

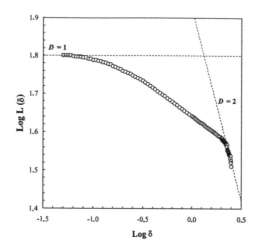

FIGURE 22.5 Illustration of the behavior of $L(\delta)$ vs. δ in a log–log plot, obtained from a simulated path of 10,000 steps resulting from a correlated random walk (thick dashed line), and from a 3D swimming path of *D. pulex*. Correlated random walks have been simulated following Keiyu et al. (1994).

and thus $D \to 2$. Such behavior is also found in a log–log plot of $L(\delta)$ vs. δ obtained from a 3D path of *D. pulex* (Figure 22.5). This path exhibits the same transition from $D = 1$ at small scales to $D = 2$ at larger scales. One must nevertheless note here that the swimming path of *D. pulex* clearly exhibits a linear (i.e., fractal) behavior over intermediate scales, contrary to the Turchin's path generated by a correlated random walk. Moreover, we raised here against Turchin's arguments that while both random walks and correlated random walks have been widely used to model animal behavior (e.g., Johnson and Milne, 1992; Tiselius et al., 1993; Wiens et al., 1993; Gautestad and Mysterud, 1993; Keiyu et al., 1994; Turchin, 1996), they mimic animal paths very poorly (Bergman et al., 2000) because they incorporate an unrealistic arbitrary distribution for the angles between successive steps. A reasonable alternative might be the inclusion of a turning angle distribution, which enables the explicit computation of the effect of persistence in the direction of travel on the expected magnitude of net displacement of the animal over time (e.g., Wu et al., 2000).

On the other hand, because different scales are often associated with different driving processes (e.g., Wiens, 1989; Seuront and Lagadeuc, 1997) the fractal dimension may have the desirable feature of being constant only over a finite, instead of an infinite, range of measurement scales. It is then useful for (1) identifying characteristic scales of variability and (2) comparing movements of organisms that may respond, for example, to the patchy structure of their environment at different absolute scales. Changes in the value of D with scale may indicate that a new set of environmental or behavioral processes is controlling movement behavior (e.g., decreased influence of patch barriers or the effect of home range behavior). Thus, the scale dependence of the fractal dimension over finite ranges of scales may carry more information, both in terms of driving processes and sampling limitation, than its scale independence over a hypothetical infinite range of scales. Alternatively, although the point of slope change may indicate the operational scale of different generative processes, it may simply reflect the limited spatial resolution of the data being analyzed (Hamilton et al., 1992; Kenkel and Walker, 1993; Gautestad and Mysterud, 1994). However, as previously shown from *D. pulex* trajectory, the effect of spatial resolution in the data will manifest as gradual changes of the fractal dimensions toward $D \to 1$ or $D \to 2$, and cannot be confused with a transition zone between two different scaling regions. What is critical for a proper interpretation of fractal dimensions is then a way to identify the range of scales over which fractal dimension is invariant.

22.4.2   Toward an Objective Identification of Scaling Range

When we are dealing with exact fractals (e.g., Koch snowflakes, the Sierpinski carpet; see Schroeder [1991] for further details), there are no difficulties in calculating a fractal dimension. The log–log plots

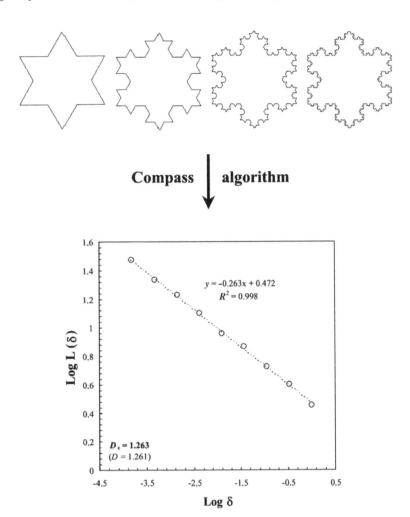

FIGURE 22.6 Illustration of the behavior of $L(\delta)$ vs. δ in a log–log plot (compass method), obtained from the "Koch snowflake." The Koch snowflake is a theoretical fractal object obtained by dividing a given segment into four subsegments three times smaller between two steps of the generation process. The expected fractal dimension is then $D =$ log 4/log 3 = 1.261 (Mandelbrot, 1983), and cannot be regarded as being significantly different from the empirical compass dimension D_c = 1.268 given in parentheses ($p < 0.01$). Note the unambiguous linearity of the behavior of log $L(\delta)$ vs. log δ over the whole range of available scales, when compared with Figure 22.5.

are very linear and we always recover an *a priori* known result (Figure 22.6 and Figure 22.7). Conversely, when we are dealing with objects or observables from nature whose properties are not known *a priori* (e.g., coastlines, swimming paths), complications begin to arise. In such cases, many analyses have implicitly made an assumption of linearity in the log–log plot (Crist et al., 1992; With, 1994a; Erlandson and Kostylev, 1995; Dowling et al., 2000) and as a result, the scaling region was estimated subjectively. However, as stated above, the apparent scaling can be simply the result of the generic property of the quantity to increase or decrease monotonically as the scale goes to zero irrespective of the geometry of the object. Consequently, one must question the validity of fitting a straight line over the whole range of available scales. We therefore propose here an objective, statistically sound procedure for testing the existence of scaling properties in animal paths.

First, consider a regression window of a varying width that ranges from a minimum of five data points (the least number of data points to ensure the statistical relevance of a regression analysis) to the entire data set. The smallest windows are slid along the entire data set at 0.01 cm (half body length of recorded *Daphnia* individuals) increments, with the whole procedure iterated ($n - 4$) times, where n is the total number of available data points. Within each window and for each width, we estimate the coefficient of

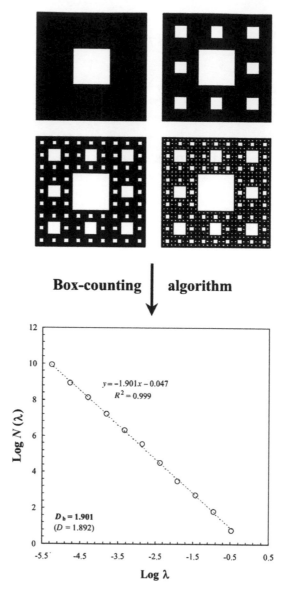

FIGURE 22.7 Illustration of the behavior of $N(\lambda)$ vs. λ in a log–log plot (box-counting method), obtained from the "Sierpinski carpet." The Sierpinski carpet is a theoretical fractal object obtained by dividing a given square into eight subsquares three times smaller between two steps of the generation process. The expected fractal dimension is then $D = \log 8/\log 3 = 1.892$, and cannot be distinguished from the empirical box dimension $D_b = 1.901$ given in parentheses ($p < 0.01$). Note the unambiguous linearity of the behavior of $\log N(\lambda)$ vs. $\log \lambda$ over the whole range of available scales, when compared with Figure 22.5.

determination (r^2) and the sum of the squared residuals (SSR) for the regression. Finally, we use the values of δ (Equation 22.1) and λ (Equation 22.2), which maximize the coefficient of determination and minimize the total sum of the squared residuals (Seuront and Lagadeuc, 1997) to define the scaling range and to estimate the related fractal dimensions (Figure 22.8A). Hereafter, this first optimization procedure will be referred to as the "$R^2 - SSR$" criterion.

Second, one may note that Equations 22.1 and 22.3 can be rewritten, respectively, as

$$d \log L(\delta) / d \log \delta = 1 - D_c \qquad (22.4)$$

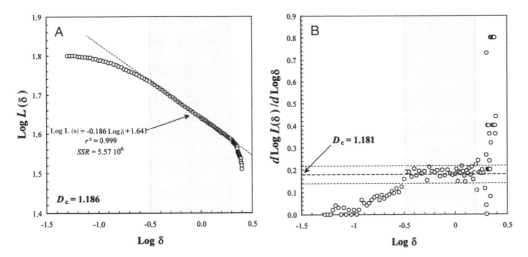

FIGURE 22.8 Illustration of the $R^2 - SSR$ (A) and the zero-slope (B) optimization criteria from a 3D swimming pathway of *D. pulex*. In both cases, the shaded areas indicate the scaling ranges. The dotted lines indicate the best regression fit obtained using the $R^2 - SSR$ criterion (A), and the 5% confidence interval of the flat behavior of $d \log L(\delta)/d \log \delta$ vs. $\log \delta$ (dashed line) found using the zero-slope criterion (B).

and

$$d \log N(\lambda) \, / \, d \log \lambda = -D_b \qquad (22.5)$$

Then if scaling exists, it will manifest as a zero-slope line in plots of both the differentials $d \log L(\delta)/d \log \delta$ vs. $\log \delta$, and $d \log N(\lambda)/d \log \lambda$ vs. $\log \lambda$ (Figure 22.8B). Equations 22.4 and 22.5 can thus be rewritten as:

$$d\big[d \log L(\delta) \, / \, d \log \delta\big] \, / \, d \log \delta = 0 \qquad (22.6)$$

or equivalently $\big[d^2 \log L(\delta) \, / \, d^2 \log \delta\big] = 0$, and

$$d\big[d \log N(\lambda) \, / \, d \log \lambda\big] \, / \, d \log \lambda = 0 \qquad (22.7)$$

or $\big[d^2 \log N(\lambda) \, / \, d^2 \log \lambda\big] = 0$. To ensure the statistical relevance of this procedure, we use a sliding regression window similar to the one described in the $R^2 - SSR$ procedure. The significance of the differences between the slope of each regression and the expected zero-slope line is directly tested using standard statistical analysis; see Zar (1996). The scaling range will then be defined as the scales that satisfy both Equations 22.6 and 22.7, and the $R^2 - SSR$ criterion. Finally the intersection of the range of scales exhibiting a zero-slope line with the *y*-axis provides the compass (see Equation 22.4) and box-counting (see Equation 22.5) dimensions. Hereafter this procedure will be referred to as the "zero-slope" criterion.

Because these procedures may lead to slightly different results in the estimates of the scaling ranges and the related fractal dimensions, we strongly recommend the inclusion in the scaling range of only the scales for which the above two criteria are fully satisfied. In that way, we ensure that the plateau exhibited by the data points in Figure 22.8B is indeed a manifestation of scaling, and not the result of a random nonfractal structure. The implementation of the "$R^2 - SSR$" and "zero-slope" procedures is illustrated using the 3D swimming paths of *D. pulex* (Figure 22.9). It can here be seen that the $R^2 - SSR$ and the zero-slope criteria lead to slight differences in the estimated scaling ranges (Figure 22.9A, B). In particular, the largest limits of the scaling range are systematically larger ($p < 0.05$, Wilcoxon–Mann–Whitney *U*-test) when estimated using the zero-slope optimization criterion (Figure 22.9A). The estimates of their lowest limits (Figure 22.9B) cannot be statistically distinguished ($p > 0.05$). These differences did not imply any significant discrepancies between the related compass and box dimensions (Figure 22.9C). Hereafter, these two optimization criteria will nevertheless be systematically used to estimate both compass

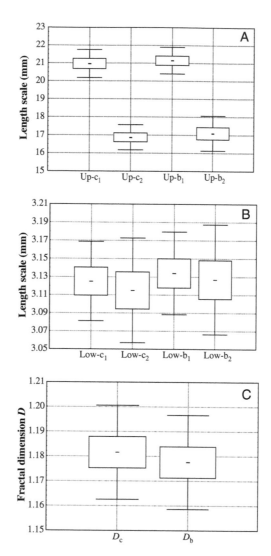

FIGURE 22.9 Upper (A) and lower (B) limits of the scaling ranges, and related compass and box dimensions (C) obtained from 3D *D. pulex* swimming pathways using the $R^2 - SSR$ and the zero-slope optimization criteria. The letters "c" and "b" refer to the compass and box-counting methods, and the subscripts "1" and "2" to the $R^2 - SSR$ and the zero-slope optimization criteria, respectively. The mean $\bar{x}$ is given within the box (–), the box limits represent $\bar{x} \pm SE$, and the outer limits $\bar{x} \pm SD$.

and box dimensions. As an example, the absence of such objective criteria in the choice of their scaling range could explain the spurious fractal dimensions ($D_c < 1$) obtained from the 2D behavior of barramundi fish larvae (Dowling et al., 2001, see their figure 1). Moreover, because it appeared from previous arguments that $D_c \geq 1$ for 2D data, we suggest here that such a result may come from the inclusion of spurious data points in the regression analysis.

22.4.3 Robustness of Fractal Dimension Estimating Algorithms

Before addressing ecological interpretations of fractal dimensions estimates, further potential limitations of fractal analysis, intrinsically related to both the compass and the box-counting methods, must be dealt with. It has been shown that (1) the values $L(\delta) = N\delta$ (i.e., compass method, Equation 22.1) may vary depending on the starting position along the curve (Sugihara and May, 1990), (2) slight reorientations

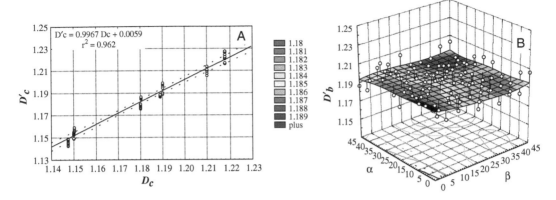

FIGURE 22.10 Distribution of the compass dimensions D'_c obtained from ten subpathways with randomly chosen starting positions and the same length within each original 3D swimming pathway of *D. pulex*, and compared to the compass dimensions D_c obtained from the original nine swimming pathways (A). Box dimensions D'_b obtained from a 3D swimming path of *D. pulex* using different values of the angles α and β controlling the position of the 3D box-counting grid in the $x - y$ and $x - z$ planes, respectively (B).

of the overlying grid can produce different values of $N(\lambda)$ (i.e., box-counting method, Equation 22.3; Appleby, 1996), and (3) the values of box dimensions might be positively correlated to a path's length (Erlandson and Kostylev, 1995). Consequently, the behaviors of Equations 22.1 and 22.3 will be biased, as will the subsequent compass and box dimension estimates.

To address the first two issues, distributions of D_c and D_b (estimated from 3D swimming paths of *D. pulex*) depending on starting point or grid orientation, respectively, are needed. First, we obtained a distribution of the compass dimension D_c by repeatedly starting the compass procedure at different, randomly chosen, positions. The resulting compass dimensions D'_c, estimated from ten random starting positions for each of the nine swimming paths available, do not show significant differences ($p > 0.05$) to the compass dimensions D_c estimated using the first point of the paths as a starting point for the compass algorithm (Figure 22.10A). Second, we obtained a distribution of the box dimension D_b from random replicates of the grid placements in the box-counting algorithm. We rotated the initial 3D orthogonal grid in 5° increments from $\alpha = 0°$ to $\alpha = 45°$ in the $x - y$ plane and from $\beta = 0°$ to $\beta = 45°$ in the $x - z$ plane. The resulting box dimension estimates D'_b fluctuate smoothly around the mean, and did not show any significant tendency related to the variations of α and β (Figure 22.10B; Friedman test, $p > 0.05$; Siegel and Castellan, 1988).

Finally, the limitation of the box-counting method raised by Erlandson and Kostylev (1995), that the values of box dimensions might be positively correlated to a path's length, has been briefly addressed by comparing the box dimensions obtained from our nine swimming paths of different length (see Table 22.2), and within each data set between ten randomly chosen subsets of decreasing length. The resulting box

TABLE 22.2

Duration and Number of Data Points Available from Nine Swimming Paths of *Daphnia pulex* Used to Illustrate Scale-Dependence and Scale-Independence Concepts

Path	N	Duration
1	864	1 min 26 s
2	2413	4 min 01 s
3	1892	3 min 09 s
4	1785	2 min 58 s
5	1733	2 min 53 s
6	2277	3 min 48 s
7	1912	3 min 11 s
8	1460	2 min 25 s
9	1479	2 min 28 s

dimensions D_b, however, did not show any significant differences between the nine available 3D paths, nor between the ten subsets taken within each available 3D path (covariance analysis, F-test, $p > 0.05$).

These different results then demonstrate the robustness of the compass and the box-counting algorithms to quantify the structure of *Daphnia*'s paths in three dimensions. Further, one may also note here that the compass dimension D_c ($D_c = 1.182 \pm 0.018: \bar{x} \pm SD$) and the box dimension D_b ($D_b = 1.178 \pm 0.018: \bar{x} \pm SD$) estimated in the different ways presented in this section cannot be regarded as being significantly different (Wilcoxon–Mann–Whitney U-test, $p > 0.05$). Although these results suggest that these two techniques can be used interchangeably to characterize *Daphnia*'s paths, we cannot claim their universality. On the contrary, we stress the need to address this question very carefully in the framework of any behavioral studies to ensure the reliability of the results.

22.4.4   Two-Dimensional vs. Three-Dimensional Fractal Dimension Estimates

The ability to characterize 3D paths based on 2D projections of these paths is an attractive proposition, as the reduction in complexity of both the data-gathering equipment and the analysis procedures is significant. However, the reliability of conclusions based on such a procedure is not clear. Therefore, we now investigate the consequences of extrapolating fractal dimensions estimated in a 2D framework to three dimensions by first testing the validity of the extrapolation procedures proposed in the literature (Morse et al., 1985; Shorrocks et al., 1991; Gunnarsson, 1992). Second, we investigate the potential for disparity among the fractal dimensions estimated from the three orthogonal 2D projections of a 3D swimming path and, in doing so, demonstrate the necessity of 3D isotropy of a swimming path as a prerequisite for extrapolating 2D fractal information into 3D space.

The philosophy behind the extrapolation of 2D fractal estimates to 3D is as follows: Morse et al. (1985) described a box-counting method for estimating the fractal dimension of vegetation habitats ($2 \leq D_3 \leq 3$, where the subscript 3 indicates a fractal object embedded in a 3D space). Consider now the problem of estimating the fractal dimension of a tree branch. In principle, a 3D grid system could be superimposed on the branch and the size of "counting cubes" varied. Such a procedure is impossible to implement in the field, however, at least given present technical limitations. Morse et al. (1985) simplified the problem by obtaining a 2D photographic image of the habitat, the fractal dimension of which was determined using the box-counting method ($1 \leq D_2 \leq 2$, where the subscript 2 indicates a fractal object embedded in a 2D space). Following Mandelbrot (1983), they determined heuristic lower ($D_{3\,min} = D_2 + 1$) and upper ($D_{3\,max} = 2D_2$) limits of the "habitat" fractal dimension under the assumption that the photograph is a randomly placed orthogonal plane. This procedure has subsequently been used to estimate the fractal dimensions of various habitats (e.g., Shorrocks et al., 1991; Gunnarsson, 1992). However, we stress here, on the basis of both simple theoretical and empirical arguments that the use of this procedure to characterize 3D animal paths is at best questionable and, at worst, meaningless.

First, the limits of the extrapolated 3D fractal dimension D_3 are not constant; instead, they increase with increasing values of the 2D fractal dimension D_2. The disparity between the upper and lower limits range from 4.76 to 31.03% for values of the 2D fractal dimension, D_2, bounded between 1.10 and 1.90, respectively (Table 22.3). Moreover, for values of D_2 greater than 1.5, the upper limit of the extrapolated fractal dimension D_3 is beyond the maximum space-filling limit $D_3 = 3$. Consider now a very complex

TABLE 22.3

Standard Procedure to Extrapolate 2D Fractal Dimensions (D_2) to 3D Fractal Dimensions (D_3)

D_2	$D_2 + 1$	$2D_2$	$1R$	$\%D_3 > 3$	
1.10	2.10	2.20	4.76	—	
1.20	2.20	2.40	9.09	—	
1.30	2.30	2.60	13.04	—	*Note:* $D_2 + 1$ and $2D_2$ are the lower and the upper limits,
1.40	2.40	2.80	16.67	—	respectively, of the extrapolated 3D fractal dimen-
1.50	2.50	3.00	20.00	—	sions; IR is the percentage of increase between the
1.60	2.60	3.20	23.08	33.33	lowest and highest limits of the extrapolated 3D
1.70	2.70	3.40	25.93	57.14	fractal dimensions; and $\%D_3 > 3$ is the percentage of
1.80	2.80	3.60	28.57	75.00	extrapolated 3D fractal dimensions exceeding the
1.90	2.90	3.80	31.03	88.89	space-filling limit $D_3 = d = 3$.

swimming path recorded in a 2D space such that $D_2 \to 2$. In this case, the percentage of D_3 values found beyond the space-filling limit, $D_3 = 3$, unrealistically tends toward 100% (Table 22.3). This can further be illustrated using the compass dimensions estimated from 2D swimming paths of barramundi fish larvae, which drops from 1.8 prior metamorphosis to 1.1 following metamorphosis (Dowling et al., 2000). Extrapolating the fishes' behavior to three dimensions will lead to reasonable values of D_3 for fish before metamorphosis, $D_3 \in [2.10 - 2.20]$. However, for post-metamorphosis fishes, 75% of the D_3 values are beyond the space-filling limit, i.e., $D_3 \in [2.80 - 3.60]$, and cannot therefore be considered legitimate. The validity of this extrapolating procedure is then highly questionable, at least in the framework of behavioral studies. One should nevertheless note here that such extrapolation can only be considered in the case of 2D records of 3D isotropic swimming paths.

Second, investigations of the fractal dimensions estimated from orthogonal 2D projections of 3D swimming paths of *D. pulex* on the $x - y$, $x - z$, and $y - z$ planes illustrate further problems. For example, the fractal dimensions estimated from the $x - y$, $x - z$ and $y - z$ projections of the same 3D path (D_{2xy}, D_{2xz} and D_{2yz}, respectively) are always significantly different (Kruskal–Wallis test, $p < 0.05$). More specifically, the dimensions D_{2xz} and D_{2yz} (side views) cannot be distinguished, and are both significantly higher than the dimension D_{2xy} (top view) (Jonckheere test, $p > 0.05$ and $p < 0.05$, respectively). This suggests that the complexity of the vertical components of the *D. pulex* swimming path is higher than that of its horizontal components, suggesting that the vertical swimming behavior of *D. pulex* is more complex than the horizontal ones. On the other hand, one may note that the average of D_{2xy}, D_{2yz}, and D_{2yz} is not significantly different from D_3 ($p > 0.05$), due to the intrinsic 3D integrative properties of Equations 22.1 and 22.3. Finally, as expected following the results presented in the previous paragraph, the 3D extrapolations of the 2D fractal dimensions D_{2xy}, D_{2yz}, and D_{2yz} are always significantly higher than the actual 3D fractal dimensions. This has been systematically verified for both compass and box dimensions, estimated in two and three dimensions (Table 22.4). Consequently, it appears that a 2D fractal dimension is not sufficient to characterize 3D swimming behavior if the swimming path is not isotropic.

TABLE 22.4

Fractal Dimensions Obtained Using Box-Counting and Compass Algorithms from 3D Pathways (D_3) and Their Three 2D Projections on the $x - y$, $x - z$, and $y - z$ Planes, Shown with the Lowest and Highest Limits of the 3D Dimensions Extrapolated from the 2D Ones

d	D	N	Mean	SD	SE
Box Dimensions D_b					
3	D_3	9	1.178	0.055	0.018
2	D_{2xy}	9	1.114	0.046	0.015
3_e	$D_{2xy} + 1$	9	2.114	1.046	1.015
3_e	$2D_{2xy}$	9	2.228	0.092	0.03
2	D_{2xz}	9	1.214	0.051	0.017
3_e	$D_{xz} + 1$	9	2.214	1.051	1.017
3_e	$2D_{2xz}$	9	2.428	0.102	0.034
2	D_{2yz}	9	1.199	0.067	0.022
3_e	$D_{2yz} + 1$	9	2.199	1.067	1.022
3_e	$2D_{2yz}$	9	2.398	0.134	0.044
Compass Dimensions D_c					
3	D_3	9	1.182	0.057	0.018
2	D_{2xy}	9	1.123	0.048	0.016
3_e	$D_{2xy} + 1$	9	2.123	1.048	1.016
3_e	$2D_{2xy}$	9	2.246	0.096	0.032
2	D_{2xz}	9	1.221	0.052	0.017
3_e	$D_{xz} + 1$	9	2.221	1.052	1.017
3_e	$2D_{2xz}$	9	2.442	0.104	0.035
2	D_{2yz}	9	1.225	0.062	0.020
3_e	$D_{2yz} + 1$	9	2.225	1.062	1.020
3_e	$2D_{2yz}$	9	2.45	0.124	0.041

Note: d is the Euclidean dimension of the considered space; D_3 is the 3D fractal dimension; D_{2xy}, D_{2xz}, and D_{2yz} are the fractal dimensions estimated from the projections of the 3D path on the $x - y$, $x - z$, and $y - z$ planes, respectively; and $D_{2ij} + 1$ and $2D_{2ij}$ are the lowest and highest limits of the 3D fractal dimensions extrapolated from the fractal dimensions D_{2ij}, estimated from paths in the $i - j$ plane.

We also note the seemingly paradoxical result that fractal dimensions estimated from 3D paths are always significantly smaller than 2, the expected lower bound of values for objects embedded in a 3D space. Indeed, following basic fractal theory, an object embedded in a d-dimensional space should have a fractal dimension bounded between $d - 1$ and d (e.g., Mandelbrot, 1983; Schroeder, 1991). In view of this, a linear succession of spaced dust particles will have a dimension bounded between 0 and 1 as they occupy a fraction of the available space greater than a single point (dimension 0), and lower than a line (dimension 1). Similarly, a convoluted curve, a coastline, for example, will occupy a fraction of space between a line (dimension 1) and a surface (dimension 2), while the dimension of a tree will be bounded between 2 (a surface) and 3 (a volume). Now consider again the case of movement paths. The path of an ant foraging on a flat surface occupies a fraction of 2D space. Its dimension is then bounded between 1 (a perfectly linear path) and 2 (a plane-filling path). Similarly, the swimming path of *D. pulex* is obviously embedded in a 3D space, the volume of water (cf. Figure 22.1). However, it does not present a 3D branching structure as does a tree, and each change of direction occurs within a 2D space. Therefore, even in 3D space, a zooplankton swimming path, or the flying path of a foraging bee, will intrinsically remain a convoluted 2D curve. The fractal dimensions of swimming paths are then bounded between a one-dimensional space (i.e., a line, $D = 1$) and a 2D space (i.e., a surface, $D = 2$). The practical consequence of this specific property of swimming paths is to call into question the validity of previous reports of fractal dimensions that fall beyond the $1 \leq D \leq 2$ limits discussed above for both 2D — $D_c < 1$ (Dowling et al., 2000) and $D_c > 2$ (Bascompte and Vilà, 1997) — and 3D ($D_c > 2$; Coughlin et al., 1992) analyses. As suggested above, these discrepancies might result from of the lack of some objective procedures to identify the scaling ranges and the subsequent fractal dimensions of movement paths. Considering this, we note that all of the fractal dimensions estimated from *D. pulex* paths were always consistently significantly higher than 1 (linear movement, $p < 0.01$) and lower than 2 (Brownian motion, $p < 0.01$).

22.5 Comparing Zooplankton Behavior with the Structure of Their Surrounding Environment

In light of the growing awareness of the scaling nature of marine ecosystems, in both their physical and biological aspects (e.g., Pascual et al., 1995; Seuront and Lagadeuc, 1997, 1998, 2001; Seuront et al., 1996a, b, 1999, 2002; Seuront and Schmitt, 2001; Lovejoy et al., 2001), it is becoming increasingly necessary to find a way to compare the composition of zooplankton swimming behaviors in relation to phytoplankton distributions. In particular, considering the remote sensing ability of zooplankton, their behavior could be strongly influenced by the distribution of their phytoplanktonic prey. Ultimately, knowledge of the precise nature of zooplankton swimming behavior could then be a way to infer the spatial distribution of prey. However, due primarily to technological limitations, it is not yet possible to obtain 3D microscale (i.e., scales smaller than 1 m) distributions of phytoplankton cells *in situ*. On the other hand, it is currently possible to obtain prolonged, simultaneous one-dimensional records (i.e., vertical profiles and time series) of physical (shear, temperature, salinity) and biological (*in vivo* fluorescence, backscatter) parameters at scales below 1 m (see, e.g., Wolk et al., 2001). From such records, one may expect a one-dimensional fractal dimension of phytoplankton distribution $D = 0.67$ (Seuront and Lagadeuc, 1997; Seuront et al., 2002). In the present study, we found a 3D fractal dimension $D = 1.18$ (Table 22.4) for *D. pulex* swimming behavior. Unfortunately, a direct comparison of these two dimensions is not possible because they characterize two processes embedded in different dimensions (Roy et al., 1987; Huang and Turcotte, 1989). We nevertheless propose here a more fundamental framework, the fractal codimension, which makes possible comparisons of the structure of processes embedded in different d-dimensional spaces. The fractal codimension c has been defined as (Seuront, 1998; Seuront et al., 1999):

$$c = d - D \tag{22.8}$$

where d is the Euclidean dimension of the embedding space and D the fractal dimension of a given process. The fractal codimension, then, measures the fraction of the dimensional space occupied by the process of interest, and is bounded between $c = 0$ and $c = 1$ for "standard" processes characterized by a fractal dimension D such as $d - 1 \leq D \leq d$. Because the fractal dimension D of swimming paths is intrinsically bounded between $1 \leq D \leq 2$, whatever the value of the embedded dimension d, for more generality we can consider a fractal codimension bounded between $c = 0$ and $c = d$. However, in such a framework, comparisons of codimensions estimated from processes embedded in different d-dimensional space are unfeasible without *a priori* knowledge of the embedding dimension d. The fractal codimension subsequently provides only a relative measure of sparseness. We therefore introduce here the "path codimension" c':

$$c' = c/d \qquad (22.9)$$

as an absolute measure of sparseness. c' is bounded between $c' = 0$ for space-filling processes, and $c' = 1$ for processes so sparse that their fractal dimension is nil, whatever the values of the original embedding dimensions d may be. Returning to the example above, the path codimensions of phytoplankton distribution and *D. pulex* behavior are $c' = 0.33$ and $c' = 0.61$, respectively. Thus, the swimming behavior of *D. pulex* appears to be less complex, or less space filling, than the distribution of its phytoplankton fodder. In particular, this result fully agrees with studies demonstrating the differences in motility between predators and prey (e.g., Tiselius et al., 1993, 1997; Seuront and Lagadeuc, 2001). In general, the path codimension provides a method of comparing the complexity of two interrelated processes, each of which may be embedded in a different dimensional space.

22.6 Conclusion

In this chapter, we have discussed the basic concepts and methods related to the fractal framework, and subsequently attempted to address the major issues related to the applicability of fractal analysis in behavioral ecology. We have tried to clarify some problematic aspects of behavioral fractal analysis, and we propose that the following recommendations are fundamental requirements for improving the robustness of fractal analyses, which in turn ensures that their interpretation will be meaningful:

1. The key component of fractal analysis is not that the fractal dimension D is a scale-independent parameter. Alternatively, we argue that the potential scale dependence of fractal dimensions over finite ranges of scales may contain more information, both in terms of driving processes and sampling limitation, than its scale dependence over a hypothetical infinite range of scales. To ensure the relevance of fractal analysis, the key issue is related to a proper estimation of the scaling range.

2. Considering the lack of objective criteria for testing the existence of scaling properties in animal paths, we present two complementary, easy to implement, robust, and statistically sound procedures to identify scaling properties and estimate fractal dimensions. More generally, we strongly recommend use of a combination of two optimization criteria to identify a scaling range.

3. We address the major objections proposed against the use of fractal analyses in ecology and demonstrate, using a series of simple testing procedures, their robustness in estimating the fractal dimensions of animal paths.

4. In an investigation of the 2D and 3D fractal properties of paths, we emphasize some intrinsic geometric properties of movement paths, and stress the need to ensure their 3D isotropy. This can be done only by comparing the fractal dimension of the three 2D projections of a 3D path.

5. We introduce a new metric, the "path codimension," which can be used to compare the absolute sparseness of related processes that are embedded in different d-dimensional spaces.

However, the main purpose of the chapter has been to confirm the ability of fractal methods to provide both qualitative and quantitative characterizations of animal paths in 2D and 3D environments. Subsequently, considering the ubiquitous geometrical nature of animal movements, we believe that our approach can be generalized to the behavior of all moving organisms. However, we emphasize that the conclusions drawn here, essentially on the robustness of fractal dimension estimates, should not be generalized to other fractal objects such as the structure of landscape heterogeneity or vegetation branching processes without preliminary, careful investigations of the properties of the algorithms involved.

We have illustrated all the concepts discussed above by applying this fractal framework to the study of 3D swimming behavior of the water flea *Daphnia pulex*. The subsequent results, while only at their preliminary stage, nevertheless have generated several salient implications for our understanding of structures and functions in marine ecosystems:

1. Because individual swimming behavior is the underlying mechanism generating population-level behaviors such as horizontal and vertical migration, the horizontal–vertical disparity in *D. pulex* swimming behavior may reflect an adaptive reminiscence of diel vertical migration as a predator avoidance strategy (Loose, 1992; Loose et al., 1992). The difference in fractal behavior identified in the horizontal and the vertical planes may also be suggested as a basis to investigate the predation risk associated with differential swimming behavior related to mating, feeding, or predator avoidance strategies. Fractal evaluation of 2D coordinates of 3D swimming paths is thus particularly useful in this context. Depending on the relative velocities of predator and prey, a swimming behavior characterized by a high fractal dimension may imply a high encounter probability with predators, relative to a more linear swimming path. In addition, a high fractal dimension may also imply a higher encounter probability with prey, depending on the foraging strategy employed (e.g., an ambush or saltatory search predator will have a higher fractal dimension swimming path with an increased level of prey).

2. Individual behavior affects the outcome of predator–prey interactions, especially in the pelagic environment, where prey movement is important both as a cue to predators (Brewer and Coughlin, 1995) and a determinant of encounter rate (Gerritsen and Strickler, 1977). Moreover, the distribution of prey organisms is very important for predators, as recently investigated numerically (Seuront, 2001; Seuront et al., 2001), because food availability changes depending on the fractal dimension. Low fractal dimensions indicate a smooth and predictable distribution of particles gathered in small numbers of patches, whereas high fractal dimensions indicate rough, fragmented, space-filling, and less predictable distributions. Therefore, when a predator can remotely detect its surroundings, prey distributions with low dimension should be more efficient. In contrast, when a predator has no remote detection ability, prey distributions with high dimension should be preferable, because available food quantity or encounter rate becomes proportional to the searched volume as fractal dimension increases. The comparison of the fractal nature of plankton swimming behavior and plankton distributions will then increase our understanding of zooplankton trophodynamics. For example, Johnson et al. (1992b) discussed the interaction between animal movement characteristics and the patch-boundary features in a "microlandscape." They argued that such interactions have important spatial consequences on gene flow, population dynamics, and other ecological processes in the community (see also Wiens et al., 1995). In the ocean, which is increasingly regarded as a "seascape" considering the growing awareness of the heterogeneous nature of microscale processes, behavioral studies would be of prime interest to improve our understanding of the functioning of marine ecosystems from a bottom-up view (Seuront, 2001; Seuront et al., 2001). Although such information is not yet available, we believe that the quickly advancing technology (e.g., Wolk et al., 2002; Franks and Jaffe, 2001) will ensure the achievement of this goal in the near future.

3. Individual feeding rate may be linked to swimming behavior; in most zooplankton, some of the same appendages are used for both behaviors. Considering the actual evidence for prey switching behavior (e.g., Kiørboe et al., 1996; Caparroy et al., 1998), fractal analysis may be suggested as a diagnostic framework to access the kind of prey zooplankton preferentially feed on in a plurispecific prey assemblage. On the other hand, swimming behavior differs among species (Tiselius and Jonsson, 1990) and among development stages within a species (van Duren and Videler, 1995). Attempts at modeling the grazing pressure resulting from both mono- or plurispecific zooplankton assemblages could then benefit from an incorporation of potential differences in swimming path complexity. For example, in a comparison of path tortuosity in

three species of grasshopper, With (1994a) found that the path fractal dimension of the largest species was smaller than those of the two smaller species. She suggested that this reflects the fact that smaller species interact with the habitat at a finer scale of resolution than do larger species. In a second study, With (1994b) found differences in the ways that gomphocerine grasshopper nymphs and adults interacted with the microlandscape. Similarly, the knowledge of the precise nature of the swimming behavior has been suggested as a way to infer the spatial distribution of foragers (Turchin, 1991; Seuront and Lagadeuc, 2001).

4. Toxic chemicals (whether natural, such as cyanobacterial toxins, or anthropogenic, such as pesticides) can have indirect effects on the entire pelagic community via effects on individual swimming behavior (Dodson et al., 1995). The demonstrated sensitivity of fractal analysis may then provide an efficient framework to use the swimming behavior of *Daphnia*, or some other zooplankton organisms, as a "living toxicometer."

5. Finally, we stress here that an important consequence of the fractal nature of zooplankton swimming behavior, illustrated here using 3D *D. pulex* paths, is its deviation from Brownian motion. Fractional Brownian motion models (Mandelbrot, 1983; Schroeder, 1991) have been suggested to characterize the movement of organisms (Frontier, 1987). However, Wiens and Milne (1989), examining beetle movements in natural fractal landscapes, found that observed beetle movements deviated from the modeled (fractional Brownian) ones. A follow-up study by Johnson et al. (1992a) found that beetle movements reflect a combination of ordinary (random) and anomalous diffusions. The latter may simply reflect intrinsic departures from randomness, or result from barrier avoidance and utilization of corridors in natural landscapes. An extensive discussion of the anomalous diffusion of a copepod in a heterogeneous environment can be found elsewhere (Marguerit et al., 1998; Schmitt and Seuront, 2001). Future modeling attempts of zooplankton swimming behavior may have to take into account the nonrandomness (i.e., fractal) of organisms' movements, and the persistence of the direction of travel (cf. Figure 22.3A), as recently suggested by Wu et al. (2000) and Schmitt and Seuront (2001).

This chapter has highlighted areas concerning the valid applications of fractal analysis so that it can be used to faithfully represent the ecological effect of plankton behavior as is found in aquatic systems. This precision should be a fundamental requirement for integrated behavioral components in plankton models (Seuront, 2001; Seuront et al., 2001; Ginot et al., 2002; Yamazaki and Kamykowski, Chapter 35, this volume; Souissi and Bernard, Chapter 23, this volume) in order for their results to be ecologically relevant. Whatever the case, the understanding of zooplankton ecology from the behavioral bottom-up approach is still in its infancy.

References

Abraham, E.R., 2001. The fractal branching of an arborescent sponge. *Mar. Biol.*, 138, 503–510.

Appleby, S., 1996. Multifractal characterization of the distribution pattern of the human population. *Geogr. Anal.*, 28, 147–160.

Bascompte, J. and Vilà, C., 1997. Fractals and search paths in mammals. *Land. Ecol.*, 12, 213–221.

Batschelet, E., 1981. *Circular Statistics in Biology.* Academic Press, New York.

Bell, W.J., 1991. *Searching Behavior, the Behavioral Ecology of Finding Resources.* Chapman & Hall, New York.

Bergman, C.M., Schaefer, J.A., and Luttich, S.N., 2000. Caribou movement as a correlated random walk. *Oecologia*, 123, 364–374.

Boinski, S. and Garber, P.A., 2000. *On the Move. How and Why Animals Travel in Groups.* University of Chicago Press, Chicago.

Bradbury, R.H. and Reichelt, R.E., 1983. Fractal dimension of a coral reef at ecological scales. *Mar. Ecol. Prog. Ser.*, 10, 169–171.

Bradbury, R.H., Reichelt, R.E., and Green, D.G., 1984. Fractals in ecology: methods and interpretation. *Mar. Ecol. Prog. Ser.*, 14, 295–296.

Brewer, M.C., 1996. *Daphnia* Swimming Behavior and Its Role in Predator–Prey Interactions. Ph.D. thesis, University of Wisconsin–Milwaukee, 155 pp.

Brewer, M.C., 1998. Mating behaviours of *Daphnia pulicaria*, a cyclic parthenogen: comparisons with copepods. *Philos. Trans. R. Soc. Lond. B*, 353, 805–815.

Brewer, M.C. and Coughlin, J.N., 1995. Virtual plankton: a novel approach to the investigation of aquatic predator–prey interactions. *Mar. Freshw. Behav.*, 26, 91–100.

Buchanan, C., Goldberg, B., and McCartney, R., 1982. A laboratory method for studying zooplankton swimming behaviors. *Hydrobiologia*, 94, 77–89.

Bundy, M.H., Gross, T.F., Coughlin, D.J., and Strickler, J.R., 1993. Quantifying copepod searching efficiency using swimming pattern and perceptive ability. *Bull. Mar. Sci.*, 53, 15–28.

Burlando, B., 1990. The fractal dimension of taxonomic systems. *J. Theor. Biol.*, 146, 99–114.

Burlando, B., 1993. The fractal dimension geometry of evolution. *J. Theor. Biol.*, 163, 161–172.

Buskey, E.J., 1984. Swimming pattern as an indicator of the roles of copepod sensory systems in the recognition of food. *Mar. Biol.*, 79, 165–175.

Buskey, E.J., 1997. Behavioral components of feeding selectivity of the heterotrophic dinoflagellate *Prorocentrum pellucidum*. *Mar. Ecol. Prog. Ser.*, 153, 776–789.

Buskey, E.J. and Stoecker, D.K., 1988. Locomotory patterns of the planktonic ciliate *Favella* sp.: adaptations for remaining within food patches. *Bull. Mar. Sci.*, 43, 783–796.

Buskey, E.J., Mills, L., and Swift, E., 1983. The effects of dinoflagellate bioluminscence on the swimming behavior of a marine copepod. *Limnol. Oceanogr.*, 28, 575–579.

Buskey, E.J., Coulter, C., and Strom, S., 1993. Locomotory patterns of microzooplankton: potential effects on food selectivity of larval fish. *Bull. Mar. Sci.*, 53, 29–43.

Buskey, E.J., Peterson, J.O., and Ambler, J.W., 1996. The swarming behavior of the copepod *Dioithona oculata*: *in situ* and laboratory studies. *Limnol. Oceanogr.*, 41, 513–521.

Caparroy, P., Pérez, M.T., and Carlotti, F., 1998. Feeding behaviour of *Centropages typicus* in calm and turbulent conditions. *Mar. Ecol. Prog. Ser.*, 168, 109–118.

Charoy, C.P., Janssen, C.R., Persoone, G., and Clément, P., 1995. The swimming behaviour of *Brachiomus calyciflorus* (rotifer) under toxic stress. I. The use of automated trajectometry for determining sub-lethal effects of chemicals. *Aquat. Toxicol.*, 32, 271–282.

Cody, M.L., 1971. Finch flocks in the Mohave desert. *Theor. Popul. Biol.*, 163, 161–172.

Costello, J.H., Strickler, J.R., Marrasé, C., Trager, G., Zeller, R., and Freise, A.J., 1990. Grazing in a turbulent environment: behavioral response of a calanoid copepod, *Centropages hamatus*. *Proc. Natl. Acad. Sci. U.S.A.*, 87, 1648–1652.

Coughlin, D.J., Strickler, J.R., and Sanderson, B., 1992. Swimming and search behaviour in clownfish, *Amphiprion perideraion*, larvae. *Anim. Behav.*, 44, 427–440.

Cowles, T.J., Desiderio, R.A., and Carr, M.E., 1998. Small-scale planktonic structure: persistence and trophic consequences. *Oceanography*, 11, 4–9.

Crist, T.O., Guertin, D.S., Wiens, J.A., and Milne, B.T., 1992. Animal movement in heterogeneous landscapes: an experiment with *Elodes* beetles in shortgrass prairie. *Functional Ecol.*, 6, 536–544.

Dodson, S. and Ramcharan, C., 1991. Size specific swimming behavior of *Daphnia pulex*. *J. Plankton Res.*, 13, 1367–1379.

Dodson, S.I., Hanazato, T., and Gorski, P.R., 1995. Behavioral responses of *Daphnia pulex* exposed to carbaryl and *Chaoborus* kairomone. *Environ. Toxicol. Chem.*, 14, 43–50.

Dodson, S.I., Ryan, S., Tollrian, R., and Lampert, W., 1997. Individual swimming behavior of *Daphnia*: effects of food, light and container size in four clones. *J. Plankton Res.*, 19, 1537–1552.

Dowling, N.A., Hall, S.J., and Mitchell, J.G., 2001. Foraging kinematics of barramundi during early stages of development. *J. Fish Biol.*, 57, 337–353.

Erlandson, J. and Kostylev, V., 1995. Trail following, speed and fractal dimension of movement in a marine prosobranch, *Littorina littorea*, during a mating and a nonmating season. *Mar. Biol.*, 122, 87–94.

Fielding, A., 1992. Applications of fractal geometry to biology. *Comput. Appl. Biosci.*, 8, 359–366.

Fisher, R., Bellwood, D.R., and Job, S.D., 2000. Development of swimming abilities in reef fish larvae. *Mar. Ecol. Prog. Ser.*, 202, 163–173.

Franks, P.J.S. and Jaffe, J.S., 2001. Microscale distributions of phytoplankton: initial results from a two-dimensional imaging fluorometer, OSST. *Mar. Ecol. Prog. Ser.*, 220, 59–72.

Frontier, S., 1985. Diversity and structure in aquatic ecosystems. *Oceanogr. Mar. Biol. Annu. Rev.*, 23, 253–312.

Frontier, S., 1987. Applications of fractal theory to ecology, in Legendre, P. and Legendre, L., Eds., *Developments in Numerical Ecology*. Springer, Berlin, 335–378.

Frontier, S., 1994. Species diversity as a fractal property of biomass, in Novak, M.M., Ed., *Fractals in the Natural and Applied Sciences*. Elsevier, New York, 119–127.

Gardner, R.H., O'Neill, R.V., Turner, M.G., and Dale, V.H., 1989. Quantifying scale-dependent effects of animal movement with simple percolation models. *Land. Ecol.*, 3, 217–227.

Gautestad, A.O. and Mysterud, I., 1993. Physical and biological mechanisms in animal movement processes. *J. Appl. Ecol.*, 30, 523–535.

Gee, J.M. and Warwick, R.M., 1994a. Body-size distribution in a marine metazoan community and the fractal dimensions of macroalgae. *J. Exp. Mar. Biol. Ecol.*, 178, 247–259.

Gee, J.M. and Warwick, R.M., 1994b. Metazoan community structure in relation to the fractal dimensions of marine macroalgae. *Mar. Ecol. Prog. Ser.*, 103, 141–150.

Gerritsen, J. and Strickler, J.R., 1977. Encounter probabilities and community structure in zooplankton: a mathematical model. *J. Fish. Res. Board Can.*, 34, 73–82.

Ginot, V., Le Page, C., and Souissi, S., 2002. A multi-agent architecture to enhance end-user individual based modelling. *Ecol. Model.* 157, 23–41.

Goldberger, A.L., Bhargava, V., Wesr, B.J., and Mandell, A.J., 1985. On a mechanism of cardiac electrical stability. The fractal hypothesis. *Biophys. J.*, 48, 525–528.

Gries, T., Jöhnk, K., Fields, D., and Strickler, J.R., 1999. Size and structure of "footprints" produced by *Daphnia*: impact of animal size and density gradients. *J. Plankton Res.*, 21, 509–523.

Gunnarsson, B., 1992. Fractal dimension of plants and body size distribution in spiders. *Funct. Ecol.*, 6, 636–641.

Hamilton, S.K., Melack, J.M., Goodchild, M.F., and Lewis, W.M., 1992. Estimation of the fractal dimension of terrain from lake size distributions, in Carling, P.A. and Petts, G.E., Eds., *Lowland Floodplain Rivers: Geomorphological Perspectives*. John Wiley & Sons, New York, 145–163.

Hastings, H.M. and Sugihara, G., 1993. *Fractals. A User's Guide for the Natural Sciences*. Oxford University Press, Oxford.

Huang, J. and Turcotte, D.L., 1989. Fractal mapping of digitized images: application to the topography of Arizona and comparisons with synthetic images. *J. Geophys. Res.*, 94, 7491–7495.

Hwang, J.S. and Strickler, J.R., 1994. Effects of periodic turbulent events upon escape responses of a calanoid copepod, *Centropages hamatus*. *Bull. Plankton Soc. Jpn.*, 41, 117–130.

Hwang, J.S., Costello, J.H., and Strickler, J.R., 1994. Copepod grazing in turbulent flow: elevated foraging behavior and habituation to escape responses. *J. Plankton Res.*, 16, 421–431.

Johnson, A.R. and Milne, B.T., 1992. Diffusion in fractal landscapes: simulations and experimental studies of tenebrionid beetle movements. *Ecology*, 73, 1968–1983.

Johnson, A.R., Milne, B.T., and Wiens, J.A., 1992a. Diffusion in fractal landscapes: simulations and experimental studies of tenebrionid beetle movements. *Ecology*, 73, 1968–1983

Johnson, A.R., Wiens, J.A., Milne, B.T., and Crist, T.O., 1992b. Animal movements and population dynamics in heterogeneous landscapes. *Land. Ecol.*, 7, 63–75.

Jonsson, P.R. and Johansson, M., 1997. Swimming behaviour, patch exploitation and dispersal capacity of a marine benthic ciliate in flume flow. *J. Exp. Mar. Biol. Ecol.*, 215, 135–153.

Kareiva, P.M. and Shigesada, N., 1983. Analyzing insect movement as a correlated random walk. *Oecologia*, 56, 234–238.

Keiyu, A.Y., Yamazaki, H., and Strickler, R., 1994. A new modelling approach for zooplankton behaviour. *Deep-Sea Res. I*, 41, 171–184.

Kenkel, N.C. and Walker, D.J., 1993. Fractals and ecology. *Abstr. Bot.*, 17, 53–70.

Kiørboe, T., Saiz, E., and Viitasalo, M., 1996. Prey switching behaviour in the planktonic copepod *Acartia tonsa*. *Mar. Ecol. Prog. Ser.*, 143, 65–75.

Lampert, W., 1987. Feeding and nutrition in *Daphnia*, in Peters, R.H. and De Bernardi, R., Eds., *Daphnia*. *Mem. Del. Inst. Ital. Di Idrobiol.*, 143–192.

Larson, P. and Kleiven, O.T., 1996. Food search and swimming speed in *Daphnia*, in Lenz, P.H., Hartline, D.K., Purcell, J.E., and Macmillan, D.L., Eds., *Zooplankton: Sensory Ecology and Physiology*. Gordon & Breach, Amsterdam, 375–387.

Leising, A.W., 2001. Copepod foraging in patchy habitats and thin layers using a 2-D individual based model. *Mar. Ecol. Prog. Ser.*, 216, 167–179.

Leising, A.W. and Franks, P.J.S., 2000. Copepod vertical distribution within a spatially variable food source: a simple foraging-strategy model. *J. Plankton Res.*, 22, 999–1024.

Li, X., Passow, U., and Logan, B.E., 1998. Fractal dimension of small (15–200 μm) particles in Eastern Pacific coastal waters. *Deep-Sea Res. I*, 45, 115–132.

Loehle, C., 1983. The fractal dimension and ecology. *Specul. Sci. Tech.*, 6, 131–142.

Loehle, C., 1990. Home range: a fractal approach. *Land. Ecol.*, 5, 39–52.

Longley, P.A. and Batty, M., 1989. On the fractal measurement of geographical boundaries. *Geogr. Anal.*, 21, 47–67.

Loose, C.J., 1992. Daphnia diel vertical migration behavior: response to vertebrate predator abundance. *Arch. Hydrobiol. Beih.*, 39, 29–36.

Loose, C.J., Von Elert, E., and Dawidowicz, P., 1992. Chemically-induced diel vertical migration in *Daphnia*: a new bioassay for kairomones exuded by fish. *Arch. Hydrobiol.*, 126, 329–337.

Lovejoy, S., Currie, W.J.S., Tessier, Y., Claereboudt, M.R., Bourget, E., Roff, J.C., and Schertzer, D., 2001. Universal multifractals and ocean patchiness: phytoplankton, physical fields and coastal heterogeneity. *J. Plankton Res.*, 23, 117–141.

Mach, J., Mas, F., and Sagués, F., 1994. Laplacian multifractality of the growth probability distribution in electrodeposition. *Europhys. Lett.*, 25, 217–276.

Maly, E.J., van Leeuwen, H.C., and Blais, J., 1994. Some aspects of size and swimming behavior in two species of *Diaptomus* (Copepoda: Calanoida). *Verh. Int. Verein. Limnol.*, 25, 2432–2435.

Mandelbrot, B., 1977. *Fractals. Form, Chance and Dimension*. W.H. Freeman, London.

Mandelbrot, B., 1983. *The Fractal Geometry of Nature*. W.H. Freeman, New York.

Marguerit, C., Schertzer, D., Schmitt, F., and Lovejoy, S., 1998. Copepod diffusion within multifractal phytoplankton fields. *J. Mar. Syst.*, 16, 69–83.

Marrasé, C., Coastello, J.H., Granata, T., and Strickler, J.R., 1990. Grazing in a turbulent environment: behavioral response of a calanoid copepod, *Centropages hamatus*. *Proc. Natl. Acad. Sci. U.S.A.*, 87, 1652–1657.

McCulloch, C.E. and Cain, M.L., 1989. Analyzing discrete movement data as a correlated random walk. *Ecology*, 70, 383–388.

Mitchell, J.G. and Furhman, J.A., 1989. Centimeter scale vertical heterogeneity and chlorophyll *a*. *Mar. Ecol. Prog. Ser.*, 54, 141–148.

Morse, D.H., 1980. *Behavioral Mechanisms in Ecology*. Harvard University Press, Cambridge, MA.

Morse, D.R., Lawton, J.H., Dodson, M.M., and Williamson, M.H., 1985. Fractal dimension of vegetation and the distribution of arthropod body lengths. *Nature*, 314, 731–733.

O'Keefe, T.C., Brewer, M.C., and Dodson, S.I., 1998. Swimming behavior of *Daphnia*: its role in determining predation risk. *J. Plankton Res.*, 20, 973–984.

Paffenhöffer, G.A., Strickler, J.R., Lewis, K.D., and Richman, S., 1996. Motion behavior of nauplii and early copepodid stages of marine planktonic copepods. *J. Plankton Res.*, 18, 1699–1715.

Pascual, M., Ascioti, F.A., and Caswell, H., 1995. Intermittency in the plankton: a multifractal analysis of zooplankton biomass variability. *J. Plankton Res.*, 17, 1209–1232.

Peitgen, H.O., Jürgensand, H., and Saupe, D., 1992. *Fractals for the Classroom*. Springer, New York.

Pietronero, L. and Labini, F.S., 2000. Fractal universe. *Physica A*, 280, 125–130.

Porter, K.G., Gerritsen, J., and Orcutt, J.D., 1982. The effect of food concentration on swimming patterns, feeding behavior, ingestion, assimilation and respiration in *Daphnia*. *Limnol. Oceanogr.*, 27, 935–949.

Price, H.J., 1989. Feeding mechanisms of krill in response to algal patches: a mesocosm study. *Limnol. Oceanogr.*, 34, 649–659.

Provata, A. and Almirantis, Y., 2000. Fractal cantor patterns in the sequence structure of DNA. *Fractals*, 8, 15–27.

Pyke, G.H., 1981. Optimal foraging in hummingbirds: rule of movement between inflorescences. *Anim. Behav.*, 29, 882–896.

Pyke, J.H., 1984. Optimal foraging theory: a critical review. *Annu. Rev. Ecol. Syst.*, 15, 523–575.

Ramcharan, C.W. and Sprules, W.G., 1991. Predator-induced behavioral defense and its ecological consequences for two calanoid copepods. *Oecologia*, 86, 276–286.

Ringelberg, J., 1964. The positively photoactic reaction of *Daphnia magna* Strauss: a contribution to the understanding of diurnal vertical migration. *Neth. J. Sea Res.*, 2, 319–406.

Roy, A.G., Gravel, G., and Gauthier, C., 1987. Measuring the dimension of surfaces: a review and appraisal of different methods, in *Proceedings, 8th International Symposium on Computer-Assisted Cartography (Auto-Carto 8)*, Baltimore, MD, 68–77.

Saiz, E., 1994. Observations of the free-swimming behavior of *Acartia tonsa*: effects of food concentration and turbulent water motion. *Limnol. Oceanogr.*, 39, 1566–1578.

Saiz, E. and Alcaraz, M., 1992. Free-swimming behaviour of *Acartia clausi* (Copepoda: Calanoida) under turbulence water movement. *Mar. Ecol. Prog. Ser.*, 80, 229–236.

Schmitt, F. and Seuront, L., 2001. Multifractal random walk in copepod behavior. *Physica A*, 301, 375–396.

Schmitt, F. and Seuront, L., 2002. Multifractal anormal diffusion in swimming behavior of marine organisms, in Pomeau, Y. and Ribotta, A., Eds., *Proc. 5th Rencontre du Non-linéaire*, Paris, Institut Poincarré. Non-linéaire publications, Paris, 237–242.

Schroeder, M., 1991. *Fractals, Chaos, Power Laws*. W.H. Freeman, New York.

Seuront, L., 1998. Fractals and multifractals: new tools to characterize space–time heterogeneity in marine ecology. *Océanis*, 24, 123–158.

Seuront, L., 2001. Microscale processes in the ocean: why are they so important for ecosystem functioning? *La Mer*, 39, 1–8.

Seuront, L. and Lagadeuc, Y., 1997. Characterisation of space–time variability in stratified and mixed coastal waters (Baie des Chaleurs, Quebec, Canada): application of fractal theory. *Mar. Ecol. Prog. Ser.*, 159, 81–95.

Seuront, L. and Lagadeuc, Y., 1998. Spatio-temporal structure of tidally mixed coastal waters: variability and heterogeneity. *J. Plankton Res.*, 20, 1387–1401.

Seuront, L. and Lagadeuc, Y., 2001. Multiscale patchiness of the calanoid copepod *Temora longicornis* in a turbulent coastal sea. *J. Plankton Res.*, 23, 1137–1145.

Seuront, L. and Schmitt, F., 2001. Describing intermittent processes in the ocean: univariate and bivariate multiscaling procedures, in Müller, P. and Garrett, C., Eds., *Stirring and Mixing in a Stratified Ocean*. SOEST, University of Hawaii, 12, 131–144.

Seuront, L., Schmitt, F., Schertzer, D., Lagadeuc, Y., and Lovejoy, S., 1996a. Multifractal intermittency of Eulerian and Lagrangian turbulence of ocean temperature and plankton fields. *Nonlinear Proc. Geophys.*, 3, 236–246.

Seuront, L., Schmitt, F., Lagadeuc, Y., Schertzer, D., Lovejoy, S., and Frontier, S., 1996b. Multifractal analysis of phytoplankton biomass and temperature in the ocean. *Geophys. Res. Lett.*, 23, 3591–3594.

Seuront, L., Schmitt, F., Lagadeuc, Y., Schertzer, D., and Lovejoy, S., 1999. Universal multifractal analysis as a tool to characterize multiscale intermittent patterns: example of phytoplankton distribution in turbulent coastal waters. *J. Plankton Res.*, 21, 877–922.

Seuront, L., Schmitt, F., and Lagadeuc, Y., 2001. Turbulence intermittency, phytoplankton patchiness and encounter rates in plankton: where do we go from here? *Deep-Sea Res. I*, 48, 1199–1215.

Seuront, L., Gentilhomme, V., and Lagadeuc, Y., 2002. Small-scale nutrient patches in tidally mixed coastal waters. *Mar. Ecol. Prog. Ser.* 232, 29–44.

Shlesinger, M.F. and West, B.J., 1991. Complex fractal dimension of the bronchial tree. *Phys. Rev. Lett.*, 67, 2106–2108.

Shorrocks, B., Marsters, J., Ward, I., and Evennett, P.J., 1991. The fractal dimension of lichens and the distribution of arthropod body lengths. *Funct. Ecol.*, 5, 457–460

Siegel, S. and Castellan, N.J., 1988. *Nonparametric Statistics*. McGraw-Hill, New York.

Siniff, D.B. and Jenssen, C.R., 1969. A simulation model of animal movement patterns. *Adv. Ecol. Res.*, 6, 185–219.

Squires, K.D. and Yamazaki, H., 1995. Preferential concentration of marine particle in isotropic turbulence. *Deep-Sea Res.*, 42, 1989–2004.

Stevens, D.W. and Krebs, J.R., 1986. *Foraging Theory*. Princeton University Press, Princeton, NJ.

Strickler, J.R., 1970. Über das Schimmverhalten von Cyclopoiden bei Verminderungen des Bestrahlungsstarke. *Schweiz. Z. Hydrobiol.*, 32, 150–180.

Strickler, J.R., 1985. Feeding currents in calanoid copepods: two new hypotheses. *Symp. Soc. Exp. Biol.*, 39, 459–485.

Strickler, J.R., 1992. Calanoid copepods, feeding currents, and the role of gravity. *Science*, 218, 158–160.

Strickler, J.R., 1998. Observing free-swimming copepods mating. *Philos. Trans. R. Soc. Lond. B*, 353, 671–680.

Suchman, C.L., 2000. Escape behavior of *Acartia hudsonica* copepods during interactions with scyphomedusae. *J. Plankton Res.*, 22, 2307–2323.

Sugihara, G. and May, R.M., 1990. Applications of fractals in ecology. *TREE*, 5, 79–86.

Tang, S., Ma, Y., and Sebastine, I.M., 2001. The fractal nature of *Escherichia coli* biological flocs. *Colloids Surf. B Interfaces*, 20, 211–218.

Tarafdar, S., Franz, A., Schulzky, C., and Hoffman, K.H., 2001. Modelling porous structures by repeated Sierpinski carpets. *Physica A*, 292, 1–8.

Tiselius, P., 1992. Behavior of *Acartia tonsa* in patchy food environments. *Limnol. Oceanogr.*, 37, 1640–1651.

Tiselius, P. and Jonsson, P.R., 1990. Foraging behaviour of six calanoid copepods: observations and hydrodynamic analysis. *Mar. Ecol. Prog. Ser.*, 66, 23–33.

Tiselius, P., Jonsson, P.R., and Verity, P.G., 1993. A model evaluation of the impact of food patchiness on foraging strategy and predation risk in zooplankton. *Bull. Mar. Sci.*, 53, 247–264.

Tiselius, P., Jonsson, P.R., Karrtvedt, S., Olsen, E.M., and Jordstad, T., 1997. Effects of copepod foraging behavior on predation risk: an experimental study of the predatory copepod *Pareuchaeta norvegica* feeding on *Acartia clausi* and *A. tonsa* (Copepoda). *Limnol. Oceanogr.*, 42, 164–170.

Titelman, J., 2001. Swimming and escape behavior of copepod nauplii: implications for predator–prey interactions among copepods. *Mar. Ecol. Prog. Ser.*, 213, 203–213.

Tsonis, A.A. and Elsner, J.B., 1995. Testing for scaling in natural forms and observables. *J. Stat. Phys.*, 81, 869–880.

Turchin, P., 1991. Translating foraging movements in heterogeneous environments into the spatial distribution of foragers. *Ecology*, 72, 1253–1266.

Turchin, P., 1996. Fractal analysis of movement: a critique. *Ecology*, 77, 2086–2090.

Turchin, P., 1998. *Quantitative Analysis of Movement*. Sinauer Associates, Sunderland, MA.

Turchin, P., Odendaal, F.J., and Rausher, M.D., 1991. Quantifying insect movement in the field. *Environ. Entomol.*, 20, 955–963.

Turner, M.G. and Gardner, R.H., 1991. Quantitative methods in landscape ecology: an introduction, in Turner, M.G. and Gardner, R.H., Eds., *Quantitative Methods in Landscape Ecology*. Springer-Verlag, New York, 3–14.

Van Ballemberghe, V., 1983. Extraterritorial movements and dispersal of wolves in southcentral Alaska. *J. Mamm.*, 64, 168–171.

van Duren, L.A. and Videler, J.J., 1995. Swimming behaviour of development stages of the calanoid copepod *Temora longicornis* at different food concentrations. *Mar. Ecol. Prog. Ser.*, 126, 153–161.

van Duren, L.A. and Videler, J.J., 1996. The trade-off between feeding mate seeking and predator–prey avoidance in copepods: behavioural responses to chemical cues. *J. Plankton Res.*, 18, 805–818.

Van Leeuwen, H.C. and Maly, E.J., 1991. Changes in swimming behavior of male *Diaptomus leptotus* (Copepoda: Calanoida) in response to gravid females. *Limnol. Oceanogr.*, 36, 1188–1195.

Voss, R.F., 1988. Fractals in nature: from characterization to simulation, in Peitgen, H.O. and Saupe, D., Eds., *The Science of Fractal Images*. Springer, New York, 21–70.

Weissburg, M.J., Doall, M.H., and Yen, J., 1998. Following the invisible trail: kinematic analysis of mate tracking in the copepod *Temora longicornis*. *Philos. Trans. R. Soc. Long. B*, 353, 701–712.

Wiens, J.A., 1989. Spatial scaling in ecology. *Funct. Ecol.*, 3, 385–397.

Wiens, J.A. and Milne, B.T., 1989. Scaling of "landscapes" in landscape ecology, or landscape ecology from a beetle's perspective. *Land. Ecol.*, 3, 87–96.

Wiens, J.A., Crist, T.O., and Milne, B.T., 1993. On quantifying insect movements. *Environ. Entomol.*, 22, 709–715.

Wiens, J.A., Crist, T.O., With, K.A., and Milne, B.T., 1995. Fractal patterns of insect movement in microlandscape mosaics. *Ecology*, 76, 663–666.

Williamson, C.E., 1981. Foraging behavior of a freshwater copepod: frequency changes in looping behavior at high and low prey densities. *Oecologia*, 50, 332–336.

Wilson, R.S. and Greaves, J.O.B., 1979. The evolution of the bugsystem: recent progress in the analysis of bio-behavioral data, in Jacoff, F.S., Ed., Advances in Marine Environmental Research. Proc. Symp. U.S. EPA (EPA-600/9-79-035), 252–272.

With, K.A., 1994a. Using fractal analysis to assess how species perceive landscape structure. *Land. Ecol.*, 9, 25–36

With, K.A., 1994b. Ontogenetic shifts in how grasshoppers interact with landscape structure: an analysis of movement patterns. *Funct. Ecol.*, 8, 477–485.

Wolk, F., Yamazaki, H., Seuront, L., and Lueck, R.G., 2002. A new free-fall profiler for measuring biophysical microstructure. *J. Atmos. Ocean. Tech*, 19, 780–793.

Wong, C.K., Ramcharan, C.W., and Sprules, W.G., 1986. Behavioral responses of a herbivorous copepod to the presence of other zooplankton. *Can. J. Zool.*, 64, 1422–1425.

Wu, K.K.S., Lahav, O., and Rees, M.J., 1999. The large-scale smoothness of the Universe. *Nature*, 397, 225–230.

Wu, H., Li, B.L., Springer, T.A., and Neill, W.H., 2000. Modelling animal movement as a persistent random walk in two dimensions: expected magnitude of the net displacement. *Ecol. Model.*, 132, 115–124.

Yamazaki, H. and Squires, K.D., 1996. Comparison of oceanic turbulence and copepod swimming. *Mar. Ecol. Prog. Ser.*, 144, 299–301.

Yen, J. and Strickler, J.R., 1996. Advertisement and concealment in the plankton: what makes a copepod hydrodynamically conspicuous? *Invert. Biol.*, 115, 191–205.

Yen, J., Weissburg, M.J., and Doall, M.H., 1998. The fluid physics of signal perception by a mate-tracking copepod. *Philos. Trans. R. Soc. Lond. B*, 353, 787–804.

Young, S. and Getty, C., 1987. Visually guided feeding behavior in the filter feeding cladoceran, *Daphnia magna. Anim. Behav.*, 35, 541–548.

Young, S. and Taylor, V.A., 1990. Swimming tracks in swarms of two cladoceran species. *Anim. Behav.*, 39, 10–16.

Zar, J.H., 1996. *Biostatistical Analysis.* Prentice-Hall International, Englewood Cliffs, NJ.

23

Identification of Interactions in Copepod Populations Using a Qualitative Study of Stage-Structured Population Models

Sami Souissi and Olivier Bernard

CONTENTS

23.1 Introduction 361
23.2 Method 363
    23.2.1 Transient Behavior of a Stage-Structured Population Model 363
        23.2.1.1 Basic Structural Transition Rules 363
        23.2.1.2 Structural Transition Rules for Predation 364
    23.2.2 Identification of Extrema in the Time-Series Abundance Data 365
    23.2.3 Age and Stage-Structured Simulation Model 366
23.3 Results 367
    23.3.1 Analysis of Transition Rules with Simulated Data 367
    23.3.2 Identification of Predation Interactions in Copepod Development Experiments 368
        23.3.2.1 Cannibalism in the Development of *A. clausi* 368
        23.3.2.2 Predation of *A. clausi* Adults on the First Naupliar Stages of
                E. acutifrons 371
    23.3.3 Identification of *in Situ* Interactions between Copepods in a Eutrophic Inlet
        of the Inland Sea of Japan 372
23.4 Discussion 374
Appendix 23.A: Sketch of the Mathematical Proof of the Basic Structural Transition Rules 375
Appendix 23.B: Sketch of the Mathematical Proof of the Transition Rules with Predation 376
Acknowledgments 376
References 377

23.1 Introduction

A succession of life stages characterizes the development of many organisms. Stages interact with resources, competitors, and predators,[1,2] so that different stages may play different roles in the food web.[3,4] In marine ecosystems, copepods are the food of larvae of many species of commercially important fishes. Copepod development is characterized by 13 morphologically well-defined stages, successively: egg, six naupliar, five copepodite, and one adult stage. In many cases, constant mortality rates of developmental stages cannot explain the observed temporal patterns of copepod populations. In consequence, predation is assumed to be the main reason for the temporal variations of mortality between stages[5–9] although direct observation remains difficult.

To identify predation effects, some ecosystem models try to reproduce the effects of interactions between copepods and predators.[10] However, difficulty in calibrating and validating such complex models with noisy data makes them difficult to use.[11] These models generally use submodels describing the dynamics of the copepod populations built for various development conditions.[12–18] The effects of some external factors, i.e., food and temperature, on the development of copepods have been heavily studied and integrated in the modeling.[6,7,19–24] Finally, the correct estimation of copepod production necessitates a special emphasis on their life representation,[25] and particularly the patterns of mortality in developmental stages.[26] Consequently, the models that describe the population dynamics involve a considerable number of assumptions that limit their generalization (parameter values, mathematical expression of recruitment, maturation and mortality processes). Among these demographic processes, mortality in developmental stages is crucial.[25,27,28] Moreover, an accurate estimation of the different sources of mortality in copepods is made difficult by the complex interactions between copepod instars and their predators. From these predators, late developmental stages of several copepods are omnivorous and thus can exert a high predation on eggs and early naupliar stages (including their own offspring), which may control their production.[6,7,28] Modeling quantitatively pelagic food webs including all these possible interactions could be laborious in terms of mathematical expressions, parameter identification, and model validation. We thus need a qualitative method capable of capturing the main properties of copepod development based on stage-structured models.

Bernard and Gouzé[29–31] developed a qualitative method for describing the dynamics of a class of nonlinear differential systems. The method, based on the description of the succession with time of the state-variable extrema (minima and maxima), is independent of parameter values and of the precise formulation of the mathematical functions used in the model. To apply it one has first to identify the extrema of each time series variable and to compare them with the theoretical succession issued from the mathematical analysis of the model. Bernard and Gouzé[29] proposed their method as a first step to test whether the structure of a given model is consistent with experimental observations. Bernard and Souissi[32] presented the mathematical basis of the application of these methods to a simplified model of stage-structured populations based on ordinary differential equations.

In this chapter we apply and extend this method to a stage-structured population model with time-delay differential equations. Because of the great number of variables in such a system, we focus on the dynamics of one developmental stage. For a copepod population isolated in nonlimiting growth conditions, the number of individuals in each stage is always determined by an input term that is the recruitment to the stage and an output term represented by the sum of maturation and mortality. These terms are usually represented by mathematical functions reproducing the different parameterizations of the demographic processes. It turns out that these functions are generally monotonic (i.e., increasing or decreasing). Nonetheless, these models are always based on the same mathematical structure, and we will show that a qualitative analysis of this structure can help to identify a perturbation in the development of the population. We then derive the rules on the possible extrema (maximum or minimum) for this stage in relation to the trend of the preceding stage. The structural transition rules depend only on the model structure, and simply express the delay in the appearance of two extrema for two successive stages: the peak of abundance of each developmental stage arises during the decreasing period of the abundance of the previous stage. These basic structural transition rules seem evident to people studying development of stage-structured populations. However, in some cases, these rules can be violated, suggesting an unexpected situation. We show that such situations are rich in information and that they can be explained by other transition rules resulting from another model including an interaction with an exogenous variable. For added simplicity we focus on predation, but the analysis is the same in principle for other disturbances occurring throughout development (food limitation, competition, advection). To illustrate the method we first show that we can identify the interactions in a population simulated by a standard numerical model. We show that the rules for predation can be used to discriminate between different interaction scenarios. Then we apply the analysis to a set of experimental data issued from a mixed culture of the copepods *Acartia clausi* and *Euterpina acutifrons*. Finally, the analysis is applied to *in situ* data of development of *Centropages abdominalis* in the Sea of Japan.

23.2 Method

23.2.1 Transient Behavior of a Stage-Structured Population Model

23.2.1.1 Basic Structural Transition Rules — Various recently developed stage-structured population models have the same formalism.[12–15,17,18,33] They assume that all individuals within a stage (or age class) are functionally identical.[33] The abundance [$N_i(t)$] of a given stage i at time t changes through the recruitment [$R_i(t - \tau)$] of new individuals according a time delay τ, maturation [$M_i(t)$] from the stage i to its successor $i + 1$, and death [$D_i(t)$]. This can be expressed by the differential equation:

$$\frac{dN_i}{dt}(t) = R_i(t - \tau) - M_i(t) - D_i(t) \tag{23.1}$$

If we define τ_{i-1} as the mean delay required by individuals in stage $(i - 1)$ to mature to stage i, Equation 23.1 can lead to the simplified equation:

$$\frac{dN_i}{dt}(t) = f_i(N_{i-1}(t - \tau_{i-1})) - g_i(N_i(t)) \tag{23.2}$$

where the quantities $f_i(N_{i-1}(t - \tau_{i-1}))$ and $g_i(N_i(t))$ represent, respectively, the input and output processes in the dynamics of stage i. The function f_i stands for the recruitment to stage i, which depends on the abundance of the preceding stage $(i - 1)$ at time $(t - \tau_{i-1})$. The function g_i is the sum of mortality in plus maturation from stage i, which both depend on the abundance of stage i at time t.

The dynamics of the system is characterized by functions f_i and g_i and their parameters. For example, in the development model of *E. acutifrons*,[14] where the functions f_i and g_i were linear, we used constant values for transfer, mortality, and egg-laying rates. In other models, the values of parameters change with stages, and can also be functions of other variables inherent to development (weight, growth) or external factors (temperature, food).[15,16] If we consider only the analytical formulation of these models (ignoring the parameterization used and the parameter values), it belongs to the general representation of system 23.2. In this study we focus on the general properties of a stage-structured population model regardless of the analytical formulation of the model and the parameter values.

It is worth noting that we have chosen a simplified model to make the mathematical analyses tractable. However, later on, we validate our conclusion on a more complicated simulation model for which analytical computations are more complicated.

The dynamics of each stage N_i expressed by the ordinary differential equation (Equation 23.2) only depends on the variable itself and on the preceding one, so that the system has a so-called loop structure.[29]

For such systems, we can apply the analysis exposed in Reference 29. It proves that the theoretically allowed successions of extrema for the state variables are derived only from the model structure. This theoretical succession of extrema can then be compared to the succession of extrema observed in the data. This analysis allows us to verify, without simulations and independent of the parameter values, the consistency of the constraints imposed by the model structure and the actual data.

The mathematical analysis is based on two hypotheses. First we assume that the functions f_i and g_i are increasing. Second, we use an estimate of the maturation time τ_{i-1}. In fact, we will see in the sequel that we only need a reasonable interval for τ_{i-1}.

Without running simulations, this study determines the possible appearance order of the extrema for the variable N_i in relation with the trend of the variable N_{i-1}. These criteria are shown in Figure 23.1 (see Appendix 23.A for the sketch of the mathematical proof), and lead to the following rule:

Consider the system (Equation 23.2) with the functions f_i increasing. Suppose that N_i has an extremum (minimum or maximum) at time t_0. If the variable N_{i-1} was decreasing at time $t_0 - \tau_{i-1}$, then the extremum of N_i is necessarily a maximum. On the contrary, if the variable N_{i-1} was increasing at $t_0 - \tau_{i-1}$, then the extremum of N_i is a minimum.

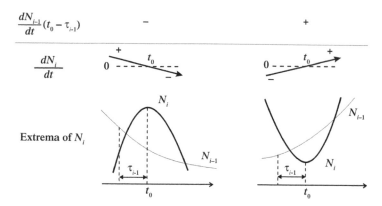

FIGURE 23.1 The basic structural transition rules representing the appearance of extrema (minima and maxima) of variable N_i at time t_0 in relation with the trend of the variable N_{i-1} in the system (Equation 23.2) corresponding to the development of a developmental stage i without external perturbation. The sign of the expression on the left is indicated by the symbols + and − (see Appendix 23.A for the sketch of the mathematical proof). τ_{i-1} represents the delay in development for individuals in stage $i-1$ for maturation in stage i.

We first notice that the rule is independent of the output term represented by the function g_i (see Appendix 23.A). The development of a single cohort follows exactly this rule. For example, a pulse of eggs produces a pulse of juveniles, which produces a pulse of adults. The maximum of adults follows in time the maximum of juveniles.

23.2.1.2 Structural Transition Rules for Predation

23.2.1.2 **Structural Transition Rules for Predation** — Controlled experiments have been performed[19,24,34,35] to understand the effects of external factors on the development of copepod populations. In the sea, the presence of several populations makes such a study more complicated, because of interactions between species (predation, competition) and spatial heterogeneity (migration, patchiness). Predation depends not only on the abundance of the prey but also on the abundance of the predator. Note that the function g_i (Equation 23.2) depends only on the abundance of the stage i, and cannot include the predation process. Suppose now that predators of concentration y feed on stage i; the corresponding equation becomes

$$\frac{dN_i}{dt}(t) = f_i(N_{i-1}(t - \tau_{i-1})) - g_i(N_i(t)) - h_i(y(t), N_i(t)) \tag{23.3}$$

where the function h_i represents the predation process. A qualitative study similar to that of Equation 23.2 (for mathematical proof see Appendix 23.B) relates the appearance of the extrema for variable N_i to trends of both variables N_{i-1} and y. The new criteria (shown in Figure 23.2) will be called the structural transition rules for predation (predation rules). These rules can be summarized in the following two points:

1. If $N_{i-1}(t_0 - \tau_{i-1})$ and $y(t_0)$ have opposite trends (one is increasing, the other decreasing), the previous rules of the extrema appearing for the variable N_i (Figure 23.1) obtained for Equation 23.2 remain valid for Equation 23.3.
2. If $N_{i-1}(t_0 - \tau_{i-1})$ and $y(t_0)$ have the same trend, N_i can have either a minimum or a maximum. New possible qualitative situations for the variable N_i, not permitted before (Equation 23.2), now become possible.

Identification of interactions will be based on the second rule, which makes nonintuitive behavior possible, e.g., the appearance of a peak of stage i before a peak of stage $i-1$.

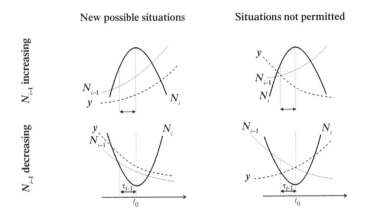

FIGURE 23.2 Schematic representation of the structural transition rules for predation corresponding to the system (Equation 23.3) where the development of stage i is perturbed by an exogenous factor y. The new criteria of appearance of extrema (minima and maxima) of variable N_i when they are not allowed by the basic transition rules (Figure 23.1) depend on the trend of the exogenous variable y (see Appendix 23.B for mathematical proof). τ_{i-1} represents the delay in development for individuals in stage $i - 1$ for maturation in stage i.

23.2.2 Identification of Extrema in the Time-Series Abundance Data

Precise identification of extrema in time series data is a crucial step in our study. The accurate determination of the critical events depends on the quality of data and on the sampling rate. Unfortunately, the time series of copepod instars abundance are generally noisy and the intervals between observations irregular. The identification of the trends in time series data by examining directly the original data can be very ambiguous. Thus, it is necessary to use a systematic procedure for identifying the trends of time series and pointing out the significant extrema. According to the objective of this study, the required procedure should satisfy the following conditions:

1. The method should smooth the original data.
2. The smoothing should conserve the transition rules developed from Equations 23.2 and 23.3, so that we can apply directly these rules to the smoothed data.

Bernard and Souissi[32] demonstrated that for a monotonic system (i.e., with monotonic functions in the model), the signal filtered by a continuous moving average still satisfies the same transition rules (see their Lemma 5 in their section 2.3). For a given time window θ, the moving average $\overline{x}(t)$, of a function $x(t)$, during the interval $[t - \theta/2, t + \theta/2]$ is calculated using the following equation:

$$\overline{x}(t) = \frac{1}{\theta} \int_{t-\theta/2}^{t+\theta/2} x(z)dz \qquad (23.4)$$

We adopted this simple low-pass filter to smooth the data. For a complete time series the filtered signal is obtained using Equation 23.5 (for $t > \theta/2$). This equation is simply obtained by decomposing the integral in Equation 23.4 between $(t - \theta/2)$ and $(t + \theta/2)$ into two integrals between $(t - \theta/2)$ and 0 on the one part and between 0 and $(t + \theta/2)$ on the other part:

$$\overline{x}(t) = \frac{1}{\theta}\left(I_N(t + \theta/2) - I_N(t - \theta/2)\right) \qquad (23.5)$$

with

$$I_N(t) = \int_0^t x(z)dz$$

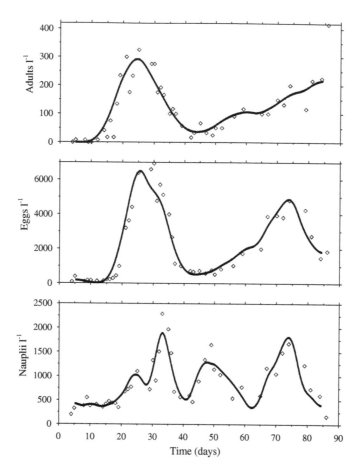

FIGURE 23.3 Time variation of developmental stages adults, eggs, and first naupliar stages (N1–N3) (diamonds) of *A. clausi* in the mixed culture *E. acutifrons–A. clausi.*[36] Continuous lines represent the data processing by moving average for the value of the time window θ equal to 1.5 days.

To obtain the filtered signal for any value of the time t and of the parameter θ, we first integrate the original data using the simple trapezoidal method to obtain at the sampling instants t_i the estimate of $I_N(t_i)$. Then, the value for any time t of $I_N(t)$ is obtained by a cubic spine interpolation (using the csaps Matlab function) between the sampling instant $I_N(t_i)$. From the interpolated functions we compute the moving average using Equation 23.5. The interpolation between data values is necessary to regularize the time series and to estimate the location of each extremum, which is generally not exactly at a sampling instant. Figure 23.3 gives an example of application of the moving average to the time series of abundance of adults, eggs, and small nauplii (N1—N3) of *A. clausi.*[36]

The extrema are detected after the smoothing steps by an automatic procedure that analyzes the changing signs in the derivative of the filtered signal. Some extrema are nevertheless difficult to localize because they are too flat. We consider therefore the significant extrema, i.e., for which the second derivative $\ddot{x}$ is larger than a threshold value. To account for both uncertainties on the extremum time t_0 and on the maturation time τ_{i-1} we apply the tests for all the values of $\tau \in [0.8\tau_{i-1}, 1.2\tau_{i-1}]$. In practice we consider 50 values of τ taken regularly from $0.8\tau_{i-1}$ to $1.2\tau_{i-1}$. For each of these values of the delay we determine if the measured extrema are compatible with the considered rules.

23.2.3   Age and Stage-Structured Simulation Model

Before applying the developed method to real data from experiments, we begin the result section with numerical experiments. The aim of this discussion is to show how these transition rules can be used to

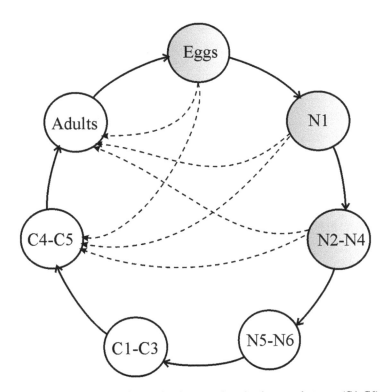

FIGURE 23.4 Conceptual scheme of the interaction between late developmental stages (C4–C6) and early stages (eggs–N4) of the copepod *A. omorii*. The life cycle is characterized by seven stages, eggs, N1, N2–N4, N5–N6, C1–C3, C4–C5, and adults. Thick continuous arrows indicate transfer from one stage to another and reproduction. Thick discontinuous arrows indicate cannibalism.

test different interaction hypotheses. We use an age-within stage model commonly used for copepod population simulations.[14,15,18,25,37,38] The model was parameterized and calibrated for the species *A. omorii*.[38] Uye and Liang[39] showed that cannibalism is common in the development of this species. Thus, instead of introducing an external predator in the simulation, we considered only an endogenous disturbance due to cannibalism as shown in Figure 23.4. Moreover, to focus on this perturbation and for added simplicity the simulation was run at constant temperature 20°C and nonlimiting food conditions. At 20°C the first naupliar stage has an average duration of 0.55 days. This value was obtained by using the expressions of the effect of temperature on stage duration of *A. omorii* given in Liang and Uye.[7] To obtain a long time series the simulation was run over 1000 days.

23.3 Results

23.3.1 Analysis of Transition Rules with Simulated Data

The model used here has a complex structure; however, when used for a constant temperature and nonlimiting food conditions, only cannibalism can control the dynamics of the system. For the set of predation parameters used here (Table 23.1), the model produced cyclic dynamics due to predator–prey interactions. Figure 23.5 shows three phase diagrams for the omnivorous stages C4–C6 (predators) and the three preyed stages (eggs, N1, and N2–N4).

Figure 23.6 shows the time evolution of the abundance of eggs, N1 and C4–C5, and adults during the first 240 days of simulation. We compare this set of data with the established transition rules given earlier. To apply the transition rules we considered a maturation time τ_1 (0.5 days) of stage N1. The maxima of stage N1 identified with continuous gray bands transgress the basic structural transition rules. Then the succession of these extrema patterns, represented by a major peak, is consistent with the basic

TABLE 23.1

Values of Capture Efficiency Parameters Used to Simulate the Development of *A. omorii* (see Figure 23.4) for a Constant Temperature (20°C) and Nonlimiting Food Conditions[a]

	Predators	
Preys	**C4–C5**	**Adults**
Eggs	0.05	0.12
N1	0.10	0.25
N2–N4	0.05	0.10

[a] All other parameters of the simulation were identical to those used by Souissi and Uye (in revision).

TABLE 23.2

Use of the Predation Rules to Test the Different Hypotheses of Predation of Stage 2 by Predators 1 and 2

Description of the Extrema (minima and maxima)	**Percentage (%)**
Extrema significant (used in the analysis)	100
H0: no interaction	19.84
H1: predation by predator 1	100
H2: predation by predator 2	8.8

Note: The extrema that transgressed the classical rules are used to calculate the percentages of cases that verify the hypotheses of predation by the predator 1 and predator 2, respectively.

transition rule (Figure 23.1), and a minor peak transgressing these rules is repeated until the end of the simulation. This dynamics leads to a trajectory of Figure 23.5B.

The minor peaks of N1 occurred during an increasing phase of omnivorous stages C4–C5 (Figure 23.6C) and adults (Figure 23.6D). Hence, these maxima are adequate for structural transition rules for predation. Some extrema remain compatible with the transition rules without predation. On the other hand, the predation effect is obtained for high predator density. The extrema of all other stages (no predation) were compatible with the structural transition rules without predation.

We have shown here that our method works well for data from a simulation model. In the next section we apply the analysis to experimental data subject to noise and uncertainty (Table 23.2).

23.3.2 Identification of Predation Interactions in Copepod Development Experiments

Yassen[36] performed various experiments to study the interactions between two copepod species. For one of these experiments, the harpacticoid *E. acutifrons* was reared together with another omnivorous calanoid *A. clausi*. The experiment was done in a 8-l tank, with nonlimiting food conditions and constant temperature. The small volume of the tank increased the encounter rate between the late copepodite stages of *Acartia*, which are omnivorous,[40] and the other developmental stages. We can thus expect an interaction of the two species corresponding to predation of *E. acutifrons* small instars.

23.3.2.1 Cannibalism in the Development of A. clausi — For sake of clarity we focus on a nonfeeding stage, which is thus less sensitive to resource fluctuations: the small nauplii of *A. clausi*. Of course, the method can apply to any stage, but the conclusion on the possible interactions will then be less clear and several possibilities must be considered. We applied then the analysis of extrema in the time series of small nauplii abundance of *A. clausi*. We used the signal processed by continuous moving average (Figure 23.3). To take into account both uncertainties of the extrema position and of the maturation time, various values of the delay parameter τ_i were considered. These values are varied continuously between 80 and 120% of the estimated value of $\tau_i = 1.3$ days.[41,42] The results of the analysis

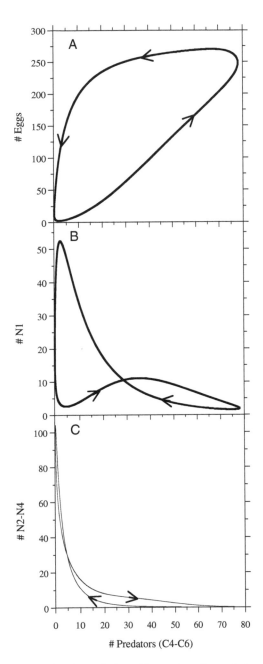

FIGURE 23.5 Diagram phases representing the evolution of the number of preyed stages — eggs (A), N1 (B), N2–N4 (C) — as a function of the total number of predators (C4–C6). The simulation of developmental stages was run for 1000 days using the system of Figure 23.4 and the set of interaction parameters given in Table 23.1. Arrows show the sense of the trajectory according to the temporal evolution of the state variable (only the portion when the system is at equilibrium was used).

are shown in Table 23.3. Figure 23.7B shows that five extrema of small nauplii of *A. clausi* transgress the classical rules. The hypothesis of predation by their adults was verified in all extrema cases (Figure 23.7C). The number of *A. clausi* adults is decreasing during the minima and increasing during the maxima of this stage (Figure 23.7C), which makes these extrema conform to the predation rules by the adults. In this experiment, we can therefore state the hypothesis of cannibalism in the development of *A. clausi*.

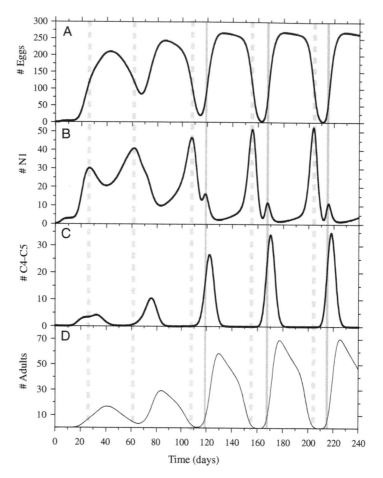

FIGURE 23.6 Simulation of the system of Figure 23.4 by using the set of parameter values given in Table 23.1. The time evolution of abundance of eggs (A), N1 (B), C4–C5 (C) and adults (D) during the first 240 days of simulation. For analyzing the extrema of N1 the timeseries B to D are delayed by 0.5 days (value of τ_1). Only maxima of stage N1 are illustrated, those in accord with the basic structural transition rules (shown in Figure 23.1) are represented by discontinuous gray bands. The continuous gray bands represent the maxima of N1 transgressing the basic structural transition rules. The adequacy of these later extrema with the predation rules (shown in Figure 23.2) is verified with the trends of their predators C4–C5 and adults.

TABLE 23.3

Use of the Predation Rules to Test the Hypothesis of Predation of Small Developmental Stages (Nauplii) in a Mixed Culture of Two Copepods *E. acutifrons* and *A. clausi* by the Adults of *A. clausi*

Description of the Extrema (minima and maxima)	Application to Small Nauplii (%)	
	A. clausi	*E. acutifrons*
Extrema significant (used in the analysis)	70	100
Extrema transgressed the classical rules	71.43	62.5
Hypothesis of predation by adults	100	100
Hypothesis of predation by larger nauplii	20	20

Note: The extrema transgressing the classical rules are used to calculate the percentage of cases that verify the hypothesis of predation by *Acartia* adults.

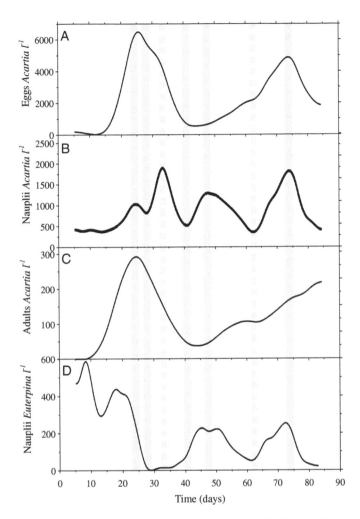

FIGURE 23.7 Data processing from moving average θ = 1.5 days for eggs (A), first naupliar stages N1—N3 (B), and adults (C) of *A. clausi* and large naupliar stages N4—N6 (D) of *E. acutifrons*. Data are obtained from the mixed culture of two copepod species *E. acutifrons*–*A. clausi*.[36] The minima and maxima of small nauplii transgressing the basic rules are shown by continuous gray strips. The discontinuous gray strips represent the maximum and minimum adequate with the basic structural transition rule. The time series of nauplii and adults (B to C) are delayed τ = 1.3 days.[41,42,56]

The hypothesis of predation by omnivorous stages (i.e., adults of *A. clausi*) is the most logical one. This hypothesis of interaction was strengthened after the application of our qualitative method. Then the same technique was applied to test additional interaction hypotheses, such as a less plausible hypothesis of the interaction between large naupliar stages (N4–N6) of *E. acutifrons* and the first naupliar stages (N1–N3) of *A. clausi*. Figure 23.7D shows the time evolution of the continuous moving average of the abundance of large nauplii of *E. acutifrons*. The hypothesis of interaction between large nauplii of *E. acutifrons* and small nauplii of *A. clausi* was rejected in 80% of the cases (Table 23.3).

23.3.2.2 Predation of A. clausi Adults on the First Naupliar Stages of E. acutifrons —

Because the adults of *A. clausi* are omnivorous[36,43,44] we can expect predation of naupliar stages of *E. acutifrons* by adults of *A. clausi*. Therefore, we applied the method of extrema analysis to the data processed with moving average for the first naupliar stages (N1—N3) of *E. acutifrons*. Figure 23.8B shows that five extrema over eight of small nauplii transgress the classical rules. The hypothesis of predation by *A. clausi* adults is also verified in 100% of cases (Table 23.3).

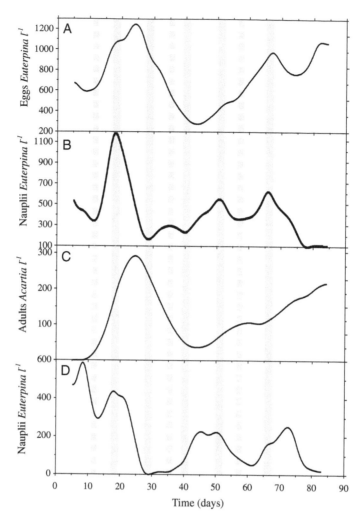

FIGURE 23.8 Data processing from moving average θ (= 1.5 days) for eggs (A), first naupliar stages N1—N3 (B), large naupliar stages N4—N6 (D) of *E. acutifrons* and adults (C) of *A. clausi*. Data are obtained from the mixed culture of two copepod species *E. acutifrons–A. clausi*.[36] The minima and maxima of small nauplii transgressing the basic transition rules are shown by continuous gray strips. The discontinuous gray strips represent the maximum and minimum following the basic structural transition rule. The time series of nauplii and adults (B to C) are delayed τ = 1.2 days.[56]

We also tested the hypothesis of interaction between large nauplii (N4–N6) and small nauplii (N1–N3) of *E. acutifrons*. This hypothesis was rejected for 80% of extrema studied (Table 23.3).

23.3.3   Identification of *in Situ* Interactions between Copepods in a Eutrophic Inlet of the Inland Sea of Japan

The population dynamics and the production of the dominant planktonic copepods (*Centropages abdominalis, A. omorii, Paracalanus* sp., and *Pseudodiaptomus marinus*) in an eutrophic inlet of the Inland Sea of Japan were studied.[6,7,45,46] In this study, the high sampling frequency (3 to 5 day intervals) and the counting of individuals in each developmental stage guarantee the applicability of our method. The position of the sampling station in the center of the Fukuyama Harbor (see fig. 1 in Liang et al.[6]) in a shallow depth justifies neglecting the physical transport effects in the demographic processes. In this region, true carnivores such as fish larvae, chaetognaths, medusae, and ctenophores, were absent or extremely rare, at least for the period from January to June.[6,7] During this period the plankton

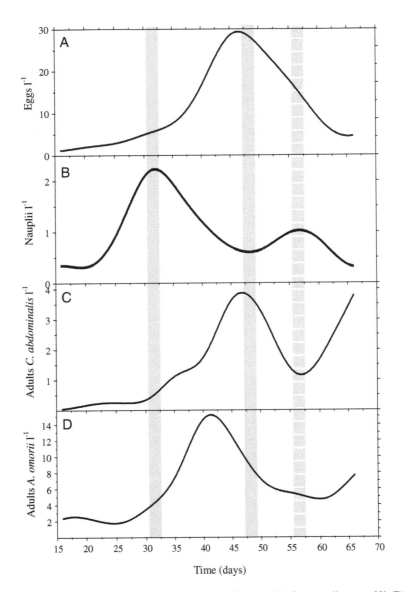

FIGURE 23.9 Data processing from moving average θ (= 4 days) for eggs (A), first naupliar stage N1 (B), adults (C) of *C. abdominalis* and adults (D) of *A. omorii*. The data are obtained from the *in situ* development of the populations in the Fukuyama Harbor region in Japan.[6,7] The maximum and minimum of N1 transgressing the basic transition rules are shown by continuous gray strips. The discontinuous gray strip represents the maximum adequate with the basic structural transition rule. The time series of N1 and adults are delayed τ = 2.8 days.[6]

was largely dominated by the copepods *C. abdominalis* and *A. omorii*. Liang et al.[6] suggested that predation including cannibalism by these copepods also might be the main source of the high mortality observed in their early development. Thus, we focus on the interaction between these populations developing simultaneously.

Figure 23.9 shows a clear maximum of stage N1 of *C. abdominalis*, occurring during an increasing phase of egg concentration (with $\tau_1 = 2.8$ days; cf. Table 23.3), which transgresses the classical rules. The abundance of the stages adult and C5 (not shown in Figure 23.9B) of both species *C. abdominalis* and *A. omorii* are increasing during the maximum of N1. This is in agreement with the predation rules if one of these species is the predator. The adults and C5 of these species are omnivorous,[6] so they represent the potential predators of the stage N1.

23.4 Discussion

The development of copepods has been extensively studied through simulation models[12–18] and through models devoted to estimation of demographic parameters.[5,8,11,47–50]

Equation 23.2 is a simplified representation of the development of a stage-structured population. In our analysis, the dynamics of variables is expressed by the functions *f* and *g* regardless of either their parameter values or their shapes. The classical structural transition rules depend only on the model structure and on development time. The qualitative feature of the succession of extrema shown in Figure 23.1 has been observed in many copepod species,[14,51,52] and other crustacean species such as crabs[53] and insects.[54] When several stages are developing together, the rules become sometimes less intuitive (in particular for the minimum), but from an extensive study of data they are verified.[32]

We must note that for these rules the knowledge of the dynamics of all developmental stages of the population is not required. Some instars can therefore exhibit more complex dynamics, which may not be described by Equation 23.2 (subject to other interactions such as predation, food limitation, and so forth).

The classical rules derived from a model of an isolated population are respected for most extrema in the time series. Nevertheless, for *in situ* data, which may be submitted to a wider range of perturbations, following of these transition rules is remarkable and shows that they are very robust in nature. Even if these classical rules are dominant in copepod development, and represent the main properties of cohorts, they may be violated. For example, it is intuitively very difficult to admit that a peak of one developmental stage may occur before the peak of the preceding stage. Rigler and Cooley[50] have mentioned some "anomalies" in their data. For the third generation of *Skistodiaptomus oregonensis*, they showed that the stages N3, C3, and C5 each had a mean pulse time preceding that of the previous instar. The Rigler–Cooley method[50] is based on visual identification of cohorts, so the use of such data would obviously lead to absurd conclusions. Such events may be due to sampling errors, particularly when the frequency of observations is very low. In this case, the singular events can be eliminated by smoothing the original data.[50] The unexpected extrema presented in Rigler and Cooley[50] represent a significant qualitative trend, which persists after smoothing the original data.

The presence of these extrema, incompatible with a classical development scheme, implies that the general structure system (Equation 23.2) is not an adequate representation of the data, so that an exogenous interaction has to be considered to explain the observations.

The introduction of a predation hypothesis in the development of a stage *i* generates new structural transition rules different from the previous ones, which explains the occurrence of new extrema. The potential predators (or external factors) can therefore be identified by comparing their trends during the observed extrema. The qualitative information on the trends of the external factors (increasing or decreasing) during the extremum situation is required. It is worth noting that with this approach the knowledge of the dynamics of the predator and of stage *i* − 1 is not necessary.

Several authors have shown the importance of predation in explaining the patterns of mortality in stage-structured populations.[5,8,9,40,50] Predation was usually suspected, but there was no method developed for its direct identification. The principal difficulty results from the different sources of mortality. The term of loss (*h*) in Equation 23.3 can represent mortality due to other biotic or abiotic factors. For example, natural disease (parasitism, intoxication, etc.), starvation, sedimentation (for eggs), and physical transport are other causes of mortality. To reduce the effects of these factors, we used naupliar stages for identification of the unexpected extrema. Indeed, the first naupliar stages are more vulnerable to predation, and generally are nonfeeding stages (at least the stage N1), which permits us to exclude the hypothesis of mortality induced by food limitation. Even though the objective of this chapter is to show the possibility of explaining these nonintuitive extrema, the method can be applied to other developmental stages and other external factors.

Inverse methods have been developed to quantify the mortality in each stage and, more generally, to estimate the principal demographic parameters (e.g., mortality rate and birth rate). Aksnes et al.[55] have presented a review of these estimation techniques. However, each technique requires a model and a number of assumptions (homogeneity in mortality between stages, instantaneous mortality equal zero, zero recruitment, and so forth). In contrast, our qualitative method requires fewer hypotheses on the model and demands the estimation of only one parameter: the stage duration. The maturation time is related to temperature for nonfeeding stages and to both food and temperature for feeding stages,[37] and it can be precisely estimated. Moreover, our method focuses on the study of one developmental stage, which means that no information on the modeling of the other developmental stages is necessary. This point is fundamental, because the different phases of the life cycle play different roles in the food web[1] and some stages can have more complex dynamics.

Because they are based on a quantitative approach, the inverse methods require higher-quality data to obtain an estimate of the abundance derivative. Finally, according to the difference in size between stages, the abundance of early development stages is often underestimated relative to those of later copepodite stages.[56] The qualitative information (e.g., appearance of a maximum or a minimum) is not sensitive to this sampling bias, whereas it can affect the estimations provided by the inverse methods.

But, in fact, the main difference between qualitative methods and inverse methods lies in their respective goals. The estimation of the mortality rate[8,11] does not necessarily explain the cause of mortality. Even if our method is qualitative (we do not quantify the loss due to mortality), the mortality source represented by the exogenous factor (the term y in Equation 23.3) is tested in the analysis and a set of potential mortality causes is rejected.

The widespread occurrence of stage-structured populations enlarges the field of application of this methodology to various domains of ecology: terrestrial (insects, amphibians), freshwater, and marine (crabs, shrimps). These populations have a very simple qualitative dynamic behavior when their development is enclosed. The simplicity of this transient behavior summarized by two classical rules makes any perturbation of this dynamics easy to detect. Thus, the transgression of the classical rules can be used as a signal of perturbation of the development of stages. This analysis shows therefore how interesting these stage-structured populations are — because of the simplicity of their endogenous dynamics they can be considered a biological indicator of the interaction with other components of the food web. The transient behavior of stage-structured populations represents a kind of memory of the interactions that have modified the dynamics of the population. Hay et al.[40] considered that a transient stress at one time in a life cycle may have a large effect on other parts of the food web and on the whole community's future development.

Appendix 23.A: Sketch of the Mathematical Proof of the Basic Structural Transition Rules

We consider the system (Equation 23.2) describing the dynamics of variable N_i. Suppose that N_i has an extremum (minimum or maximum) at t_0, so

$$\frac{dN_i}{dt}(t_0) = 0$$

By differentiation along time of the dynamic equation of stage i (Equation 23.2), we obtain

$$\frac{d^2 N_i}{dt^2}(t_0) = \frac{df_i}{dN_{i-1}} \frac{dN_{i-1}}{dt}(t_0 - \tau_{i-1}) - \frac{dg_i}{dN_i} \frac{dN_i}{dt}(t_0)$$

So, that for an extremum of N_i at t_0:

$$\frac{d^2 N_i}{dt^2}(t_0) = \frac{df_i}{dN_{i-1}} \frac{dN_{i-1}}{dt}(t_0 - \tau_{i-1})$$

The function f_i is increasing and then $\dfrac{df_i}{dN_{i-1}} > 0$.

Thus, when $\frac{dN_{i-1}}{dt}(t_0 - \tau_{i-1}) > 0$, we have $\frac{d^2N_i}{dt^2}(t_0) > 0$. In this case $\frac{dN_i}{dt}$ is increasing for t_0. $\frac{dN_i}{dt}$ is then negative for some $t < t_0$, and positive for some $t > t_0$, which implies that N_i is decreasing before t_0 and increasing after. N_i admits thus a minimum at t_0.

Similarly, when $\frac{dN_{i-1}}{dt}(t_0 - \tau_{i-1}) < 0$, we have $\frac{d^2N_i}{dt^2}(t_0) < 0$, so that N_i has a maximum at t_0.

Appendix 23.B: Sketch of the Mathematical Proof of the Transition Rules with Predation

We will consider the system (Equation 23.3) describing the dynamics of variable N_i. Suppose that N_i has an extremum (minimum or maximum) at t_0, and thus ($\frac{dN_i}{dt}(t_0) = 0$).

By differentiation along time of the dynamic equation of stage i (Equation 23.3), we obtain

$$\frac{d^2N_i}{dt^2}(t_0) = \frac{df_i}{dN_{i-1}}\frac{dN_{i-1}}{dt}(t_0 - \tau_{i-1}) - \frac{dg_i}{dN_i}\frac{dN_i}{dt}(t_0) - \frac{\partial h_i}{\partial N_i}\frac{dN_i}{dt}(t_0) - \frac{\partial h_i}{\partial y}\frac{dy}{dt}(t_0)$$

So, that for an extremum of N_i at t_0:

$$\frac{d^2N_i}{dt^2}(t_0) = \frac{df_i}{dN_{i-1}}\frac{dN_{i-1}}{dt}(t_0 - \tau_{i-1}) - \frac{\partial h_i}{\partial y}\frac{dy}{dt}(t_0)$$

In the previous system (Equation 23.2) $\frac{d^2N_i}{dt^2}(t_0)$ and $\frac{dN_{i-1}}{dt}(t_0 - \tau_{i-1})$ have the same sign (see Appendix 23.A), but with predation the sign of $\frac{d^2N_i}{dt^2}(t_0)$ depends on both signs of $\frac{dN_{i-1}}{dt}(t_0 - \tau_{i-1})$ and $\frac{dy}{dt}(t_0)$. This means that the succession of extrema in stage i depends not only on the trend of the previous stage but also on the trend of the predator.

The functions f_i and h_i are increasing and then $\frac{df_i}{dN_{i-1}}, \frac{\partial h_i}{\partial y} > 0$.

Thus, when $\frac{dN_{i-1}}{dt}(t_0 - \tau_{i-1}) > 0$ and $\frac{dy}{dt}(t_0) < 0$, we have $\frac{d^2N_i}{dt^2}(t_0) > 0$, so that N_i has a minimum at t_0. Similarly when $\frac{dN_{i-1}}{dt}(t_0 - \tau_{i-1}) < 0$ and $\frac{dy}{dt}(t_0) > 0$ we have $\frac{d^2N_i}{dt^2}(t_0) < 0$, so that N_i has a maximum at t_0.

When $\frac{dN_{i-1}}{dt}(t_0 - \tau_{i-1})$ and $\frac{dy}{dt}(t_0)$ have the same sign, new possible situations not permitted before become possible. The new criteria, called the structural transition rules for predation, are summarized in Figure 23.2.

Acknowledgments

The authors thank Antoine Sciandra and Jean-Luc Gouzé for their comments. This work is a contribution to ELICO (Ecosystèmes Littoraux Côtiers), COMORE project, and ECOREG (Ecosystem Complexity Research Group). We thank Shin-Ichi Uye for providing the data used to produce Figure 23.9. We also thank Konstantinos Ghertsos and Peter G. Strutton for help with the English.

References

1. Osenberg, C.W., Mittelbach, G.G., and Wainwright, P.C., Two-stage life histories in fish: the interaction between juvenile competition and adult performance, *Ecology*, 73, 255, 1992.
2. Mittelbach, G.G. and Osenberg, C.W., Stage-structured interactions in bluegill: consequences of adult resource variation, *Ecology*, 74, 2381, 1993.
3. Uye, S.-I. and Kayano, Y., Predatory feeding behavior of *Tortanus* (Copepoda: Calanoida): life-stage differences and the predation impact on small planktonic crustaceans, *J. Crust. Biol.*, 14, 473, 1994.
4. Stuart, V. and Pillar, S.C., Diel grazing patterns of all ontogenetic stages of *Euphausia lucens* and *in situ* predation rates on copepods in the southern Benguela upwelling region, *Mar. Ecol. Prog. Ser.*, 64, 227, 1990.
5. Hairston, N.G.J. and Twombly, S., Obtaining life table data from cohort analyses: a critic of current methods, *Limnol. Oceanogr.*, 30, 886, 1985.
6. Liang, D., Uye, S.-I., and Onbé, T., Population dynamics and production of the planktonic copepods in a eutrophic inlet of the Inland Sea of Japan. I. *Centropages abdominalis*, *Mar. Biol.*, 124, 527, 1996.
7. Liang, D. and Uye, S.-I., Population dynamics and production of the planktonic copepods in a eutrophic inlet of the Inland Sea of Japan. II. *Acartia omorii*, *Mar. Biol.*, 125, 109, 1996.
8. Ohman, M.D. and Wood, S.N., Mortality estimation for planktonic copepods: *Pseudocalanus newmani* in a temperate fjord, *Limnol. Oceanogr.*, 41, 126, 1996.
9. Saunders, J.F. and Lewis, W.M.J., A perspective on the use of cohort analysis to obtain demographic data for copepods, *Limnol. Oceanogr.*, 32, 511, 1987.
10. Jones, R. and Henderson, E.W., The dynamics of energy transfer in marine food chains, *S. Afr. J. Mar. Sci.*, 5, 447, 1987.
11. Wood, S.N. and Nisbet, R.M., *Estimation of Mortality Rates in Stage-Structured Populations*, Springer-Verlag, Berlin, 1991, 101.
12. Wroblewski, J.S., A simulation of the distribution of *Acartia clausi* during Oregon Upwelling, August 1973, *J. Plankton Res.*, 2, 43, 1980.
13. Davis, C.S., Predatory control of copepod seasonal cycles on Georges Bank, *Mar. Biol.*, 82, 31, 1984.
14. Sciandra, A., Study and modelling of development of *Euterpina acutifrons* (Copepoda, Harpacticoida), *J. Plankton Res.*, 8, 1149, 1986.
15. Carlotti, F. and Sciandra, A., Population dynamics model of *Euterpina acutifrons* (Copepoda: Harpacticoida) coupling individual growth and larval development, *Mar. Ecol. Prog. Ser.*, 56, 225, 1989.
16. van den Bosch, F. and Gabriel, W., A model of growth and development in copepods, *Limnol. Oceanogr.*, 39, 1528, 1994.
17. Gaedke, U., Population dynamics of the calanoid copepods *Eurytemora affinis* and *Acartia tonsa* in the Ems-Dollart-Estuary: a numerical simulation, *Arch. Hydrobiol.*, 118, 185, 1990.
18. Souissi, S. and Nival, P., Modeling of population dynamics of interacting species: effect of exploitation, *Environ. Model. Assess.*, 2, 55, 1997.
19. Klein Breteler, W.C.M., Fransz, H.G., and Gonzalez, S.R., Growth and development of four calanoid copepod species under experimental and natural conditions, *Neth. J. Sea Res.*, 16, 195, 1982.
20. Klein Breteler, W.C.M. and Gonzalez, S.R., Culture and development of *Temora longicornis* (Copepoda, Calanoida) at different conditions of temperature and food, *Syllogeus*, 58, 71, 1986.
21. Klein Breteler, W.C.M., Gonzalez, S.R., and Schogt, N., Development of *Pseudocalanus elongatus* (Copepoda, Calanoida) cultured at different temperature and food conditions, *Mar. Ecol. Prog. Ser.*, 119, 99, 1995.
22. Uye, S.-I., Fecundity studies of neritic calanoid copepods *Acartia clausi* Giesbrecht and *A. steueri* Smirnov: a simple empirical model of daily egg production, *J. Exp. Mar. Biol. Ecol.*, 50, 255, 1981.
23. Vidal, J., Physioecology of zooplankton. I. Effects of phytoplankton concentration, temperature, and body size on the growth rate of *Calanus pacificus* and *Pseudocalanus* sp., *Mar. Biol.*, 56, 111, 1980.
24. Vidal, J., Physioecology of zooplankton. II. Effects of phytoplankton concentration, temperature, and body size on the development and molting rates of *Calanus pacificus* and *Pseudocalanus* sp., *Mar. Biol.*, 56, 135, 1980.
25. Souissi, S. and Ban, S., The consequences of individual variability in moulting probability and the aggregation of stages for modelling copepod population dynamics, *J. Plankton Res.*, 23, 1279, 2001.

26. Eiane, K., Aksnes, D.L., Ohman, M.D., Wood, S., and Martinussen, M.B., Stage-specific mortality of *Calanus* spp. under different predation regimes, *Limnol. Oceanogr.*, 47, 636, 2002.

27. Ohman, M.D. and Wood, S.N., The inevitability of mortality, *ICES J. Mar. Sci.*, 52, 517, 1995.

28. Ohman, M.D. and Hirche, H.-J., Density-dependent mortality in an oceanic copepod population, *Nature*, 412, 638, 2001.

29. Bernard, O. and Gouzé, J.L., Transient behaviour of biological loop models, with application to the Droop model, *Math. Biosci.*, 127, 19, 1995.

30. Bernard, O. and Gouzé, J.L., Robust validation of uncertain models, *Eur. Conf. Control*, 2–4, 1261, 1995.

31. Bernard, O. and Gouzé, J.L., Global qualitative behavior of a class of nonlinear biological systems: application to the qualitative validation of phytoplankton growth models, *Artif. Intel.*, 136, 29, 2002.

32. Bernard, O. and Souissi, S., Qualitative behavior of stage-structured populations: application to structural validation, *J. Math. Biol.*, 37, 291, 1998.

33. Gurney, W.S.C., Nisbet, R.M., and Lawton, J.H., The systematic formulation of tractable single-species population models incorporating age structure, *J. Anim. Ecol.*, 52, 479, 1983.

34. Ban, S., Effect of temperature and food concentration on post-embryonic development, egg production and adult body size of calanoid copepod *Eurytemora affinis*, *J. Plankton Res.*, 16, 721, 1994.

35. Escribano, R., Irribarren, C., and Rodriguez, L., Influence of food quantity and temperature on development and growth of the marine copepod *Calanus chilensis* from northern Chile, *Mar. Biol.*, 128, 281, 1997.

36. Yassen, S.T., Compétition entre trois espèces de copépodes planctoniques en élevage : *Euterpina acutifrons, Temora stylifera, Acartia clausi.* Étude ecophysiologique. Thèse Doctorat-Univ. P. et M. Curie Paris VI, 1984, 161.

37. Souissi, S., Carlotti, F., and Nival, P., Food and temperature-dependent function of moulting rate in copepods. An example of parameterization for population dynamics models, *J. Plankton Res.*, 19, 1331, 1997.

38. Souissi, S. and Uye, S.I., Simulation of the interaction between two copepods, *Centropages abdominalis* and *Acartia omorii* in an eutrophic inlet of the Inland Sea of Japan: role of predation and cannibalism, *J. Plankton Res.*, in revision.

39. Uye, S.-I. and Liang, D., Copepods attain high abundance, biomass and production in the absence of large predators but suffer cannibalistic loss, *J. Mar. Syst.*, 15, 495, 1998.

40. Hay, S.J., Evans, J.T. and Gamble, J.C., Birth, growth and death rates for enclosed populations of calanoid copepods, *J. Plankton Res.*, 10, 431, 1988.

41. Klein Breteler, W.C.M. and Schogt, N., Development of *Acartia clausi* (Copepoda, Calanoida) cultured at different conditions of temperature and food, in *Ecology and Morphology of Copepods*, Ferrari, F.D. and Bradley, B.P., Eds., Kluwer Academic, Dordrecht, 1994, 469.

42. Landry, M.R., Population dynamics and production of a planktonic marine copepod, *Acartia clausii*, in a small temperate lagoon on San Juan Island, Washington, *Int. Rev. Ges. Hydrobiol.*, 63, 77, 1978.

43. Gaudy, R., Feeding four species of pelagic copepods under experimental conditions, *Mar. Biol.*, 25, 125, 1974.

44. Paffenhöfer, G.A. and Knowles, S.C., Omnivorousness in marine planktonic copepods, *J. Plankton Res.*, 2, 355, 1980.

45. Liang, D. and Uye, S.-I., Population dynamics and production of the planktonic copepods in a eutrophic inlet of the Inland Sea of Japan. III. *Paracalanus* sp., *Mar. Biol.*, 127, 219, 1996.

46. Liang, D. and Uye, S.-I., Population dynamics and production of the planktonic copepods in a eutrophic inlet of the Inland Sea of Japan. IV. *Pseudodiaptomus marinus*, the egg-carrying calanoid, *Mar. Biol.*, 128, 415, 1997.

47. Caswell, H. and Twombly, S., Estimation of stage-specific demographic parameters for zooplankton populations: methods based on stage-classified matrix projection models, in *Estimation and Analysis of Insect Populations*, McDonald, L.L., Ed., Springer, New York, 1989, 93.

48. Parslow, J. and Sonntag, N.C., Technique of system identification applied to estimating copepod population parameters, *J. Plankton Res.*, 1, 137, 1979.

49. Jellison, R., Dana, G.L. and Melack, J.M., Zooplankton cohort analysis using systems identification techniques, *J. Plankton Res.*, 17, 2093, 1995.

50. Rigler, F.H. and Cooley, J.M., The use of field data to derive population statistics of multivoltine copepods, *Limnol. Oceanogr.*, 19, 636, 1974.

51. Green, E.P., Harris, R.P., and Duncan, A., The production and ingestion of faecal pellets by nauplii of marine calanoid copepods, *J. Plankton Res.*, 14, 1631, 1992.
52. Peterson, W.T. and Painting, S.J., Developmental rates of the copepods *Calanus australis* and *Calanoides carinatus* in the laboratory, with discussion of methods used for calculation of development time, *J. Plankton Res.*, 12, 283, 1990.
53. Blaszkowski, C. and Moreira, G.S., Combined effects of temperature and salinity on the survival and duration of larval stages of *Pagurus criniticornis* (Dana) (Crustacea, Paguridae), *J. Exp. Mar. Biol. Ecol.*, 103, 77, 1986.
54. Manly, B.F.G., *Stage-Structured Populations Sampling, Analysis and Simulation*, Chapman & Hall, London, 1990, 187.
55. Aksnes, D.L., Miller, C.B., Ohman, M.D., and Wood, S.N., Estimation techniques used in studies of copepod population dynamics: a review of underlying assumptions, *Sarsia*, 82, 279, 1997.
56. Miller, C.B. and Tande, K.S., Stage duration estimation for *Calanus* populations, a modelling study, *Mar. Ecol. Prog. Ser.*, 102, 15, 1993.

Section III

Simulation

24

The Importance of Spatial Scale in the Modeling of Aquatic Ecosystems

Donald L. DeAngelis, Wolf M. Mooij, and Alberto Basset

CONTENTS

24.1 Introduction 383
24.2 Spatial Scales in Aquatic Ecosystems: Origins and Effects 384
    24.2.1 General Considerations of Spatial Scale in Models 384
    24.2.2 Hierarchy of Spatial Scales 385
24.3 A Brief Survey of Spatial Scales in Models 386
    24.3.1 Spatial Resolution Chosen to Reflect the Scale of Variation of Abiotic Conditions 386
        24.3.1.1 Lower Trophic Level Biomass Dynamics: Marine Models 387
        24.3.1.2 Lower Trophic Level Biomass Dynamics: Freshwater Models 388
        24.3.1.3 Fish Movement and Dynamics: Marine Systems 388
        24.3.1.4 Fish Movement and Dynamics: Streams 389
    24.3.2 Spatial Resolution Chosen to Reflect Scale of Variation of Biotic Conditions 389
    24.3.3 Spatial Scale Imposed by Individual Organism Size 390
    24.3.4 Spatial Scale Imposed by Emergent Patterns 392
24.4 Discussion 393
Acknowledgment 395
References 395

24.1 Introduction

Although the earliest mathematical models in ecology were, by necessity, simple and avoided spatial complexity, ecologists gradually realized that, without taking into account spatial heterogeneity, models could not explain many of the empirical observations of ecology. A current view is that "the problem of pattern and scale is the central problem in ecology, unifying population biology and ecosystems science, and marrying basic and applied ecology."[64]

The topic of scale in aquatic systems (marine and freshwater) has received a great deal of attention. Several books[42,53,70,80] emphasize spatial scale in, respectively, phytoplankton communities, the geography of fisheries, fish recruitment, and coral communities. Edited volumes have been devoted to patterns and processes in aquatic systems [38,118] and a classic ecological text is largely concerned with modeling such patterns.[86] In addition, overview papers and volumes have dealt with such important topics as effects of scale in doing research in lakes,[35] spatial scale and species diversity,[65,75,134] and the broad topic of spatial scale in general.[92]

The modeling of aquatic systems is a rapidly evolving field. The way spatial scale is included in models continues to evolve as well, as modelers learn which techniques are successful and which are not. Our primary goal here is to observe how ecologists develop models that deal with spatial scale. Spatially explicit models are now common in aquatic ecology and other branches of ecology and the models span a broad range of types of systems. In each case, the construction of a system model requires making choices about the spatial extent and spatial resolution that reflect the peculiarities of the specific systems being modeled. But the choices must also be based on underlying logic concerning what spatial scales are appropriate. After introducing some of the definitions and issues involved in spatial scale, we examine this logic through a number of examples.

24.2   Spatial Scales in Aquatic Ecosystems: Origins and Effects

24.2.1   General Considerations of Spatial Scale in Models

What is meant by spatial scale and why is it important in modeling? Powell[97] defined *scale* as synonymous with "scale of variability," and Levin and Pacala[67] noted that the scale of a process is the range over which it varies according to some criterion. Unless there is variation of a process or pattern in space, the topic of scale does not arise. Spatial scales come from a variety of sources. One means of classification is first to divide the sources of spatial scale into physical and biological factors.

Physical factors can be divided into physical structures, or groups of structures, and disturbances. Scales associated with physical structures can be relatively permanent in form, such as the size of a bay, the width of a water channel, or the size of a crevice that provides shelter to an animal. Other structures are features of water movement maintained by energy flows. In ocean systems these include ocean currents and fronts, upwellings, internal waves that can concentrate planktonic organisms, Langmuir cells,[43] eddies and mixing at many scales,[37,102] gradients in temperature, dissolved oxygen, or nutrient concentration, and microscale turbulence that affects nutrient uptake by phytoplankton.[80] The effects of these scales on marine ecosystems and fish populations have been discussed in detail by numerous authors.[38,42,50,53,70,80] Similar sets of spatial scales can be listed for freshwater lakes, rivers, and streams.

Physical disturbances also act at characteristic spatial scales. An El Niño event has a scale of oceanic basin dimensions,[99] while a hurricane may affect an area of coastal ocean,[96] a storm may wash out a patch of macroalgae,[120] and heavy rainfall may cause a scour event in a stream.[79]

Spatial scales are also associated with biotic factors, such as the sizes of organisms, or parts of modular organisms, such as clonal plants.[130] Because their range of influence can extend beyond their sizes, organisms can create larger spatial scales. For example, the effect of an individual fish on other fishes in a school or on prey organisms is related to its range of perception. The territories and home ranges of individuals are also spatial scales. These are aspects of "ecological neighborhood," defined by Addicott et al.[1] as "the region within which an organism is active or has some influence during the appropriate period of time." Beyond individual organisms, there are spatial scales related to aggregations of organisms, such as plankton patches, seagrass beds, or coral reefs, and to the spatial structure of populations and metapopulations. Scales may be associated with outbreaks of organisms, as well, such as eruptions of echinoderms and other populations.[33,109] Finally, spatial scales may emerge from the interactions of biotic populations. Phytoplankton and zooplankton have been hypothesized to interact to create patterns of patches with scales on the order of kilometers.[68,73,85,86]

Spatial scale has also been defined in a slightly different way that is sometimes more convenient to use when discussing observations or models. Wiens[135] and Schneider[112] defined "outer" and "inner" scales. The outer scale is the whole spatial area over which one observes or models a particular phenomenon, whereas the inner scale, or grain, is the minimum area resolvable by measurement. A spatial model typically prescribes some extent, which is the size of the spatial arena over which the simulation is performed, and some size of the spatial cells making up the model's spatial extent, which is limit of resolution of the model. By choosing an inner scale, or spatial cell size in a model, the modeler is, in effect, assuming that the variables in the model exhibit no spatial variation, or are homogeneous, within the cell. The model is capable of describing any characteristic spatial scales of structures and processes that lie between these inner and outer scales. How spatial scales are chosen in models is the subject of the next section.

24.2.2 Hierarchy of Spatial Scales

A number of questions revolve around the issue of spatial scale. A key question confronts modelers: What are the best scales of resolution and extent to use in modeling a particular phenomenon? Discussing this question from a broad viewpoint, Levin[64] stated that "there is no single 'correct' scale on which to describe populations or ecosystems." Levin concluded, as a generalization, that "the problem is not to choose the correct scale of description, but rather to recognize that change is taking place on many scales at the same time, and that it is the interaction among phenomena on different scales that must occupy the attention." This is an insight that was recognized as well by the proponents of a hierarchical view of ecosystems[3,5,87] and is reflected in some models.[31]

Any pattern observed in nature is the result of processes acting on many spatial and temporal scales, from detailed chemistry at the atomic level to evolutionary events like continental drift. Scientific investigations are typically focused narrowly on specific aspects of any phenomenon, which reduces the number of scales that have to be considered at once. An investigation of the viability of a particular salmon population in a given home river, for example, might focus primarily on spawning and the downstream movement of smolts. A key idea of the hierarchical concept of ecological systems is that, often, spatial scales that are not immediately above or below the one of immediate interest can be ignored in modeling a phenomenon.[77,87] Thus, for the salmon study, the large spatial scales of the evolutionary history or the total life cycle (e.g., survival in the ocean) can be ignored for the limited aims of the study. If the study is concerned only with demographic aspects, then smaller scales, such as the small-scale random movements of individual salmon, may also be ignored. Such elimination of spatial scales not directly related to the problem of interest is an aspect of simplifying a study into solvable subunits that is typical in science.

Even when effects from higher-level scales can be ignored and effects from lower-level scales filtered out, the question remains: What spatial scales do we need to consider in dealing with a particular problem? While Levin[64] noted that there is no "right" scale, it is still highly important to understand how scale is chosen in particular models. This is something that is usually addressed in the conceptual model of that system. Consider a specific question, for example: What is the likely fraction of a pulse of salmon smolts swimming downstream to survive passage through a long reservoir? This specific question means that only aspects of the reservoir (e.g., predators and other sources of mortality) and the smolts (their sizes, speed, etc.) relevant to their survival will occur in the model. The conceptual model is a picture of how the system works, specifically how the smolts will interact with their predators. This model is based on general theoretical ideas and specific knowledge and data concerning that system. Within that conceptual model are notions of what the important spatial scales are. These include the most obvious scale, the size of the reservoir, as the outer scale or extent. The spatial scale of the reservoir determines how long smolts are exposed to predators in that reservoir. What is the inner scale, or scale of resolution? Taking the entire reservoir as the scale of resolution is unreasonable, as it would imply the smolt prey and their predators are well mixed in the whole reservoir, which is not likely to be the case. If the modeler were attempting to describe individual predation interactions in detail, then the range of perception of predators might form a natural scale of resolution. If less detail is wanted, it can be argued that the larger scale of the mean home range size over which individual predators move in search of prey forms a reasonable scale of resolution.[26,93] Other spatial scales may exist in such a system, however. One is the spatial size of the "pulse" of smolts, as these often move downstream in spatially distinct patches, smaller than the size of the reservoir. Another might be the scale of spatial heterogeneity within the reservoir. Either of these may influence the scale of resolution needed. There is no automatic way to decide on the conceptual model or the spatial scales needed to describe the system sufficiently well to address the question. This is a matter of the knowledge and insight of the ecologist. After the conceptual model is developed, it is translated into a quantitative model to compute the loss of smolts in the reservoir.

Another question regarding spatial modeling concerns whether space in a system actually has to be modeled explicitly to correctly incorporate the effects of spatial scales on the dynamics. We know in fact that the effects of space can be usefully included in many models that are not spatially explicit. Biological processes such as growth, respiration, and feeding rates are well known as functions of body size.[14,91,111] The effects of organism body size on such processes as feeding rates can be built into functional responses of models that do not include space in any explicit way (e.g., Reference 140). In addition, spatial scale is implicit in many models that are not spatial simulations. Such models include spatially implicit models, such as those of Roughgarden and Iwasa[105] and Roughgarden et al.[106] for bivalves, which do not model the explicit spatial dynamics of the system, but consciously focus on small spatial domains that are open to larval settlement. Many metapopulation models[41] are spatially implicit, with spatial scales implicit in the colonization and extinction probabilities, and predator–prey interaction models can be formulated to take into account the scales of aggregation of predators and prey.[21] Thus, many models that are not simulation models per se contain spatial scales of one sort or another (see Roughgarden[104] for a discussion of spatially implicit models).

As adaptable as spatially implicit models are, they cannot easily represent many aspects of spatial pattern, especially the patterns of spatially complex systems. To describe patterns in space one usually must use simulation models and must consider carefully how to structure the model, particularly with respect to the two scales defined by Wiens[135] and Schneider:[112] the spatial extent, or area covered by the model, and the spatial resolution of the model. If, for example, one is modeling a whole population over some time period, then the spatial extent should cover the entire spatial range of the population during that time. But it is possible also for a model to represent only a small representative part of a population or community (e.g., a unit area of the pelagic zone to represent a phytoplankton community), so the extent is often assigned somewhat arbitrarily for the particular purposes of the model. Spatial resolution, or environmental grain of a system, refers to the smallest spatial units (e.g., spatial grid cells), which depend on the level of spatial detail that is included in an explicit way. The grid cells must be small enough to represent essential spatial details. For example, in modeling a population in which foraging by a consumer occurs, one might assume that an important aspect of the model is how effective a particular searching strategy is at locating prey. Then it will be necessary to have a spatial resolution fine enough to represent the distribution of prey concentrations fairly precisely. If there are no other processes in the model that require finer resolution than this, then the cell size will be chosen to represent the prey distribution at precise enough resolution for the model to be useful.

The general issues concerning the types and vast numbers of spatial scales associated with any aquatic system must be considered in all spatial modeling efforts. However, given Levin's[64] point that there are no "right" scales, there is no prescribed path for choosing the inner and outer scales of a spatial model. One way to gain insight, however, is to review a broad spectrum of models and try to identify common patterns. We do this in the following section.

24.3   A Brief Survey of Spatial Scales in Models

This section reviews a sampling of published spatially explicit aquatic (both marine and freshwater) models. The purpose is to gain insight into how spatial scales are chosen in such models. Several types of models that are in common use in modeling ecological systems in space (see Czaran[22] in particular for definitions): (1) State variable models based on reaction-diffusion formulations; (2) spatially explicit metapopulation models; (3) cellular automata models; (4) individual-based simulation models; and (5) spatially implicit models, including some types of metapopulation models that incorporate spatial scales in some parameter values, but that, unlike the first four types, do not model the system explicitly in space.

It is more convenient, however, to organize our survey not around model types, but around the underlying causes of the spatial scales. We have identified four main types of reasons for the choice of this scale. The scale of resolution may be chosen (1) to relate to the scale of spatial variation in the background abiotic conditions, (2) to relate to the spatial variation in background biotic conditions, (3) to relate to the individual body sizes or ranges of influence of individuals being modeled, or (4) to relate to emergent spatial scales caused by processes in the model.

24.3.1   Spatial Resolution Chosen to Reflect the Scale of Variation of Abiotic Conditions

Physical scales are frequently dominant in aquatic ecosystems. There are at least a few model types in which biotic dynamics are simulated on a background of varying abiotic conditions. Here we consider models that (1) simulate lower trophic level biomass dynamics (including, sometimes, fish biomass), and (2) that focus on fish movements and population dynamics in inhomogeneous aquatic environments. In both model types the abiotic conditions vary spatially and the biological variables generally follow them. To capture the variation, the models are often formulated on a spatial grid. Even when they are initially formulated as spatially continuous, they must be solved by numerical methods, so a spatial cell size must be set to evaluate the model. This cell size represents the knowledge or intuition of the modeler about what spatial resolution is needed to represent the key structures and processes.

24.3.1.1   *Lower Trophic Level Biomass Dynamics: Marine Models* — An early systems

ecology model of lower trophic level production is that of O'Brien and Wroblewski,[83] who simulated the lower trophic levels in a marine upwelling over the continental shelf off western Florida. The upwelling system was modeled as a two-dimensional region, with the extent of 200 km along an axis perpendicular to the coastline and up to 200 m deep. To capture the variations in magnitude and direction of the upwelling, this region was divided into 3362 internally homogeneous rectangular cells, each 2.5 m vertically and 5 km horizontally, reflecting the fact that characteristic vertical scales tend to be orders of magnitude smaller than horizontal scales in most marine models. Within each cell a set of biological processes was represented; light- and nutrient-limited primary production of phytoplankton, nutrient uptake and excretion, consumption of phytoplankton by zooplankton, planktivory by fish, natural mortality, and decomposition, as well as movement by advection (both sinking and movement with current) and diffusion across the cell walls. Nutrients were connected with the outside through interchange with sediments and the external waters. O'Brien and Wroblewski used the model to explore the factors controlling the dynamics of the system, showing that spatial variation in phytoplankton and zooplankton stocks resulted from the spatial variation in nutrient upwelling.

The model of O'Brien and Wroblewski[83] showed the spatial modeling approach could be a powerful tool for ecological research and was a precursor of many subsequent models developed to simulate regions of the seas or oceans. We cannot review them all or any in detail, but we will attempt to give some rationale for the scales of resolution used. Tett et al.[121] simulated the vertical distribution of phytoplankton in the shelf near the Scilly Isles, and the Sound of Jura, both near Scotland. As in much oceanic modeling, the authors used a one-dimensional spatial model, considering only the vertical scale. The scale of resolution was set to allow simulation of mixing. For the Scilly Isles, where the mixing zone was narrow, a 5 m resolution was used, whereas, for the Sound of Jura, where mixing was greater, the resolution was 10 m. Predictions were compared with field data and conclusions drawn for why the model application to the Scilly Isles appeared to work better than to Sound of Jura. Computers are fast enough now to easily allow finer structure models than that of Tett et al. as well as to take into account factors, omitted by those authors, such as self-shading. Oguz et al.[84] modeled plankton productivity and N cycling in the Black Sea by dividing a 150-m-thick water column into 50 vertical layers, which was able to resolve the sharp vertical variations that occur during certain times of the year. Another vertical layer model is that of Chapelle,[15] who modeled P and N cycles in lagoon. That model used five 1-m vertical layers to capture the variation in the water column. Three additional cells, of vertical size 0.01, 0.01, and 0.08 m, were used to represent the much steeper gradient of nutrient concentration in the sediment.

In other models horizontal variations are at least as important as the vertical ones. A classic example is the model of Narragansett Bay,[82] which contains a detailed ecological model (phytoplankton, zooplankton, meroplankton, fish, benthos, and nutrients) on top of a hydrologic model. To obtain sufficient resolution for the hydrologic model, the authors used 324 square cells of about 0.86 km², but only one vertical layer was assumed. In addition, the square cells were aggregated into eight larger spatial elements, in which driving variables that change less rapidly in space, such as temperature, tidal mixing, sewage input, etc., could be individually set. Models of this general type have been applied to many coastal waters, as well as freshwater bodies. Kishi et al.[58] simulated effects of nutrients released by aquaculture in Mikame Bay, Japan. They used a grid of 2.25 ha cells to describe oxygen demand, dissolved oxygen, accumulated organic matter, as affected by currents and tides. Pastres et al.[89] modeled eutrophication in a Venice lagoon in three dimensions, with cells 1 ha horizontally and 1 m in depth to capture the variation in nutrients, spatial irregularities, and scale of eddy diffusivity. Periáñez et al.[90] modeled suspended matter in an estuarine system on a grid of 1-ha cells. Tsanis et al.[124] also justified a 100-m scale of resolution for their model of phosphorus and suspended solids dynamics in a 250-ha area of Cootes Paradise marsh (Ontario) in relation to eddy diffusivity. Because the marsh was shallow and well mixed vertically, only a single layer was modeled.

Sometimes natural features of a water body are taken as the compartments of the model. Wulff and Stigebrandt,[139] for example, modeled fluxes of nutrients (N, P, Si, and humics) in the Baltic Sea. In this case, the three basins of that sea, each thousands of square kilometers and relatively mixed internally, formed the natural cells of the model. Beyer[10] modeled eutrophication of a Danish 40-km fiord ecosystem, using six or ten well-mixed surface segments, as well as a deep segment below one of the surface segments, to obtain a good representation of water flow.

In rare cases the opportunity has been taken to compare models using different assumptions on spatial resolution. Baretta et al.[6] discussed two models of North Sea eutrophication; ERSEM and MIKE21. ERSEM contains a highly detailed ecosystem model (70 compartments), but has low spatial resolution (10 surface cells and 5 deep cells for the entire North Sea). MIKE21 has a more rudimentary ecosystem model (6 compartments), but has a much finer scale of 340-km^2 compartments (although no depth differentiation). Simulations of the two models showed that MIKE21 is more appropriate for showing specific spatial locations where eutrophication might occur, as it was much better at describing the advection-dispersion, tidal, and meteorologically induced flows, but the greater ecological complexity of ERSEM might be more important for following long-term trends. Eventually, increased computer speeds may lessen the need for such trade-offs between ecological and spatial resolution in models.

Not all models attempt to simulate the dynamics of linked spatial cells; some merely take into account the spatial variability that affects processes across geographic domains. Wroblewski et al.[138] modeled plankton dynamics in the North Atlantic, using a 1° longitudinal and latitudinal grid. Both light and nutrient, nitrogen, varied geographically, and the scale was chosen to represent the resolution over which variation can be significant, but no horizontal mixing was assumed. In this case, the spatially distributed model was simply the sum of local models, each weighted by the area of the zone covered. The spatial resolution was chosen to reflect the important differences in abiotic conditions across the modeled region.

24.3.1.2   Lower Trophic Level Biomass Dynamics: Freshwater Models — Many large lakes

have been modeled spatially; an early example is the phytoplankton–zooplankton–nutrient model of western Lake Erie,[28] which included mass transport. The system was simulated spatially using seven irregular boxes of scale of roughly 1600 km^2, on the basis that there was relative homogeneity of conditions within the cells. A partial differential equation in terms of finite differences described mass transport.

A similar rationale of using spatially homogeneous compartments motivated the models used to describe Lake Baikal. Silow et al.[114] reviewed several such models that describe seasonal biomass dynamics in that lake. Their own model, including a pollutant, divided the lake into ten regions that differed significantly in environmental conditions. Each region had three vertical layers; 0 to 50 m, 50 to 250 m, and >250 m, although the bottom layers were aggregated into a single compartment.

24.3.1.3   Fish Movement and Dynamics: Marine Systems — Higher trophic levels, such

as fish and krill, have also been simulated in spatial models as functions of spatially varying abiotic conditions. Such models may allow active movement of adults, as well as passive movement of eggs and larvae. Power[98] simulated the transport of anchovy eggs in coastal California waters. More than 500 square grid cells of 1370 km^2 were used, a scale allowing mean geostrophic flow, wind-driven Ekman transport, and eddy diffusivity to be simulated reasonably well. The model predicted substantial movement of eggs from high initial density to low initial density regions. Reyes et al.[101] simulated the migration of fish in the Laguna de Terminos, Mexico. This estuarine lagoon was represented by 14 areas of 100 km^2 cells each, and the fish density was a state variable, with growth variables in each cell fluctuating through the year due to local conditions. The spatial scale resolved relates to the scale of changes in habitat quality. Model fish migrated when conditions in one deteriorated, as described by a finite-difference equation. Daan[23] used 186 square cells (of approximate size 2100 km^2) to simulate the effects of closed fishing on some North Sea stocks, as this resolution allowed realistic description of spatial variation in fishing. Sekine et al.[113] modeled the effects of coastal development on a fish population in Shijuki Bay, Japan. They modeled several plankton types, and fish in five spatial cells, varying from 1,640,000 to 102,500,000 m^3, differentiated environmentally, where the fish could move between the cells. The effect of thermal preferences on the migration patterns of large fish (bluefin tuna in the Gulf of Maine) was simulated by Humston et al.,[46] using a kinetic model for individual fish movement (see also Reference 27) in response to ambient temperature, and satellite data on surface water temperatures at a spatial resolution of 1 km^2. Using a corresponding spatial cell size of 1 km^2, the model was able to produce reasonable agreement with observed bluefin tuna spatial distributions.

Fish movement and predator–prey dynamics have been combined in some models. Ault et al.[4] modeled the interaction of spotted seatrout and pink shrimp in Biscayne Bay, Florida, using a finite-element hydrologic model[131,132] to simulate water currents and salinity. The mean spacing between grid nodes was about 500 m, but this resolution was finer or coarser in different regions of the bay, depending on the sharpness of the gradients. The model was able to show important relationships between spawning time and larval transport and settlement, as well as the effects of changing habitat quality on growth of the spotted seatrout.

24.3.1.4 *Fish Movement and Dynamics: Streams* — Variations in abiotic factors influence the modeling of stream systems. Longitudinal scales of resolution are the most conspicuous for streams. Clark and Rose[19] in modeling competition between brook char and rainbow trout in a stream, divided a 600 m reach of stream into cells for pools (2 m), riffles (1.6 m), and runs (0.4 m), as these posed different habitat conditions for the fish. Perhaps less conspicuous are the vertical scales in streams. An important scale in many streams is the boundary layer on the bottom, created by bottom roughness. Although this bottom layer may be only several centimeters thick, it is significant in that there is almost no water flow and because nutrients and detritus may be temporarily stored there before moving farther downstream. Movement of nutrients between this zone and the free-flowing water is important to stream production and has been included in stream models (e.g., Reference 25).

24.3.2 Spatial Resolution Chosen to Reflect Scale of Variation of Biotic Conditions

The habitat of marine and aquatic organisms is shaped not only by the spatial scales of physical structures and processes, but also by the scales created by biota themselves. Coral reefs and beds of seagrasses and algae are among the more conspicuous biotic structures, but important forms of transient patchiness, such as patches of zooplankton, krill, and other prey populations, also have spatial scales that affect their predators.

Coral reefs have attracted great attention from marine biologists because of the biotic diversity they harbor, e.g., 100 families of fish have coral reef representatives.[108] Coral reefs are relatively permanent physical features on ecological timescales of years. The characteristic sizes of reefs vary from small patch reefs that have a spatial scale of tens of meters, to continuous reefs having scales of hundreds of meters, and finally to larger spatial groupings, from clusters in local areas to larger geographic complexes.[117] This typically defines three spatial scales useful for modeling populations.[16] The largest scale, or metapopulation scale, must include the regional groupings of reefs that incorporate the whole population, and that are connected by long-distance larval transport. The smallest scale is the local area of the order of patch reef or part of a larger reef, on which individuals interact. A mesoscale, between these two scales, can also be defined. (Of course, even a small patch reef is itself a complex topographic structure that has microhabitats and important refuges from predation of the order of a meter or less in scale.) Often it is of interest to incorporate two or more of these scales into a model to study the effects of regional processes, such as dispersal, on local species richness.[29] Forrester et al.[34] have developed a model for a subpopulation on the mesoscale. On this scale the adult population is relatively closed (that is, there is relatively little movement of adults in and out of the system) but it is open to settlement from fish larvae external to the system modeled. Their mesoscale model explicitly simulates the location of 625 reefs that are in close enough spatial proximity for adults to migrate among them according to rules derived from field studies. Adults may also suffer mortality and obtain recruits from settling larvae.

The model is used to study the relationships among larval settlement, mortality, and population size at this mesoscale. An interesting theoretical concept being examined by this modeling is that of "scale transitions," or changes in the effective dynamics when one looks at a system at two different scales. In this case, the authors used the model to determine whether the density-dependent relationship between recruitment and adult abundance that is often observed at the local scale is also evident at a larger spatial scale.

Organisms must also respond to an environment of much more transient spatial patches, such as their prey organisms. Zooplankton are usually patchily distributed. In models of the growth and survival of larval fish, it may be essential to capture this patchiness. Letcher and Rice[63] modeled the growth and movement of larval fish in a spatial model. Larval fish may not be able to survive on average densities of zooplankton[80] and depend on high-density zooplankton patches. The fish in Letcher and Rice's model may move hundreds of meters in a day, encountering high-density patches of zooplankton prey as they move. Letcher and Rice, using their model, showed that encounters with patches of a scale of the order of a hundred meters, which the fish may stay in for hours, is critical for larval growth and survival. More generally, areas of habitat especially favorable for survival of larval fish may range from hundreds of meters[63] to thousands of square miles in extent,[70,98] and these scales have been included in models. Likewise, zooplankton may depend on high-density phytoplankton patches, which in turn are influenced by patches of nutrient.[126] At microscales, zooplankters may excrete tiny nutrient patches that algal cells can use.[119]

Other mobile prey species, such as krill, are also patchily distributed. A complex cellular automata model, consisting of passively drifting krill, foraging penguins, and a small fishing fleet, was developed by Marin and Delgado.[74] An area of greater than 800,000 km² was modeled, with a cell size of 85.75 km². The cell size is fine enough both to sufficiently resolve the ocean currents that carry the krill and the changing spatial pattern of krill abundance. The complex spatial patterns of marine prey and predators have also been studied by Fauchald et al.,[32] who found that murres and their prey, capelins, exhibit spatial structure on three scales. They found overlapping structures of capelin and murres at a scale of >90,000 km², possibly in part reflecting common habitat preferences. They also found overlapping patterns on a scale of ~2500 km², possibly reflecting aggregative movement of murres toward high capelin densities. They found spatial structures of both murres and capelin at the ~9 km² scale, as well. In this case, there was no significant overlap between the structures, but this may be due to the ability of the capelin to elude the birds for some time at this scale. Fauchald[31] developed a hierarchical approach for modeling foraging by predators that can take into account each of these different scales. The author assumed a fractal landscape, with high-quality patches of prey at small spatial scales nested within low-quality patches at larger scales, there being several steps in this hierarchy. In such an environment the forager could optimize its success at encountering prey by adjusting its search radius according to its current level of prey encounters.

24.3.3 Spatial Scale Imposed by Individual Organism Size

A vast array of models has been developed to simulate local dynamics of marine and aquatic organisms, including the interactions of organisms with their neighbors, both for sessile organisms such as corals and aquatic macrophytes and for fish, tunicates, krill, and other mobile organisms. In these cases the models contain spatial scales related to the size of individuals, their ranges of influence, and dispersal.

It is important to note that even models that do not simulate dynamics in space explicitly usually have parameters that incorporate information on spatial scale. Models of population growth or interactions between populations contain parameters for growth rates, foraging rates, respiration, and so forth, all of which depend on the sizes of the organisms being modeled.[14,91,111] The sizes of the parameter values in such models reflect the spatial units and sizes of the organisms being modeled. Yodzis and Innis[140] showed how to make the parameters scale for organisms of arbitrary size, an approach followed by other modelers as well (e.g., References 7 and 8).

Here, however, we restrict ourselves to the uses of spatial scales of resolution in a spatially explicit model. Models of fish schooling exemplify the importance of local scales in determining larger-scale patterns. Huth and Wissel[47] simply assumed the basic scale to be the body length of the fish. They assumed a minimum nearest neighbor distances ranging from 0.5 to 2 body lengths, and assigned a set of behavioral rules regarding searching and the attraction, repulsion, and parallel orientation of fish with their neighbors. Reuter and Breckling[100] elaborated somewhat on this basic model, allowing a fish to be influenced by all other visible fish in the school, weighted by their distance. A fish length of 30 cm was assumed, a unit of movement of 50 cm, and a basic unit cell size of 1 cm^2, in a two-dimensional representation were used. A number of scale lengths were incorporated into Reuter and Breckling's model, including a sight range of 500 cm, a distance of maximum parallel orientation of 50 cm, and a distance of 8 cm, below which only repulsion between fishes occurs. Both Huth and Wissel's and Reuter and Breckling's models were able to produce some of the characteristics of real schools of fish, although the latter's assumption that each fish takes into account the orientations of all other visible fish seems to be important for holding the school together.

Even when relatively stationary, fish may space themselves according to fairly consistent rules. In their individual-based model of rainbow trout and brook char in a stream, Clark and Rose,[19] in addition to dividing the stream in pool, riffle, and run segments, further divided these into 3600 cm^2 cells that represent the feeding territories of the individual fish. Van Winkle et al.[129] included several spatial scales in their individual-based model of sympatric populations of brown and rainbow trout: the mean length of locations providing cover (15 cm), the size-dependent area of trout feeding stations (from about 100 cm^2 to more than 1 m^2), the mean area of a redd (0.3 m^2), and the minimum usable depth as a function of trout length (0.25 × trout length). This illustrates that complex models may include a variety of important spatial scales. Still larger scales, relating to home range size of fish, may be relevant for some situations. Petersen and DeAngelis[93] and DeAngelis and Petersen[26] modeled predation on salmon smolts migrating downstream through a reservoir. As discussed in an earlier section of this chapter, in that case the home range of the predators was relevant in determining the effect of predators on a pulse of smolts. Because the home range of key predators was estimated to be about 2 km along the axis of flow of the reservoir, the length of cells was chosen to be 2 km, with the other two dimensions the width and depth of the reservoir.

In cases in which a dispersing population is modeled, the spatial scale of resolution in a model is often chosen to represent the characteristic dispersal distance in a given time step.[24,48] Lee and DeAngelis[62] developed an individual-based model for the spatiotemporal dynamics of unionid mussels, in particular modeling their spread in a river system due to transport of glochidia by host fish. The river was divided into 500-m longitudinal cells to resolve the distribution of dispersal steps, which ranged up to 5 km.

Spatial dispersal is also key to reproduction of free-spawning benthic organisms, such as bivalves, echinoderms, sponges, hydrozoans, and ascinians. If densities of reproducing organisms are too low, an Allee-type effect may depress reproduction enough to create a vicious cycle of population decline.[56,69,94] Claereboudt[18] has modeled the effects of the spatial distribution of sessile benthic spawners on reproduction, using three-dimensional spatial cells with either 25 or 400 cm^2 horizontal dimensions and 10 cm vertical layers, representing the approximate height of the turbulent boundary layer, to simulate the probability of fertilization.

An increasing number of models simulate the dynamics of sessile marine organisms explicitly in space. A convenient method for modeling many sessile populations has been the cellular automata (CA) approach, or variations on it. Unlike individual-based models, which focus on modeling individuals and their movement, CA models focus on spatial locations on a grid and the state of each given location through time. Typically, a grid cell is in one of two states: presence or absence of an organism (although more than just these binary states can be assigned in more complex models). In the CA approach the size of grid cells is chosen to approximate the size of an organism or its immediate range of influence. In the case of clonal or colonial organisms, each cell may represent a part of an organism. An early example of such a model was Maguire and Porter's[72] model of growth and competition between species of coral. A 25 × 25 grid of 1-m^2 cells was used. Coexistence of six species of different competitive abilities was studied under processes of larval settlement, growth, overtopping, and disturbance-induced death. Karlson and colleagues[51–55] have extended these studies to examine a number of issues concerning competition of sessile invertebrates, using 40 × 40 grids, with each cell the site for a part of a clonal organism. CA models have also been applied to the settlement and competition of intertidal species, such as mussels.[13,88,103,136] In these cases, the size of adult mussels usually form a scale of resolution. In Robles and Desharnais'[103] model the width of the area between the mean low and mean high tides forms an important additional spatial scale influencing the dynamics.

Disturbance scales may also be important. Petraitis and Latham[95] suggest that scour disturbances of 10 to 100 m^2 or larger may allow a stable barnacle–mussel bed to switch to another stable state, that of the macroalga *Ascophyllum*, or vice versa. Such models have also been applied to the spread of the crown-of-thorns sea star in a reef system.[128] In that case, individual reefs, organized onto 16 columns and 100 rows, formed the spatial scale of resolution.

The CA approach has also been used for aquatic macrophytes, in which case the size of the cell is chosen for purposes other than to fit a single individual or component, and is usually much larger than the size of an individual. For example, Chiarello and Barrat-Segretain[17] simulated the recolonization of cleared patches in an estuarine channel using 9-m^2 cells. This scale was large enough to encompass distinct patches of macrophytes, but small enough to represent fairly detailed patterns in the channel. Wortmann et al.[137] modeled the vegetative spread of *Zostera* using a cell size of 1.2 m^2, this size chosen such that the macrophyte biomass in a given cell is roughly homogeneous. Collins and Wlonsinski,[20] in a general model for aquatic macrophytes, used a similar cell size with the same motivation. Bearlin et al.[9] also modeled *Zostera*, but at a much finer scale of resolution. Their aim was to capture the dynamics of each individual module of the clonal plant (section of rhizome that usually bears a vegetative shoot and root bundle), instead of the larger-scale dynamics of the other models. A 1-m^2 patch constituted the outer scale, and each module, in one of three size classes, occupied a 2-cm^2 cell. Thus there were 5000 spatial cells in the model.

Models of aquatic grazers feeding on bottom resources must utilize appropriate spatial scales that reflect the patch sizes that might be utilized by the grazer in one feeding bout as well as the typical scale of its choice of movements to new patches. Blaine and DeAngelis[11] modeled the grazing of snails on periphyton using a grid of 100-cm^2 cells. In this individual-based model the snails were allowed to graze the periphyton in a given cell down to uneconomical levels before moving to another cell. More general models of this type[7,8] have been used to explore the effects of different functional responses and different strategies for when to leave a patch on the dynamics of the grazer-resource system.

24.3.4   Spatial Scale Imposed by Emergent Patterns

Most of the spatial scales used in models described above were set based on input information, such as the spatial variation in environmental conditions or the sizes of the organisms being modeled. However, there are also situations in which the processes simulated in a model interact to create spatial variation in the output variables. In these cases the modeler has to guess what the scale of variation will be and choose an appropriate grid size to capture the variation.

Simple abiotic processes can give rise to spatial scales of variation. For example, sedimentation on a floodplain is due to an interplay between the advection by water currents and rate of sedimentation, which together imply a spatial scale of variation in the depth of the sediment that is deposited. Ulbrich et al.[127] found that a cell length of 5 m along the direction of flow was sufficient to capture the spatial variation in sediment depth. In a biological example, Webster[133] modeled the detritus dynamics in streams, in which the benthic macroinvertebrate and microbial communities processed allochthonous detrital inputs into coarse and fine particulate organic matter, which also underwent stream transport and settlement. The author used 100-m longitudinal segments to describe the spatial patterns of these components along the stream. A class of stream models termed *nutrient spiraling* models[81] was developed to estimate, among other things, nutrient uptake lengths, or the mean distance traveled by a nutrient atom along the stream before being taken up by stream biota. Mulholland and DeAngelis[78] similarly modeled the effects of stream periphyton on nutrients moving downstream, with various rates of transition between the free-flowing water and the periphyton in the boundary layer. They found that a cell length of 0.1 m was easily sufficient to describe the spatial distribution of first-uptake distances of a pulse of nutrients to the system, and that typical nutrient uptake lengths (mean distance traveled by a nutrient ion before being taken up) were between 20 and 40 m in the streams modeled.

One of the classic models of theoretical ecology is that for persistence of an algal patch in a water mass favorable for growth, but surrounded by unfavorable habitat. Kierstad and Slobodkin[57] and Skellam[115] calculated the minimum size of the water mass for algal growth to offset the losses of diffusion. They showed, using a simple model, that the critical size for a patch was about 100 m, below which diffusion would wipe out growth effects. This is an example of an emergent scale. It is also a simple example of a more general question of the relationship between the size of a patch of a population or community and its persistence, resilience, or other forms of stability (e.g., Reference 30).

Much more complex biological processes are also involved in the emergence of scale lengths in ecological systems. Among the most important are those formed from Turing instabilities,[68,85,86,125] in which spatial patterns spontaneously develop in an initially uniform distribution of prey and predators. Malchow[73] studied this phenomenon in detail, using a model of phytoplankton–zooplankton interaction developed and parameterized by Scheffer,[110] adding realistic values of diffusion of the phytoplankton and zooplankton and including such factors as phytoplankton self-shading. Malchow modeled the interaction in three spatial dimensions in a region of 6400-km^2 horizontal cells with 100 m in depth. Under the realistic assumption that zooplankton diffusion was greater than that of phytoplankton, diffusive instabilities could develop in the model, leading to a cyclically varying spatial sinusoidal pattern. The horizontal scale of the pattern was roughly 20 km. Spatial resolution of 1 km horizontally and 5 m in depth was sufficient to describe the structure.

24.4 Discussion

The examples above help to answer some questions regarding the choice of scales of resolution in spatial models. The spatial size of the cells must be fine enough to represent the spatial pattern of quantities of interest at the level of detail needed to address the questions the model is attempting to answer. These quantities may be either input or output variables.

An idea of the sort of spatial extents and scales of resolution found in aquatic models can be gained from a simple plot (Figure 24.1). This shows a few of the model types described above, ranging from models of spawning and competition by benthic invertebrates to models of variation in primary productivity across oceanic scales.

The choice of a particular scale of resolution implies an assumption that model variables are relatively constant at finer scales or that effects of variability at finer scales are unimportant. How do we know that variation at finer scales is not important? An example may illustrate how to think about this. A model of a fish population on a coral reef needs to include the presence of refuges for the fish from predators. Such refuges may be much smaller than 1 m^2. It may not be necessary to model the reef at such a small scale of resolution, however. It is possible that the precise spatial locations or configurations of refuges are unimportant, if these are more or less uniformly distributed. It is then sufficient simply to know the average number of refuges, and use this density as a fish-carrying capacity parameter in a model that uses larger spatial cells, say, 100 m^2. The 100-m^2 spatial cell would contain the information about the refuges in a spatially *implicit* way. In some cases it is possible to go even further and make a model completely spatially implicit, without any division of space into cells. If there is no spatial variability involved whose spatial configuration is needed for addressing the model question, a spatially implicit model is sufficient. Roughgarden[104] shows how neighborhood models of plants may be stripped down to spatially implicit models. Along similar lines, Adler and Mosquera[2] have shown that spatial structure, formerly thought to be necessary to explain the coexistence of an arbitrary number of species competing exploitatively,[122] may not be necessary. Thus spatial models may not be needed for describing the coexistence properties of such communities.

As a generalization, we might say that the more theoretical the question that is being addressed, the more incentive there is to make it spatially implicit, as Slatkin and Anderson[116] and Gavrikov[36] did with Maguire and Porter's[72] model of coral competition. When models attempt to describe particular geographic environments, spatial configuration is often found to be important and provides motivation to model the system spatially with whatever scale of resolution is needed to capture relevant spatial pattern.

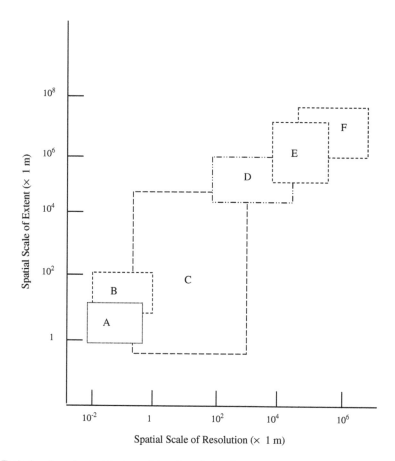

FIGURE 24.1 Typical scales of extent (outer scale) and resolution (inner scale) of several general model types used for describing phenomena in aquatic systems. These are models of (A) spawning and competition among sessile benthic organisms, (B) fish schooling behavior, (C) fish movement and dynamics in streams, (D) lower trophic level dynamics in bays, lakes, and wetlands, (E) lower trophic level dynamics and of fish movement and dynamics in coastal oceans, and (F) variations in primary productivity on an oceanic scale.

However, it is often the case even in more generic models, such as the stream fish model by Van Winkle et al.[129] (see above), that the spatial complexity is too great to attempt to reduce them to spatially implicit models. Even in theoretical models, such as that of Ruxton[107] for foraging animals, the spatially explicit model yields results that are not captured by a spatially implicit model.

Levin[64] raised another issue concerning spatial scale of models. He noted that, although it may be advantageous to develop models at fine spatial resolution, where individual-based mechanisms can be simulated, the models' predictive capabilities may be limited:

> At very fine spatial and temporal scales, stochastic phenomena (or deterministically driven chaos) may make the systems of interest unpredictable. Thus we focus attention on larger spatial regions, longer time scales, or statistical ensembles, for which macroscopic statistical behaviors are more regular. In physics this tradeoff is well studied, and goes to the heart of the problem of measurement. At fine scales, quantum mechanical laws must replace classical mechanical laws; laws become statistical in character, dealing only with probabilities of occupancy.[64]

Thus general laws are macroscopic ones, derived by averaging up to large enough scales, as one averages over atoms to obtain the laws of fluid mechanics.[67] This is a good general approach, but it is essential to take seriously the remark of Levin et al.:[66] "To derive robust statements about these systems, it is essential to understand what detail at the local level affects the broader scale patterns, and what is noise." In biological systems, not all details, even at the finest scales, constitute noise. As Morowitz[76] remarked, "In steam-engine thermodynamics we do not require very detailed knowledge of the system; however, in biology the case is entirely different as phenomena depend on molecular detail. A misplaced methyl group can eventually kill a whale." In the same sense, populations frequently are governed more by the detailed irregularity of a landscape rather than by its average values, just as the biota in a stream may be affected as much by a 1-h spate as by the average yearly conditions in a stream.

The choice of a proper spatial scale of resolution for a given model, therefore, must be approached carefully, as it is closely related to the problem of what can safely be averaged over. The best approach may be not only to look at a given problem from the particular scale that the modeler believes is most useful for addressing the problem, but also to follow this by looking at the same problem from both a somewhat coarser scale (scaling up) and a somewhat finer scale (scaling down). Grimm[39] recommends this hierarchical type of approach. He notes that Brandl et al.[12] and Jeltsch et al.,[49] in modeling rabies, first used a model that explained in a coarse fashion the wavelike dispersal pattern of rabies, and then scaled down to a finer resolution of detail to simulate the individual dispersal events that triggered outbreaks in front of the main wave. An example of scaling up might be to follow a spatially explicit model by one that is spatially implicit, as Roughgarden[104] did. The scaled up version may be amenable analysis that could not be performed on the spatially explicit form.

But scaling up is more typically a problem for field studies than it is for modeling. A long-standing problem has been how to extrapolate the results of aquatic studies that are small in spatial scale, such as microcosms or enclosures, to whole water bodies (e.g., Reference 35). Modeling provides one way to estimate how well a small-scale study can be scaled up to a lake or other relevant water body. MacNally[71] used models that explicitly simulated enclosures with areas of 0.25, 1, and 4 m^3, as well as a completely open system, to elucidate some of the factors that change across scales.

Such conscious application of hierarchical ideas is not yet common in modeling, but is receiving increased attention,[31,44,45,59–61,77,123] as are innovative techniques to use information at various spatial scales to calibrate and validate models.[40] The future of modeling of aquatic systems will probably not consist of the development of ever more complex models, but the skillful application of hierarchical concepts to probe a system from a number of different spatial scales.

Acknowledgment

D.L.D. was supported by the U.S. Geological Survey's Florida Integrated Science Centers. This is publication 3193 of the NIOO-KNAW Netherlands Institute of Ecology, Centre for Limnology.

References

1. Addicott, J.F., Aho, J.M., Antolin, M.F., Padilla, D.K, Richardson, J.S., and Soluk, D.A., Ecological neighborhoods: scaling environmental patterns, *Oikos*, 49, 340–346, 1987.
2. Adler, F.R. and Mosquera, J., Is space necessary? Interference competition and limits to biodiversity, *Ecology* 81, 3226, 2000.
3. Allen, T.F.H. and Starr, T.B., *Hierarchy*. University of Chicago Press, Chicago, 1982.
4. Ault, J.S., Luo, J., Smith, S.G., Serafy, J.E., Wang, J.D., Humston, R., and Diaz. G.A. A spatial dynamic multistock production model, *Can. J. Fish. Aquat. Sci*, 56(S1), 4, 1999.
5. Ball, G.L., and Gimblett, R., Spatial dynamic emergent hierarchies simulation and assessment system, *Ecol. Modelling*, 62, 107, 1992.
6. Baretta, J.W., Ruardij, P., Vested, H.J., and Baretta-Becker, J.G., Eutrophication modelling in the North Sea: two different approaches, *Ecol. Modelling*, 75/76, 471, 1994.
7. Basset, A., DeAngelis, D.L., and Diffendorfer, J.E., Stability and functional response of grazers on a landscape, *Ecol. Modelling*, 101, 153, 1997.

8. Basset, A.M., Fedele, A.M., and DeAngelis, D.L., Optimal exploitation of spatially distributed trophic resources and population stability, *Ecol. Modelling*, 151, 245, 2002.

9. Bearlin, A.R., Burgman, M.A., and Regan, H.M., A stochastic model of seagrass (*Zostera muelleri*) in Port Phillip Bay, Victoria, Australia, *Ecol. Modelling*, 118, 131, 1999.

10. Beyer, J.E., *Aquatic Ecosystems: An Operational Research Approach*. University of Washington Press, Seattle, 1981.

11. Blaine, T.W. and DeAngelis, D.L., The effects of spatial scale on predator–prey functional response, *Ecol. Modelling*, 95, 319, 1997.

12. Brandl, R., Jeltsch, F., Grimm, V., Mueller, M., and Kummer, G., Modelle zu lokalen and regionalen Aspekten der Tollwutausbreitung, *Z. Oekol. Nat.*, 3, 207–216, 1994.

13. Burrows, M.T. and Hawkins, S.J., Modeling patch dynamics on rocky shores using deterministic cellular automata, *Mar. Ecol. Prog. Ser.*, 167, 1, 1998.

14. Calder, W.A., III, *Size, Function, and Life History*. Harvard University Press, Cambridge, MA, 1984.

15. Chapelle, A., A preliminary model of nutrient cycling in sediments of a Mediterranean lagoon, *Ecol. Modelling*, 80, 131, 1995.

16. Chesson, P., Spatial scales in the study of reef fishes: a theoretical perspective, *Aust. J. Ecol.*, 23, 209, 1998.

17. Chiarello, E. and Barrat-Segretain, M.H., Recolonization of cleared patches by macrophytes: modelling with point processes and random mosaics, *Ecol. Modelling*, 96, 61, 1997.

18. Claereboudt, M., Fertilization success in spatially distributed populations of benthic free-spawners: a simulation model, *Ecol. Modelling* 121, 221, 1999.

19. Clark, M.E. and Rose, K.A., Individual-based model of stream-resident rainbow trout and brook char: model description, corroboration, and effects of sympatry and spawning season duration, *Ecol. Modelling*, 94, 157, 1997.

20. Collins, C.D. and Wlonsinski, J.H., A macrophyte submodel for aquatic ecosystems, *Aquat. Bot.*, 33, 191, 1989.

21. Cosner, G.C., DeAngelis, D.L., Ault, J.S., and Olson, D.B., Effects of spatial grouping on the functional response of predators, *Theor. Popul. Biol.*, 56, 65, 1999.

22. Czaran, T., *Spatiotemporal Models of Population and Community Dynamics*. Chapman & Hall, London, 1998.

23. Daan, N., Simulation study of effect of closed areas to all fishing, with particular reference to the North Sea ecosystem, in *Large Marine Ecosystems*, Sherman, K., Alexander, L.M., and Gold, B.D., Eds., AAAS Press, Washington, D.C., 1993.

24. Dawson, P.J. and Green, D.G., Viability of population in a landscape, *Ecol. Modelling*, 85, 165, 1996.

25. DeAngelis, D.L., Loreau, M., Neergaard, D., Mulholland, P.J., and Marzolf, E.R., Modeling nutrient-periphyton dynamics in streams: the importance of transient storage zones, *Ecol. Modelling*, 80, 149, 1995.

26. DeAngelis, D.L. and Petersen, J.H., Importance of the predator's ecological neighborhood in modeling predation on migrating prey, *Oikos*, 94, 315, 2001.

27. DeAngelis, D.L. and Yeh, G.T., An introduction to modeling migratory behavior in fishes, in *Mechanisms of Migration in Fishes*, McCleave, J.D., Arnold, G.P., Dodson, J.J., and Neill, W.H., Eds., Plenum Press, New York, 1984, 445.

28. DiToro, D.M., O'Connor, D.J., Thomann, R.V., and Mancini, J.L., Phytoplankton-zooplankton-nutrient interaction model for western Lake Erie, in *Systems Analysis in Ecology*, Vol. III, Patten, B., Ed., Academic Press, New York, 1975.

29. Doherty, P.J., Tropical territorial damselfishes: is density limited by aggression or recruitment, *Ecology*, 64, 176, 1983.

30. Donalson, D.D. and Nisbet, R.M., Population dynamics and spatial scale: effects of system size on population persistence, *Ecology*, 80, 2492, 1999.

31. Fauchald, P., Foraging in a hierarchical patch system, *Am. Nat.*, 153, 603, 1999.

32. Fauchald, P., Erikstad, K.E., and Skarsfjord, H., Scale-dependent predator–prey interactions: the hierarchical spatial distribution of seabirds and prey, *Ecology*, 81, 773, 2000.

33. Foreman, R.E., Benthic community modification and recovery following intensive grazing by *Strongylocentrotus droebachiensis*, *Helgol. Wiss. Meeresunters.*, 30, 468, 1977.

34. Forrester, G.E., Vance, R.R., and Steele, M.A., Population dynamics of reef fishes at large scales: using a simulation model to make predictions from small-scale data, in *Coral Reef Fishes. Dynamics and Diversity in a Complex Ecosystem*, Sale, P., Ed., Academic Press, New York, 2001.

35. Frost, T.M., DeAngelis, D.L., Bartell, S.M., Hall, D.J., and Hurlbert, S.H., Scale in the design and interpretation of aquatic community research, in *Complex Interactions in Lake Communities*, Carpenter, S.R., Ed., Springer-Verlag, New York, 1987, 229.

36. Gavrikov, V.L., A model of collisions of growing organisms: a further development, *Ecol. Modelling*, 79, 59, 1995.

37. Gervais, F., Opitz, D., and Behrendt, H., Influence of small-scale turbulence and large-scale mixing on phytoplankton primary production, *Hydrobiologia,* 342/343, 95, 1997.

38. Giller, P.S., Hildrew, A.G., and Raffaelli, D.G., Eds., *Aquatic Ecology: Scale, Pattern and Process*. 34th Symposium of the British Ecological Society, Blackwell Scientific, Oxford, U.K., 1994.

39. Grimm, V., Ten years of individual-based modelling in ecology: What have we learned and what could we learn in the future? *Ecol. Modelling*, 115, 129, 1999.

40. Grimm, V., Frank, K., Jeltsch, F., Brandl, R., Uchmański, J., and Wissel, C., Pattern-oriented modeling in population ecology, *Sci. Total Environ.*, 183, 151, 1996.

41. Hanski, I., *Metapopulation Ecology*. Oxford University Press, Oxford, 1999.

42. Harris, G.P., *Phytoplankton Ecology: Structure, Function, and Fluctuation*. Chapman & Hall, London, 1986.

43. Harris, G.P. and Lott, J.N.A., Observations of Langmuir circulations in Lake Ontario, *Limnol. Oceanogr.*, 18, 584, 1973.

44. Harris, G.P., Algal biomass and biogeochemistry in catchments and aquatic ecosystems: scaling of processes, models and empirical tests, *Hydrobiologia*, 349, 19, 1997.

45. Higgins, S.I. and Richardson, D.M., A review of models of alien plant spread, *Ecol. Modelling*, 87, 249, 1996.

46. Humston, R., Ault, J.S., Lutcavage, M., and Olson, D.B., Schooling and migration of large pelagic fishes relative to environmental cues, *Fish. Oceanogr.,* 9, 136, 2000.

47. Huth, A. and Wissel, C., The simulation of fish schools in comparison with experimental data, *Ecol. Modelling*, 75/76, 135, 1994.

48. Ims, R.A., Movement patterns related to spatial structure, in *Mosaic Landscapes and Ecological Processes,* Hansson, L., Fahrig, L., and Merriam, G., Eds., Chapman & Hall, London, 1995, 85.

49. Jeltsch, F., Muller, M.S., Grimm, V., Wissel, C., and Brandl, R., Pattern formation triggered by rare events: lessons from the spread of rabies, *Proc. R. Soc. Lond. B*, 264, 495, 1997.

50. Jumars, P.A., *Concepts in Biological Oceanography: An Interdisciplinary Primer*. Oxford University Press, Oxford, 1993.

51. Karlson, R.H., A simulation study of growth inhibition and predator resistance in *Hydractinia echinata*, *Ecol. Modelling*, 13, 29, 1981.

52. Karlson, R.H., Competitive overgrowth interactions among sessile colonial invertebrates: a comparison of stochastic and phenotypic variation, *Ecol. Modelling*, 27, 299, 1985.

53. Karlson, R.H., *Dynamics of Coral Communities*. Kluwer Academic, Dordrecht, 1999.

54. Karlson, R.H. and Buss, L.W., Competition, disturbance and local diversity patterns of substratum-bound clonal organisms: a simulation, *Ecol. Modelling*, 23, 243, 1984.

55. Karlson, R.H. and Jackson, J.B.C., Competitive networks and community structure: a simulation study, *Ecology,* 62, 670, 1981.

56. Karlson, R.H. and Levitan, D.R., Recruitment-limitation in an open population of *Diadema antillarus*: an evaluation, *Oecologia*, 82, 40, 1990.

57. Kierstead, H. and Slobodkin, L.B., The size of water masses containing plankton bloom, *J. Mar. Res.*, 12, 141, 1953.

58. Kishi, M.J., Uchiyama, M., and Iwata, Y., Numerical simulation for quantitative management of aquaculture, *Ecol. Modelling*, 72, 21, 1994.

59. Kolasa, J., Ecological systems in hierarchical perspective: breaks in community structure and other consequences, *Ecology*, 70, 36, 1989.

60. Kotliar, N.B. and Wiens, J.A., Multiple scales of patchiness and patch structure: a hierarchical framework for the study of heterogeneity, *Oikos*, 59, 253, 1990.

61. Laval, P., Hierarchical object-oriented design of a concurrent, individual-based, model of a pelagic Tunicate bloom, *Ecol. Modelling*, 82, 265, 1995.
62. Lee, H.-L. and DeAngelis, D.L., A simulation study of the spatio temporal dynamics of the Unionid mussels, *Ecol. Modelling*, 95, 171, 1997.
63. Letcher, B.H. and Rice, J.A., Prey patchiness and larval fish growth and survival: inferences from an individual-based model, *Ecol. Modelling*, 95, 29, 1997.
64. Levin, S.A., The problem of pattern and scale in ecology, *Ecology*, 73, 1943, 1992.
65. Levin, S.A., Multiple scales and the maintenance of biodiversity, *Ecosystems*, 3, 498, 2000.
66. Levin, S.A., Grenfell, B., Hastings, A., and Perelson, A.S., Mathematical and computational challenges in population biology and ecosystems science, *Science*, 275, 334, 1997.
67. Levin, S.A. and Pacala, S., Theories of simplification and scaling of spatially distributed processes, in *Spatial Ecology*, Tilman, D. and Karieva, P., Eds., Princeton University Press, Princeton, NJ, 1997, 271.
68. Levin, S.A. and Segel, L.A., Hypothesis for origin of planktonic patchiness, *Nature*, 259, 659, 1976.
69. Levitan, D.R., The ecology of fertilization in free-spawning invertebrates, in *Ecology of Marine Invertebrate Larvae*, McEdward, L.R., Ed., CRC Press, Boca Raton, FL, 1995, 123.
70. MacCall, A.D., Dynamics Geography of Marine Fish Populations. Washington Sea Grant Program. Distributed by the University of Washington Press, Seattle, 1990.
71. MacNally, R., Scaling artefacts in confinement experiments: a simulation model, *Ecol. Modelling*, 99, 229, 1997.
72. Maguire, L.A. and Porter, J.W., A spatial model of growth and competition strategies in coral communities, *Ecol. Modelling*, 3, 249, 1977.
73. Malchow, H., Nonequilibrium structure in plankton dynamics, *Ecol. Modelling*, 75/76, 123, 1994.
74. Marin, V.H. and Delgado, L.E., A spatially explicit model of the Antarctic krill fishery off the South Shetland Islands, *Ecol. Appl.*, 11, 1235, 2001.
75. McGowan, J.A. and Walker, P.W., Pelagic diversity patterns, in *Species Diversity in Ecological Communities*, Ricklefs, R.E. and Schluter, D., Eds., University of Chicago Press, Chicago, 1993, 203.
76. Morowitz, H.J., *Energy Flow in Biology*. Academic Press, New York, 1968.
77. Mueller, F., Hierarchical approaches to ecosystem theory, *Ecol. Modelling*, 63, 215, 1992.
78. Mulholland, P.J. and DeAngelis, D.L., Surface–subsurface exchange and nutrient spiralling, in *Streams and Ground Waters*, Jones, J.B. and Mulholland, P.J., Eds., Academic Press, San Diego, 2000, 149.
79. Mulholland, P.J., Steinman, A.D., Palumbo, A.V., DeAngelis, D.L., and Flum, T.E., Influence of nutrients and grazing on the response of stream periphyton communities to a scour disturbance, *J. North Am. Benthol. Soc.*, 10, 127–142, 1991.
80. Mullin, M.M., Webs and Scales: Physical and Ecological Processes in Marine Fish Recruitment, Washington Sea Grant Program. Distributed by the University of Washington Press, Seattle, 1993.
81. Newbold, J.D., Elwood, J.W., O'Neill, R.V., and Van Winkle, W., Measuring nutrient spiralling in streams, *Can. J. Fish. Aquat. Sci.*, 38, 860, 1981.
82. Nixon, S.W. and Kremer, J.N. Narragansett Bay — the development of a composite simulation model for a New England estuary, in *Ecosystem Modeling in Theory and Practice*, Hall, C.A.S. and Day, J.W., Jr., Eds., John Wiley Interscience, New York, 1977, 622.
83. O'Brien, J.J. and Wroblewski, J.S., An ecological model of the lower marine trophic levels on the continental shelf off West Florida, Tech. Rep. Geophysical Fluid Dynamics Institute, Florida State University, 1972, 170 pp.
84. Oguz, T., Ducklow, H.W., Malanotte-Rizzoli, P., Murray, J.W., Shushkina, E.A., Vedernikov, V.I., and Unluata, U., A physical–biochemical model of plankton productivity and nitrogen cycling in the Black Sea, *Deep-Sea Res. I*, 46, 597, 1999.
85. Okubo, A., Horizontal dispersion and critical scales for phytoplankton patches, in *Spatial Pattern in Plankton Communities*, Steele, J.H., Ed., Plenum Press, New York, 1978.
86. Okubo, A., *Diffusion and Ecological Problems: Mathematical Models*. Springer-Verlag, Berlin, 1980.
87. O'Neill, R.V., DeAngelis, D.L., Waide, J.B., and Allen, T.F.H., *A Hierarchical Concept of Ecosystems*, Princeton University Press, Princeton, NJ, 1986.
88. Paine, R.T. and Levin, S.A., Intertidal landscapes: disturbances and the dynamics of pattern, *Ecol. Monogr.*, 51, 145, 1981.
89. Pastres, R., Franco, D., Pacenik, G., Solidoro, C., and Dejak, C., Using parallel computers in environmental modelling: a working example, *Ecol. Modelling*, 80, 69, 1995.

90. Periáñez, R., Abril, J.M., and García-León, M., Modelling the suspended matter distribution in an estuarine system: application to the Odiel River in southwest Spain, *Ecol. Modelling*, 87, 169, 1996.
91. Peters, R.H., *The Ecological Implications of Body Size.* Cambridge University Press, Cambridge, U.K., 1983.
92. Peterson, D.L. and Parker, V.T., *Ecological Scale: Theory and Applications.* Columbia University Press, New York, 1998.
93. Petersen, J.H. and DeAngelis, D.L., Dynamics of prey moving through a predator field: a model of migrating juvenile salmon, *Math. Biosci.*, 165, 97, 2000.
94. Petersen, C. and Levitan, D.R., The Allee effect: a barrier to repopulation of exploited seas, in *Conservation of Exploited Species*, Reynolds, J.D., Mace, G.M., Redford, K.H., and Robinson, J.G., Eds., Cambridge University Press, New York, 2001, 281.
95. Petraitis, P.S. and Latham, R.E., The importance of scale in testing the origins of alternative community states, *Ecology*, 80, 429, 1999.
96. Porter, J.W., Kosmynin, V., Patterson, K.L., Porter, K.G., Jaap, W.C., Wheaton, J.L., Hackett, K., Lybolt, M., Tsokos, C.P., Yanev, G., Marcinek, D.M., Dotten, J., Eaken, D., Patterson, M., Meier, O.W., Brill, M., and Dustan, P., Detection of coral reef change by the Florida Keys coral reef monitoring project, in *The Everglades, Florida Bay, and Coral Reefs of the Florida Keys*, Porter, J.W. and Porter, K.G., Eds., CRC Press, Boca Raton, FL, 2002, 749.
97. Powell, T.M., Physical and biological scales of variability in lakes, estuaries, and the coastal ocean, in *Perspectives in Ecological Theory*, Roughgarden, J., May, R.M., and Levin, S.A., Eds., Princeton University Press, Princeton, NJ, 1989, 157.
98. Power, J.H., A model of the drift of northern anchovy, *Engraulis mordax*, larvae in the California current, *Fish. Bull.* (U.S.), 84, 585, 1986.
99. Ray, G.C., Hayden, B.P., Bulger, A.J., Jr., and McCormick-Ray, M.G., Effects of global warming on the biodiversity of coastal-marine zones, in *Global Warming and Biological Diversity*, Peters, R.L and Lovejoy, T.E., Eds., Yale University Press, New Haven, CT, 1992, 91.
100. Reuter, H. and Breckling, B., Self-organization of fish schools: an object-oriented model, *Ecol. Modelling*, 75/76, 147, 1994.
101. Reyes, E., Sklar, F.H., and Day, J.W., A regional organism exchange model for simulating fish migration, *Ecol. Modelling*, 74, 255, 1994.
102. Rintoul, S. and Macintosh, P., Ocean circulation modelling for climate studies, in *Modelling Change in Environmental Systems*, Jakeman, A.J., Beck, M.B., and McAleer, M.J., Eds., John Wiley, Chichester, U.K., 1993, 407.
103. Robles, C. and Desharnais, R., History and current development of paradigm of predation in rocky intertidal communities, *Ecology*, 83, 1521, 2002.
104. Roughgarden, J., Production functions from ecological populations: a survey with emphasis on spatially implicit models, in *Spatial Ecology: The Role of Space in Population Dynamics and Interspecific Interactions*, Tilman, D. and Karieva, P., Eds., Princeton University Press, Princeton, NJ, 1997, 296.
105. Roughgarden, J. and Iwasa, Y., Dynamics of a metapopulation with space-limited subpopulations, *Theor. Popul. Biol.*, 29, 235, 1986.
106. Roughgarden, J., Gaines, S.D., and Possingham, H., Recruitment dynamics in complex life cycles, *Science*, 241, 1460, 1988.
107. Ruxton, G.D., Foraging in flocks: non-spatial models may neglect important costs, *Ecol. Modelling*, 82, 277, 1995.
108. Sale, P.F., Introduction, in *The Ecology of Fishes on Coral Reefs*, Sale, P.F., Ed., Academic Press, New York, 1991, 3.
109. Sapp, J., *What Is Natural? Coral Reef Crisis.* Oxford University Press, New York, 1999.
110. Scheffer, M., Fish and nutrients interplay determines algal biomass, a minimal model, *Oikos*, 62, 271, 1991.
111. Schmidt-Nielsen, K., *Scaling: Why Is Animal Size So Important?* Cambridge University Press, Cambridge, U.K., 1984.
112. Schneider, D.C., The rise of the concept of scale in ecology, *BioScience*, 51, 545, 2001.
113. Sekine, M., Nakanishi, H., and Ukita, M., A shallow-sea ecological model using an object-oriented programming language, *Ecol. Modelling*, 57, 221, 1991.
114. Silow, E.A., Garman, V.J., Stom, D.J., Rosenraukh, D.M., and Baturin, V.I., Mathematical models of the Lake Baikal ecosystem, *Ecol. Modelling*, 82, 27, 1995.

115. Skellam, J.G., The formulation and interpretation of mathematical models of diffusionary processes in population biology, in *The Mathematical Theory of the Dynamics of Biological Populations*, Bartlett, M.S. and Hiorns, R.W., Eds., Academic Press, New York, 1973, 63.

116. Slatkin, M. and Anderson, D.J., A model of competition for space, *Ecology*, 65, 1840, 1984.

117. Springer, V.G., Pacific Plate biogeography, with special reference to shorefishes, *Smithsonian Contrib. Zool.*, 367, 1, 1982.

118. Steele, J.H., Ed., *Spatial Pattern in Plankton Communities*. Plenum Press, New York, 1978.

119. Sterner, R.W., Herbivores' direct and indirect effects on algal populations, *Science*, 231, 605, 1986.

120. Syms, C. and Jones, G.P., Scale of disturbance and the structure of a temperate fish guild, *Ecology*, 80, 921, 1999.

121. Tett, P., Edwards, A., and Jones, K., A model for the growth of shelf-sea phytoplankton in summer, *Estuaries Coastal Shelf Sci.*, 23, 641, 1986.

122. Tilman, D., Competition and biodiversity in spatially structured habitats, *Ecology*, 75, 2, 1994.

123. Tischendorf, L., Modelling individual movements in heterogeneous landscapes: potential of a new approach, *Ecol. Modelling*, 103, 33, 1997.

124. Tsanis, I.K., Prescott, K.L., and Shen, H., Modelling of phosphorus and suspended solids in Cootes Paradise marsh, *Ecol. Modelling*, 114, 1, 1998.

125. Turing, A.M., On the chemical basis of morphogenesis, *Philos. Trans. R. Soc. Lond. B*, 237, 37, 1952.

126. Turpin, D.H., Parslow, J.S., and Harrison, P.J., On limiting nutrient patchiness and phytoplankton growth: a conceptual approach, *J. Plankton Res.*, 3, 421, 1981.

127. Ulbrich, K., Marsula, R., Jeltsch, F., Hofmann, H., and Wissel. C., Modelling the ecological impact of contaminated river sediments on wetlands, *Ecol. Modelling*, 94, 221, 1997.

128. Van der Laan, J.D., and Hogeweg, P., Waves of crown-of-thorns starfish outbreaks, *Coral Reefs*, 11, 207, 1992.

129. Van Winkle, W., Jager, H.I., Railsback, S.F., Holcomb, B.D., Studley, T.K., and Baldrige, J.E., Individual-based model for sympatric populations of brown and rainbow trout for instream flow assessment: model description and calibration, *Ecol. Modelling*, 110, 175, 1998.

130. Waller, D.M. and Steingraber, D.A., Branching and modular growth: theoretical models and empirical patterns, in *Population Biology and Evolution of Clonal Organisms*, Jackson, J.B.C, Buss, L.W., and Cook, R.E., Eds., Yale University Press, New Haven, CT, 1985, 225.

131. Wang, J.D., Luo, J., and Ault, J.S., Flows, salinity, and some implications on larval transport in South Biscayne Bay, Florida, *Bull. Mar. Sci.*, in press.

132. Wang, J.D., Cofer-Shabica, S.V., and Chin-Fatt, J., Finite element characteristic advection model, *J. Hyd. Eng.*, 114, 1098, 1988.

133. Webster, J.R., The role of benthic macroinvertebrates in detritus dynamics of streams: a computer simulation, *Ecol. Monogr.*, 53, 383, 1983.

134. Westoby, M., The relationship between local and regional diversity: comment, *Ecology*, 79, 1827, 1998.

135. Wiens, J.A., Spatial scaling in ecology, *Funct. Ecol.*, 3, 385, 1989.

136. Wilson, W., Nesbit, R., Ross, R., Robles, C.D., and Desharnais, R.A., Abrupt population changes along continuous gradients, *Bull. Math. Biol.*, 58, 907, 1996.

137. Wortmann, J., Hearne, J.W., and Adams, J.W., A mathematical model of an estuarine grass, *Ecol. Modelling*, 98, 137, 1997.

138. Wroblewski, J.S., Sarmiento, J.L., and Flierl, G.R., An ocean basin scale model of plankton dynamics in the North Atlantic. 1. Solutions for the climatological oceanographic conditions in May, *Global Biogeochem. Cyc.*, 2, 199, 1988.

139. Wulff, F. and Stigebrandt, A., A time-dependent budget model for nutrients in the Baltic Sea, *Global Biogeochem. Cyc.*, 3, 63, 1989.

140. Yodzis, P. and Innis, S., Body size and consumer-resource dynamics, *Am. Nat.*, 139, 1151, 1992.

25

Patterns in Models of Plankton Dynamics in a Heterogeneous Environment

Horst Malchow, Alexander B. Medvinsky, and Sergei V. Petrovskii

CONTENTS

25.1   Introduction...401
25.2   The Habitat Structure 402
25.3   The Model of Plankton–Fish Dynamics 403
     25.3.1   Parameter Set...403
     25.3.2   Rules of Fish School Motion 403
25.4   Numerical Study of Pattern Formation in a Heterogeneous Environment..............404
     25.4.1   No Fish, No Environmental Noise, Connected Habitats.............................404
     25.4.2   One Fish School, No Environmental Noise, Connected Habitats:
            Biological Pattern Control 405
     25.4.3   Environmental Noise, No Fish, Connected Habitats: Physical Pattern Control 405
     25.4.4   Environmental Noise, No Fish, Separated Habitats: Geographical Pattern Control...406
25.5   Conclusions..406
Acknowledgments..407
References...407

25.1   Introduction

The horizontal spatial distribution of plankton in the natural marine environment is highly inhomogeneous.[1–3] The data of observations show that, on a spatial scale of dozens of kilometers and more, the plankton patchy spatial distribution is mainly controlled by the inhomogeneity of underlying hydrophysical fields such as temperature, nutrients, etc.[4,5] On a scale less than 100 m, plankton patchiness is controlled by turbulence.[6,7] However, the features of the plankton heterogeneous spatial distribution are essentially different (uncorrelated to the environment) on an intermediate scale, roughly, from a 100 m to a dozen kilometers.[5–8] This distinction is usually considered as an evidence of the biology's "prevailing" against hydrodynamics on this scale.[9,10]

This problem has generated a number of hypotheses about the possible origin of the spatially heterogeneous distribution of species in nature. Several possible scenarios of pattern formation have been proposed; see References 11 and 12 for a brief summary. Using reaction-diffusion equations as a mathematical tool,[13–17] many authors attribute the formation of spatial patterns in natural populations to well-known general mechanisms, e.g., to differential-diffusive Turing[18–20] or differential-flow-induced[21–23] instabilities; see References 24 and 25. However, these theoretical results, whatever their importance in a general theoretical context, are not directly applicable to the problem of spatial pattern formation in plankton. Actually, the formation of "dissipative" Turing patterns is only possible under the limitation that the diffusivities of the interacting species are not equal. This is usually not the case in a planktonic system where the dispersal of species is due to turbulent mixing. Furthermore, and this is probably more important, the patterns appearing as a result of a Turing instability are typically stationary and regular while the spatial distribution of plankton species in a real marine community is nonstationary and irregular. The impact of a differential or shear flow may be important for the pattern formation in a benthic community as a result of tidal forward–backward water motion[26] but seems to be rather artificial concerning the pelagic plankton system. Again, the patterns appearing according to this scenario are usually highly regular, which is not realistic.

Recently, a number of papers has been published about pattern formation in a minimal phyto-plankton–zooplankton interaction model[24,25,27–30] that was originally formulated by Scheffer,[31] accounting for the effects of nutrients and planktivorous fish on alternative local equilibria of the plankton community. Routes to local chaos through seasonal oscillations of parameters have been extensively studied with several models.[32–43] Deterministic chaos in uniform parameter models and data of systems with three or more interacting plankton species have been studied as well.[44,45] The emergence of diffusion-induced spatiotemporal chaos along a linear nutrient gradient has been found by Pascual[46] as well as by Pascual and Caswell[47] in Scheffer's model without fish predation. Chaotic oscillations behind propagating diffusive fronts have been shown in a prey–predator model;[48,49] a similar phenomenon has been observed in a mathematically similar model of a chemical reactor.[50,51] Recently, it has been shown that the appearance of chaotic spatiotemporal oscillations in a prey–predator system is a somewhat more general phenomenon and must not be attributed to front propagation or to an inhomogeneity of environmental parameters.[52,53] Plankton-generated chaos in a fish population has been reported by Horwood.[54]

Other processes of spatial pattern formation after instability of spatially homogeneous species distributions have been reported, as well, e.g., bioconvection and gyrotaxis,[55–58] trapping of populations of swimming microorganisms in circulation cells,[59,60] and effects of nonuniform environmental potentials.[61,62] In this chapter we focus on the influence of fish, noise, and habitat distance on the spatiotemporal pattern formation of interacting plankton populations in a nonuniform environment. Scheffer's planktonic prey–predator system[31] is used as an example. The fish are considered as localized in schools, cruising and feeding according to defined rules.[63] The process of aggregation of individual fishes and the persistence of schools under environmental or social constraints has already been studied by many other authors[64–77] and is not considered here.

25.2   The Habitat Structure

The marine environment is not a homogeneous medium. Therefore, as a simple approach, the considered model area is divided into three habitats of sizes $S \times S/2$, $S \times S$, and $S \times S/2$ with distances l_{12} and l_{23}, respectively (Figure 25.1). The inner-habitat dynamics are identical. One can think of a reaction-diffusion metapopulation dynamics, in contrast to the standard approach,[78] which does not explicitly include the inner-habitat space.

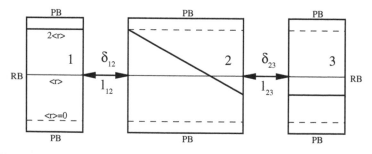

FIGURE 25.1 Model area with three habitats of different productivity r. Double mean productivity $r = 2\langle r \rangle$ in the left and low productivity $r = 0.6\langle r \rangle$ in the right habitat, connected by a linear productivity gradient in the middle. Periodic boundary (PB) conditions at lower ($x = 0$) and upper ($x = S$) border, no-flux boundary conditions (RB) at the left- ($y = 0$) and right-hand ($y = 2S + l_{12} + l_{23}$) side.

The first habitat on the left-hand side is of double mean phytoplankton productivity $2\langle r \rangle$; the third habitat on the right-hand side has 60% of $\langle r \rangle$. Both are coupled by the second with linearly decreasing productivity via coupling constants $\delta_{12} = \delta_{21}$ and $\delta_{23} = \delta_{32}$. Left and right habitats are not coupled, i.e., $\delta_{13} = \delta_{31} = 0$. The productivity gradient in the middle habitat corresponds to assumptions by Pascual.[46] This configuration and the chosen model parameters yield a fast prey–predator limit cycle in the left habitat, continuously changing into quasi-periodic and chaotic oscillations in the middle, coupled to slow limit cycle oscillations in the right habitat.

25.3 The Model of Plankton–Fish Dynamics

The inner-habitat population dynamics is described by reaction-diffusion equations whereas the inter-habitat migration is modeled as a difference term. The spatiotemporal change of two growing and interacting populations i in three habitats j at time t and horizontal spatial position (x,y) is modeled by

$$\frac{\partial X_{ij}}{\partial t} = \phi_{ij}(X_{1j}, X_{2j}) + d_{ij}\,\Delta X_{ij} + \sum_{k=1}^{3} \delta_{ik}(X_{ik} - X_{ij}); i = 1,2; j = 1,2,3 \tag{25.1}$$

Here, ϕ_{ij} stands for growth and interactions of population i in habitat j and d_{ij} for its diffusivity. For the Scheffer model of the prey–predator dynamics of phytoplankton X_{1j} and zooplankton X_{2j}, one finds with dimensionless quantities[46,63,79]

$$\phi_{1j} = r(x,y)X_{1j}\left(1 - X_{1j}\right) - \frac{aX_{1j}}{1 + bX_{1j}}X_{2j} \tag{25.2}$$

$$\phi_{2j} = \frac{aX_{1j}}{1 + bX_{1j}}X_{2j} - m(x,y,t)X_{2j} - \frac{g^2 X_{2j}^2}{1 + h^2 X_{2j}^2}f_j; \qquad j = 1,2,3 \tag{25.3}$$

The dynamics of the top predator, i.e., the planktivorous fish f_j, is not modeled by another partial differential equation but by a set of certain rules; see Section 25.3.2.

25.3.1 Parameter Set

The following set of model parameters has been chosen for the simulations described in Section 25.4; see References 46, 63, and 79:

$$\langle r \rangle = 1,\ a = b = 5,\ g = h = 10,\ \langle m_2 \rangle = 0.6,\ f_1 = f_2 = 0.5,\ f_3 = 0 \tag{25.4}$$

$$S = 100,\ x \in [0,S],\ y \in \left[0, 2S + l_{12} + l_{23}\right],\ d_{1j} = d_{2j} = 5 \times 10^{-2}; \quad j = 1,2,3 \tag{25.5}$$

The fish parameters are discussed in the following subsection.

25.3.2 Rules of Fish School Motion

The present mathematical formulation assumes fish to be a continuously distributed species, which is certainly wrong on larger scales. Furthermore, it is rather difficult to incorporate the behavioral strategies of fish. Therefore, it is more appropriate to look for a discrete model of fish dynamics; i.e., fish are considered as localized in a number of schools with specific characteristics. These schools are treated as superindividuals.[80] They feed on zooplankton and move on the numerical grid for the integration of the plankton-dynamic reaction-diffusion equations, according to the following rules:

1. The fish schools feed on zooplankton down to its protective minimal density and then move.
2. The fish schools might even have to move before reaching the minimal food density because of a maximum residence time, which can be due to protection against higher predation or security of the oxygen demand.
3. Fish schools memorize and prefer the previous direction of motion. Therefore, the new direction is randomly chosen within an "angle of vision" of ±90° left and right of the previous direction with some decreasing weight.
4. At the reflecting northern and southern boundaries the fish schools obey some mixed physical and biological laws of reflection.
5. Fish schools act independently of each other. They do not change their specific characteristics of size, speed, and maximum residence time.

The rules of motion posed are as simple but also as realistic as possible, following related reports; see References 81 through 83. In previous papers,[84,85] it has been shown that the path of a fish school obeying the above rules can have certain fractal and multifractal properties.

One of the current challenges of modelers is to find appropriate interfaces between the different types of models. Hydrophysics and low trophic levels are modeled with standard tools like differential, difference, and integral equations. Therefore, these methods are often called equation based. However, higher trophic levels like fish or even a number of zooplankton species show distinctive behavioral patterns, which cannot be incorporated in equations, but rather in rules. That is why these methods, such as cellular automata,[86] intelligent agents,[87,88] and active Brownian particles,[89] are called rule based.

As mentioned above, the fish school moves simply on the numerical grid here. Recently developed software for the grid-oriented connection of rule- and equation-based dynamics has been used. This grid connection bears a number of problems, related to the matching of characteristic times and lengths of population dynamics and such more technical conditions as the Courant–Friedrichs–Lewy (CFL) criterion for the stability of explicit numerical integration schemes for partial differential equations.[90] An improved version is in preparation. Other approximations for discrete-continuum couplings are reported in recent publications.[91–93]

25.4 Numerical Study of Pattern Formation in a Heterogeneous Environment

The motion of fish according to the defined rules will be restricted to the left and left half of the middle habitat with highest plankton abundance. Environmental noise will be incorporated following an idea by Steele and Henderson:[94] The value of m will be chosen randomly at each point and each unit time step from a truncated normal distribution between $I = \pm10\%$ and 15% of $\overline{m}$, i.e., $m(x,y,t) = \overline{m}[1 + I - rndm(2I)]$ with $rndm(z)$ as a random number between 0 and z.

Starting from spatially uniform initial conditions, we now examine whether fish and/or environmental noise and/or habitat distance can substantially perturb the plankton dynamics in the three habitats and whether they can cause transitions between homogeneous, periodic, and aperiodic spatiotemporal structures.

The phytoplankton patterns are displayed on a gray scale from black ($X_{1j} = 0$) to white ($X_{1j} = 1$). Fish will appear as a white spot.

25.4.1 No Fish, No Environmental Noise, Connected Habitats

First, the pattern formation according to the three-habitat spatial structure of the environment is studied. Fish and environmental noise are set aside. Two snapshots of the spatiotemporal dynamics after a long-term simulation are presented in Figure 25.2.

The densities in the left habitat oscillate rather quickly throughout the simulation. The diffusively coupled limit cycles along the gradient in the middle habitat generate a transition from periodic oscillations near the left border of the habitat to quasi-periodic in the middle part and to chaotic oscillations near the right border,[46] which couple to the slowly oscillating right habitat. The slow oscillator is too weak to fight the chaotic forcing from the left border. Finally, chaos prevails in the right half of the model area.

FIGURE 25.2 Rapid spatially uniform prey–predator oscillations in the left habitat and transition from plane to chaotic waves in the middle and right habitats. No fish, no environmental noise, $t = 1950, 3875$.

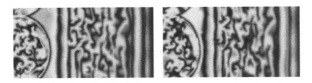

FIGURE 25.3 Fish-induced pattern formation in the left habitat. One fish school, no environmental noise, $t = 1475, 2950$.

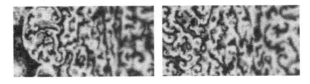

FIGURE 25.4 Noise-induced pattern formation in the left habitat. No fish, 15% environmental noise, $t = 2950, 3825$.

25.4.2 One Fish School, No Environmental Noise, Connected Habitats: Biological Pattern Control

Now, the left habitat and the left half of the middle "are stocked with fish," i.e.,

$$f_2 = \begin{cases} f_1 > 0 & \text{if } y \in \left[S/2 + l_{12}, S + l_{12} \right] \\ 0 & \text{otherwise} \end{cases} \tag{25.6}$$

The influence of one fish school is considered (Figure 25.3).

The feeding of fish leads to local perturbations of the quick oscillator in the left habitat. The perturbed site at the left model boundary acts as excitation center for a target pattern wave, however, the "inner" wave fronts are destroyed by the feeding fish and spirals are rapidly formed, invading the whole left habitat as well as the regularly oscillating part of the middle. The right half of the model area shows the same scenario of pattern formation as in Section 25.4.1. Finally, one has the left area filled with spiral plankton waves, coupled to chaotic waves on the right-hand side.

The fish induces the spatiotemporal plankton structure in the left half of the model area. External noise does not alter the dynamics; it only accelerates the pattern formation process and blurs the unrealistic spiral waves. The pronounced structures fade away and look much more realistic. The effects of fish and noise on the pattern-forming process are not distinguishable. Therefore, we investigate now whether fish is a necessary source of plankton pattern generation or whether some noise might be sufficient.

25.4.3 Environmental Noise, No Fish, Connected Habitats: Physical Pattern Control

Keeping the fish out and starting with a weak 10% noise intensity, one finds structures very similar to those in Section 25.4.1 without noise. The patterns remain qualitatively the same, however, the noise supports the expansion of the wavy and chaotic part toward the left-hand side and the borders between the areas become blurred. A slightly higher noise intensity of 15% changes the results (Figure 25.4).

The wavy and chaotic region on the right-hand side "wins the fight" against the left-hand regular structures and invades the whole space. This corresponds to a pronounced noise-induced transition[95] from one spatiotemporally structured dynamic state to another. This transition can be also seen in the

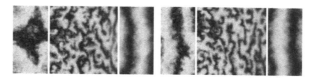

FIGURE 25.5 Suppression of irregular pattern formation in the left and right habitat. No fish, 15% environmental noise, coupling parameters $\delta_{12} = \delta_{23} = 5 \times 10^{-3}$, $t = 2500, 5000$.

local power spectra, which have been processed for the left habitat close to the left reflecting boundary, using the software package SANTIS.[96]

A very weak noise intensity of only 5% changes the scale of the power spectrum drastically; however, at an intensity of 10% some leading frequencies can be clearly distinguished (see Reference 79).

The increase to 15% lets the periodicity disappear and a nonperiodic system dynamics remains. This is another proof of the noise-induced transition from periodical to aperiodical local behavior in the left half of the model area after crossing a critical value of the external noise intensity.

The further enhancement of noise up to 25% does not change the result qualitatively. Breakthrough of the right-hand side structures only occurs earlier. However, the final spatiotemporal dynamics looks very much like the real turbulent plankton world.

25.4.4 Environmental Noise, No Fish, Separated Habitats: Geographical Pattern Control

Maintaining the same conditions as in Section 25.4.3, but separating the habitats, prevents the left and right habitat from being swamped by chaotic waves. However, the coupling along the opposite habitat borders is strong enough, i.e., the distances are not large enough, to disturb the spatially uniform oscillations in both outer habitats, and plane waves are generated, blurred by the noise (Figure 25.5).

Larger distances would, of course, decouple the dynamics. The left habitat would exhibit fast spatially uniform oscillations, the right habitat slow oscillations, whereas the middle would behave like Pascual's model system.[46] On the other hand, stronger noise and/or cruising fish would reestablish the structures found in the foregoing subsections.

25.5 Conclusions

A conceptual coupled biomass- and rule-based model of plankton–fish dynamics has been investigated for temporal, spatial, and spatiotemporal dissipative pattern formation in a spatially structured and noisy environment. Environmental heterogeneity has been incorporated by considering three diffusively coupled habitats of varying phytoplankton productivity and noisy zooplankton mortality. Inner-habitat growth, interaction, and transport of plankton have been modeled by reaction-diffusion equations, i.e., continuous in space and time. Inter-habitat exchange has been treated as proportional to the density difference. The fish have been assumed to be localized in a school, obeying certain defined behavioral rules of feeding and moving, which essentially depend on the local zooplankton density and the specific maximum residence time. The school itself has been treated as a static superindividual; i.e., it has no inner dynamics such as age or size structure. The predefined spatial structure of the model area has induced a certain spatiotemporal structure or "prepattern" in plankton and it has been investigated whether fish and/or noise and/or habitat distance would change this prepattern.

In the connected system, it has turned out that the chaotic waves of the middle habitat always prevail against the slow population oscillations on the right-hand side. The considered single fish school induces a transition from oscillatory to wavy behavior in plankton, regardless of the noise intensity. This is "biologically controlled" pattern formation.

Leaving fish aside, it has been shown that a certain supercritical noise intensity is necessary to induce a similar final dynamic pattern; i.e., the existence of a noise-induced transition between different spatiotemporal structures has been demonstrated. This is "physically controlled" pattern formation.

The dominance of biological or physical control of natural plankton patchiness is difficult to distinguish. However, biology plays its part. In the presented model simulations, noise has not only induced and accelerated pattern formation, but it has also been necessary to blur distinct artificial population structures like target patterns or spirals and plane waves and to generate more realistic fuzzy patterns.

In the system of three separated habitats, one finds "geographically controlled" patterns. Increasing distance, i.e., decreasing coupling strength, leads to decoupled dynamics, and vice versa.

Acknowledgments

The authors acknowledge helpful discussions with E. Kriksunov (Moscow). This work has been partially supported by Deutsche Forschungsgemeinschaft, Grant 436 RUS 113/631.

References

1. Fasham, M., The statistical and mathematical analysis of plankton patchiness, *Oceanogr. Mar. Biol. Annu. Rev.*, 16, 43, 1978.
2. Mackas, D. and Boyd, C.M., Spectral analysis of zooplankton spatial heterogeneity, *Science*, 204, 62, 1979.
3. Greene, C., Widder, E., Youngbluth, M., Tamse, A., and Johnson, G., The migration behavior, fine structure, and bioluminescent activity of krill sound-scattering layer, *Limnol. Oceanogr.*, 37, 650 1992.
4. Denman, K., Covariability of chlorophyll and temperature in the sea, *Deep-Sea Res.*, 23, 539, 1976.
5. Weber, L., El-Sayed, S., and Hampton, I., The variance spectra of phytoplankton, krill and water temperature in the Antarctic ocean south of Africa, *Deep-Sea Res.*, 33, 1327, 1986.
6. Platt, T., Local phytoplankton abundance and turbulence, *Deep-Sea Res.*, 19, 183, 1972.
7. Powell, T., Richerson, P., Dillon, T., Agee, B., Dozier, B., Godden, D., and Myrup, L., Spatial scales of current speed and phytoplankton biomass fluctuations in Lake Tahoe, *Science*, 189, 1088, 1975.
8. Seuront, L., Schmitt, F., Lagadeuc, Y., Schertzer, D., and Lovejoy, S., Universal multifractal analysis as a tool to characterize multiscale intermittent patterns: example of phytoplankton distribution in turbulent coastal waters, *J. Plankton Res.*, 21, 877, 1975.
9. Levin, S., Physical and biological scales and the modelling of predator–prey interactions in large marine ecosystems, in *Large Marine Ecosystems: Patterns, Processes and Yields*, Sherman, K., Alexander, L., and Gold, B., Eds., American Association for the Advancement of Science, Washington, D.C., 1990, 179.
10. Powell, T., Physical and biological scales of variability in lakes, estuaries and the coastal ocean, in *Ecological Time Series*, Powell, T. and Steele, J., Eds., Chapman & Hall, New York, 1995, 119.
11. Malchow, H., Nonequilibrium spatio-temporal patterns in models of nonlinear plankton dynamics, *Freshw. Biol.*, 45, 239, 2000.
12. Malchow, H., Petrovskii, S., and Medvinsky, A., Pattern formation in models of plankton dynamics: a synthesis, *Oceanol. Acta*, 24(5), 479, 2001.
13. Vinogradov, M. and Menshutkin, V., Modeling open-sea systems, in *The Sea: Ideas and Observations on Progress in the Study of the Seas*, Goldberg, E., Ed., Vol. 6, Wiley, New York, 1977, 891.
14. Okubo, A., *Diffusion and Ecological Problems: Mathematical Models*, Biomathematics Texts, Vol. 10, Springer-Verlag, Berlin, 1980.
15. Okubo, A. and Levin, S., *Diffusion and Ecological Problems: Modern Perspectives,* Interdisciplinary Applied Mathematics, Vol. 14, Springer-Verlag, Berlin, 2001.
16. Murray, J., *Mathematical Biology,* Biomathematics Texts, Vol. 19, Springer-Verlag, Berlin, 1989.
17. Shigesada, N. and Kawasaki, K., *Biological Invasions: Theory and Practice,* Oxford University Press, Oxford, 1997.
18. Turing, A., On the chemical basis of morphogenesis, *Philos. Trans. R. Soc. Lond. B*, 237, 37, 1952.
19. Segel, L. and Jackson, J., Dissipative structure: an explanation and an ecological example, *J. Theor. Biol.*, 37, 545, 1972.
20. Levin, S. and Segel, L., Hypothesis for origin of planktonic patchiness, *Nature*, 259, 659, 1976.

21. Rovinsky, A. and Menzinger, M., Chemical instability induced by a differential flow, *Phys. Rev. Lett.*, 69, 1193, 1992.
22. Menzinger, M. and Rovinsky, A., The differential flow instabilities, in *Chemical Waves and Patterns*, Kapral, R. and Showalter, K., Eds., *Understanding Chemical Reactivity*, Vol. 10, Kluwer, Dordrecht, 1995, 365.
23. Klausmeier, C., Regular and irregular patterns in semiarid vegetation, *Science*, 284, 1826, 1999.
24. Malchow, H., Flow- and locomotion-induced pattern formation in nonlinear population dynamics, *Ecol. Modelling*, 82, 257, 1995.
25. Malchow, H., Motional instabilities in predator–prey systems, *J. Theor. Biol.*, 204, 639, 2000.
26. Malchow, H. and Shigesada, N., Nonequilibrium plankton community structures in an ecohydrodynamic model system, *Nonlinear Processes Geophys.*, 1(1), 3, 1994.
27. Malchow, H., Spatio-temporal pattern formation in nonlinear nonequilibrium plankton dynamics, *Proc. R. Soc. Lond. B*, 251, 103, 1993.
28. Malchow, H., Nonequilibrium structures in plankton dynamics, *Ecol. Modelling*, 75/76, 123, 1994.
29. Malchow, H., Nonlinear plankton dynamics and pattern formation in an ecohydrodynamic model system, *J. Mar. Syst.*, 7(2–4), 193, 1996.
30. Malchow, H., Flux-induced instabilities in ionic and population-dynamical interaction systems, *Z. Phys. Chem.*, 204, 95, 1998.
31. Scheffer, M., Fish and nutrients interplay determines algal biomass: a minimal model, *OIKOS*, 62, 271, 1991.
32. Doveri, F., Scheffer, M., Rinaldi, S., Muratori, S., and Kuznetsov, Y., Seasonality and chaos in a plankton-fish model, *Theor. Popul. Biol.*, 43, 159, 1993.
33. Kuznetsov, Y., Muratori, S., and Rinaldi, S., Bifurcations and chaos in a periodic predator–prey model, *Int. J. Bifurcation Chaos*, 2, 117, 1992.
34. Popova, E., Fasham, M., Osipov, A., and Ryabchenko, V., Chaotic behaviour of an ocean ecosystem model under seasonal external forcing, *J. Plankton Res.*, 19, 1495, 1997.
35. Rinaldi, S. and Muratori, S., Conditioned chaos in seasonally perturbed predator–prey models, *Ecol. Modelling*, 69, 79, 1993.
36. Rinaldi, S., Muratori, S., and Kuznetsov, Y., Multiple attractors, catastrophes and chaos in seasonally perturbed predator–prey communities, *Bull. Math. Biol.*, 55, 15, 1993.
37. Ryabchenko, V., Fasham, M., Kagan, B., and Popova, E., What causes short-term oscillations in ecosystem models of the ocean mixed layer? *J. Mar. Syst.*, 13, 33, 1997.
38. Scheffer, M., Rinaldi, S., Kuznetsov, Y., and van Nes, E., Seasonal dynamics of daphnia and algae explained as a periodically forced predator–prey system, *OIKOS*, 80, 519, 1997.
39. Scheffer, M., *Ecology of Shallow Lakes,* Population and Community Biology Series, Vol. 22, Chapman & Hall, London, 1998.
40. Steffen, E. and Malchow, H., Chaotic behaviour of a model plankton community in a heterogeneous environment, in *Selforganisation of Complex Structures: From Individual to Collective Dynamics*, Schweitzer, F., Ed., Gordon & Breach, London, 1996, 331.
41. Steffen, E. and Malchow, H., Multiple equilibria, periodicity, and quasiperiodicity in a model plankton community, *Senckenbergiana Mar.*, 27, 137, 1996.
42. Steffen, E., Malchow, H., and Medvinsky, A., Effects of seasonal perturbation on a model plankton community, *Environ. Modeling Assess.*, 2, 43, 1997.
43. Truscott, J., Environmental forcing of simple plankton models, *J. Plankton Res.*, 17, 2207, 1995.
44. Ascioti, F., Beltrami, E., Carroll, T., and Wirick, C., Is there chaos in plankton dynamics? *J. Plankton Res.*, 15, 603, 1993.
45. Scheffer, M., Should we expect strange attractors behind plankton dynamics — and if so, should we bother? *J. Plankton Res.*, 13, 1291, 1991.
46. Pascual, M., Diffusion-induced chaos in a spatial predator–prey system, *Proc. R. Soc. Lond. B*, 251, 1, 1993.
47. Pascual, M. and Caswell, H., Environmental heterogeneity and biological pattern in a chaotic predator–prey system, *J. Theor. Biol.*, 185, 1, 1997.
48. Sherratt, J., Lewis, M., and Fowler, A., Ecological chaos in the wake of invasion, *Proc. Natl. Acad. Sci. U.S.A.*, 92, 2524, 1995.
49. Sherratt, J., Eagan, B., and Lewis, M., Oscillations and chaos behind predator–prey invasion: mathematical artefact or ecological reality? *Philos. Trans. R. Soc. Lond. B*, 352, 21, 1997.

50. Davidson, F., Chaotic wakes and other wave-induced behavior in a system of reaction-diffusion equations, *Int. J. Bifurcation Chaos*, 8, 1303, 1998.
51. Merkin, J., Petrov, V., Scott, S., and Showalter, K., Wave-induced chemical chaos, *Phys. Rev. Lett.*, 76, 546, 1996.
52. Petrovskii, S. and Malchow, H., A minimal model of pattern formation in a prey–predator system, *Math. Comput. Modelling*, 29, 49, 1999.
53. Petrovskii, S. and Malchow, H., Wave of chaos: new mechanism of pattern formation in spatio-temporal population dynamics, *Theor. Popul. Biol.*, 59(2), 157, 2001.
54. Horwood, J., Plankton generated chaos in the modelled dynamics of haddock, *Philos. Trans. R. Soc. Lond. B*, 350, 109, 1995.
55. Pedley, T. and Kessler, J., Hydrodynamic phenomena in suspensions of swimming microorganisms, *Annu. Rev. Fluid Mech.* 24, 313, 1992.
56. Platt, J., "Bioconvection patterns" in cultures of free-swimming organisms, *Science*, 133, 1766, 1961.
57. Timm, U. and Okubo, A., Gyrotaxis: a plume model for self-focusing micro-organisms, *Bull. Math. Biol.*, 56, 187, 1994.
58. Winet, H. and Jahn, T., On the origin of bioconvective fluid instabilities in *Tetrahymena* culture systems, *Biorheology*, 9, 87, 1972.
59. Leibovich, S., Spatial aggregation arising from convective processes, in *Patch Dynamics*, Levin, S., Powell, T., and Steele, J., Eds., Lecture Notes in Biomathematics, Vol. 96, Springer-Verlag, Berlin, 1993, 110.
60. Stommel, H., Trajectories of small bodies sinking slowly through convection cells, *J. Mar. Res.*, 8, 24, 1948.
61. Malchow, H., Spatial patterning of interacting and dispersing populations, *Mem. Fac. Sci. Kyoto Univ. (Ser. Biol.)*, 13, 83, 1988.
62. Shigesada, N. and Teramoto, E., A consideration on the theory of environmental density (in Japanese), *Jpn. J. Ecol.*, 28, 1, 1978.
63. Malchow, H., Radtke, B., Kallache, M., Medvinsky, A., Tikhonov, D., and Petrovskii, S., Spatio-temporal pattern formation in coupled models of plankton dynamics and fish school motion, *Nonlinear Anal. Real World Appl.*, 1, 53, 2000.
64. Blake, R., *Fish Locomotion*, Cambridge University Press, Cambridge, U.K., 1983.
65. Cushing, D., *Marine Ecology and Fisheries*, Cambridge University Press, Cambridge, U.K., 1975.
66. Flierl, G., Grünbaum, D., Levin, S., and Olson, D., From individuals to aggregations: the interplay between behavior and physics, *J. Theor. Biol.*, 196, 397, 1998.
67. Grünbaum, D. and Okubo, A., Modelling social animal aggregations, in *Frontiers in Mathematical Biology*, Levin, S., Ed., Lecture Notes in Biomathematics, Vol. 100, Springer-Verlag, Berlin, 1994, 296.
68. Gueron, S., Levin, S., and Rubenstein, D., The dynamics of herds: from individuals to aggregations, *J. Theor. Biol.*, 182, 85, 1996.
69. Huth, A. and Wissel, C., The simulation of fish schools in comparison with experimental data, *Ecol. Modelling*, 75/76, 135, 1994.
70. Niwa, H.-S., Newtonian dynamical approach to fish schooling, *J. Theor. Biol.*, 181, 47, 1996.
71. Okubo, A., Dynamical aspects of animal grouping: swarms, schools, flocks, and herds, *Adv. Biophys.*, 22, 1, 1986.
72. Radakov, D., *Schooling in the Ecology of Fish*, Wiley, New York, 1973.
73. Reuter, H. and Breckling, B., Self-organization of fish schools: an object-oriented model, *Ecol. Modelling*, 75/76, 147, 1994.
74. Romey, W., Individual differences make a difference in the trajectories of simulated schools of fish, *Ecol. Modelling*, 92, 65, 1996.
75. Steele, J., Ed., *Fisheries Mathematics*, Academic Press, London, 1977.
76. Stöcker, S., Models for tuna school formation, *Math. Biosci.*, 156, 167, 1999.
77. Weihs, D., Warum schwimmen Fische in Schwärmen? *Naturw. Rdsch.*, 27(2), 70, 1974.
78. Hanski, I., *Metapopulation Ecology*, Oxford Series in Ecology and Evolution, Oxford University Press, New York, 1999.
79. Malchow, H., Petrovskii, S., and Medvinsky, A., Numerical study of plankton–fish dynamics in a spatially structured and noisy environment, *Ecol. Modelling*, 149, 247, 2002.
80. Scheffer, M., Baveco, J., DeAngelis, D., Rose, K., and van Nes, E., Super-individuals as simple solution for modelling large populations on an individual basis, *Ecol. Modelling*, 80, 161, 1995.

81. Ebenhöh, W., A model of the dynamics of plankton patchiness, *Modeling Identification Control*, 1, 69, 1980.

82. Fernö, A., Pitcher, T., Melle, W., Nøttestad, L., Mackinson, S., Hollingworth, C., and Misund, O., The challenge of the herring in the Norwegian Sea: making optimal collective spatial decisions, *Sarsia*, 83, 149, 1998.

83. Misund, O., Vilhjálmsson, H., Hjálti i Jákupsstovu, S., Røttingen, I., Belikov, S., Asthorson, O., Blindheim, J., Jónsson, J., Krysov, A., Malmberg, S., and Sveinbjørnsson, S., Distribution, migration and abundance of Norwegian spring spawning herring in relation to the temperature and zooplankton biomass in the Norwegian sea as recorded by coordinated surveys in spring and summer 1996, *Sarsia*, 83, 117, 1998.

84. Tikhonov, D., Enderlein, J., Malchow, H., and Medvinsky, A., Chaos and fractals in fish school motion, *Chaos Solitons Fractals*, 12(2), 277, 2001.

85. Tikhonov, D. and Malchow, H., Chaos and fractals in fish school motion, II, *Chaos Solitons Fractals*, 16(2), 277, 2003.

86. Wolfram, S., *Cellular Automata and Complexity. Collected Papers*, Addison-Wesley, Reading, MA, 1994.

87. Bond, A. and Gasser, L., Eds., *Readings in Distributed Artificial Intelligence*, Morgan Kaufmann, San Mateo, CA, 1988.

88. Wooldridge, M. and Jennings, N., Intelligent agents — theory and practice, *Knowledge Eng. Rev.*, 10(2), 115, 1995.

89. Schimansky-Geier, L., Mieth, M., Rosé, H., and Malchow, H., Structure formation by active Brownian particles, *Phys. Lett. A*, 207, 140, 1995.

90. Courant, R., Friedrichs, K., and Lewy, H., On the partial difference equations of mathematical physics, *IBM J.*, 215, March 1967.

91. Savill, N. and Hogeweg, P., Modelling morphogenesis: from single cells to crawling slugs, *J. Theor. Biol.*, 184, 229, 1997.

92. Pitcairn, A., Chaplain, M., Weijer, C., and Anderson, A., A discrete-continuum mathematical model of *Dictyostelium* aggregation, *Eur. Commun. Math. Theor. Biol.*, 2, 6, 2000.

93. Schofield, P., Chaplain, M., and Hubbard, S., Mathematical modelling of the spatio-temporal dynamics of host–parasitoid systems, *Eur. Commun. Math. Theor. Biol.*, 2, 12, 2000.

94. Steele, J. and Henderson, E., A simple model for plankton patchiness, *J. Plankton Res.*, 14, 1397, 1992.

95. Horsthemke, W. and Lefever, R., *Noise-Induced Transitions. Theory and Applications in Physics, Chemistry, and Biology*, Springer Series in Synergetics, Vol. 15, Springer-Verlag, Berlin, 1984.

96. Vandenhouten, R., Goebbels, G., Rasche, M., and Tegtmeier, H., SANTIS — a tool for Signal ANalysis and TIme Series processing, version 1.1, User Manual, Biomedical Systems Analysis, Institute of Physiology, RWTH Aachen, Germany, 1996.

26

Seeing the Forest for the Trees, and Vice Versa: Pattern-Oriented Ecological Modeling

Volker Grimm and Uta Berger

CONTENTS

26.1 Introduction...411
26.2 Why Patterns, and What Are Patterns?..412
26.3 The Protocol of Pattern-Oriented Modeling...413
    26.3.1 Formulate the Question or Problem...414
    26.3.2 Assemble Hypotheses about the Essential Processes and Structures.......................414
    26.3.3 Assemble Patterns..414
    26.3.4 Choose State Variables, Parameters, and Scales...415
    26.3.5 Construct the Model...415
    26.3.6 Analyze, Test, and Revise the Model..416
    26.3.7 Use Patterns for Parameterization...416
    26.3.8 Search for Independent Predictions...416
26.4 Examples...417
    26.4.1 Independent Predictions: The Beech Forest Model BEFORE................................417
    26.4.2 Parameterization of a Mangrove Forest Model..419
    26.4.3 Habitat Selection of Stream Trout..422
26.5 Pattern-Oriented Modeling of Aquatic Systems...423
26.6 Discussion...424
References...425

26.1 Introduction

Continuous critical reflection of the question posed by Hall[1] — "What constitutes a good model and by whose criteria?" — is part and parcel of a sound practice of ecological modeling. Attempts have therefore been made to formulate general modeling paradigms and to distinguish between different categories of models: analytically tractable mathematical models on the one hand, and simulation models that have to be run on computers on the other. In the 1960s and 1970s, a paradigm was formulated[2–4] stating that analytically tractable models are preferable because simulation models are too complex to be understood and too case specific to be of general significance. And indeed, this paradigm was useful, in particular because the simulation models of that time used the same language as analytical models, i.e., differential or difference equations, and therefore were fundamentally no different from analytical models.

However, for about 15 years now computers have been so powerful that a new kind of simulation model caught on in ecology, which may be dubbed "bottom-up simulation models" as they start with the entities at the "bottom" level of ecological systems (i.e., individuals, local spatial units). Individual-based models[5,6] belong to this category, as do grid-based models[7–11] and neighborhood models.[11–13] They have added a new dimension to the old dichotomy of analytical and simulation models, as for the first time the oversimplifying assumptions of mathematical models, which usually were made merely for analytical tractability, can be relinquished. However, it became so simple to include all kinds of empirical knowledge in these new simulation models that, for example, many individual-based models seem unnecessarily complex.[14] What is still lacking are guidelines on how to find the appropriate level of resolution: What aspects of a real system should be included in a model, and what not?

As a powerful guideline, Grimm[15] and Grimm et al.[16] propose orientation toward patterns observed in natural systems. Explaining these patterns may by itself be the objective of a model (e.g., species–area relationship,[17] the wavelike spread of rabies,[10] patterns in the size distributions of populations[13,18,19]); alternatively — if the model has other objectives — the patterns may be used to decide on model structure and to make the model testable. Grimm et al.[16] use three example models to show how natural patterns help decide what aspects of real systems are to be described in a coarse, aggregated way, and what aspects have to be taken into account in more detail.

Since the publication of Grimm et al.,[16] new applications of the "pattern-oriented modeling" (POM) approach have enabled it to be broadened and refined. Thulke et al.[20] show how a POM designed for basic ecological questions is refined step-by-step toward specific applied problems. Wiegand et al.[21–23] demonstrate how patterns may be used not only to decide on model structure and resolution, but also to determine parameter values that would otherwise be unknown. A number of case studies[21,24–27] show that in most cases the usage of multiple "weak" patterns is more fruitful than focusing on one single "strong" pattern. And, perhaps most importantly, the pattern-oriented approach leads to "structurally realistic" models. This means that they allow for predictions that are independently testable without concerning the aspects of the real system used to develop or validate the model.

The objective of this chapter is to summarize all these aspects in a new, comprehensive formulation of the POM approach. Example models demonstrate the different aspects and benefits of the approach. We also discuss some points that are specific to the application of POM for problems in aquatic ecology. This chapter is aimed at not only modelers but aquatic ecologists in general, because if POM is to succeed it is crucial that empirical researchers understand the role of patterns for modeling.

26.2   Why Patterns, and What Are Patterns?

Most analytical models of classical theoretical ecology focus on logical relationships. For example, what would happen if the *per capita* growth rate of a population were positive and constant? The answer is unlimited exponential growth. The logical conclusion of this is that there must be some mechanism by which the *per capita* growth rate becomes zero, which leads to the concept of the density-dependent growth of populations. Logical considerations of this kind are indispensable to ecology; they help develop general concepts and identify underlying general principles. However, logical models are not sufficient for systems analysis because they usually do not leave the realm of logical possibilities. They are so general that they do not apply to any real system.

What we need are models that provide not only logical possibilities (which cannot be sorted out because the models are not testable), but real descriptions of actual ecological phenomena. Therefore, modeling requires us to design models according to what we observe in real ecological systems. This does not, however, mean naively trying to mimic nature with as much detail as possible, because including everything in a model is impossible.[28] The only fully realistic model of nature can be nature itself.[29] Instead, modeling means trying to take into account solely the "essential" aspects of the system. But how are we to know what aspects are "essential" — and what does "essential" mean in the first place?

First of all, whether something is essential depends on the problem or question for which the model is designed. For example, if we want to model the population dynamics of Alpine marmots (*Marmota marmota*), which are territorial and socially breeding mammals living in mountains, the very location of the burrows where they hibernate is not very likely to be essential for their population dynamics. On the other hand, the observation that the territories of the marmots are not randomly scattered over the landscape but occur in clusters is likely to be essential.[30,31] Random distribution would be nothing more than we would expect to observe by chance, whereas the clusters constitute a pattern. A pattern is anything beyond random variation and therefore indicates specific mechanisms (essential underlying processes and structures) responsible for the pattern. These mechanisms may be abiotic (the clusters may indicate suitable habitat) or biotic (marmot groups in isolated territories may not be able to survive in the long run, and only clusters of territories provide a metapopulation structure allowing for long-term persistence). Patterns observed (at a certain level of observation) and which are characteristic of the system are likely to be indicators of underlying essential processes and structures. Non-essential factors probably do not leave any clearly identifiable traces in the structure and dynamics of the system.

Undoubtedly, the attempt to identify patterns and, by "decoding" these patterns, to ascertain the essential properties of an observed system, is nothing but the basic research program of any science, and the natural sciences in particular. Physics and other natural sciences provide numerous examples of patterns that provide the key to the essence of physical systems: classical mechanics (Kepler's laws), quantum mechanics (atomic spectra), cosmology (red shift), molecular genetics (Chargaff's rule), and mass extinctions (the iridium layer at the Cretaceous boundary).

In ecology, however, this basic pattern-oriented research program seems to be less acknowledged. And even if a pattern were reproduced by a model, this was usually not explicitly perceived as a modeling strategy, and so the full potential of the pattern-oriented approach was not used. A good example of this is the well-known cycles of snowshoe hares and the lynx in Canada. These cycles of population abundance are certainly a pattern, but it is relatively easy to reproduce cycles with all different kinds of mechanisms (see preface of Czárán[11]). Therefore, none of the existing models explaining the pattern was able to outdo the others. However, until recently, another pattern in the hare–lynx cycles had been overlooked: the period length of the cycles is almost constant, whereas the amplitudes of the peaks vary chaotically.[32] This observation could only be reproduced by a model with a specific structure (a food chain of vegetation, hare, and lynx) and a specific, previously ignored mechanism: in times of low hare abundance the lynx may switch to other prey (presumably squirrels).

This example is particularly revealing because of the frequent complaint that there are so few clear patterns in ecology. Two or more seemingly weak patterns (constant period *and* chaotic amplitudes) may provide an even more fruitful key to the essence of a system than one single strong pattern. This is because it is usually harder to reproduce patterns in different aspects of the system simultaneously than to reproduce just one pattern regarding one aspect (see "multicriteria assessment" of models[33]). The whole set of patterns that can be identified in a system constitutes a kind of "ecological fingerprint," which is vital not only for identifying the system,[34] but also to trace the essential processes and structures of a system.

26.3 The Protocol of Pattern-Oriented Modeling

It should be noted that POM modeling as described below is not genuinely new per se. Many modelers apply this method intuitively (e.g., References 32, and 35 through 42). There have also been attempts to describe the usage of patterns as a general strategy (e.g., References 43 and 44), but most of this work is concerned with selecting the most appropriate analytical models reproducing certain population census time series. In contrast, POM as it is presented here is concerned with bottom-up models. DeAngelis and Mooij[45] independently developed a notion of bottom-up models that is very similar to POM (they refer to "mechanistically rich" models).

What is new about POM is the attempt to make the usage of patterns explicit and to integrate this usage into a general protocol of ecological modeling. This protocol, however, can only describe the general tasks of modeling and their sequence. Modeling cannot be formalized into a simple recipe because modeling is a creative process whose details depend not only on the system studied and the question asked, but also largely on the modeler's skills, experience, and background (for more detailed descriptions of the modeling process, see References 46 through 48). Note that the following sequence of tasks has to be cycled through numerous times. Modeling is a cyclic process,[20] which is repeated until no further improvements can be made given the empirical knowledge available, or until there are no resources (time, money) left to continue the process. Not all the tasks described below are genuinely "pattern-oriented," but we still include them because an isolated description of the genuine pattern-oriented aspect of ecological modeling would not be useful.

26.3.1   Formulate the Question or Problem

Modeling requires deciding what aspects of the real system to take into account and at what resolution. Without a clear question or problem to be tackled with the model, these decisions could not be made and therefore everything would have to included, leading to a hopelessly complex model. Thus, the modeling strategy "first model the system, then answer questions using the model" cannot work. However, if we start with an explicitly stated question or problem, we can ask whether we believe each known element and process to be essential for the question or problem.

26.3.2   Assemble Hypotheses about the Essential Processes and Structures

Certainly, every answer to the question whether something is considered essential is nothing but a hypothesis that may be true or false. But modeling means starting with hypotheses. The ultimate objective of a model is to check whether the hypotheses are useful and whether they are sufficient to explain and predict the phenomena observed. One important consequence of this is that if we are unable to formulate at least a minimum set of such hypotheses, we cannot build a model. But what would be the basis of such hypotheses?

Besides hard data, qualitative empirical knowledge is decisive. Every field ecologist or natural resource manager who is familiar with the system in question knows far more than is or can be expressed in hard data. Often, this qualitative knowledge is latent and will only be expressed if the right questions are asked. And often this knowledge can be expressed in "if–then" rules. A forest manager who has seen a hundred times or more how a canopy gap in a beech forest is closed is very likely to be able to formulate empirical rules of the following kind: either the neighboring canopy trees will close the gap, or one of the smaller and younger trees of the lower canopy will grow into the gap and close it. Although it may not be possible to predict exactly what will happen in one particular gap, it may be possible to estimate the probabilities of the alternative outcomes. One of the main advantages of modern bottom-up simulation models is that qualitative empirical rules can easily be represented in the simulation programs without any mathematical constraints.

Another important source of hypotheses is theory: even if no data are available on a certain process, general theoretical principles might help to formulate test hypotheses. Or, even if there is no such principle, one might assume extreme scenarios, such as a constant, linear, or random relationship between variables. All in all, assembling hypotheses about the essential processes and structures of the system is a crucial step of modeling that usually has to be repeated many times.

26.3.3   Assemble Patterns

In addition to the hypotheses that may reflect empirical knowledge or theoretical principles, patterns are used in POM to decide on the model's structure and resolution. For example, in natural beech forests local stands that are at different developmental stages have typical percentages of cover in different vertical layers. This observation suggests not only considering the horizontal but also the vertical structure of the beech forest. To this end, a model structure is chosen that distinguishes between vertical layers (see Section 26.4.1 example). This does not mean that the typical layers of a beech forest are hardwired into the model, but that a model structure is provided that allows whether these typical layers will or will not emerge to be tested.

This is the general idea of POM: if we decide to use a pattern for model construction because we believe this pattern contains information about essential structures and processes, we have to provide a model structure that in principle allows the pattern observed to emerge. Whether it does emerge depends on the hypotheses we have built into the model. Examples of how patterns determine and constrain model structure include spatial patterns that require a spatially explicit model; temporal patterns not only in abundance but also in population structure, which require a structured population model; different life history strategies in different biotic or abiotic environments, which require describing the life cycle of individuals; in benthic marine systems, the different settlement of larvae at different altitudes, which requires including topography into the spatially explicit model;[49] and different behavior in different habitats, which requires including habitat quality;[27] etc.

When considering patterns that might be used to structure and, later on, to test the model, it is important not to exclusively focus on "strong" patterns that are strikingly different from random variation and therefore seem to be strong indicators of underlying processes. As the above-mentioned example of population cycles shows, individual strong patterns may not be sufficient to narrow down an appropriate model structure. A combination of seemingly "weak" patterns may be much more powerful to find the right model. Multiple patterns concerning different aspects of a system reduce the degrees of freedom in model structure. We are all familiar with this effect of additional information: it is virtually impossible to identify a person if we know only his or her age, but if we know the person's age, sex, profession, nationality, etc., the chances of finding the right person increases with each additional piece of information.

Therefore, trying to assemble characteristic patterns of an ecological system is similar to trying to describe an individual so that it can be identified. We have to ask ourselves what distinguishes this system from other, similar or neighboring systems. What makes us identify the system? What is the system's "ecological fingerprint"? Searching for patterns means thinking in terms of the entire system instead of focusing exclusively on its parts. The title of this chapter refers to a situation frequently encountered in ecology: the focus is so much on the elements (the "trees") that we fail to see the system (the "forest") or system-level patterns. Good modeling and good ecology require us to focus on both the elements and the system at the same time (hence, the "vice versa" in the title).

26.3.4 Choose State Variables, Parameters, and Scales

Once the hypotheses on essential structures and processes and the pattern characteristics of the system have been assembled (and note that this constellation will have to be revised every time a new model version has been implemented and analyzed), the next task is to decide on the state variables describing the state of the system (i.e., the structure), and on the parameters that quantify when, how, and how fast the state variables change (i.e., the dynamics). And since POM explicitly deals with patterns observed in the real system that are linked to certain spatial and temporal scales, the spatial extension and resolution of the model and the time horizon and temporal resolution have to be defined. To avoid getting bogged down in long lists of variables and parameters, it makes sense to use simple graphical representations of the model's elements, e.g., simplified Forrester diagrams[46] or "influence diagrams,"[10,50] where boxes delineate structural elements or processes and arrows indicate influence, e.g., process *A* has an influence on structure *B*. Influence diagrams are also useful for aggregating blocks of processes to keep initial model versions simple and thus manageable.

26.3.5 Construct the Model

First of all, the order in which processes occur has to be defined (this does not apply to "event-driven" models, where the model entities and events determine the order by themselves). To visualize their sequence, flowcharts are useful tools. The next step is to implement the model. Although it is possible to develop useful analytical models that are pattern oriented (but that are usually only formulated analytically, whereas the results are obtained numerically, i.e., by simulation), in most cases pattern-oriented models will be simulation models that have to be implemented as computer programs. Describing the implementation of simulation models in detail would go beyond the scope of this chapter. However, the quality of the process of implementation is decisive for the quality and efficiency of the modeling project. For general issues regarding simulation models and their implementation, see Haefner;[46] for software considerations that are particular to individual-based (and grid-based) models, see Ropella et al.[51] Two aspects of implementing a simulation model that are of particular significance are careful protocols to test subunits of the program (functions and procedures) independently, for example, using independent implementations in spread-sheets,[51] and the implementation of a graphical user interface that visualizes the state variables and allows both the developers of the model and peers to perform controlled experiments with the model ("visual debugging"[52]).

26.3.6 Analyze, Test, and Revise the Model

Nonmodelers often believe that the formulation of a model is the most difficult part of modeling. However, formulating and implementing *some sort of* model is not a problem. The tricky part is building a model that produces meaningful results, and this requires finding ways of assessing the quality of the model output so that we can rank different versions of the model. Modeling requires a kind of currency to compare different model versions and parameterizations. And this is the point where the orientation toward patterns pays off: the patterns provide the currency. A comparison of observed and simulated patterns allows the potential of different model versions to reproduce what we observe in reality to be assessed.

It will not always be easy to identify clear "signals" in the output of the model. This is because the output is the summary result of all model processes, which may mask the effect of individual subunits of the models. Therefore, even if the aim of modeling is to construct structurally realistic models, model analysis requires courageous and forceful modifications of the model structure leading to model versions that are deliberately unrealistic and "do not occur in nature."[53] The appropriate attitude for analyzing models is that of experimenters.[14] The objective of the experiments performed with the model is to identify those mechanisms that are responsible for the patterns observed. Only these mechanisms will be kept in the model.

26.3.7 Use Patterns for Parameterization

Model output depends both quantitatively and qualitatively on the values of the model parameters. Therefore, the model parameters need to be known. Yet in most models of real systems, only a minority of parameters are known precisely. For other parameters it may be possible to specify biologically meaningful ranges. But if these ranges are too broad for too many parameters, the model's output may be too uncertain to narrow down an appropriate model structure and to answer the original question. Now, one new, very important aspect of POM is that patterns can be used to determine parameters indirectly. We will not describe this method in detail here (see References 21 through 23; see also the mangrove forest example below). This aspect is, however, closely related to the pattern-oriented approach because it will work only with structurally realistic models.

Hanski[54,55] uses indirect parameterization to determine parameters of real metapopulations that were otherwise unknown. His simple but structurally realistic model (the "Incidence Function Model") includes, for example, the position and size of habitat patches, and a simple relationship between the patch size and extinction risk of a population inhabiting that patch. Then, the model output is fitted to empirical presence–absence data (occupancy pattern) of the real network of patches. Similarly, Wiegand et al.[23] use specific patterns in the census time series of brown bears (including information about family structure) to narrow down the uncertainty of the demographic parameters of their model.

Wiegand et al.[21] describe the general strategy of parameterizing models for conservation biology — where data are scarce as a rule — by utilizing patterns. Wiegand et al.[22] apply four observed patterns for parameterization, one after the other. They show that after the use of each pattern the uncertainty about parameters is reduced and that the resulting parameter set leads to much less uncertain results at the system level than the original parameterization, which was based on educated guesses. Thus, error propagation was no longer a severe problem after the pattern-oriented parameterization. Other examples are given in DeAngelis and Mooij.[45]

26.3.8 Search for Independent Predictions

If a model reproduces patterns observed in nature, this is a success because the patterns were not hardwired into the model; instead, only a model structure was provided that in principle allowed for these patterns to emerge. However, even if a model reproduces a pattern, it is not possible to deduce logically that the model mechanisms responsible for the pattern match those in the real world.[29] The population cycles mentioned above are an example of this: different model mechanisms generate cycles. The problem in this case is that cycles are just a single pattern, which is rather easy to reproduce. Therefore, patterns reproduced by a model cannot prove that the model is "correct." Instead, we have to use multiple patterns to gradually increase confidence in the model. These may be additional properties of the time series, besides being cyclic,[44] or, preferably, additional patterns regarding the structure of the ecological system. The more patterns a model is capable of reproducing simultaneously, the higher the confidence that the model is structurally realistic and can therefore be used to serve its original objective.

The strongest evidence of structural realism is independent predictions — predictions of system properties that were utilized during neither model construction nor parameterization; i.e., these properties (or even patterns) were not used to narrow down model structure or parameter values. The idea of independent predictions is that if we used multiple patterns to build, parameterize, and test a model, we ought to obtain a model that reflects the key structures of the real system, including their hierarchical organization, in just the right way. If this is so, it should be possible to analyze additional model structures and processes that previously were not at the focus of attention. An example of this is the beech forest model BEFORE presented as an example in Section 26.4.1. The structural realism of a model implies a richness in model structure that allows for new, additional ways of looking at the model. Structural and mechanistic[45] richness is a key property of pattern-oriented models based on multiple patterns.

26.4 Examples

Here we briefly discuss example models that highlight different aspects of POM. We cannot, however, give a full description of the background of the models, of the models themselves, or of the results, all of which are described in the cited literature.

26.4.1 Independent Predictions: The Beech Forest Model BEFORE

In this example we demonstrate how a model that is constructed with regard to more than just one pattern will be so rich in structure and mechanisms that it will enable independent predictions. The model BEFORE[56,57] was designed to model the spatiotemporal dynamics of natural mid-European beech (*Fagus silvatica*) forests on large spatial and temporal scales. These forests would be the dominating type of ecosystem in large areas of Central Europe, but except in Bohemia and some parts of the Balkans, no natural forests exist anymore. For management and conservation, two questions are of particular interest: How large has a beech forest to be to develop its characteristic spatiotemporal dynamics? And how can the "naturalness" of a certain managed beech forest be assessed?

The structure of BEFORE was determined by the objective of the model and by two patterns. The objective — spatiotemporal dynamics on large spatial and temporal scales (hundreds of hectares and centuries) — suggested a model structure much coarser than, for example, models of forest stand dynamics, which are used in management and which typically are concerned with hectares and a decade or two. Moreover, the model should focus exclusively on the dominating species, the beech; other species and any kinds of spatial heterogeneity of the environment were ignored.

Two patterns were used to narrow down the model structure. First, natural beech forests show a mosaic pattern of small areas (0.1 to 2 ha) in certain developmental stages of the local stands (e.g., "mature" stands with closed canopy and almost no understory, or "decaying" stand with open canopy, scattered canopy trees, and growing cohorts of juvenile trees). To allow this mosaic pattern to emerge in the model, space was divided into grid cells considerably smaller than the typical mosaic patches. The cells were of the size of one very large, old canopy tree (about 14×14 m^2, or about 1/50 ha).

Second, the developmental stages were characterized by typical covers in different vertical layers of a local stand. Therefore, BEFORE distinguishes between four vertical height classes (Figure 26.1). In the lower two classes, only percentage cover was considered as a state variable (100% cover means that no light can penetrate to lower layers), whereas in the upper two classes (lower and upper canopy), individuals were distinguished (by age in the lower canopy layer, and by age and canopy size in the upper canopy layer).

Growth and mortality within a cell were described by empirical rules (e.g., "if light is reduced by the higher layers by more than 70%, then mortality increases by 20%"). Regarding the interactions between

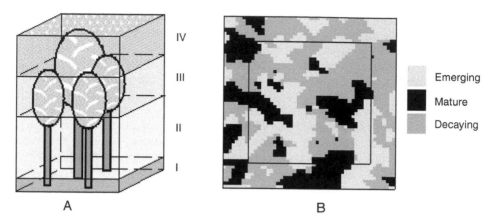

FIGURE 26.1 Visualization of the model structure and output of BEFORE. (A) Vertical discretization of the model forest. Within a grid cell of 14.29 m² (which corresponds to the maximum canopy area of an old beech tree), four vertical layers are distinguished. I: seedlings; II: juvenile beech; III: lower canopy; IV: canopy. Note that the tree icons are only for visualization and do not "exist" in this form in the model. (B) The model reproduces a typical mosaic of small areas that may be in the three different developmental stages: "emerging," "mature," or "decaying." A model forest of 54 × 54 cells (≈ 60 ha) is presented. For evaluating the model, only the delineated inner area is used (38 × 38 cells ≈ 29.5 ha.) (Modified from Reference 26.)

trees in neighboring cells, the mosaic pattern indicated there must be some interactions. If the dynamics in the cell were completely independent of each other, no pattern could emerge. Even so, it took several cycles of model formulation and thorough testing of the model assumptions until neighbor interactions were identified: damage by wind-thrown trees in the neighborhood, increasing susceptibility to wind-fall due to canopy gaps in the neighborhood, and an increased input of light falling through canopy gaps in the neighborhood.

BEFORE was able to reproduce the observed spatiotemporal dynamics on both the local and regional scale and to highlight the mechanisms responsible for these dynamics.[56] However, one of the purposes of BEFORE was to establish criteria or indicators that would allow assessment of how close a certain forest is to the structure of a natural forest; i.e., what are good indicators of "naturalness"? Although the mosaic of small forest areas at different developmental stages proved to be a good indicator in theory, in practice it will often be difficult and too time-consuming to assess the naturalness of a forest in this way (S. Winter, personal communication). Therefore, Rademacher et al.[26] looked for other indicators. BEFORE turned out to be rich enough in structure for other aspects of the forest to be studied that were not even mentioned during model development and parameterization. These aspects were the spatial distribution of very old and/or very large individual trees, and the local and regional age structure of the upper canopy.

Regarding both aspects, BEFORE made — without additionally fine-tuning the parameters — predictions that matched the sparse and scattered empirical information about these aspects that could be found in the literature about natural forests or very old forest reserves (Figure 26.2). The very large "giant" individuals exhibit a typical spatial distribution that is almost independent of even large storm events, and the upper canopy shows a typical age structure (neighboring canopy trees usually have an age difference of about 60 years) at both the local and regional scale.

It is encouraging to see that a conceptually simple model constructed following the pattern-oriented approach is able to produce independent predictions that match empirical observations. However, despite its conceptual simplicity, the implementation of BEFORE is rather complex (including about 100 if–then rules). Yet the model proved to be robust with regard to most of the model parameters. It seems that in structurally realistic models, which reflect the hierarchical structure of real systems, just counting the number of parameters, rules, or state variables is not an ideal indicator of the model's actual complexity. In a way, the same is true for real ecological systems: if we ignore all the details and peculiarities that certainly exist, a natural beech forest produces very robust and rather simple spatiotemporal dynamics.

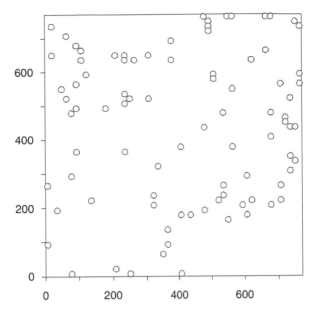

FIGURE 26.2 Spatial distribution of trees older than 300 years on an area of 770 m² as predicted by the beech forest model BEFORE. (Modified from Reference 26.)

And this is one of the tasks of bottom-up modeling: to reveal how simple and predictable patterns emerge from the seemingly chaotic interaction of a large number of different entities and events.

26.4.2 Parameterization of a Mangrove Forest Model

This example demonstrates indirect parameterization: a pattern at the system level is used to determine lower-level parameters that would otherwise be unknown. The example model is KiWi, a mangrove forest model that is designed both to tackle basic questions (e.g., the emergence of zonation patterns) and applied ones (the sustainable use of mangrove forests).[58] In contrast to the beech forest model above, the model could not be based on empirical rules because such empirical knowledge does not exist for mangroves. Therefore, the grid-based approach could not be used. Moreover, an essential characteristic of mangrove trees is their plastic allometry. Depending on nutrient availability, the inundation regime, freshwater input, and the resulting pore water salinity, the number, length, and circumference of the prop roots of one of the three dominating species in the region under consideration (northern Brazil), *Rhizophora mangle*, may change. Furthermore, the maximum height of adult mangrove trees varies depending on the environmental conditions. If mangrove forest dynamics are to be investigated with a simulation model, the model must consequently consider not only the influence of the abiotic factors on the growth rate and the mortality of the trees but also the change in the area required by the individuals, since this may change their strength in neighborhood competition.

For this purpose, the mangrove model KiWi was developed, which is based on the field of neighborhood (FON) approach to the individual-based modeling of plant populations.[12,58] In its first version, it describes a three-species forest parameterized by the growth functions of the mangroves occurring in America: *Avicennia germinans, R. mangle,* and *Laguncularia racemosa*.[59] The model defines a tree through its stem position, its stem diameter, and a zone of influence (ZOI) within which the tree competes for light, nutrients, and space with its neighbors. The biological significance of the ZOI (for example, projected root or crown area) was initially not explicitly defined. It is nevertheless plausible that the ZOI must increase with the size of the tree. This is described by the relationship:

$$R = a \cdot \left(\frac{dbh}{2}\right)^b \qquad (26.1)$$

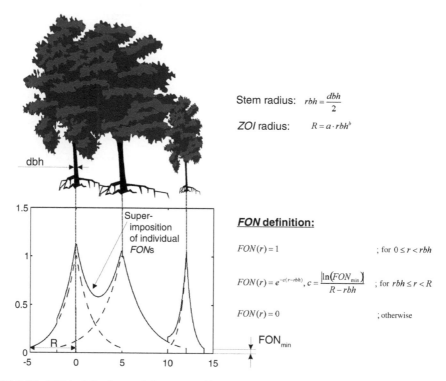

Stem radius: $rbh = \dfrac{dbh}{2}$

ZOI radius: $R = a \cdot rbh^{b}$

FON definition:

$FON(r) = 1$; for $0 \le r < rbh$

$FON(r) = e^{-c(r - rbh)}, \; c = \dfrac{\left|\ln\left(FON_{min}\right)\right|}{R - rbh}$; for $rbh \le r < R$

$FON(r) = 0$; otherwise

FIGURE 26.3 The ZOI is defined around the stem position and grows with the size of the individual. It marks the area within which the individual influences its neighbors (or environment). The competition strength of the individual is described by its FON, which is defined on the ZOI. The superimposition of the FONs marks the competition strength exerted by the individuals at any position (x,y). This value can for example be used to decide whether the establishment of seedlings is possible at a certain location (x,y).

where R is the radius of the ZOI and *dbh* is the stem diameter at breast height. The competition strength exerted by a tree on a certain position is described through a scalar neighborhood field (FON), which is defined on the ZOI (Figure 26.3). The FON is scaled to 1 within the stem and falls exponentially to a minimum (>0) at the boundary of the ZOI. The FON exponential form is directly justified if the neighborhood competition is interpreted as competition for light.[60] However, the species-specific mechanism could also be exerted through root competition, competition of the branches, or a combination of these factors.

For the parameterization of KiWi it is therefore important to check whether an exponential FON is suitable to describe the intra- and interspecific competition of the mangrove species considered and to find a way of determining the parameters a and b (Equation 26.1). The empirically determined relationships between stem and crown diameter obtained for *R. mangle* and *A. germinans* by Cintrón and Schaeffer-Novelli[61] seem suitable for this purpose:

$$\text{\textit{Rhizophora mangle}} \qquad R_{Crown} = 7.113 \cdot \left(\frac{dbh}{2}\right)^{0.6540} \qquad\qquad (26.2)$$

$$\text{\textit{Avicennia germinans}} \qquad R_{Crown} = 5.600 \cdot \left(\frac{dbh}{2}\right)^{0.2994} \qquad\qquad (26.3)$$

Unfortunately, there is no direct information about the competition processes of mangrove trees older than saplings. Thus it remains unclear whether the crown size really determines the competition processes of neighboring trees. However, the very robust ecological *pattern* presented in Figure 26.4 can be used to test the suitability of this assumption. The points mark density-biomass data obtained from different mangrove forests in Latin America.[62] The figure is reminiscent of the frequently examined self-thinning

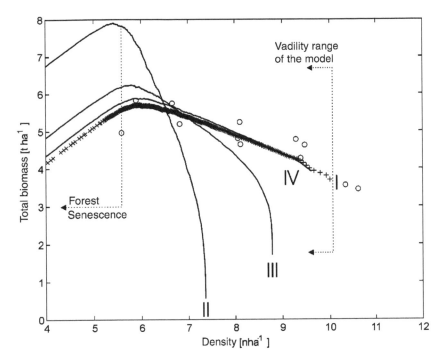

FIGURE 26.4 Biomass-density trajectories of different simulated mangrove forests. The points mark empirical data (see Table 26.1) of different mangroves obtained by Fromard et al.[62] Curve I and curve II were simulated for *R. mangle* and *A. germinans* assuming crown–stem diameter relations as FON parameters measured by Cintrón and Schaeffer-Novelli[61] (Equations 26.2 and 26.3). Curves III and IV are simulated trajectories for *A. germinans* with $b = 0.5$ and $b = 0.6$, respectively, as FON exponent (Equation 26.1).

TABLE 26.1

Empirical Data of the Mangrove Forests Considered in Figure 26.4

Type	Density (N ha^{-1})	Total Biomass (t ha^{-1} dry wt)	Species Composition	Rel. Density (%)
Pioneer stage	41111 ± 3928	31.5 ± 2.9	L:A	99.5:0.5
Young stage	11944 ± 1064	71.8 ± 17.7	L:A	75.5:24.5
Pioneer (1 year)	31111 ± 12669	35.1 ± 14.5	L:A:R	40.7:55.7:3.6
Mature (coast)	917 ± 29	180.0 ± 4.4	L:A:R:others	13.6:60.9:20.0:5.5
Mature (coast)	780 ± 154	315.0 ± 39.0	L:A:R	3.8:14.7:81.4
Adult (riverine)	3310 ± 1066	188.6 ± 80.0	L:A:R:others	50.5:37.7:1.8:10.3
Adult (riverine)	3167 ± 2106	122.2 ± 76.4	A:R:others:*Pterocarpus*	1.5:10.7:16.4:71.4
Senescent	267 ± 64	143.2 ± 15.5	L:A:R:*Aerostichum*	1.4:81.7:5.6:11.2

Source: Modified from Fromard, F. et al., *Oecologia*, 115, 39, 1998.

curves of plant cohorts.[19] For our purpose it is interesting that all the data lie on the same curve irrespective of which species dominates the forest (Table 26.1). Consequently, all trajectories of mixed forests and monospecific forests adhere to the same pattern; hence so should a simulated mangrove forest.

Curve I in Figure 26.4 shows the trajectory for a simulated *R. mangle* forest. The FON parameters of the individuals were defined according to crown parameters (Equation 22.2). The biomass was calculated to be $BIOM = 128.2 * dbh^{2.6}$ (biomass in g).[62] Obviously the parameters chosen are suitable for reproducing the empirical findings. This supports the assumption that light competition within the crown is a main factor driving the neighborhood competition processes of this species.

For *A. germinans*, however, this is not the case. Curve II presents the simulated trajectory for this species based on FON assumptions according to Equation 26.3. The biomass was calculated to be $BIOM = 0.14 * dbh^{2.4}$ (biomass in kg).[62] The resulting trajectory is much steeper than the empirical trajectory, although as b increases (curve III: $b = 0.5$; curve IV: $b = 0.6$) the concordance improves. Thus, a reasonable parameter range can be indirectly determined.

In conclusion, it can be stated that the exponential shape of the FON is suitable for describing tree-to-tree competition in mangrove trees. For *R. mangle,* the FON parameters a and b can be defined according to the stem–crown diameter relationship. For *A. germinans* a reasonable range of these parameters can be found by the pattern of the density–biomass curve, although it still has to be checked whether the values chosen can be empirically explained through root competition. The parameterization of *L. racemosa* is probably possible following the same procedure.

It should be noted that many of the assumptions of the FON approach were made on a phenomeno-logical level. They were based on observations and patterns at the local scale of individual mangrove trees (high variability of the area required by trees depending on environmental factors and local competition with neighbor trees). Therefore, the system level pattern of Figure 26.4 was not only needed for parameterization but also to prove that all these phenomenological descriptions at the local and individual level are sufficient to capture the essence of the processes they are supposed to describe. And indeed, the FON approach is not only able to reproduce the overall pattern of Figure 26.4 but also linear self-thinning trajectories in general,[58] patterns in size distributions,[19] and patterns in the age-related decline of forest productivity (Berger et al., unpublished manuscript).

26.4.3   Habitat Selection of Stream Trout

Here we demonstrate how a set of six seemingly "weak" patterns can be used to narrow down model structure. Railsback[25] and Railsback and Harvey[27] were led to this approach because of two reasons: first, their model of the habitat choice of trout in streams was based on many assumptions and parameters not precisely known and so it seemed unwise to focus on the numerical values of the model's individual output variables. Instead, the model was supposed to reproduce overall patterns in the behavior of stream trout. Second, regarding several submodels describing the behavior of rainbow trout, different assump-tions were made that were based on standard models in the literature, or assumptions that seemed more reasonable to the authors. Without using patterns, it would have been impossible to decide which of the competing submodels (or "theories" regarding the behavior of the individuals; Grimm and Railsback, unpublished manuscript) was more appropriate.

The behavior modeled by Railsback and Harvey[27] was habitat choice in reaction to changes in river flow, temperature, and trout density.[25] These factors affect food availability and mortality risks. The model fish move to the habitat that provides the highest fitness (according to a fitness criterion introduced by Railsback et al.[63]). The model was tested with six patterns taken from fishery literature:

1. The dominant fish acquires the best feeding site; if removed the next-dominant fish moves into the best site and the other fish move up the hierarchy.
2. During flood flows, adult trout move to the stream margins where stream velocities are low.
3. Juvenile trout use higher velocities competing with larger trout.
4. Juvenile trout use faster and shallower habitats in the presence of predatory fish.
5. Trout use higher velocities on average when temperatures are higher (metabolic rates increase with temperature, so more food is needed to avoid starvation).
6. When general food availability is decreased, trout shift to habitats with higher mortality risks but also higher food intake.

None of these patterns seems to be very "strong" and reproducing them with simple hypotheses about the behavior of the trout seems easy. However, it proved to be much more difficult to reproduce all these six patterns *simultaneously.* Three different hypotheses for how trout select habitat were tested. Only one of these (maximizing predicted survival and growth to reproductive size over an upcoming time horizon) led to results that matched all six patterns simultaneously, whereas the other two hypotheses (one of them being a kind of "standard" in modeling habitat selection[63]) were only able to match two or three patterns.

The entire trout model, which is rather complex and includes a detailed description of the habitat and the growth, behavior, and mortality of the fish, should be considered as a kind of "virtual laboratory" (Grimm and Railsback, unpublished manuscript). Sample individuals are fed into this laboratory to test hypotheses or theories about their behavior. However, this will not work if the virtual laboratory itself is not designed properly. Thus, to test the parts of a system (the individuals, or local units), we need a good model of the entire system, but to get a good model of the entire system, we need good models of its parts. This chicken-and-egg problem highlights that neither an exclusive system-level approach to modeling nor an exclusive "bottom-up" approach focusing solely on the parts is sufficient. Grimm[14] briefly discusses this mutual relationship of top-down (or holistic) and bottom-up (or reductionistic) approaches. Auyang[64] presents a thorough discussion of this issue and recommends an approach called *synthetic microanalysis*, which integrates synthesis at the system level with analysis of the individuals making up the system.[25] Auyang independently arrives at the same conclusion as we do within the framework of POM: examining neither only the parts nor only a whole system is adequate (see Section 26.3.3).

26.5 Pattern-Oriented Modeling of Aquatic Systems

Most of the early models that were explicitly built following the pattern-oriented approach were about terrestrial systems. However, in principle there is no difference in POM between terrestrial and aquatic systems (as is also reflected in the examples above). The reason is obvious: the research program of detecting patterns and trying to find the underlying mechanisms is generic and therefore independent of the subject of a discipline.

Many patterns in aquatic and, in particular, marine ecological systems gave rise to important concepts and models, such as zonation patterns in the intertidal,[65] patchiness in the distribution of plankton,[66,67] and the relationship between disturbance and diversity in coral reefs.[68] Sometimes the lack of a clear static pattern may also constitute the pattern, for example, the high spatiotemporal dynamics in the distribution and abundance of macrozoobenthos in mudflats that are strongly affected by disturbance events, e.g., the Wadden Sea.[49]

Modern bottom-up simulation models, in particular spatially explicit individual-based models, have become established in marine ecology and have in recent years become a "*de facto* tool in large-scale efforts studying the interactions of marine organisms with their environment" (Reference 69, p. 411). Reviewing models of marine systems, which are inherently pattern-oriented, would go beyond the scope of this chapter. Nevertheless, such a review would be worthwhile and there are certainly many interesting models that have been built with regard to certain patterns, e.g., Hermann et al.[70] and Walter et al.[71]

Marine systems are much more determined by physical processes than most terrestrial systems. Therefore, individual-based models "have, by necessity, focused on *explicitly* coupling the biological and ecological formulations of hydrodynamic models of varying degrees of three-dimensional and temporal complexity" (Reference 69, p. 411; this review gives an overview of spatially explicit individual-based models of marine populations). However, the physical processes are usually modeled by oceanographers, who often seem to consider it natural — or even necessary — for both the hydrodynamic and the ecological parts of a model to be formulated in the same language, i.e., mathematical equations. For example, Fennel and Neumann,[72] in their useful overview of coupling biology and oceanography in models, state that "it is widely accepted that biological models should be as simple as reasonable and as complex only as necessary," which is certainly true. Yet in their next sentence they equate biological models with equations: "this implies that the answer to the question: 'which are the right model equations' depends ... on the problem under consideration" (Reference 72, p. 236). We believe that, on the biological side of marine ecological models, bottom-up simulation models, which consider by means of computer programs individual, local units, behavioral rules, and stochastic events, are frequently more powerful tools than mathematical equations. For example, in the stream trout model described above, the flow regime of the river is modeled with a standard hydrological model, but the biological model is individual-based. Verduin and Backhaus[73] present an even closer integration of a physical and biological model describing the interaction of near-bottom flow with a seagrass that locally dissipates the energy of flow.

One particular problem of pelagic systems seems to be that here patterns are notoriously harder to detect than in many benthic and terrestrial systems. Therefore, the search for multiple patterns is even more important here. And if there is no pattern at all, there can be no science because "change without pattern is beyond science."[74]

26.6 Discussion

In this chapter we have described the rationale and the tasks of pattern-oriented ecological modeling. Although many modelers and theorists use patterns to design and test their models (see references in Section 26.3), it seems worthwhile to explicitly formulate how patterns can be used in ecological modeling in general and in bottom-up simulation models in particular, because in the latter models there is, in contrast to analytical models, no built-in limitation of model complexity. We have tried to show that patterns provide guidelines for deciding model structure and resolution, they provide a currency for testing all the versions of a model, they allow low-level parameters to be determined, and they lead, if multiple patterns are used, to structurally realistic models that allow independent predictions.

This is not to say that POM is free of limitations. First of all, care has to be taken with the patterns used for POM. The human mind is inclined to perceive patterns all the time, even if they do not exist. It is therefore important to quantify, if possible, the significance of the patterns. It may also be that the data that display a certain pattern are flawed because of defective designs or changes in the sampling protocol. DeAngelis and Mooij[45] discuss an example where a flawed census time series was used as a pattern for parameterization.

Second, it must never be naively inferred that in reality the same mechanisms are responsible for the patterns as in the model. The pattern produced by a model may be correct, but the mechanism may still be completely wrong. Therefore, care has to be taken to search for as many patterns as possible to gradually increase confidence in the model pattern by pattern. The highest degree of confidence that can be achieved is independent predictions that indicate that the model captures the essential elements and processes of the hierarchical organization of the actual system. But even if a model allows for independent predictions that match observations, the model is still a model and will never be able to represent all the relevant aspects of the real system at the same time. However, modeling and models have a momentum of their own,[15] which harbors the risk that modelers will stop sufficiently distinguishing between the real system and their model. It must therefore be borne in mind that although "structural realism,"[21] which we referred to repeatedly in this chapter, is a very useful metaphor, it may also be dangerous if it is taken too literally.

A further critical point of POM is the question of emergence: Do the patterns in the model really emerge or are they hardwired into the model by choosing appropriate model structures ("imposed behavior"[75])? One could, for example, argue that in the beech forest model BEFORE the vertical layers are hardwired into the system precisely because we provided the model's vertical structure. Did we not simply impose the existence of layers, causing the model structure to reflect our own bias about the system? Real emergence would mean specifying only the properties of the entities, i.e., the individual beech trees, and then seeing if vertical layers emerge. One could, for example, think of a beech forest model based on the FON approach[58] (see Section 26.4.2). However, even with the FON approach, we would have to specify rules that describe the high shade tolerance of beech trees. And this rule has a "real" basis: young beech trees that reach into the lower canopy are able to survive several decades of overshading before they proceed to the upper canopy. Therefore, we necessarily seem to have to "impose" this rule.

A good way to test whether patterns have been imposed is to try model rules that are obviously absurd, such as a young beech dies whenever it is shaded; or it never dies but always waits until the canopy opens again. If these absurd rules still lead to similar patterns as the serious rules, then we probably did hardwire the pattern into the model structure.

If we provide no hierarchical model structure at all, this will seldom lead to useful results. This was also observed in "agent-based modeling," which is used in many disciplines such as sociology, economy, engineering, and/or general complex systems theory. The "agent" is a generalization of the "individual" in the individual-based models of ecology. Agent-based models are usually designed according to the principle of emergence;[75] i.e., they do not provide model structures that bear the risk of hardwired model outcome. However, as Railsback[25] reports, it has been acknowledged now in the field of agent-based modeling that the strategy of just providing entities with properties and then letting the model run may be good fun but rarely leads to really interesting results telling us something about the real world. Therefore, "getting results"[25] is a main problem in the field of agent-based modeling, and Railsback[25] showed by referring to the trout model by Railsback and Harvey[27] (see Section 26.4.3 above) that the pattern-oriented approach to the narrow-down model structure is also a powerful tool for agent-based modeling in general. Thus it seems that the somewhat naive modeling strategy of just throwing a bunch of entities (individuals, agents) with certain properties into a certain environment and then seeing what happens is not especially useful if problems and questions of the real world are concerned. Instead, models have to be designed carefully with regard to their objective, the knowledge available, and the patterns we observe and which are potential indicators of underlying essential structures and processes.

POM is certainly not a miracle cure for all the problems of ecological modeling in general and bottom-up modeling in particular. But it still seems to be an approach worth using. After all, POM means nothing more than applying the general research approach of science: searching to reveal the mechanism underlying the patterns we observe.

References

1. Hall, C.A.S., What constitutes a good model and by whose criteria? *Ecol. Modelling*, 43, 125, 1988.
2. Holling, C.S., The strategy of building models of complex ecological systems, in *The Strategy of Building Models of Complex Ecological Systems,* Watt, K.E.F., Ed., Academic Press, New York, 1966, 195.
3. Levins, R., The strategy of model building in population biology, *Am. Sci.,* 54, 421, 1966.
4. May, R.M., *Stability and Complexity in Model Ecosystems*, Princeton University Press, Princeton, NJ, 1973.
5. Huston, M., DeAngelis, D., and Post, W., New computer models unify ecological theory, *BioScience,* 38, 682, 1988.
6. DeAngelis, D.L. and Gross, L.J., *Individual-Based Models and Approaches in Ecology*, Chapman & Hall, New York, 1992.
7. Green, D.G., Simulated effects of fire, dispersal and spatial pattern on competition within forest mosaics, *Vegetatio,* 82, 163, 1989.
8. Czárán, T. and Bartha, S., Spatiotemporal dynamic models of plant populations and communities, *Trends Ecol. Evol.,* 7, 38, 1992.
9. Ratz, A., Long-term spatial patterns created by fire: a model oriented toward boreal forests, *Int. J. Wildland Fire.,* 5, 25, 1995.
10. Jeltsch, F., Milton, S.J., Dean, W.R.J., and van Rooyen, N., Tree spacing and coexistence in semiarid savannas, *J. Ecol.,* 84, 583, 1996.
11. Czárán, T., *Spatiotemporal Models of Population and Community Dynamics*, Chapman & Hall, London, 1997.
12. Berger, U., Hildenbrandt, H., and Grimm, V., Towards a standard for the individual-based modeling of plant populations: self-thinning and the field-of-neighborhood approach, *Nat. Resour. Modeling*, 15, 39, 2002.
13. Bauer, S., Wyszomirski, T., Berger, U., Hildenbrandt, H., and Grimm, V., Asymmetric competition as a natural outcome of neighbour interactions among plants: results from the field-of-neighbourhood modelling approach, *Plant Ecol.,* in press.
14. Grimm, V., Ten years of individual-based modelling in ecology: what have we learned, and what could we learn in the future? *Ecol. Modelling*, 115, 129, 1999.
15. Grimm, V., Mathematical models and understanding in ecology, *Ecol. Modelling*, 75/76, 641, 1994.
16. Grimm, V., Frank, K., Jeltsch, F., Brandl, R., Uchmanski, J., and Wissel, C., Pattern-oriented modelling in population ecology, *Sci. Total Environ.,* 183, 151, 1996.

17. Wissel, C. and Maier, B., A stochastic model for the species-area relationship, *J. Biogeogr.*, 19, 355, 1992.
18. Wyszomirski, T., Wyszomirska, I., and Jarzyna, I., Simple mechanisms of size distribution dynamics in crowded and uncrowded virtual monocultures, *Ecol. Modelling*, 115, 253, 1999.
19. Berger, U. and Hildenbrandt, H., The strength of competition among individual trees and the biomass-density trajectories of the cohort, *Plant Ecol*, 167, 89, 2003.
20. Thulke, H., Müllser, M.S., Grimm, V., Tischendorf, L., Wissel, C., and Jeltsch, F., From pattern to practice: a scaling-down strategy for spatially explicit modelling illustrated by the spread and control of rabies, *Ecol. Modelling*, 117, 179, 1999.
21. Wiegand, T., Jeltsch, F., Hanski, I., and Grimm, V., Using pattern-oriented modeling for revealing hidden information: a key for reconciling ecological theory and conservation practice, *Oikos*, 100, 209, 2003.
22. Wiegand, T., Revilla, E., and Knauer, F., Dealing with uncertainty in spatially explicit population models, *Biodiversity and Conservation*, in press.
23. Wiegand, T., Naves, J., Stephan, T., and Fernandez, A., Assessing the risk of extinction for the brown bear (*Ursus arctos*) in the Cordillera Cantabrica, Spain, *Ecol. Monogr.*, 68, 539, 1998.
24. Huth, A. and Wissel, C., The simulation of the movement of fish schools, *J. Theor. Biol.*, 156, 365, 1992.
25. Railsback, S.F., Getting "results": the pattern-oriented approach to analyzing natural systems with individual-based models, *Nat. Resour. Modeling*, 14, 465, 2001.
26. Rademacher, C., Neuert, C., Grundmann, V., Wissel, C., and Grimm, V., Was charakterisiert Buchenur-wälder? Untersuchungen der Altersstruktur des Kronendachs und der räumlichen Verteilung der Baumriesen in einem Modellwald mit Hilfe des Simulationsmodells BEFORE, *Forstwiss. Centralbl.*, 120, 288, 2001.
27. Railsback, S.F. and Harvey, B.C., Analysis of habitat selection rules using an individual-based model, *Ecology*, 83, 1817, 2002.
28. Wissel, C., Aims and limits of ecological modelling exemplified by island theory, *Ecol. Modelling*, 63, 1, 1992.
29. Levin, S.A., The problem of pattern and scale in ecology, *Ecology*, 73, 1943, 1992.
30. Dorndorf, N., Zur Populationsdynamik des Alpenmurmeltiers: Modellierung, Gefährdungsanalyse und Bedeutung des Sozialverhaltens für die Überlebensfähigkeit, Doctoral thesis, Philipps-Universität Marburg, 1999.
31. Grimm, V., Dorndorf, N., Arnold, W., Frey-Roos, F., Wissel, C., and Wyszomirski, T., Modelling the role of social behavior in the persistence of the alpine marmot *Marmota marmota*, *Oikos*, 102, 124, 2003.
32. Blasius, B., Huppert, A., and Stone, L., Complex dynamics and phase synchronization in spatially extended ecological systems, *Nature*, 399, 354, 1999.
33. Ford, E.D., *Scientific Method for Ecological Research*, Cambridge University Press, Cambridge, U.K., 2000.
34. Jax, K., Jones, C.G., and Pickett, S.T.A., The self-identity of ecological units, *Oikos*, 82, 253, 1998.
35. Casagrandi, R. and Gatto, M., A mesoscale approach to extinction risk in fragmented habitats, *Nature*, 400, 560, 1999.
36. Claessen, D., De Roos, A.M., and Persson, L., Dwarfs and giants: cannibalism and competition in size-structured populations, *Am. Nat.*, 155, 219, 2000.
37. Doak, D.F. and Morris, W., Detecting population-level consequences of ongoing environmental change without long-term monitoring, *Ecology*, 80, 1537, 1999.
38. Elliot, J.A., Irish, A.E., Reynolds, C.S., and Tett, P., Modelling freshwater phytoplankton communities: an exercise in validation, *Ecol. Modelling*, 128, 19, 2000.
39. Harrison, G.W., Comparing predator–prey models to Luckinbill's experiment with *Didinium* and *Paramaecium*, *Ecology*, 76, 357, 1995.
40. Johst, K. and Brandl, R., The effect of dispersal on local population dynamics, *Ecol. Modelling*, 104, 87, 1997.
41. Lewellen, R.H. and Vessey, S.H., The effect of density dependence and weather on population size of a polyvoltine species, *Ecol. Monogr.*, 68, 571, 1998.
42. Wood, S.N., Obtaining birth and death rate patterns from structured population trajectories, *Ecol. Monogr.*, 64, 23, 1994.
43. Hilborn, R. and Mangel, M., *The Ecological Detective: Confronting Models with Data*, Princeton University Press, Princeton, NJ, 1997.

44. Kendall, B.E., Briggs, C.J., Murdoch, W.W., Turchin, P., Ellner, S.P., McCauley, E., Nisbet, R.M., and Wood, S.N., Why do populations cycle? A synthesis of statistical and mechanistic modeling approaches, *Ecology,* 80, 1789, 1999.

45. DeAngelis, D.L. and Mooij, W.M., In praise of mechanistically rich models, in *Understanding Ecosystems: The Role of Quantitative Models in Observation, Synthesis and Prediction* (Cary Conference IX), Canham, C. and Lauenroth, W., Eds., Princeton University Press, Princeton, NJ, in press.

46. Haefner, J.W., *Modeling Biological Systems: Principles and Applications*, Chapman & Hall, New York, 1996.

47. Starfield, A.M., Smith, K.A., and Bleloch, A.L., *How to Model It: Problem Solving for the Computer Age*, McGraw-Hill, New York, 1990.

48. Starfield, A.M. and Bleloch, A.L., *Building Models for Conservation and Wildlife Management*, Collier Macmillan, London, 1986.

49. Grimm, V., Günther, C.-P., Dittmann, S., and Hildenbrandt, H., Grid-based modelling of macrozoobenthos in the intertidal of the Wadden Sea: potentials and limitations, in *The Wadden Sea Ecosystem: Stability Properties and Mechanisms*, Dittmann, S., Ed., Springer, Berlin, 1999, 207.

50. Brang, P., Courbaud, B., Fischer, A., Kissling-Naf, I., Pettenella, D., Schönenberger, W., Spörk, W., and Grimm, V., Developing indicators for sustainable management of mountain forests using a modelling approach, *Forest Policy and Economics*, 4, 113, 2002.

51. Ropella, G.E.P., Railsback, S.F., and Jackson, S.K., Software engineering considerations for individual-based models, *Nat. Resour. Modeling,* 15, 5, 2002.

52. Grimm, V., Visual debugging: a way of analyzing, understanding, and communicating bottom-up simulation models in ecology, *Nat. Resour. Modelling*, 15, 23, 2002.

53. Kaiser, H., The dynamics of populations as result of the properties of individual animals, *Fortschr. Zool.*, 25, 109, 1979.

54. Hanski, I., *Metapopulation Ecology*, Oxford University Press, Oxford, 1999.

55. Hanski, I., A practical model of metapopulation dynamics, *J. Anim. Ecol.*, 63, 151, 1994.

56. Neuert, C., Rademacher, C., Grundmann, V., Wissel, C., and Grimm, V., Struktur und Dynamik von Buchenurwäldern: Ergebnisse des regelbasierten Modells BEFORE, *Natursch. Landschaftspl.*, 33, 173, 2001.

57. Neuert, C., Die Dynamik räumlicher Strukturen in naturnahen Buchenwäldern Mitteleuropas, Doctoral thesis, Universität Marburg, 1999.

58. Berger, U. and Hildenbrandt, H., A new approach to spatially explicit modelling of forest dynamics: spacing, ageing and neighourhood competition of mangrove trees, *Ecol. Modelling*, 132, 287, 2000.

59. Chen, R.G. and Twilley, R.R., A gap dynamic model of mangrove forest development along gradients of soil salinity and nutrient resources, *J. Ecol.*, 86, 37, 1998.

60. Grote, R. and Reiter, I.M., Competition-dependent modelling of foliage and branch biomass in forest stands, submitted.

61. Cintrón, G. and Schaeffer-Novelli, Y., Caracteristicas y desarrollo estructural de los manglares de norte y sur america, *Cienc. Interam.,* 25, 4, 1985.

62. Fromard, F., Puig, H., Mougin, E., Marty, G., Betoulle, J.L., and Cadamuro, L., Structure, above-ground biomass and dynamics of mangrove ecosystems: new data from French Guinea, *Oecologia,* 115, 39, 1998.

63. Railsback, S.F., Lamberson, R.H., Harvey, B.C., and Duffy, W.E., Movement rules for individual-based models of stream fish, *Ecol. Modelling*, 123, 73, 1999.

64. Auyang, S.Y., *Foundations of Complex System Theories in Economics, Evolutionary Biology, and Statistical Physics*, Cambridge University Press, New York, 1998.

65. Raffaelli, D. and Hawkins, S., *Intertidal Ecology*, Chapman & Hall, London, 1996.

66. Okubo, A., *Diffusion and Ecological Problems: Mathematical Models*, Springer, Berlin, 1980.

67. Abraham, E.R., The generation of plankton patchiness by turbulent stirring, *Nature*, 391, 577, 1998.

68. Connell, J.H., Diversity in tropical rain forests and coral reefs, *Science,* 199, 1302, 1978.

69. Werner, F.E., Quinlan, J.A., Lough, R.G., and Lynch, D.R., Spatially explicit individual-based modeling of marine populations: a review of advances in the 1990s, *Sarsia*, 86, 411, 2001.

70. Hermann, A.J., Hinckley, S., Megrey, B.A., and Stabeno, P.J., Interannual variability of the early life history of walleye pollock near Shelikof Strait as inferred from a spatially explicit, individual-based model, *Fish. Oceanogr.*, 5, 39, 1996.

71. Walter, E.E., Scandol, J.P., and Healey, M.C., A reappraisal of the ocean migration patterns of Fraser River sockeye salmon (*Oncorhynchus nerka*) by individual-based modelling, *Can. J. Fish. Aquat. Sci.*, 54, 847, 1997.

72. Fennel, W. and Neumann, T., Coupling biology and oceanography in models, *Ambio*, 30, 232, 2001.

73. Verduin, J.J. and Backhaus, J.O., Dynamics of plant-flow interactions of the seagrass *Amphibolis antartica*: field observations and model simulations, *Estuarine Coastal Shelf. S.*, 50, 185, 2000.

74. Zeide, B., Quality as a characteristic of ecological models, *Ecol. Modelling*, 55, 161, 1991.

75. Railsback, S.F., Concepts from complex adaptive systems as a framework for individual-based modelling, *Ecol. Modelling*, 139, 47, 2001.

27

Spatial Dynamics of a Benthic Community: Applying Multiple Models to a Single System

Douglas D. Donalson, Robert A. Desharnais, Carlos D. Robles, and Roger M. Nisbet

CONTENTS

27.1 Introduction 429
27.2 Four Model Classes 430
27.3 Modeling Mussels in the Intertidal Zone 432
27.4 Examples of Model Verification and Validation 437
27.5 Discussion 441
References 443

27.1 Introduction

Understanding the effects of spatial processes on population dynamics is essential to the advancement of the science of ecological modeling. Inclusion of spatial processes in a model can significantly alter the dynamic predictions made by nonspatial models. While models that allow for spatial interactions can provide insights into ecological dynamics, there are difficulties as well. It can be difficult to deduce the relative importance of underlying mechanisms causing the altered dynamics from simulation results without the tools of mathematical analysis. Also, confirming that the computer code driving the model is working is a nontrivial task. As additional complexity is added to increase realism of the model, these tasks increase in difficulty. Using different model types to represent the dynamic(s) under study can provide both additional ways to test that models are working correctly and better understanding of the various mechanisms contributing to the system dynamics.

When choosing a model for use in a research project, it is important to make sure that the model class can represent all of the important dynamic interactions in the system. A model class can loosely be defined as a group of models that all have the same underlying structure. Within this grouping, models based on differential equations are a model class, as are all deterministic models. Model classes are often described along Holling's[1] gradient with end points *strategic* (very general), such as ordinary differential equation models (ODE), and *tactical* (very specific), such as agent-based models (ABMs). However, this is not always an accurate way to partition different model classes. For example, the ODE and ABM in Donalson and Nisbet[2] are both strategic models. Instead, we look at the relative ease of design and analysis against the assumptions inherent in the structure of the model class. We use a sliding scale of simplicity to complexity of model class. At one end of the scale are models that are simple to design and analyze but have very restrictive assumptions; on the other end are models that are difficult to design and analyze but remove constraints of simple model classes. Although there are many published examples of comparisons between two model classes, usually a population-based model to an individual-based model,[3-5] there are relatively few that compare more than two model classes representing the same system dynamics (but see References 2 and 6 through 8 for examples).

A key to understanding the difference between model classes is recognition of implicit model assumptions. All models involve explicit and implicit assumptions. Explicit assumptions are the mechanisms chosen by the researcher to simulate the various dynamics involved in the system under study. These are, for the most part, independent of the class of model used. Examples of these are logistic vs. exponential growth, and Type I, II, or III functional responses.[9] Implicit model assumptions are those that characterize the model class. A good starting point for a discussion of implicit model assumptions are the Lotka-Volterra predator–prey equations.[10] This pair of equations is a well-known example of the ODE class of models. The two most significant implicit assumptions of the Lotka-Volterra models (and in fact, any ODE model) are first, that the system is "well mixed," that all individuals are identical and all individuals experience identical conditions;[11] and, second, that there are enough individuals that demographic stochasticity has minimal effect on the population dynamics. In natural systems, these two assumptions are usually mutually exclusive. Spatial scales small enough to guarantee a homogeneous environment can only host small populations and are therefore potentially subject to strong demographic stochasticity. Spatial scales that can maintain populations large enough to minimize demographic stochasticity are seldom homogeneous. The combination of these two assumptions in a single model can be restrictive.

How then can we determine what type of model class should be used for what situation?[12] Choosing the model class at the beginning of the project on the basis of projected system dynamics is often not a good option. This is particularly true when dealing with potentially complex spatial processes where the interactions in time and space at different scales can cause the dynamics to deviate from linear predictions.[2,13–15] In contrast, using a complex spatial model to analyze basic system behavior such as potential equilibria and sensitivity analysis is seldom practical because of the slower speed of the simulations and the stochasticity inherent in the results.[16] Therefore, we need some way to evaluate different models against the system we wish to simulate to ensure we are using the correct tool(s) for the job.

In this chapter we describe four complementary model classes and their interrelations. We then introduce four models of an intertidal predator–prey system, each of which represents one of the four model classes. We use results from our testing of the four models to demonstrate advantages of our multiple model approach. The focus of this chapter is on model comparisons; future work will attempt to match model results to natural systems.

27.2 Four Model Classes

The properties of the ODE class of models are well known. Aside from the implicit assumptions previously discussed, there are two important properties of ODEs for the purposes of this chapter. First, there are a wide range of analytical and numerical tools available that allow detailed analysis of the dynamics of the system. Second, simulations of ODE-based models are fast when compared to their more complex counterparts. Because of this, ODE models provide an excellent baseline for modeling projects.

The stochastic birth–death (SBD) model class removes the restriction that demographic stochasticity is negligible while still requiring that all spatial processes be "well mixed" and that all individuals are identical. We spend a bit more time on SBD models because they are powerful but seldom-used additions to the modeler's tool kit. SBD models are generally defined using an underlying deterministic model. To construct an SBD, two changes are made to the deterministic model. The first is that the state variables are constrained to integer values (but see the SBD model later in this chapter for a specific exception to this rule). This represents a more realistic view of populations since changes in population size occur through loss or gain of individuals, not pieces of individuals. Second, the terms of the deterministic equation are redefined as the probabilities of an increase or decrease in the population per unit time. This represents the unpredictability in natural populations of exact times between individual births and deaths. The random distribution applied to these probabilities is usually either the Poisson (number of state changes in a fixed time interval) or the negative exponential (time to next state change). These two distributions are used because they are memoryless; that is, the next state is dependent only on the present state and no other history is required. This choice of distribution matches the memoryless assumption present in ODE-based deterministic models.

Detailed examples of designing more complex SBD models are given in Donalson and Nisbet,[2] Donalson,[7] Renshaw,[17] Nisbet and Gurney,[18] and in the second part of this chapter. However, a brief example is useful here. We look at a simple birth/death process whose deterministic representation is the ODE $dN(t)/dt = bN(t) - \delta N(t)$, where $N(t)$ is the total number of individuals at time t, b is the per capita growth rate, and δ is the per capita death rate. There are two possible state transitions in this process, $N \rightarrow N + 1$ and $N \rightarrow N - 1$. Because we are now dealing with a stochastic system, we redefine the rates as a set of probabilities. In a very small time interval $[t, t + \Delta t]$, the chance that there is a birth, $[N \rightarrow N + 1]$, is $bN(t) \Delta t$ and the chance of a death, $[N \rightarrow N - 1]$, is $\delta N(t) \Delta t$. Here we require that Δt be small enough that the chance of multiple state changes within the time interval be negligibly small. Alternately, instead of solving for ΔN, we can solve for Δt, the time to the next state change. The probability distribution of possible times to next event is then memoryless with mean $1/bN$ for $N \rightarrow N + 1$ and $1/\delta N$ for $N \rightarrow N - 1$. Law and Kelton[19] provide the formula for the memoryless distribution of possible times given a mean time as $-\ln(U[0,1]) * (\text{mean})$, where $U[0,1]$ is a uniform random number generated between 0 and 1. In this example the stochastic time to next state transition is then the shorter of $-\ln(U[0,1]) * (1/bN)$ or $-\ln(U[0,1]) * (1/\delta N)$. Note that we can now simply advance from one state change to the next by calculating the two possible times to the next state change, advancing time to the shorter of the two times, and then executing the appropriate state change. This formulation is possible because these probabilities are independent of each other and of state changes at previous times. This is called a *continuous time* or *event-driven* model and is used in the construction of two of the four models described in this chapter. SBD models are only slightly more difficult to construct than deterministic models, but, because their predictions are probabilistic, one must also deal with statistical properties in their analysis.

Cellular automata (CA) are a class of models that relax the requirement that space be "well mixed" and allow explicitly for the presence, in some form, of individual entities. CA models can be either deterministic or stochastic. We describe the stochastic version. This class of models uses many of the same ideas as SBD models and is relatively easy to construct. Space is now included in the model explicitly as a grid of cells, and, unlike the ODE and SBD models, the focus of state changes is the individual cell as opposed to the population. The probability transitions described for the SBD model are applied to each cell with the added feature that the transition probabilities for the next state of a cell are not only dependent upon its present state but also a function of the states of some set of its neighbors. One way to handle simultaneously updating the states of multiple cells is to "double buffer" space. Double buffering space means that there are actually two alternate cellular grids. One grid represents the present state and the other the next state. The next state for each cell in the present grid is recorded in the next state grid. When this is completed, the next state grid becomes the new present state grid, and the grid that was the present becomes the next state. Analyses of CA models are at a level of difficulty above that of SBD models because the effects of nonrandom spatial patterns may have to be included in analysis of the dynamics.

There are implicit assumptions regarding space and individuals in the CA class of models. A cellular array consists of a fixed number of locations. This means that, by default, in the special case of one individual per lattice site, the system has a carrying capacity that is equal to the number of cells. Because space consists of an array of cells, there is also fine-gained quantization of space. Thus, there is a minimum spatial scale at which interactions can occur. Individuals are represented in their simplest form, just a set of discrete states.

Agent-based models (sometimes also referred to as individual-based models) are a class of model that can, potentially, remove all implicit assumptions from the model structure.[20-22] This class of model can be conveniently broken down into two parts: the agents and the infrastructure. The agents are typically the entities that comprise the populations of the state variables in a deterministic model. The infrastructure comprises the implementation of space and time.

There are two ways space can be represented in an ABM. Discrete space is similar to that of the CA model; however, the focus of interaction is now the individual, not the cell, so the cell now just represents some aspect of environmental space. Space can also be represented as continuous, where the position of an individual is represented by a real-valued X, Y coordinate pair (or triplet in the came of three-dimensional simulations). Time can be implemented as either discrete or continuous. In the case of continuous time, state transitions for individuals are inserted into or removed from a time-ordered event schedule, and, like the SBD, time is advanced from one event to the next. Because state transitions are now tied to the individual agent, the per capita rate as opposed to the total rate is used to calculate the time to state transition.

ABMs can be quite difficult to code and verify. In addition, given a result, deducing the contributions of the various underlying mechanisms is not trivial. In contrast, if we believe that individual behavior, in all its forms, is a contributing factor to population dynamics, this class of model must become part of the modeler's tool kit.[3,7,11,20]

Each of the aforementioned class of models has strengths and weaknesses. On the gradient described earlier, as we move in the direction of more complex model classes, the ability to represent more complex interactions increases. However, the analysis of results and, in particular, the ability to isolate the relative strength of the contributions of the underlying mechanisms to the dynamic outcomes becomes far more difficult. Comparing and contrasting the results of different model classes can mitigate many of these problems. As discussed previously, each model class has some advantage of "realism" gained from removing an implicit assumption, with the negative trade-off of new technical challenges and more difficult analysis of results. We advocate multiple model analysis for two purposes, model *verification* and results *validation*, in a similar (but not identical) manner to Rykiel.[23] Model verification is confirming that the code/equations that comprise the structure of the model are working as intended. We use the term *validation*, in its most general sense, to describe the analysis of the dynamic results once verification is completed. In general, a more complex model can be configured to match the implicit assumptions of a simpler model for the purposes of verification. Comparing the results of a simple model to a complex counterpart can validate or invalidate its use of implicit assumptions associated with the simpler model. Using simpler models to factor out various mechanisms can also help decompose complex results into the individual contributions of different mechanisms.

We now introduce an experimental system of mussels and their predators. In our present research, we are using the four previously described model classes to assist in our understanding of the dynamics of this system. In this chapter we use these models to demonstrate comparing and contrasting different model classes. We will introduce some of the preliminary results from this work to help demonstrate the multiple-model approach.

27.3 Modeling Mussels in the Intertidal Zone

The mussel *Mytilus californianus* is a dominant species of the intertidal zones of the North American continent. This species is found in narrow bands in shore sites of moderate to high wave exposure. The predators of *M. californianus* are the seastar, *Pisaster ochraceus*, in the Pacific Northwest,[24,25] and the spiny lobster, *Panulirus interruptus*, in Southern California.[26,27]

The ecology of mussel communities has been studied for over five decades. Early experiments suggested that the lower limits for *M. californianus* are set by predation and the upper limits are set by physical factors such as intolerance to desiccation.[24,28] Thus, mussels experience a spatial refuge from predation at the upper intertidal zone. Paine[25] observed that below the upper intertidal zone there were patches of very large mussels that escape predation. This fact and the observation that seastars eat mussels smaller than the maximum available size suggested that mussels reach a certain size and become resistant to predation; this represents a "size refuge" hypothesis. More recent experiments emphasized supply-side effects,[29] suggesting that variation in recruitment rates is a source of variation in the adult populations, and recruitment produces feedback effects in the processes of competition and predation (see discussion in Robles and Desharnais[30]).

Later studies contradict the hypotheses of spatial and size refuges. In Southern California, time-lapse photography has revealed that spiny lobsters enter the upper intertidal zone at night and to consume mussels.[26,31] Similar behavior was observed in the Pacific Northwest;[32] seastars move with the tides and are found foraging above lower boundaries of mussel beds. Also, experiments have shown that concentrations of adult mussels that occur above lower shore levels of the most wave-exposed locations appear to result as much from the elevated rates of recruitment in these locations as from the impact of hydrodynamic stresses and tidal exposure on predator foraging.[33] It appears that the refuge hypothesis is an oversimplification of a more complex situation.

Mussel growth depends on the flow of water providing food, resulting in higher growth rates for mussels located lower in the intertidal zone and on wave-exposed shores.[34] The probability of being attacked by a predator decreases when a mussel is surrounded by larger mussels.[25,32,35,36] Thus, the rates of production and mortality in any specific location depend on the location of a mussel in the gradients of tidal height and wave exposure and on the size and density of surrounding mussels. These observations suggest the need for a new theoretical synthesis that will study rates of recruitment, growth, and predation mortality as a dynamic spatially explicit process.

We have developed a multiple-model approach to the study of predation in benthic communities. Four classes of models have been developed and analyzed, one from each previously discussed model class: (1) an analytical "mean field" approximation consisting of ODEs, (2) an SBD version of the mean field ODE model, (3) a CA model, and (4) an ABM. A set of model parameter values will be common to all four model types. Comparison and cross-validation can be made among models to take advantage of the strengths that each has to offer.

Our ODE model is based on the work of Nisbet et al.[8] where "space" is made up of a large number of very small "patches" that can be occupied by at most one mussel (algae in their model) and predators ("grazers" in their model) that move randomly among patches. Prey biomass grows in size in each patch until a predator grazes a patch to size zero. In Nisbet et al.,[8] as soon as grazers remove the algae from a patch there is algal regrowth. In our model each patch is either empty or occupied by a mussel. This model is given by a pair of differential equations:

$$\frac{\partial n(a,t)}{\partial t} = -\frac{\partial n(a,t)}{\partial a} - \mu(a,t)n(a,t), \tag{27.1}$$

$$\frac{dP(t)}{dt} = I - e_P(t)P(t), \tag{27.2}$$

where $n(a, t)$ is the density of prey of age a at time t and $P(t)$ is the density of predators. The first equation is the McKendrick–von Foerster model[37,38] for aging and death in an age-structured population where $\mu(a, t)$ is mortality rate for prey of age a at time t. We assume prey settle at a constant rate σ into empty patches, but the overall recruitment of prey decreases linearly until all available space is occupied at a maximum density K. This yields $n(0, t) = \sigma[1 - N(t)/K]$ as the boundary condition for new prey, where $N(t)$ is the density of prey of all ages. In this open system, predators immigrate at the constant rate I and emigrate at the per capita rate $e_P(t)$, which may depend on the age and size structure of the prey population.

Prey size plays an important role in the predator–prey dynamics. We let $s(a)$ represent the size of a prey of age a and describe growth using the von Bertalanffy[39] formulation $s(a) = s_\infty - (s_\infty - s_0)e^{-\beta a}$, where β is the growth rate, s_∞ is the maximum size, and s_0 is the size of a newly settled recruit. We assume each prey's vulnerability to predation depends on its size and the density and size of prey in some spatial neighborhood of radius h surrounding the individual. In our mean field approximation, we assume that the size of the neighborhood $h = \infty$, and define $S(t) = \int s(a)n(a,t)da$ as the mean size of prey weighted by prey density. We write the mean field approximation for the per capita mortality rate of prey as $\mu(a,t) = \mu_0 + \theta P(t)e^{-cS(t)}$, which is independent of prey age but decreases exponentially with the weighted mean size of prey, $S(t)$. The parameter μ_0 is the mortality rate due to causes other than predation, θ is the predator attack rate, and c is a measure of how quickly resistance to predation increases with prey size. For predators, we assume that their emigration rate from the system is inversely proportional to the per capita rate of prey consumption; we use $e_P(t) = e_0[S(t)\theta e^{-cS(t)}]^{-1}$, where e_0 is the constant of proportionality relating prey consumption to predator emigration. Defining $N(t) = \int n(a,t)da$ as the overall prey density and taking the time derivative of $S(t)$, we can replace Equation 27.1 with a pair of ODEs

TABLE 27.1

ODE and SBD Models and Parameters

ODE Model

State Variables	Equations
$S(t) \equiv$ weighted mean prey size (mm prey area^{-1})	$\dfrac{dS}{dt} = s_0\sigma + \left(s_\infty\beta - s_0\sigma K^{-1}\right)N(t) - \left(\beta + \mu_0 + \theta P(t)\, e^{-cS(t)}\right)S(t)$
$N(t) \equiv$ prey density (prey area^{-1})	$\dfrac{dN}{dt} = \sigma - \left(K^{-1}\sigma + \mu_0 + \theta P(t)e^{-cS(t)}\right)N(t)$
$P(t) \equiv$ predator density (predator area^{-1})	$\dfrac{dP}{dt} = I - e_0\left(\theta S(t)e^{-cS(t)}\right)^{-1}P(t)$

Stochastic Birth–Death Model

Transition	Transition Rate	State Changes
Prey recruitment	$A\sigma\!\left(\dfrac{K - N(t)}{K}\right)$	$N(t + \Delta t) = N(t) + A^{-1}$ $S(t + \Delta t) = S(t) + s_0 A^{-1} + \Delta S$
Prey death	$A\left(\mu_0 + \theta P(t)e^{-cS(t)}\right)N(t)$	$N(t + \Delta t) = N(t) - A^{-1}$ $S(t + \Delta t) = \left(S(t) + \Delta S\right)\left(AN(t) - 1\right)/\left(AN(t)\right)$
Predator immigration	AI	$P(t + \Delta t) = P(t) + A^{-1}$ $S(t + \Delta t) = S(t) + \Delta S$
Predator emigration	$Ae_0\left(\theta S(t)e^{-cS(t)}\right)^{-1}P(t)$	$P(t + \Delta t) = P(t) - A^{-1}$ $S(t + \Delta t) = S(t) + \Delta S$
Prey growth	$\Delta S = \left(N(t)s_\infty - S(t)\right)\left(1 - e^{-\beta\Delta t}\right)$	

Model Parameters

Symbol	Definition	Default Values and Units
area	Unit area (one "cell")	25 cm²
A	Total system area	4×10^4 units of area
t	Time	1 day
a	Age	1 day
σ	Prey recruitment rate	1 prey area^{-1} day^{-1}
s_0	Size of newly settled prey	1 mm
s_∞	Maximum prey size	200 mm
β	Decrease in prey growth rate with size	0.0004 day^{-1}
μ_0	Background *per capita* prey mortality rate	0.0001 day^{-1}
K	Maximum prey density	1 prey area^{-1}
θ	Predator attack coefficient	1.0 unit area predator^{-1} day^{-1}
c	Resistance to predation with prey size	0.04 unit area mm^{-1}
I	Predator immigration rate	0.01 predator area^{-1} day^{-1}
e_0	Predator emigration constant	5.0 mm predator^{-1} day^{-2}

yielding the system of three ODEs given in Table 27.1. (Details of this derivation are available in Desharnais et al., in preparation.) This system of equations is our ODE mean field approximation model.

The SBD model is a probabilistic version of the ODE model. Prey and predator densities are treated as discrete system variables. The dynamics are modeled as series of four possible transitions: the recruitment or death of a prey or the immigration or emigration of a predator. The time interval Δt between transitions is a continuous random variable chosen from a negative exponential probability distribution with an expected value that is the inverse of the transition rates. The transition rates follow from the ODE model.

Since it is a model of demographic stochasticity, the SBD forces us to consider the scale of the process being modeled. Transitions will occur much more frequently in a large system than in a small one. An additional parameter A must be used to represent the total size of the area being modeled. Since our system variables $N(t)$ and $P(t)$ are densities, the total numbers of prey and predators are $AN(t)$ and $AP(t)$, respectively. These quantities increase or decrease by one, or, equivalently, the variables $N(t)$ and $P(t)$ may increase or decrease by an amount A^{-1} at each transition. The transition rates are just the rates from the ODE model multiplied by A. The transitions and their rates are given in Table 27.1.

Prey growth presents a complication. Because size is a continuous variable, we must increment the system variable $S(t)$ by an amount that represents the growth of all prey during the time interval Δt. Fortunately, the von Bertalanffy growth model has the convenient property that if $s_i(t)$ is the size of an individual i at time t, then the average size $\bar{s}(t)$ of all individuals follows the same growth equation as for a single individual: $\bar{s}(t + \Delta t) = s_\infty - (s_\infty - \bar{s}(t))e^{-\beta\Delta t}$. Since $S(t)$ is mean size weighted by density, $S(t) = \bar{s}(t)N(t)$, and we obtain the expression for the increment ΔS due to growth given in Table 27.1. For a prey recruitment transition, we also add the size of the new recruit divided by the total area, $s_0 A^{-1}$. Unlike the ABM, individuals are not tracked in an SBD, so for a prey death transition, we decrease $(S(t) + \Delta S)$ by an amount equivalent to the loss of one prey of average size; this amounts to multiplying by $(AN(t) - 1)/(AN(t))$.

Implementation of the SBD is fairly straightforward. A random time to the next transition is computed for each of the four possible transition types by generating a uniform random number, taking the negative of the natural logarithm, and dividing by the transition rate. The smallest of the four transition times determines Δt and the type of transition that occurs. Time is advanced to $t + \Delta t$ and the state variables are modified according to the transition rules. This process is repeated until a final simulation time is reached.

In the CA model space is represented explicitly as a rectangular grid of cells, where each cell has one unit of area and A is the total number of cells. Each cell represents a potential site of occupation by an individual prey. Since only one prey can occupy a cell at any time, we explicitly assume that there is no intraspecific competition for space among growing prey: an easily overlooked implicit assumption of the ODE and SBD models.

The recruitment, growth, and death of prey are modeled as transitions among "cell states." Each cell is denoted by its x,y coordinates in the grid and the state of a cell is represented by a discrete variable $S_{xy}(t)$. We use $S_{xy}(t) = 0$ to indicate an empty cell, $S_{xy}(t) = s_0$ for a cell with a new recruit, and $S_{xy}(t) = s_\infty$ for a cell with a prey at the maximum size. The size range $[s_0, s_\infty]$ is divided into m discrete increments, where m is chosen large enough to approximate continuous growth. Time is advanced in discrete steps Δt and all cell transitions occur simultaneously. Prey recruitment is represented as a transition from $S_{xy}(t) = 0$ to $S_{xy}(t + \Delta t) = s_0$. Prey growth is represented as a transition from $S_{xy}(t) = s$ to $S_{xy}(t + \Delta t) = s + \Delta s$, where $\Delta s \geq 0$. Prey death is represented as a transition from $S_{xy}(t) = s$ to $S_{xy}(t + \Delta t) = 0$ for $s > 0$. An additional "global variable" $P(t)$ represents the density of predators.

At each time step, cell transitions occur with probabilities that are derived using the rates specified in the ODE model (Table 27.1). For example, if Δt is sufficiently small, the probability that an empty cell will receive a recruit is given by $\sigma\Delta t$. Similarly, the probability that an occupied cell x,y becomes empty due to prey death is given by

$$\left(\mu_0 + \theta P(t)e^{-cS_{xy,h}(t)}\right)\Delta t,$$

where

$$S_{xy,h}(t) = (2h + 1)^{-2} \sum_{i=x-h}^{x+h} \sum_{j=y-h}^{y+h} S_{ij}(t)$$

is the mean of the prey sizes in a neighborhood of $(2h + 1) \times (2h + 1)$ cells centered at x,y. (At the sides and corners of the lattice, $S_{xy,h}(t)$ is the mean of the subset of neighboring cells within the system.) If

$h = 0$, then $S_{xy,0}(t) = S_{xy}(t)$ and the neighborhood is the cell x,y itself. When $h = \infty$, then $S_{xy,\infty}(t) = S(t)$ and we have the mean field approximation of the ODE and SBD models.

Prey growth is also treated as a stochastic process. For a prey of size s, the expected increase in size λ_s is given by the differential form of the von Bertalanffy model: $\lambda_s = \beta(s_\infty - s)\Delta t$. The expectation λ_s is then used as the mean parameter in a Poisson probability function to determine the actual increase in size Δs for each prey. In rare cases when $s + \Delta s > s_\infty$, we set $\Delta s = s_\infty - s$.

Predator immigration and emigration are treated as a random birth–death process. The number of predators that enter the system at each time step is chosen from a Poisson distribution with a mean value $AI\Delta t$. Each predator in the system at time t either leaves the system with probability $e_P(t)\Delta t$ or remains with probability $1 - e_P(t)\Delta t$. Assuming each predator makes its decision independently, we use a binomial distribution to determine the loss of predators. In the ODE model we assume that the per capita emigration rate $e_P(t)$ is inversely proportional to the predator's per capita consumption of prey. The same assumption is used in the CA model, except the actual loss of prey biomass density to predators during the previous interval, $S_P(t)$, is computed by summing the sizes of the individual prey consumed by predators and dividing by the system area A. The emigration rate is then computed using $e_P(t) = e_0[P(t)\Delta t/S_P(t)]$, where the term in square brackets is the inverse of the per capita rate of prey consumption by predators. In rare cases when $S_P(t) = 0$, the probability that predators leave the system is set equal to one.

In our ABM model, space is represented as a rectangle of area A. The boundaries are reflecting in the sense that predator movements are modeled as discrete jumps, but predators are not allowed to jump outside the system boundaries. (However, there is predator immigration into and emigration from the system across the boundaries.) The default space is continuous as opposed to the more common grid-based approach. Individuals are allowed to reside at any X,Y coordinate as opposed to the center of a grid cell.

State changes in the ABM model are tied to the individual. Each agent has a number of potential changes, such as feeding, death, or emigration. The times to each of these potential state changes are calculated using the per capita rates from the SBD model. All the times (with their associated individual and type of state change) are inserted into a time-ordered list called the event schedule. The next event (state change) is the one with the shortest time, which is also the event at the top of the list. As new individuals are added to the system their events are added to the event schedule, and, when an individual is removed from the system, any remaining events tied to that individual are removed from the event schedule. Finally, a predicted event associated with one individual can be altered by a change in another individual, thereby altering a previously scheduled event.

Mussels recruit at a constant rate σ, multiplied by the total area of the system A. New recruits are given a random X and Y pair of coordinates constrained to be within the system area and so that the initial diameter of the recruit will not overlap an edge. They recruit at size s_0 and are not allowed to overlap another individual. By default, if the coordinates of a new recruit overlap those of another individual, that recruitment is aborted and a new recruitment event is scheduled. Mussels are represented as circles and size is equivalent to the circle diameter. Note that, unlike the CA model, many small individuals can be packed into an area the size of one grid cell. Because of this, a K value was added to the ABM to allow comparison to the other three models.

Mussels grow in the same manner as described for the SBD model. Because the ABM model uses continuous space, spatial competition must be included in the model. In the CA model, a grid cell has a diameter that is equal to the diameter of a full-grown mussel. As a result no two individuals can overlap. At most, two neighbors will just touch when both are at maximum size. With continuous space, this is not the case. Therefore, as a first cut at modeling spatial competition, we introduce the rule that no two individuals are allowed to overlap. At the point where two neighboring individuals just touch, they quit growing. The time at which this event occurs is computed explicitly from the von Bertalanffy equations and added to the event schedule. They remain at this size until either the individual dies, or its neighbor dies. Upon the death of a constraining neighbor, an individual is allowed to resume growing. Death occurs from either background mortality as described in the CA model or from predation, as described later. Time to death from background mortality is chosen from a negative exponential distribution with mean $1/\mu_0$ and added to the event schedule at the time a mussel is recruited.

Predatory seastars enter the system at a constant rate IA and are placed at random within the system boundaries. Unlike the mussels, overlap is allowed. All predators are assumed to be the same size and there is no predator growth. When the predator enters the system it is provided with a time in the future when it will leave the system using the equation $\dfrac{e_0}{\theta S(t)} e^{cS(t)}$ and calculated using the present system state. At that time, the individual is removed from the system, regardless of its location in space.

Seastars move randomly about the system in a series of jumps. Two random numbers $U[0,1]$ are generated. The first number is multiplied by the maximum distance that the predator is allowed to jump. (Note that a maximum jump distance close to the dimension of the system simulates a well-mixed system, whereas a short jump distance allows for the possibility of local interactions affecting the system dynamics.) The second random number is multiplied by 2π to generate the direction in which the predator will move. The predator is then moved from its present position at the specified angle to the previously calculated distance. If this moves the predator off the map then a new movement vector is chosen.

Each time a predator lands at a new location it searches an area the size of one unit of space. All mussels with centers that lie within this area are subject to attack. As with the CA model, the chance of a successful attack is a combination of the predator's intrinsic ability to feed θ, modified by the protection provided by the mussel's own size and those of its neighbors, $e^{-cS(t)}$.

To match the CA model, it was necessary to add an additional configuration option to the ABM model. When an individual colonizes a grid cell in the CA model, it by default reserves the entire area of the grid cell for its use. This is in spite of the fact that at size s_0 it takes up only a very small part of the grid cell. To match this assumption, a grid-based cell array was overlaid on space. The rule that all individuals must "snap to" the center of each grid cell (the default discrete space requirement) and that only one mussel could occupy each grid cell was then an option that could be applied to the system.

27.4 Examples of Model Verification and Validation

We now compare and contrast some basic results from each of our four models. Our analysis follows the general theme of this chapter. We first confirm that all models have the same basic behavior when all are configured with the same assumptions. We then look at some of the implicit assumptions inherent within each model and test them for robustness against the explicit results of a more complex model. We start by describing the dynamics expected from a well-mixed large population using the ODE model.

The ODE model provides the baseline for our model analysis. The predicted dynamics of this model are summarized graphically in Figure 27.1. Setting $dP/dt = dN/dt = 0$, we can solve for the equilibria P^* and N^* in terms of S. Substituting these expressions into the equation for dS/dt and setting $dS/dt = 0$, one can show that the model equilibria are the real roots of the function $f(S) = s_0\sigma - (\beta + \mu_0)S - \kappa S^2 e^{-2cS} + \dfrac{K^{-1}\sigma(\beta s_\infty - \sigma s_0)}{K^{-1}\sigma + \mu_0 + \kappa S e^{-2cS}}$, $0 \leq S \leq s_\infty$, where $\kappa = I\theta^2/e_0$. This function is plotted in Figure 27.1A for the parameter values in Table 27.1. In this case the model has two stable equilibria and one unstable equilibrium. At the stable equilibrium S_1^*, prey density and sizes are kept low by high levels of predation. At the stable equilibrium S_3^*, there is a high density of large prey that are resistant to predation. The unstable equilibrium S_2^* lies on a boundary separating the basins of attraction of the two stable equilibria. When predation rates are low, only the upper equilibrium S_3^* exists (Figure 27.1B). When predation rates are high, only the lower equilibrium S_1^* exists (Figure 27.1C). For biologically feasible parameter values such as those in Table 27.1, there is no evidence of exotic dynamics such as limit cycles or chaos.

Verification of the three stochastic models is demonstrated in Figure 27.2. In this case all three of the stochastic models were configured to match the implicit assumptions of the ODE model. We tested three elements here. The first is that the three stochastic models all converge to the same two stable equilibrium values as the ODE. The second is that all three stochastic models move to the correct stable equilibrium when they are started in that equilibrium's basin of attraction. Lastly, it is useful to note that, qualitatively, all the transient responses are similar.

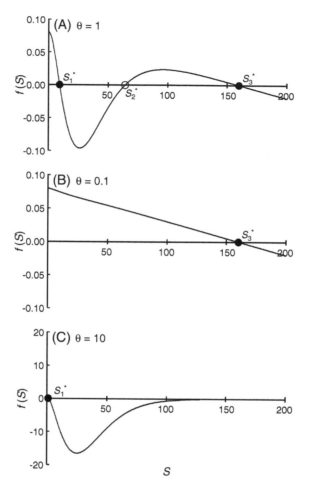

FIGURE 27.1 The ODE model predictions over a range of values of the predator attack coefficient, θ. The remaining parameters have the values given in Table 27.1. S_1^* is the lower stable equilibrium, S_2^* the middle unstable equilibrium, and S_3^* the upper stable equilibrium. (A) Two stable equilibria are present when the predator attack rate is at an intermediate value ($\theta = 1$). (B) When the predator attack rate is low ($\theta = 0.1$), only the upper stable equilibrium S_3^* exists. (C) When the predator attack rate is high ($\theta = 10$), only the lower stable equilibrium S_1^* exists.

There are two parts to our ODE validation. The first, testing the implicit assumption that demographic stochasticity can be discounted at a scale reasonable for our research, is demonstrated in Figure 27.3. Figure 27.3A shows ten realizations of the SBD model run at a system size that allows a total of 400 individuals ($A = 400$). Each run was started just above the unstable equilibrium (S_2^*) with a different random seed. As expected, the result is no longer deterministic, with three of the ten runs stabilizing at the lower equilibrium. Figure 27.3B shows the probability that the time series will stabilize at the upper equilibrium for a range of initial conditions about the unstable equilibrium for three different system sizes. The probability converges rapidly toward the prediction of the ODE model even at the relatively small system sizes used in this experiment. This indicates that demographic stochasticity should not have a significant effect on the steady state response at the default system size of $A = 40,000$ individuals (but see Donalson[7] and Harrison[40] for counter examples.)

The CA model demonstrates a breakdown in the ODE assumptions when space is not homogeneous and when individuals are not well mixed (Figure 27.4). Figure 27.4A shows the mean field approximation for the neighborhood effect at the lower equilibrium. There is little evidence of any heterogeneity when the size protection for each mussel is calculated using the entire population ($h = \infty$). The same is true at the upper equilibrium (Figure 27.5B), with the exception that there is more variation in size among mussels. In contrast, when a mussel is protected by only its eight immediate neighbors ($h = 1$),

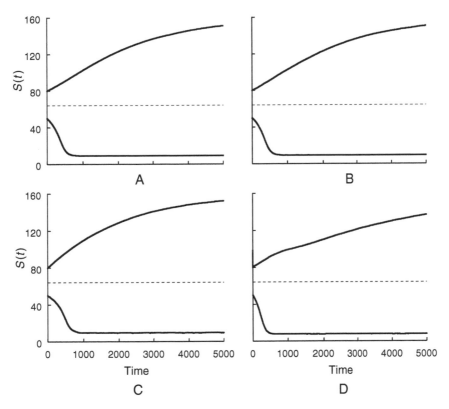

FIGURE 27.2 Two time series plots for each type of model. One plot starts at $S(0) = 50$ and the other at $S(0) = 80$. The dashed line is S_2^*. Time runs through $t = 5000$ days. All models were configured to match the implicit assumptions of the ODE model and used the default parameters from Table 27.1.

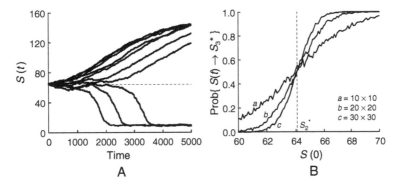

FIGURE 27.3 Results from the SBD model: (A) ODE, (B), SBD, (C) CA, and (D) ABM. (A) Ten realizations of the SBD with a 20×20 system started at $S(0) = 65$, which is just above the unstable equilibrium of $S_2^* = 64.1$ (dashed line). Three realizations go to the lower equilibrium and seven go to the upper equilibrium. (B) The effect of initial conditions and system size on the equilibrium; 1000 simulations were run at each of 100 initial conditions between $S(0) = 60$ and $S(0) = 70$ and the proportion that went to the upper equilibrium was computed. This experiment was repeated for three system sizes: $(a)10 \times 10$, $(b)20 \times 20$, and $(c)30 \times 30$. The vertical dashed line locates the unstable equilibrium S_2^*.

occasionally a mussel escapes predation long enough to become too large for seastars to handle it easily and it becomes the center of an "island" of protected individuals (Figure 27.4C). These islands can increase slowly in size and merge until the system resembles the upper equilibrium in Figure 27.4B. On the other hand, at the parameter values in Table 27.1, the ODE prediction of lower and upper stable equilibria holds well for moderate neighborhood sizes of $h \geq 2$. Figures 27.4D and E show an interesting

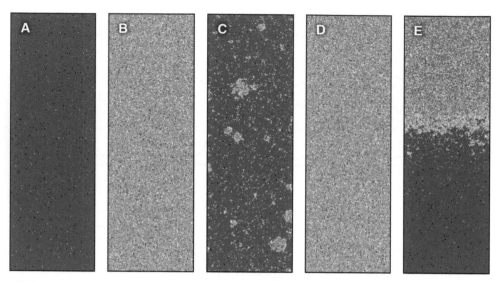

FIGURE 27.4 **(Color figure follows p. 332.)** Two possible spatial effects are observed in the CA model when the mean field assumptions are relaxed. Each plot shows the size distribution of mussels over space, where mussel sizes are color-coded from blue for the smallest through red for the largest. The size distributions over space are shown for the mean field approximation ($h = \infty$) at (A) the lower and (B) the upper equilibria. (C) When the neighborhood size is small ($h = 1$) and the seastar attack rate is reduced ($\theta = 0.8$), "islands" of larger individuals develop at the lower equilibrium. (D) When a smooth linear gradient in the seastar attack rate is imposed from $\theta = 0$ at the upper boundary of the lattice to $\theta = 2$ at the lower boundary, and the mean field approximation ($h = \infty$) is used, the lattice equilibrates at the upper equilibrium. (E) When the same gradient in θ is combined with a small neighborhood effect ($h = 1$), the result is a sharp transition in prey cover similar to what is seen in many natural mussel beds. This predation gradient is what might be expected from differences in foraging times at different tidal heights.

change in system dynamics when a continuous gradient is applied to space in the CA model and neighborhood size is small. In reality, seastars have more time to forage in areas that are submerged for longer periods during the tidal cycle. A simple representation of this is to assume a constant gradient of decreasing predator foraging rate starting in deep water and moving to the high tide mark. In Figures 27.4 D through E a vertical gradient in the predation rate was applied to space, with $\theta = 2$ at the bottom of the area, and θ decreased linearly to zero at the top. In Figure 27.2D the mean field approximation ($h = \infty$) was used and the area was initially empty of mussels. In this case, predators no longer keep mussels in check at the lower equilibrium and the system is eventually filled with large mussels. Figure 27.2E uses the same gradient in predation rates but with neighborhood size, $h = 1$. Instead of a continuous change in the mussel sizes and density across space, we find a sharp transition from high density/large size to low density/small size. Given this result, and that we believe that there are potentially both neighborhood effects from size protection and gradients in the predation rates in the natural system, it is clear that neither the ODE nor SBD representations of our system alone will suffice for our modeling effort.

An implicit assumption of the CA model is that only one "individual" can occupy a single cell, regardless of the size of the individual. Figure 27.5 demonstrates the effect on the size distribution of individuals when this assumption of fine-grained spatial quantization, implicit in the CA model, is relaxed. Figure 27.5A shows a spatial plot of the ABM model at the upper equilibrium, configured to match the grid-based spatial assumptions of our CA model. Figure 27.5C is the size distribution of the spatial plot in Figure 27.5A. Figures 27.5B and D show the same results for the default continuous spatial configuration of the ABM model. There is a strong shift in the distribution of sizes between the two configurations. This is caused by a combination of continuous space and the requirement that individuals not overlap. Individuals can recruit into spaces that would be reserved for a single individual in a grid-based model. As the smaller individuals grow, they contact each other at a smaller size and stop growing.[11] Interestingly, although the size distributions are significantly different between forms of

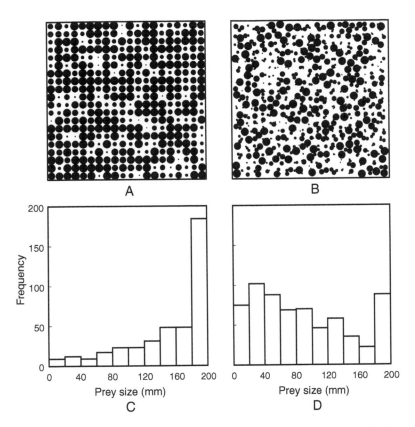

FIGURE 27.5 Results from the IBM model demonstrate the effects of using continuous space and allowing competition for space among mussels. The system size was 20×20 and the default parameters of Table 27.1 were used. The diameters of the filled circles are proportional to the sizes of the mussels. (A) The spatial distribution of mussels at the upper equilibrium for the mean field approximation when each prey is confined to a grid point. (B) The spatial distribution of mussels at the upper equilibrium for the mean field approximation when space is continuous and mussels grow until they touch another individual. In this case, a recruitment limit was put on the total number of mussels allowed in the system because given the small size of recruits and the high immigration rate, without this the system would quickly be swamped by hundreds of thousands of individuals. The total number of mussels in the system was capped at 150% of the number predicted by the ODE approximation. The ODE prediction of a lower and upper stable equilibrium also holds when the assumption of grid-based space is relaxed. The histograms in (C) and (D) show the size distribution of mussels for the plots in (A) and (B).

the model, qualitatively, the dynamics between the two are similar. In particular, both forms of model exhibit two stable equilibria at low and high prey biomass.

27.5 Discussion

It is certainly no small task to develop and analyze four different models of a single system. Why go to all this effort? The simplest route is to apply Occam's razor, that is, choose the simplest model. However, there is a caveat to that: it must be the simplest model that adequately represents the system. Therefore, for a simple model to be accepted, its results must be robust when its implicit assumptions are relaxed, otherwise the results are simply an artifact of the model class. On the other hand, a stand-alone complex model is very difficult both to test and to analyze.[3,41,42] Without the support of simpler models, the results can be problematic. However, when different models are combined, the results may be quite robust. There are two main lessons that we have learned from implementing a multiple-model approach. The first is that implicit assumptions in simple models are really explicit. The second is that building different models of the same process forces clearer understanding of the dynamics being represented.

We started with the assertion that models are comprised of two parts: the explicit assumptions, those defined by the researcher, and the implicit assumptions, those set by the model class. In reality, an implicit assumption in a simple model class becomes an explicit assumption in a more complex model class. For all but the most complex model (the ABM model), we were able to configure a more complex model to match the implicit assumptions of a simpler model. Another way to look at this is that the set of all possible dynamics of a more complex model class is a superset of the set of all possible dynamics from a simpler model class. There has been much discussion in the ecological community on the possible errors introduced into complex models by introducing additional parameters that must be specified.[41] The argued solution has been to use a simpler model or model class. However, as was demonstrated above, the simpler model still contains an implicit representation of that parameter. Instead of an educated guess, the value is dictated solely by the assumptions implicit in that model class. It is reasonable to assume that setting a parameter implicitly may actually add greater error to a result that using an explicit but imperfect measurement of that parameter in a more complex model.

Perhaps the best reason for using multiple model classes to explore a system is that it forces researchers to understand the dynamics that they are attempting to simulate far better than if only a single model class were used. For example, imagine a model that includes logistic growth. Its representation in an ODE model is straightforward. Now imagine the same logistic growth function represented in an ABM. From the view of the individual, what causes the suppression in population growth? Is it caused by the local or global population? Perhaps the phenomenon is triggered by lack of a resource such as food. Where in the life cycle does the density dependence occur? Mating could decrease, pregnancies could not come to term, or it could manifest itself in an increased death rate in a juvenile age class. These different possibilities can potentially have different effects on the dynamics of the population.

Within our research, comparing the continuous space ABM to the CA model forced detailed questions about how the individuals interact with one another and with space and led to the idea of implicit competition. Grid-based models by definition have a minimum scale for model interactions, the size of a single cell. An interesting consequence of this assumption occurs when using models where individuals have size as one of their attributes and only one individual is allowed per cell. When an individual colonizes a cell it preempts the entire area of the cell, even if it physically occupies only a very small portion of that area. This built-in spatial competition only became obvious to us when we were comparing the CA model to the ABM model.

As our research on the seastar/mussel dynamics has progressed, each of our four models has found its own niche. The ODE and SBD models will continue to be used as the deterministic and stochastic baselines for the rest of the model work. In particular we will be exploring different functional forms for phenomena such as size-dependent predation and neighborhood effects.

The CA model will be used to explore further the hypothesis that the abrupt lower boundary of mussel beds is a manifestation of two stable equilibria over smooth environmental gradients. The multiple equilibria hypothesis has already proved robust against demographic stochasticity (SBD), neighborhood effects (CA), and continuous space (ABM). Within the intertidal environment, we find fine-grained spatial heterogeneity such as boulders underlying continuous gradients such as tidal height and wave exposure. One question that will be explored with the CA model is at what spatial scale will this spatial heterogeneity have an effect on the dynamics.

There are many assumptions within the ABM configuration. They occur either because this initial version of the ABM model is designed to match the simpler models or because it is necessary to have some starting point in any modeling project. Perhaps two of the most obvious assumptions are the emigration and immigration of the predators and the spatial competition among the prey. In the natural system, predators enter and leave the system across a boundary. This implies two things. First, the immigration rate is dependent on the system perimeter as opposed to its area. Second, only individuals close to a boundary can choose to emigrate. A later phase of this research will explore the effect of these assumptions.

The spatial competition presently used in this model is very restrictive. However, it was deemed the best starting point for the modeling task. There are other possibilities for modeling spatial competition among prey. Our current implementation permits only a monolayer of mussels. The other extreme is to allow complete overlap of the mussels. We have already developed a version of the ABM that has this assumption. However, it is clear that there is spatial competition within mussel beds. This and the fact that the other three models all use the no overlap assumption (the K value in the ODE and SBD models and the fixed number of cells in the CA model) argue against this configuration. A third possibility is to use layered space with a three-dimensional spatial model. This is a nontrivial task and will be attempted in a later stage of the research.

In this chapter we introduced the concept of applying multiple-model classes to the analysis of a single system. The argument for this approach is that each model class brings with it both strengths and liabilities. Complex models are better suited for exploring interactions in space and time. However, they are difficult to verify and analyze and tend to be slow. Simple models are easier to analyze and faster than their more complex counterparts, but they often require use of implicit assumptions that can significantly alter the resulting system dynamics.[2,3,7,11,20,41] Comparing and contrasting the results of different model classes configured to simulate the same system can allow the researcher to tailor each model class to specific tasks that are optimized for that class. Although the multiple-model approach takes a greater investment in time over the short term, it provides more complete and robust results over the long term.

References

1. Holling, C.S., The strategy of building models of complex ecological systems, in *Systems Analysis in Ecology*, Watt, K.E.F., Ed., Academic Press, New York, 1966.
2. Donalson, D.D. and Nisbet, R.M., Population dynamics and spatial scale: effects of system size on population persistence, *Ecology*, 80, 2492–2507, 1999.
3. Grimm, V., Ten years of individual-based modelling in ecology: what we have learned, and what could we learn in the future? *Ecol. Modelling*, 115, 129–148, 1999.
4. Parunak, V., Savit, R., and Riolo, R., Agent-Based Modeling vs. Equation-Based Modeling: A Case Study and Users Guide, Proceedings of Workshop on Multi-Agent Systems and Agent-Based Simulation, (MABS'98), Springer, 1998. Center for Electronic Commerce Paper CEC-0 115.
5. Wilson, W.G., Resolving discrepancies between deterministic population models and individual-based simulations, *Am. Nat.*, 151, 116–134, 1998.
6. Wilson, W.G., Lotka's game in predator-prey theory: linking populations to individuals, *Theoret. Pop. Biol.*, 50, 368–393, 1996.
7. Donalson, D.D., Modeling Complex Interactions: Theoretical Ecology meets Pacman, Ph.D. dissertation, University of California at Santa Barbara, Santa Barbara, 2000.
8. Nisbet, R.M., Diehl, S., Cooper, S.D., Wilson, W.G., Donalson, D.D., and Kratz, K., Primary productivity gradients and short-term population dynamics in open systems, *Ecol. Monographs* 67, 535–553, 1997.
9. Holling, C.S., The components of predation as revealed by a study of small mammal predation of the European pine sawfly, *Can. Entomol.*, 91, 293–320, 1959.
10. Volterra, V., Fluctuations in the abundance of a species considered mathematically, *Nature*, 118, 558–560, 1926.
11. Huston, M., DeAngelis, D., and Post, W., New computer models unify ecological theory. *BioScience*, 38, 682–691, 1988.
12. Axelrod, R., *The Complexity of Cooperation: Agent-Based Models of Competition and Collaboration*, Princeton University Press, Princeton, NJ, 1997.
13. McCauley, E., Wilson, W.G., and De Roos, A.M., Dynamics of age-structured and spatially structured predator-prey interactions individual-based models and population-level formulations, *Am. Nat.*, 142, 412–442, 1993.
14. Wolff, W.F., Microinteractive predator–prey interactions, in *Research Reports in Physics: Ecodynamics: Contributions to Theoretical Ecology*, Wolff, W.F., Soeder, C.-J., and Drepper, F., Eds., Springer-Verlag, New York, 1988, 285–308.
15. De Roos, A.M., McCauley, E., and Wilson, W.G., Mobility versus density-limited predator-prey dynamics on different spatial scales, *Proc. Royal Soc. London B Biol. Sci.*, 246, 117–122, 1991.

16. Fahse, L., Wissel, C., and Grimm, V., Reconciling classical and individual-based approaches in theoretical population ecology: a protocol for extracting population parameters from individual-based models, *Am. Nat.*, 152, 838–852, 1998.

17. Renshaw, E., *Modeling Biological Populations in Space and Time*, Cambridge University Press, New York, 1991.

18. Nisbet, R.M. and Gurney, W.S.C., *Modelling Fluctuating Populations*, Wiley, Chichester, U.K., 1982.

19. Law, A.M. and Kelton, W.D., *Simulation Modeling and Analysis*, McGraw-Hill, New York, 1991.

20. Deangelis, D. and Gross, L., Eds., *Individual-Based Models and Approaches in Ecology: Populations, Communities and Ecosystems*, Chapman & Hall, New York, 1992.

21. Judson, O., The rise of individual-based modeling in ecology, *Trends Ecol. Evol.*, 9, 9–14, 1994.

22. McGlade, J.M., Individual-based models in ecology, in *Advanced Ecological Theory: Principles and Applications,* McGlade, J.M., Ed., Blackwell Science, Oxford, 1999, chap. 1.

23. Rykiel, E.J., Testing ecological models: the meaning of validation, *Ecol. Modelling*, 90, 229–244, 1996.

24. Paine, R.T., Intertidal community structure: experimental studies on the relationship between a dominant competitor and its principal predator, *Oecologia*, 15, 93–120, 1974.

25. Paine, R.T., Size limited predation: an observational and experimental approach with the *Mytilus-Pisaster* interaction, *Ecology*, 57, 858–873, 1976.

26. Robles, C.D., Predator foraging characteristics and prey population structure on a sheltered shore, *Ecology*, 68, 1502–1514, 1987.

27. Robles, C.D. and Robb, J., Varied carnivore effects and the prevalence of intertidal algal turfs, *J. Exper. Mar. Biol. Ecol.*, 166, 65–91, 1993.

28. Connell, J.H., Community interactions on marine rocky intertidal shores, *Ann. Rev. Ecol. Syst.*, 3, 169–192, 1972.

29. Lewin, R., Supply-side ecology, *Science*, 234, 25–27, 1986.

30. Robles, C.D. and Desharnais, R.A., History and current development of a paradigm of predation in rocky intertidal communities, *Ecology*, 83, 1521–1536, 2002.

31. Manly, B.F.J., Comments on design and analysis of multiple-choice feeding-preference experiments, *Oecologia*, 93, 149–152, 1993.

32. Robles, C., Sweetnam, D.A., and Eminike, J., Lobster predation on mussels: shore-level differences in prey vulnerability and predator preference, *Ecology*, 71, 1564–1577, 1990.

33. Robles, C.D., Alvarado, M.A., and Desharnais, R.A., The shifting balance of marine predation in regimes of hydrodynamic stress, *Oecologia*, 128, 142–152, 2001.

34. Dahlhoff, E.P. and Menge, B.A., Influence of phytoplankton concentration and wave exposure on the ecophysiology of *Mytilus californianus*, *Mar. Ecol. Prog. Ser.*, 144, 97–107, 1996.

35. McClintock, J.B. and Robnett, T.J., Jr., Size selective predation by the asteroid *Pisaster ochraceous* on the bivalve *Mytilus californianus*, a cost benefit analysis, *Mar. Ecol.*, 7, 321–332, 1986.

36. Robles, C., Sherwood-Stephens, R., and Alvarado, M., Responses of a key intertidal predator to varying recruitment of its prey, *Ecology*, 76, 565–579, 1995.

37. McKendrick, A.G., Applications of mathematics to medical problems, *Proc. Edinburgh Math. Soc.*, 44, 98–130, 1926.

38. Foerster, H. von, Some remarks on changing populations, in *The Kinetics of Cellular Proliferation*, Stohlman, F., Ed., Grune and Stratton, 1959, 382–407.

39. Bertalanffy, L. von, A quantitative theory of organic growth, *Hum. Biol.*, 10, 181–213, 1938.

40. Harrison, G.W., Comparing predator-prey models to Luckinbill's experiment with *Didinium* and *Paramecium*, *Ecology*, 76, 357–374, 1995.

41. Levin, S., Grenfell, A.B., Hastings, A., and Perelson, A.S., Mathematical and computational challenges in population biology and ecosystem science, *Science*, 275, 334–343, 1997.

42. Grimm, V., Wyszomirski, T., Aikman, D., and Uchmanski, J., Individual-based modeling and ecological theory: synthesis of a workshop, *Ecol. Modelling*, 115, 275–282, 1999.

28

The Effects of Langmuir Circulation on Buoyant Particles

Eric D. Skyllingstad

CONTENTS

28.1 Introduction...445
28.2 Simulation Setup..446
28.3 Flow and Particle Patterns..447
28.4 Discussion..449
References..452

28.1 Introduction

Linear structures in sea surface foam, commonly referred to as "windrows," have long interested ocean-ographers and mariners. The first rigorous study of circulations associated with windrows was presented by Langmuir (1938) who noticed that seaweed would form into regular rows whenever the winds exceeded ~3 to 5 m s^{-1}. Because of Langmuir's work, we now refer to these circulations as Langmuir circulation (LC). LC are important for establishing the ocean surface mixed layer and dispersing heat, salinity, nutrients, and pollution through the upper water column (see review by Barstow, 1983).

Very little research has focused on the importance of LC for biological systems. Observations (e.g., Weller et al., 1985) show that LC can generate vertical velocities of up to ~0.2 m s^{-1} and generate surface upwelling regions with scales of tens to hundreds of meters depending on forcing conditions. Passive effects of Langmuir circulation are thought to affect the sinking rates of *Sargassum* by providing an initial downward transport until the *Sargassum* buoyancy chambers collapse due to increased pressure. Johnson and Richardson (1977) show that vertical velocities of ~0.02 m s^{-1} are sufficient to cause this transport. Research on phytoplankton (Yamazaki and Kamykowski, 1991) and medusa (Larson, 1992) also indicates that particle transport by LC probably has an effect on the behavior of swimming upper-ocean organisms. In the case of medusa, Larson hypothesized that LC clustering of *Linuche unguiculata* improved reproduction success and available food supply. Because LC generates strong surface convergence, it seems natural that organisms could be affected by these motions and perhaps adapt in a way to exploit the concentrating effects produced by LC. With this in mind, Woodcock (1944) suggested that Portuguese man-of-war (*Physalia*) have adapted to cross regions of upwelling produced by LC.

Stommel (1948) determined using an idealized LC roll structure that sinking particles, such as phytoplankton, could be trapped in the upwelling portion of the LC leading to a locally higher concen-tration of particles. More recently, Pershing et al. (2001) tested this idea by tracking the motion of particles in a prescribed cellular circulation pattern. They found that positively buoyant particles, and not sinking particles, tended to collect in downwelling regions with a vertical distribution dependent on the strength of the circulation and the buoyancy rise velocity of the particles.

In this chapter, the work of Pershing et al. (2001) is extended by replacing their simple prescribed cellular circulation with results from a large-eddy simulation (LES) turbulence model of the ocean surface boundary layer. LES models resolve the largest scales of turbulence in the ocean, for example, eddy scales that encompass some fraction of the mixed layer depth. Model resolution can range from 0.5 to 1 m, yielding accurate physical structures from 2 to 5 m in scale. LC in the model is produced by including a term in the equations of motion that accounts for the interaction of surface waves with upper-ocean currents. This term, known as the vortex force, was derived by Craik and Leibovich (1976) and is defined as

$$V_s \times \omega$$

where V_s is the surface wave Stokes drift and ω is the current vorticity, defined as $\nabla \times v$, and v is the current. As shown in Skyllingstad and Denbo (1995), the horizontal scale of LC is largely dependent on the depth of the mixed layer, although variations in the vertical profile of V_s can also affect the degree of LC organization and spacing of the convergence zones. In general, LC will occur for even small wave amplitudes.

Simulations are presented here for typical oceanic conditions with a wind- and wave-driven mixed layer. Lagrangian particles with different rise velocities are used to examine how buoyant organisms and material might behave in these conditions. These experiments differ from previous Lagrangian particle experiments (e.g., McWilliams et al., 1996; Skyllingstad, 2001) that examined floating particles confined to the surface. The chapter is organized as follows. In Section 28.2, the simulation initial conditions are described along with the procedure for tracking particles. Section 28.3 describes the results of the simulations, focusing on the role of particle buoyancy and the stabilizing effects of solar radiation. A brief conclusion is given in Section 28.4.

28.2 Simulation Setup

The LES model used in these experiments is described in Skyllingstad et al. (1999, 2000) and is based on the incompressible Navier–Stokes equations. Verifications of the model accuracy in simulating mixed layer turbulence have been presented in Skyllingstad and Denbo (1995) and Skyllingstad et al. (1999). Domain size in the simulations is set to 128×128 m in the horizontal and 20 m in the vertical with 1 m resolution. Temperature in the initial setup is set to 29°C to a depth of 10 m, decreasing to 26°C at a depth of 50 m (below the model bottom). Setting the temperature constant in the upper 30 m is equivalent to prescribing a boundary layer depth of 30 m, which should force LC with similar scales once the boundary layer is fully turbulent. Salinity is set to a constant value of 34 psu. Periodic lateral boundaries are applied in the LES with a radiative boundary condition at the model bottom (allowing for the transmittance of internal waves out of the model). The model is forced with wind stresses of 0.1 and 0.4 N m⁻², and surface wave Stokes drift forcing for a monochromatic wave 2 m high and with a 30 m wavelength. Wind forcing in this case can be scaled using the friction velocity, $u_* = \sqrt{\tau / \rho}$, yielding 0.01 and 0.02 m s⁻¹ for the two prescribed wind stress values.

After a 2.5-h spin up, a 150×150 array of particles was initialized as a sheet oriented across the flow at location $x = 32$ m, from the surface to a depth of 30 m. Particle buoyancy was implemented by applying a rise velocity of 0.0, 0.01, and 0.02 m s⁻¹. The effects of breaking waves on particles were crudely parameterized by eliminating the rise velocity between the surface and 1 m depth. This adjustment prevents particles from becoming trapped at the surface where turbulence velocities reduce to zero in the model. Rise velocities were selected based primarily on vertical velocity ranges thought to exist in LC and produced in the simulations. Actual velocities of swimming phytoplankton, for example, range from ~200 μm s⁻¹ for dinoflagellate *Gyrodinium dorsum* to ~0.014 m s⁻¹ for ciliate *Mesodinium rubrum* (Yamazaki and Kamykowski, 1991). Vertical rise velocities for *Sargassum* near the surface are ~0.02 m s⁻¹ (Johnson and Richardson, 1977).

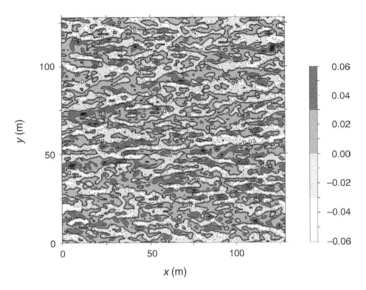

FIGURE 28.1 Vertical velocity (m s⁻¹) at a depth of 2 m after 2 h.

28.3 Flow and Particle Patterns

The structure of the fully developed boundary layer is revealed by a horizontal cross-section plot of the vertical velocity as shown in Figure 28.1. Although coherence does appear in the velocity fields, linear structures do not stand out as clearly as windrows observed over the ocean. This is because the vertical velocity represents both turbulent motions that have short time and length scales and the more persistent LC that sets surface convergence. The longer timescale of LC allows for the surface accumulation of material as long as the rise velocity of the particles is significant relative to the turbulence strength. A plot of the Lagrangian particles without a rise velocity demonstrates the importance of particle buoyancy (Figure 28.2). After 1 min, the effects of vertical motions are evident in the vertical cross section, which shows regions of particle motion extended to the bottom of the mixed layer at 30 m. Particles near the surface are advected downstream more rapidly than those deeper in the water column as shown by the plane view of the particle locations. At 39 min later, the vertical distribution of particles has an almost random quality with only a hint of structure shown by slightly higher density particle clouds between $y = 20$ to 40 m and 80 to 100 m. Horizontal placement of particles also shows little organization, except for higher concentrations near the bottom of the water column (i.e., between 0 to 20 m and ~100 to 128 m) that are a remnant of the original sheet structure.

Adding a vertical rise velocity of 0.01 m s⁻¹ causes a significant change in the behavior of the particles as shown in Figure 28.3. At 10 min, the vertical location of the particles shows considerable structure with particles starting to form linear structures that extend through the depth of the mixed layer. These structures appear to reach an equilibrium as shown by Figure 28.3B and C with a large concentration of material located in the 1-m wave-breaking zone near the model surface. Horizontal locations of particles in this case show linear structures that are beginning to look like windrows, although there are still many particles between the windrow clusters.

Increasing the rise velocity further to 0.02 m s⁻¹ produces a pattern with even fewer particles in the mixed layer interior (Figure 28.4). In this case, most of the particles are trapped in the wave-breaking zone and have converged into well-defined windrows. Also apparent in the horizontal view (Figure 28.4D) is a domain scale convergence pattern as shown by the relative absence of windrows in the center of the domain. The large-scale pattern is typical of LC simulations, which tend to produce eddy scales that are some multiple of the domain size.

We also performed a second set of simulations using a higher wind stress value, 0.4 N m⁻², to see if increased turbulence from shear instability could disrupt the LC-induced patterns. Results showing

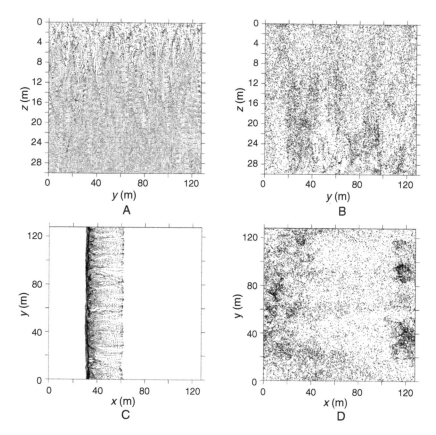

FIGURE 28.2 Total particle positions integrated along the x axis after (A) 1 min and (B) 30 min and integrated along the z axis after (C) 1 min and (D) 30 min, for particle rise velocity of zero. Particles are introduced in the simulation at hour 3.5.

particles after 30 min for rise velocities of 0.01 and 0.02 m s^{-1} are presented in Figure 28.5. As this plot shows, a rise velocity of 0.01 m s^{-1} with stronger winds is overwhelmed by increased turbulence so that only a diffuse plume of particles is produced under the largest-scale circulation upwelling center. Surface convergence in this case is also much weaker in comparison with the 0.1 N m^{-2} wind example. Increasing the rise velocity to 0.02 m s^{-1}, yields more concentrated convergence of particles, but still shows considerably more scatter relative to the weak wind cases.

A final test was performed using sinking particles with a vertical sinking velocity of –0.01 m s^{-1} and a wind stress of 0.1 N m^{-2}. Results from this test show a gradual reduction in particle concentration throughout the mixed layer, with the highest concentration centered near the mixed layer bottom in the LC return flow. This result disagrees with Stommel (1948), probably because of the more complex circulation generated in the LES model in comparison with that used by Stommel. In Stommel's model, sinking particles became trapped in a retention zone created as the particles dropped out of the circulation in the surface divergence zone. As the particles descended, they eventually reencountered upward velocities that were able to counteract the downward buoyancy force. Stommel notes that the particle paths in the idealized circulation pattern used in his study would be greatly affected by turbulent motions, which would displace sinking particles away from the retention.

In our simulation, sinking particles did initially become trapped in the upward-moving zones of the LC. However, random particle motion from small-scale turbulence circulations quickly transported particles away from these regions and particles gradually sank below the mixed layer. Finally, Stommel noted that the retention zone idea would also apply for rising particles in the upward moving sections of the flow as validated here.

Particle behavior in the simulations is similar to two-dimensional experiments performed by Colbo and Li (1999). In their experiments, values of the dispersion coefficient were estimated from the time

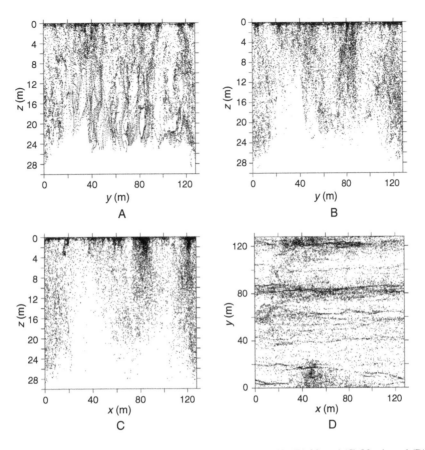

FIGURE 28.3 Integrated particle positions vertical cross sections after (A) 10, (B) 20, and (C) 30 min and (D) horizontal cross section at 30 min for particle rise velocity of 0.01 m s⁻¹.

history of the particle spatial variance. They found that neutrally buoyant particles exhibited the highest dispersion coefficients, in agreement with observations and results presented here. Buoyant particles had much smaller dispersion coefficients because of concentrations in the LC surface convergence zones.

28.4 Discussion

The simulations presented in this chapter demonstrate some basic principles regarding the effects of LC and mixed layer turbulence on particle behavior. Results show that, without buoyancy or rise velocity, particles are rapidly dispersed throughout the water column, agreeing with two-dimensional results from Colbo and Li (1999). Flow structures, such as surface convergence zones and regions of upwelling, cause only minor concentrations of particles relative to a random field.

Adding a rise velocity similar in scale to the surface friction velocity (in this case, about 0.01 and 0.02 m s⁻¹) produces a significant change in the particle behavior. Now, instead of particles drifting in a random way, LC has forced clouds of particles below surface convergence regions. Increasing the rise velocity to 0.02 m s⁻¹ exceeds the strength of the LC-induced motions and the concentration of particles in the mixed layer slowly decreases as the windrow regions collect particles near the surface. With stronger surface winds and friction velocity increased to 0.02 m s⁻¹, particle concentrations become less organized, particularly in the rise velocity of 0.01 m s⁻¹ case where particles appear as diffuse plumes. These results suggest that particle collection by LC might be related to the strength of the winds; however, a more extensive set of experiments is needed to develop a more robust parameterization of particle collection. Specifically, a full range of wind and wave parameters is needed to establish the behavior of

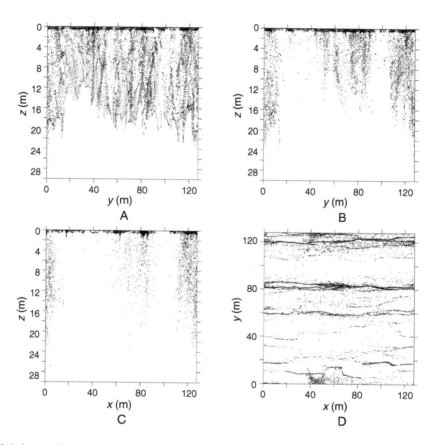

FIGURE 28.4 Same as Figure 28.3, but for rise velocity of 0.02 m s⁻¹.

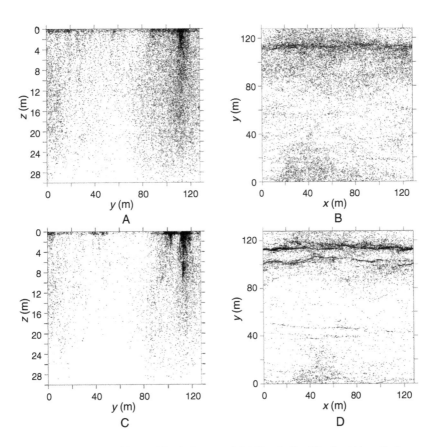

FIGURE 28.5 Integrated particle positions with wind stress of 0.4 N m^{-2} after 30 min, (A) vertical cross section and (B) horizontal cross section with rise velocity of 0.01 m s^{-1} and (C) vertical cross section and (D) horizontal cross section with rise velocity of 0.02 m s^{-1}.

buoyant particles in the LC-dominated boundary layer. With respect to swimming organisms, it would appear that swimming speeds roughly equivalent to u_* are needed to overcome the convergence and collection effects of LC.

References

Barstow, S.F., The ecology of Langmuir circulation: a review, *Mar. Environ. Res.*, 9, 211, 1983.

Colbo, K. and M. Li, Parameterizing particle dispersion in Langmuir circulation, *J. Geophys.*, 104, 26,059, 1999.

Craik, A.D.D. and S. Leibovich, A rational model for Langmuir circulations, *J. Fluid Mech.*, 73, 401, 1976.

Johnson, D.L. and P.L. Richardson, On the wind-induced sinking of *Sargassum*, *J. Exp. Mar. Biol. Ecol.*, 28, 255, 1977.

Langmuir, I., Surface motion of water induced by wind, *Science*, 87, 119, 1938.

Larson, R.J., Riding Langmuir circulation: a review, *Mar. Environ. Res.*, 9, 211, 1992.

McWilliams, J.C., P.P. Sullivan, and C.-H. Moeng, Langmuir turbulence in the ocean, *J. Fluid Mech.*, 334, 1, 1997.

Pershing, A.J., P.H. Wiebe, J.P. Manning, and N.J. Copley, Evidence for vertical circulation cells in the well-mixed area of Georges Bank and their biological implications, *Deep-Sea Res. II*, 48, 283, 2001.

Skyllingstad, E.D., Scales of Langmuir circulation generated using a large-eddy simulation model, *Spill Sci. Technol.*, 6, 239, 2001.

Skyllingstad, E.D. and D.W. Denbo, An ocean large eddy simulation of Langmuir circulations and convection in the surface mixed layer, *J. Geophys. Res.*, 100, 8501, 1995.

Skyllingstad, E.D., W.D. Smyth, J.N. Moum, and H. Wijesekera, Upper ocean turbulence during a westerly wind burst: a comparison of large-eddy simulation results and microstructure measurements, *J. Phys. Oceanogr.*, 29, 5, 1999.

Skyllingstad, E.D., W.D. Smyth, and G.B. Crawford, Resonant wind-driven mixing in the ocean boundary layer, *J. Phys. Oceanogr.*, 30, 1866, 2000.

Smith, J.A., Observations of wind, waves, and the mixed layer: the scaling of surface motion, in *The Wind-Driven Air-Sea Interface: Electromagnetic and Acoustic Sensing, Wave Dynamics, and Turbulent Fluxes*, University of New South Wales, Sydney, Australia, 1999, 231.

Stommel, H., Trajectories of small bodies sinking slowly through convection cells, *J. Mar. Res.*, 8, 24, 1948.

Weller, R.A., J.P. Dean, J. Marra, J.F. Price, E.A. Francis, and D.C. Boardman, Three-dimensional flow in the upper ocean, *Science*, 227, 1552, 1985.

Woodcock, A.H., A theory of surface water motion deduced from the wind-induced motion of the *Physalia*, *J. Mar. Res.*, 5, 196, 1944.

Yamazaki, H. and D. Kamykowski, The vertical trajectories of motile phytoplankton in a wind-mixed water column, *Deep-Sea Res.*, 38, 219, 1991.

29

Modeling of Turbulent Intermittency: Multifractal Stochastic Processes and Their Simulation

François G. Schmitt

CONTENTS

29.1   Introduction 453
29.2   Multiplicative Cascades to Describe Intermittency 454
        29.2.1 Scaling and Intermittency for Velocity Fluctuations in Turbulence 454
        29.2.2 The Multiplicative Cascade Model and Its Main Properties 454
                29.2.2.1 Scaling Properties of Multiplicative Cascades 455
                29.2.2.2 Correlation Properties of Multiplicative Cascades 456
29.3   Simulation of Discrete Cascades 457
29.4   Simulation of Continuous Cascades 459
        29.4.1 Theory: Continuous Multiplicative Cascades 459
        29.4.2 A Causal Lognormal Stochastic Equation and Its Properties 460
        29.4.3 Simulation of a Continuous Lognormal Multifractal 462
29.5   Conclusion 463
Appendix 29.A 464
        Characteristic Functions 464
        Infinitely Divisible Distributions 464
        Stable Stochastic Integrals 465
References 466

29.1 Introduction

Geophysical fields generally possess a high variability at all scales. This is particularly true in oceanography, where physical, chemical, and biological fields display highly heterogeneous patterns at all scales, from millimeters to thousands of kilometers and from milliseconds to years. At small scales, the origin of this variability is mainly oceanic turbulence, and in this framework the study of aquatic ecosystems is deeply linked to the turbulent variability of the medium.

For more than 50 years, turbulence has been known to lead to intermittency, i.e., large fluctuations having a spatial and temporal structure.[1] In the Kolmogorov 1941 framework,[2,3] these turbulent fluctuations have scaling properties, stating that from an outer scale L to the inner Kolmogorov scale η there is no characteristic scale. Such scaling intermittency is now generated using multiplicative cascades. An energy cascade from large to small scales to describe the dynamics of the velocity field in fully developed turbulence was first proposed by Richardson[4] in 1922. Multiplicative cascades lead to multifractal fields[5] possessing scale invariance of intermittent fluctuations.

For oceanic turbulence, many fields of interest in aquatic ecosystem studies have already been shown to possess intermittent multifractal fluctuations: temperature,[6-8] phytoplankton[6-9] and zooplankton densities,[8,10,11] nutrient density,[12] salinity,[8,9] as well as the atmospheric wind forcing[13–15] and the sea state.[16] This shows the relevance of this approach to studies on marine ecosystems. It has many theoretical consequences, such as possible modifications of predator–prey contact rates.[17] Here we do not focus on such consequences, but rather on the numerical modeling of the fields. Indeed, for many applications such as the grazing of a copepod in a heterogeneous phytoplankton field,[18] it is interesting to be able to simulate such fields. Here we propose to describe procedures and algorithms to simulate discrete and continuous (in scale) multiplicative cascades. In the following, Section 29.2 recalls the theory; Section 29.3 is devoted to simulation of discrete cascades, and Section 29.4 to simulation of continuous cascades.

29.2 Multiplicative Cascades to Describe Intermittency

29.2.1 Scaling and Intermittency for Velocity Fluctuations in Turbulence

In fully developed turbulence (corresponding to large Reynolds number flows) there is a range of scales for which advective terms of the Navier–Stokes equations are dominant compared to the dissipative term. Following an intuitive idea originally stated by Richardson[4] in 1922, for these scales, forming the so-called "inertial range," there is a cascade of energy from large to small scales. This inspired Kolmogorov to formulate his famous law in 1941, assuming that for the inertial range, the statistics of velocity fluctuations $\Delta V_\ell = V(x + \ell) - V(x)$ at scale ℓ are locally isotropic, and depend only on the small-scale homogeneous dissipation ε and on the scale ℓ. This gives:[2]

$$< \left| \Delta V_\ell \right| > \approx \varepsilon^{1/3} \ell^{1/3} \tag{29.1}$$

which corresponds to the famous –5/3 spectrum in Fourier space:[3]

$$E_V(k) \approx \varepsilon^{2/3} k^{-5/3} \tag{29.2}$$

This law is now ubiquitous for three-dimensional (3D) fully developed turbulence. It has been experimentally verified for oceanic velocity fluctuations on many occasions (see, e.g., References 19 and 20).

It was later discovered that one point of Kolmogorov's hypothesis — the fact that the dissipation at small scale was a smooth homogeneous field — was not verified: the small-scale dissipation, in fact, experimentally displayed intermittent fluctuations, and this intermittency was increasing with the Reynolds number.[1] To take this into account, Kolmogorov and Obukhov proposed a new approach in 1962, assuming that the small-scale dissipation is a random variable, with a lognormal probability distribution function (pdf).[21,22] This hypothesis was rather arbitrary, and no real justification was proposed: Kolmogorov simply indicated "it is natural to suppose that …" (Reference 21, p. 83). Soon after this, several experimental studies showed that the small-scale dissipation is a random field with a specific spatial structure with long-range correlations:[23,24]

$$< \varepsilon(x)\varepsilon(x + r) > \approx r^{-\mu} \tag{29.3}$$

with an experimental value of μ around 0.4. Yaglom[25] then proposed a multiplicative cascade model compatible with all these experimental facts about intermittency of dissipation: the model depends on the Reynolds number, produces large fluctuations for the small-scale dissipation, and moreover generates a random variable having lognormal statistics and long-range power law correlations as given by Equation 29.3. The lognormal hypothesis was in fact introduced to be compatible with Kolmogorov's hypothesis, but it is not a necessary hypothesis and can be relaxed and replaced with any positive random variables. This is discussed below.

29.2.2 The Multiplicative Cascade Model and Its Main Properties

The discrete multiplicative cascade model presented here is an adaptation from Yaglom's original proposal;[25] it is still at the basis of most cascade models currently used to generate intermittency in turbulence. This is basically a discrete model in scale, but it can be densified. The term *discrete in scale*

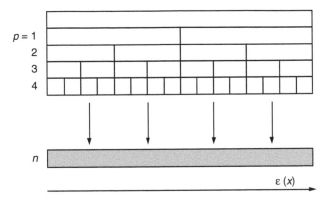

FIGURE 29.1 A schematic illustration of the cascade process leading, after n steps, to the resulting field ε.

refers to the fact that the scale ratio of mother to daughter structures is a finite number, strictly larger than 1. This model is multiplicative, and embedded in a recursive manner. The multiplicative hypothesis generates large fluctuations, and the embedding generates long-range correlations, which give spatially to these large fluctuations their intermittent character.

29.2.2.1 Scaling Properties of Multiplicative Cascades —

The eddy cascade is symbolized by cells, with each cell associated with a random variable W_i. All these random variables are assumed positive, independent, and to possess the same law, with the following conservative property: $\langle W \rangle = 1$. The larger scale corresponds to a unique cell of size L. We introduce a scale ratio $\lambda_1 > 1$ (usually for discrete models $\lambda_1 = 2$) and take for convenience $L = \ell_0 \lambda_1^n$. Because the model is discrete, the next scale involved corresponds to λ_1 cells, each of size $L / \lambda_1 = \ell_0 \lambda_1^{n-1}$. This is iterated, and at step p ($1 \leq p \leq n$) there are λ_1^p cells, each of size $L / \lambda_1^p = \ell_0 \lambda_1^{n-p}$. There are n cascade steps, and at step n there are λ_1^n cells, each of size $L / \lambda_1^n = \ell_0$, the dissipation scale (Figure 29.1). To reach the dissipation scale, all intermediary scales have been involved, corresponding to a property of localness of interactions in turbulent cascades. Finally, at each point the dissipation is written as the product of n random variables:

$$\varepsilon(x) = \prod_{p=1}^{n} W_{p,x} \tag{29.4}$$

where $W_{p,x}$ is the random variable corresponding to position x and level p in the cascade. For two different levels of the cascade, the random variables are assumed independent, so that taking the moment of order $q > 0$ of Equation 29.4 gives

$$< \varepsilon^q > = < \left(\prod_{p=1}^{n} W_{p,x} \right)^q > = \prod_{p=1}^{n} < \left(W_{p,x} \right)^q > = < W^q >^n \tag{29.5}$$

Introducing the scale ratio $\lambda = (L / \ell_0)^n = \lambda_1^n$ between the outer scale and the dissipation scale, Equation 29.5 gives finally the moments of the small-scale dissipation ε, denoted ε_λ to note its scale ratio (and hence Reynolds number) dependence:

$$< \left(\varepsilon_\lambda \right)^q > = \lambda^{K(q)} \tag{29.6}$$

where the moment-function $K(q) = \log_{\lambda_1} < W^q >$ is introduced. The conservative property $\langle W \rangle = 1$ corresponds to $K(1) = 0$ and also $\langle \varepsilon_\lambda \rangle = 1$.

Equation 29.6 is a generic scaling property of multifractal fields obtained through multiplicative cascades. Depending on the model chosen, a different form of the pdf of W can be assumed. A lognormal pdf for W corresponds to a quadratic expression $K(q) = \frac{\mu}{2}(q^2 - q)$, with $\mu = K(2)$ proportional to the variance of $\log W$ (the exact expression is given in the next section). It can also be noticed that $K(q)$ is

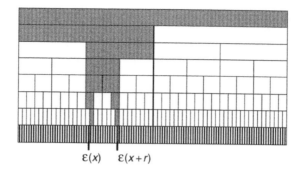

$$\varepsilon(x) \qquad \varepsilon(x+r)$$

FIGURE 29.2 Illustration of two paths leading to the points $\varepsilon(x)$ and $\varepsilon(x + r)$; they have a common path before a bifurcation.

(up to a $\log \lambda_1$ factor) the second Laplace characteristic function of the random variable $\log W$ (see appendix to this chapter), showing that it is a convex function.

29.2.2.2   *Correlation Properties of Multiplicative Cascades* — We now consider the long-range correlations generated by this cascade model. The correlation $< \varepsilon(x)\varepsilon(x+r) >$ can be decomposed as a product of random variables (see Reference 25). For this, one must estimate at what point $\varepsilon(x)$ and $\varepsilon(x + r)$ have a common ancestor. In fact, the distance r corresponds, on average, to m steps, with $\lambda_1^m \approx r$, so that the $n - m$ first steps are common to the two random variables, whereas there is a "bifurcation" at step m, and the two paths separate (Figure 29.2). In the product of the two random variables, there are thus $n - m$ identical variables and m different and independent. This is written:

$$< \varepsilon(x)\varepsilon(x+r) >=< \prod_{p=1}^{n} W_{p,x} \prod_{p'=1}^{n} W_{p',x+r} >= \prod_{p=1}^{n-m} < W_{p,x}^2 > \prod_{p=n-m+1}^{n} < W_{p,x} > \prod_{p'=n-m+1}^{n} < W_{p',x+r} > \quad (29.7)$$

$$=< W^2 >^{n-m} < W >^{2m} \quad (29.8)$$

Finally, introducing $K(q)$ and recalling that $\lambda_1^n = \lambda$ and $\lambda_1^m \approx r$ yields

$$< \varepsilon(x)\varepsilon(x+r) >\approx \lambda^{K(2)} r^{-K(2)} \quad (29.9)$$

For a given cascade process, the total scale ratio is fixed, so that this relation corresponds to a long-range power law correlation with exponent $\mu = K(2)$.

On the other hand, the logarithm of the cascade process also possesses interesting correlation properties. Let us define the generators γ and g of, respectively, the cascade process and the weight random variable:

$$\begin{cases} \gamma = \log \varepsilon \\ g = \log W \end{cases} \quad (29.10)$$

When it is defined,* we are interested in the autocorrelation function of γ:

$$C_\gamma(r) =< \gamma(x)\gamma(x+r) > \quad (29.11)$$

As before, we may consider the path leading to $\gamma(x)$ and $\gamma(x + r)$: for the last m steps (with $\lambda_1^m \approx r$), the paths are different, whereas before this bifurcation the path is common. The main difference with the calculation leading to Equation 29.9 is that in Equation 29.11 γ is not the product but the sum of random variables. A simple calculation shows that $<\gamma(x)\gamma(x + r)>$ involves $n^2 - n + m$ terms of the form $<g>^2$ and $n - m$ terms $<g^2>$. Let us note $G = <g> = <\log W>$ and $\sigma^2 = <(g - G)^2> = <g^2> - G^2$. We then have

---

* The autocorrelation function requires the existence of second-order moments to be defined. For log-Lévy cascades the generator is a Lévy-stable process, so that its autocorrelation function, a second-order moment, diverges.

$$C_\gamma(r) \approx < g >^2 (n^2 - n + m) + < g^2 > (n - m) \tag{29.12}$$

$$\approx G^2 n^2 + \sigma^2 (n - m) \tag{29.13}$$

$$\approx C - \frac{\sigma^2}{\log \lambda_1} \log r \tag{29.14}$$

where we introduced the constant $C = G^2 n^2 + \sigma^2 n$. This corresponds to a logarithmic decay of the autocorrelation function of the generator.[26,27] Its Fourier transform gives the power spectrum (when defined) of the singularity process $\gamma(x) = \log \varepsilon(x)$:

$$E_\gamma(k) \approx k^{-1} \tag{29.15}$$

which corresponds to an exactly $1/f$ noise.[5,28] Properties 29.6, 29.9, 29.14, and 29.15 may be used to check numerical simulations.

29.3 Simulation of Discrete Cascades

We perform here numerical simulations of a discrete multifractal cascade, choosing for simplicity $\lambda_1 = 2$. The choice of the pdf of W determines the model to be used. Several simple discrete models may be found in the literature, but in fact the potential choice is infinite: to normalize the cascade we need to take $<W> = 1$. We need to take a pdf such that at least some moments converge; we restrict ourselves also to strictly positive random variables. The scaling properties of ε are then mainly characterized by

$$K(q) = \log_2 < W^q > \tag{29.16}$$

We take here as a generic example the lognormal cascade, for which $K(q)$ is known to be quadratic. The properties $K(1) = 0$ and $\mu = K(2)$ fully determine the moment function, which is written:

$$K(q) = \frac{\mu}{2}\left(q^2 - q\right) \tag{29.17}$$

Let us chose $W = e^g$ with g Gaussian: $g = \sigma g_0 + G$, where σ and G are constants to determine and g_0 is a centered ($<g_0> = 0$) and unitary Gaussian variable:

$$< e^{q g_0} > = e^{\frac{1}{2} q^2} \tag{29.18}$$

A simple calculation then gives

$$< W^q > = \exp\left\{\tfrac{1}{2} \sigma^2 q^2 + qG\right\} \tag{29.19}$$

Equations 29.16 and 29.17 provide the constants σ and G:

$$\mu \log 2 = \sigma^2$$
$$-\frac{\mu}{2} \log 2 = G \tag{29.20}$$

so that finally the discrete lognormal cascade of parameter μ is generated by the weight:

$$W = \exp\left\{\sqrt{\mu \log 2}\, g_0 - \frac{\mu}{2} \log 2\right\} \tag{29.21}$$

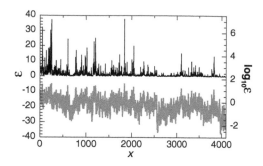

FIGURE 29.3 A sample of a lognormal discrete cascade. The generator γ and its exponential ε are shown. The generator is a correlated Gaussian noise, and ε is an intermittent multifractal field.

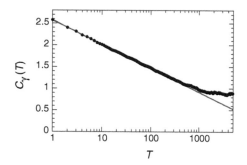

FIGURE 29.4 The autocorrelation function of the generator, estimated from 100 realizations of a discrete lognormal cascade with $\mu = 0.25$. As expected, this shows a logarithmic decrease over the whole scaling range.

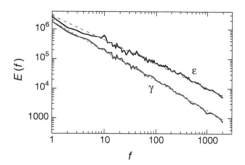

FIGURE 29.5 The Fourier power spectrum of a discrete multifractal field and of its generator. As expected, over the whole scaling range, the generator displays an exactly k^{-1} power spectrum, and the multifractal field a $k^{-1+\mu}$.

This can be generated by the following algorithm:

1. Choose n and μ. The total scale ratio will be $\lambda = 2^n$.
2. Initialize all values of E_i ($i = 1 \ldots \lambda$) to 1.
3. Cascade steps. For $p = 1$ to n:

   Separate E_i in 2^p cells, each of size 2^{n-p}. For each cell, chose a new realization of a random variable W_i according to Equation 29.21, and multiply the value of each point of the cell by W_i.

This may be repeated for several realizations N_R so that the total number of points of the simulation will be λN_R with a scaling between the smallest scale $\ell_0 = 1$ and the scale $L = \lambda \ell_0 = \lambda$. This is illustrated below with $n = 12$, $N_R = 100$, and $\mu = 0.25$. Figure 29.3 shows the first realization (4096 points) of the series $\varepsilon(x)$ and $\gamma(x)$: it is seen that $\varepsilon(x)$ is highly intermittent while $\gamma(x)$ is a correlated noise. This

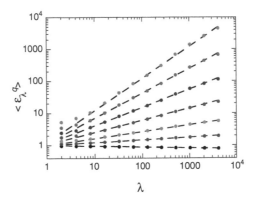

FIGURE 29.6 The scaling of the moments of order 0.5, 1.5, 2, 2.5, 3, 3.5, and 4 (from below to above) of the discrete multifractal simulation. The different straight lines indicate the accuracy of the scaling property. The slopes of these straight lines give estimates of $K(q)$.

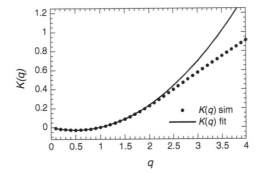

FIGURE 29.7 The $K(q)$ function estimated from 100 realizations of the discrete multifractal simulation, compared to the theoretical curve given by Equation 29.17. There is an excellent agreement until a critical moment corresponding to the maximum moment that can be accurately estimated with 100 realizations.

correlation is illustrated in Figure 29.4, showing $C_\gamma(r)$: the logarithmic decay is extremely well verified for more than three orders of magnitude, with a slope of about 0.24. This is very close to the theoretical value, which is $\sigma^2/\log 2 = \mu$. Figure 29.5 represents the power spectrum of γ and of ε, which are both scaling, with theoretical slopes of -1 and $-1 + \mu$, respectively (Equation 29.15 and the Fourier transform of Equation 29.9). Figure 29.6 is a direct test of the scaling of the resulting field, for various orders of moments. Figure 29.7 represents the moment function $K(q)$ estimated from 100 realizations of the numerical simulation, compared to the theoretical expression. There is excellent agreement until a critical moment $q_s \cong 2.4$, above which the estimated function is linear; this corresponds to a maximum order of moment that can be estimated with 100 realizations of a multifractal field of maximum scaling ratio 2^{12}. There are not enough realizations for an accurate estimation of the scaling exponent corresponding to larger moments.

29.4 Simulation of Continuous Cascades

29.4.1 Theory: Continuous Multiplicative Cascades

The cascade formalism was historically developed for discrete-in-scale cascades, having a fixed scale ratio (typically 2) between the scale of a structure and of the daughter.[25,29,30] It is in fact clear that scale densification is needed to obtain a realistic description, since there is no physical justification for such characteristic scale ratio.

Densification corresponds to taking $\lambda_1 \rightarrow 1^+$, while keeping $\lambda = \lambda_1{}^n$ fixed ($n \rightarrow \infty$).[31,32] This leads to weights W belonging to log-infinitely divisible (ID) probability laws (see the appendix, and also Feller,[33] on ID distributions). Continuous models have been discussed in Reference 5, introducing the "log-Lévy" model. ID distributions in the context of multiplicative cascades have been discussed by Saito,[34] Novikov,[35] and She and Waymire.[36] There are infinitely many potential log-ID continuous cascade models, as there is no bound to the number of ID probability laws. Among these, some log-ID models have received more attention. These are the log-Poisson model[36–38] and the log-stable models, also called universal multifractals,[5,28,31,39,40] characterized by the Lévy index α ($0 \leq \alpha \leq 2$), including for $\alpha = 2$ the lognormal model.[21,22,25]

Continuous multifractal processes may be generated using the procedure described in the previous section, with a value of λ_1 fixed and close to 1, as done in Reference 32. On the other hand, it is interesting to dispose of a stochastic equation corresponding to the limit process, instead of relying on an algorithmic procedure. This was proposed in Reference 41, where stochastic equations generating log-ID continuous multifractal processes have been provided, involving ID random measures. We do not here discuss the general log-ID case, but only the log-stable case, which corresponds to simpler expressions. For ID processes the equation providing a log-stable continuous multifractal process is the following:[5,41]

$$\varepsilon_\Lambda(x) = \Lambda^{-c} \exp\left\{ c^{1/\alpha} \int_{A(x)} |u - x|^{-1/\alpha} dL_\alpha(u) \right\} \tag{29.22}$$

where $\Lambda = X/\tau$ is the total scale ratio corresponding to the cascade process, $0 \leq c \leq 1$ is a parameter, $A(x) = \left[x - \dfrac{X}{2}, x - \dfrac{\tau}{2} \right] \cup \left[x + \dfrac{\tau}{2}, x + \dfrac{X}{2} \right]$, and $L_\alpha(u)$ is a log-stable measure of index α (see the appendix). This process is characterized by the log-stable moment function:[5]

$$K(q) = c\left(q^\alpha - q\right) \tag{29.23}$$

It is also of interest to have an expression for a causal process, where the position is time and the past does not depend on the future. In particular, this is important for prediction of multifractal time series. We discuss this below in the context of lognormal processes.

29.4.2   A Causal Lognormal Stochastic Equation and Its Properties

For a causal lognormal process, Equation 29.22 becomes:[41]

$$\varepsilon_\Lambda(t) = \Lambda^{-\mu/2} \exp\left\{ \mu^{1/2} \int_{t-T}^{t-\tau} (t - u)^{-1/2} dB(u) \right\} \tag{29.24}$$

where $B(u)$ is a Brownian motion and the process is developed in time over a scale ratio of $\Lambda = T/\tau$. We show here that this generates a stochastic process with the same properties as discrete cascades.

We have the following moments:

$$\left\langle \left(\varepsilon_\Lambda\right)^q \right\rangle = \Lambda^{-q\mu/2} \left\langle \exp\left\{ q\mu^{1/2} \int_{t-T}^{t-\tau} (t - u)^{-1/2} dB(u) \right\} \right\rangle \tag{29.25}$$

$$= \Lambda^{-q\mu/2} \exp\frac{\mu q^2}{2} \int_{t-T}^{t-\tau} \frac{du}{t - u} \tag{29.26}$$

$$= \Lambda^{-q\mu/2} \exp\frac{\mu q^2}{2} \log \Lambda \tag{29.27}$$

giving Equation 29.17.

We then consider the correlation function

$$C_\varepsilon(r) = <\varepsilon(t)\varepsilon(t+r)> \tag{29.28}$$

for $\tau \le r \le T - \tau$. The stochastic integrals are split in order to separate the overlapping integration domains, corresponding to independent random variables:

$$C_\varepsilon(r) = \Lambda^{-\mu} < e^{\sqrt{\mu}\int_{t-T}^{t+r-T}(t-u)^{-1/2}dB(u)} >< e^{\sqrt{\mu}\int_{t-\tau}^{t+r-\tau}(t+r-u)^{-1/2}dB(u)} >< e^{\sqrt{\mu}\int_{t+r-T}^{t-\tau}[(t-u)^{-1/2}+(t+r-u)^{-1/2}]dB(u)} > \tag{29.29}$$

$$= \Lambda^{-\mu}\exp\left\{\frac{\mu}{2}\int_{t-T}^{t+r-T}\frac{du}{t-u}\right\}\exp\left\{\frac{\mu}{2}\int_{t-\tau}^{t+r-\tau}\frac{du}{t+r-u}\right\}I_1 \tag{29.30}$$

$$= \Lambda^{-\mu}\left(\frac{T}{T-r}\right)^{\mu/2}\left(\frac{\tau+r}{\tau}\right)^{\mu/2}I_1 \tag{29.31}$$

where I_1 is the last integral to evaluate:

$$I_1 = \exp\left\{\frac{\mu}{2}\int_{t+r-T}^{t-\tau}\left[(t-u)^{-1/2}+(t+r-u)^{-1/2}\right]^2 du\right\} \tag{29.32}$$

$$= \exp\left\{\frac{\mu}{2}\int_{t+r-T}^{t-\tau}\frac{du}{t-u}\right\}\exp\left\{\frac{\mu}{2}\int_{t+r-T}^{t-\tau}\frac{du}{t+r-u}\right\}I_2 \tag{29.33}$$

$$= \left(\frac{T-r}{\tau}\right)^{\mu/2}\left(\frac{T}{\tau+r}\right)^{\mu/2}I_2 \tag{29.34}$$

with finally:

$$I_2 = \exp\left\{\mu\int_{t+r-T}^{t-\tau}\frac{du}{\sqrt{(t-u)(t+r-u)}}du\right\} \tag{29.35}$$

$$= \left(\frac{\sqrt{T-r}+\sqrt{T}}{\sqrt{\tau}+\sqrt{r+\tau}}\right)^{2\mu} \tag{29.36}$$

where we have used the identity

$$\int\frac{dx}{\sqrt{x(x+t)}} = 2\ln\left(\sqrt{x}+\sqrt{x+t}\right) \tag{29.37}$$

Finally, Equations 29.31, 29.34, and 29.36 give

$$C_\varepsilon(r) = \left(\frac{\sqrt{\Lambda-r^*}+\sqrt{\Lambda}}{1+\sqrt{r^*+1}}\right)^{2\mu} \tag{29.38}$$

with $r^* = r/\tau$. Whenever $1 \ll r^* \ll \Lambda$ this gives finally

$$C_\varepsilon(r) \approx (4\Lambda)^\mu\left(\frac{r}{\tau}\right)^{-\mu} \tag{29.39}$$

which has the same power law behavior as Equation 29.9 for discrete cascades.

On the other hand, we may introduce the singularity process $\gamma(t) = \log \varepsilon(t)$ and estimate its autocorrelation function for $\tau \le r \le T - \tau$:

$$C_\gamma(r) = < \gamma(t)\gamma(t + r) > \tag{29.40}$$

$$= (\tfrac{\mu}{2}\log\Lambda)^2 + \mu < \int_{t-T}^{t-\tau} (t - u)^{-1/2}\, dB(u) \int_{t+r-T}^{t+r-\tau} (t + r - u)^{-1/2}\, dB(u) > \tag{29.41}$$

where, as before, the stochastic integrals can be split in different nonoverlapping domains. Then, using Equation 29.A18 of the appendix, we have

$$C_\gamma(r) = (\tfrac{\mu}{2}\log\Lambda)^2 + \mu \int_{t+r-T}^{t-\tau} (t - u)^{-1/2}(t + r - u)^{-1/2}\, du \tag{29.42}$$

and using again Equation 29.37 this gives finally

$$C_\gamma(r) = (\tfrac{\mu}{2}\log\Lambda)^2 + 2\mu \log\left(\frac{\sqrt{\Lambda - r^*} + \sqrt{\Lambda}}{1 + \sqrt{r^* + 1}} \right) \tag{29.43}$$

$$\approx C_0 - \mu\log\frac{r}{\tau} \tag{29.44}$$

where $C_0 = (\tfrac{\mu}{2}\log\Lambda)^2 + \mu\log(4\Lambda)$. The last line used, as before, is the assumption $1 \ll r^* \ll \Lambda$. Equation 29.44 is very close to Equation 29.14, which was obtained for discrete cascades. The stochastic integral given by Equation 29.24 thus shows the adequate multiscaling properties (Equation 29.27), and autocorrelation properties for the process (Equation 29.39), and its generator (Equation 29.44). This proves that Equation 29.24 is a stochastic evolution equation generating causal lognormal multifractal processes. The generalization to the full log-stable family is straightforward (but the demonstrations for the correlations may not be as tractable as for Gaussian processes). We have given here all the steps of the calculations for correlation properties, because they have not been published before in this form.

29.4.3 Simulation of a Continuous Lognormal Multifractal

Equation 29.24 can be directly used for stochastic simulations: because the parameters Λ and μ are given, the simulation of this process needs only the discretization of the stochastic integral. An algorithm to perform this can be the following:

1. Choose Λ, μ, and the total number N of points to simulate; there is no upper bound to $N \ge \Lambda$.
2. Simulate $(g_j)_{j=1-\Lambda..N}$: $N + \Lambda$ realizations of a Gaussian white noise of variance 1 and 0 mean.
3. Integration. For $i = 1$ to n:

$$\text{Estimate } \varepsilon(t) \text{ as } \varepsilon_i = \Lambda^{-\mu/2} \exp\left\{ \sqrt{\mu} \sum_{j=i-\Lambda}^{i-1} \frac{g_j}{\sqrt{i-j}} \right\}$$

Figure 29.8 shows a sample of 5000 points simulated with this algorithm, with the parameters $\Lambda = 5000$ and $\mu = 0.25$. The process and its generator have the same visual shape as for discrete cascades. We have checked the multiscaling and correlation properties of this stochastic simulation: they have the same shape as for discrete cascades (Figure 29.3 to Figure 29.7) and we do not reproduce them here.

Equation 29.24 and the associated algorithm are not only more realistic than discrete cascades, but also can be used sequentially to generate a causal multifractal process with a scale ratio Λ, and no limit on the number of points for the simulation. This equation also generates a new family of stochastic

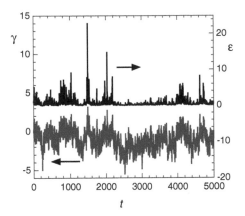

FIGURE 29.8 A sample of the generator and the multifractal process simulated with a continuous and causal multifractal process. As expected, the generator is a correlated noise while the multifractal process is highly intermittent.

processes having interesting mathematical properties that are here only beginning to be explored. Other methods to simulate universal multifractals are given in Reference 42.

29.5 Conclusion

We have considered the simulation of discrete and continuous multiplicative cascades. In both cases the theory has been briefly recalled, and a complete algorithm provided in the simple but generic case of lognormal cascades. For discrete cascades, the simulation is done using a simple iterative procedure, whereas for continuous cascades a stochastic evolution equation has been provided for the log-stable family, which can be easily discretized for simulation purposes. This discretization is in fact linked to the numerical implementation of a continuous (in scale and in time or space) evolution equation, and must not be confused with discrete cascades corresponding to a discreteness in scale ratio.

The field that is generated with this type of process or equation is intermittent, scale invariant, and characterized by multiple-scaling properties. The large fluctuations produced have a very specific structure, with long-range power law correlations. Such properties are fully compatible with experimental results in fully developed turbulence.

The need to be able to simulate "synthetic" multifractal fields or time series may arise in different contexts in oceanography. It may be useful to be able to reproduce turbulent activity to test inversion methods for acoustic propagation or radiative transfer, as well as for multiagent models or to simulate planktonic heterogeneity or contact rates in ecosystem models. In many of these applications, a preliminary data analysis is of course necessary to estimate various parameters such as the scaling range, the cascade model closer to the data, and its parameters.

The simulation procedures and methods proposed here can be directly applied to reproduce and simulate synthetic dissipation fields, which is directly involved, e.g., to estimate planktonic contact rates[17] or diffusivities of various quantities.[43] But we must stress here that this approach applies to fields resulting directly from a multiplicative cascade, which are called variously in the literature: "conservative multifractals," "multifractal measures," or "multifractal fields." On the other hand, there are also fields, such as the velocity or temperature in turbulence, with slightly different properties: their fluctuations (or structure functions), and not the field itself, have multiple scaling properties. These also received a variety of names: "nonconservative multifractals," "multifractal functions," or "multiaffine fields." These fields cannot be simulated using the procedures described here; some adaptations are necessary, which are not addressed in this chapter because the theory to perform such simulations (especially for continuous cascades) is not yet well established.

Appendix 29.A

We review here some useful properties of characteristic functions, infinitely divisible distributions, Gaussian and Lévy-stable random variables, and stable stochastic integrals. Useful references for these topics are Feller,[33] Samorodnitsky and Taqqu,[44] and Janicki and Weron.[45]

Characteristic Functions

Let X be a random variable with probability density $p(x)$. The usual characteristic function is the Fourier characteristic function $\varphi(k)$, defined as

$$\varphi(k) = < e^{ikx} > = \int_{-\infty}^{+\infty} e^{ikx} p(x)dx \qquad (29.A1)$$

Since when studying multiplicative cascades real moments are introduced, we use here Laplace characteristic functions, assuming that the positive tail of the probability density decreases rapidly enough for the integral to converge:

$$f(q) = < e^{qx} > = \int_{-\infty}^{+\infty} e^{qx} p(x)dx \qquad (29.A2)$$

We thus have $f(q) = \varphi(-iq)$, and the tables given for Fourier characteristic functions (e.g., in Reference 33) provide Laplace characteristic functions, as well. For example, for a Gaussian random variable of variance σ and 0 mean

$$f(q) = e^{\frac{\sigma^2}{2}q^2} \qquad (29.A3)$$

Infinitely Divisible Distributions

Infinite divisibility is a property that has no simple expression for probability densities or distributions. Instead, this property is very simply characterized using characteristic functions. A probability distribution is said to be infinitely divisible (ID) if its characteristic function f has the following property: for any n integer, there exists a characteristic function f_n such that

$$f_n^{\,n} = f \qquad (29.A4)$$

In such a case, we can introduce the second characteristic function $\Psi = \log f$:

$$\Psi(q) = \log < e^{qx} > \qquad (29.A5)$$

If $\Psi(q)$ is a characteristic function, whenever, for any $a > 0$, $a \Psi(q)$ is still a characteristic function, the distribution associated to $\Psi(q)$ is ID.

Let us give two examples. A Poisson distribution with mean α has a second characteristic function of

$$\Psi(q) = \alpha\left(e^q - 1\right) \qquad (29.A6)$$

Lévy-stable random variables can be defined as follows: let $\left(X_i\right)_{i=1..n}$ be independent random variables of the same law, and n any integer. Then the law is stable if there exist $a_n > 0$ and b_n such that

$$a_n \sum_{i=1}^{n} X_i - b_n$$

has the same law as the X_i. Strictly stable random variables are the ones for which $b_n = 0$, corresponding to centered ($\langle X \rangle = 0$) variables. Using the second characteristic function, it is easily seen that Lévy random variables, for which

$$\Psi(q) = \Gamma^{\alpha} q^{\alpha} \tag{29.A7}$$

are stable with $a_n = n^{1/\alpha}$. The parameter $\Gamma > 0$ is called the scale or dispersion parameter; it plays the same role as the variance for Gaussian random variables: for $\alpha = 2$, we recover Gaussian random variables with $\sigma^2 = \Gamma^2/2$. We note that Lévy-stable can take high values with hyperbolic probability, so that for the Laplace characteristic function to converge, we must consider only skewed Lévy variables, for which the hyperbolic tail corresponds to negative values only (see References 5, 28, 40, and 42). In this case there is no problem of convergence of moments of order $q > 0$.

The moments of order q of ε are linked to the second characteristic function of its generator:

$$K(q) = \log_{\lambda} \langle \varepsilon_{\lambda}^{q} \rangle = \log_{\lambda} \langle e^{q\gamma} \rangle = \frac{\Psi_{\gamma}(q)}{\log \lambda} \tag{29.A8}$$

since characteristic functions add for independent random variables. It can be seen that

$$\varepsilon = \prod_{p=1}^{n} W_p = e^{\sum_{p=1}^{n} g_p} \tag{29.A9}$$

gives

$$\Psi_{\gamma}(q) = n \Psi_{g}(q) \tag{29.A10}$$

showing that for the densification to be consistent, it is necessary to have infinitely divisible distributions for the logarithms of the weights W (and hence for the generators).

Stable Stochastic Integrals

The rescaling property of stable random variables can be used to define stable stochastic integrals. Let us consider again

$$Y = \sum_{i=1}^{n} X_i$$

where $(X_i)_{i=1..n}$ are independent strictly stable random variables having the same law, with parameters α and Γ_X. Then we have $\Psi_Y(q) = n\Psi_X(q) = n\Gamma_X^{\alpha} q^{\alpha}$ showing that

$$\Gamma_Y = n^{1/\alpha} \Gamma_X \tag{29.A11}$$

This property can be used to build stable random measures and in a consistent way stable stochastic integrals.

For an interval of width dx, $M(dx)$ is defined as a strictly stable random measure, i.e., a strictly stable random variable of scale parameter:

$$\Gamma_M = (dx)^{1/\alpha} \tag{29.A12}$$

It has thus the following second characteristic function:

$$\Psi_M(q) = \log \langle e^{qM(dx)} \rangle = \left(\Gamma_M q\right)^{\alpha} = dx q^{\alpha} \tag{29.A13}$$

For a positive valued function F, such that

$$\int_{a}^{b} F^{\alpha}(x)dx$$

exists, the stochastic integral

$$I = \int_a^b F(x)M(dx) \tag{29.A14}$$

is defined as the strictly stable random variable of scale parameter

$$\Gamma_I = \left(\int_a^b F^\alpha(x)dx \right)^{1/\alpha} \tag{29.A15}$$

Its second characteristic function is then the following:

$$\Psi_I(q) = \log < e^{q\int_a^b F(x)M(dx)} >= \left(\int_a^b F^\alpha(x)dx \right) q^\alpha \tag{29.A16}$$

This fully characterizes stable stochastic integrals. In particular, the Gaussian stochastic integral

$$I = \int_a^b F(x)B(dx)$$

is still a Gaussian random variable with the variance

$$\sigma_I^2 = \int_a^b F^2(x)dx \tag{29.A17}$$

Gaussian stochastic integrals have also the property:

$$< \int_{E_1} F(x)B(dx) \int_{E_2} G(x)B(dx) >= \int_{E_1 \cap E_2} F(x)G(x)dx \tag{29.A18}$$

Indeed, if $E_1 \cap E_2 = \varnothing$, the two integrals are independent random variables and whenever $E_1 \cap E_2 \neq \varnothing$ only the intersection contributes to the correlation.

References

1. Batchelor, G.K. and Townsend, A.A., The nature of turbulent motion at large wave-numbers, *Proc. R. Soc.*, A99, 238, 1949.
2. Kolmogorov, A.N., The local structure of turbulence in incompressible viscous fluid for very large Reynolds numbers, *Dokl. Akad. Nauk SSSR*, 30, 299, 1941.
3. Obukhov, A.M., Spectral energy distribution in a turbulent flow, *C. R. Acad. Sci. URSS*, 32, 22, 1941.
4. Richardson, L.F., *Weather Prediction by Numerical Process*, Cambridge University Press, Cambridge, U.K., 1922.
5. Schertzer, D. and Lovejoy, S., Physical modeling and analysis of rain and clouds by anisotropic scaling multiplicative processes, *J. Geophys. Res.*, 92, 9693, 1987.
6. Seuront, L. et al., Multifractal analysis of phytoplankton biomass and temperature variability in the ocean, *Geophys. Res. Lett.*, 23, 3591, 1996.
7. Seuront, L. et al., Multifractal analysis of Eulerian and Lagrangian variability of oceanic turbulent temperature and plankton fields, *Nonlinear Proc. Geophys.*, 3, 236, 1996.
8. Lovejoy, S. et al., Universal multifractals and ocean patchiness: phytoplankton, physical fields and coastal heterogeneity, *J. Plankton Res.*, 23, 117, 2001.
9. Seuront, L. et al., Universal multifractal analysis as a tool to characterize multiscale intermittent patterns: example of phytoplankton distribution in turbulent coastal waters, *J. Plankton Res.*, 21, 877, 1999.

10. Pascual, M., Ascioti, F.A., and Caswell, H., Intermittency in the plankton: a multifractal analysis of zooplankton biomass variability, *J. Plankton Res.*, 17, 1209, 1995.

11. Seuront, L. and Lagadeuc, Y., Multiscale patchiness of the calanoid copepod *Temora longicornis* in a turbulent coastal sea, *J. Plankton Res.*, 23, 1137, 2001.

12. Seuront, L., Gentilhomme, V., and Lagadeuc, Y., Small-scale patches in tidally mixed coastal waters, *Mar. Ecol. Prog. Ser.*, 232, 29, 2002.

13. Schmitt, F. et al., Empirical study of multifractal phase transitions in atmospheric turbulence, *Nonlinear Proc. Geophys.*, 1(2/3), 95, 1994.

14. Schmitt, F. et al., Multifractal temperature and flux of temperature variance in fully developed turbulence, *Europhys. Lett.*, 34, 195, 1996.

15. Kurien, S. et al., Scaling structure of the velocity statistics in atmospheric boundary layers, *Phys. Rev. E*, 61, 407, 2000.

16. Kerman, B., A multifractal equivalent for the Beaufort scale in sea-state, *Geophys. Res. Lett.*, 20, 8, 1991.

17. Seuront, L., Schmitt, F., and Lagadeuc, Y., Turbulence intermittency, small-scale phytoplankton patchiness and encounter rates in plankton: where do we go from here? *Deep-Sea Res. I*, 48, 1199, 2001.

18. Marguerit, C. et al., Copepod diffusion within multifractal phytoplankton fields, *J. Mar. Syst.*, 16, 69, 1998.

19. Grant, H.L., Stewart, R.W., and Moilliet, A., Turbulent spectra from a tidal channel, *J. Fluid Mech.*, 12, 241, 1962.

20. Phillips, O.M., The Kolmogorov spectrum and its oceanic cousins: a review, *Proc. R. Soc. Lond. A*, 434, 125, 1991.

21. Kolmogorov, A.N., A refinement of previous hypotheses concerning the local structure of turbulence in a viscous incompressible fluid at high Reynolds number, *J. Fluid Mech.*, 13, 82, 1962.

22. Obukhov, A.M., Some specific features of atmospheric turbulence, *J. Fluid. Mech.*, 13, 77, 1962.

23. Gurvich, A.S. and Zubkovski, S.L., *Izv. Akad. Nauk. SSSR, Ser. Geofiz.*, 1856, 1963.

24. Pond, S. and Stewart, R.W., *Izv. Akad. Nauk. SSSR, Ser. Fis. Atmos. Okeana*, 1, 914, 1965.

25. Yaglom, A.M., The influence of fluctuations in energy dissipation on the shape of turbulent characteristics in the inertial interval, *Sov. Phys. Dokl.*, 11, 26, 1966.

26. Arneodo, A. et al., Analysis of random cascades using space-scale correlation functions, *Phys. Rev. Lett*, 80, 708, 1998.

27. Arneodo, A., Bacry, E., and Muzy, J.-F., Random cascades on wavelet dyadic tress, *J. Math. Phys.*, 39, 4142, 1998.

28. Schertzer, D. and Lovejoy, S., Nonlinear variability in geophysics: multifractal analysis and simulation, in *Fractals: Physical Origin and Consequences*, Pietronero, L., Ed., Plenum Press, New York, 1989, 49.

29. Mandelbrot, B.B., Intermittent turbulence in self-similar cascades: divergence of high moments and dimension of the carrier, *J. Fluid Mech.*, 62, 305, 1974.

30. Schertzer, D. and Lovejoy S., On the dimension of atmospheric motions, in *Turbulence and Chaotic Phenomena in Fluids*, Tatsumi, T., Ed., Elsevier, North-Holland, Amsterdam, 1984, 505.

31. Schertzer, D., Lovejoy, S., and Schmitt, F., Structures in turbulence and multifractal universality, in *Small-Scale Structures in 3D Hydro and MHD Turbulence*, Meneguzzi, M., Pouquet, A., and Sulem, P.-L., Eds., Springer Verlag, New York, 1995, 137.

32. Schmitt, F., Vannitsem, S., and Barbosa, A., Modeling of rainfall time series using two-state renewal processes and multifractals, *J. Geophys. Res.*, 103, D18, 23181, 1998.

33. Feller, W., *An Introduction to Probability Theory and Its Applications*, Vol. 1 and 2, Wiley, New York, 1971.

34. Saito, Y., Log-gamma distribution model of intermittency in turbulence, *J. Phys. Soc. Jpn.*, 61, 403, 1992.

35. Novikov, E.A., Infinitely divisible distributions in turbulence, *Phys. Rev. E*, 50, R3303, 1994.

36. She, Z.-S. and Waymire, E., Quantized energy cascade and log-Poisson statistics in fully developed turbulence, *Phys. Rev. Lett.*, 74, 262, 1995.

37. She, Z.-S. and Leveque, E., Universal scaling laws in fully developed turbulence, *Phys. Rev. Lett.*, 72, 336, 1994.

38. Dubrulle, B., Intermittency in fully developed turbulence: log-Poisson statistics and generalized scale covariance, *Phys. Rev. Lett.*, 73, 959, 1994.

39. Kida, S., Log-stable distribution and intermittency of turbulence, *J. Phys. Soc. Jpn.*, 60, 5, 1991.

40. Schertzer, D. et al., Multifractal cascade dynamics and turbulent intermittency, *Fractals*, 5, 427, 1997.

41. Schmitt, F. and Marsan, D., Stochastic equations for continuous multiplicative cascades in turbulence. *Eur. Phys. J. B*, 20, 3, 2001.

42. Pecknold, S., Lovejoy, S., and Schertzer, D., Stratified multifractal magnetization and surface geomagnetic fields. II. Multifractal analysis and simulations, *Geophys. J. Int.*, 145, 127, 2001.
43. Kantha, L.H. and Clayson, C.A., *Small Scale Processes in Geophysical Flows*, International Geophysics Series, Vol. 67, Academic Press, New York, 888 pp.
44. Samorodnitsky, G. and Taqqu, M.S., *Stable Non-Gaussian Random Processes: Stochastic Models with Infinite Variance*, Chapman & Hall, New York, 1994, 632 pp.
45. Janicki, A. and Weron, A., *Simulation and Chaotic Behavior of α-Stable Stochastic Processes*, Marcel Dekker, New York, 1994, 355 pp.

30

An Application of the Lognormal Theory to Moderate Reynolds Number Turbulent Structures

Hidekatsu Yamazaki and Kyle D. Squires

CONTENTS

30.1   Introduction...469
30.2   Lognormal Theory 470
30.3   Simulations 471
30.4   Discussion...474
      30.4.1   Surface Turbulent Layer 475
      30.4.2   Subsurface Stratified Layer 477
Acknowledgments..477
References..478

30.1 Introduction

Kolmogorov (1941) proposed one of the most successful theories in the area of turbulence, namely, the existence of an inertial subrange. Successively, Kolmogorov (1962) revised the original theory to take the variability of the dissipation rate in space into account. The process of this refinement introduced a lognormal model to describe the distribution of dissipation rates. The inertial subrange theory requires an energy cascade process, whose length scale is much larger than that of the viscous dominating scale. Thus, the types of flows to which the theory applies occur at high Reynolds numbers. Geophysical flows provide an example in that they typically occur at high Reynolds numbers because the generation mechanism is usually much larger than the viscous dominating scale. In fact, the first evidence of the existence of an inertial subrange came from observations of a high Reynolds number oceanic turbulent flow (Grant et al. 1962). Gurvich and Yaglom (1967) further developed the lognormal theory that described the probability distribution of the locally averaged dissipation rates. In their work, the theory was also intended for high Reynolds number flows to simplify the development (see also Monin and Ozmidov, 1985).

Although both the inertial subrange and lognormal theories successfully describe high Reynolds number turbulence, an important question arises: To what degree are these theories appropriate to turbulence occurring over a moderate Reynolds number range, whose power spectrum does not attain an inertial subrange? Clearly, the inertial subrange theory is out of the question; i.e., there is a limited range of scales at moderate Reynolds numbers. However, is it possible that the dissipation rate in moderate Reynolds number turbulence obeys the lognormal theory?

Relevant to the present chapter is that turbulence generated at laboratory scales in many facilities does not attain high Reynolds numbers; thus, energy spectra do not typically exhibit an inertial subrange. Microorganisms, such as zooplankton in the ocean, may be transported in the water column by a large-scale flow that is clearly occurring at high Reynolds numbers, but the immediate flow field surrounding

0-8493-1344-9/04/$0.00+$1.50
© 2004 by CRC Press LLC

the individual organism in a seasonal thermocline is another example of moderate Reynolds number turbulence (Yamazaki et al., 2002). The lognormal theory provides a simple statistical representation of the flow, as well as yielding a tool to predict the local properties of the strain field. If lognormality holds at moderate Reynolds numbers, it would enable one to predict the probability of the strain field in many flows of practical interest.

Turbulence dissipation rates reported in the literature are normally values averaged over a scale of a few meters. On the other hand, a relevant scale for the encounter rate of predator/prey is normally much shorter than 1 m. It is important to note that the volume-averaged dissipation rate associated with this length scale will not be identical to that obtained for the original domain since the dissipation rate for this length scale is an additional random variable that obeys a different probability density function from the mother domain. The lognormal theory assists in understanding the local properties of velocity strains.

Direct numerical simulation (DNS) is well suited for investigating the applicability of the lognormal theory at moderate Reynolds numbers. A significant advantage of DNS relevant to this study is that all components of the strain rate can be directly computed and the dissipation rate can be calculated as a function of position and time. DNS studies, e.g., Jiménez et al. (1993), show that the strain field of turbulence is dominated by filament-like structures. These coherent structures are crucial to understanding flow dynamics. Yamazaki (1993) proposed that planktonic organisms may make use of these structures to find mates and prey/predator. Presented in this chapter is a demonstration that the lognormal theory is consistent with the strain properties associated with the filament structures, at least, for moderate Reynolds numbers.

30.2 Lognormal Theory

A complete discussion of the lognormal theory can be found in Gurvich and Yaglom (1967). The theory can be developed by considering a domain, Q, with energy-containing eddies of size, L, where Q is proportional to L^3. The volume-averaged dissipation rate over Q is denoted $\langle \varepsilon \rangle$ and is defined as

$$\langle \varepsilon \rangle = Q^{-1} \int_Q \varepsilon(x) dx \qquad (30.1)$$

where $\varepsilon(x)$ is the local dissipation rate. The original domain, Q, is successively divided into subdomains denoted q_i, whose length scale is l_i. This successive division process is referred to as a breakage process. The average dissipation in a volume q_i is then

$$\varepsilon_i = q_i^{-1} \int_{q_i} \varepsilon(x) dx \qquad (30.2)$$

The dissipation rate ε_i is a random variable representing the average within q_i. The breakage coefficient, α, is defined as a ratio of two successive ε_i:

$$\alpha_i = \varepsilon_i / \varepsilon_{i-1} \quad \text{for} \quad i = 1,....,N_b \qquad (30.3)$$

where N_b is the number of breakage processes. In the original lognormal theory, the ratio of length scales l_{i-1} and l_i for two successive breakages is a constant, $\lambda_b = l_i/l_{i-1}$. At the N_b breakage, the volume averaged dissipation rate in a single cell, ε_r, for the averaging scale $r = l_{N_b}$ can be expressed in terms of $\langle \varepsilon \rangle$ by

$$\log \varepsilon_r = \log \langle \varepsilon \rangle + \sum_{i=1}^{N_b} \log \alpha_i \qquad (30.4)$$

where r might be considered as an encounter rate length scale, such as perception distance/reaction distance. Gurvich and Yaglom (1967) assumed that the random variable $\log \alpha_i$ follows a normal distribution. One drawback of the Gurvich and Yaglom theory is that, if α is lognormal, the maximum value of α is infinity. Yamazaki (1990) argues that the maximum value of α cannot exceed λ_b^3 and proposes the B-model, which assumes a beta probability density function for α. The B-model predicts high-order statistics of velocity well.

An important question arises in the above development: Is the assumption of high Reynolds number required in the lognormal theory? There are two constraints: α_i is mutually independent and N_b is large. However, in practice, the first condition is not so strict, and the second requirement may be as small as 2 or 3 (Mood et al., 1974). In other words, the sum of a few random variables, e.g., log α_i, tends to approach a normal distribution as the central limit theorem predicts. Therefore, there is no explicit requirement for the existence of an inertial subrange to satisfy these conditions. Hence, it may be reasonable to expect that the lognormal theory might be applicable to turbulence occurring at modest Reynolds numbers in which there is no inertial subrange.

It should be noted that, while Gaussian statistics is an approximation, increasingly less accurate for the higher-order moments as shown by Novikov (1971) and Jiménez (2000), the lognormal theory has provided a reasonable model for some applications (e.g., see Arneodo et al., 1998). The practical advantages offered via assumption of Gaussian statistics outweigh the inaccuracies in many instances, e.g., as applied to positive-value statistics such as temperature and rainfall. In this chapter, we emphasize the practical aspects of application of the lognormal theory for analyzing the dissipation rate for turbulent flows at moderate Reynolds numbers, bearing in mind the limitations of the theory as shown by other investigators.

30.3 Simulations

We have simulated isotropic turbulence using DNS of the incompressible Navier–Stokes equations (Rogallo, 1981). A statistically stationary flow was achieved by artificially forcing all nonzero wavenumbers within a spherical shell of radius K_F (Eswaran and Pope, 1988). For the simulations presented here, $K_F = 2\sqrt{2}$, corresponding to 92 forced modes. The small-scale resolution is measured by the parameter $k_{max}\eta$, where η is the Kolmogorov length scale and k_{max} is the highest resolved wavenumber. The value of η is obtained from $(\nu^3/\varepsilon)^{1/4}$ where ν is the kinematic viscosity of the fluid. In this study, $k_{max}\eta$ was approximately 2. Several preliminary computations were performed to ensure the adequacy of the numerical parameters and to test the data reduction used to acquire the dissipation rate. Most of the results presented in this chapter are from simulations performed using 64^3 collocation points, corresponding to a Taylor-microscale Reynolds number $Re_\lambda = 29$ (Case C64). Although a single simulation (sampled over time) should be sufficient for testing the hypothesis that the lognormal theory is applicable to a moderate Reynolds number flow, simulations performed at higher resolution were desired to give some confidence that conclusions from this study were relatively free of resolution effects and not adversely influenced by the scheme used to maintain a statistically stationary state. Therefore, calculations were also performed at a higher resolution 96^3 (Case C96) and used to confirm the trends observed at the lower resolution, in which there is less separation between the peaks of the energy and dissipation spectra (Figure 30.1). The Taylor-microscale Reynolds number for the higher-resolution flow is 42. The calculations were run using a fixed time step, chosen so that the Courant number remained approximately 0.40. The flow was allowed to evolve to a statistically stationary state; flow-field statistics were then acquired over a total time period T (Figure 30.2). Flow fields were saved every eddy turnover time $T_e = L_f/u'$, in which L_f is the longitudinal integral time scale and u' is the root-mean-square velocity, for subsequent postprocessing of the dissipation rate.

For each grid resolution, an ensemble of ten velocity fields was processed to determine the minimum averaging scale at which lognormality was satisfied as well as to calculate breakage coefficients. Each velocity field was subdivided into smaller volumes, and the dissipation rate within a given subdomain was calculated by integrating over the grid point values within a given volume. B-spline integration (de Boor, 1978) was used for calculation of the dissipation rate within subvolumes to faithfully follow the definition of local averaging given in Equation 30.2. Note that Wang et al. (1996) averaged grid point dissipation rates arithmetically.

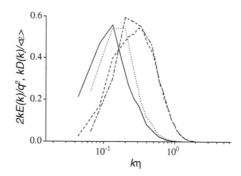

FIGURE 30.1 Three-dimensional energy and dissipation spectra. Case C64: dotted line is energy and chain dot line is dissipation; Case C96: solid line is energy and dashed line is dissipation.

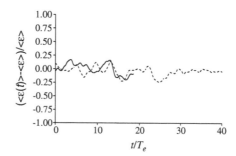

FIGURE 30.2 Temporal variation of the volume-averaged dissipation rate. Case C64, solid line; Case C96, dashed line.

Lognormality for the compiled data is tested by making use of the Kolmogorov–Smirnov test (KS test) at a 5% significance level. The KS test is a powerful tool to distinguish if the samples are drawn from a hypothesized distribution; however, the target distribution must be free from the estimation of parameters or without parameters involved in the distribution (Mood et al., 1974). In other words, if the hypothesized distribution contains some parameters, e.g., the mean and the variance, the KS test is not, rigorously speaking, applicable. As usual, in the practical application of statistical theories, since no other simple test is available to determine if the samples come from the hypothesized distribution, the KS test is employed in this work, albeit with the limitations described above.

If the theory is applicable to the present simulations, locally averaged ε_r should be lognormal, but no information is given in the theory on how the instantaneous dissipation rate, $\varepsilon(x)$, distributes. Figure 30.3 shows the quantile–quantile plot (qq-plot) of instantaneous dissipation rates, equivalent to grid-level dissipation rates, for Case C64. The distribution is clearly different from a lognormal distribution. Yeung and Pope (1989) and Wang et al (1996) also show a similar distribution for the grid-level dissipation rates, but at higher Reynolds numbers, $Re_\lambda = 93$ in Yeung and Pope and $Re_\lambda = 151$ in Wang et al. There is of course no *a priori* knowledge of the probability distribution of the instantaneous dissipation rates and, hence, it should not seem surprising that the grid-level values do not distribute as lognormal. The lognormal theory is only applicable to a locally averaged quantity; therefore it is necessary to consider a locally averaged dissipation rate, ε_r.

The grid-level dissipation rate exhibits features remarkably similar to instantaneous dissipation rates observed in geophysical data (Yamazaki and Lueck, 1990). Stewart et al. (1970) measured the velocity in the atmospheric boundary layer over the ocean. They attributed the departure from lognormality to be caused by a limited cascade process with an insufficient Reynolds number. They presumed that to satisfy the lognormal theory, it was necessary for the turbulence Reynolds number to be very high. Because there was no local averaging applied to their data, the reported values were essentially the same as the grid-level dissipation rates in the present DNS. They also argued that the departure from lognormality at the low end of the distribution was caused by instrument noise. The DNS results, however,

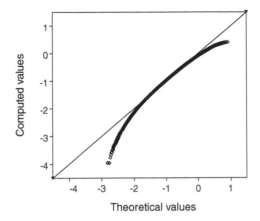

FIGURE 30.3 The quantile–quantile plot of grid-level dissipation rate and prediction from lognormal distribution for Case C64.

do not suffer from analogous problems. Small-scale resolution of the velocity field has been carefully maintained. Therefore, the concave nature of the grid-level dissipation rates (the instantaneous values) is possibly a more universal characteristic of the kinetic energy dissipation rate. If one is interested in extremely high values of the local dissipation rate, the lognormal theory provides an upper bound for the estimate. In other words, the actual value should be smaller than the predicted value. On the other hand, if one is interested in extremely low values, the lognormal theory overpredicts the values compared to the actual dissipation rate.

To investigate what averaging scale satisfies the lognormal theory, we have computed the local average of dissipation rates with varying averaging scales for each of the ten fields, as well as compiled all data. Lognormality is tested for these compiled data sets. The minimum averaging scale for lognormality to hold in terms of the Kolmogorov scale for the two cases are similar, 9.5 for Case C64 and 10.2 for Case C96.

Because statistics may change from one realization (i.e., velocity field) to the next, lognormality of the dissipation rate for each of the ten different fields has also been examined. Shown in Table 30.1 are the numbers of individual fields passing lognormality for Case C64. The minimum averaging scale for the entire ensemble of ten fields is 9.5, but there are several individual fields satisfying lognormality at smaller averaging scales. Although one field at $r/\eta = 7.9$ failed the KS test, all individual fields follow lognormality for an averaging scale as small as 6.3. This is roughly 30% smaller than that obtained using the entire ensemble.

TABLE 30.1

Number of Individual Fields Passing KS Test for C64 Case

No. of Cells for Local Averaging	r / η	No. of Fields Passing KS Test
10^3	9.5	10
11^3	8.6	10
12^3	7.9	9
13^3	7.3	10
15^3	6.3	10
16^3	5.9	8
20^3	4.7	6
25^3	3.8	3
27^3	3.5	3
30^3	3.2	1
32^3	3.0	1

To consider why individual cases can satisfy lognormality at smaller averaging scales compared to the entire ensemble of ten fields, we consider the nature of the KS test. The test statistic is the maximum difference between the observed cumulative distribution function and the hypothesized cumulative distribution function. The critical value for the test statistic is defined as, $d = d_\gamma / \sqrt{n}$, where d_γ is the critical value at a certain significance level γ, and n is the number of samples. When the test statistic exceeds d, the hypothesis that the samples come from the proposed probability density function is rejected at the specified significance level. For a significance level of 5%, as used in this study, the value of d_γ is 1.36. As the number of samples increases, the test value decreases. Thus, the test is more difficult to pass for larger sample sizes. As we have mentioned earlier, the KS test is developed for a parameter-free distribution. However, we are using an estimated mean and variance for the hypothesized lognormal distribution, so we are violating the assumptions for the KS test. Therefore, the observed minimum averaging scale difference between the ten-field case and single-field cases is, most likely, due to the violation of the KS test assumption. Unfortunately, we do not have any other simple way to test the hypothesized distribution. Practically speaking, the observed dissipation rate is very close to a lognormal distribution even at the smallest averaging scale obtained from the single-field case.

It is further interesting to note that one field satisfies lognormality at an averaging scale $r/\eta = 3.0$. This is almost identical to the minimum averaging scale for oceanic data (Yamazaki and Lueck, 1990). Despite the difference in the nature of the data source, the minimum averaging scales obtained from the present moderate Reynolds number flow calculated using DNS, which are roughly between 5 and 10, are remarkably close to the geophysically observed values. Making use of a laboratory air-tunnel experiment, van Atta and Yeh (1975) report 36η as the length scale that assures statistical independence between successive observations. The sample independence length scale should be larger than the corresponding minimum averaging scale for lognormality. The laboratory experiment also provides a similar minimum averaging scale to the present simulation results. Recently, Benzi et al. (1995, 1996) show velocity scale similarity as small as 4η using both wind-tunnel experiments and direct numerical simulations, and propose a new scaling notion: extended self-similarity (ESS). These observations are consistent with each other, showing that the lognormal theory is fairly robust at moderate Reynolds numbers.

How the breakage coefficient distributes is an important issue in the lognormal theory. However, no previous investigation has been made to examine the appropriate distribution of this coefficient. Yamazaki (1990) proposed the Beta distribution and developed the B-model. The minimum averaging scale at which lognormality holds for each individual field has been used as a child domain length scale, i.e., $l_c = 6.32$. The corresponding mother domain for $\lambda = 5$, which is the recommended value, is then $l_m = 31.6$. Thus, the entire volume is subdivided into 15^3 cells for the child domain and 3^3 cells of the mother domain. The breakage coefficient, α, is tested against both the Beta distribution (the B-model) and the lognormal distribution (the Gurvich and Yaglom model). As shown in Figure 30.4, the Beta distribution predicts the observed statistics well. The lognormal distribution, on the other hand, exhibits a poor fit to the data.

30.4 Discussion

Although the lognormal theory is not developed from a vigorous fluid mechanical point of view, the theory seems to work remarkably well even if the flow occurs at moderate Reynolds numbers, which lack an inertial subrange. Therefore, it offers the possibility of a practical tool for predicting locally averaged dissipation rates at spatial scales larger than 10η and the minimum averaging scales as small as three times η. The theory can be extended to smaller averaging scales bearing in mind that the theory overpredicts high value of dissipation rates.

A perception distance of larval fish may be taken as the local averaging scale of dissipation rate in order to predict the upper band for encounter rate with prey. Another example is that an ambient flow field around a single organism can be extrapolated from the average dissipation rate of a turbulent water column. Incze et al. (2001) observed that several copepod species avoided high turbulent water column when the dissipation rate exceeded 10^{-6} W kg^{-1} and they interpreted this observed feature via the

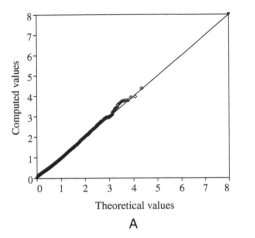
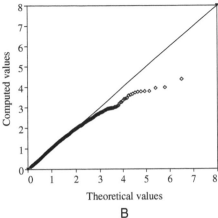

FIGURE 30.4 (A) The qq-plot of the breakage coefficient for Case C64 and $\lambda = 5$; the Beta distribution is assumed. (B) The qq-plot of the breakage coefficient for Case C64 and $\lambda = 5$; the lognormal distribution is assumed.

behavioral response of the organisms to the flow field. The majority moved from the surface to a stratified intermediate water column where the dissipation rate was reduced to 10^{-8} W kg^{-1} or less. Haury et al. (1990) also observed that a shift in the community structure of zooplankton took place when the average dissipation rate of the water column exceeded 10^{-6} W kg^{-1}. The Kolmogorov scale associated with 10^{-6} W kg^{-1} is 10^{-3} m, roughly the size of a copepod. Is this the reason the community structure of zooplankton is responding to the turbulence level at 10^{-6} W kg^{-1}? According to the universal spectrum for oceanic turbulence, the peak in the shear spectrum takes place at no higher than 30 cycles m^{-1} at this dissipation rate (Gregg, 1987; Oakey, 2001). At η scale, the kinetic energy is virtually exhausted. The dissipation rates reported in the literature are normally based on at least 1-m scale averaging, but the highly intermittent nature of instantaneous dissipation rates is masked (Yamazaki et al., 2002).

Clearly, the average dissipation rate does not describe the ambient flow field for a single organism. To provide an estimate of the representative ambient flow field around a single plankter, we make use of the lognormal theory. We assume that the plankter is a sphere whose radius is 1 mm. Based on the observed evidence (Haury et al., 1990; Incze et al., 2001), we consider the following scenario: the assumed organism moves from *a surface turbulent layer* whose dissipation rate is 10^{-6} W kg^{-1} and whose thickness, L_1, is 10 m to *a subsurface stratified layer* whose dissipation rate is 10^{-8} W kg^{-1} and whose thickness, L_2, is 1 m. Then we consider two levels of averaging scales for the lognormal theory: $r_1 = 10\eta$ and $r_2 = 3\eta$.

For low-order moments, such as mean and variance, any lognormal models give nearly identical predictions; thus, we make use of the Gurvich and Yaglom model with the intermittency coefficient $\mu = 0.25$ (Yamazaki et al., 2002). The model provides the following relationship for the local average dissipation rate ε_r and the domain average dissipation rate $<\varepsilon>$:

$$m_r = \log<\varepsilon> - 0.125 \log(Lr^{-1}) \tag{30.5}$$

$$\sigma_r^2 = 0.25 \log(Lr^{-1}) \tag{30.6}$$

where m_r is the mean and σ_r^2 the variance of log ε_r.

30.4.1   Surface Turbulent Layer

In this layer, we use the following values:

$<\varepsilon> = 10^{-6}$ W kg^{-1}

$L_1 = 10$ m

Thus,

$$\eta = 1.0 \times 10^{-3} \text{ m}$$
$$r_1 = 10\eta = 1.0 \times 10^{-2} \text{ m}$$
$$r_2 = 3\eta = 3.0 \times 10^{-3} \text{ m}$$

The turbulence rms velocity q may be expressed in terms of L and $<\varepsilon>$ (Tennekes and Lumley, 1972):

$$q = (<\varepsilon>L)^{1/3} \tag{30.7}$$

These values lead to $q = 2.15 \times 10^{-2} \text{ m s}^{-1}$. The Reynolds number based on L is

$$\text{Re} = (qL)/\nu \tag{30.8}$$

and is related to the Taylor scale Reynolds number as followed (Levich, 1987).

$$\text{Re}_\lambda \approx (8\text{Re})^{1/2} \tag{30.9}$$

For the specific example considered here, $\text{Re} = 2.15 \times 10^5$ and $\text{Re}_\lambda = 1311$. Since $\log \varepsilon_r$ distribute as normal, the following z value distributes as a standard normal distribution:

$$z = \frac{\log \varepsilon_r - m_r}{\sigma_r} \tag{30.10}$$

For a given L/r, the probability that local ε_r exceeds the global mean, $<\varepsilon>$, can be assessed by taking $\log<\varepsilon> = \log \varepsilon_r$ in Equation 30.10 (Figure 30.5). For r_1 and r_2, the probability is 0.256 and 0.238, respectively. Hence, nearly 75% of spatial volume is occupied by the local average dissipation rate less than $<\varepsilon>$. Large values are taking place in less than 25% of the total volume.

To estimate an extreme value of the local average dissipation rate for each averaging scale r_1 and r_2, we suppose that the extreme values take place at a probability that is equivalent to the volume occupancy of the assumed organisms. As a typical number of copepod observed in field, we assume ten individuals per liter. The volume occupied by organisms is $4.19 \times 10^{-2} \text{ m}^3$ and the corresponding probability, p_r, is 4.19×10^{-5}. This probability is equivalent to an extreme event that takes place for less than 0.15 s in

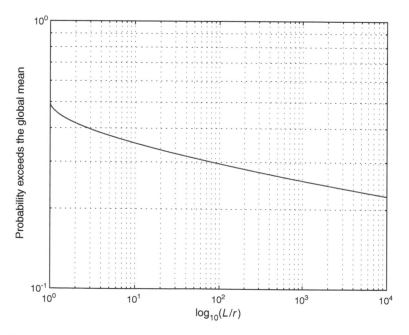

FIGURE 30.5 Probability exceeds the global mean against $\log_{10}(L/r)$.

1 h. The lognormal theory provides $\varepsilon_r = 7.5 \times 10^{-5}$ W kg^{-1} and 1.6×10^{-4} W kg^{-1} for r_1 and r_2, respectively. When we equate this dissipation rate with the isotropic formula ($\varepsilon = 7.5 s_m^2$), the mean cross stream turbulence shear, s_m, is 3.3 and 4.6 s^{-1} for each case. These are substantial values, although the volume occupation of such high values is low. Where do these high strain rates take place? Unfortunately, the lognormal theory does not predict the actual flow structures. Thus, we relate the lognormal theory to the coherent structure studies with DNS.

Numerical simulations show the strain field of turbulence is dominated by a filament-like structure (Vincent and Meneguzzi, 1991; Jiménez et al., 1993). Jiménez (1998) shows that the mean radius of filament R is roughly 5η and a maximum azimuthal velocity u_θ is roughly q. A maximum vorticity ω_{max} is $3(q/R)$. The volume fraction of filament p_f is related to the Taylor scale Reynolds number:

$$p_f = 4 \, Re_\lambda^{-2} \tag{30.11}$$

For our case, p_f is 2.33×10^{-6} so that the actual volume occupied by the filament in 10^3 m^3 is 2.33×10^{-3} m^3. Thus, if we assume the cross section of the filament is a circle whose radius is 5η and that the remaining length scale of a "typical" filament is the same as the Taylor microscale, then there are roughly 250 filaments for this particular volume. According to the development above, the maximum dissipation rate associated with the filament is 6.13×10^{-4} W kg^{-1}. The lognormal theory predicts that the local dissipation rate based on p_f is 1.7×10^{-4} and 2.5×10^{-4} W kg^{-1} for r_1 and r_2. The maximum dissipation rate for the filament should be larger than the local average value; thus two independent assessments for the local shear values are consistent.

30.4.2   Subsurface Stratified Layer

We use the following values for this layer:

$\langle \varepsilon \rangle = 10^{-8}$ W kg^{-1}

$L_2 = 1$ m

Thus,

$\eta = 3.16 \times 10^{-3}$ m

$r_1 = 10\eta = 3.16 \times 10^{-2}$ m

$r_2 = 3\eta = 9.48 \times 10^{-3}$ m

The probability that local average values exceed the global mean is 0.32 and 0.29 for each averaging scale. Thus, nearly 70% of space is occupied by the local dissipation rate that is below the global mean. Based on the same argument for extreme values, the volume occupancy ratio by the organism, $p_f = 4.19 \times 10^{-5}$, provides $\varepsilon_r = 2.4 \times 10^{-7}$ W kg^{-1} and 3.9×10^{-7} W kg^{-1} for r_1 and r_2, respectively. The mean cross stream turbulence shear, s_m, is 0.18 and 0.23 s^{-1} for each case. The number of filaments expected in 1 m^3 in this case is smaller, roughly three, and the maximum dissipation rate occurring within the filament is 6.16×10^{-7} W kg^{-1}. The lognormal theory predicts that the dissipation rates associated with the filament occupancy ratio are 1.5×10^{-7} and 2.3×10^{-7} W kg^{-1}.

Zooplankton in the surface mixed layer may be reacting to the intermittent high shear that can be argued quantitatively from the lognormal theory as presented in the chapter. The local quantities should be used to investigate the effects of turbulence on individual microscale organism behaviors.

Acknowledgments

We are indebted to A. Abib for his patient work running the simulation codes. This work was supported by Grant-in-Aid for Scientific Research C-10640421.

References

Arneodo, A., Manneville, S., and Muzy, J.F., Toward log-normal statistics in high Reynolds number turbulence, *Eur. Phys. J. B*, 1, 129, 1998.

Benzi, R., Ciliberto, S., Baudet, C., and Chavarria, G.R., On scaling of three-dimensional homogenous and isotropic turbulence, *Physica D*, 80, 385, 1995.

Benzi, R., Struglia, M.V., and Tripiccione, R., Extended self-similarity in numerical simulations of three-dimensional anisotropic turbulence, *Phys. Rev. E*, 53, R5565, 1996.

de Boor, C., *A Practical Guide to Spline*, Springer-Verlag, New York, 1978.

Eswaran, V. and Pope, S.B., An examination of forcing in direct numerical simulations of turbulence, *Comput. Fluid*, 16, 257, 1988.

Grant, H.L., Stewart, R.W., and Moilliet, A., Turbulence spectra from a tidal channel, *J. Fluid Mech.*, 12, 241, 1962.

Gregg, M.C., Diapycnal mixing in the thermocline: a review, *J. Geophys. Res.*, 92, 5249, 1987.

Gurvich, A.S. and Yaglom, A.M., Breakdown of eddies and probability distributions for small-scale turbulence, boundary layers and turbulence, *Phys. Fluids*, 10, 59, 1967.

Haury, L.R., Yamazaki, H., and Itsweire, E.C., Effects of turbulent shear flow on zooplankton distribution, *Deep-Sea Res.*, 37, 447, 1990.

Incze, L.S., Hebert, D., Wolff, N., Oakey, N., and Dye, D., Changes in copepod distributions associated with increased turbulence from wind stress, *Mar. Ecol. Prog. Ser.*, 213, 229, 2001.

Jiménez, J., Small scale intermittency in turbulence, *Eur. J. Mech. B Fluids*, 17, 405, 1998.

Jiménez, J., Intermittency and cascades, *J. Fluid Mech.*, 409, 99, 2000.

Jiménez, J., Wray, A.A., Saffman, P.G., and Rogallo, R.S., The structure of intense vorticity in isotropic turbulence, *J. Fluid Mech.*, 255, 65, 1993.

Kolmogorov, A.N., The local structure of turbulence in incompressible viscous fluid for very large Reynolds numbers, *Dokl. Akad. Nauk SSSR*, 30, 299, 1941.

Kolmogorov, A.N., A refinement of previous hypotheses concerning the local structure of turbulence in a viscous incompressible fluid at high Reynolds number, *J. Fluid Mech.*, 13, 82, 1962.

Levich, E., Certain problems in the theory of developed hydrodynamical turbulence, *Phys. Rep.*, 151, 129, 1987.

Monin, A.S. and Ozmidov, R.V., *Turbulence in the Ocean*, D. Reidel, Dordrecht, the Netherlands, 1985.

Mood, A.M., Graybill, F.A., and Boes, D.C., *Introduction to the Theory of Statistics*, 3rd ed., McGraw-Hill, New York, 1974.

Novikov, E.A., Intermittency and scale similarity in the structure of a turbulent flow, *Prikl. Mat. Mech.*, 35, 266, 1971.

Oakey, N.S., Turbulence sensors, in *Encyclopedia of Ocean Sciences*, J.H. Steele, S.A. Thorpe, and K.K. Turekian, Eds., Academic Press, San Diego, 2001, 3063.

Rogallo, R.S., Numerical experiments in homogeneous turbulence, Tech. Rep. NASA TM 81315, NASA Ames Research Center, 1981.

Stewart, R.W., Wilson, J.R., and Burling, R.W., Some statistical properties of small-scale turbulence in an atmospheric boundary layer, *J. Fluid Mech.*, 41, 141, 1970.

Tennekes, H. and Lumley, J.L., *A First Course in Turbulence*, MIT Press, Cambridge, MA, 1972.

van Atta, C.W. and Yeh, T.T., Evidence for scale similarity of internal intermittency in turbulent flows at large Reynolds numbers, *J. Fluid Mech.*, 71, 417, 1975.

Vincent, A. and Meneguzzi, M., The spatial structure and statistical properties of homogenous turbulence, *J. Fluid Mech.*, 225, 1, 1991.

Wang, L.P., Chen, S., Brasseur, J.G., and Wyngaard, J.C., Examination of hypotheses in the Kolmogorov refined turbulence theory through high-resolution simulations. Part 1: Velocity field, *J. Fluid Mech.*, 309, 113, 1996.

Yamazaki, H., Breakage models: lognormality and intermittency, *J. Fluid Mech.*, 219, 181, 1990.

Yamazaki, H., Lagrangian study of planktonic organisms: perspectives, *Bull. Mar. Sci.*, 53, 265, 1993.

Yamazaki, H. and Lueck, R.G., Why oceanic dissipation rates are not lognormal, *J. Phys. Oceanogr.*, 20, 1907, 1990.

Yamazaki, H., Mackas, D., and Denman, K., Coupling small scale physical processes to biology: towards a Lagrangian approach, in *The Sea, Biological-Physical Interactions in the Ocean*, A.R. Robinson, J.J. McCarthy, and B.J. Rothschild, Eds., John Wiley & Sons, New York, 2002, 51–112.

Yeung, P.K. and Pope, S.B., Lagrangian statistics from direct numerical simulations of isotropic turbulence, *J. Fluid Mech.*, 207, 531, 1989.

31

Numerical Simulation of the Flow Field at the Scale Size of an Individual Copepod

Houshuo Jiang

CONTENTS

31.1 Introduction...479
31.2 Dynamic Coupling...481
    31.2.1 Navier–Stokes Equations Governing the Flow Field around
        a Free-Swimming Copepod...481
    31.2.2 Dynamic Equation of a Free-Swimming Copepod's Body 482
    31.2.3 A Simple Example for the Dynamic Coupling 483
31.3 Numerical Simulation..484
    31.3.1 Methods 484
    31.3.2 Results..486
        31.3.2.1 Comparison with an Observational Result.......................................486
        31.3.2.2 Swimming Behavior and Flow Geometry...487
        31.3.2.3 Swimming Behavior and Feeding Efficiency....................................488
31.4 A Future Application 489
31.5 Concluding Remarks 489
Acknowledgments..490
References...490

31.1 Introduction

Calanoid copepods are generally negatively bouyant.[1] Strickler[1] hypothesized that the reason these planktonic animals, living in a nutritionally dilute environment,[2] are negatively buoyant can be found in an analysis of the forces acting on them. His hypothesis was based on hours of minute observations of registering the paths of algae and of free-swimming calanoid copepods on film. When a copepod swims horizontally, the effect of negative buoyancy or excess weight is counterbalanced by the creation of a feeding current. In other words, were the copepods neutrally buoyant, basically conserving energy by not having to swim constantly, they would not encounter as many algae as they need for survival in these nutritionally dilute waters. Intuitively, Strickler[1] suggested that the configuration of forces acting on a free-swimming copepod determines the copepod's body orientation and swimming velocity. He further drew diagrams of different configurations of forces for several different copepod species. Along the same line, Emlet and Strathmann[3] argued that the drag on the main body of a copepod — along with the excess weight — also plays an important role in setting up the flow field around the copepod. Their argument emphasized the role of the copepod's swimming behavior (including the body orientation and swimming direction and speed) and morphology (including the morphology of the main body and the morphology and motion pattern of the cephalic appendages). Both are the determining factors of drag forces. Observational evidence supports their argument. For example, the experiments done by Emlet[4] revealed the difference in flow geometry between tethered and free-swimming ciliated larvae. Although the planktonic organisms that Emlet studied are larvae of the bivalve *Crassostrea gigas* and the gastropod *Calliostoma ligatum*, he has pointed out that the results may apply generally to other small, self-propelled organisms. For copepods, the observations by Bundy and Paffenhöfer[5] showed large differences in flow geometry between tethered copepods and free-swimming copepods, and between different copepod species. The differences in flow geometry are due to three facts. (1) For free-swimming copepods, flow field velocity and geometry are controlled by the balance of forces, i.e., drag, negative buoyancy, and thrust obtained by the appendages from the water. (2) For tethered copepods, tethering will alter the balance of these forces. (3) Different species may have different configurations of the balance of forces.

The advances in understanding the creation of copepod (or other zooplankton) feeding currents should be at least partially credited to an innovative technical breakthrough in high-speed micro-cinematography.[6–9] With this technical breakthrough researchers were able to take high-speed movies of live zooplankton. From the early 1980s on, miles of film have documented a vivid world in which zooplankters swim, feed, and breed. By watching and carefully studying the movies of live copepod feeding frame by frame, researchers[1,8,10,11] have found that calanoid copepods are "suspension-feeders." "They [the copepods] capture and handle the food particles not passively according to size and shape but, in most cases, actively using sensory inputs for detection, motivation to capture, and ingestion."[11] The technical breakthrough also contributed directly to another important finding, that many calanoid copepods create feeding currents. Moreover, it is feasible to use this technique to measure the feeding currents. In the past two decades, many new observations have provided qualitative and quantitative information about the feeding currents.[1,5,9,12–30] In some studies, the three-dimensional structure of the feeding currents, including velocity magnitudes and some other flow properties, has been measured.[5,17,19,22,23,25,28,29] They are particularly useful for developing theoretical and numerical studies.

The successful work done by zooplankton biologists has inevitably stimulated research interests of some physicists and fluid dynamicists. However, accompanying theoretical studies have not been satis-factory. Most of the studies only chose some simple solutions based on the Stokes flow model to fit data obtained from observations, ignoring the fact that these simple solutions were not able to reproduce even the simplest features of a feeding current (but these features may be important for a copepod's feeding or sensing). The failure of these theoretical studies stems from the fact that they did not take into account the fundamental mechanisms underlying the creation of feeding currents (or generally speaking, the creation of the water flow around a free-swimming copepod).

Recently, Jiang et al.[31] simulated the feeding current created by a tethered copepod. They did this through a computational fluid dynamics (CFD) model based on the idea that a copepod exerts propulsive forces on the surrounding water to create the feeding current by beating its cephalic appendages. The simulated feeding current was shown to be quite comparable with an observation by Yen and Strickler.[28] Then, through coupling the Navier–Stokes equations with the dynamic equation of an idealized body of a copepod, a hydrodynamic model[32] was proposed to calculate the flow field around a free-swimming copepod in steady motion. Following this hydrodynamic model, Jiang et al.[33] developed a CFD simulation framework to simulate the flow field around a free-swimming copepod in steady motion with realistic body shape. The parameter inputs for this simulation framework are the swimming behavior, morphology, and excess weight of a copepod. (Apparently, numerous original observations and published results have contributed to the documentation and validation of the parameter inputs for the simulation framework, and it is impossible to name them all here.) It is now clear that the dynamic coupling between a copepod's swimming motion and the copepod's surrounding water determines the flow field around the copepod. The importance of considering free-swimming copepods as self-propelled bodies is highlighted. For steady motion, this means a free-swimming copepod must gain thrust (equal in magnitude but opposite in direction to the vector sum of the propulsive forces that the copapod exerts on the surrounding water through its appendages, i.e., the reacting force of the total propulsive forces) from the surrounding water to counterbalance the drag force by water and the excess weight. The propulsive forces, which determine how many forces and where they are exerted by the copepod on the surrounding water, are apparently an important factor in determining the flow field around the copepod. The morphology and swimming motion of the copepod's body is another factor in determining the flow field and actually controls the boundary condition at the body–fluid interface. Theoretical and numerical studies have demonstrated that the feeding currents can be reproduced from first principles, namely, Newton's laws of motion.

In this chapter, first the basic ideas underlying the above-mentioned theoretical[32] and numerical[33] studies are generalized. Then, some results are reviewed. To validate the hydrodynamic model and numerical simulation framework, the flow field around a backward-swimming copepod, obtained from an observation done in the Strickler laboratory at the Great Lakes WATER (Wisconsin Aquatic Technology and Environmental Research) Institute of the University of Wisconsin–Milwaukee, is compared with the counterpart results obtained from the hydrodynamic model and numerical simulation framework. In addition, a future application of the numerical simulation method in the study of the on/off or time-dependent feeding current is outlined.

31.2 Dynamic Coupling

31.2.1 Navier–Stokes Equations Governing the Flow Field around a Free-Swimming Copepod

Consider a free-swimming copepod in a water column and assume the water is otherwise quiescent in the absence of the copepod. The equations governing the flow-velocity vector field $\mathbf{u}(\mathbf{x}, t)$ around the copepod are the Navier–Stokes equations and the continuity equation:

$$\rho\frac{\partial \mathbf{u}}{\partial t} + \rho\mathbf{u}\cdot\nabla\mathbf{u} = -\nabla p + \mu\nabla^2\mathbf{u} + \mathbf{f}_a(\mathbf{x}, t) \tag{31.1}$$

$$\nabla\cdot\mathbf{u} = 0 \tag{31.2}$$

where ρ is the density of the water, μ is the dynamic viscosity, and p is the flow pressure field. The boundary conditions of Equations 31.1 and 31.2 are the no-slip boundary condition on the surface of the main body (i.e., the body excluding the beating appendages, denoted as Ω_{mb}):

$$\mathbf{u} = \mathbf{V}_{\text{swimming}}, \text{ at } \Omega_{mb} \tag{31.3}$$

and the boundary condition at infinity:

$$\mathbf{u} \to 0, \text{ at infinity} \tag{31.4}$$

The first term on the left-hand side of Equation 31.1 is identified as the inertial acceleration forces, the second term as the inertial convective forces, the first term on the right-hand side of Equation 31.1 as the pressure forces, and the second term as the viscous forces. The force field $\mathbf{f}_a(\mathbf{x}, t)$ (force per unit volume) is discussed in the following two paragraphs.

The force field $\mathbf{f}_a(\mathbf{x}, t)$ approximates the mean effect of the beating movement of the copepod's cephalic appendages. This approximation is made probable by taking into account two characteristics of the beating movement of the cephalic appendages. (1) The beating movement is operated at a high frequency. (2) The beating movement is performed in certain asymmetric patterns. (For a detailed analysis, see Jiang et al.[32]) By doing so, we avoid dealing with the difficulty resulting from the highly time-dependent moving boundary conditions imposed by these cephalic appendages; however, the mean effect of the beating movement of these appendages can still be included in the governing equations. In fact, the mean effect of the beating movement of the cephalic appendages, as represented by the force field $\mathbf{f}_a(\mathbf{x}, t)$, is the propulsive forces exerted by the copepod on the water. (Note that thrust is the reacting force of the vector sum of the propulsive forces.) The above-described two characteristics of the beating movement suggest that both the resistive and the reactive types of forces are likely to contribute to the thrust generation. Moreover, it is noteworthy that the thrust is not simply generated due to the pressure gradient resulting from the ventrally positioned feeding current.

Because the cephalic appendages are spatially distributed (for many species, ventrally to the copepod), $\mathbf{f}_a$ is interpreted as a spatially distributed force field, i.e., a function of space $\mathbf{x}$. On the other hand, since a copepod may beat its cephalic appendages intermittently,[12,14,20,34,35] $\mathbf{f}_a$ may also be a function of time t, reflecting the long timescale variation of the mean effect of the beating movement. Cowles and Strickler[12] provide an example of the intermittent beating, where the studied copepod (*Centropages typicus*) beat its appendages for 1 s (mean) and then stopped beating (thereby sank) for 4 s when it was in filtered seawater. In this situation, the feeding current created by the copepod is termed the "on/off" feeding current. To simulate this on/off feeding current mathematically, $\mathbf{f}_a$ has to be turned on for 1 s and then turned off (i.e., set to zero) for 4 s. It is noted that immediately adjacent to the beating appendages there exist short timescale variations in the feeding current, due to the high-frequency characteristic of the beating movement. However, they have been safely eliminated from the governing equations.[32]

The magnitudes of the terms in Equation 31.1 can be conveniently estimated by performing scale analysis of (or scaling) the equation. In scale analysis, we specify typical expected values of the following quantities: (1) the magnitudes of the field variables, (2) the magnitudes of fluctuations in the field variables, and (3) the characteristic length and time scales on which these fluctuations occur. Inspecting the flow field around a copepod we find a characteristic length L related to the body size of the copepod, a characteristic velocity U determined by the copepod's swimming behavior, and a characteristic time T that is either imposed by the force field $\mathbf{f}_a$ or simply defined as L/U (the convective timescale). Then, we nondimensionalize Equation 31.1 by scaling time by T, distance by L, $\mathbf{u}$ by U, and pressure by $\mu U/L$. Substituting the nondimensional variables (denoted by primes) $t' = t/T$, $\mathbf{x}' = \mathbf{x}/L$, $\mathbf{u}' = \mathbf{u}/U$, and $p' = p/(\mu U/L)$, Equation 31.1 becomes

$$\beta \frac{\partial \mathbf{u}'}{\partial t'} + \mathrm{Re}\,\mathbf{u}' \cdot \nabla' \mathbf{u}' = -\nabla' p' + \nabla'^2 \mathbf{u}' + \frac{L^2}{\mu U} \mathbf{f}_a \tag{31.5}$$

Two important nondimensional numbers appear in Equation 31.5: the frequency parameter $\beta = L^2/(\nu T)$ and the Reynolds number $\mathrm{Re} = UL/\nu$, where $\nu = \mu/\rho$ is the kinematic viscosity of the fluid. The frequency parameter β measures the relative importance between the inertial acceleration forces and the viscous forces. For the previously mentioned situation of a copepod (*C. typicus*) beating its cephalic appendages intermittently, $\beta \sim 0.6$ if we choose $L = 2.0 \times 10^{-3}$ m, $\nu = 1.350 \times 10^{-6}$ m$^2 \cdot$ s^{-1}, and $T = 5.0$ s (period of the intermittent beating). This indicates that the inertial acceleration forces cannot be neglected in comparison with the viscous forces and that the on/off feeding current so created is intrinsically unsteady. In the absence of the force field $\mathbf{f}_a$ or if the force field is time independent, T may be defined as L/U in which case β reduces to Re. In this situation, a steady flow will be achieved after a period of time for initial adjustment. (This may be termed the "time" boundary layer.)

The Reynolds number Re represents the magnitude of the inertial convective forces relative to the viscous forces. When $\mathrm{Re} \ll 1$, the inertial convective forces (and the inertial acceleration forces if $\mathbf{f}_a$ is time independent) are small compared with the viscous forces and therefore may be neglected. Usually, the Reynolds number of the flow field around a free-swimming copepod does not satisfy the condition of $\mathrm{Re} \ll 1$ but is of the order $\mathrm{Re} \sim 1$; this means that the viscous forces are as important as the inertial forces. In some situations Re can be up to several hundreds, where the inertial forces dominate over the viscous forces outside the boundary layer around the copepod.

31.2.2 Dynamic Equation of a Free-Swimming Copepod's Body

The dynamic equation of a free-swimming copepod's main body can be approximately written as

$$\left(m + m_a \right) \frac{d\mathbf{u}_c}{dt} = \mathbf{W}_{\text{excess}} + \mathbf{F} + \mathbf{T} \tag{31.6}$$

where m is the mass of the copepod, m_a is the added mass, and $\mathbf{u}_c$ is the instantaneous velocity of the copepod's body. $\mathbf{W}_{\text{excess}}$ is the copepod's excess weight and can be calculated according to the formula:

$$\mathbf{W}_{\text{excess}} = \Delta\rho \, \Omega_{\text{copepod}} \, \mathbf{g} \tag{31.7}$$

where $\Delta\rho$ is the copepod's excess density relative to seawater, $\Omega_{copepod}$ is the body volume of the copepod, and $\mathbf{g}$ is the acceleration due to gravity. $\mathbf{F}$ is the drag force exerted by the flow field on the copepod's main body and calculated as

$$F = \int_{\Omega mb} \mathbf{n} \cdot 2\mu S d\Omega - \int_{\Omega mb} p\mathbf{n} d\Omega \qquad (31.8)$$

where $\mathbf{n}$ is the outward unit vector normal to the surface element $d\Omega$ and

$$S_{ij} = \frac{1}{2}\left(\frac{\partial u_i}{\partial x_j} + \frac{\partial u_j}{\partial x_i}\right) \qquad (31.9)$$

with u and p calculated from Equations 31.1 through 31.4. The thrust $\mathbf{T}$ that the copepod gains from the water is calculated from the integral

$$\mathbf{T} = -\int_{\mathbf{x}} \mathbf{f}_a(\mathbf{x},t)d\mathbf{x} \qquad (31.10)$$

For simplicity, the equations of moments are not considered.

In the studies of some intrinsically unsteady and highly time-dependent problems such as the jumping reaction, the full Equation 31.6 has to be used. However, Equation 31.6 can be greatly simplified for some other swimming behaviors. For example, when a copepod is in steady motion, i.e., either hovering at the same position ($\mathbf{V}_{swimming} = 0$) or swimming at a constant velocity ($\mathbf{V}_{swimming} = $ constant), Equation 31.6 becomes

$$-\mathbf{T} = \mathbf{W}_{excess} + \mathbf{F} \qquad (31.11)$$

which means that the copepod must gain thrust (i.e., $\mathbf{T}$) from the surrounding water to counterbalance the drag force by water and the excess weight. When a copepod stops beating its cephalic appendages, so that the thrust $\mathbf{T} = 0$, it sinks freely due to the excess weight. In the final steady state, the drag force resulting from sinking balances the excess weight, i.e., Equation 31.6 reduces to

$$\mathbf{W}_{excess} + \mathbf{F} = 0 \qquad (31.12)$$

The copepod's terminal velocity of sinking ($\mathbf{V}_{terminal}$) can be determined from this equation.

Comparing a copepod's swimming velocity with its terminal velocity of sinking can qualitatively determine the property of the flow field created by the copepod. If $|\mathbf{V}_{swimming}| \ll |\mathbf{V}_{terminal}|$, then $|\mathbf{F}| \ll |\mathbf{W}_{excess}|$. From Equation 31.11, one can see that the thrust, $\mathbf{T}$, that the copepod gains from the water is mainly used to counterbalance the excess weight. In this situation, the flow field around the copepod looks like the flow generated by a force monopole; from a biological point of view, the copepod creates a wide and cone-shaped feeding current. On the other hand, if $|\mathbf{V}_{swimming}| \gg |\mathbf{V}_{terminal}|$, then $|\mathbf{F}| \gg |\mathbf{W}_{excess}|$. From Equation 31.11, one can see that the thrust, $\mathbf{T}$, is mainly used to counterbalance the drag forces resulting from swimming. In this situation, the flow field looks like the flow generated by a force dipole, as the copepod exerts both drag forces and the propulsive forces on the water; from a biological point of view, the copepod does not create a feeding current, and the flow geometry is cylindrical, narrow, and long. (For a detailed analysis, see Jiang et al.[32]) In this sense, the terminal velocity of sinking is the most natural scaling of the swimming velocity of different copepod species.

31.2.3 A Simple Example for the Dynamic Coupling

Equations 31.1 through 31.4 together with Equations 31.6 through 31.10 are a set of equations describing the dynamic coupling between a copepod's swimming motion and the water surrounding the copepod. Fully solving these equations is not easy and needs very sophisticated computational techniques. To comprehend the dynamic coupling, which couples the flow generation process with the swimming behaviors of copepods, Jiang et al.[32] provided a simple example. They used the Stokes approximation (or inertia-free approximation) to simplify Equation 31.1. Assuming steady motion for the copepod, they used Equation 31.11 as the dynamic equation for the copepod's body. For the model copepod's morphology,

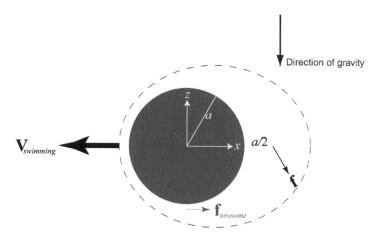

FIGURE 31.1 Schematic illustration of the model copepod consisting of a spherical body of radius a and a point force **f** (representing the mean effect of a single appendage) located outside the spherical body at a distance of $a/2$ away from the surface of the spherical body. The positive z-direction is opposite to the direction of gravity. The application point for the point force **f** is placed on the positive x-axis. The whole system translates at a constant velocity $V_{swimming}$ through the water. $f_{urosome}$ represents the effects of the beating of the urosome; however it is neglected in the work. (Note that this is not a free body diagram.) (From Jiang, H. et al., *J. Plankton Res.*, 24, 167, 2002. With permission.)

they used an idealized morphology, consisting of a spherical body and a single appendage represented by a point force outside the spherical body (Figure 31.1). With these simplifications, the equations for the dynamic coupling become

$$-\nabla p + \mu \nabla^2 \mathbf{u} + \mathbf{f}\delta(\mathbf{x} - \mathbf{x}_0) = 0 \qquad (31.13)$$

$$\mathbf{W}_{excess} + \mathbf{F} - \mathbf{f} = 0 \qquad (31.14)$$

with a known formula relating the drag force **F** to the point force **f** together with some morphological parameters, and with suitable boundary conditions. This simple hydrodynamic model can be analytically solved and in general can be used to calculate the flow field created by the model copepod (as shown in Figure 31.1) with arbitrary steady motion. Using this model, the authors showed how the flow geometry varies with different swimming behaviors.

Essentially, the net force exerted by a steady-swimming copepod on the surrounding water must be equal to the copepod's excess weight in spite of the copepod's swimming behavior. This is because the copepod is self-propelled. Concerning the spatial decay of the velocity field around the copepod, this indicates that the velocity field should decay in the far field to the velocity field generated by a point force of magnitude of the copepod's excess weight in an infinite domain (which is termed the *point force model*). Fortunately, the simple hydrodynamic model is able to reproduce this important property in velocity decay. Figure 31.2 clearly shows that the velocity magnitudes for different swimming behaviors (e.g., hovering, forward swimming fast or slowly) decay to the velocity field generated by the point force model. It should be pointed out that the Stokes solution of the flow around a translating sphere cannot reproduce this velocity-decay property because the translating sphere is actually not self-propelled but towed; i.e., additional forces are applied to the surrounding water. A correct hydrodynamic model for free-swimming copepods must consider the dynamic coupling at the very beginning. In other words, the model copepod must be a self-propelled body.

31.3 Numerical Simulation

31.3.1 Methods

The simple hydrodynamic model described in Section 31.2.3 takes advantage of two strong assumptions: (1) assuming a spherical body shape with a single appendage and (2) neglecting inertial effects. However,

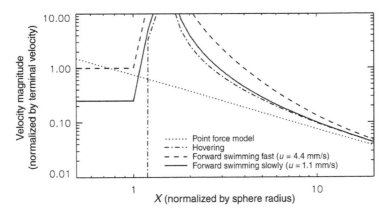

FIGURE 31.2 Velocity decay along the line $y = 0$, $z = 0$ (see Figure 31.1 for definition of the coordinate system) for different swimming behaviors. The velocity magnitudes have been normalized by the terminal velocity of sinking of the spherical copepod (4.4 mm · s^{-1} for the present case). (From Jiang, H. et al., *J. Plankton Res.*, 24, 167, 2002. With permission.)

a real copepod is unlikely to be spherical and has many appendages; the Reynolds number associated with the flow field around a free-swimming copepod usually does not satisfy the condition of Re << 1 required by the inertia-free approximation. (On the contrary, the assumption of steady motion is suitable, because most calanoid copepods are in steady motion in most of their time.) To release the above-mentioned two strong assumptions, Jiang et al.[33] developed a framework of numerical simulation to solve the coupling between the steady Navier–Stokes equations and the dynamic equation of a copepod's body in steady motion:

$$\rho \mathbf{u} \cdot \nabla \mathbf{u} = -\nabla p + \mu \nabla^2 \mathbf{u} + \mathbf{f}_a \qquad (31.15)$$

$$\mathbf{W}_{excess} + \mathbf{F} - \int_{\mathbf{x}} \mathbf{f}_a(\mathbf{x}) d\mathbf{x} = 0 \qquad (31.16)$$

with suitable boundary conditions. In general, this framework can be used to solve for the flow field around a model copepod with a realistic body shape — for example, the body morphology shown in Figure 31.3 — and in arbitrary steady motion, such as hovering, sinking, and steady swimming with various body orientations.

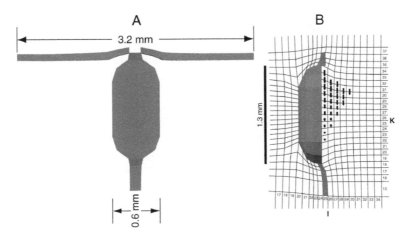

FIGURE 31.3 Morphology of the model copepod: (A) ventral view and (B) lateral view with a ventrally distributed force field modeling the mean effect of the beating movement of the cephalic appendages. (From Jiang, H. et al., *J. Plankton Res.*, 24, 191, 2002. With permission.)

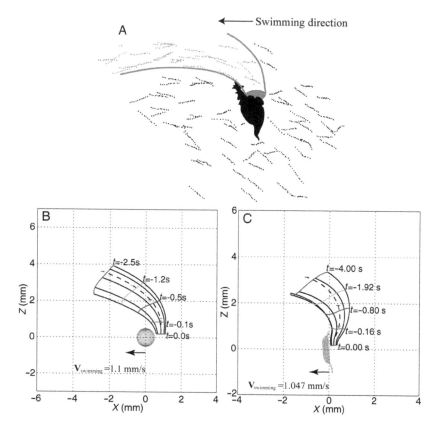

FIGURE 31.4 Comparison between results from observation, theoretical analysis, and numerical simulation, respectively. (A) Trajectories of suspended particles as seen from the copepod's point of view. The copepod (*D. minutus*) was observed swimming backward slowly. Total time of observation was 2 s. Note that those particles between the two lines would intersect the copepod's capture area. (B) From theoretical analysis,[32] lateral view of the streamtube through the capture area of a spherical copepod swimming backward (in negative *x*-direction) at a speed of 1.1 mm · s[-1]. (C) From numerical simulation,[33] lateral view of the streamtube through the capture area of a model copepod swimming backward (in negative *x*-direction) at a speed of 1.047 mm · s[-1]. Note that the frame of reference is fixed on the copepod.

31.3.2   Results

31.3.2.1   Comparison with an Observational Result — The results obtained from the theoretical[32] and numerical[33] studies can be at least qualitatively compared with observations on free-swimming copepods. An example is given. From a video clip taken in the Strickler laboratory, the flow field around a backward-swimming *Diaptomus minutus* was visualized by constructing trajectories of suspended particles around the copepod (Figure 31.4A). It can be seen that particles that would finally intersect the copepod's capture area come from a cone-shaped region behind and above the copepod, i.e., the region between the two lines as shown in Figure 31.4A. This kind of flow geometry is similar to that obtained (for a similar scenario) from both theoretical analysis (Figure 31.4B) and numerical simulation (Figure 31.4C). In all three plots, the copepod was shown to create a wide and cone-shaped feeding current, as the copepod's swimming velocity was much less than its terminal velocity of sinking. However, quantitative comparison point by point still challenges experimental biologists to obtain an accurate measurement of the three-dimensional velocity vector field around a free-swimming copepod.

Previous observational studies[5,19] have documented characteristics of the three-dimensional flow field around free-swimming copepods. Many of the flow characteristics have been reproduced in the theoretical[32] and numerical[33] studies. Note that no previous hydrodynamic analysis has had the capability of reproducing these flow characteristics.

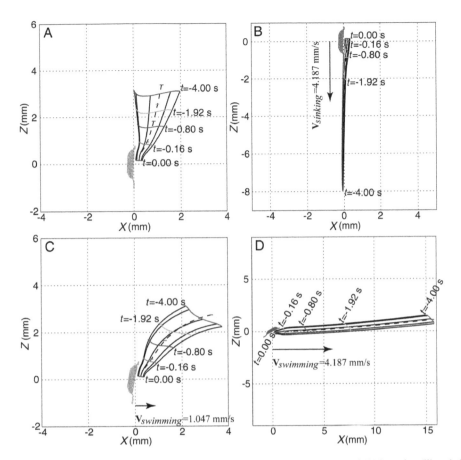

FIGURE 31.5 Lateral view of the streamtube through the capture area of a model copepod (A) hovering (like a helicopter) in the water, (B) sinking freely with the anterior pointing upward, at its terminal velocity (4.187 mm · s⁻¹ and along its body axis in the present case), (C) swimming forward (in positive *x*-direction) at a speed of 1.047 mm · s⁻¹, and (D) swimming forward (in positive *x*-direction) at a speed of 4.187 mm · s⁻¹. Note that the frame of reference is fixed on the copepod. (From Jiang, H. et al., *J. Plankton Res.*, 24, 191, 2002. With permission.)

31.3.2.2 Swimming Behavior and Flow Geometry — An important conclusion drawn from

the numerical simulation study[33] is that the geometry of the flow field around a free-swimming copepod varies significantly with different swimming behaviors. The geometry of the flow field around a copepod can be visualized by constructing a streamtube through the capture area of the copepod.[31] The streamtube associated with a copepod swimming slowly (i.e., swimming at a speed at least several times slower than the copepod's terminal velocity of sinking, termed the slow-swimming behavior) resembles the streamtube of a copepod hovering in the water. In both situations, the cone-shaped and wide streamtube transports water to the capture area of the copepod, and the copepod creates a feeding current (Figure 31.5A, C). Conversely, when a copepod swims at a speed equal to or greater than the terminal velocity (termed the fast-swimming behavior), the streamtube through the capture area is cylindrical, long, and narrow, and the flow field created is not a feeding current (Figure 31.5D). In addition, when a copepod sinks freely, the flow comes from below relative to the copepod and the streamtube through the capture area is much narrower and longer than hovering and swimming slowly, but shorter than swimming fast (see Figure 31.5B). Again, the flow field around a free-sinking copepod does not resemble a feeding current. A theoretical analysis[32] has explained such dependence of flow geometry on swimming behaviors. Although no observational evidence can be found in the literature for copepods to support this conclusion, supportive evidence can be found in the literature for other organisms. For example, Emlet[4] found that tethered bivalve larvae in still water and tethered polychaete larvae created flow fields in which particles followed curved trajectories, whereas particles followed straighter trajectories around free-swimming polychaete larvae and bivalve larvae tethered in flowing water.

The dependence of flow geometry on swimming behaviors is reflected in sensory modes (mechanoreception and/or chemoreception) adopted by copepods in detecting prey and food particles. This is because the sensory modes depend largely on the flow geometry. Using a three-dimensional alga-tracking, chemical advection–diffusion model, Jiang et al.[36] showed that a copepod's swimming behavior can place a constraint on its chemoreception. When it hovers or swims slowly, a copepod can use chemoreception to remotely detect individual algae entrained by the flow field around itself. A free-sinking copepod may also be able to use chemoreception to detect algal particles. In contrast, a fast-swimming copepod is not able to rely on chemoreception to remotely detect individual algae. As pointed out in the very beginning of this chapter, an advantage for copepods of negative buoyancy is the creation of a strong feeding current, thereby increasing the number of encounters. Here, a further advantage for negatively buoyant copepods to hover or to swim slowly is to create a strong feeding current, which allows deformation of the active space around an entrained alga, thereby conducive to the early warning system of deformed active space.[1,8] What really matters is how much food the copepod realizes is going by. However, systematic studies relating swimming behaviors to mechanoreception are still needed.

31.3.2.3 Swimming Behavior and Feeding Efficiency — The results obtained from the numerical study[33] also reveal the dependence of feeding efficiency on swimming behaviors. Without considering sensory inputs, feeding efficiency is simply measured by a ratio between the volumetric flux through a copepod's capture area and the power input by the copepod in creating the flow field around itself. Figure 31.6 clearly shows that the ratio is a function of swimming behaviors (including swimming velocity and direction). The behaviors of hovering or swimming slowly are energetically more efficient in terms of relative capture volume per energy expended than the behaviors of swimming fast. That is, for the same amount of energy expended, a hovering or slow-swimming copepod (which creates a feeding current) is able to scan more water than a fast-swimming copepod. The adaptive advantage for calanoid copepods may be from this very dependence of feeding efficiency on swimming behaviors — many calanoid copepods create a feeding current because the feeding mode of creating a feeding current is energetically more efficient. Even though hovering/slow-swimming behaviors are energetically more efficient (i.e., with a larger ratio of volumetric-flux to power input), a hovering or slow-swimming copepod does not scan more volume of water than a fast-swimming copepod does in a given period of time. In fact, the volumetric flux calculated for a hovering or slow-swimming copepod is less than that calculated for a fast-swimming copepod, provided the two have the same body size and excess density. This contradicts previous understanding of this problem. However, the new understanding is based on considering free-swimming copepods self-propelled and is, therefore, more convincing.

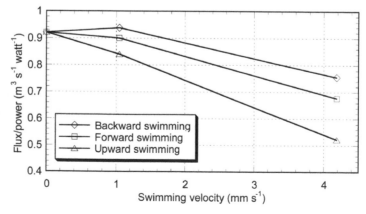

FIGURE 31.6 Ratio between the volumetric flux through the capture area and the power input as a function of swimming behaviors (swimming velocity and direction). (Drawn from the data first reported in Jiang et al.[33])

31.4 A Future Application

Some copepods create a feeding current all the time. They do have to stop sometimes — to groom their mouthparts, to jump to a new position within the water column, or to escape from a perceived danger. However, at all other times their mouthparts create the feeding current. Those are the copepods considered in Section 31.3. Other copepods beat the mouthparts for a period of time and then stop for a few seconds. (Sometimes the beating activity stops for less than a second, but there is a clear stop and the feeding current stops.) The on/off feeding current so created has been well documented.[12,14,20,34,35,]

Quite possibly, the copepods creating the on/off feeding current expend more energy because they have to accelerate the water every time they start. However, the on/off feeding current may enable the copepods to better detect prey and food particles via chemoreception and/or mechanoreception. With these sensory inputs, the on/off feeding current mode may be energetically more efficient (more food is captured even if it costs more energy). (Note that sensory inputs are neglected in quantifying the feeding efficiency in Section 31.3.2.3.) This kind of temporal partitioning of feeding activity ("on" or "off") was also observed to be dependent on food concentrations,[12] and on/off feeders can survive in lower food concentrations. Small-scale turbulence in aquatic systems can increase the perceived concentration of prey to predators.[37] In response to the higher encounter rates (i.e., higher perceived concentration of prey) due to small-scale turbulence of suitable intensity, copepods (*Centropages hamatus*) were observed to increase feeding activity as if they were experiencing altered prey concentrations.[20] Several questions arise: (1) What is the advantage of the on/off feeding current? (2) How does the on/off feeding current affect the transmission of chemical and/or mechanical signals associated with approaching prey and food particles? (3) What kinds of combination between the frequency of the on/off feeding current (more precisely the time duration for both "on" and "off" activities) and the frequency at which the copepod encounters a food particle (depending on food concentration and intensity of small-scale turbulence) will enable the copepod to maximize its feeding efficiency?

These questions can be answered by performing numerical experiments using the previously described framework of numerical simulation. Here, the time-dependent terms in Equation 31.1 have to be considered. An easy start is to simulate the on/off feeding current created by a tethered copepod. Therefore, there is no need to consider Equation 31.6. Thereafter, the simulation will be extended to free-swimming copepods.

Another problem is the possible dependence of the on/off feeding current on a copepod's body size, i.e., the size effect. From the frequency parameter β (defined in Section 31.2.1 after Equation 31.5), one can see that the effect of a given pattern of the on/off activity (i.e., given period and temporal partitioning between "on" and "off") is more prominent for a copepod of a larger body size. In other words, copepods of small body size are less likely to gain any possible benefit from creating an on/off feeding current. This size effect can also be investigated by performing numerical experiments.

31.5 Concluding Remarks

Owing to their huge size, weather phenomena are very difficult to reproduce in the laboratory. With advances in theories of atmospheric sciences, especially atmospheric dynamics, and with the emergence of high-speed computers and the development of sophisticated computational techniques, numerical experiments have been widely used in research on atmospheric sciences. Today, numerical weather prediction has become an everyday practice using the data flow from the global observational network. As an analogy, it is not easy to study or measure the copepod feeding currents in the laboratory, due to copepods' small size. Postprocessing the data is also time-consuming. However, experimental studies have already accumulated a huge database for copepod research. On the ground of this database and with mathematical formalism describing some processes taking place at the scale of individual copepods, a first set of numerical experiments[31–33,36,38] has been carried out. It is probable that in the near future numerical experiments on computers will become a powerful tool for studying zooplankton and their interactions with the environment. The key point is that numerical studies should be combined with experimental studies in an interactive way in which one complements the other.

Acknowledgments

The postdoctoral scholarship award to the author by the Woods Hole Oceanographic Institution (WHOI) is gratefully acknowledged. The author would like to thank Professor T. R. Osborn and Professor C. Meneveau for their guidance, encouragement, and many hours of discussion. The author acknowledges Professor J. R. Strickler for discussion on the on/off feeding current and for allowing the author to use his observational data in the chapter. The author thanks Dr. K. G. Foote and an anonymous reviewer for very helpful comments on the manuscript. This is Contribution Number 10765 from WHOI.

References

1. Strickler, J.R., Calanoid copepods, feeding currents, and the role of gravity, *Science*, 218, 158, 1982.
2. Conover, R.J., Zooplankton — life in a nutritionally dilute environment, *Am. Zool.*, 8, 107, 1968.
3. Emlet, R.B. and Strathmann, R.R., Gravity, drag, and feeding currents of small zooplankton, *Science*, 228, 1016, 1985.
4. Emlet, R.B., Flow fields around ciliated larvae: effects of natural and artificial tethers, *Mar. Ecol. Prog. Ser.*, 63, 211, 1990.
5. Bundy, M.H. and Paffenhöfer, G.-A., Analysis of flow fields associated with freely swimming calanoid copepods, *Mar. Ecol. Prog. Ser.*, 133, 99, 1996.
6. Barber, R.T. and Hilting, A.K., Achievements in biological oceanography, in *50 Years of Ocean Discovery, National Science Foundation 1950–2000*, Ocean Studies Board, National Research Council, 2000, 11–21.
7. Strickler, J.R., Observation of swimming performances of planktonic copepods, *Limnol. Oceanogr.*, 22, 165, 1977.
8. Alcaraz, M., Paffenhöfer, G.-A., and Strickler, J.R., Catching the algae: a first account of visual observations on filter-feeding calanoids, in *Evolution and Ecology of Zooplankton Communities*, Kerfoot, W.C., Ed., Am. Soc. Limnol. Oceanogr. Spec. Symp., Vol. 3, University Press of New England, Hanover, NH, 1980, 241.
9. Strickler, J.R., Feeding currents in calanoid copepods: two new hypotheses, in Physiological Adaptations of Marine Animals, Lavarack, M.S., Ed., *Symp. Soc. Exp. Biol.*, 23, 459, 1985.
10. Koehl, M.A.R. and Strickler, J.R., Copepod feeding currents: food capture at low Reynolds number, *Limnol. Oceanogr.*, 26, 1062, 1981.
11. Paffenhöfer, G.-A., Strickler, J.R., and Alcaraz, M., Suspension-feeding by herbivorous calanoid copepods: a cinematographic study, *Mar. Biol.*, 67, 193, 1982.
12. Cowles, T.J. and Strickler, J.R., Characterization of feeding activity patterns in the planktonic copepod *Centropages typicus* Kroyer under various food conditions, *Limnol. Oceanogr.*, 28, 105, 1983.
13. Price, H.J., Paffenhöfer, G.-A., and Strickler, J.R., Modes of cell capture in calanoid copepods, *Limnol. Oceanogr.*, 28, 116, 1983.
14. Strickler, J.R., Sticky water: a selective force in copepod evolution, in *Trophic Interactions within Aquatic Ecosystems*, Meyers, D.G. and Strickler, J.R., Eds., American Association for the Advancement of Science, Washington, D.C., 1984, 187–239.
15. Vanderploeg, H.A. and Paffenhöfer, G.-A., Modes of algal capture by the freshwater copepod *Diaptomus sicilis* and their relation to food-size selection, *Limnol. Oceanogr.*, 30, 871, 1985.
16. Price, H.J. and Paffenhöfer, G.-A., Capture of small cells by the copepod *Eucalanus elongates*, *Limnol. Oceanogr.*, 31, 189, 1986.
17. Paffenhöfer, G.-A. and Lewis, K.D., Perceptive performance and feeding behavior of calanoid copepods, *J. Plankton Res.*, 12, 933, 1990.
18. Jonsson, P.R. and Tiselius, P., Feeding behaviour, prey detection and capture efficiency of the copepod *Acartia tonsa* feeding on planktonic ciliates, *Mar. Ecol. Prog. Ser.*, 60, 35, 1990.

19. Tiselius, P. and Jonsson, P.R., Foraging behaviour of six calanoid copepods: observations and hydro-dynamic analysis, *Mar. Ecol. Prog. Ser.*, 66, 23, 1990.

20. Costello, J.H., Strickler, J.R., Marrasé, C., Trager, G., Zeller, R., and Freise, A.J., Grazing in a turbulent environment: behavioral response of a calanoid copepod, *Centropages hamatus*, *Proc. Natl. Acad. Sci. U.S.A.*, 87, 1648, 1990.

21. Marrasé, C., Costello, J.H., Granata, T., and Strickler, J.R., Grazing in a turbulent environment: energy dissipation, encounter rates, and efficacy of feeding currents in *Centropages hamatus*, *Proc. Natl. Acad. Sci. U.S.A.*, 87, 1653, 1990.

22. Yen, J., Sanderson, B., Strickler, J.R., and Okubo, A., Feeding currents and energy dissipation by *Euchaeta rimana*, a subtropical pelagic copepod, *Limnol. Oceanogr.*, 36, 362, 1991.

23. Yen, J. and Fields, D.M., Escape responses of *Acartia hudsonica* nauplii from the flow field of *Temora longicornis*, *Arch. Hydro. Beih.*, 36, 123, 1992.

24. Bundy, M.H., Gross, T.F., Coughlin, D.J., and Strickler, J.R., Quantifying copepod searching efficiency using swimming pattern and perceptive ability, *Bull. Mar. Sci.*, 53, 15, 1993.

25. Fields, D.M. and Yen, J., Outer limits and inner structure: the 3-dimensional flow fields of *Pleuromamma xiphias*, *Bull. Mar. Sci.*, 53, 84, 1993.

26. Kiørboe, T. and Saiz, E., Planktivorous feeding in calm and turbulent environments, with emphasis on copepods, *Mar. Ecol. Prog. Ser.*, 122, 135, 1995.

27. Saiz, E. and Kiørboe, T., Predatory and suspension-feeding of the copepod *Acartia-tonsa* in turbulent environments, *Mar. Ecol. Prog. Ser.*, 122, 147, 1995.

28. Yen, J. and Strickler, J.R., Advertisement and concealment in the plankton: what makes a copepod hydrodynamically conspicuous? *Invert. Biol.*, 115, 191, 1996.

29. Fields, D.M. and Yen, J., Implications of the feeding current structure of *Euchaeta rimana*, a carnivorous pelagic copepod, on the spatial orientation of their prey, *J. Plankton Res.*, 19, 79, 1997.

30. Kiørboe, T., Saiz, E., and Visser, A.W., Hydrodynamic signal perception in the copepod *Acartia tonsa*, *Mar. Ecol. Prog. Ser.*, 179, 97, 1999.

31. Jiang, H., Meneveau, C., and Osborn, T.R., Numerical study of the feeding current around a copepod, *J. Plankton Res.*, 21, 1391, 1999.

32. Jiang, H., Osborn, T.R., and Meneveau, C., The flow field around a freely swimming copepod in steady motion. Part I: Theoretical analysis, *J. Plankton Res.*, 24, 167, 2002.

33. Jiang, H., Meneveau, C., and Osborn, T.R., The flow field around a freely swimming copepod in steady motion. Part II: Numerical simulation, *J. Plankton Res.*, 24, 191, 2002.

34. Price, H.J. and Paffenhöfer, G.-A., Perception of food availability by calanoid copepods, *Arch. Hydro-biol. Beih. Erg. Limnol.*, 21, 115, 1985.

35. Price, H.J. and Paffenhöfer, G.-A., Effects of concentration on the feeding of a marine copepod in algal monocultures and mixtures, *J. Plankton Res.*, 8, 119, 1986.

36. Jiang, H., Osborn, T.R., and Meneveau, C., Chemoreception and the deformation of the active space in freely swimming copepods: a numerical study, *J. Plankton Res.*, 24, 495, 2002.

37. Rothschild, B.J. and Osborn, T.R., Small-scale turbulence and plankton contact rates, *J. Plankton Res.*, 10, 465, 1988.

38. Jiang, H., Osborn, T.R., and Meneveau, C., Hydrodynamic interaction between two copepods: a numerical study, *J. Plankton Res.*, 24, 235, 2002.

32

Can Turbulence Reduce the Energy Costs of Hovering for Planktonic Organisms?

Hidekatsu Yamazaki, Kyle D. Squires, and J. Rudi Strickler

CONTENTS

32.1 Introduction 493
32.2 Methods 494
    32.2.1 Flow Fields 494
        32.2.1.1 Direct Numerical Simulation 494
        32.2.1.2 Random Flow Simulation 496
    32.2.2 Planktonic Swimming Models 497
        32.2.2.1 Equation of Motion 497
        32.2.2.2 Velocity-Based Swimming Model 498
        32.2.2.3 Strain-Based Swimming Model 498
32.3 Results 499
    32.3.1 Generated Flow Fields 499
    32.3.2 Statistics 500
    32.3.3 Mean Swim Velocity V_{s2} 501
    32.3.4 Ambient Flow Field U_{f2} 502
    32.3.5 Particle Rising/Sinking Velocity ΔV_2 502
32.4 Discussion 503
Appendix 32.A: From Nondimensional Numbers to Dimensional Numbers 504
Acknowledgments 504
References 504

32.1 Introduction

Turbulence is one of the complex physical processes that play major roles in shaping the environment of planktonic organisms in oceans and lakes (Mann and Lazier, 1998; Yamazaki et al., 2002). It produces fluid motions at all spatial and temporal scales, from millimeters to kilometers and from milliseconds to days (Tennekes and Lumley, 1972). Several investigators have recognized the importance of this turbulent environment, which surrounds the planktonic organisms and enhances the encounter rates between predators and prey (i.e., Rothschild and Osborn, 1988; Costello et al., 1990; Marrasé et al., 1990; Browman, 1996; Osborn and Scotti, 1996; Yamazaki, 1996; Sundby, 1996; Strickler and Costello, 1996; Browman and Skiftesvik, 1996). Because of the interplay between turbulence and group properties of various components of the pelagic food web, both biological and physical variables exhibit spatial heterogeneity (Denman and Gargett, 1995).

Some years ago, rapid advances in computing power made it possible to solve the Navier–Stokes equations directly (called direct numerical simulation, DNS). It enhanced our understanding of the underlying structures of turbulent flows a great deal. One notable feature of the turbulence is the organized structures exhibited in both the velocity and the strain fields (e.g., Vincent and Meneguzzi, 1991). Yamazaki (1993) proposed that those organized structures could provide helpful information for planktonic organisms. Because these structures are of similar temporal and spatial scales as the plankton (Strickler, 1985) they could act, for example, as "landmarks" in the flow field. At that time, Yamazaki (1993) hypothesized that these organized structures help zooplankters find mates and detect prey and predators. However, to test his hypothesis we would need to know a large repertoire of behavioral responses of different zooplankters to different flow fields and then model the behaviors numerically — a daunting task even in these days of advanced computing techniques.

In this study, we concentrate on a more testable hypothesis. We ask the hypothetical question of whether zooplankters could use the information contained in a turbulent flow field to save the swimming cost. Because most planktonic organisms are negatively buoyant, they sink at their terminal speed unless they actively swim against gravity (e.g., Strickler, 1982). If an organism were to perceive the organized structures, especially the up-flowing ones, or some characteristic parameters of them, it may behave in such a way that it could result in reducing the cost of swimming against gravity. The organisms could hover at a preferred depth with a reduced energy output, similar to birds soaring in a thermal plume. We, therefore, formulate our hypothesis as follows:

> A planktonic organism can reduce the cost of hovering by making use of the local flow structures of turbulence.

To simplify the computations we assumed that the sizes of the organisms are of the order of 1 mm and we approximated their shapes by spheres. Although former studies (e.g., Yen et al., 1991; Jiang et al., 2002) quantified flows around a single copepod, the interaction between turbulence and the biologically generated flows is a considerably complicated problem. Here we assumed that the turbulence flow field is not altered by the presence of the organism. Also, we assumed that the organisms respond to the local flow structures with fixed action patterns (Lorenz, 1935). Our "experimental" design was to subject the "organisms" to two types of "flow," either a turbulent flow or a kinematically correct random flow field.

The first type of flow, the turbulent flow with its coherent structures, has been constructed using the DNS technique of Rogallo (1981). The second type of flow, referred to as the random flow simulation (RFS), has been constructed by observing that the kinematic condition is satisfied while the velocity field maintains the continuity condition. We, therefore, generated a flow field with a prescribed spectral shape, which satisfies the continuity constraint but is not a solution of the Navier–Stokes equations, and, therefore, does not possess the nonlinear interactions inherent in turbulent flows. In this case, we did not need to compute the Navier–Stokes equations, but time-advanced the flow field keeping the continuity condition.

We then introduced "organisms" to these two flow fields and computed their motions in response to local flow conditions. Among the literature on the reactions of zooplankters to flow fields, there are two schools of thought (e.g., Kiørboe and Visser, 1999). Real organisms could react to the velocity field and changes therein (e.g., Hwang, 1991; Hwang et al., 1994; Hwang and Strickler, 1994, 2001), or they could react to the strain field and changes therein (e.g., Strickler and Bal, 1973; Strickler, 1975; Zaret, 1980). Thus, we constructed two different planktonic swimming models, one, referred to as the velocity-based swimming model (VBS), and the other, referred to as the strain-based swimming model (SBS).

In the following section, first we present how we computed DNS and RFS, and then discuss the construction of the two swimming models. The results and discussion are presented in the last section.

32.2 Methods

32.2.1 Flow Fields

32.2.1.1 Direct Numerical Simulation — The incompressible Navier–Stokes equations are solved making use of the pseudo-spectral method of Rogallo (1981). The dependent variables are represented as Fourier series expansions using 48^3 collocation points on a periodically reproduced cubic domain.

TABLE 32.1

Flow Field Parameters Used in the DNS Model

ν	4.49×10^{-3}	τ_e	4.51	λ	0.629
L_f	1.138	η	5.40×10^{-2}	N	48
L_g	0.575	τ_η	0.649	τ_F	1.12
$q^2/2$	9.56×10^{-2}	v_η	8.33×10^{-2}	σ_F^2	2.835×10^{-5}
ε	1.08×10^{-2}	$k_{max}\eta$	1.22	K_F	$2\sqrt{2}$
u_{rms}	0.252	Re_λ	35	N_F	92

Note: The number shows nondimensional values based on the terminal sinking speed.

The flow is made statistically stationary by adding a stirring force to the low wavenumber components of the velocity using the scheme developed by Eswaran and Pope (1988). In this method, an artificial forcing term is specified as a complex, vector-valued Uhlenbeck–Ornstein (UO) stochastic process. Three input parameters are necessary to specify the stirring force: the radius k_F of forced modes, the amplitude of the forcing σ_F^2, and the integral timescale τ_F. The radius k_F determines the number of modes that are subjected to forcing. Following Yeung and Pope (1989), a radius $k_F = 2\sqrt{2}$ is used, corresponding to 92 forced modes. This value is recommended by Yeung and Pope (1989) as a compromise between contamination of a large range of low wavenumber modes by the forcing or having too few modes subjected to forcing, which may result in an unacceptable variability in flow field statistics. The forcing amplitude σ_F^2 is chosen so that the resulting flow has desired values of the Taylor microscale Reynolds number, $Re_\lambda = u_{rms}\lambda / \nu$ and Kolmogorov length scale, $\eta = (\nu^3 / \varepsilon)^{1/4}$. For an isotropic turbulence the Taylor microscale λ is related to the mean dissipation rate ε as

$$\lambda = \left(\frac{15\nu u'^2}{\varepsilon} \right)^{1/2} \tag{32.1}$$

where ν is the fluid kinematic viscosity, and $u_{rms} = \sqrt{q^2 / 3}$ is the root-mean-square (rms) velocity fluctuation. Twice the turbulent kinetic energy, q^2, and the mean dissipation rate, ε, are related to the three-dimensional energy spectrum, $E(k)$, by

$$q^2 = \int_0^{k_{max}} E(k)dk, \qquad \varepsilon = 2\nu \int_0^{k_{max}} k^2 E(k)dk \tag{32.2}$$

where k_{max} is the maximum resolved wave number in the simulation and is dependent on the de-aliasing scheme used for the nonlinear terms. In the pseudo-spectral method of Rogallo (1981), $k_{max} = \sqrt{2}N / 3$. The forcing timescale τ_F, which is prescribed as a fraction of the eddy turnover time, $\tau_e = L_f/u_{rms}$, so that the ratio τ_F/τ_e, is in the range 0.1 to 1.0 (e.g., Yeung and Pope, 1989), where the longitudinal integral length scale of the turbulence, obtained from integration of the two-point correlation, is denoted L_f.

Small-scale resolution of the turbulence is characterized by the parameter $k_{max}\eta$. A value $k_{max}\,\eta > 1$ is sufficient for adequate resolution of lower-order statistics (Eswaran and Pope, 1988). From an initial condition the flow evolves to a stationary condition. In the statistically stationary portion of the calculation, $k_{max}\eta$ and the Taylor-microscale Reynolds number are 0.63 and 35, respectively. The forcing parameters and relevant flow field statistics are given in Table 32.1. In this table, L_g is the transverse integral length scale. The other statistical quantities shown are the Kolmogorov time and velocity scales, which are defined as $\tau_\eta = (\nu/\varepsilon)^{1/2}$ and $v_\eta = (\nu\varepsilon)^{1/4}$, respectively. We have checked the isotropy of the simulated flow by comparing the distribution of turbulent kinetic energy along the coordinate axes of mutually orthogonal solenoidal unit vectors, $(e_1(k), e_2(k))$, defined as

$$e_1(k) = \frac{k \wedge z}{|k \wedge z|} \tag{32.3}$$

$$e_2(k) = \frac{k \wedge e_1(k)}{|k \wedge e_1(k)|} \tag{32.4}$$

where $\wedge$ is the vector product operator and $\mathbf{z}$ is a unit vector along one of the Cartesian coordinate axes (e.g., Curry et al., 1984). The fluid kinetic energy decomposed along $\mathbf{e}_1$ and $\mathbf{e}_2$ is obtained from the following scalar product:

$$\varphi_1(t) = <\left|\mathbf{e}_1 \cdot \hat{\mathbf{u}}(\mathbf{k},t)\right|^2 > \tag{32.5}$$

$$\varphi_2(t) = <\left|\mathbf{e}_2 \cdot \hat{\mathbf{u}}(\mathbf{k},t)\right|^2 > \tag{32.6}$$

where the angle brackets denote a volume average and $\hat{\mathbf{u}}(\mathbf{k},t)$ is the Fourier coefficient at wavenumber $\mathbf{k}$. For an isotropic turbulence $\varphi_1 = \varphi_2$ and, therefore, a measure of departure from isotropy is given by

$$I = \sqrt{\varphi_1 / \varphi_2} \tag{32.7}$$

For the calculations presented in this chapter, $I = 1.05$. As also shown in Table 32.1, $L_f/L_g = 1.98$ is only slightly smaller than the isotropic value of 2. The discrepancy from the isotropic value arises from statistical errors in computing the time averages as well as from the imposition of periodic boundary conditions.

32.2.1.2 *Random Flow Simulation* — Random flow fields were generated as a linear combination
of Fourier modes. The construction of the random flows is very similar to the method used by Rogallo (1981) to prescribe initial conditions for DNS of homogeneous turbulence. The Fourier coefficients of the velocity field $\mathbf{u}(\mathbf{k},t)$ are given by

$$\hat{u}_i(\mathbf{k},\omega) = \exp(j\omega t)\left[\phi_1(\mathbf{k})\mathbf{e}_i^{(1)} + \phi_2(\mathbf{k})\mathbf{e}_i^{(2)}\right], \quad i = 1,2,3 \tag{32.8}$$

where $\mathbf{e}_i^{(1)}$ and $\mathbf{e}_i^{(2)}$ are unit basis vectors defined along mutually orthogonal axes perpendicular to the wavenumber $\mathbf{k}$, j is the square root of -1, and ω is the phase. The components ϕ_1 and ϕ_2 are constructed as projections of the velocity field onto the $\mathbf{e}^{(1)}$ and $\mathbf{e}^{(2)}$ basis to ensure that the velocity field is divergence free. The prescription of ϕ_1 and ϕ_2 takes the form:

$$\phi_1(k) = \frac{1}{k_\perp}\left[k_2\frac{a(k)kk_2 + b(k)k_1k_3}{kk_\perp} - k_1\frac{b(k)k_2k_3 - a(k)kk_1}{kk_\perp}\right] \tag{32.9}$$

$$\phi_2(k) = b(k) \tag{32.10}$$

where $k = \left(k_1^2 + k_2^2 + k_3^2\right)^{1/2}$ is the magnitude of the wavenumber vector $\mathbf{k}$ and $k_\perp = \left(k_1^2 + k_2^2\right)^{1/2}$. The above expression holds for $k \neq 0$ and, when either of these wavenumbers is zero, analogous expressions can be derived. The complex-valued coefficients $a(k)$ and $b(k)$ are given by

$$a(k) = \left(\frac{E(k)}{4\pi k^2}\right)^{1/2}\exp(j\theta_1)\cos\psi \tag{32.11}$$

$$b(k) = \left(\frac{E(k)}{4\pi k^2}\right)^{1/2}\exp(j\theta_2)\sin\psi \tag{32.12}$$

where $E(k)$ is the three-dimensional energy spectrum, and θ_1, θ_2, and ψ are uniformly distributed random numbers on the interval $(0,2\pi)$. The coefficients $a(k)$ and $b(k)$ are random in phase and are subject to the constraint that the energy in each mode is required to have the expected value:

$$\left\langle u_n^2\right\rangle = \int\left\langle\hat{\mathbf{u}}\cdot\hat{\mathbf{u}}^*\right\rangle dA = E(k) \tag{32.13}$$

The spectrum $E(k)$ in Equation 32.13 used to obtain the Fourier modes is a curve-fit to the DNS result. The three-dimensional energy spectrum $E(k)$ from the DNS is shown in Figure 32.1 and is compared with the experimental measurements from Comte-Bellot and Corrsin (1971). When normalized using

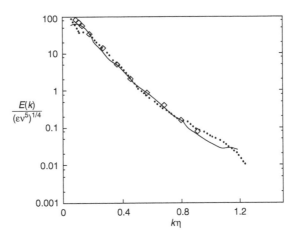

FIGURE 32.1 Radial energy spectrum. Turbulent flow computed by DNS (solid line); random flow (dotted line); Comte-Bellot and Corrsin (1971) (open boxes).

the Kolmogorov scales the experimental measurements and the DNS results are in good agreement. Finally, the temporal frequency ω in the random flows is chosen from a Gaussian distribution with a mean $\bar{\omega} = \varepsilon^{1/3} k^{2/3}$ and standard deviation $\sigma_k = \varepsilon^{1/3} k^{2/3}$ (Fung et al., 1992).

32.2.2 Planktonic Swimming Models

32.2.2.1 Equation of Motion —

Although planktonic organisms come in many sizes and shapes, we have assumed the target organisms have a spherical ridged body with a diameter of 1 mm. We also assumed that the swimming speeds of the organism are of the same order of magnitude as the terminal sinking velocities. Any heavier-than-water "organism" requires a swimming speed that at least exceeds the sinking speed in order to maintain a preferred depth within the water column. The specific density of planktonic organisms is, in general, slightly larger than that of the surrounding water (Hutchinson, 1967). In this study, we assumed the organism has a 3% larger density than the ambient water. The equation of motion for a small rigid sphere moving through a nonuniform flow field can be used by neglecting the Besset history terms and Faxen correction terms (Maxey and Riley, 1983). The equation is written as follows:

$$\frac{d\mathbf{v}}{dt} = \alpha\left[\mathbf{u} - \mathbf{v} + \mathbf{w}_s + \mathbf{v}_s\right] + \frac{3}{2} R \frac{D\mathbf{u}}{Dt} \tag{32.14}$$

where $\mathbf{v}$ is the particle velocity, $\mathbf{u}$ is the fluid velocity along the particle path, $\mathbf{w}_s$ is the Stokes settling velocity, and $\mathbf{v}_s$ is swimming velocity of the organism. Gravity is considered to act along the y-axis (the index number 2). The parameters in Equation 32.14 are the Stokes settling velocity, $\mathbf{w}_s$, the inertia parameter α, and mass ratio R,

$$\mathbf{w}_s = \frac{\left(m_p - m_f\right)g}{6\pi r\mu} \tag{32.15}$$

$$\alpha = \frac{6\pi r\mu}{m_p + \frac{1}{2}m_f} \tag{32.16}$$

$$R = \frac{m_f}{m_p + \frac{1}{2}m_f} \tag{32.17}$$

where r is the particle radius, μ is the fluid viscosity, m_p is the particle mass, m_f is the displaced fluid mass, and g is acceleration due to gravity. For the case of pure water with $2r = 10^{-3}$ m, the settling velocity is 1.6×10^{-2} m s^{-1}.

Equation 32.14 is time-advanced using a second-order Runga-Kutta scheme with the modification for particles having small inertia (Lamirini, 1987). Particles are introduced into the calculation following the development of a statistically stationary condition in the fluid. A total of 24,000 particles are then randomly seeded throughout the computational box.

We now proceed to model the swimming velocity vector, v_s, of the planktonic organism. We need to make a large assumption for the construction of the model; namely, the organism makes a behavioral decision based on the local flow structure with a given simple rule. Our objective of the swimming is to make use of the existing flow field in order to conserve the hovering swimming energy; thus the effectiveness of the model is measured with the amount of the upward lifting. Flow structures exhibit coherency due to either the kinematic condition (continuity) or the dynamic condition (the nonlinearity of the Navier–Stokes equations). The RFS focuses on former reason, and the DNS enables us to look at the nonlinear effects. Recent DNS studies show the organized structures in the strain rate field (e.g., Vincent and Meneguzzi, 1991; Jiménez, 1992; Jiménez et al., 1993, Jiménez and Wray, 1998). Thus, in this study, two different flow structures are used to model the swimming patterns. The first is based on the local velocity vectors, and the other makes use of the local strain tensor.

32.2.2.2   *Velocity-Based Swimming Model* — In the velocity-based swimming model (VBS), the swimming velocity is aligned with the local fluid velocity. The motivation of this model is to move the particle faster than its own swimming speed. In regions where the local velocity is "upward," i.e., against gravity, this scenario possesses the advantage that the particle is assisted in its upward motion. In regions of "down-flow," i.e., the fluid velocity has its vertical component along the gravity direction, a swimming velocity with the opposite sense to the local flow increases the probability that the particle will travel to an upflow region. Mathematically, the velocity-based model may be written as

$$v_{s,i} = b_i \cos\theta_i \quad i = 1,2,3 \tag{32.18}$$

where $v_{s,i}$ is the ith component of the swimming velocity and $\cos\theta_i = u_i / |u_i|$ is the direction cosine of the fluid velocity along the particle path. The factor b_i is a set of random numbers, scaled such that the average speed is a prescribed value. The swimming velocities in the velocity-based model are sampled from a uniform distribution with orientation sensitive to the local flow. Note that in "downward" flowing regions the sign of the vertical component calculated from Equation 32.18 is reversed to maintain continually upward swimming.

32.2.2.3   *Strain-Based Swimming Model* — For the strain-based swimming model (SBS), we related the swimming pattern to the coherent structures exhibited in the strain field. From previous investigations (Siggia, 1981; Kerr, 1985; Vincent and Meneguzzi, 1991; Jiménez and Wray, 1998) vortex tubes are the dominant small-scale structure in many classes of turbulent flow, and numerical results show that the vorticity vector is preferentially aligned with the eigenvector corresponding to the intermediate value of the strain rate tensor. We imagined that a vortex tube is like a thermal plume in the atmosphere; thus we forced the particle to move along the direction of the tube. Lund and Rogers (1994) proposed a nondimensional parameter to characterize the intermediate eigenvector with the following parameter, s^*,

$$s^* = \frac{-3\sqrt{6}\alpha\beta\gamma}{\left(\alpha^2 + \beta^2 + \gamma^2\right)^{3/2}} \tag{32.19}$$

where the quantities α, β, and γ are the minimum, intermediate, and maximum eigenvalues, respectively. For incompressible flow, $\alpha + \beta + \gamma = 0$, and α must always be positive, γ must always be negative, and β can be either positive or negative. The parameter s^* is defined such that it ranges between ± 1.

The components of the strain-based swimming model take the form:

$$v_{s,i} = bs * \cos\theta_i \tag{32.20}$$

where the factor b is used to scale the average magnitude of the swimming vector to a desired value, and the scale is adjusted to the same level as in the velocity-based model to facilitate comparison of the results obtained using the two models. Another parameter $\cos\theta_i$ is the direction cosine of the intermediate principal axis of the strain rate tensor. As shown by Equation 32.20, the magnitude of the swimming vector is proportional to $s*$ while the orientation is dictated by the intermediate principal axis of the strain rate. There are numerous possible choices for the orientation of the swimming vector. We use the direction cosine of the intermediate strain principal axis that has a tendency for alignment with the vorticity. The direction cosine is biased against gravity, but the sense of the swimming velocity can be either aligned along or against gravity, as $s*$ takes on both positive and negative values.

Finally, for both the strain- and velocity-based models, it is assumed that the particle maintains its swimming speed for a prescribed period, τ_s. In the calculations presented in this work, the swimming period τ_s is equal to the Kolmogorov timescale. Calculations using different τ_s show a negligible dependence on the swimming period.

32.3 Results

32.3.1 Generated Flow Fields

Both DNS and RFS show good agreement between the three-dimensional spectra of the flow fields. As we mentioned earlier, the isotropy index, I, which is defined in Equation 32.7, is 1.05, and the ratio of the longitudinal and the transversal correlation length scale is 1.98. These values suggest that turbulence generated in DNS is close to an isotropic state. Because the power spectra for DNS and RFS are of the same level, the kinetic energy of the generated flow field is also of the same level. The difference between DNS and RFS occurs in the phase of each Fourier component.

The state of the strain field can be characterized by the $s*$ defined in Equation 32.19. The parameter $s*$ is basically the product of three eigenvalues of the strain rate. The sign of this value indicates whether the flow is axisymmetric expanding or contracting. Shown in Figure 32.2 is the distribution of $s*$ from both the DNS and the RFS considered in this work. As shown in Figure 32.1, both the turbulent and random flows have essentially the same energy spectrum. It is clear from Figure 32.2, however, that the small-scale features of the flow as quantified using $s*$ are very different. The values obtained from DNS are mostly positive, whereas the corresponding values from RFS are only slightly skewed toward positive. This means that the isotropic turbulence has a preferred strain state, i.e., axisymmetric expansion. But the same tendency in RFS is quite weak; thus this tendency is due mostly to the dynamic condition of flow stemming from the nonlinear dynamics. However, it is important to note that the positive skewness is not entirely due to the dynamic condition, because the $s*$ from RFS is not totally uniform.

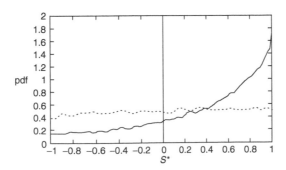

FIGURE 32.2 Probability density of $s*$. Turbulent flow computed by DNS (solid line); random flow (dotted line).

32.3.2   Statistics

A total of 24,000 particles were seeded into the flow fields. The particles were divided into six groups containing 4000 particles each. For each group the orientation of gravity was changed in two opposite directions for each three coordinates. The different orientations of gravity prevented any bias from a single direction, and improved the convergence of the ensemble averages using the triplex averaging procedure (Wang and Maxey, 1993). The six orientations of the gravity vector, $\mathbf{g} / |\mathbf{g}|$, in the simulations are $(0,-1,0)$, $(0,1,0)$, $(-1,0,0)$, $(1,0,0)$, $(0,0,-1)$, and $(0,0,1)$, and are denoted as $\mathbf{e}^{(m)}$ with $m = 1,2,\ldots,6$.

We now consider the mean and rms values of several quantities in order to examine the difference between the response of the particles in the turbulent and random flows. The change in the rising/settling velocity relative to Stokes terminal velocity was calculated from its definition:

$$\Delta v_i^m = \left\langle v_i - w_{s,i} \right\rangle^{(m)} \quad i = 1,2,3 \tag{32.21}$$

where (m) indicates six different orientations, and the angle bracket denotes the averaging operator for all particles. The rms fluctuating rising/settling velocity relative to the settling velocity is

$$\sigma_{v_i}^{(m)} = \left[\left\langle \left(v_i - w_{s,i} \right)^2 \right\rangle - \left(\Delta v_i \right)^2 \right]^{1/2(m)} \tag{32.22}$$

The mean, $V_{s,i}^{(m)}$, and rms swimming velocity, $\sigma_{s,i}^{(m)}$, are defined as

$$V_{s,i}^{(m)} = \left\langle v_{s,i} \right\rangle^{(m)} \tag{32.23}$$

$$\sigma_{si}^{(m)} = \left[\left\langle \left(v_{s,i} - V_{s,i} \right)^2 \right\rangle \right]^{1/2(m)} \tag{32.24}$$

The mean, $U_{f,i}^{(m)}$, and the rms fluid velocity, $\sigma_{f,i}^{(m)}$, along the particle path are

$$U_{f,i}^{(m)} = \left\langle u_i \right\rangle^{(m)} \tag{32.25}$$

$$\sigma_{fi}^{(m)} = \left[\left\langle \left(u_i - U_{f,i} \right)^2 \right\rangle \right]^{1/2(m)} \tag{32.26}$$

The mean rising/settling velocity and the rms particle fluctuating velocity along the direction of gravity are

$$\Delta V_2 = \frac{1}{6} \sum_{m=1}^{6} \Delta v_i^{(m)} \cdot \mathbf{e}^{(m)} \tag{32.27}$$

$$\sigma_{v2} = \frac{1}{6} \left(\sigma_{v2}^1 + \sigma_{v2}^2 + \sigma_{v1}^3 + \sigma_{v1}^4 + \sigma_{v3}^5 + \sigma_{v3}^6 \right) \tag{32.28}$$

In the direction orthogonal to gravity, the mean rising/settling velocity and the rms particle fluctuating velocity are

$$\Delta V_1 = \frac{1}{12} \left(\Delta v_1^1 + \Delta v_3^1 + \Delta v_1^2 + \Delta v_3^2 + \Delta v_2^3 + \Delta v_3^3 + \Delta v_2^4 + \Delta v_3^4 + \Delta v_1^5 + \Delta v_2^5 + \Delta v_1^6 + \Delta v_2^6 \right) \tag{32.29}$$

$$\sigma_{v1} = \frac{1}{12} \left(\sigma_{v1}^1 + \sigma_{v3}^1 + \sigma_{v1}^2 + \sigma_{v3}^2 + \sigma_{v2}^3 + \sigma_{v3}^3 + \sigma_{v2}^4 + \sigma_{v3}^4 + \sigma_{v1}^5 + \sigma_{v2}^5 + \sigma_{v1}^6 + \sigma_{v2}^6 \right) \tag{32.30}$$

Finally, the third level of averaging is done over time. The initial transients are discarded and only the statistically stationary portion of particle motion is included in the time average. The statistical error $e_{s,d}$ for an averaged quantity (e.g., ΔV_2, ΔV_1, etc.) is estimated by

$$e_{s,d} = 2\sigma \left(\Im / T \right)^{1/2} \tag{32.31}$$

TABLE 32.2

Mean and rms Particle Velocity, Swimming Velocity, and Fluid Velocity along the Particle Trajectory in the Direction Parallel to Gravity

Run	Flow Type	Swimming Model	ΔV_2	V_{s2}	U_{f2}	σ_2	σ_{s2}	σ_{f2}
TS	DNS	SBS	−0.921	−0.638	−0.283	1.90	0.97	1.78
RS	RFS	SBS	−0.242	−0.099	−0.141	1.93	1.18	1.73
TV	DNS	VBS	−0.399	−0.781	0.382	1.49	0.68	1.51
RV	RFS	VBS	−0.576	−0.836	0.255	1.69	0.66	1.64

Note: Numbers are normalized by the terminal velocity.

TABLE 32.3

Mean and rms Particle Velocity, Swimming Velocity, and Fluid Velocity along the Particle Trajectory in the Direction Orthogonal to Gravity

Run	Flow Type	Swimming Model	ΔV_1	V_{s1}	U_{f1}	σ_1	σ_{s1}	σ_{f1}
TS	DNS	SBS	−0.006	-2.36×10^{-4}	−0.006	2.02	1.16	1.78
RS	RFS	SBS	−0.038	-3.28×10^{-2}	−0.005	1.92	1.19	1.72
TV	DNS	VBS	−0.013	-6.24×10^{-4}	−0.011	1.65	1.12	1.68
RV	RFS	VBS	0.002	-4.47×10^{-3}	−0.004	1.76	1.11	1.73

Note: Numbers are normalized by the terminal velocity.

where σ is the rms particle fluctuating velocity, $\Im = \int_0^\infty \rho(\tau)d\tau$ is the integral timescale computed from an autocorrelation coefficient $\rho(\tau)$, and T is the total integration time (e.g., Tennekes and Lumley, 1972). The statistical error for the mean swimming velocity is $10^{-4} v_\eta$ and the error in the average rising/settling velocity is $10^{-3} v_\eta$.

32.3.3 Mean Swim Velocity V_{s2}

The VBS model always forces the particle to take the swimming velocity vector against gravity, so the swimming vector in the direction parallel to gravity is always a negative value. The negative value is against gravity. The mean and the standard deviation are nearly the same for this single sign random variable, V_{s2} (Table 32.2). However, the swimming vector for the SBS model can point either against gravity or with gravity, although more values are biased toward against gravity. The mean swimming vector in the direction of gravity is negative for the DNS case, but almost zero for the RFS case. No bias in horizontal direction was observed in computed statistics (Table 32.3).

The scaling parameter b for both models is adjusted so that the mean value of the swimming vector is 1.75 times the particle settling velocity. If a particle swims upward against gravity in still water, the particle moves upward at the speed of 0.75 times the settling velocity. Because particles are responding to the flow structures according to the proposed swimming models, the direction of the swimming vector is not always against gravity. The velocity-based model for both RFS and DNS show the upward component of the swimming vector is roughly 80% of the settling velocity.

The SBS model shows a significant difference in the swimming vector for two different flow fields (Figure 32.3). The RFS does not show much bias toward a single direction, whereas the DNS shows about 60% of the settling velocity in the direction against gravity. The reason the RFS shows nearly close to zero values is that the flow structure generated by RFS does not exhibit much coherency in the strain field. The magnitude of the coherency in the strain field for DNS is quite different from RFS. This reflects the difference in both swimming velocity components. The strain-based model spends 36% of the total velocity vector in the direction against gravity. For the vertical velocity component, the velocity-based model forces more effort in the vertical component.

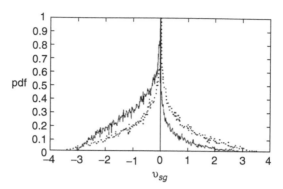

FIGURE 32.3 Probability density function of the particle swimming velocity along the direction of gravity, $v_{sg} = \frac{1}{6}\sum_{m=1}^{6} \mathbf{v}_s \cdot \mathbf{e}^{(m)}$. The particle swimming is based on SBS swimming. Turbulent flow (solid line); random flow (dotted line).

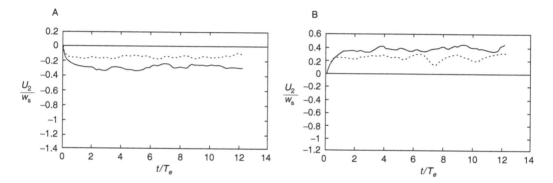

FIGURE 32.4 Time development of gravity component of mean fluid velocity. (A) SBS model; (B) VBS model. Turbulent flow (solid line); random flow (dotted line).

32.3.4   Ambient Flow Field U_{f2}

A significant difference is obtained in the ambient flow field that each particle experiences. The velocity-based model "sees" more downward flow field than upward flow field in both RFS and DNS (Figure 32.4). But the strain-based model sees the upward flow field more often than the downward flow field. We were puzzled about the velocity-based model results at the beginning, because we were supposed to force particles to swim in an upward direction in the upward flow field. Actually this acts in a negative fashion. Because the flow field does not have any preferred orientation, other than the gravity direction, the probability of seeing the upward flow field and the downward flow field is the same. But we are forcing particles to run through the upward portion faster than the downward portion. As a result, the particles spend more time in the downward portion of flow.

On the other hand, the strain-based model spends more time in the upward portion of flow. This is because the coherency in the flow structure is indeed identified more effectively from the strain field. In other words, by detecting a certain feature of the strain field *a modeled plankter can detect an upward orientation of flow structure effectively*. Interesting enough, the total amounts of the upward flow component in both RFS and DNS are not much different. Hence, although the coherency due to the kinetic condition is small, the flow field still exhibits a certain level of coherency, which is useful to identify the upward portion of flow.

32.3.5   Particle Rising/Sinking Velocity ΔV_2

The combination of the previous swimming effort and the flow field makes a large difference in the upward movements (Figure 32.5). The strain-based model for DNS shows nearly the same magnitude

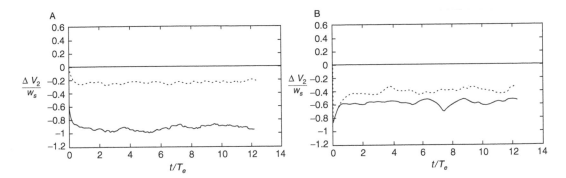

FIGURE 32.5 Time development of gravity component of mean particle rise velocity relative to terminal velocity; see Equation 32.21. (A) SBS model; (B) VBS model. Turbulent flow (solid line); random flow (dotted line).

in ΔV_2 with the settling velocity. In other words, organisms are moving upward against gravity at about twice as fast as the terminal sinking speed. The strain-based model for RFS is about 120% of the settling velocity. The velocity-based model shows almost the same amount of the vertical rising velocity for both DNS and RFS. The particles are moving against gravity at roughly 150% of the settling velocity.

Because these values are the mean of each random variable, we must consider the variability exhibited by each random variable. The standard deviation of each case varies from the minimum 1.49 for the velocity-based model with DNS to the maximum 1.93 for the strain-based model with RFS. Some fraction of particles is actually going with the direction of gravity, because they are trapped in a downward flow domain. What this implies is that for a given short duration of time a particle may not move against gravity, but if it waits long enough the net result will be always be against gravity.

As we forced the models to maintain 1.75 times the settling velocity for the average swimming speed, the average rise speed should exceed 75% of the settling velocity if the model is taking advantage of the existing flow structures. Both velocity-based models do not achieve this minimum threshold. However, the strain-based model for DNS shows a rise speed that exceeds the minimum threshold.

32.4 Discussion

Our hypothesis was that a planktonic organism could reduce the cost of hovering by making use of the local flow structures of turbulence. We tested this hypothesis with numerical simulations, two for the flow fields, and two for swimming behaviors. Our results based on combining flow fields with swimming behaviors show that in turbulent waters negatively buoyant zooplankters reacting to the local strain field could conserve energy while maintaining their position within the water column. For this case the mean rise speed exceeds the corresponding one in a still water case and our hypothesis proves to be true, at least theoretically.

The organisms employing this strategy would need a sensory system, which would allow them to perceive the direction of gravity and the principal component of the flow strain field. For crustacean zooplankton, for example, both sensory inputs would be perceived via the mechanoreceptors on their antennules and other body parts (e.g., Strickler and Bal, 1973; Strickler, 1975, 1982; Kiørboe and Visser, 1999). Because these, and many other reports and their experiments, show that crustacean zooplankton (e.g., calanoid and cyclopoid copepods) do perceive gravity and the local strain field, one would wonder whether these organisms may take advantage of turbulence to counteract their sinking speed due to their negative buoyancy. For the moment, we recognize that such an adaptive strategy can *theoretically* exist. Further investigation with live organisms may confirm whether at least some zooplankters employ this strategy.

Appendix 32.A: From Nondimensional Numbers to Dimensional Numbers

DNS is usually performed in a nondimensional fashion. The numbers shown in this study are also nondimensionalized using the terminal sinking speed of the target organism as a reference scale. When one specifies the dimensional and the viscosity of fluid, the nondimensional numbers automatically scale to the dimensional equivalences. Suppose a surface mixing layer whose dissipation rate (ε_*) is 10^{-6} W Kg^{-1} (m^2 s^{-3}). And we assume the viscosity of the sea water (ν_*) is 10^{-6} m^2 s^{-1}. The corresponding Kolmogorov scale, η_*, is 10^{-3} m. These dimensional numbers are expressed in terms of two scale ratios that are the length scale, l_*, and t_*. Hence, the following relations are derived:

$$\nu_* = \nu\, l_*^2\, t_*^{-1}$$

$$\varepsilon_* = \varepsilon\, l_*^2\, t_*^{-3}$$

Thus, l_* is 0.0186 and t_* is 1.55. For instance, the dimensional velocity scale q_* is 5.25×10^{-3} m s^{-1}. In this study, we assumed that the diameter of the organism is 10^{-3} m and the density difference between the organism and the water is 3%. These conditions set the dimensional terminal velocity (w_*) and the value is 1.6×10^{-2} m s^{-1}.

Acknowledgments

We express our appreciation to A. Abib for his patient work running the numerical codes. We also thank L.R. Haury for his useful comments. This work was supported by NSF Grant OCE-9409073, ONR Grant. N00014-92-1653. H.Y. was also supported by a Grant-in-Aid for Scientific Research (C) 10045026.

References

Browman, H.I., Predator–prey interactions in the sea: commentaries on the role of turbulence. *Mar. Ecol. Prog. Ser.,* 139, 301, 1996.

Browman, H.I. and Skiftesvik, A.B., Effects of turbulence on the predation cycle of fish larvae: comments on some of the issues. *Mar. Ecol. Prog. Ser.,* 139, 309, 1996.

Comte-Bellot, G. and Corrsin, S., Simple Eulerian time correlation of full- and narrow-band velocity signals in grid-generated, isotropic turbulence. *J. Fluid Mech.,* 48, 273, 1971.

Costello, J.H., Marrasé, C., Strickler, J.R., Zeller, R., Freise, A.J., and Trager, G., Grazing in a turbulent environment: I. Behavioral response of a calanoid copepod, *Centropages hamatus. Proc. Nat. Acad. Sci. U.S.A.,* 87, 1648, 1990.

Curry, J.H., Herring, J.R., Loncaric, J., and Orszag, S.A., Order and disorder in two- and three-dimensional Bènard convection. *J. Fluid Mech.,* 147, 1, 1984.

Denman, K.L. and Gargett, A.E., Biological–physical interaction in the upper ocean: the role of vertical and small scale transport processes. *Annu. Rev. Fluid Mech.,* 27, 225, 1995.

Eswaran, V. and Pope, S.B., An examination of forcing in direct numerical simulations of turbulence. *Comput. Fluids,* 16, 257, 1988.

Fung, J.C.H., Hunt, J.C.R., Malik, N.A., and Perkins, R.J., Kinematic simulation of homogeneous turbulence by unsteady random Fourier modes. *J. Fluid Mech.,* 236, 281, 1992.

Hutchinson, G.E., *A Treatise on Limnology,* Vol. 2. Wiley, New York, 1967, 1116 pp.

Hwang, J.S., Behavioral Responses and Their Role in Prey/Predator Interactions of a Copepod. Ph.D. thesis, Boston University, Boston, 1991, 165 pp.

Hwang, J.S. and Strickler, J.R., Effects of periodic turbulent events upon escape responses of a calanoid copepod, *Centropages hamatus. Bull. Plankton Soc. Jpn.,* 41, 117, 1994.

Hwang, J.S. and Strickler, J.R., Can copepods differentiate prey from predator hydromechanically? *Zool. Stud.,* 40, 1, 2001.

Hwang, J.S., Costello, J.H., and Strickler, J.R., Copepod grazing in turbulent flow: elevated foraging behavior and habituation of escape responses. *J. Plankton Res.*, 16, 421, 1994

Jiang, H., Osborn, T.R., and Meneveau, C., The flow around a freely swimming copepod in stead motion: Part I. Theoretical analysis. *J. Plankton Res.*, 24, 167, 2002.

Jiménez, J., Kinematic alignment effects in turbulent flows. *Phys. Fluids A*, 4, 652, 1992.

Jiménez, J. and Wray, A.A., On the characteristics of vortex filaments in isotropic turbulence. *J. Fluid Mech.*, 373, 255, 1998.

Jiménez, J., Wray, A.A., Saffman, P.G., and Rogallo, R.S., The structure of intense vorticity in isotropic turbulence. *J. Fluid Mech.*, 255, 65, 1993.

Kerr, R.M., Higher order derivative correlation and the alignment of small scale structures in isotropic numerical turbulence. *J. Fluid Mech.*, 153, 31, 1985.

Kiørboe, T. and Visser, A.W., Predator and prey perception in copepods due to hydromechanical signals. *Mar. Ecol. Prog. Ser.*, 179, 81, 1999.

Lamirini, A., Mise en oeuvre d'un traitement Lagrangien des particules dans un modèle numèrique d'ècoulement turbulent diphasique à inclusions disperses. Project de fin d'ètudes Ecole Nationale des Ponts et Chaussèes, L.N.H., Chatou, France, 1987.

Lorenz, K., Der Kumpan in der Umwelt des Vogels. *J. Ornithol.*, 83, 137, 1935.

Lund, T.S. and Rogers, M.M., An improved measure of strain state probability in turbulent flows. *Phys. Fluids*, 6, 1838, 1994.

Mann, K.H. and Lazier, J.R.N., *Dynamics of Marine Ecosystems: Biological–Physical Interactions in the Ocean*, 2nd ed. Blackwell Science, Cambridge, MA, 1998, 394 pp.

Marrasé, C., Costello, J.H., Granata, T., and Strickler, J.R., Grazing in a turbulent environment: II. Energy dissipation, encounter rates and efficiency of feeding currents in *Centropages hamatu*. *Proc. Natl. Acad. Sci., U.S.A.*, 87, 1653, 1990.

Maxey, M.R. and Riley, J.J., Equation of motion for a small rigid sphere in a nonuniform flow. *Phys. Fluids*, 26, 883, 1983.

Osborn, T. and Scotti, A., Effect of turbulence on predator–prey contact rates: where do we go from here? *Mar. Ecol. Prog. Ser.*, 139, 302, 1996.

Rogallo, R.S., Numerical experiments in homogeneous turbulence, Tech. Rep. NASA TM81315, NASA Ames Research Center, 1981.

Rothschild, B.J. and Osborn, T.R., Small-scale turbulence and planktonic contact rate. *J. Plankton Res.*, 10, 465, 1988.

Siggia, E.D., Numerical study of small scale intermittency in three dimensional turbulence. *J. Fluid Mech.*, 107, 375, 1981.

Strickler, J.R., Swimming of planktonic *Cyclops* species (Copepoda, Crustacea): pattern, movements and their control, in *Swimming and Flying in Nature*, Vol. 2, Wu, T.Y.T., Brokaw, C.J., and Brennen, C., Eds., Plenum Press, New York, 1975, 599–613.

Strickler, J.R., Calanoid copepods, feeding currents and the role of gravity. *Science*, 218, 158, 1982.

Strickler, J.R., Feeding currents in calanoid copepods: two new hypotheses, in *Physiological Adaptations of Marine Animals*, Laverack, M.S., Ed., *Symp. Soc. Exp. Biol.*, 39, 459, 1985.

Strickler, J.R. and Bal, A.K., Setae of the first antennae of the copepod *Cyclops scutifer* (Sars): their structure and importance. *Proc. Natl. Acad. Sci. U.S.A.*, 70, 2656, 1973.

Strickler, J.R. and Costello, J.H., Calanoid copepod behavior in turbulent flows. *Mar. Ecol. Prog. Ser.*, 139, 307, 1996.

Sundby, S., Turbulence-induced contact rates in plankton: the issue of scales. *Mar. Ecol. Prog. Ser.*, 139, 305, 1996.

Tennekes, H. and Lumley, J.L., *A First Course in Turbulence*. MIT Press, Cambridge, MA, 1972.

Vincent, A. and Meneguzzi, M., The spatial structure and statistical properties of homogenous turbulence. *J. Fluid Mech.*, 225, 1, 1991.

Wang, L-P. and Maxey, M.R., Settling velocity and concentration distribution of heavy particles in homogeneous isotropic turbulence. *J. Fluid Mech.*, 256, 27, 1993.

Yamazaki, H., Lagrangian study of planktonic organisms: perspectives. *Bull. Mar. Sci.*, 53, 265, 1993.

Yamazaki, H., Turbulence problems for planktonic organisms. *Mar. Ecol. Prog. Ser.*, 139, 304, 1996.

Yamazaki, H., Mackas, D., and Denman, K., Coupling small scale physical processes to biology: towards a Lagrangian approach, in *The Sea, Biological–Physical Interactions in the Ocean*, A.R. Robinson, J.J. McCarthy, and B.J. Rothschild, Eds., John Wiley & Sons, New York, 2002, 51–111.

Yen, J., Sanderson, B., Strickler, J.R., and Okubo, A., Feeding currents and energy dissipation by *Euchaeta rimana*, a subtropical pelagic copepod. *Limnol. Oceanogr.*, 36, 362, 1991.

Yeung, P.K. and Pope, S.B., Lagrangian statistics from direct numerical simulations of isotropic turbulence. *J. Fluid Mech.*, 207, 531, 1989.

Zaret, R.E., Zooplankters and Their Interactions with Water, with Each Other, and with Their Predators. Ph.D. thesis, Johns Hopkins University, Baltimore, MD, 1980, 166 pp.

33

Utilizing Different Levels of Adaptation in Individual-Based Modeling

Geir Huse and Jarl Giske

CONTENTS

33.1 Introduction 507
33.2 Individual-Based Modeling 508
33.3 Methodology 509
    33.3.1 The Attribute Vector 509
    33.3.2 The Strategy Vector 509
    33.3.3 Criteria for Evaluating Adaptation 510
33.4 Three Levels of Adaptation 510
    33.4.1 Fixed Genetic Strategies 510
    33.4.2 Phenotypic Plasticity 511
    33.4.3 Individual Learning 512
        33.4.3.1 Artificial Neural Networks 513
        33.4.3.2 Supervised Learning 514
        33.4.3.3 Reinforcement Learning 514
33.5 Case Studies 515
    33.5.1 Overwintering Diapause in Daphnids 515
    33.5.2 Diel Vertical Migration in a Salmonid 515
    33.5.3 Antipredator Responses 516
33.6 Discussion 517
    33.6.1 Pros and Cons of the Different Levels of Adaptation 517
    33.6.2 The Baldwin Effect 517
33.7 Conclusions 518
Acknowledgments 518
References 518

33.1 Introduction

The most obvious trait of any biological organism is its functionality. It is so universal that its presence was almost unnoticed until the early 19th century. Charles Darwin's (1859) fundamental contribution to biology was to propose natural selection and adaptation as the explanation for the great variety of seemingly intelligent design in Nature. In biology, the most common use of the word *adaptation* is the modification of a trait by natural selection during evolution. For example the white camouflage color of hares in winter and their gray-brown fur in summer can be referred to as adaptations. Here we refer to adaptation in a wider sense: as any process that allows adjustments to the environment. The high importance of adaptation for biological units can be seen from the fact that there are three separate ways

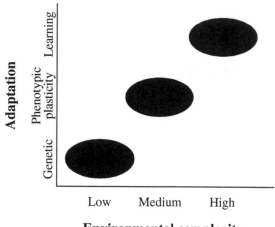

FIGURE 33.1 Different levels of adaptation related to environmental complexity.

that adaptation takes place in biology (Figure 33.1). All three types of adaptation may partly determine individual behavior; whichever is dominant may depend on the current circumstances. We refer to them as different levels of adaptation, which reflects the different timescales that they work on. Although most models of populations and organisms do not specify any process of adaptation, or assume the dominance of a particular kind of adaptation, the idea of adaptation underlies most, if not all, theories and models in ecology. If not for genetic adaptation, one could not assume that a parameter value measured for some individuals also would apply to other members of the population. If not for phenotypic plasticity, one could not assume life history adjustment resulting from climate changes. And if not for learning and memory, one could not assume group dynamics and territoriality.

Although an unconscious use of the power of adaptation in biological theory and in individual-based modeling often is sufficient for many purposes, it is also obvious that conscious attention to the effect of adaptations would improve our ability to mimic natural organisms and their responses to environmental variation (Railsback, 2001). The purpose of this chapter is to illustrate how different levels of adaptation, from fixed strategies to phenotypic plasticity and learning, can be utilized in individual-based modeling of life history and behavioral strategies, to improve models of aquatic populations.

33.2 Individual-Based Modeling

It is not evident from the name, but individual-based modeling (IBM) is a tool for studying group and population processes. For young practitioners of individual-based models, it can be useful to recall that before IBM there were population-based models. This tradition goes back to Lotka and Volterra, and has roots in Euler (1760) and beyond. Before computers became standard equipment, there was a huge need to simplify problems. Thus populations were modeled as consisting of N identical individuals. Alfred Lotka (1925) and Vito Volterra (1926) developed the Lotka–Volterra equations, which enabled simulations of the dynamics of two (or more) interacting populations. These approaches thus described the dynamics of the population by the average members, and dynamics usually occurred only in the temporal dimension. The population approach is still a useful exercise, and the methods developed by Lotka, Volterra, and others have led to the development of structured population models where the population is divided according to age, stage, or a physiological criterion (Fisher, 1930; Metz and Diekman, 1986; Tuljapurkar and Caswell, 1997). Such structured models have proved successful for many applications in ecology and fisheries science. Their advantage, in addition to being more flexible than population models, is that the use of differential equations or matrix models allows analytical solutions (Huston et al., 1988; Caswell and John, 1992).

One of the first applications of the individual-based approach was to explore causes of recruitment variability to commercial fish stocks (DeAngelis et al., 1979; Beyer and Laurence, 1980). This issue has prevailed and been studied empirically in fishery science since Hjort (1914). By IBM it is possible to simulate the individual variability in survival and spatial distribution of early life stages of fish cohorts. This is important for population dynamics, because the survivors tend to differ from average individuals at earlier stages (Crowder et al., 1992). Studies of fish early life history have therefore become one of the major topics for IBM applications (Grimm, 1999). Although the IBM approach was initiated in the late 1970s, it is only since the influential review of Huston et al. (1988) that it has been applied extensively in ecology. The so far most valued ability of IBM is clearly the disintegration of the population into individuals, and the reintegration of individual events into population processes. However, while the approach benefits from considering the population impact of local conditions on individual physiology, there has been little attention to the effect of adaptation to local conditions through differences in individual performance. On this matter IBM has the advantage over population models in that they have the same basic unit as natural selection. This allows a fairly straightforward implementation of adaptation in individual-based models. The IBM approach is not as formalized as the analytic life history models and the Lotka–Volterra models, and as the name implies it is an approach rather than a specific set of equations. Still, IBM has some common structural elements, and below we provide a description of attribute and strategy vectors that are used for bookkeeping of individual characters and implementation of adaptation, respectively.

33.3 Methodology

33.3.1 The Attribute Vector

Individuals can be specified in individual-based models by using an attribute vector $\mathbf{A_i}$ (Chambers, 1993), which contains all the states αm_i used to specify an individual i such as age, weight, sex, hormone levels, and spatial coordinates (x_i, y_i, z_i) at time t:

$$\mathbf{A_i} = (\alpha 1_i, \alpha 2_i, \alpha 3_i, \ldots, \alpha m_i, x_i, y_i, z_i, t) \tag{33.1}$$

Even though the individual-based structure is appealing, it is often difficult to simulate populations on a truly individual basis due to the great abundances involved. This can be solved using the super-individual approach (Scheffer et al., 1995). A super-individual represents many identical individuals and the number of such identical siblings (n_s) thus becomes an attribute of the super-individual:

$$\mathbf{A_s} = (\alpha 1_s, \alpha 2_s, \alpha 3_s, \ldots, \alpha m_s, x_s, y_s, z_s, n_s, t) \tag{33.2}$$

where $\mathbf{A_s}$ is the attribute vector of super-individual s. Mortality operates on the super-individual, and the number of siblings of each super individual is thus decreased in proportion to the mortality rate (Scheffer et al., 1995).

33.3.2 The Strategy Vector

In addition to possessing states, real individuals have adaptive traits, such as life history and behavioral strategies that specify how they should live their lives. The previous lack of IBM studies involving life history strategies and behavior of individuals is in part due to a lack of appropriate techniques for implementing these features. However, adaptive traits can be modeled by introducing a strategy vector $\mathbf{S_i}$ (Huse, 2001; Huse et al., 2002a) that specifies the adaptive traits, such as life history traits or behavior, of an individual:

$$\mathbf{S_i} = (\beta 1_i, \beta 2_i, \beta 3_i, \ldots, \beta m_i) \tag{33.3}$$

where βm_i is the adaptive trait m of individual i. The strategy vector may be considered analogous to a biological chromosome as in the genetic algorithm (Holland, 1992), but may also be updated during the individual's life as a way to simulate learning. In both cases the trait values are modified iteratively in search of the best strategy vectors, and both these approaches for establishing trait values will be discussed below. The combination of attribute and strategy vectors thus enables most relevant characteristics of individuals to be implemented in individual-based models. The classification based on attribute and strategy vectors can be used to describe IBM verbally even though the actual programming implementation is not vector based, as, for example, in object-oriented programming (Maley and Caswell, 1993).

33.3.3   Criteria for Evaluating Adaptation

In nature, adaptation is evaluated by natural selection (Darwin, 1859). To allow analysis of behavioral and life history traits, criteria mimicking the process of natural selection have been constructed and implemented in models as Darwinian fitness measures. The argument behind the application of a fitness measure is that a particular feature or trait under investigation has become optimally adapted over evolutionary time under the given constraints (Stearns and Schmid-Hempel, 1987; Parker and Maynard Smith, 1990). Although there is an ongoing debate of what is the most appropriate definition of fitness (Stearns, 1992; Giske et al., 1993; Mylius and Diekmann, 1995; Roff, 2002), there seems to be a consensus about the importance of including aspects of survival and fecundity (growth) into the fitness definition. The common fitness measures based on this assumption are the instantaneous rate of increase r and the net reproductive rate R_0 (Roff, 1992). An alternative approach is to use endogenous fitness rather than an explicit fitness criterion (Mitchell and Forrest, 1995; Menczer and Belew, 1996; Huse, 1998; Strand et al., 2002). In endogenous systems, no fitness measure to maximize is provided, but rather fitness emerges by interactions between the organism and its environment. Individuals grow and die according to probability functions and Monte Carlo simulations (Judson, 1994). Criteria are set for reproduction and the fittest organisms will, by definition, be those that are able to reproduce more, relative to the other individuals under the set conditions. Such models operate similarly to the way in which evolution works: through "adaptation execution" rather than by "fitness maximization" (Wright, 1994).

33.4   Three Levels of Adaptation

33.4.1   Fixed Genetic Strategies

Traditionally, evolutionary adaptations have been implemented in individual-based models in two ways. First, all parameter values that are results of experiments on the modeled organism, such as physiological and morphological parameters, reflect the results of natural selection. In such cases modelers may not always be conscious about their application of adaptation, but application of measured parameters is a shortcut to the adapted state without performing the selection process over again. Second, rules of behavior implemented in IBMs are often derived from evolutionary considerations. The major source of such rules is life history theory, and we take the much-used "Gilliam's rule" as an example. This rule states that juvenile fish should seek the habitat where the mortality risk per growth rate (μ/g) is minimal. Werner and Gilliam (1984) showed that for a specific life history pattern of fish, a given growth and mortality regime, and under the assumption that fitness could be modeled by the net reproductive rate (R_0), this was the optimal policy for pre-reproductive organisms. This rule has been applied in many studies of fish spatial distribution, but also in studies of many kinds of organisms for which it was not developed. Aksnes and Giske (1990) developed other life history–based rules for other aquatic life histories.

    Although Gilliam's rule and other rules for optimal behavior yield both analytical and intuitive understanding of the major trade-offs for individuals in a given environment, they cannot be said to be truly individual, but rather population rules. The rule is derived from typical aspects of the life cycles, not from individual characteristics. Also, when using such life history–based rules, all individuals tend to make the same decisions. An alternative approach to deriving rules from some theory could therefore be to evolve them. The genetic algorithm (GA) developed by Holland (1992) is a technique that applies evolution by natural selection in computer programs to find optimal solutions to a problem by representing solutions in "genetic code." It involves: a numeric genetic code, selection of the best combination of numbers in consecutive generations using a fitness criterion, and mutations and recombinations to produce

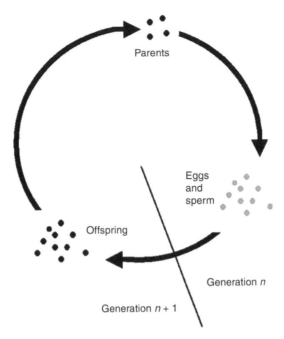

FIGURE 33.2 The genetic algorithm. Of a large initial population of newborn offspring, a smaller number will survive and become the parents for next generation. These survivors produce sexual products in proportion to their accumulated reserves. These sexual products form the eggs for next generation, possibly after recombinations and mutations of the parent strategy vectors. This process is repeated in an individual-based model over a number of generations. In each generation, strategy vectors that make the offspring more likely to become parents will increase in abundance, and hence the individual genomes and the whole gene pool of the offspring population will become better adapted to the local conditions.

new variation (Figure 33.2). The GA has successfully been applied to problems within a large number of fields, such as engineering, physics, economics, medicine, artificial life, and biology (Goldberg, 1989; Mitchell, 1996). For recipes on applications of GAs in IBMs, see Huse et al. (2002a).

The simplest application of GAs to evolve adaptations in an IBM is through life history switch genes. In classical models of zooplankton dynamics, the seasonal dynamics of the population was taken care of by programming codes, such as

$$\text{IF (JulianDay = 280) THEN descend to overwintering depth} \qquad (33.4)$$

The programmer knew that zooplankton in spring and summer fed in surface waters, and overwintered at greater depths during the winter season. Hence, the programmer used the first level of adaptation by driving the model through observations and fixed preset values. Alternatively, the GA could be used to search for the optimal date for seasonal migration, and the pseudocode could have been as

$$\text{IF (JulianDay = character } \beta m) \text{ THEN descend to overwintering depth} \qquad (33.5)$$

where now character βm (Equation 33.3) is the character value of character number m in the strategy vector of an individual. Individuals with the same character values at character m would then descend to overwintering depths the same day. After some generations of adaptation, the gene pool of the population would consist of one single character value of character m, or a series of values that on average gave their bearers the same fitness. Another modeling technique related to the GA is genetic programming (Koza, 1992), where computer code rather than allele values are evolved. GP can be used instead of the GA for most of the cases discussed in this chapter.

33.4.2 Phenotypic Plasticity

Above we discussed how to adapt fixed strategies for maximizing survival, growth, and reproduction using the GA. When the environment varies in a fairly predictable fashion such as the seasonal changes

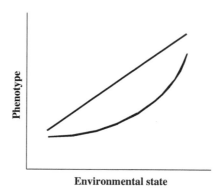

FIGURE 33.3 Reaction norms emerge as different phenotypes are expressed under different environmental states. The relationship between phenotype and environment is shown for two different genotypes.

in day length or interannual climatic variation, fixed strategies (Equation 33.5) tend to yield for flexible strategies that are dependent on the state of the environment. We now come to the second level of adaptation in biology: phenotypic plasticity.

Phenotypic plasticity is defined as "a change in the average phenotype expressed by a genotype in different macro environments" (Via, 1987). Thus one genotype adapts by expressing different phenotypes under different environmental states. This does not imply that an individual is equally well adapted to all environmental states, but rather that the individual resources are allocated to give the best attainable life history for the particular environment (Lessells, 1991). The relationship emerging between the expressed phenotype for an environmental state and the environmental state is referred to as the reaction norm. This concept is illustrated in Figure 33.3.

To implement phenotypic plasticity in models it is necessary to include environment state (E) as a variable in the model. To model, for example, the linear reaction norm seen in Figure 33.3, one needs intercept and slope characters embedded on the strategy vector. In this case the phenotype is $\beta_1 \cdot E + \beta_2$. To allow more complex reaction norms the complexity of the equation needs to be increased, along with the number of adapted variables. The variables are estimated using the GA as discussed above. In some cases several environmental features affect a trait. One way to model this would simply be to add more terms to the reaction norm. A different approach would be to use an artificial neural network (ANN), as discussed below.

State dependency is a phenomenon similar to phenotypic plasticity. From the perspective of the organism, phenotypic plasticity is the genes' changes in strategy under a variable external environment, whereas state dependency is the genes' changes in strategy under a variable internal environment. But from the perspective of a gene, the nearest external environment is the rest of the genome and the organism. The organism is the survival tool for its genes (Dawkins, 1982; Keller, 1999). State-dependent behavior or life history decisions appear because the genes code for flexible strategies. State-dependent decisions are usually modeled by stochastic dynamic programming (Houston and McNamara, 1999; Clark and Mangel, 2000). This method, however, is backward running, and therefore not always easy to combine with other individual-based methods. An alternative would then be to use the ANN and a GA, which also can solve state-dependent problems (Huse et al., 1999). The ANN is explained in the next section.

33.4.3 Individual Learning

The final and perhaps most refined level of adaptation is learning. Learning can be defined as "any process in an animal in which behavior becomes consistently modified as a result of experience" (Lawrence, 1989). As opposed to the other two kinds of adaptation discussed previously, learning is not passed on to offspring. Instead, learning is an independent process for each individual, although it may be facilitated by parents. Still the capacity for learning is evolved. Learning is particularly efficient in complex or unpredictable environments where changes take place at a small timescale. Learning requires a higher mental capacity than the other means of adaptation, and in general it is more important for

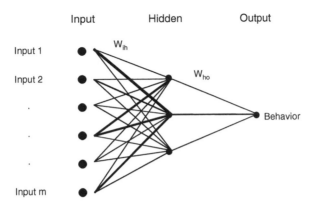

FIGURE 33.4 A schematic outline of an ANN. The connection points of the lines are referred to as nodes.

marine mammals (Rendell and Whitehead, 2001) than for fish, and rather restricted in invertebrates. The evolutionary role of learning is very obvious in mammals, where parents usually teach their offspring to find food and to avoid predators. For good introductions to learning and modeling learning, respectively, see Schmajuk (1997) and Ballard (1997).

Learning can be implemented in models in different ways and below we address some basic methods for doing this. Simulation models of learning can be divided into supervised learning paradigms and reinforcement learning. Common to many of the models is that they are based on ANNs.

33.4.3.1 *Artificial Neural Networks* — The ANN is a computing method inspired by a conceptual model of how the human brain functions. Neurons in the brain are interconnected by synapses; similarly layers of nodes in an ANN are linked together and pass signals between each other (Figure 33.4). Thus, the ANN is made to mimic the decision process in a biological organism, with multiple sensory inputs, a complex and hidden brain, a decision and a muscle output. This paradigm was initiated by McCulloch and Pitts (1943) and their theoretical outline of the two-state neuron. Since then, there has been considerable development in ANNs, and today a wide variety of ANN architectures are available (Rosenblatt, 1958; Rummelhart et al., 1986; Montana and Davis, 1989). In Figure 33.4 a three-layer feedforward ANN is illustrated.

The connection between a series of stimuli and the decision in an ANN goes from the input layer through one or several layers of hidden nodes. Each hidden node adds and weighs the input from a series of input nodes:

$$N_h = \sum_{i=1}^{m} W_{ih} \cdot I_i \tag{33.6}$$

where I_i is input data i, W_{ih} is the connection weight between input data i and hidden node h (i.e., the relative influence of input data i for hidden node h), N_h is the sum of the weighted input data of hidden node h, and m is the number of input nodes connected to hidden node h. At the hidden node, values are transformed to the [0,1] range using the standard sigmoid transformation:

$$TN_h = \frac{1}{(1 + e^{-(N_h + B_h)})} \tag{33.7}$$

where TN_h is the transformed node value and B_h is the bias (van Rooij et al., 1996) of hidden node h. The bias B_h is similar to an intercept value in a regression model. The transformation in the ANN introduces nonlinearity, which allows the network to solve complex problems. The output O is calculated by adding together the sums of the transformed hidden node values multiplied by the output weights (W_{ho}):

$$O = \sum_{h=1}^{n} W_{ho} \cdot TN_h \tag{33.8}$$

where *n* is the number of hidden nodes. Finally the output is transformed using an equation similar to Equation 33.7. The ANN is adapted by defining all connection weights and biases as separate characters on the strategy vector, and then continuously select for the best ANNs using the GA or learning techniques. With an initial random set of connection weights in each modeled individual in the first generation, there is a potential risk that no phenotype will be able to perform sensible decisions. This can be overcome by introducing the first few generations in the GA to a simpler and friendlier environment, where the pressure to perform is weaker than for later generations.

33.4.3.2 *Supervised Learning* —
Supervised learning is a method where a training set of known input–response pairings is used to produce generalizing capabilities in an ANN. This corresponds to having an omniscient teacher, hence the name. In supervised learning, the output produced by the ANN is compared to the correct response and the weights are modified according to the discrepancy in an iterative procedure. This procedure is repeated for the entire training set. The trained network can then be used to generate predictions for cases outside of the training set. One of the simplest ANNs is the so-called perceptron (Rosenblatt, 1958). This network consists of an input and an output layer and can be trained using the Widrow–Hoff or delta learning rule (Ballard, 1997). In this case the output is the sum of the input to the network multiplied by the weights (W) between the input and the output node, and the weights are then changed by

$$W = W_{old} + d \cdot L \cdot I \tag{33.9}$$

where L is the learning rate, I is the input, and d is then defined as $O_{target} - O$. Thus, the discrepancy between the observed and predicted output is used to modify the weights. This procedure is a simple way to implement associative learning into models, for example, classical conditioning. In classical conditioning, a conditioned response toward a conditioned stimulus, which does not elicit response, is learned as the conditioned stimulus is presented in conjunction with an unconditioned stimulus that elicits response (Schmajuk, 1997). In the classic case of Pavlov's dog, the bell ringing before the food is presented is the conditioned stimulus and the food is the unconditioned stimulus. After several pairings of the two stimuli, the conditioned stimulus is able to produce a conditioned response similar to an unconditioned response. This is referred to as stimulus substitution (Schmajuk, 1997). Classical conditioning can be simulated by presenting the stimuli as input to the ANN and then correcting the weights through iterations to reproduce conditioned responses similar to the observed (unconditioned) responses using Equation 33.9.

There are some limitations to what problems the perceptron can solve, and for more complex problems, backpropagation is a better technique (Rummelhart et al., 1986). Implicit in this approach is the use of a hidden layer, as seen in Figure 33.4, in addition to the input and output layers of the perceptron. This along with a sigmoid transfer function and a generalized delta rule that is propagated backward into the ANN allows solution of complex problems (Ballard, 1997). Thus errors are computed for each unit in the hidden and output layers and the weights are modified correspondingly.

33.4.3.3 *Reinforcement Learning* —
While the supervised learning paradigm assumes the presence of an omniscient supervisor able to tell the network the correct response, one often does not know the correct answer. Instead, it may be possible to evaluate actions in a less rigorous manner. Such situations are suited for reinforcement learning where actions are associated with rewards and punishment for "good" and "bad" behaviors, respectively (Ackley and Littman, 1992). Thus as opposed to supervised learning where the correct output is known during training, the ANN has to discover the desired output. This process allows individuals to produce increasingly more favorable behaviors as they explore and learn about their environment. A behavior network can be updated by calculating a reinforcement signal based on the fitness consequence of the previous action, so-called delayed rewards. The weights of the ANN are then modified according to a reinforcement learning procedure that resembles backpropagation, to produce increasingly "better" behavior. Recipes for applying reinforcement learning can be found in Ballard (1997), and for an application of reinforcement learning to simulation of movement behavior in a spatial lattice, see Ackley and Littman (1992).

33.5 Case Studies

The discussion above has focused on presenting and categorizing adaptation and adaptive modeling techniques. Below some cases are provided where we discuss the different levels of adaptation and model implementation. The cases focus on diapause in zooplankton, vertical migration in salmonids, and antipredator responses.

33.5.1 Overwintering Diapause in Daphnids

In autumn, many life-forms shift focus from growth to overwintering survival. This applies to boreal trees as well as to boreal bears. We illustrate this with two examples, starting with hibernation decision in a freshwater herbivorous zooplankter. *Daphnia* enters diapause in fall in temperate lakes by producing two resting eggs, which overwinter in the lake sediments. Next spring the resting eggs hatch and grow into adult females that commence asexual reproduction at maturity. The timing of the onset of diapause is important because the alternative is to produce several female offspring that reach maturity and can reproduce themselves again. On the other hand *Daphnia* strains that remain in the water column through winter risk extinction due to the long generation time, low fecundity, a potentially high predation pressure, as well as environmental hazards.

The simplest modeling strategy for this phenomenon would be to obtain the most probable date for hibernation from a field study. In many cases this would also be the most appropriate method, as this date reflects the actual adaptation that has taken place in this or a similar lake, and as it allows the modeler to concentrate on other dynamic aspects of the model. But if one were to model the timing, the first approach would be to assume a Julian day for descent and code this on the strategy vector. However, if interannual variation in the environment is great, this fixed strategy might be inferior to a more flexible strategy taking the state of the environment into account. By assuming that temperature is a key variable in this respect, the decision variable (D) could be expressed, for example, as a power function of the temperature T at some early stage:

$$D = \beta_1 T^{\beta_2} \tag{33.10}$$

The corresponding strategy vector is $S_i = (\beta_1, \beta_2)$. A similar approach has been used in a model of the life cycle of the boreal marine copepod *Calanus finmarchicus* (Fiksen, 2000). Fiksen equipped each individual in the population with three evolvable characters: (1) the day of the year (i.e., the day length) at which a resting stage V copepodid should wake from overwintering diapause, (2) the day of year when it should shift allocation pattern from somatic growth to production of fat reserves for overwintering, and (3) the fat/somatic tissue ratio required to initialize overwintering diapause. Thus, he used two static characters and one state-dependent character.

If in addition to the interannual variation in temperature, additional factors such as the predation pressure or density of conspecifics, or both, are important, the problem of the *Daphnia* becomes a lot more complex. To model this, one could build on Equation 33.5, and include more factors in a similar fashion. This kind of model would soon become rather complicated, especially if the different factors interfere with each other in affecting the survival of *Daphnia*. An alternative way of doing this would be to use a formalized structure such as the ANN. For example, the problem could be solved sequentially so that each day the *Daphnia* would be presented with the relevant information and perform the decision whether to enter diapause or continue producing regular offspring. This example illustrates how increasing environmental complexity needs to be approached with increasing model complexity in order to solve the important trade-offs.

33.5.2 Diel Vertical Migration in a Salmonid

Vertical migration is a classic theme in ecology occurring in a wide range of aquatic organisms. Diel vertical migration (DVM) comes about as an adaptation to the daily light cycle and usually moves organisms between deep waters during the day and shallower waters at night (Figure 33.5). This behavior

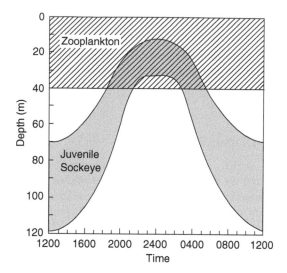

FIGURE 33.5 The diel vertical distribution of salmonids in a lake. (From Clark, C.W. and Levy, D.A., *Am. Nat.*, 131:271–290, 1988. With permission.)

is driven by the trade-off between feeding and predation risk. Below we develop a model for vertical migration in a small freshwater fish, for example, a small salmonid (Clark and Levy, 1988). The fish feed on zooplankton and are eaten by visually foraging fish.

The surface light level goes through tremendous changes during the diel cycle, peaking at midday and going to virtually zero at night, except for regions with a midnight sun. A simple way of performing DVM would thus be to adapt to a constant light level and adopt this throughout the day. Thus the fish would always stay at or move toward the depth corresponding to this light level, and the best isolume will be estimated iteratively using the GA. Movement would be constrained by the surface and the bottom, and at night the fish would tend to occupy the surface layer.

The isolume strategy is simple, but also rather inflexible, and might leave the fish more exposed to predation than optimal, for example, when it is satiated. A state-dependent strategy involving stomach fullness (SF) would therefore improve the flexibility of the model:

$$D = \beta_1 + \beta_2 \cdot SF^{\beta_3} \qquad (33.11)$$

where D is the decision variable (depth), and β_1 is the isolume level, and β_2 and β_3 represent the depth dependence on stomach fullness. The strategy vector would then be $S_i = (\beta_1, \beta_2, \beta_3)$. In addition to the state dependence, there could be other factors influencing depth selection, such as the distribution of food resources, conspecifics, or the temperature. Again it can be profitable to use ANNs if several factors are to be taken into account. See Strand et al. (2002) for an application of ANNs and GAs in an IBM of vertical migration in the mesopelagic fish *Maurolicus muelleri*.

33.5.3 Antipredator Responses

A number of aquatic species are able to learn antipredator responses. Even flatworms with a very simple nervous system are able to learn avoidance of predators by associating it with odors of injured conspecifics (Wisenden and Millard, 2001). This is a case of classical conditioning where the response to chemical cues released by injured conspecifics is evolved. Once this smell is associated with a strange odor, the strange odor is responded to as the presence of a predator. In this case the predator odor becomes the conditioned stimulus, and the smell of injured conspecifics is the unconditioned stimulus. Similarly, young stages of daphnids can alter their morphology on detection of chemicals associated with digestive processes of their predator, the phantom midge *Chaoborus* sp. (Krueger and Dodson, 1981; Larsson and Dodson, 1993). The alteration reduces the likelihood of *Daphnia* being ingested by the predator. Because the morphological alteration is irreversible and costs energy, it can be fruitful to study at what predator densities or *Daphnia* densities or time of year it should be induced. Alterations in morphology can be predicted using an ANN with the factors mentioned above as input data along with equations for the costs and benefits associated with the alterations. So here we see examples of adaptation by phenotypic plasticity toward a specific and evolutionary well-known predator, and through learning toward any new predator. This is seen in fish species as well. For example, gobies in western Norway do not have inherited aversion against cod, their main predator, but rather learn to be averse through combined visual–olfactory associations (Utne-Palm, 2001). Mathis et al. (1996) further showed that such predator aversions could be transferred culturally from experienced to naive individuals, also across species barriers.

33.6 Discussion

33.6.1 Pros and Cons of the Different Levels of Adaptation

There has been no thorough account of the relative profitability of the different levels of adaptation (Frank, 1996), although some preliminary efforts have been attempted (Boyd and Richerson, 1985; Holland, 1992). The lack of a lucid theory for understanding the profitability of the different levels of adaptation makes it difficult to provide recommendations for when to use the different approaches in IBM. However, there is a general tendency for the level of adaptation to increase with the complexity of the environment (Figure 33.1). Thus in a constant environment a fixed strategy is favorable as no time or energy is wasted on failed explorations (Frank, 1996). In a seasonally variable environment or if the environment varies between rather predictable states, phenotypic plasticity would be the adaptive level of choice. For short-term variation caused, for example, by the tidal cycle, learning might be the most profitable way of exploiting the environment. In general, it is useful to consider the kind of adaptation exerted in nature and the capabilities of the target species when deciding on what level of adaptation to use in a model. Each level of adaptation has its pros and cons. Genetic adaptation, through fixed or flexible phenotypes, is slow, inflexible, and irreversible in the organism. Mental adaptation, through learning, is quick and reversible, but requires individual association and explorations by trial and error. Hence, very dangerous or very rare events are not open to learning. Learning is costly to the organism in several ways. It costs time, lost opportunities, and sometimes enhanced mortality risk to make the observations and associations that can be learned. It also costs energy to transfer them into storage, and keep them stored in the neural system. The storage capacity itself is also under adaptive evolution, competing for resources with all other anatomical structures (Fisher, 1930). The capacity for learning is therefore strongly dependent on the evolutionary lineage, and learning ability is, for example, much greater in mammals than in fish. Finally, with a limited storage capacity, different elements of learning compete for available memory. This explains also why organisms should be quick to forget as soon as the learned association is not of high fitness value any more (Healy, 1992).

Social learning is widespread in primates and other mammals. Similar cultural exchange between age groups is seen in herring where young individuals seem to learn features of their migration pattern from schooling with older ones (McQuinn, 1997; Fernö et al., 1998). However, the collective dynamics of the herring schools are important for the information exchange among herring cohorts, and in years when the recruiting year class is especially abundant, the information exchange seems to be interrupted (Huse et al., 2002b). This shows that social learning can be vulnerable to interference at large population sizes.

33.6.2 The Baldwin Effect

James Mark Baldwin (1896) proposed a mechanistic connection between individual learning and genetic evolution. In a variable environment, the genome can improve its survival by including genes that code for alternative phenotypes. Hence, phenotypic plasticity is a genomic bet-hedging strategy in variable environments. But in a complex fitness landscape, where events seldom repeat, the organisms cannot possess a genetic preparedness for all possible circumstances. It is therefore in the interest of genes in a genome to include new members that code for mental structures facilitating learning. Hence, individual learning is adaptive if learners reproduce their genetic codes more frequent than nonlearners. This is the standard explanation for all types of phenotypic adaptation. However, Baldwin suggested an additional mechanism: in variable environments, new external factors can appear, impacting genetic fitness of the organism. Individuals with genomes that contain genes coding for a capacity to learn how to handle new situations can more easily survive and reproduce their genome. One can say that learning smooths the peaks and valleys in the fitness landscape, allowing the genes in the learner to be fit in more environments. And while the genes are kept alive through individual learning, any mutation in the genome leading toward genetic adaptation toward the changed environment will be facilitated. So although random favorable mutations are equally likely in learners and nonlearners, learners are more likely to remain alive and reproducing until the favorable mutation has occurred. As a result, Baldwin said, adaptive genetic evolution is more likely in learners, and hence the genetic trait of learning is maintained and spread.

The Baldwin effect moves the evolutionary level of selection upward from single genes and genomes to organisms and groups. The effect of the third level of adaptation is that social information transfer on the level of groups, as well as nonsocial individual learning, affects both the behavior of the organism and the survival of its genes. In fish, this is clearly seen in the potential strength of imitation-learning in fish feeding (Baird et al., 1991), in predator avoidance (Mathis et al., 1996; Krause et al., 2000), and mating behavior (Dugatkin and Godin, 1993), but is also likely to affect long-range migrations (McQuinn, 1997; Fernö et al., 1998). Hence, to study population processes in fish and other Baldwinian creatures, consideration of all three levels of adaptation may be warranted.

33.7 Conclusions

Above we have presented an approach for thinking about adaptation as well as a conceptual framework for utilizing different levels of adaptation in IBM. In behavioral ecology it has been commonplace to disregard the level of adaptation involved and simply assume that optimal behavior is achieved. Adaptation in aquatic animals spans from fixed genetic strategies via phenotypic plasticity, to individual and social learning, and IBM can be used to address all these levels of adaptation. The cases discussed above illustrate the importance of including adaptation in IBM, as well as being specific about the level of adaptation.

Acknowledgments

G.H. was supported by the European Commission. J.G. was partly supported by the Research Council of Norway. We thank Sami Souissi for providing valuable comments on a former version of this chapter.

References

Ackley, D. and Littman, M., 1992. Interactions between learning and evolution, in Langton, C., Taylor, C., Farmer, J., and Rasmussen, S., Eds., *Artificial Life III*. Reading, MA: Addison-Wesley, 487–509.

Aksnes, D.L. and Giske, J., 1990. Habitat profitability in pelagic environments. *Mar. Ecol. Prog. Ser.*, 64:209–215.

Baird, T.A., Ryer, C.H., and Olla, B.L., 1991. Social enhancement of foraging on an ephemeral food source in juvenile walleye pollock, *Theragra chalcogramma*. *Environ. Biol. Fish.*, 31:307–311.

Baldwin, J.M., 1896. A new factor in evolution. *Am. Nat.*, 30:441–451.

Ballard, D.H., 1997. *An Introduction to Natural Computation*. Cambridge, MA: MIT Press, 307 pp.

Beyer, J.E. and Laurence, G.C., 1980. A stochastic model of larval fish growth. *Ecol. Modelling*, 8:109–132.

Boyd, R. and Richerson, P.J., 1985. *Culture and the Evolutionary Process*. Chicago: University of Chicago Press, 331 pp.

Caswell, H. and John, A.M., 1992. From the individual to population in demographic models, in DeAngelis, D.L. and Gross, L.J., Eds., *Individual-Based Models and Approaches in Ecology*. New York: Chapman & Hall, 36–66.

Chambers, C.R., 1993. Phenotypic variability in fish populations and its representation in individual-based models. *Trans. Am. Fish. Soc.*, 122:404–414.

Clark, C.W. and Levy, D.A., 1988. Diel vertical migration by juvenile sockeye salmon and the antipredation window. *Am. Nat.*, 131:271–290.

Clark, C.W. and Mangel, M., 2000. *Dynamic State Variable Models in Ecology. Methods and Applications*. New York: Oxford University Press, 289 pp.

Crowder, L.B., Rice, J.A., Miller, T.J., and Marschall, E.A., 1992. Empirical and theoretical approaches to size-based interactions and recruitment variability in fishes, in DeAngelis, D.L. and Gross, L.J., Eds., *Individual-Based Models and Approaches in Ecology*. New York: Chapman & Hall, 237–255.

Darwin, C., 1859. *On the Origin of Species by Means of Natural Selection, or the Preservation of Favoured Races in the Struggle for Life*. London: J. Murray, 502 pp.

Dawkins, R., 1982. *The Extended Phenotype*. Oxford: Oxford University Press, 318 pp.

DeAngelis, D.L., Cox, D.C., and Coutant, C.C., 1979. Cannibalism and size dispersal in young-of-the-year largemouth bass: experiments and model. *Ecol. Modelling*, 8:133–148.

Dugatkin, L.A. and Godin, J.G.J., 1993. Female mate copying in the guppy (*Poecilia-Reticulata*) — age-dependent effects. *Behav. Ecol.*, 4:289–292.

Euler, L., 1760. Recherches générales sur la mortalité: la multiplication du genre humain. *Mem. Acad. Sci. Berlin*, 16:144–164.

Fernö, A., Pitcher, T.J., Melle, W., Nøttestad, L., Mackinson, S., Hollingworth, C., and Misund, O.A., 1998. The challenge of the herring in the Norwegian Sea: making optimal collective spatial decisions. *Sarsia*, 83:149–167.

Fiksen, Ø., 2000. The adaptive timing of diapause — a search for evolutionarily robust strategies in *Calanus finmarchicus*. *ICES J. Mar. Sci.*, 57:1825–1833.

Fisher, R.A., 1930. *The Genetical Theory of Natural Selection*. Oxford: Clarendon Press, 318 pp.

Frank, S.A., 1996. The design of natural and artificial systems, in Rose, M.R. and Lauder, G.V., Eds., *Adaptation*. San Diego: Academic Press, 451–505.

Giske, J., Aksnes, D.L., and Førland, B., 1993. Variable generation time and Darwinian fitness measures. *Evol. Ecol.*, 7:233–239.

Goldberg, D.E., 1989. *Genetic Algorithms in Search, Optimization and Machine Learning*. Reading, MA: Addison-Wesley, 412 pp.

Grimm, V., 1999. Ten years of individual-based modelling in ecology: what have we learned and what could we learn in the future? *Ecol. Modelling*, 115:129–148.

Healy, S., 1992. Optimal memory: toward an evolutionary ecology of animal cognition? *Trends Ecol. Evol.*, 7:399–400.

Hjort, J., 1914. Fluctuations in the great fisheries of northern Europe reviewed in the light of biological research. *Rapp. P.-V. Réun. Cons. Int. Explor. Mer*, 20:1–28.

Holland, J.H., 1992. *Adaptation in Natural and Artificial Systems*. Cambridge, MA: MIT Press, 211 pp.

Houston, A.I. and McNamara, J.M., 1999. *Models of Adaptive Behaviour. An Approach Based on State*. Cambridge, U.K.: Cambridge University Press, 378 pp.

Huse, G., 1998. Life History Strategies and Spatial Dynamics of the Barents Sea Capelin (*Mallotus villosus*), thesis, Bergen: University of Bergen, 37 pp.

Huse, G. 2001. Modelling habitat choice in fish using adapted random walk. *Sarsia*, 86:477–483.

Huse, G., Strand, E., and Giske, J., 1999. Implementing behaviour in individual-based models using neural networks and genetic algorithms. *Evol. Ecol.*, 13:469–483.

Huse, G., Giske, J., and Salvanes, A.G.V., 2002a. Individual-based models, in Hart, P.J.B. and Reynolds, J., Eds., *Handbook of Fish and Fisheries*. Oxford: Blackwell Science, 228–248.

Huse, G., Railsback, S.F., and Fernö, A., 2002b. Modelling changes in migration pattern of herring: collective behaviour and numerical domination. *J. Fish Biol.*, 60:571–582.

Huston, M., DeAngelis, D., and Post, W., 1988. New computer models unify ecological theory. *BioScience*, 38:682–691.

Judson, O.P., 1994. The rise of individual-based model in ecology. *Trends Ecol. Evol.*, 9:9–14.

Keller, L., 1999. *Levels of Selection in Evolution*. Princeton, NJ: Princeton University Press, 318 pp.

Koza, J.R., 1992. *Genetic Programming.* Cambridge, MA: MIT Press, 840 pp.

Krause, J., Hoare, D., Krause, S., Hemelrijk, C.K., and Rubenstein, D.I., 2000. Leadership in fish shoals. *Fish Fish.,* 1:82–89.

Krueger, D.A. and Dodson, S.I., 1981. Embryological induction and predation ecology in *Daphnia pulex. Limnol. Oceanogr.,* 26:212–223.

Larsson, P. and Dodson, S., 1993. Invited review: chemical communication in planktonic animals. *Arch. Hydrobiol.,* 129:129–155.

Lawrence, E., 1989. *Henderson's* Dictionary of *Biological Terms.* Harlow, U.K.: Longman Scientific & Technical, 637 pp.

Lessells, C.M., 1991. The evolution of life histories, in Krebs, J.R. and Davies, N.B., Eds., *Behavioural Ecology.* Oxford: Blackwell Scientific, 32–68.

Lotka, A.J., 1925. *Elements of Physical Biology.* Baltimore: Williams & Wilkins, 465 pp.

Maley, C.C. and Caswell, H., 1993. Implementing *i*-state configuration models for population dynamics: an object-oriented programming approach. *Ecol. Modelling,* 68:75–89.

Mathis, A., Chivers, D.P., and Smith, R.J.F., 1996. Cultural transmission of predator recognition: intraspecific and interspecific learning. *Anim. Behav.,* 51:185–201.

McCulloch, W.S. and Pitts, W.H., 1943. A logical calculus of the ideas immanent in nervous activity. *Bull. Math. Biophys.,* 5:115–133.

McQuinn, I.H., 1997. Metapopulations and the Atlantic herring. *Rev. Fish Biol. Fish.,* 7:297–329.

Menczer, F. and Belew, R.K., 1996. From complex environments to complex behaviors. *Adaptive Behav.,* 4:317–363.

Metz, J.A.J. and Diekman, O., 1986. *The Dynamics of Physiologically Structured Populations.* Berlin: Springer-Verlag, 511 pp.

Mitchell, M., 1996. *An Introduction to Genetic Algorithms.* Cambridge, MA: MIT Press, 205 pp.

Mitchell, M. and Forrest, S., 1995. Genetic algorithms and artificial life, in Langton, C.G., Ed., *Artificial Life: An Overview.* Cambridge, MA: MIT Press, 267–289.

Montana, J. and Davis, L., 1989. Training feedforward networks using genetic algorithms, in Sridharan, N.S., Ed., *Eleventh International Joint Conference on Artificial Intelligence.* San Mateo, CA: Morgan Kaufman, 762–767.

Mylius, S.D. and Diekmann, O., 1995. On evolutionary stable life histories, optimization and the need to be specific on density dependence. *Oikos,* 74:218–224.

Parker, G.A. and Maynard Smith, J., 1990. Optimality theory in evolutionary biology. *Nature,* 348:27–33.

Railsback, S.F., 2001. Concepts from complex adaptive systems as a framework for individual-based modelling. *Ecol. Modelling,* 139:47–62.

Rendell, L. and Whitehead, H., 2001. Culture in whales and dolphins. *Behav. Brain Sci.,* 24:309–382.

Roff, D.A., 1992. *The Evolution of Life Histories.* New York: Chapman & Hall, 548 pp.

Roff, D.A., 2002. *Life History Evolution.* Sunderland, MA: Sinauer Associates, Inc., 527 pp.

Rosenblatt, F., 1958. The perceptron: a probabilistic model for information storage and organization in the brain. *Psychol. Rev.,* 65:386–408.

Rummelhart, D.E., Hinton, G.E., and Williams, R.J., 1986. Learning representations by back-propagating errors. *Nature,* 323:533–536.

Scheffer, M., Baveco, J.M., DeAngelis, D.L., Rose, K.A., and van Nes, E.H., 1995. Super-individuals a simple solution for modelling large populations on an individual basis. *Ecol. Modelling,* 80:161–170.

Schmajuk, N.A., 1997. *Animal Learning and Cognition: A Neural Network Approach.* Cambridge, U.K.: Cambridge University Press, 352 pp.

Stearns, S.C., 1992. *The Evolution of Life Histories.* Oxford: Oxford University Press, 320 pp.

Stearns, S.C. and Schmid-Hempel, P., 1987. Evolutionary insight should not be wasted. *Oikos,* 49:118–125.

Strand, E., Huse, G., and Giske, J., 2002. Artificial evolution of life history and behavior. *Am. Nat.,* 159:624–644.

Tuljapurkar, S. and Caswell, H., 1997. *Structured-Population Models in Marine, Terrestrial, and Freshwater Systems.* New York: Chapman & Hall, 643 pp.

Utne-Palm, A.C., 2001. Response of naïve two-spotted gobies *Gobiusculus flavescens* to visual and chemical stimuli of their natural predator, cod *Gadus morhua. Mar. Ecol. Prog. Ser.,* 218:267–274.

van Rooij, A.J.F., Jain, L.C., and Johnson, R.P., 1996. *Neural Network Training Using Genetic Algorithms.* Bunke, H. and Wang, P.S.P., Eds., Singapore: World Scientific, 130 pp.

Via, S., 1987. Genetic constraints on the evolution of phenotypic plasticity, in Loeschcke, V., Ed., *Genetic Constraints on Adaptive Evolution.* Berlin: Springer-Verlag, 47–71.

Volterra, V., 1926. Variazioni e fluttuazioni del numero d'individui in specie animali conviventi. *Atti R. Mem. Cl. Sci. Fis. Mat. Nat.,* 6:31–113.

Werner, E.E. and Gilliam, J.F., 1984. The ontogenetic niche of reproduction and species interactions in size-structured populations. *Annu. Rev. Ecol. Syst.,* 15:393–425.

Wisenden, B.D. and Millard, M.C., 2001. Aquatic flatworms use chemical cues from injured conspecifics to assess predation risk and to associate risk with novel cues. *Anim. Behav.,* 62:761–766.

Wright, R., 1994. *The Moral Animal. Why We Are the Way We Are.* London: Abacus, 466 pp.

34

Using Multiagent Systems to Develop Individual-Based Models for Copepods: Consequences of Individual Behavior and Spatial Heterogeneity on the Emerging Properties at the Population Scale

Sami Souissi, Vincent Ginot, Laurent Seuront, and Shin-Ichi Uye

CONTENTS

34.1 Introduction 523
34.2 Development of the Model 524
    34.2.1 Main Characteristics of the *Mobidyc* Platform 524
    34.2.2 Life Cycle Representation 525
    34.2.3 Model Architecture 525
        34.2.3.1 Modeling Survival, Growth, and Molting of Individuals 527
        34.2.3.2 Reproduction 528
        34.2.3.3 Predation 528
            34.2.3.3.1 Effect of Body Weight and Temperature on the Ingestion Rate.... 528
            34.2.3.3.2 The Predation Task 528
34.3 Results 528
    34.3.1 Simulation of the Mortality of a Single Stage: Role of the Demographic Noise 528
    34.3.2 Simulation without Predation 530
    34.3.3 Effect of Cannibalism and Spatial Representation 530
    34.3.4 Effect of Predation on the Spatiotemporal Dynamics of the Population 530
34.4 Discussion 533
    34.4.1 Toward an Improvement of Individual-Based Models 534
        34.4.1.1 A Common Methodology for IBMs 534
        34.4.1.2 Common Tools for IBMs 535
        34.4.1.3 Common Languages and Common Descriptions for IBMs 536
    34.4.2 IBMs and Population Dynamics of Zooplankton 536
34.5 Conclusion 538
Appendix 34.A 539
Acknowledgments 540
References 540

34.1 Introduction

In marine ecosystems the production of planktonic copepods supports most food webs, directly affecting higher trophic levels (including pelagic fish populations) and the biological pump of carbon into the deep ocean.[1] Traditionally, in ecological models dealing with pelagic processes, the lowest trophic levels have been represented with functional groups. Except when size is treated explicitly, these models assume that all individuals within a trophic compartment are identical.[2] Hence, the prevailing approach in observing and modeling pelagic populations is to characterize average processes (such as feeding, birth, growth, migration, and mortality) that are acting on total abundances or on spatially distributed fields of concentrations.[3–5] However, several studies have shown that zooplankters, including copepods, possess individual traits and behaviors;[6–8] consequently they act and react individually. Furthermore, the individuals may encounter different patterns of food and physical parameters, which may increase the inter-individual variability.[9] At the population level the individual variability on the molting rate of copepods was observed from experimental work.[10,11] The new emerging individual-based modeling (IBM) techniques are appropriate to cope with such variability.[12] The success of this approach can be attributed to the fact that IBMs make more realistic assumptions than the more traditional state variable models.

The IBM is a "bottom-up" approach, and may be opposed to the classical state variable modeling, which is a "top-down" approach.[13,14] Uchmański and Grimm[15] defined individual-based models as models that should include discrete individuals with their complete life cycles as well as differences between individuals, and the dynamics of the resources (in general with spatial heterogeneity) over which the individuals compete. A corollary to this definition, which is a fundamental aspect of IBM, is the explicit representation of individual variability.[16] The differences among individuals affect the dynamics of the population. The interactions between individuals and/or their environment lead to global properties at the population level and/or at large spatial scales. This property referred to as "emergence" represents another fundamental concept of the IBM approach.[17]

The last decade (1990s) has been characterized by an exponential increase in the number of publications using IBMs in ecology.[18] Certainly the increase in computer power and the development of object computer languages and associated tools as simulators have significantly contributed to the rise of the IBM.[19] Nevertheless, the creation and execution of an IBM is still a delicate operation that requires substantial computational investment.[20] Therefore, there is often a distance, if not a gap, between modelers and end users, which may have negative consequences. Even if some problems still exist, caused by the complexity of individual-based models,[12] the recent progress in developing specialized platforms and tools has increased the accessibility of this approach to a broader range of scientists who may or may not have previous modeling experience.[14]

In this chapter we use a recently developed platform based on multiagent systems and specialized in the development of population dynamics models.[20] This platform, called *Mobidyc*, is designed to make the end users true masters of their models without the assistance of computer experts. This new platform has been used to introduce an educational example that shows how IBM development can lead to new perspectives on our understanding of aquatic ecosystems structures and functions. First, *Mobidyc* has been used to build easily and without any hard-coding a model of the whole life cycle of the copepod *Centropages abdominalis*. We subsequently introduced (1) cannibalism, (2) predation, and (3) explicit spatial patterns to study their consequences on *C. abdominalis* at the population level. The results obtained here can be extended to other populations and/or organisms. In particular, within this platform the user can focus on model development and running experiments rather than hard-coding the model, which needs qualifications in computer programming that the biologist has not yet attained.

34.2 Development of the Model

34.2.1 Main Characteristics of the *Mobidyc* Platform

In the multiagent system (MAS) paradigm, agents perform their individual tasks with their own operational autonomy and reaction to their environment, but they influence the global behavior of the simulated system through local interactions.[21] *Mobidyc* is an all-agents architecture that focuses more on what each agent does than on what it actually is. The advantage of this approach is that all the different elementary tasks that form the behavior of individuals can be clustered into a low number of classes of activities, e.g., locate, select, translate, compute, end, and workflow control.[20] *Mobidyc* provides three kinds of agents:

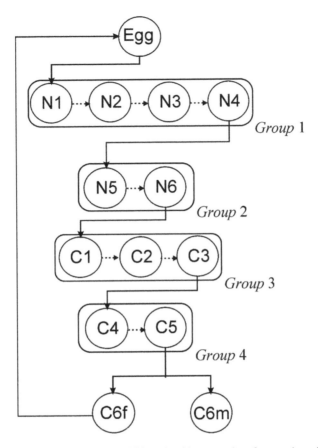

FIGURE 34.1 Schematic representation of a copepod life cycle with aggregation of post-embryonic stages into four groups: N1–N4, N5–N6, C1–C3, and C4–C5. Each stage is represented by a *Mobidyc* agent containing a dictionary of attributes and an ordered dictionary of tasks.

1. Animats that represent the typical individuals
2. Cells that represent a discretization of the space
3. Nonlocated agents that are optional and may provide general scenarios for all other agents or compute the results that the user wishes to save

34.2.2 Life Cycle Representation

For a cohort starting with eggs, and from an individual point of view, each individual egg has to follow the schematic template of Figure 34.1. To reach the adult stage (C6), an egg should pass successively through the six larval (naupliar) stages and the five juvenile (copepodite) stages. The success of this biological trajectory depends on the survival throughout all stages. Instead of using detailed representations of developmental stages, the life cycle of the copepod was simplified by aggregating post-embryonic developmental stages into four groups (Figure 34.1). The feeding mode of later developmental stages is omnivorous, and hence they were supposed to eat early stages (eggs to N4). So this life cycle representation aggregates similar stages from an ecological point of view (i.e., N1 to N4 accessible to predation), avoiding any disadvantages resulting from an over-aggregation of stages.[11,22]

34.2.3 Model Architecture

A basic animat corresponding to the classical individuals of IBMs already contains three attributes: age, location, and number. The last attribute is set to one to represent an individual; however, for other models

TABLE 34.1

List of Attributes Used to Define the Life Cycle of the Copepod *Centropages abdominalis*

Attribute	Owner	Definition
Age	All stages	Age (days)
Location	All stages	Position in the grid of cells
Number	All stages	Number of individuals per agent
Rand	All stages	Intermediate, used to assign a random real number between 0 and 1
State	All stages	Intermediate, used to define the state of survival
pSurv	All stages	Probability of survival in the stage
Duration	Egg to C4–C5	Stage duration
Weight	All stages	Body carbon weight (µg C)
stateSex	C4–C5	Intermediate, used to define the type of sex in the stage C5
pFemale	C4–C5	Sex ratio of the population (proportion of females)

TABLE 34.2

Definition of the Tasks Used to Develop the Life Cycle of the Copepod *Centropages abdominalis*

Agent Task	Definition	Type	Setting
Grow older	Compute the age of the agent	Predefined	my_age:=my_age + simulator_timeStep
StateSurvive	Determine the state of the agent for survival	Predefined (ModifyAttributes)	my_rand : = Random_real [0,1[my_state : = my_pSurv – my_rand
CondMortality	Conditional mortality	Predefined	if my_state < 0 then I am dead
TempDuration	Effect of temperature on the duration on the stage	Predefined (ModifyAttributes)	my_duration : = Random real [Dmi, Dmax[
Growth	Effect of age on the growth of weight for groups of stages	Predefined (ModifyAttributes)	my_weight : = W0*exp(g*my_age)
Hatching/Molting	Transfer from one stage to the next	Predefined (Metamorphosis)	If my_age > my_duration
DefineSex	Define the sex of agent after molting in the stage C4–C5	Predefined (ModifyAttributes)	my_rand : = Random_real [0,1[my_stateSex : = my_pFemale – my_rand
MoltingM/MoltingF	Transfer from C4–C5 to C6m/from C4–C5 to C6f	Predefined (Metamorphosis)	If ((my_stateSex<0) AND (my_age > my_duration)) /((my_stateSex>=0) AND (my_age > my_duration))
Reproduction	Reproduction of C6F into eggs	Predefined (Reproduction)	Number of offspring per individual equal to my_fecund

Note: The predefined task *ModifyAttributes* is very useful for developing several models as the example shown here because it is designed to compute mathematical calculations involving agent attributes.

the number can be an integer for a group of identical individuals (i.e., clutch of eggs) or a real number to represent a concentration of a subpopulation. In our example, seven additional attributes are needed to complete the definition of the model without predation (Table 34.1). Souissi and Uye[23] focused on a period of 5 months where zooplankton was largely dominated by the copepods *C. abdominalis* and *Acartia omorii*.[24] For the same period the average values for temperature and chlorophyll concentrations were 12.25°C (SD = 2.73°C) and 4.50 µg/l (SD = 2.86 µg/l), respectively. Hence, for the sake of simplicity, the simulations will be done using the constant values of 12°C and 5 µg/l for temperature and chlorophyll, respectively. The definition and the settings of the different tasks used for the first version of the model (without predation) are shown in Table 34.2.

Most agent behaviors require a computation of mathematical relations to update their attribute values. In *Mobidyc*, this is done through the task *ModifyAttributes*. An example of its use is demonstrated in Figure 34.2 with the setting of the copepod survival.

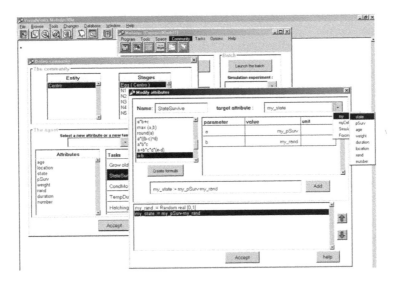

FIGURE 34.2 Snapshot of the task *ModifyAttributes* to define the state of survival in the egg stage. The mathematical expression can be chosen from the list on the left or can be created using a simple mathematical grammar syntax. A parameter value can be constant (fixed numerical value), an attribute value, or a function of an attribute value. For an animat, expressions can involve attributes of the running animat (i.e., my_*pSurv*), of its cell (myCell_*attributeName*), of one of the simulator characteristics (i.e., Simulator_*timeStep*), or of any nonlocated agent (i.e., Forcing_*temp*). One task using *ModifyAttributes* can contain a series of mathematical instructions. The window in the second plane is used to define the community from the principal menu of *Mobidyc* interface (the interface in the third plane and the VisualWorks environment in the background).

TABLE 34.3

Parameter Values Used in the Simulation

Stages	D (days)	W_{min} ($\mu gC/ind$)	W_{max} ($\mu gC/ind$)	g ($\mu gC/d$)	$pSurv$
Egg	2.16	0.027	0.027	0	0.850
N1–N4	8.65	0.015	0.077	0.19	0.900
N5–N6	4.57	0.109	0.162	0.09	0.925
C1–C3	8.22	0.233	0.882	0.16	0.925
C4–C5	5.41	1.567	3.394	0.14	0.950
C6m		4.187	5.117	0	0.975
C6f		6.062	7.409	0	0.975

34.2.3.1   *Modeling Survival, Growth, and Molting of Individuals* — The survival (or death)
of any individual for a time step is given by the sign of the difference between its probability of survival (*pSurv*) and a random number drawn from an uniform distribution between 0 and 1. For example, for a high value of survival (e.g., *pSurv* = 0.95), the probability of drawing a random number greater than *pSurv* is low; consequently, the individual has a greater chance of surviving than dying during the simulation. The individuals surviving in the stage should pass successively from one stage to another until they reach the adult stage. For constant conditions of food and temperature the probability of transfer from one stage to another is a simple function of age.[10] The individual variability of molting explains the shape of the probability distribution of molting rate at the population level.[11] Using the empirical expressions proposed by Liang et al.,[25] we compute the stage duration (*D*) and the weight (*W*) of developmental stages (Table 34.3) for the considered temperature (12°C). For aggregated stages (e.g., N1 to N4), the individual body mass varies between the value of the first developmental stage in the group (W_{min}) and the value of the last developmental stage in the group (W_{max}) (Table 34.3). We assume that the individuals newly recruited in the group grow exponentially with a constant growth rate *g* (Table 34.3). For adult stages the individual weight interval [W_{min}–W_{max}] is simply obtained by considering 10% of deviation around the computed value

of weight. To introduce individual variability in stage duration a similar deviation (10%) is also considered for the mean stage duration (*D*). For these parameters (*W* of C6 and *D*) the value for each individual is taken randomly in the corresponding interval. The sex of individuals may be identified at stage C5.[25] Because only adult stages are involved in the reproduction process, for the sake of simplicity, the sex of individuals is only separated in the adult stage with a sex-ratio 1:1. In stage C4–C5 two tasks of meta-morphosis are defined, one corresponding to the molting into males (*MoltingM*) and the other task represents the molting into the female stage (*MoltingF*).

34.2.3.2   Reproduction — The empirical expression developed by Liang et al.[26] for *C. abdominalis* fecundity is used. For nonlimiting food conditions and at temperature of 12°C, the daily egg production rate computed with Liang's model equals 85 eggs/female. Consequently, during one time step of simulation (0.5 days) the number of eggs produced by one female is taken randomly between 38 and 47, which represents 10% of dispersion.

34.2.3.3   Predation — The expressions given here are used for representing both cannibalism of older developmental stages and for predation by an external predator that we will introduce later. In all cases the vulnerable individuals belong to early developmental stages: from eggs to N4.

34.2.3.3.1   Effect of Body Weight and Temperature on the Ingestion Rate.   The general allometric relationship between the maximal specific ingestion rate (I_{max}: day^{-1}) and the body mass (μg C) given by Moloney and Field[27] was used:

$$I_{max} = 0.56W^{-0.25} \tag{34.1}$$

The effect of temperature on ingestion rate is introduced by assuming an exponential effect with Q_{10} set to 3, a common value for zooplankton equal to 3.[23] The combination of the effects of body mass and temperature leads to the following expression:

$$I_{(W,T)} = \Delta t \times I_{max} . W . \left(e^{0.1\log(Q_{10})(T-T_r)} \right) \tag{34.2}$$

where $I_{(W,T)}$ is the maximal ingestion ($\mu gC \cdot ind^{-1}$) during a simulation time step ($\Delta t = 0.5d$), *W* is the individual body mass ($\mu gC \cdot ind^{-1}$), and T_r is the reference temperature (20°C).

34.2.3.3.2   The Predation Task.   We have to build a user-defined task that we have chosen to call *PredateOneStage* by using the proper primitives (Figure 34.3). The predator should first locate the prey agents according to its spatial perception. In our example we limit the predator perception to the cell where it is located, so the starting primitive is *MyCell*. But a more general primitive, called *MyNeighborhood*, returns the adjacent cells of the running animat or cell according to a radius, and can also be used as a starting primitive.

In a second step, the predator should select the prey. This action is represented by two successive primitives *FromCellToAnimat* giving all animats on the cell and *SelectOnName* giving only the reduced list of stages accessible to predation. Then the predator agent eats the preys randomly until its satiety is reached or when the list of prey is empty. The primitive *ModifyAttributes* computes the stocking of ingested matter with an efficiency of 70%. This quantity is stocked on a new attribute called *ing*.

We can now add this new task to any developmental stage or other new entity. For example, for C6 stages we need only to compute the maximal ingestion using the mathematical expression of *ModifyAttributes* and then to adequately parameterize the predation task. We will show in Section 34.3.4 how the same task of predation can be parameterized differently for another predator.

34.3   Results

34.3.1   Simulation of the Mortality of a Single Stage: Role of the Demographic Noise

Several studies have focused on the effects of demographic stochasticity on the extinction risk of populations.[28] Kendall and Fox[29] suggest that demographic variation should be interpreted as individual

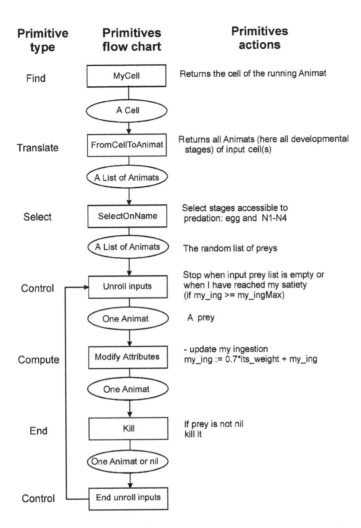

Primitive type	Primitives flow chart	Primitives actions
Find	MyCell	Returns the cell of the running Animat
	A Cell	
Translate	FromCellToAnimat	Returns all Animats (here all developmental stages) of input cell(s)
	A List of Animats	
Select	SelectOnName	Select stages accessible to predation: egg and N1-N4
	A List of Animats	The random list of preys
Control	Unroll inputs	Stop when input prey list is empty or when I have reached my satiety (if my_ing >= my_ingMax)
	One Animat	A prey
Compute	Modify Attributes	- update my ingestion my_ing := 0.7*its_weight + my_ing
	One Animat	
End	Kill	If prey is not nil kill it
	One Animat or nil	
Control	End unroll inputs	

FIGURE 34.3 An example of building a task with primitives, here a task of predation of omnivorous stages (C6m and C6f) on early developmental stages (eggs, N1–N4). Seven primitives (represented with boxes) corresponding to six types are strung out to build this task. The output arguments of each primitive (represented with ellipses) represent the input argument of the next primitive.

variation in demographic traits, rather than as sampling error. One individual may do well and reproduce, a predator may eat another, while another may fail to reproduce because it did not find any mate. All these random variations are ignored when we consider the average birth or death rates of a population.[30,31] To illustrate the role of demographic noise the basic example of a single stage is presented. The probability of survival (*pSurv*) varied from 0.975 to 0.800 with 0.025 increments. For each *pSurv* value 20 runs were realized, to reach a significant representation of the variability between runs. Figure 34.4A shows the results of these simulations for the four selected values of *pSurv*: 0.975, 0.925, 0.875, and 0.825. For fairly large numbers of individuals, this probabilistic interpretation of *pSurv* is similar to a deterministic model (Figure 34.4). However, if only a small number of individuals remain (e.g., fewer than 30 individuals) significant deviations from the deterministic mean may occur.

Each set of 20 simulations is used to estimate the constant mortality rate (z, day^{-1}) by fitting the exponential model ($N_t = N_0 e^{-zt}$). The different fits for each set of simulations are shown in Figure 34.4A with continuous lines. The effects of varying time interval windows used to fit the model are negligible; hence the average value of z is used to establish a linear relationship between *pSurv* and z (Figure 34.4B). The importance of taking the individual probability of survival is emphasized, especially when the number of individuals is low.

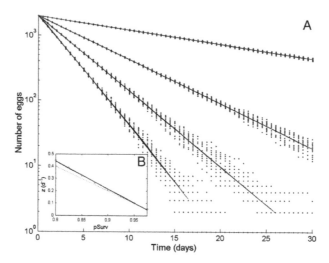

FIGURE 34.4 (A) Simulations of the decrease of individuals in a single stage with four different values of probability of survival (0.975, 0.925, 0.875, and 0.825). The symbols represent the simulated points for 20 repetitions and continuous lines are the fit of Equation 34.4 for several time intervals (see text for more details). (B) The continuous line represents the linear regression *pSurv* used in the model and the fitted value of ($z = -2.269pSurv + 2.257$; $r^2 = 0.999$). The discontinuous line represents the theoretical line of equation $z = 2\,(1 - pSurv)$.

34.3.2   Simulation without Predation

Now consider a local population of the copepod *C. abdominalis*, all simulations starting with four females and four males. The model correctly simulates the succession of the developmental stages with respect to their stage durations and other demographic parameters such as reproduction and mortality (Figure 34.5). The stochastic representation of the stage survivals (Table 34.2) shows the visible differences between the five runs realized with the same set of parameters (Figure 34.5). For longer simulation durations the population should increase exponentially, which is a general property of stage-structured populations simulated under nonlimited conditions.

34.3.3   Effect of Cannibalism and Spatial Representation

As a first step, the predation task *PredateOneStage* is activated for adult stages (i.e., C6m and C6f) without modifying the other parameters of the model. With no spatial representation (i.e., all individuals in the same cell), all runs resulted in the extinction of the population. There is no viability for the *Centropages* population under these conditions. This occurred because without spatial representation, the encounter between prey and predators is maximal as there is no refuge for the prey.

   To study the effects of spatial representation we consider a grid of hexagonal cells (each cell has six neighbors) and add the predefined task *Move* for all developmental stages except eggs. This task can be parameterized to reproduce a random walk according to a radius of action, given in number of cells, or following a gradient of an attribute of the cell. For example, if cells represent a resource, agents can move following an increasing (or decreasing) gradient of the resource. Here, all agents move randomly with a radius of action of one cell. Figure 34.6 shows the results of simulation of the population (five runs) for a grid of 5 × 5 cells. The introduction of the spatial representation increased the time duration of the simulation before the extinction of the population. There was no recruitment for the second generation for all runs (Figure 34.6). By increasing the spatial grid to 10 × 10 cells, the viability of the population becomes possible (Figure 34.7). However, two runs of five produced extinction of the population.

34.3.4   Effect of Predation on the Spatiotemporal Dynamics of the Population

The early developmental stages of copepods are also the prey of many other zooplankters, particularly fish larvae.[32] These visual predators can locate prey patches or individual prey according to their visual

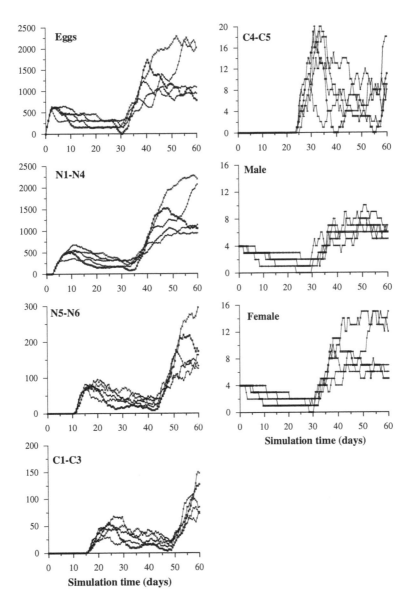

FIGURE 34.5 Simulation of the development of the population of *C. abdominalis* during 2 months under constant conditions of food (5 μg chlorophyll/l) and temperature (12°C). Five runs (represented with different symbols) are realized with the same initial conditions of a mixture of eight adults with a sex-ratio 1:1.

field before attacking.[33] In this simulation an additional predator (i.e., small fish larvae) eating only the first developmental stages is added. Figure 34.8A shows the initial conditions with four males (gray triangles), four females (black triangles), and six predators (squares). Each predator had only one developmental stage and is able to move in its search area (Figure 34.8A) to the cell containing the highest density of prey. The user-defined task *PredateOneStage* is now added to the new predator (agent) and parameterized as previously done for adult stages. The maximal ingestion per predator and per day (*Imax*) was set to the arbitrary value of 1 μgC. In order to compute the densities of prey, a simple task was assembled and added to each cell. The representation of this agent (predator) and the cells is detailed according to the syntax (macrolanguage) proposed by Houssin et al.[34] in the appendix to this chapter.

Figure 34.8 shows the spatiotemporal evolution of the system. The spatial heterogeneity of prey distribution can be explained by the random walk of copepod females (with a one-cell range). Here, the order of the tasks of adult females is crucial and may modify the results of the simulation. We used the

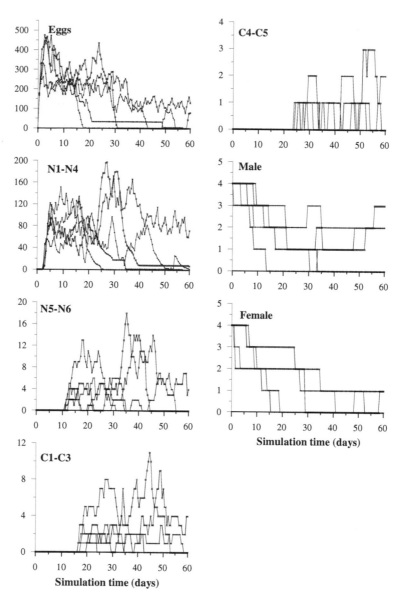

FIGURE 34.6 Simulation of the development of *C. abdominalis* over 2 months. Space is represented by a grid of 5 × 5 hexagonal cells. All postembryonic developmental stages move with a radius = 1 cell. The predation task shown in Figure 34.3 was activated for the stage C6. The other parameters of the simulation are identical to those of the runs obtained in Figure 34.5.

following task order, *Move → PredateOneStage → Reproduction*, which gives an additional chance to the laid eggs in the parental cell to escape from parental predation. Figure 34.8B shows an isolated local maximum (low area of the grid) of prey density after 5 days of simulation. This maximum corresponds to the reproduction of the nearest female (Figure 34.8B). This "patch" of prey is located in the search area of two predators that moved simultaneously to the same cell. The model is defined in a sequential way so the first predator, taken randomly, will modify the state of the system (i.e., number of prey). With such a simple representation of the feeding behavior of the predators, a school of predators is obtained after 8.5 days of simulation. Figure 34.8C shows an intersection of search areas of all predators composed by three cells and containing one female. All predators locate the same patch and move to the same cell (Figure 34.8D). The predators with simple deterministic behavior stay within the school

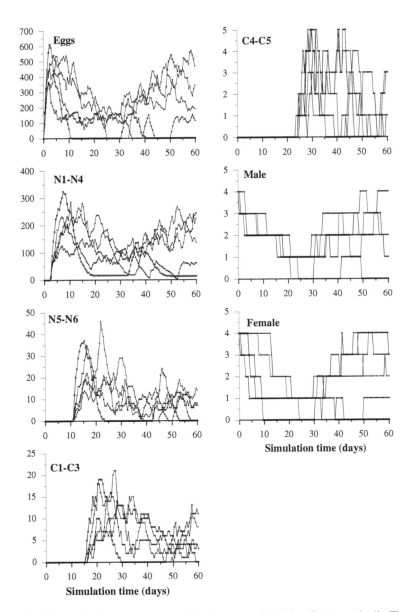

FIGURE 34.7 Simulation of the development of *C. abdominalis* in a grid of 10 × 10 hexagonal cells. The other conditions and model parameters are identical to those of Figure 34.6.

until 13 days of simulation (Figure 34.8E). The school then locates itself to a poorer area and the patch of prey outside their search area. The predators have no advantage to stay in the school because the prey density does not sustain their bioenergetic demand and then the school is separated (Figure 34.8F). All these behavioral and spatial patterns result from the interaction between individuals and can be regarded as the emerging properties of the system.

34.4 Discussion

Because IBM-related approaches are sometimes nonconvergent, noncompatible, and/or nonaccessible to a wide range of computer- and noncomputer-oriented scientists, we stress here the need to agree on a scientifically sound and common framework. This may speed and facilitate direct exchanges among

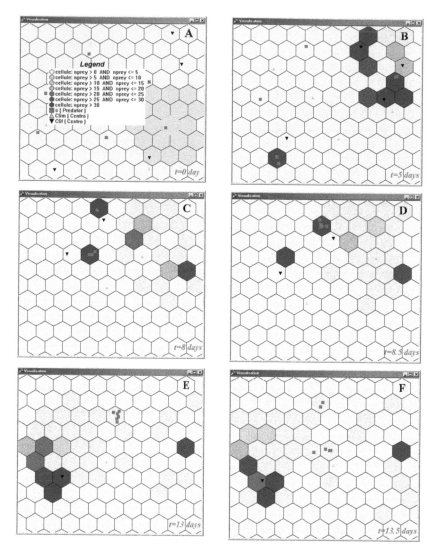

FIGURE 34.8 Effect of predation on the spatiotemporal dynamics of the development of the population of *C. abdominalis*. (A) Initial condition of the simulation, cells count the vulnerable stages to predation (eggs and N1–N4), which are then represented with a color scale. The legend shows only stages exerting a predation C6m (gray triangle), C6f (black triangle), and the predator (square). The gray cells show the search area of one predator corresponding to the radius of action = 2 cells. (B) Spatial structure after 5 days of simulation. (C, D) Two successive time steps representing the formation of a school of predators. (E, F) Two time steps representing the separation of the school of predators.

IBM-oriented scientists. Particularly, this goal may be achieved only via establishment of a common procedure requesting clear definition of the link between IBMs, theory, common methodology, tools, and languages, as well as clarification of the connections between IBMs and population dynamics.

34.4.1 Toward an Improvement of Individual-Based Models

34.4.1.1 *A Common Methodology for IBMs* — As the populations are represented by a collection of interacting individuals and not by continuous densities, the development of IBMs should change our way of modeling biological systems.[35,36] This new way of modeling, referred to as the "bottom-up" approach, presents a great potential for challenging theoretical and practical ecological studies.[15,19] However, a recent review covering the last decade[18] clearly shows that the potential offered

by this approach is still far from being fully exploited. Additional experiments with IBMs should be conducted to try to relate them more directly to theory. Levin and Durrett[37] have argued that phenomenological approaches are deficient in that they lack attention to underlying processes, whereas IBMs may obscure the essential interactions in a sea of details. They nevertheless recognize the advantages of IBMs and suggest that the gap be bridged between the mathematical formalism of classical approaches and the detail of IBMs. On the other hand, recent papers based on simple mathematical models of population dynamics have suggested the use of stochastic IBM approaches, which are more appropriate for considering environmental variability and individual behavior.[9] In general, IBMs are used because a fundamental significance of individual variability is suspected, i.e., significance with respect to persistence and other dynamic properties (resilience, resistance) of the population.[12] However, in their present state, IBMs have several drawbacks and should be developed according to a common methodology. Some recent studies have focused on comparative studies between classical and individual-based models. Wilson[38] showed that the IBM representation of a spatially distributed predator–prey population leads to strong spatial structuring in contrast to predictions from its representative analytical formulation. Sumpter and Broomhead[39] approximate the emerging properties of an IBM at the population level by a logistic equation. They further demonstrate how this approximation is obtained mathematically and how the parameters of the exponential logistic equation can be written in terms of the parameters of the individual-based model. Recently, Berec[40] showed that a class of spatially explicit IBMs could be analyzed mathematically by relating them to partial differential equations. In these examples the IBM formulation is close to the analytical model at the population level. From these basic examples, it can be seen that the outcome of IBMs can and cannot be comparable to the results provided by more standard modeling approaches. This is an extremely strong argument to demonstrate the ability of IBMs to provide new insights into *a priori* known results regarding patterns and processes in both space and time.

The development and use of IBMs is continuously increasing, and the development of a new framework for their analysis and comparison is recommended. Only a few studies have proposed such techniques capable of capturing generalities to gain insight beyond a specific simulation model.[13,41] Recently, Railsback[17] proposed using the main concepts from a new research field of complex adaptive systems (CAS) as a theoretical foundation for IBMs. CAS study the emergence of complex behaviors even in systems based on relatively simple interactions between few individuals. The proposed framework converges toward Grimm's recommendations proposing that IBM modelers concentrate more on the process of modeling itself rather than merely on the entities that are modeled.[18] All these recent papers have aimed at improving the great potential of IBMs in contributing significantly to basic and applied ecology recognize the necessity of improving computer implementation and reproducibility of models.

34.4.1.2 Common Tools for IBMs

— Unfortunately, most published IBMs are not readily available to the whole scientific community and still look like "black boxes" of bygone years. Although Grimm[18] presented a review of IBMs published in the last decade on the basis of 50 papers, he simply classified these models according to a set of criteria (see his table 1). One fundamental difficulty in the evaluation and comparison of IBMs in the literature arises from the absence of any theoretical formalism, such as differential equations, where one can express, conserve, and compare one model to another, or to export it to another modeling tool. This problem of readability, which was briefly raised by Judson,[19] amplifies the suspicion with which the results of an individual-based model may be viewed. On the other hand, the number of platforms and tools dedicated to developing object-oriented modeling has increased although their accessibility to noncomputer scientists remains limited. A good example of a multiagent software platform for the simulation of complex adaptive systems is the *Swarm* software.[42] Lorek and Sonnenschein[14] revised these tools and focused on the conception of a modeling tool (WESP-TOOL) more accessible to end users in developing and analyzing IBMs for metapopulations. This imposed a better separation of the components of the model from those of the platform, and the result was more reliable as the platform controls more components involved in the execution of the model. These efforts try to clarify the "black box" or the model in simulators, but even if scientists develop most of these models they often remain inaccessible to experts in population dynamics. These systems require ecologists to develop some computer programming skills, although the software provided makes the process of programming easier.[43]

In the specific framework of *Mobidyc*, we have focused on the concept of constructing several IBMs with increasing levels of complexity and using them to study different questions. All these steps were developed without hard coding. Moreover, the use of primitives allowed for the decomposition of each task into some elementary tasks. With this approach, the end-user modeler can easily control all steps of model implementation. Consequently, he or she can test easily several algorithms and/or parameterizations involved in the model. This concept is fundamentally different from building a unique and complete model, which may be more realistic but too complex to allow easy validation of all the steps used to build it. In such a situation it is difficult to compare this model with others because the code sources or the architecture are not easily accessible. On the contrary, the examples developed here with *Mobidyc* can be made by any end user after familiarization with the *Mobidyc* architecture. This platform is fully adapted to end users and should be enriched and strengthened in the future by incorporating feedback from users, for example, adding new primitives and enriching the library of examples. The main objective is to dissociate IBMs from the simulator carrying out their execution.

34.4.1.3   *Common Languages and Common Descriptions for IBMs* — Up until now, the

biggest weakness of modeling tools is that there is no link between the natural language in which the researcher describes the model and the actual programming language in which this model is written. Mamedov and Udalov[43] proposed the CENOCON software to provide a means to run conceptual ecological models where programming is not needed at all. Even though this software takes a text file with descriptions of organisms and space, giving more flexibility to the running of IBMs, the main structure of the simulator and the proposed rules of interactions between species are still imposed by the authors. For example, a user should understand the whole complex mechanisms behind this simulator (see their fig. 1). This situation may slow the process of standardization of IBMs as a useful tool for ecological studies. Lorek and Sonnenschein[14] proposed a WESP-DL specification language, which is completely defined and stable. However, this macrolanguage requires a minimum of syntax to structure the program. Alternatively, the macrocommands used here with the primitive *ModifyAttributes* (see Figure 34.2) provide a closer description because the mathematical expressions are associated with the agent's attributes. In both cases the descriptive language is nevertheless not separated from the platform or the simulator. The use of component programming (through primitives in this example) may facilitate the design of a domain-specific language in which these models could be written independently of the platform. An example shown in Appendix 34.A was inspired by Houssin et al.[34] showing us the expert language used to describe a predator searching for prey and cell counting for those prey. Using primitives improves task description (i.e., example of Figure 34.3) and offers a promising framework for developing macrolanguages of model description, which are closer to expert languages (see Appendix 34.A). The aim of this framework is to make model development independent of platforms. The model can be built interactively using the platform and then translated to this language. Furthermore, this textual language can be compiled with any adapted compiler without needing an interface, gaining in execution and performance.[34]

34.4.2   IBMs and Population Dynamics of Zooplankton

With the exception of a few examples of object-oriented models developed for zooplankton species[44–46] representing "true IBMs," the classical state variable approach has been common practice in the modeling of zooplankton populations.[47,48] Moreover, the term IBM is sometimes used whereas the average values are modeled.[2,49] The later examples cannot be considered "true IBMs" according to the previously given definition in this chapter because resources used by individuals typically are not explicit. As a matter of fact, these models assume that individuals are born, produce offspring, and die practically at the same moment, as well as assuming that all individuals are identical.[50] The introduction of physical variability and advection can produce different spatial and temporal patterns.[51] However, the same equations and parameterization are used for all individuals. The only possible difference between individuals comes from the set of environmental factors generated by physical models. With such coupling in physical–biological models, physical forcing drives most of the biological patterns.

Uchmański[50] considered that the use of IBMs only where the description of individual life cycles includes stochastic elements that allow us to explore the importance of individual variation as opposed to population dynamics. The behavioral interactions between individuals represented in IBMs can also be responsible for global population dynamics.[39]

In this chapter illustrative applications are developed to show the potential of IBMs. The simulations shown here confirm the earlier conclusions on their role for both small-sized populations and local scales.[41] Mooij and Boersma[46] showed that most demographic parameters of *Daphnia* populations simulated using an object-oriented model varied as a function of the number of modeled individuals until a threshold value was reached (around 100 individuals) above which the simulated parameters remained constant. In our first example, the variation in the probability of survival in a single stage showed an increase of dispersion between runs for low numbers of eggs (Figure 34.5). This is fully consistent with our above-stated results. The examples developed in this chapter are specifically adapted to local populations, which encompass most ecological situations in aquatic ecology, at least in zooplankton ecology.[52,53] The introduction of both individual behavior, even though simply represented, and spatial representation, pointed out the need for careful consideration of the links between spatial scales and their consequences on population dynamics. To introduce the potential effect of spatial dispersion (e.g., due to diffusion or advection or any other processes) we considered a large grid of 50×50 cells. Figure 34.9 shows the effects of initial density of adults and the introduction of advection on the temporal evolution of eggs. Without advection, each female produced very local populations and consequently fully controlled the development of its production. This situation leads to a stable global population with a low abundance (Figure 34.9A, C, and E). On the other hand, the introduction of dispersion of eggs and N1 to N4 increased the size of the population (Figure 34.9B, D, and F). One must also note here the increase in variability between runs for the same initial conditions. In this way, any extrapolation conducted on both *in situ* biotic and abiotic processes and based, for example, on averaging or least squares procedures, is very suspicious if not irrelevant. It is thus of major interest to take into account possible interactions between individuals, local populations, and physical forcing. Here again spatially explicit IBMs can be helpful to study these between-scale transfers as well as the interaction between physical and biological processes.

Recently, Seuront and Lagadeuc[54] continuously sampled the population of the calanoid copepod *Temora longicornis*, and assessed the high heterogeneity observed using multiscaling analysis techniques. The observed patterns of spatial distributions of copepods[55] at several scales gave new direction to the development of new sampling techniques adapted to the small scales characterizing plankton development. By taking into account the possible complex interactions between external (turbulence, advection, food patchiness) and endogenous (feeding behavior, vertical migrations) mechanisms leading to patchy properties of plankton distribution,[52,53] IBMs will also be useful for improving studies concerned with small-scale processes. Other sampling techniques (plankton nets) give estimations for seasonal (or larger spatial scales) patterns. However, small-scale variability is excluded, increasing the difficulties in interpreting the results when either models of population dynamics or inverse techniques are used. Souissi and Uye[23] developed a model representing the predator–prey interactions involving several developmental stages of the copepods, *C. abdominalis* and *Acartia omorii*. An age-within-stage model based on differential equations and the previously published parameterization of demographic processes was used.[10,11,48] The model simulated interesting patterns of the temporal evolution of the predation rate.[23] The different numerical simulations confirmed the role of predation and cannibalism in controlling these copepod populations, as argued empirically by Uye and Liang.[24] The modeling exercise allowed some biases in sampling strategies to be revealed, leading to an underestimation of early developmental stages. The data used in Souissi and Uye[23] and published earlier in Liang et al.[25] are not adapted to be used for validation of IBMs such as the models developed here. However, it is interesting to mention that models without spatial dimension (0D) give one deterministic run for a fixed combination of the number of parameters. Stage-structured population models, even those based on mathematical representations, generally ordinary differential equations (ODEs), still remain difficult to validate.[56] In addition, these 0D models assume a homogeneous distribution of copepods. In other words, irrespective of volume, these models represent processes in the same way (1 liter, 1 cubic meter, other volume). Most other cases coupling copepod development with physical models have led to one-, two-, or three-dimensional

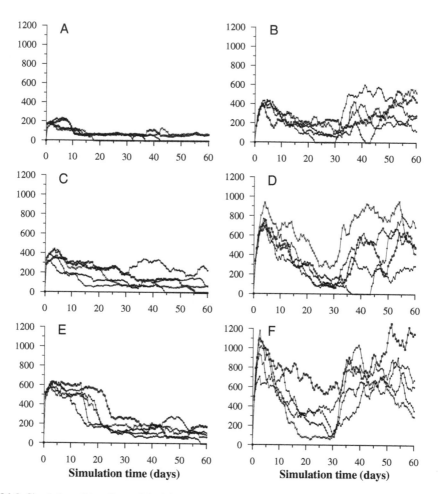

FIGURE 34.9 Simulation of the effects of the initial density of predators and the advection represented by a nondirectional random walk on the time evolution of eggs in a large grid of 50 × 50 hexagonal cells. The other conditions and model parameters are identical to those of Figure 34.5 to Figure 34.7. The figures on the left (A, C, E) correspond to simulation without advection; the produced eggs stay in the cell and the stage N1–N4 move randomly with a radius = 1 cell. The figures on the right (B, D, F) represent a dispersion of prey (eggs and N1–N4) introduced by considering a random walk with a maximal radius = 3 cells. The initial conditions were modified: 4 males and 4 females (A, B), 8 males and 8 females (C, D), and 12 males and 12 females (E, F).

simulations where the representation of copepods was very simple.[51,57,58] In most simulations, copepods follow the hydrodynamic fields without expressing possible feedback on physical processes.

Another major advantage of IBMs based on object-oriented modeling is the possibility of synchronous (pseudo-parallelism) representation; for example if a prey is met by two predators only one predator can capture it. This modeling approach considers a population as a consequence of individual interactions, so this advantage should be explored to study the emerging properties using several behavioral scenarios. In particular, the framework of complex adaptive systems proposed by Railsback[17] seems to be well adapted to develop IBMs for copepod populations.

34.5 Conclusion

We suggest at this point that the development of individual-based approaches in aquatic ecology should be regarded as the basis of our future understanding of patterns and processes observed in the ocean. Generally speaking, in aquatic ecology an increasingly synoptic viewpoint should also be established taking into account the rapidly developing multiscale technology (e.g., References 59 and 60). More specifically, in plankton ecology the combination of both endogenous (physiology and behavior) and exogenous (e.g., turbulence) processes via IBMs will undoubtedly ensure the reconciliation of the observed discrepancies between experimental or field data and the more classical standard modeling approaches.

Appendix 34.A

Expert language proposed to describe the predator and the cells used in the simulations of Figure 34.8. This macrolanguage used to describe the models developed with *Mobidyc* is closer to an expert language. It is inspired by the example developed by Houssin et al.[34] Reserved words or system-generated locutions are in italics. Comments (both user-defined or system-generated) are delimited with { }.

{ Predator definition }
[*Entity*]
   Name : Predator ;
[Stages]
   s
[*Attributes*]
   location, age, number ;   {default attributes}
   Imax = 1;   {maximal ingestion rate}
   Ing = 0;   {stock of ingested preys}
[*Tasks*]
   Grow older   {predefined: *GrowOlder*}
      {*my*_age : = *my*_age + *Simulator_timeStep*}
   Move   {predefined: *Move*}
   {*Reach a cell according to the following parameters and options: include the initial cell in the search area; range of vision: 2 cells; go upward a cell gradient of:* nprey; *cell final selection: the nearest one*}
   PredOneStage   {user-defined task}
      InitIng   {predefined: *ModifyAttributes*}
         *my*_ing : = 0
      My Cell -> *aCell*   {primitive: *MyCell*}
      From Cells to Animats -> *aListOfAnimats*   {primitive: *FromCellsToAnimats*}
         {*returns all animats living on input cells*}
      Select on name -> *aListOfAnimats*   {primitive: *SelectOnName*}
         {*Select objects (animats) of input list that match name:* egg (Centro); N1–N4 (Centro)}
      To unroll a list -> *aAnimat*   {primitive: *ToUnrollAList*}
      {*Returns one by one the objects (animats) of input list, and stops when list is over or when my*_ing > = *my*_imax}
         Eat the prey -> *aAnimat*   {primitive: *ModifyAttributes*}
            *my*_ing : = *my*_ing + *its*_weight ;
         Kill the prey   {primitive: *Kill*}
      End – To unroll a list   {primitive: *EndToUnrollAList*}
  { Cell definition }
  [*Entity*]
    OtherEntities

[*Stages*]
    Cell
[*Attributes*]
    nprey   {number of prey}
[*Tasks*]
    Count in a Cell   {user-defined task}
        Me -> *aCell*   {primitive: *Me*}
        All Animats -> *aListOfAnimats*   {primitive: *AllAnimats*}
            {returns all living animats}
        SelectOnName -> *aListOfAnimats*   {primitive: *SelectOnName*}
           {*Select objects (animats) of input list that match name:* egg (Centro); N1–N4 (Centro)}
        Count -> *aListOfAnimats*   {primitive: *Count*}
           {*Counts input items and puts the result in the attribute: nprey*}

Acknowledgments

The MOBIDYC platform (version 2.0) is freely available in French, English, and German and a tutorial (in French) is provided. It uses the Smalltalk Visual Works 7 environment developed by the Cincom Company. This environment is free for noncommercial use and runs on almost all platforms: http://www.avignon.inra.fr/mobidyc. This project aims to create a network group, which can participate in the improvement and development of these tools and also enrich the library of models. The models developed here can be downloaded (or obtained from S.S.) and immediately used. For example, biologists with *in situ* or experimental data who may like to test their own hypotheses can use this model as a starting point to adapt it to their own situation and obtain quick simulations. This work is a contribution to ELICO (Ecosystèmes Littoraux et Côtiers) and the contribution 4 of ECOREG (Ecosystem Complexity Research Group). We are very grateful to Volker Grimm and an anonymous referee for their helpful and constructive comments. We also thank Konstantinos Ghertsos and Peter G. Strutton for help with the English.

References

1. Ohman, M.D. and Hirche, H.-J., Density-dependent mortality in an oceanic copepod population, *Nature*, 412, 638, 2001.
2. Batchelder, H.P., Edwards, C.A., and Powell, T.M., Individual-based models of copepod populations in coastal upwelling regions: implications of physiologically and environmentally influenced diel vertical migration on demographic success and nearshore retention, *Prog. Oceanogr.*, 53, 307, 2002.
3. Gaedke, U., Population dynamics of the calanoid copepods *Eurytemora affinis* and *Acartia tonsa* in the Ems-Dollart-Estuary: a numerical simulation, *Arch. Hydrobiol.*, 118, 185, 1990.
4. Plagányi, É., Hutchings, L., Field, J.G., and Verheye, H.M., A model of copepod population dynamics in the Southern Benguela upwelling region, *J. Plankton Res.*, 21, 1691, 1999.
5. Robinson, C.L.K. and Ware, D.M., Modelling pelagic fish and plankton trophodynamics off southwestern Vancouver Island, British Columbia, *Can. J. Fish. Aquat. Sci.*, 51, 1737, 1994.
6. Bundy, M.H., Gross, T.F., Coughlin, D.J., and Strickler, J.R., Quantifying copepod searching efficiency using swimming pattern and perceptive ability, *Bull. Mar. Sci.*, 53, 15, 1993.
7. Schmitt, F. and Seuront, L., Multifractal random walk in copepod behavior, *Physica A*, 301, 375, 2001.
8. Doall, M.H., Colin, S.P., Strickler, J.R., and Yen, J., Locating a mate in 3D: the case of *Temora longicornis*, *Philos. Trans. R. Soc. Lond. B*, 353, 681, 1998.
9. Pitchford, J.W. and Brindley, J., Prey patchiness, predator survival and fish recruitment, *Bull. Math. Biol.*, 63, 527, 2001.

10. Souissi, S., Carlotti, F., and Nival, P., Food and temperature-dependent function of moulting rate in copepods: an example of parameterization for population dynamics models, *J. Plankton Res.*, 19, 1331, 1997.

11. Souissi, S. and Ban, S., The consequences of individual variability in moulting probability and the aggregation of stages for modelling copepod population dynamics, *J. Plankton Res.*, 23, 1279, 2001.

12. Grimm, V., Wyszomirski, T., Aikman, D., and Uchmański, J., Individual-based modelling and ecological theory: synthesis of a workshop, *Ecol. Modelling*, 115, 275, 1999.

13. Fahse, L., Wissel, C., and Grimm, V., Reconciling classical and individual-based approaches in theoretical population ecology: a protocol for extracting population parameters from individual-based models, *Am. Nat.*, 152, 838, 1998.

14. Lorek, H. and Sonnenschein, M., Modelling and simulation software to support individual-based ecological modelling, *Ecol. Modelling*, 115, 199, 1999.

15. Uchmański, J. and Grimm, V., Individual-based modelling in ecology: what makes the difference? *TREE*, 11, 437, 1996.

16. Grimm, V. and Uchmański, J., Individual variability and population regulation: a model of the significance of within-generation density dependence, *Oecologia*, 131, 196, 2002.

17. Railsback, S.F., Concepts from complex adaptive systems as a framework for individual-based modelling, *Ecol. Modelling*, 139, 47, 2001.

18. Grimm, V., Ten years of individual-based modelling in ecology: what have we learned and what could we learn in the future? *Ecol. Modelling*, 115, 129, 1999.

19. Judson, O.P., The rise of individual-based model in ecology, *TREE*, 9, 9, 1994.

20. Ginot, V., Le Page, C., and Souissi, S., A multi-agents architecture to enhance end-user individual based modelling, *Ecol. Modelling*, 157, 23, 2002.

21. Ferber, J., *Multi-Agents Systems. An Introduction to Distributed Artificial Intelligence*, Addison-Wesley, Reading, MA, 1999, 509.

22. Souissi, S., Modélisation du cycle de vie d'un poisson: conséquences pour la gestion des ressources exploitées. Application à l'étude de l'interaction entre les populations de copépodes et la population d'anchois. Ph.D. thesis, Université Pierre et Marie Curie Paris VI, Paris, 1998, 416 pp.

23. Souissi, S. and Uye, S.I., Simulation of the interaction between two copepods, *Centropages abdominalis* and *Acartia omorii* in an eutrophic inlet of the Inland Sea of Japan: role of predation and cannibalism, *J. Plankton Res.*, in revision.

24. Uye, S.-I. and Liang, D., Copepods attain high abundance, biomass and production in the absence of large predators but suffer cannibalistic loss, *J. Mar. Syst.*, 15, 495, 1998.

25. Liang, D., Uye, S.-I., and Onbé, T., Population dynamics and production of the planktonic copepods in a eutrophic inlet of the Inland Sea of Japan. I. *Centropages abdominalis*, *Mar. Biol.*, 124, 527, 1996.

26. Liang, D., Uye, S.-I., and Onbé, T., Production and loss of eggs in the calanoid copepod *Centropages abdominalis* Sato in Fukuyama Harbor, the inland sea of Japan, *Bull. Plankton Soc. Jpn.*, 41, 131, 1994.

27. Moloney, C.L. and Field, J.G., General allometric equations for rates of nutrient uptake, ingestion, and respiration in plankton organisms, *Limnol. Oceanogr.*, 34, 1290, 1989.

28. Jager, H.I., Individual variation in life history characteristics can influence extinction risk, *Ecol. Modelling*, 144, 61, 2001.

29. Kendall, B.E. and Fox, G., Variation among individuals and reduced demographic stochasticity, *Cons. Biol.*, 16, 109, 2002.

30. Burgman, M.A., Ferson, S., and Akcakaya, H.R., *Risk Assessment in Conservation Biology*, Chapman & Hall, London, 1993.

31. Wissel, C., Stephan, T., and Zaschke, S.-H., Modelling extinction and survival of small populations, in *Minimum Animal Populations*, Remmert, H., Ed., Springer, Berlin, 1994, 67.

32. Schmitt, P.D., Prey size selectivity and feeding rate of larvae of the northern anchovy, *Engraulis mordax* Girard, *CalCOFI Rep.*, 27, 153, 1986.

33. Caparroy, P., Thygesen, U.H., and Visser, A.W., Modelling the attack success of planktonic predators: patterns and mechanisms of prey size selectivity, *J. Plankton Res.*, 22, 1871, 2000.

34. Houssin, D., Bornhofen, S., Souissi, S., and Ginot, V., Entre programmation par composants et langages d'experts: Mobidyc, une plate forme multi-agents orientée utilisateur pour la dynamique des populations, *Tech. Sci. Inf.*, 21, 525, 2002.

35. Lomnicki, A., Population ecology from the individual perspective, in *Individual-Based Models and Approaches in Ecology*, DeAngelis, L. and Gross, L.J., Eds., Chapman & Hall, New York, 1992, 3.

36. Schmitz, O.J., Combining field experiments and individual-based modeling to identify the dynamically relevant organizational scale in a field system, *Oikos*, 89, 471, 2000.
37. Levin, S.A. and Durrett, R., From individuals to epidemics, *Philos. Trans. R. Soc. Lond. B*, 351, 1615, 1996.
38. Wilson, W.G., Lotka's game in predator–prey theory: linking populations to individuals, *Theor. Pop. Biol.*, 50, 368, 1996.
39. Sumpter, D.J. and Broomhead, D.S., Relating individual behaviour to population dynamics, *Proc. R. Soc. Lond. B*, 268, 925, 2001.
40. Berec, L., Techniques of spatially explicit individual-based models: construction, simulation, and mean-field analysis, *Ecol. Modelling*, 150, 55, 2002.
41. Wilson, W.G., Resolving discrepancies between deterministic population models and individual-based simulations, *Am. Nat.*, 151, 116, 1998.
42. Minar, H., Burkhart, R., Langton, C., and Askenazi, M., The swarm simulation system: a toolkit for building multiagent simulations, Santa Fe Institute Working Paper 96-06-042, 1996.
43. Mamedov, A. and Udalov, S., A computer tool to develop individual-based models for simulation of population interactions, *Ecol. Modelling*, 147, 53, 2002.
44. Laval, P., Hierarchical object-oriented design of a concurrent, individual-based, model of a pelagic tunicate bloom, *Ecol. Modelling*, 82, 265, 1995.
45. Laval, P., A virtual mesocosm with artificial salps for exploring the conditions of swarm development in the pelagic tunicate *Salpa fusiformis*, *Mar. Ecol. Prog. Ser.*, 154, 1, 1997.
46. Mooij, W.M. and Boersma, M., An object-oriented simulation framework for individual-based simulations (OSIRIS): *Daphnia* population dynamics as an example, *Ecol. Modelling*, 93, 139, 1996.
47. Sciandra, A., Study and modelling of development of *Euterpina acutifrons* (Copepoda, Harpacticoida). *J. Plankton Res.*, 8, 1149, 1986.
48. Souissi, S. and Nival, P., Modeling of population dynamics of interacting species: effect of exploitation, *Environ. Model. Assess.*, 2, 55, 1997.
49. Carlotti, F. and Wolf, K.-U., A Lagrangian ensemble model of *Calanus finmarchicus* coupled with a 1-D ecosystem model, *Fish. Oceanogr.*, 7, 191, 1998.
50. Uchmański, J., What promotes persistence of a single population: an individual-based model, *Ecol. Modelling*, 115, 227, 1999.
51. Miller, C.B., Lynch, D.R., Carlotti, F., Gentleman, W., and Lewis, C.V.W., Coupling of an individual-based population dynamic model of *Calanus finmarchicus* to a circulation model for the Georges Bank region, *Fish. Oceanogr.*, 7, 219, 1998.
52. Abraham, E.R., The generation of plankton patchiness by turbulent stirring, *Nature*, 391, 577, 1998.
53. Folt, C.L. and Burns, C.W., Biological drivers of zooplankton patchiness, *TREE*, 14, 300, 1999.
54. Seuront, L. and Lagadeuc, Y., Multiscale patchiness of the calanoid copepod *Temora longicornis* in a turbulent coastal sea, *J. Plankton Res.*, 23, 1137, 2001.
55. Tsuda, A., Sugisaki, H., and Kimura, S., Mosaic horizontal distributions of three species of copepods in the subarctic Pacific during spring, *Mar. Biol.*, 137, 683, 2000.
56. Bernard, O. and Souissi, S., Qualitative behavior of stage-structured populations: application to structural validation, *J. Math. Biol.*, 37, 291, 1998.
57. Fennel, W., Modeling of copepods with links to circulation models, *J. Plankton Res.*, 23, 1217, 2001.
58. Pedersen, O.P., Tande, K.S., and Slagstad, D., A model study of demography and spatial distribution of *Calanus finmarchicus* at the Norwegian coast, *Deep Sea Res. II*, 48, 567, 2001.
59. Franks, P.J.S. and Jaffe, J.S., Microscale distributions of phytoplankton: initial results from a two-dimensional imaging fluorometer, OSST, *Mar. Ecol. Prog. Ser.*, 220, 59, 2001.
60. Wolk, F., Yamazaki, H., Seuront, L., and Lueck, R. G., A new free-fall profiler for measuring biophysical microstructure, *J. Atmos. Ocean. Tech.*, 19, 780, 2002.

35

Modeling Planktonic Behavior as a Complex Adaptive System

Atsuko K. Yamazaki and Daniel Kamykowski

CONTENTS

35.1 Introduction...543
35.2 Copepod Model 544
35.3 Dinoflagellate Model 545
    35.3.1 Decision-Making Mechanism...547
    35.3.2 Simulation..549
        35.3.2.1 With Different External Nitrate Conditions 550
        35.3.2.2 With Different Threshold Settings...553
35.4 Discussion..554
References..556

35.1 Introduction

Many observational studies have shown that plankton act as "intelligent" organisms in the sense that they cope with their environment. Plankton move autonomously to accommodate various changes in their environment, such as temperature, nutrients, and turbulence. They alter their behavior or behavior patterns in response to environmental variation, but such behavior switching is not described by simple linear relationships with a few environmental factors. Dinoflagellates, for example, often are described as phytoplankton, which exhibit a regular diel migration pattern characterized by near-surface aggregations during daylight hours. On the other hand, many observational studies have shown that the diel migration can be irregular, in response to changes in the physical/chemical environment (Kamykowski, 1981; Cullen and Horrigan, 1981). Laboratory observations on zooplankton have pointed out relations between small-scale turbulence and changes in their behavior (Castello et al., 1990; Saiz et al., 1992). Strickler (1985) has shown that copepods choose paths according to their environment in his microscopic observational studies. These observational results documenting behavioral responses to environmental factors led us to model plankton behavior as adaptive to the physical/chemical environment.

Many models have adapted stochastic or mathematical methods to describe relationships between plankton and their surroundings. These modeling techniques tend to treat an organism as a passive element, and deal solely with the nonadaptive aspect of plankton behavior. However, plankton behavior in reality is sufficiently adaptive to require a more detailed consideration of behavior in plankton modeling. We need to integrate a more realistic behavior scheme into plankton modeling to better represent the autonomous and adaptive aspects of plankton behavior in relation to environmental changes.

Adaptive behavior modeling involves important elements that are closely related to each other: parallelism, distributed processing, and representation (Langton, 1988; Wilson, 1991). When an organism makes a decision to select the next behavioral step, it simultaneously evaluates many factors, such as properties in its surroundings in relation to its internal state. This checking process or data processing takes place in parallel fashion, and this process is highly distributed inside the organism. No single component is solely responsible for the final decision. This decision-making scheme works as a combination of small influences that are nonlinearly interacting with each other, and organism behavior emerges from such interactions.

Adaptive behavior modeling involves another important element: the concept of bottom-up control. A change in a component in a system's basic mechanisms or in an interaction among its lower-level components leads to a remarkable difference in properties, which we can observe macroscopically. A bottom-up scheme is not linear since properties in the upper levels often affect the states of properties in the lower level by feedback to the lower levels. Designating ecosystems as prototypical examples of *complex adaptive systems* (CAS), Levin (1998) notes that macroscopic properties of a system emerge from interaction among components in the system and may feed back to influence the subsequent development of their interaction. Kaufman (1993) states that interactions between individuals in a complex system influence the system's further development. Complexity is the collective behavior of many basic but interacting units, and their interactions lead to coherent collective phenomena, which can be described only at levels higher than those of the individual units, but should emerge from the interaction among them.

Levin (1998) also notes that any ecosystem can be characterized by flows, which are identified as one of four basic properties of any CAS by Holland (1992). Flows of nutrients, energy, materials, and information characterize the interconnections between parts in the systems. As a result, the flows transform an unordered collection of lower-level components into an integrated system that has a macroscopic order. The basic mechanisms in the system are interspersed with the interconnections of the flows. In the system, behavioral selection is not predetermined, but emerges from the interaction among the flows and connections. Railsback (2001) describes these characteristics of CAS as the concepts that should be applied to individual-based modeling.

In this chapter, we assume that behavioral changes of a plankter emerge from its internal state changes. We describe plankter behavior as an instance of macroscopic properties and the plankter's internal states as lower-level units of an integrated system. As noted above, observational studies have shown that external changes induce internal changes (Kamykowski, 1981; Cullen and Horrigan, 1981). We integrate environmental changes into a system for behavior selection. We define the flows of biochemicals inside the organism as well as the flows between its environment and its internal components. Then we illustrate interrelations among lower-level units and environmental factors associated with biochemicals to construct the behavior selection scheme. We present two models in which the properties of CAS are implemented for plankton behavior selection. One model of copepod swimming behavior based on a bottom-up scheme is briefly described; the other model of dinoflagellate vertical movement based on changes in biochemical fluxes is examined in more detail.

35.2 Copepod Model

Inspired by the work of Beer (1990) and Maes (1991) on the simulation of adaptive behavior, Keiyu et al. (1994) constructed a behavior model with a decision-making scheme for an individual copepod. The model has two mechanisms, motion and decision making, which influence each other. Their interactions determine the plankter's next move. In the model, an organism's three-dimensional motion is represented as discrete steps in a Lagrangian coordinate system. The motion mechanism computes an outcome of momentum produced by the plankter's fin and antenna motions for each step. This motion scheme itself is completely mechanical and does not involve autonomous elements. The plankter's behavioral choices are determined by the second scheme, the decision-making mechanism. This mechanism adopts the bottom-up and feedback characteristics of a complex adaptive system.

The decision-making mechanism is implemented in a heterogeneous network (Figure 35.1) in which nodes represent the database and the switches for behavior selections. A switch node is toggled according to the status of the conditions associated with the node. The conditions evaluate signals sent by the database and switch nodes. The decision for the next step is made in the interaction among the nodes,

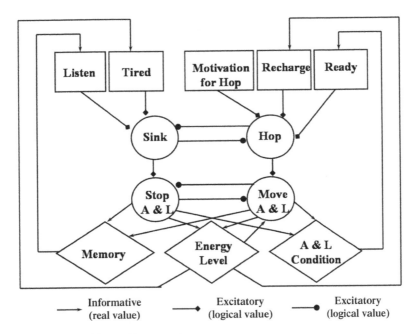

FIGURE 35.1 Zooplankton decision-making network.

and its decision is changed into values sent to the motion mechanism. These values change parameter values in the motion mechanisms, and then the model outputs the plankter's next step. In turn, the databases in the decision-making mechanism are updated by values set by the motion mechanisms.

The movements obtained from this copepod model were represented in computer simulation. Even though the adaptive part of the model was limited, the comparison of the simulation results to observational studies for swimming organisms suggested the need for a scheme that incorporates the adaptive nature of copepod behavior in order to model the organism's movements in response to changes in its surroundings (Keiyu et al., 1994). This modeling effort also showed that consideration of the relations among internal states can be a good candidate for simulating the adaptive nature of planter behavior. In particular, it revealed that the network technique exemplifies the characteristics of the bottom-up approach for plankton behavior modeling. We can simulate plankter behavior as outcomes from interactions among internal and external components in the system by connecting factors associated with behavior. The network can be a very simple or a more complex one based on the number of linking factor nodes in the basic model. In this process, we can examine how each factor node participates in decision making, by changing the parameters associated with the node. Similarly, another important characteristic of the bottom-up approach, possible feedback to the interactions, can be implemented by simply adding a connection between two nodes.

35.3 Dinoflagellate Model

After the positive experience with the copepod model, we constructed an adaptive model for an individual dinoflagellate, assuming that the organism is monitoring its internal biochemical changes to determine its next move. The results of this model are discussed in detail to more completely demonstrate the CAS associations of the adaptive behavioral approach. The behavior of dinoflagellates is a lower-level component of ecosystem operation, one whose characteristics can be described by flows of energy, nutrients, and information. We also assume that the adaptive nature of organism behavior resides in these flows and that we can replicate a portion of that behavior by describing the flows and the interrelations and conditions among the flows. The effects of different external environmental and internal cellular factors on dinoflagellate behavior have been examined in many studies (Liu et al., 2001). External factors include

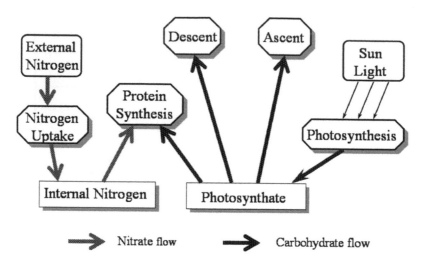

FIGURE 35.2 Biochemical flow diagram: for the dinoflagellate model.

light intensity, temperature, salinity, oxygen, and nutrients. Some of these external influences may affect photosynthetic responses or the internal balance among different biochemical pools. Some researchers suspect that dinoflagellate migration is induced by changes in its internal pools of carbohydrate or lipid, which are products of photosynthesis and of the nitrogen that dinoflagellates absorb from the seawater (Cullen, 1985; Cullen et al., 1985; Kamykowski, et al., 1992, 1998; MacIntyre, et al., 1997).

In this model, we focused on external biochemical fluxes from the environment to the organism as well as on biochemical flows within the organism. We also limited our concern to ascent and descent, although dinoflagellates can move in other directions. We chose light intensity and nutrients in the ocean as the main external factors influencing the vertical moves. The light intensity can change according to the amount of daylight and the vertical location of a dinoflagellate in the water column. The light intensity decreases exponentially with depth. The amount of nitrate absorbed by the dinoflagellate depends on the nitrate concentration, which is usually less available from the surface to a certain depth below the surface and then increases rapidly with depth. We focus on photosynthate (specifically carbohydrate here), which is produced through photosynthesis and which is consumed as an energy source for such activities as vertical movement and as an energy and carbon source for protein production, and on nitrogen compounds (mainly NO_3), which are absorbed from the seawater and which are used as a nitrogen source for protein production in support of cell growth (Lalli and Parsons, 1993).

The flows of biochemicals are carried out as described in Figure 35.2. This diagram shows the biochemical fluxes chosen to directly affect behavioral switching. Other activities related to the consumption of biochemicals that can be considered insignificant for behavior selection in terms of the amount of the biochemicals, such as respiration, are treated as a part of these nodes and are not clearly laid out in the diagram.

The carbohydrate is produced through photosynthesis and is accumulated in the photosynthate pool to be consumed mainly for such actions as descent, ascent, and protein synthesis. This flow of energy inside the dinoflagellate cell is expressed through directional links among the nodes "Photosynthesis," "Photosynthate," "Protein Synthesis," "Descent," and "Ascent." The links from the external nitrate node to the internal NO_3 pool node and from the pool node to the nutrient uptake node model the flow of nitrate. The consumption of nitrate for protein production is marked as the link from the pool node to the protein production node. These substance fluxes and changes in the pools themselves do not directly decide the cell's vertical movements.

The function to determine the rate of carbohydrate (photosynthate) fixation through photosynthesis is taken from Lalli and Parsons (1993), and the amount of photosynthate produced by the cell per minute (P_{id}) is defined as Equation 35.1 for the light intensity, I_{id}. Recall that the value of I_{id} depends not only on the time of the day but also on dinoflagellate vertical location.

$$P_{id} = P_{max} (I_{id} - I_c)/(K_I + I_{id} - I_c) \tag{35.1}$$

$$P_{id} = 0 \text{ if } I_{id} < I_c$$

where P_{max} is the maximum photosynthetic rate, I_c is a compensation light intensity, and K_I is a half saturation constant.

The uptake of nitrate from the seawater is defined using the Michaelis–Menten equation (McCarthy, 1981). The uptake rate is a function of nitrate concentration in water at depth d (S_d) and the nitrogen uptake rate per minute at depth d (V_d) is expressed by

$$V_d = V_{max}S_d/(K_S + S_d) \tag{35.2}$$

where V_{max} is the maximum uptake rate and K_S is a half saturation constant. For the rate of protein synthesis, Droop's equation (McCarthy, 1981) is used and the protein synthetic rate at time step i (G_i) is given by Equation 35.3 as a function of N_i, which is the amount of nitrogen inside the cell at time step i.

$$G_i = G_{max}(N_i - K_N)/N_i \tag{35.3}$$

$$G_i = 0 \text{ if } N_i < K_N$$

where G_{max} is the maximum synthetic rate and K_N is a half saturation constant for protein production. Then, the amount of photosynthate in organism's internal photosynthate pool at time step i (P_i) is described as

$$P_i = P_{i-1} + P_{id} - (\alpha_{Resp} + \alpha_{Descent} + \alpha_{Ascent} + \alpha_{ProdMax} G_i/G_{max}) \tag{35.4}$$

where α_{Resp}, $\alpha_{Descent}$, and α_{Ascent} are the amounts of photosynthate used for respiration, a movement of descent, and a movement of ascent per minute, respectively, and $\alpha_{ProdMax}$ is the maximum amount of photosynthate used to synthesize protein. $\alpha_{Descent}$ is set to 0 when the cell ascends, while α_{Ascent} is set to zero if the cell descends. Assuming that internal nitrogen is used only for protein production, we have defined the consumption of the internal nitrogen as a function of the protein production rate, G_i. The amount of nitrogen in an organism's internal nitrogen pool at time step i (N_i) is given by Equation 35.5:

$$N_i = N_{i-1} + V_d - \beta G_i \tag{35.5}$$

where β is the nitrogen consumption unit for protein synthesis. β is set to $1.0 \times 10^{-8} \, \mu M \, NO_3$–N cell^{-1} min^{-1} in the models. The swimming speeds for descent and ascent are set constant according to the values in Kamykowski et al. (1992) and Kamykowski and McCollum (1986). Other specific parameter values for equations related to biochemical flows from Equations 35.1 through 35.5 are given in Yamazaki and Kamykowski (2000).

35.3.1 Decision-Making Mechanism

The decision-making mechanism was inserted into the biochemical flow by adding a set of nodes that contain rules to control the behavior switching. The heterogeneous network for behavior selection in Figure 35.3 shows the integration of the decision-making mechanism into the biochemical fluxes and interaction among nodes in the network. The nodes marked with "A" through "F," named "condition nodes," play a core role in the model because their interactions decide the next move. Each condition node contains a simple criterion with a threshold value or values, and sends a signal to nodes associated with organism's activities or another condition node to activate or deactivate a particular action if the criterion is met.

Condition node "A" monitors the value of the NO_3 pool to send a signal for activating or deactivating the action of nitrogen intake, whereas node "B" monitors the NO_3 pool and the photosynthate pool to start or stop the cell's protein synthesis activity. The carbohydrate accumulation in the photosynthesis pool is toggled by node "E" such that photosynthate is produced through the "Photosynthesis" node if node "E" sends an activate signal to the behavior node when the light intensity is greater than a value contained in node "E." Then, nodes "C" and "D" check the amount of material in the photosynthate pool. At every time step, these parallel checking routines with predetermined threshold values associated

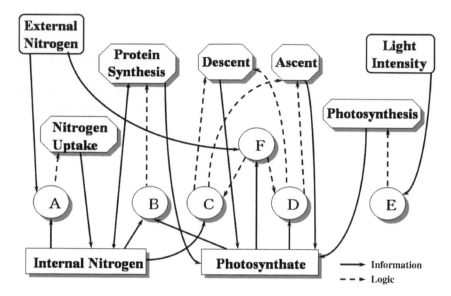

FIGURE 35.3 Decision-making network for the dinoflagellate model.

with the condition nodes make the decision whether the dinoflagellate should swim up or sink/swim down. Node "C" switches the descent or ascent node on/off depending on the amount of the internal NO_3, and node "D" acts similarly according to the amount of the accumulated carbohydrate in the photosynthate pool. Node "F" resolves any conflict between nodes "C" and "D" by turning on and off these nodes according to the values of the external nitrogen and the accumulated photosynthate. The specification of these criteria in the condition nodes can be expressed as

Node C
    IF signal from node F = ON
    THEN if value of {Internal Nitrogen} > threshold value for {Internal Nitrogen}
        SWITCH_ON {Ascent}
    ELSE {Descent}

Node D
    IF signal from node F = ON
    THEN if value of {Photosynthate} > threshold value for {Photosynthate}
        SWITCH_ON {Descent}
    ELSE {Ascent}

Node F
    IF value of {External Nitrogen} > threshold value for {External Nitrogen}
    AND if value of {Photosynthate} > threshold value for {Photosynthate}
        THEN SWITCH_ON {Node C} and SWITCH_OFF {Node D}
        ELSE SWITCH_ON {Node D} and SWITCH_OFF {Node C}

where within the braces are the corresponding threshold labels used in Figure 35.3.

At each time step, biochemical flows are updated according to the nutrient concentration and the light intensity at the organism's location in the water column. The updated values change the values of biochemical pools, which are also sent to the condition nodes and evaluated by the criteria. Then, each condition node sends its decision to its connected nodes, and as the result, the behavior of ascent or descent emerges from the interaction among the condition nodes. This checking is done at every time step, and the dinoflagellate changes its vertical location in the ocean according to a decision for ascent or descent. This location change results in changes in the light intensity and in the amount of nitrate available in the water. For example, when the dinoflagellate moves up, the light intensity increases in daytime, but the nitrate concentration decreases. In the case of a descent, the situation reverses. Therefore, the values of the internal pools will decrease or increase according to each vertical move. Changes in the internal pools will be reflected in the decision for vertical movement, as signals from the condition nodes to the behavior nodes controlling descent and ascent alternate depending on whether or not the value of the pool exceeds the threshold values defined in the condition nodes. All nodes interact with each other to yield a decision for the next move in this stepwise chain mechanism. In other words, many rules in the model are working in parallel and affect each other in one time step to make a decision for the next movement.

35.3.2 Simulation

We implemented the model to simulate different external conditions as well as different threshold settings for the condition nodes, and made graphs to represent the simulation results. The vertical movement of the dinoflagellate cell is plotted along with the light intensity at the surface in the main graphical window, while another window above the main window shows the amounts of internal nitrate, photosynthate, and protein relative to their predefined maximum values. The value of accumulated protein is reset to half when the amount reaches its predefined maximum value, so that the number of peaks in the plot of protein production indicates how advantageous the cell's movements are in terms of protein production. This halving routine at present ignores the tendency for dinoflagellate species to divide at specific times of day (Berdalet et al., 1992), often around dawn, for the sake of modeling convenience. We have added a GUI (to the left) to change the threshold values in the condition node in order to examine the effects of different threshold values (Figure 35.4).

In the simulation, the maximum value of the light intensity at the ocean surface, I_{MAX}, was set to a constant value taken from Lalli and Parsons (1993). The diel change of the light intensity at the surface (I_i) was represented as a cosine curve given by Equation 35.6 for the daytime, and $I_i = 0$ otherwise. Furthermore, the daytime is set to a period between 0500 and 1900 hours.

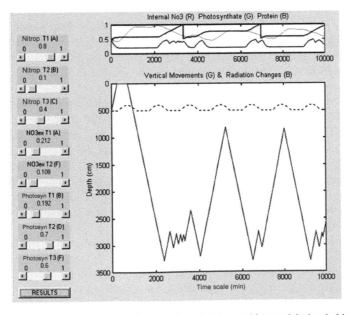

FIGURE 35.4 An example of GUI plots for the dinoflagellate decision-making model: threshold values can be varied with sliders.

$$I_i = I_{MAX} \sin \{\pi(i - SR)/DP\} \tag{35.6}$$

where SR is the time of the sunrise in minutes and DP is the length of the daytime.

Also, the light attenuation in the ocean was computed with a fixed extinction coefficient value obtained from Lalli and Parsons (1993). As defined as Equation 35.1, the amount of photosynthate produced by the cell per minute (P_{id}) is a function of the light intensity at time step i at dinoflagellate vertical location d (I_{id}). Since the light attenuates exponentially with ocean depth, the light intensity at depth d at time step i (I_{id}) can be described as an exponential function of the surface light intensity (I_i). In the simulation this external condition, I_{id}, was defined as Equation 35.7:

$$I_{id} = I_i e^{-Kd} \tag{35.7}$$

where K is an attenuation constant and d is the depth at which the cell is located.

The other external condition that affects biochemical flux inside the cell is nutrient concentration at dinoflagellate vertical location d. In the simulation, a hyperbolic tangent function given by Equation 35.8 was used with different parameter values to express the conditions of nitrate concentration in the seawater. See Yamazaki and Kamykowski (2000) for the values I_{MAX} and K used in the simulation.

$$S_d = \alpha \tanh((d - h)/1500.0) + \beta \tag{35.8}$$

where α, h, and β are constant values, d is the depth at which the cell is located, and S_d is nitrate concentration in water at a depth, d.

Sensitivity of the threshold associated with each node has been examined by using the GUI (Figure 35.4). By setting the threshold values to their maximums or zero, we can isolate the effect of each internal or external condition. We have changed the threshold values in the condition nodes for photosynthesis and nitrate uptake and examined the effects of these influx changes on modeled behavior. The simulation has shown that threshold values play an important role in changing the migration pattern in the simulation. Some of the threshold values were found to contribute more to the migration pattern of the model than others. In particular, the behavioral pattern is sensitive to small changes of the threshold values associated with the uptake of the external nitrate and the consumption of photosynthate for protein synthesis (Yamazaki and Kamykowski, 1998).

35.3.2.1 With Different External Nitrate Conditions

— Three different conditions shown in Figure 35.5 are used for nitrate concentration: NO_3 replete condition in which nitrate concentration is high through the ocean surface to the deep section (note that the deep nitrate levels exceed those normally encountered in the ocean); NO_3 moderate condition, which has a similar nitrate distribution to one observed in Haury et al. (1990); and NO_3 depleted condition in which nitrate concentration is relatively low throughout the vertical locations defined for simulation. The values of α, h, and β used for these conditions are listed in Figure 35.5.

In the simulation, we represented a cell's migration pattern and the amount of internal protein as plots in two windows (Figure 35.6 through Figure 35.8). The solid line in the larger window shows the migration pattern produced by the model. In the smaller window, the solid line shows the protein accumulated by that modeled cell. The simple halving routine for the protein creates peaks in the line so that more peaks represent more protein accumulated by the cell. The thin solid line in the same window shows the relative light intensity at the ocean surface. The dotted lines in both windows represent a regular migration pattern and the amount of protein yielded by another modeled cell which has the same biochemical flow scheme but migrates regularly such that it swims up between 0100 and 1500 hours; otherwise it descends.

In simulating these NO_3 conditions, we chose a set of threshold values that yields a migration pattern similar to the regular diel vertical migration under the moderate NO_3 condition (Figure 35.6). With the same threshold setting, the modeled cell displayed different migration patterns for the depleted and replete NO_3 conditions. For the replete NO_3 condition, the cell stayed at the ocean surface (Figure 35.7). This movement in the simulation is similar to observations in laboratory experiments by Kamykowski et al. (1992), which show that many dinoflagellate cells do not undertake organized mass descents during the night under nutrient replete conditions. For the depleted NO_3 condition, on the other hand, the cell

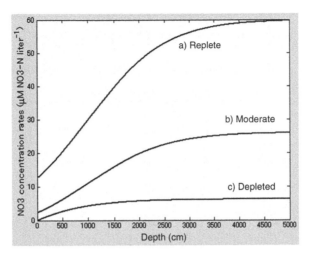

FIGURE 35.5 Three NO_3 concentration conditions expressed by hyperbolic tangent functions in the simulation: (A) The replete condition ($\alpha = 300.0$, $h = -1000.0$, $\beta = 300.0$); (B) the moderate condition ($\alpha = 150.0$, $h = -1000.0$, $\beta = 110.0$); (C) the depleted condition ($\alpha = 150.0$, $h = 1000.0$, $\beta = -88.0$).

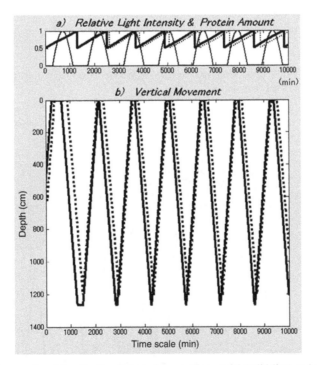

FIGURE 35.6 Simulation results of the dinoflagellate decision-making model (solid line) and regular migration model (dotted line) with the NO_3 moderate condition.

migrated irregularly and yielded more protein (Figure 35.8) than the regularly migrating cell (Figure 35.6). The irregular migration pattern induced by the decision-making mechanism for the depleted condition can be interpreted as similar to laboratory observations mentioned in Cullen (1985), where dinoflagellates tended to stay in the mid-depth when external nitrogen became depleted.

These results indicate that irregular migration patterns can be advantageous in terms of protein production under some nutrient conditions. This tendency is prominent in the case of the nitrate depleted condition in the simulation, and the results suggest that the regular diel vertical migration is not a preferable strategy for the cell when its surrounding seawater is scarce in nitrate. These simulation results

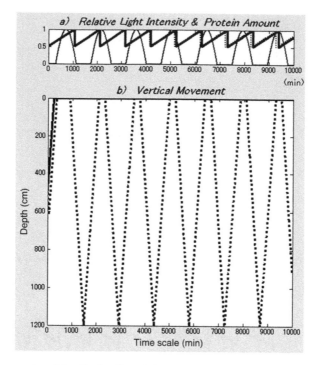

FIGURE 35.7 Simulation results of the dinoflagellate decision-making model (solid line) and regular migration model (dotted line) with the NO$_3$ replete condition.

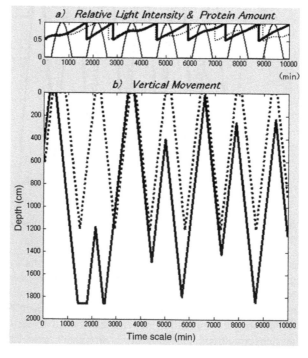

FIGURE 35.8 Simulation results of the dinoflagellate decision-making model (solid line) and regular migration model (dotted line) with the NO$_3$ depleted condition.

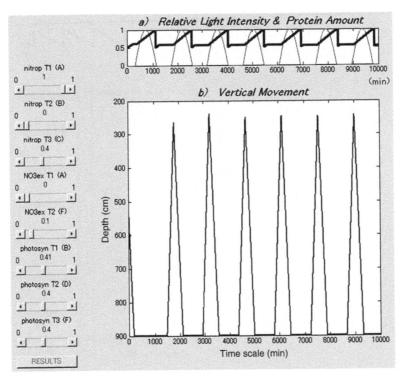

FIGURE 35.9 Regular migration of decision-making cell with 0.41 for the photosynthate criterion in Node B and 0.4 in Node B (with the NO₃ moderate condition), seven protein peaks.

also suggest that sensitivity to changes in the surrounding nutrient condition can benefit dinoflagellate cells undertaking irregular migration and that the cell adapts to environmental changes.

35.3.2.2 With Different Threshold Settings

— The shifts in the threshold value can also be considered a proxy for a set of threshold values that are more or less rapidly approached depending on how cell physiology is forced by day-to-day variability in the external availability of light and nutrients. The set of threshold values can be adjusted to make the cell behave almost in the same manner even if the external nutrient condition is different. On the other hand, with the same external conditions, the cell displays different swimming patterns when the threshold setting changes (Figure 35.9 and Figure 35.10). With the nitrate moderate condition, a cell's regular vertical movement shown in Figure 35.9 shifts to a more irregular movement shown in Figure 35.10 when changes are made in the threshold values of Nodes B and D where the amount of photosynthate is evaluated. These characteristics of the model allow the cell to behave differently in order to adapt to the external changes by changing its internal criteria. In other words, the changes in the internal criteria are converted into changes in a cell's behavioral pattern as a comprehensive illustration of the adaptive process.

In the model, we have assumed that the more protein produced, the more protein peaks in the smaller plotting window or the better the adaptation result. This assumption leads to another: if the external condition is not favorable for the cell to reproduce itself, the cell can try to change its internal criteria, which are a set of thresholds in this model, to reach the optimum result. For the same external NO₃ condition, the cell can yield more protein with one setting of the threshold values compared with another setting as shown in Figure 35.9 and Figure 39.11. The difference between the two settings resides in the threshold value for the amount of photosynthate in Node B, which controls protein synthesis by evaluating the amounts of internal nitrogen and the accumulated photosynthate inside the cell. Figure 35.9 shows that with the moderate NO₃ concentration, the cell migrates regularly with the photosynthate threshold value of 0.41 in Node B and obtained seven protein peaks. With the same NO₃ concentration, the cell produced more protein peaks (nine peaks) with the value of 0.1 for the same threshold under

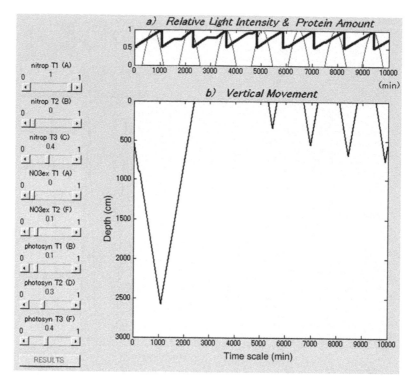

FIGURE 35.10 Irregular migration of decision-making cell with 0.1 for the photosynthate criterion in Node B and 0.3 in Node B (with the NO_3 moderate condition).

which the cell stays at the ocean surface longer, as seen in Figure 35.10. The difference between the two settings for the cell is a change in the threshold value for the criteria in Node B in this case, and the amounts of protein produced with these settings are very different as a result. This means that the cell can adapt to the external changes to yield the best result by changing its inside criteria. This may be related to enzymatic adaptations at the cellular level or to altered species composition based on a species-specific range of criteria.

In the simulation, changes in the threshold values produce the different fluxes of nutrient and energy in the cell, and the interaction of these changes resulted in different behavioral patterns. A small change of a threshold value often results in a very different movement in simulation. This matches one of the characteristics of complex systems: a small change at the lower level can produce a large difference at the upper level. An organism in which many systems are involved in the flow of chemical substances also exhibits this characteristic. The combination of such small changes can produce the best result for an organism. In this model, the combination of threshold values to obtain the most peaks of protein under a certain NO_3 concentration can be sought by moving the sliders. This process of seeking the best combination resembles an organism's trial-and-error adaptation process.

35.4 Discussion

In this chapter, we have presented models constructed by focusing on biochemical flows inside plankton as well as between organisms and their environment. The models illustrate that the plankter adaptive behavior decision-making mechanism is a complex adaptive system. Even though the copepod model had only one node to respond to external stimuli, its simulation results showed that it is important for plankton behavior modeling to contain a behavior selection mechanism and that the mechanism should consider an organism's internal condition and its relation to the environment. The dinoflagellate model, whose behavior selection mechanism involves multiple adaptive elements in relation to a cell's internal

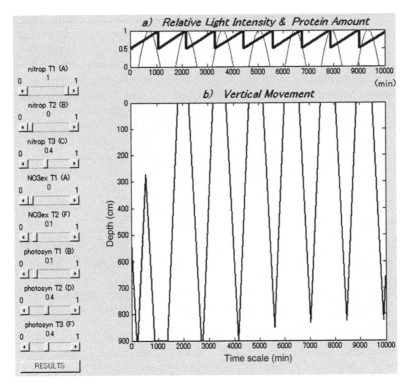

FIGURE 35.11 Migration of decision-making cell with 0.1 for the photosynthate criterion in Node B and 0.4 in Node B (with the NO_3 moderate condition), nine protein peaks.

and external conditions, demonstrated even more clearly the significance of adaptive factors in plankton behavior modeling. We note that the irregularity that emerged in the dinoflagellate simulation model arose in a target organism that is a single cell whose internal mechanism has to be greatly affected by environmental changes. Waters and Mitchell (2002) note that the adaptation of plankton occurs at the individual scale where centimeter-scale interactions between phytoplankton and the surrounding environment are important. Introducing such small-scale interaction into the simulation by using nutrient data that have more variable distributions is valuable to further examine the relationship between cell behavior and environmental factors under increasingly realistic environmental conditions.

The simulation results for the model suggest that phytoplankter behavioral changes in response to its environment may result from changes in its internal biochemical flow criteria. At the same time, the criteria themselves may vary depending on changes in the amounts of the biochemical fluxes, which in turn are affected by changes in a cell's environmental surroundings. The different migration patterns generated by varying the thresholds for internal quotas suggest either different species with different quotas or a single species that changes its quota because it is stressed. In the latter case, the threshold values can express the tolerance of an individual organism for environmental changes.

The essence of adaptive behavior, therefore, may reside in complex interactions among criteria, biochemical fluxes, and their balances as described for nutrient uptake kinetics by Morel (1987) or for plankton acclimation to light, nutrients, and temperature by Flynn et al. (1997) and Geider et al. (1998). Values for the criteria associated with variable internal plankton composition may produce various behavior patterns. Plankton may select the most favorable criteria setting and then display movements induced by the setting. Then, it is important to investigate the process of how plankton determine the best combination of criteria in terms of their goal under a certain environmental condition, and how that combination is determined in the individual cell or how it is passed on to the next generation. In fact, this adaptive behavioral mechanism probably is associated with how natural selection acts on the genetic make-up of a species population and how environmental changes act on the expression of that genetic potential.

We note that more precise observational studies on plankton are essential to refine the models, as pointed out by Jørgensen (1997) for modeling and simulating ecological systems. Particularly, biochemicals have to be measured to model the flows of the substances and to understand the relation between their changes and plankton movement. It is not easy to attain such measurements, and it is often difficult to take biological observational results into account when modeling adaptive behavior. Compared to conceptually simpler mathematical or statistical approaches to describe the movements of an individual organism, such as Franks (1997) and Blackwell (1997), the heterogeneous and bottom-up nature of our approach induces complexity in analyzing or estimating the movements in simulation. We believe that the nature of adaptive behavior resides in biochemical fluxes, and by looking at them and adopting them to describe the plankton behavioral mechanism as a complex adaptive system, we can attain a more realistic modeling of the ocean planktonic ecosystem.

References

Beer, R., 1990. *Intelligence as Adaptive Behavior: An Experiment in Computational Neuroethology.* Academic Press, New York, 213 pp.

Berdalet, E., Latasa, M., and Estrada, M., 1992. Variations in the biochemical parameters of *Heterocapsa* sp. and *Olithodiscus luteus* groen in 12:12 light:dark cycles. I. Cell cycle and nucleic acid composition. *Hydrobiologia,* 238:139–147.

Blackwell, P.G., 1997. Random diffusion models for animal movement. *Ecol. Modelling,* 100:87–102.

Castello, J.H., Marrase, C., Strickler, J.R., Zeller, R., Freise, A.J., and Trager, G., 1990. Grazing in a turbulent environment: I. Behavioral response of a calanoid copepod, *Centropages hamatus. Proc. Natl. Acad. Sci. U.S.A.,* 87:1648–1652.

Cullen, J.J., 1985. Diel vertical migration by dinoflagellates: roles of carbohydrate metabolism and behavioral flexibility, in M.A. Rankin, Ed., Migration Mechanisms and Adaptive Significance. *Contrib. Mar. Sci.,* 27:135–152.

Cullen, J.J. and Horrigan, S.G., 1981. Effects of nitrate on the diurnal vertical migration, carbohydrate metabolism and behavioral flexibility, in M.A. Rankin, Ed., Migration: Mechanisms and Adaptive Significance. *Contrib. Mar. Sci.,* 27: 135–152.

Cullen, J., Zhu, M., Davis, R.F., and Pierson, D.C., 1985. Vertical migration, carbohydrate synthesis and nocturnal uptake during growth of *Heterocapsa niei* in a laboratory column, in D.M. Anderson, A.W. White, and D.G. Baden, Eds., *Toxic Dinoflagellates.* Elsevier, New York, 189–194.

Flynn, K.J. Fasham, M.J.R., and Hipkin, C.R., 1997. Modelling the interactions between ammonium and nitrate in marine phytoplankton. *Philos. Trans. R. Soc.,* 352:1625–1645.

Franks, P., 1997. Models of harmful algal blooms. *Limnol. Oceanogr.,* 42:1273–1282.

Geider, R.J., MacIntyre, H.L., and Kana, T.M., 1998. A dynamic regulatory model of phytoplankton acclimation to light, nutrients, and temperature. *Limnol. Oceanogr.,* 43:679–694.

Haury, L.R., Yamazaki, H., and Itsweire, E.C., 1990. Effect of turbulent shear flow on zooplankton distribution. *Deep-Sea Res.,* 37:447–461.

Holland, J.H., 1992. *Adaptation in Natural and Artificial Systems.* MIT Press, Cambridge, MA, 211 pp.

Jørgensen, S.E., 1997. Ecological modelling by "Ecological Modelling." *Ecol. Modelling,* 100:5–10.

Kamykowski, D., 1981. Laboratory experiments on the diurnal vertical migration of marine dinoflagellates through temperature gradients. *Mar. Biol.,* 62:81–89.

Kamykowski, D. and McCollum, A., 1986. The temperature acclimated swimming speed of selected marine dinoflagellates. *J. Plankton Res.,* 8:275–287.

Kamykowski, D., Reed, R.E., and Kirkpatrick, G.J., 1992. Comparison of sinking velocity, swimming velocity, rotation and path characteristics among six marine dinoflagellate species. *Mar Biol.,* 113:319–328.

Kamykowski, D. Milligan, E.J., and Reed, R.E., 1998. Biochemical relationship with the orientation of the autotrophic dinoflagellate *Gymnodinium breve* under nutrient replete conditions. *Mar. Ecol. Prog. Ser.,* 167:105–117.

Kaufman, S.A., 1993. *The Origins of Order: Self-Organization and Selection in Evolution.* Oxford University Press, New York.

Keiyu, A.K., Yamazaki, H., and Strickler, J.R., 1994. A new modeling approach for zooplankton behavior. *Deep-Sea Res. II,* 41:171–184.

Lalli, C.M. and Parsons, T.R., 1993. *Biological Oceanography: An Introduction.* Pergamon Press, New York, 301 pp.

Langton, C.G., 1988, Artificial life, in C.G. Langton, Ed., *Artificial Life,* Vol. VI, *Santa Fe Institute Studies in the Sciences of Complexity.* Addison-Wesley, Reading, MA, 1–47.

Levin, S.A., 1998. *Ecosystems and the Biosphere as Complex Adaptive Systems,* Vol. 1, *Ecosystems.* Springer-Verlag, New York, 431–436.

Liu, G., Janowitz, G.S., and Kamykowski, D., 2001. A biophysical model of population dynamics of the autotrophic dinoflagellate *Gymnodinium breve. Mar. Ecol. Prog. Ser.,* 210:101–124.

MacIntyre, J.G., Cullen, J.J., and Cembella, A.D., 1997. Vertical migration, nutrition and toxicity in the dinoflagellate *Alexandrium tamarense. Mar. Ecol. Prog. Ser.,* 148:201–216.

McCarthy, J.J., 1981. The kinetics of nutrient utilization. *Can. Bull. Fish. Aquat. Sci.,* 210:211–233.

Maes, P., 1991. A bottom-up mechanism for behavior selection in an artificial creature, in J.-A. Meyer and S.W. Wilson, Eds., *From Animals to Animats: Proceedings of the First International Conference on Simulation of Adaptive Behavior.* MIT Press, Cambridge, MA, 238–246.

Morel, F.M.M., 1987. Kinetics of nutrient uptake and growth in phytoplankton. *J. Phycol.,* 23:137–150.

Railsback, S.F., 2001. Concepts from complex adaptive systems as a framework for individual-based modeling. *Ecol. Modelling,* 139: 47–62.

Saiz, E., Alcaraz, M., and Paffenhöfer, G.A., 1992. Effects of small scale turbulence on feeding rate and gross-growth efficiency of three Acartia species (Copepod: California). *J. Plankton Res.,* 14:1085–1097.

Strickler, J.R., 1985. Feeding currents in calanoid copepods: two new hypothesis, in M.S. Laverack, Ed., Physiological Adaptations of Marine Animals. *Symp. Soc. Exp. Biol.,* 39:459–458.

Waters, R. and Mitchell, J.G., 2002. Centimeter-scale spatial structure of estuarine *in vivo* fluorescence profiles. *Mar. Ecol. Prog. Ser.,* 237:51–63.

Wilson, S.W., 1991. The animat path to *AI,* in J.-A. Meyer and S.W. Wilson, Eds., *From Animals to Animats: Proceedings of the First International Conference on Simulation of Adaptive Behavior.* MIT Press, Cambridge, MA, 15–21.

Yamazaki, A.K. and Kamykowski, D., 1998. Modeling vertical movements of phytoplankton, in *Proceedings of the 17th Simulation Technology Conference,* Tokyo, June 1998, 57–60.

Yamazaki, A.K. and Kamykowski, D., 2000. A dinoflagellate adaptive behavior model: response to internal biochemical cues. *Ecol. Modelling,* 137:59–72.

36

Discrete Events-Based Lagrangian Approach as a Tool for Modeling Predator–Prey Interactions in the Plankton

Philippe Caparroy

CONTENTS

36.1 Introduction...559
36.2 Methodology 560
     36.2.1 Representation of Swimming Behavior 560
     36.2.2 Discrete Events Discrimination and Timescale of Numerical Integration.................562
     36.2.3 Vectorial Estimation of the Encounter Timescale..562
36.3 Case Study 564
     36.3.1 Simulating Gerritsen and Strickler Analytical Solution 564
     36.3.2 Effect of Varying Swimming Sequence Duration..567
36.4 Discussion...569
36.5 Conclusion 571
Acknowledgments...571
References..572

36.1 Introduction

Predation and food availability are probably the most important biological factors controlling food web structure, and hence production, in marine ecosystems (Steele and Frost, 1977; Landry, 1978; Ohman, 1988). In an effort to understand and predict the structuring role of predation, marine ecologists have put much effort into the development of predation rate models over the last half-century. Grazing or predation in the plankton is not directly performed by populations but by individuals, and interactions between predator and prey are discrete events that occur at the smallest time and space scales in the sea. Therefore, a common conceptual scheme was provided by Holling's predation cycle model (Holling, 1966). This model is currently recognized as the most realistic approach (Price, 1988), as it dissociates predatory events into component processes of search, encounter, attack, escape, pursuit, and capture, which occur at the lower end of the microscale (millimeters and seconds).

Within this succession of conditional predatory events, encounter has been the most extensively modeled, starting with the work of Gerritsen and Strickler (1977). This paper emphasized the strong coupling between swimming behavior, in terms of kinetics, and foraging strategies. Inspired by this fundamental work, a large number of theoretical studies have provided detailed mechanistic models of predator–prey interactions, including the effect of physical factors, such as small-scale turbulence, and swimming behavior (Rothschild and Osborn, 1988; Evans, 1989; Kiørboe and Saiz, 1995). Similarly, a few theoretical studies have concerned the subsequent steps of pursuit (MacKenzie et al., 1994) and attack/capture (Beyer, 1980; Heath, 1993; Caparroy et al., 2000). However, despite this intense experimental and modeling effort, it is still unclear how these processes integrate up to population-level parameters such as predation rates or grazing pressure.

In contrast to these simple models of predator–prey interactions, more complex analytical solutions incorporating conditional joint probability density function of predator and prey velocities have recently been proposed (Lewis and Pedley, 2000). This approach was specifically designed to evaluate the effect of oceanic small-scale turbulence on predator–prey encounter rates, and compared successfully with previous direct numerical simulation (DNS) of the motion of planktonic organisms in fully turbulent flow fields (Yamazaki et al., 1991). This latter Lagrangian representation of planktonic organism simultaneous displacement is a more realistic way to compute encounter rates between populations of interacting predators and prey. However, it has a high computational cost, particularly if one wishes to minimize the error associated with an *a posteriori* estimation of discrete encounter events number.

In an attempt to overcome these limitations, we propose a Lagrangian representation of trophic interactions between *n* planktonic organisms, based on the temporal discrimination of discrete predatory events. Our model has the ability to identify any forthcoming predatory event (e.g., encounters between predator and prey) in the form of its corresponding timescale, and has the potential to include:

1. A deterministic and realistic representation of swimming behavior
2. A deterministic representation of the predation processes in form of a succession of conditional discrete temporal events: encounter; attack, triggered by encounter of prey; escape, triggered by encounter of predator; scanning or searching (actively/passively) for other planktonic organisms (potential predators or prey)
3. A stochastic representation of the pelagic entities motion

36.2 Methodology

The general mechanism of our model consists of the Lagrangian simulation of the simultaneous displacements of *n* potentially interacting planktonic organisms, into a three-dimensional volume of water. According to the conceptual scheme of Holling (1966), predatory interactions can take the form of discrete temporal events such as encounter (remote perception or direct physical collision), attack, capture, and escape. Each component process of the predation rate can be individually estimated by tracking each planktonic organism and integrating the concerned events number over time.

36.2.1 Representation of Swimming Behavior

An explicit representation of planktonic organism swimming behavior is a key point in the scope of developing a mechanistic model of predatory interactions. For rheotactic planktonic organisms (e.g., copepods, chaetognaths, ciliates), processes of predator and prey perception are known to rely upon hydromechanical signals (Yen and Fields, 1992; Kiørboe et al., 1999; Jakobsen, 2001). The magnitude of these signals depends on the different components of the flow field generated by a swimming organism (Kiørboe and Visser, 1999), and their spatiotemporal distribution is simultaneously governed by the motility pattern of the organism. Previous modeling studies have essentially focused on the impact of simple behavioral strategies on individual component processes of the predation rate. Encounter success has been considered in terms of kinetics (Gerritsen and Strickler, 1977) and directionality (Gerritsen, 1980) for simple antagonist swimming strategies (continuously cruising predator vs. an ambush predator) or has been evaluated for a particular behavioral sequence (Kiørboe and Saiz, 1995): passive sinking vs. feeding current generation. However, encounter, attack, or escape reactions are discrete events that trigger behavioral responses in the form of a switch of behavioral sequence (e.g., switch from swimming to jumping sequence) and associated kinetic changes (modification of speed and direction of motion). Therefore, to simulate the outcome of the full predation cycle, modeled organisms must be given the possibility of expressing a more complex and realistic behavioral repertoire.

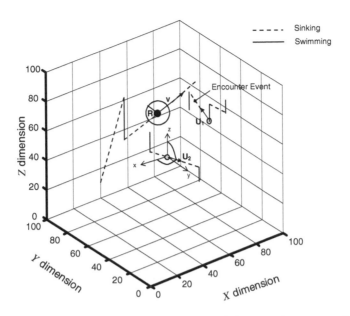

FIGURE 36.1 Conceptual scheme of the three-dimensional resolution of discrete predatory events. Predator (black sphere) with perceptive ability (perceptive sphere of radius R) and prey (gray spheres without perceptive ability) is alternating between sinking (continuous paths) and swimming (dotted paths) sequences. For the instantaneous situation considered both predator and preys are swimming. Each individual behavioral sequence is described in terms of a duration: τ (s) and a velocity vector with norm being the speed, and polar (θ) and azimuthal (φ) angles defining the direction of motion. By applying vectorial calculus to velocity vectors ($\mathbf{V}$, predator velocity vector; $\mathbf{U_n}$, prey n velocity vector) of all potentially interacting entities, it is possible to identify the time lag (numerical integration time step of the model) before the next discrete predatory event. Here encounter event between predator and prey $n = 1$ will occur at the position depicted on the graph.

To do so, we consider that the swimming behavior of a particular planktonic organism can be described by a collection of k behavioral sequences, which alternate over time according to a stochastic process. Each behavioral sequence: S_i ($i \in [1,k]$) is described in terms of an associated velocity vector $\mathbf{V}_i$ (Figure 36.1) and a duration $\tau_{behavior,i}$ (s). The norm of the velocity vector ($\|\mathbf{V}_i\|$) is the speed of the considered behavioral sequence, and its directionality is defined by the azimuthal (θ) and polar (φ) angles (radians) of the velocity vector, relative to an external frame of reference $(0,X,Y,Z)$.

As an illustration, a typical motility pattern for copepods (e.g., *Acartia tonsa*; Saiz, 1994) can be simulated as the result of the random succession of three behavioral sequences:

1. A swimming sequence of speed ~1 body length s^{-1} and random directions: $\varphi \in [0, \Pi]$ and $\theta \in [0, 2\Pi]$
2. A sinking sequence of speed ~0.01 mm s^{-1} and random directions: $\varphi = 0$, $\theta \in [0, 2\Pi]$
3. A jumping sequence of speed ~10 body length s^{-1} and random directions: $\varphi \in [0, \Pi]$ and $\theta \in [0, 2\Pi]$

During the time course of a simulation, spatial coordinates of each planktonic organism are integrated over time according to

$$X(t + dt) = X(t) + \sin \varphi \cos \theta \, \|\mathbf{V}_i\| \, dt \tag{36.1}$$

$$Y(t + dt) = Y(t) + \sin \varphi \sin \theta \, \|\mathbf{V}_i\| \, dt \tag{36.2}$$

$$Z(t + dt) = Z(t) + \cos \varphi \, \|\mathbf{V}_i\| \, dt \tag{36.3}$$

where dt (s) is the timescale of numerical integration of the model, and $\|\mathbf{V}_i\|$ is the speed of the behavioral sequence characterizing the plankter's swimming behavior at time t. Because values of φ, θ, and $\|\mathbf{V}_i\|$ are associated with the instantaneous swimming behavior (S_i) of each organism:

$$0 < dt \leq \tau_{behavior, i} \tag{36.4}$$

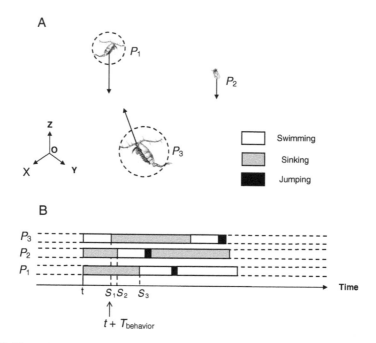

FIGURE 36.2 (A) Three planktonic organisms (P_1, P_2, P_3) with different sizes and velocity vectors (arrows) considered at time t. P_3 and P_1 have perceptive abilities (perceptive fields represented as dotted line spheres) and P_2 has no perceptive ability. (B) Horizontal bars represent how the behaviors of each organism evolve over time, as the result of alternating behavioral sequences: three behavioral sequences (with associated velocities and directions, φ, θ) alternate over time: swimming, jumping, sinking. Planktonic organism behaviors are not synchronized. To integrate motion of these three organisms over time, we must first define a behavioral timescale ($T_{behavior}$), such that no discrete behavioral transitions (S_1, S_2, S_3) are missed. In this example, $T_{behavior}$ is defined by S_1 (switch from swimming to sinking for P_1).

36.2.2   Discrete Events Discrimination and Timescale of Numerical Integration

Equation 36.4 illustrates the general mechanism of our approach, which is based on the discrimination of discrete temporal events. To simulate the three-dimensional displacements of a single planktonic organism, with swimming behavior represented as a succession of distinct behavioral sequences of fixed duration, the time step of numerical integration must be chosen so that no switches of behavioral sequences, for this particular individual, are missed.

In the more complex situation where n organisms are simultaneously tracked, with asynchronous swimming behaviors, the same constraint applies (Figure 36.2). If each individual behavior is a collection of k alternating behavioral sequences, a behavioral timescale for numerical integration of the model is given by $T_{behavior}$ (s):

$$T_{behavior} = MIN\{\tau_{behavior(1,j)}, \ \cdots \ \tau_{behavior\,(i,j)},..., \tau_{behavior\,(n,j)}\} \qquad (36.5)$$

with $j \in [1, k]$.

Once this timescale is computed, potential discrete predatory events, concerning any of the tracked organisms over this time interval, must be identified. For example, simultaneously swimming organisms might encounter each other before the end of the swimming sequences in which they are involved. Similarly, a predator and its prey, respectively, involved in an attack/escape reaction following their encounter, might capture/escape their respective protagonist before the end of the jumping sequence they are performing.

36.2.3   Vectorial Estimation of the Encounter Timescale

To identify the first potential predatory event, encounter, we compute an encounter timescale: $T_{encounter}$ (s). We consider a mixed predator–prey population of n planktonic organisms, modeled as translating spheres

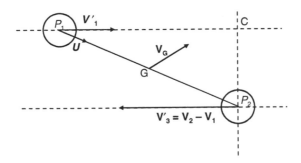

FIGURE 36.3 Two moving spherical planktonic organisms P_1 and P_2, with their associated velocity vectors $\mathbf{V'}_1$ and $\mathbf{V'}_2$ (drawing made in the plane of motion). $\mathbf{V'}_1$ and $\mathbf{V'}_2$ are velocity vectors of displacement relative to the geometric barycenter G, itself moving with velocity vector $\mathbf{V_G}$. For this particular case, $\mathbf{V_G}$ is in the same plane as $\mathbf{V'}_1$ and $\mathbf{V'}_2$. Therefore in an external frame of reference $(0,X,Y,Z)$, an organism's plane of motion is invariant with time.

of radius r (mm), with perceptive abilities. As a first approximation (e.g., Gerritsen and Strickler, 1977) perceptive field can be considered spherical, with radius R (mm), and

$$R > r \tag{36.6}$$

The encounter/perception event is defined as the contact between the spherical perceptive field of a scanning organism, and a spherical scanned organism. At any time t during the simulation, the encounter timescale is therefore the shortest time lag before the next future encounter event between at least two planktonic organisms.

To test whether two spheres P_1 and P_2, with radius r_1 and r_2 (mm) and known velocity vectors at time t, will collide at time $t + T_{\text{encounter}}$, we explicit their relative motion in a frame of reference centered at their spatial barycenter, G (Figure 36.3).

In a three-dimensional frame of reference $(0,X,Y,Z)$, the barycenter of the P_1P_2 segment G is given by

$$\mathbf{OG} = {}^1\!/_2(\mathbf{OP_1} + \mathbf{OP_2}) \tag{36.7}$$

If P_1 and P_2, respectively, are moving with velocity vectors $\mathbf{V_1}$ and $\mathbf{V_2}$ in $(0,X,Y,Z)$, G is moving with velocity vector $\mathbf{V_G}$:

$$\mathbf{V_G} = {}^1\!/_2(\mathbf{V_1} + \mathbf{V_2}) \tag{36.8}$$

If we consider relative motion between P_1, P_2 in a frame of reference with origin at G, then in this new frame of reference (moving with velocity $\mathbf{V_G}$ in $(0,X,Y,Z)$) velocity vectors for each sphere ($\mathbf{V'}_1$ and $\mathbf{V'}_2$) are given by differentiating Equation 36.7 over time:

$$\mathbf{V'}_1 = {}^1\!/_2(\mathbf{V_1} - \mathbf{V_2}) \tag{36.9}$$

$$\mathbf{V'}_2 = {}^1\!/_2(\mathbf{V_2} - \mathbf{V_1}) \tag{36.10}$$

And the velocity vector $\mathbf{V'}_3$ describing the relative displacement of P_2 with respect to P_1, still in the frame of reference with origin at G, is given by

$$\mathbf{V'_3} = \mathbf{V'_2} - \mathbf{V'_1} \tag{36.11}$$

As Equations 36.9 and 36.10 state that $\mathbf{V'}_1$ and $\mathbf{V'}_2$ are parallel, collision between P_1 and P_2 is possible only if P_2 moves toward P_1, which translates to

$$\mathbf{V'_3} \bullet \mathbf{u} < 0 \tag{36.12}$$

with $\mathbf{u}$ the unit vector, such that

$$\mathbf{P_1P_2} = \mathbf{u} \, \|\mathbf{P_1P_2}\| \tag{36.13}$$

If Equation 36.12 is verified, an encounter event might occur between the two organisms if

$$\|\mathbf{P}_2\mathbf{C}\| = \sqrt{\|\mathbf{P}_1\mathbf{P}_2\|^2 - \left[\mathbf{P}_1\mathbf{P}_2 \bullet \frac{\mathbf{V}_1'}{\|\mathbf{V}_1'\|}\right]} \leq r_1 + r_2 \tag{36.14}$$

If Equation 36.14 is verified, the time lag before the forthcoming encounter: $\tau_{\text{encounter}}$ (s), can be computed as the time necessary for both spheres to come into contact:

$$\tau_{\text{encounter}} = \frac{\|\mathbf{P}_1\mathbf{C}\| - \sqrt{(r_1 + r_2)^2 - \|\mathbf{P}_2\mathbf{C}\|^2}}{\|\mathbf{V}_3'\|} \tag{36.15}$$

For a population of n potentially interacting organisms, all possible interactions leading to an eventual encounter must be examined, by using the above principles. Finally, the encounter timescale is given by

$$T_{\text{encounter}} = \text{MIN}\{\tau_{\text{encounter } 1}, \dots, \tau_{\text{encounter } i}\} \tag{36.16}$$

with

$$i \in [1, C_n^2] \tag{36.17}$$

Similar to the encounter process, collision events might occur between organisms without perceptive ability. By analogy to the previous demonstration, a collision time scale: $T_{\text{collision}}$ (s) can be computed, if we consider that a collision event occurs as a result of direct contact between two spherical planktonic organisms.

If we are interested in a complete assessment of the predation cycle, we must also take into account possible attack and escape reactions resulting in possible capture (for the predator) and escape (for the prey) events. This can be simply performed by considering that any predator encountering a prey will initiate an attack reaction, and conversely an escape reaction for the prey. Attack and escape reactions are specialized behavioral sequences performed at high velocity for most planktonic organisms with perceptive ability (e.g., jumps performed at ~10 to 100 body length s^{-1} for copepods). The simpler representation for both behaviors could be to consider that on prey perception (encounter) the predator immediately orients toward the prey position, whereas on predator perception the prey initiates an escape reaction, which can be oriented in any direction. Once engaged in these concurrent behavioral sequences, a capture event will occur if prey, modeled as a translating sphere, collides with the predator's capture sphere volume (volume swept out by the feeding appendages for copepods or the mouth for fish larvae) during the attack sequence. Again, and similarly to the encounter event case, a capture timescale: T_{capture} (s) can be estimated using Equations 36.12 to 36.15.

Finally, having identified the different timescales for all potential discrete predatory events, the timescale of numerical integration of the model, dt (s) is given by

$$dt = \text{MIN}\{T_{\text{behavior}}, T_{\text{encounter}}, T_{\text{capture}}\} \tag{36.18}$$

Combining Equation 36.18 with Equations 36.1 to 36.3, the motion of each planktonic organism can be deduced, and the eventually associated predatory events can be integrated over time.

36.3 Case Study

36.3.1 Simulating Gerritsen and Strickler Analytical Solution

To illustrate the potential applicability of our approach, we first consider the simple case of a cruising planktonic predator with a constant spherical perceptive sphere of radius R (mm), continuously searching for prey in a finite space volume of 1 m^3. Both predator and prey behavioral repertoire is made up of a single "swimming" behavioral sequence of constant duration, τ (s), performed at constant speed and in random directions. For all individually tracked organisms, a new swimming direction is randomly generated at the start of each new swimming sequence. This is performed by randomly extracting polar

TABLE 36.1

Parameters of the Model

Parameter	Description/Values	Unit
Velocity vector		
Polar angle: ϕ	Uniformly distributed $\phi \in [0, \Pi]$	Radians
Azimuthal angle: θ	Uniformly distributed $\theta \in [0, 2\Pi]$	
Speed: v	Magnitude of the velocity vector / 0–3	mm s^{-1}
Behavioral sequence duration: τ	Constant 1–50	s
Perceptive field radius: R	Constant 3	mm
Simulation duration: T	3600	s
Prey density: N	40	l^{-1}

(ϕ), and azimuthal (θ) angles of the velocity vector from independent uniform distributions, respectively, $[0, \Pi]$ and $[0, 2\Pi]$. Other simulation parameter values are listed in Table 36.1.

At the start of a simulation, the initial Cartesian coordinates of predator and prey are randomly generated from a uniform distribution, and integration of each organism motion path starts according to the algorithm described above. Because the purpose of this first set of simulations is to estimate only the encounter process, we neglect the representation of any attack reaction on prey perception for our theoretical predator. Similarly, our theoretical prey are considered less perceptive ($R = r$) and cannot react in the form of an escape reaction, as they never encounter any predator.

Periodic boundary conditions are imposed over the cubic spatial domain. Each time a moving planktonic organism encounters a boundary of the space volume, it leaves this face of the cube and spontaneously reenters the domain on the opposite face, without interrupting its current behavioral (swimming) sequence.

For predator and prey, discrete boundary encounter events can be discriminated using the general equations (Equations 36.7 to 36.15) proposed above, simply by considering the encounter event a direct collision between the organism spherical body and a dimensionless spatial point located on a boundary.

To assess the potential extra contacts brought about by the periodic boundary conditions imposed to our spatial domain, each prey was identified by a single integer index. From the predator point of view, this procedure allowed discrimination between the first and possible subsequent multiple encounters of each individual prey.

A first set of simulations presented in Figure 36.4 considers the following scenario: three individual predators are simultaneously tracked over 1 h, while faced with motionless prey, at low (10 prey l^{-1}) and high (80 prey l^{-1}) prey densities.

Simultaneous encounter time series (Figure 36.4) clearly illustrate the ability of our numerical approach to represent discrete predatory events over time. The encounter signal at low prey density appears to be randomly distributed over time. For the high prey density case, signal frequency sharply increases, as expected.

Another important point raised by this simulation is the ability of our model to assess interindividual variability in terms of encounter signal. Here the three predators have identical speeds and swimming bout durations; therefore, this variability results only from the aleatory aspects of their respective paths and the prey spatial distribution.

To test the soundness of our numerical scheme of integration, a second set of simulations was designed to compare encounter rates generated by our Lagrangian model with the analytical model proposed by Gerritsen and Strickler (1977). This model, currently used by marine ecologists to estimate encounter rates between planktonic predators and prey, has the following null hypothesis:

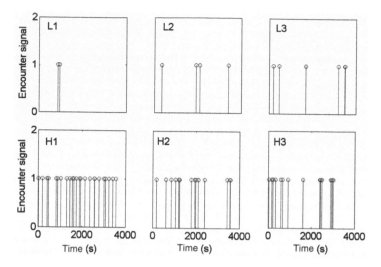

FIGURE 36.4 Simulated encounter signal as perceived over time (3600 s) by three distinct cruising predators (1, 2, and 3) swimming simultaneously in a 1 m³ space volume with constant speed $v = 1$ mm s⁻¹ at low prey density (*L*: 10 prey l⁻¹) and high prey density (*H*: 80 prey l⁻¹). In these simulations, prey are motionless, and the predator has a 3 mm perception distance.

It computes the encounter rate (*Z*: s⁻¹) of a moving predator as it swims across an infinite cloud of prey uniformly distributed in space. Predator and prey swimming directions are uniformly distributed, and speeds are constant. The expected encounter rate is given by

$$Z = \frac{\Pi R^2 N}{3} \frac{3v^2 + u^2}{v} \qquad \text{for } v \geq u \tag{36.19}$$

where *R* (mm) is predator perception distance, *N* (mm⁻³) is prey density, and *u* and *v* are, respectively, prey and predator speeds (mm s⁻¹).The same formula, with *u* and *v* interchanged, applies when $u \geq v$.

To simulate Gerritsen and Strickler's analytical model (G&S model) we consider a single predator and its prey swimming at random (Table 36.1). Simulation results presented on Figure 36.5 and Figure 36.6 are performed at a prey density of 40 prey l⁻¹, for various combinations of predator and prey speeds, and a total simulation duration of *T* = 3600 s.

For each predator–prey speed couple, a sample of 200 simulations was performed, providing a corresponding sample of 200 estimated encounter numbers. For each sample, a Poisson probability density function (Poisson pdf) was fitted to the frequency distribution of encounter numbers. This fit was performed, using the maximum likelihood method under Matlab statistical toolbox (function poissfit.m). This method allowed us to obtain estimates of the Poisson pdf parameter: λ (dimensionless), which is defined as

$$\lambda = ZT \tag{36.20}$$

where *T*(s) is simulation duration and *Z* (s⁻¹) is given by Equation 36.19. Therefore, according to Equation 36.20 and to the statistical properties of the Poisson pdf, an estimate of the expected encounter rate was simply obtained by dividing λ by the simulation duration (*T*).

In parallel, inter-encounter time lags were determined for our theoretical cruising predator, by systematically measuring the time between two successive encounters over the time course of each simulation. Each sample of inter-encounter time lags was processed by fitting a negative exponential pdf.

Results obtained through this procedure are presented in Figure 36.5. The encounter number distribution clearly reproduces the shape of a Poisson pdf, with the mean encounter number shifted toward larger values as the predator's speed increases. This Poisson pattern for the encounter rate is confirmed by the simultaneous negative exponential pattern of the inter-encounter time lag distribution. These results of our model are consistent with the hypothesis initially formulated by Gerritsen and Strickler (1977) and logically expected from a random time-dependent process.

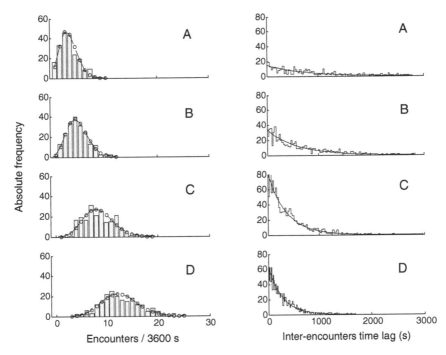

FIGURE 36.5 Simulated encounters numbers distributions (left bar plots) and inter-encounter time lag distributions (right step plots) for distinct values of predator swimming speeds v (mm s^{-1}). (A) $v = 0.5$ mm s^{-1}, (B) $v = 1$ mm s^{-1}, (C) $v = 2$ mm s^{-1}, (D) $v = 3$ mm s^{-1}. These simulations consider a theoretical cruising predator with rectilinear swimming sequences ($\tau = 50$ s duration) oriented at random (see the text for details). Predator perception distance is $R = 3$ mm, prey swimming speed is $u = 0$ mm s^{-1}. Fitted functions are Poisson pdf (left plots: circles) and negative exponential pdf (right plots: continuous line). Prey density is $N = 40$ prey l^{-1}, simulation duration is $T = 3600$ s.

Mean encounter rate, estimated as λT, once plotted against predator swimming speeds (Figure 36.6) exactly reproduces the results predicted by the G&S model (Equation 36.19) for the concerned values of R, v, and u. As predicted from the G&S model, encounter rate increases with predator's speed, and the magnitude of this effect is maximal in situations where predator and prey speeds differ greatly.

Again, as a result of the stochastic nature of our approach, the model provides additional information. Mean values are correctly reproduced (Figure 36.3), but variability of the encounter pattern is also computed. Here, in the situation of an underlying Poisson process, variance of the encounter rate is identical to the mean.

36.3.2 Effect of Varying Swimming Sequence Duration

After demonstrating the reliability of our numerical approach, we further examine the effect of varying the swimming sequence duration. This is the only parameter of our model that is ignored in the formulation of Gerritsen and Strickler (1977), and this discrepancy simply results from the fact that these authors considered a theoretical predator swimming on an infinite rectilinear straight path.

This straight-line approach is somewhat conceptually different from the situation most often observed in planktonic organisms, where the swimming path can be more realistically described as a succession of finite linear segments.

According to the previous description of the model, our conceptual scheme considers swimming plankters that randomly change direction at the end of a rectilinear swimming sequence of duration τ (s). Because we consider that our theoretical predator swims with a constant speed v (mm s^{-1}), the length L (mm) of the rectilinear swimming path is simply given by

$$L = v\,\tau \tag{36.21}$$

Therefore, increasing swimming bout duration (τ) is equivalent to an increase in bout length (L).

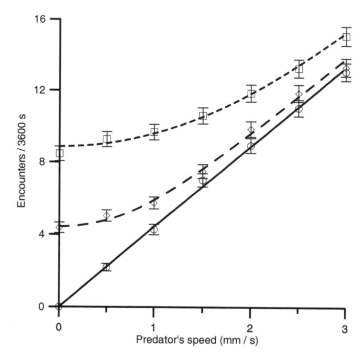

FIGURE 36.6 Simulated changes in average encounter number over 3600 s (symbols) with increasing predator speed, for three different values of prey speed: $u = 0$ mm s^{-1} (circles), $u = 1$ mm s^{-1} (diamonds), and $u = 2$ mm s^{-1} (squares). Average encounter numbers are estimated by fitting a Poisson pdf to a sample of 200 simulations for each combination of predator and prey speeds (see text for details). Vertical bar represents ±95% C.I obtained for the fit of the Poisson pdf. For $u = 0$ mm s^{-1}, encounter numbers pdf obtained from a sample of 200 simulations are presented in Figure 36.5. Lines represent the predicted values by the Gerritsen and Strickler (1977) analytical solution. Continuous line: $u = 0$ mm s^{-1}; dashed line: $u = 1$ mm s^{-1}; dotted line: $u = 2$ mm s^{-1}. Prey density is 40 prey l^{-1}. Other parameter values are given in Table 36.1.

We tested the effect of varying swimming bout duration (hence, bout length) by analyzing samples of 200 simulations for bout lengths varying between 2 and 50 s, which are representative of plankter swimming behavior, e.g., copepod (Saiz, 1994; Caparroy et al., 1998), first feeding fish larvae (MacKenzie et al., 1994; Kiørboe and MacKenzie, 1995). For each simulation, we distinguished the encounters occurring with one or more previously encountered prey, which we shall further refer to as "redundant encounters," from the encounters occurring with distinct prey (i.e., the first encounters with prey not previously encountered). The sum of both encounter rates is equal to the total encounter rate. This distinction was designed because, intuitively, it appeared reasonable to expect multiple encounters with the same prey, as soon as a segment of predator's swimming path of finite length overlapped with previously explored zones during the time course of the simulation.

Results of those simulations (Figure 36.7, left bar plots) show that with decreasing bout duration (and bout length) the total encounter pdf diverges from the initial Poisson pdf pattern predicted by G&S model, and becomes overdispersed. This large apparent increase in encounter variance is not associated with any concomitant change in average total encounter number (not shown on the graph). Therefore, in a first approximation, Gerritsen and Strickler's equation seems to correctly predict the average encounter rate, independently of the swimming bout length.

To the contrary, the redundant encounter pdf appears more sensitive to changes in swimming bout length (Figure 36.7, right bar plots). This long-tailed asymmetric distribution is skewed toward 0 with increasing bout length, and experiences both a reduction of its mean and variance. Therefore, with such a strategy, our theoretical predator decreases the probability of reencountering the same prey, one or more times. This effect is simply due to changes in the predator's global path with varying bout length. With small linear displacements the predator's perceptive field has a high probability of overlapping previously explored areas, and hence the predator has a high probability of perceiving previously encountered prey.

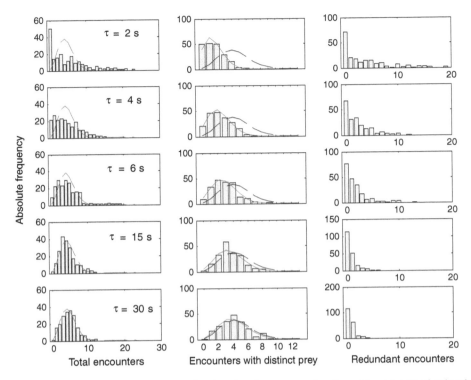

FIGURE 36.7 Simulated effect of changes in predator's swimming sequence duration τ (s) and length of swimming bout L (mm) ($L = v\tau$). For these simulations, $v = 1$ mm s^{-1}, and therefore $L = \tau$. Swimming bout durations are shown on the graphs (2 to 30 s). Right bar plots represent total encounter numbers pdf (including possible multiple encounters with the same prey). Dashed line is the predicted Poisson pdf, with the λ parameter given by Gerritsen and Strickler's analytical solution (Equation 36.20). Central bar plots represent the encounter numbers pdf with distinct prey (i.e., similar situation as if once encountered, the prey was removed from the medium, i.e., captured). The continuous line shows a fitted Poisson pdf to the simulated distributions. Left bar plots represent multiple encounter numbers (redundant encounters) with prey already encountered once. Other parameter values are given in Table 36.1.

Of more interest is the observed variation of the single encounter pdf (Figure 36.7, middle bar plots). These distributions appear to be correctly described by a fitted Poisson pdf, but show a decrease in average encounter number with decreasing bout length. As shown on the graphs, the average encounters number is decreased by up to a factor of four when bout length is 2 mm (2 s bout duration at a speed of 1 mm s^{-1}), compared to the average value predicted by the G&S model. Such a discrepancy between our numerical approach has severe consequences. The counting process we used to estimate the single encounter number with distinct prey is strictly similar to removing the prey from the medium, once it has been encountered a first time. Therefore, this variable is simply an estimate of the capture rate, assuming 100% capture efficiency for our predator, and can be compared with the G&S model predictions, as encounter rate equates to capture rate for such a capture efficiency value. The capture rate is therefore clearly overestimated by the G&S model, by up to a factor of four in the present simulations. This effect is clearly independent of the decrease in accessible prey density resulting from the removal of encountered prey over the time course of the simulation. Still for a value of $\tau = 2$ s, the average number of removed (encountered) prey over 1 h simulation is ~1 prey. At a density of 40 prey l^{-1}, in a 1 m^3 volume, this only represents a $25 \times 10^{-4}\%$ decrease in prey density.

36.4 Discussion

The main objective of this work was to demonstrate the possibility of extracting, from a classical Lagrangian representation of planktonic organism three-dimensional displacements, the inherent information related to predator–prey interactions. This goal was achieved by using a barycentric representation of predator–prey relative displacements. The numerical reliability of our approach has also been demonstrated through comparison with the Gerritsen and Strickler (1977) analytical solution.

However, despite the intensive use of G&S model in the literature, this straight-line approach does not appear appropriate for the correct simulation of the encounter rate of planktonic predators when swimming bout length tends toward small, but realistic, values.

This effect has already been observed by Lewis and Pedley (2000), and results from the overlap between volume mapped out by the predator's perception sphere as it swims along successive linear segments oriented in random directions. These authors proposed a correction factor for the G&S model (Equation 36.20) in the form of a volume fraction fv (dimensionless): ratio of the reduced volume mapped out by predator due to overlaps, to the straight line volume mapped out. However, they did not provide any analytical solution for this correction term, and further used direct numerical simulation to estimate fv, in situations where the length of the swimming bout is small compared with the radius of the perception sphere: $L < 2R$. Further studies should therefore concentrate on a precise estimation of these fraction volume and overlap ratio effects, for the range of parameter values (R, v, u, L, τ) characteristic of planktonic predator and prey motility.

Similarly to Lewis and Pedley (2000), direct numerical simulation (DNS) of planktonic contact rates have been previously used by Yamazaki et al. (1991). This approach relies on the numerical integration of planktonic equations of motion (Equations 36.1 to 36.3) using the Runge–Kutta numerical scheme of integration. A fixed time step of integration is assigned as small as possible to detect all possible contacts between displacements of spherical organisms. Because our method provides an analytical solution for the time step of numerical integration between two successive encounters, it is clear that it avoids unnecessary computations during periods where no encounters occur. However, there is a common effect of increased computational cost with the number of organisms tracked. Similarly to Yamazaki et al. (1991) or Lewis and Pedley (2000), this problem can be solved by applying the algorithm initially proposed by Sundaram and Collins (1996). The simulation domain can be divided into cubic cells of smaller size, and potential collisions between prey and predator's perception sphere are tested only in those cells situated on the predator's path before the next change of direction.

The use of DNS of planktonic organism motion was initially applied to study the effect of small-scale turbulence on predator–prey encounter rates (Yamazaki et al., 1991). By adding the effect of a turbulent flow field to the plankter swimming velocities, more realistic estimates of *in situ* encounter rates are obtained. This is because still fluid conditions, as assumed by Equation 36.20, are rare in the pelagic realm. To do so, Equations 36.1 through 36.3 can be coupled with a realistic stochastic representation of turbulent flow fields (Malik and Vassilicos, 1999). This kinematic approach has the ability to reproduce the main features of homogeneous isotropic small-scale turbulence, and the enormous advantage of lowering the computational cost of turbulent flow field simulation. It has been used successfully by Lewis and Pedley (2000), and provides a more realistic representation of turbulence-induced increases in encounter rate compared to simple analytical solutions (Rothschild and Osborn, 1988; Evans, 1989).

Small-scale turbulence affects not only encounter rates, but also postencounter processes, such as predator attack success (MacKenzie et al., 1994). It is possible within the framework of our model to assess both the capture rate of planktonic predators and the escape success of their prey. Similarly to encounter, capture is a discrete event that can be modeled as the collision between a prey and the spherical capture sphere of the predator. Equations for prey escape directions, with varying orientation and randomness have been formulated (Caparroy et al., 2000) and could be incorporated into the model. This is of particular importance in estimating the integrated effects of small-scale fluid motions on the full predation cycle.

Another advantage of the discrete events discrimination technique is that it allows for the association of complex behavioral repertoires to specific categories of tracked particles. Such a higher level of complexity in the representation of plankter swimming behavior is necessary to understand the comparative advantages of different foraging strategies. Complexity of swimming behavior might concern the number of different behavioral sequences, their length, duration, and speeds, but also the dominant directionalities associated with each sequence (Titelman, 2000).

A more exact representation of the encounter process also involves taking into account a more realistic description of predator and prey perceptive fields. Spherical perceptive fields generally considered are mathematical abstractions that allow us to obtain simple analytical solutions for the encounter rate. However, for real planktonic predators, the shape of the perceptive field is far from isotropic; rather, it is a result of the expression of microscale physical processes. For visual predators such as fish larvae, perception distance can be modeled as a complex function of water column and prey optical properties (Aksnes and Utne, 1997). For rheotactic organisms such as copepods, perception distance of predator and prey results from the magnitude and shape of the flow field associated with protagonist displacements (Kiørboe et al., 1999; Kiørboe and Visser, 1999; Visser, 2001). Because perception distance is the most sensitive parameter in Equation 36.20 (Gerritsen and Strickler, 1977), further modeling effort should concentrate on incorporating physical models of perceptive field into the expression of the encounter rate function. In the case of hydromechanical signal perception by copepods, this is not a simple task because instantaneous perception distances depend on instantaneous relative orientation between predator and prey (Visser, 2001). Therefore, averaging such processes over time and space to obtain simple analytical solutions is not reasonable. However, because our Lagrangian approach integrates predator and prey spatial coordinates and orientations over time, it is potentially more appropriate to incorporate such process models.

In a population of potentially interacting pelagic organisms (phytoplanktonic cell, ciliate, copepod, fish larvae, and so forth) predatory interactions generally concern more components than a single predator–prey couple. Because of their schizoid nature, planktonic organisms are often predator and prey at the same time, and modern optimal foraging theory (Lima and Dill, 1990) considers that feeding behavior should be driven by the conflicting objective of simultaneously maximizing feeding rates and minimizing predation risk (Tiselius et al., 1997). However, no functional response model (neither numerical, nor analytical) has tried to represent/compute the outcome of such complex interactions. Complex feeding rate models generally propose a deterministic representation of simultaneous feeding rates on mixtures of two prey (e.g., Hassel et al., 1976).

Above this number of interacting organisms, all existing models are still deterministic, but rather complex mathematical functions without evident functional/biological sense. Using our discrete events-based Lagrangian approach, it is possible to explore sophisticated scenarios where the feeding success of particular planktonic organism will depend not only on its access to a pool of prey resources, but also on the presence and encounter probability with its own predators and competitors.

36.5 Conclusion

Predation rate functions represent the energetic input that will greatly influence individual predator's survival, growth, and reproductive success; as well as prey death rate. Therefore, a correct assessment of mean predation rates and their associated variance is a key modeling process, as both statistical properties will be integrated up to population dynamics and community food web levels. However, individual-based models of planktonic communities most often represent feeding rates of individuals with the help of simple nonlinear functions of prey concentration (Ivlev, 1945; Steele, 1974; Lam and Frost, 1976; Lehman, 1976), generally fitted to experimental values obtained from group of animals, rather than individuals. Although these models describe the shape of most planktonic predator's functional response and have the advantage of mathematical simplicity; they describe a steady-state solution, which is not the rule in the sea, and remain simple correlation studies, which do not provide causal relationships for the observed predation rates.

Because our approach provides access to the interindividual variability of feeding rates, it is more adapted for the purpose of the emerging individual-based modeling techniques (see Souissi et al., Chapter 34, this volume), and should allow for the exploration of how small-scale processes of predation integrate up to higher levels of ecosystem dynamics in the sea.

Acknowledgments

The author thanks two anonymous referees for their critical comments as well as Sami Souissi, Laurent Seuront, Denis Busardo, Andy Visser, and Jérome Casas for their helpful suggestions and encouragement. Thanks are also extended to J. Seymour who kindly improved the language.

References

Aksnes, D. and Utne, A.C. 1997. A revised model of visual range in fish. *Sarsia* 82: 137–147.

Beyer, J.E. 1980. Feeding success of clupeoid fish larvae and stochastic thinking. *Dana* 1: 65–91.

Caparroy, P., Perez, M.T., and Carlotti, F. 1998. Feeding behaviour of *Centropages typicus* in calm and turbulent conditions. *Mar. Ecol. Prog. Ser.* 168: 109–118

Caparroy, P., Høgsbro, U., and Visser, A.W. 2000. Modelling the attack success of planktonic predators: patterns and mechanisms of prey size selectivity. *J. Plankton Res.* 22(10): 1871.

Evans, G.T. 1989. The encounter speed of moving predator and prey. *J. Plankton Res.* 11(2): 415–417.

Gerritsen, J. 1980. Adaptative response to encounter problems, in Kerfoot, W.C., Ed., *Evolution and Ecology of Zooplankton Communities,* University Press of New England, London, 52–61.

Gerritsen, J. and Strickler, J.R. 1977. Encounter probabilities and community structure in zooplankton: a mathematical model. *J. Fish. Res. Board Can.* 34: 73–82.

Hassel, M.P., Lawton, J.H., and Beddington, J.R. 1976. The components of arthropod predation I. The prey death rate. *J. Anim. Ecol.* 45(1): 135–164.

Heath, M.R. 1993. The role of escape reaction in determining the size distribution of prey captured by herring larvae. *Environ. Biol. Fish.,* 38: 331–344.

Holling, C.S. 1966. The functional response of invertebrate predators to prey density. *Mem. Entomol. Soc. Can.* 48: 5–78.

Ivlev, V.S. 1945. The biological productivity of waters. *Usp. Sovrem. Biol.* 19(1): 88–120.

Jakobsen, H. 2001. Escape response of planktonic protists to fluid mechanical signals. *Mar. Ecol. Prog. Ser.* 214: 67–78.

Kiørboe, T. and MacKenzie, B.R. 1995. Turbulence-enhanced prey encounter rates in larval fish: effects of spatial scale, larval behaviour and size. *J. Plankton. Res.* 17(12): 2319–2331.

Kiørboe, T. and Saiz, E. 1995. Planktivorous feeding in calm and turbulent environments with emphasis on copepod. *Mar. Ecol. Prog. Ser.* 122: 135–145.

Kiørboe, T. and Visser, A. 1999. Hydrodynamic signal perception in the copepod *Acartia tonsa. Mar. Ecol. Prog. Ser.* 179: 81–95.

Kiørboe, T., Saiz, E., and Visser, A., 1999. Predator and prey perception in copepods due to hydrodynamical signals. *Mar. Ecol. Prog. Ser.* 179: 97–111.

Lam, K.R. and Frost, W.B. 1976. Model of copepod filtering response to changes in size and concentration of food. *Limnol. Oceanogr.* 21(4): 490–500.

Landry, M.R. 1978. Population dynamics and production of a planktonic marine copepod *Acartia clausii* in a small temperate lagoon on San Juan Island, Washington. *Int. Rev. Gesamten Hydrobiol.* 63: 77–119.

Lehman, T.J. 1976. The filter-feeder as an optimal forager, and the predicted shapes of feeding curves. *Limnol. Oceanogr.* 21(4): 501–516.

Lewis, D.M. and Pedley, T.J. 2000. Planktonic contact rates in homogeneous isotropic turbulence: theoretical predictions and kinematic simulations. *J. Theor. Biol.* 205: 377–408.

Lima, S.L. and Dill, L.M. 1990. Behavioral decision made under the risk of predation: a review and prospectus. *Can. J. Zool.* 68: 619–640.

MacKenzie, B., Miller, T. et al. 1994. Evidence for a dome-shaped relationship between turbulence and larval fish ingestion rates. *Limnol. Oceanogr.* 39(8): 1790–1799.

Malik, N.A. and Vassilicos, J.C. 1999. A Lagrangian model for turbulent dispersion with turbulent-like flow structures: comparison with direct numerical simulation for two-particle statistics. *Phys. Fluids* 11: 1572–1580.

Ohman, M.D. 1988. Behavioral responses of zooplankton to predation. *Bull. Mar. Sci.* 43: 530–550.

Price, H. J., 1988. Feeding mechanisms in marine and freshwater zooplankton. *Bull. Mar. Sci.* 43(3): 327–343.

Rothschild, B.J. and Osborn, T.R. 1988. Small scale turbulence and plankton contact rates. *J. Plankton Res.* 10(3): 465–474.

Saiz, E. 1994. Observations of the free-swimming behavior of *Acartia tonsa*: effects of food concentration and turbulent water motion. *Limnol. Oceanogr.* 39(7): 1566–1578.

Souissi et al. 1993.

Steele, J.H. 1974. *The Structure of Marine Ecosystem.* Harvard University Press, Cambridge. MA.

Steele, J.H. and Frost, B.W. 1977. The structure of plankton communities. *Philos. Trans. R. Soc. Lond. Ser. B* 280(976): 485–534.

Sundaram, S. and Collins, L.R. 1996. Numerical considerations in simulating a turbulent suspension of finite volume particles. *J. Comput. Phys.*124: 337–350.

Tiselius, P., Jonsson, P.R. et al. 1997. Effects of copepod foraging behavior on predation risk: an experimental study of the predatory copepod *Paraeuchaeta norvegica* feeding on *Acartia clausi* and *A. tonsa* (Copepoda). *Limnol. Oceanogr.* 42(1): 164–170.

Titelman, J. 2000. Swimming and escape behaviour of copepod nauplii: implications for predator–prey interactions among copepods. *Mar. Ecol. Prog. Ser.* 213: 203–213.

Visser, A. 2001. Hydromechanical signals in the plankton. *Mar. Ecol. Prog. Ser.* 222: 1–24.

Yamazaki, H., Osborn, T. et al. 1991. Direct numerical simulation of planktonic contact in turbulent flow. *J. Plankton. Res.* 13(3): 629–643.

Yen, J. and Fields, D. M., 1992. Escape response of *Acartia hudsonica* (Copepoda) nauplii from the flow field of *Temora longicornis* (Copepoda). *Arch. Hydrobiol. Beih. Erg. Limnol.* 36: 123–134.

Index

A

Abbott, Denman and, studies, 31, 59, 62
Abbott and Letelier studies, 31, 55, 58, 62
Abiotic conditions, 386–389
Abraham studies, 226, 271, 336
Acartia clausi, 362, 366, 368–372
Acartia husonica, 152
Acartia omorii
    Mobidyc platform, 526, 537
    stage-structured population model, 367, 372, 373
Acartia tonsa, 153–154
Acceleration fields, 166–167, 173–174
Ackley and Littman studies, 514
Acoustic Doppler Velocimeter (ADV)
    East Sound, 33
    vertical velocity gradients, 39, 40–42
Acoustic remote sensing
    acoustic measurements, 72–73
    ambient noise, 72
    basics, 66–67, 92–93
    bubbles, 67–68, 84–88
    directional characteristics, 91
    dissolved oxygen, 73–75, 83–84
    energy time distribution, 80–81
    equalized matched-filter processing, 73
    experimental configuration, 71–72
    gases, 69, 91–92
    leaf blade, gaseous interchange, 91–92
    measurements, 73–75
    medium response, 80–81
    multipaths, 84–88
    multiscale acoustic effects, 75–83
    noise, 72, 88–91
    non-photosynthesis-related effects, 81–83
    oxygen production, 69
    photosynthesis influence, 67–69
    Posidonia oceanica, 69
    propagation channel modeling, 77–80
    reverberation, 88
    sea surface motion, 82
    signal transmission, 72
    sound propagation, 83–92
    spectral characteristics, 88–90
    test site, 69–71
    tide, 81–82
    time-frequency characteristics, 90–91
    time-varying medium impulse response, 75–76
    transducer calibration, 72–73
    Ustica experiment (September 1999), 69–75
    water temperature, 83

Adamec, Wilson and, studies, 53, 57
Adamic studies, 259
Adams and Sterner studies, 130
Adaptation levels
    artificial neural networks, 513–514
    evaluation criteria, 510, 517
    fixed genetic strategies, 510–511
    individual learning, 512–514
    phenotypic plasticity, 511–512
    pros and cons, 517
    reinforcement learning, 514
    supervised learning, 514
Addicott studies, 384
Adelaide, Australia, 272
Adler and Mosquera studies, 393
Adobe Photoshop software, 112
Adriatic sardine, 194
ADV, *see* Acoustic Doppler Velocimeter (ADV)
Advanced Very High Resolution Radiometer
        (AVHRR), 220
Agent-based models, 431–432, 442
Aggregative force field, 164–165
Aksnes and Utne studies, 571
Alec Electronics, 272
Alexander and Roughgarden studies, 230
Algae, 45, 337
Algorithms, robustness, 346–348
Alldredge studies, 162, 163
Almirantis, Provata and, studies, 336
Alpine marmots, 412
Altabet and FranHois studies, 128
Ambient flow field, 502
Ambient noise
    directional characteristics, 91
    photosynthesis effect, 88–91
    spectral characteristics, 88–90
    time-frequency characteristics, 90–91
    Ustica experiment, 72
Ambler studies, 162, 166
Analysis
    benthic community patterns, 229–240
    biogeochemical variability, 215–226
    copepod interactions, 361–375
    environmental variability, temporal scaling,
        201–211
    epiphytal and epilithic communities, 245–255
    fractal characterization, 297–318
    fractal dimension estimation, 245–255
    gelatinous organisms, 329–331
    patchiness, 297–318
    rank-size analysis, 257–276

sea fan orientation, 321–327
skipjack tuna dynamics, 183–196
stage-structured population models, 361–375
swimming behavior, 333–358
vertical phytoplankton distribution patterns,
    257–276
wavelets, 279–296
zooplankton, 333–358
Analysis of Variance (ANOVA), 229
Anchovy eggs, 388
ANCOVA, 236
Anderson, Okubo and, studies, 163, 164, 165, 172, 175
Anderson, Slatkin and, studies, 393
Ankistrodesmus sp., 337
ANNs, *see* Artificial neural networks (ANNs)
ANOVA, *see* Analysis of Variance (ANOVA)
Antarctic krill, 162
Antipredator responses, 516–517
Appleby studies, 347
Aquatic ecosystem functioning, 4–5
Aquatic systems, pattern-oriented modeling, 423–424
Archer, Mahadevan and, studies, 221
ARcInfo software, 113
Arheimer and Lidén studies, 202, 209
Ariño and Pimm studies, 210
Arithmetic method, 295
Ariz studies, 185
Armi and Haidvogel studies, 177
Arneodo studies, 471
Artificial neural networks (ANNs), 512, 513–514
Aschenwald studies, 119
Ascophyllum, 392
Assembling patterns, 414–415
Astudillo and Caddy studies, 194
Atherosclerosis, 137
Attribute vector, IBM, 509
Ault studies, 389
Australia, 272
Autocorrelations, 171–173, 229
Automated Underwater Vehicle (AUV), 44
Auyang studies, 423
AVHRR, *see* Advanced Very High Resolution
    Radiometer (AVHRR)
Avicennia germinans, 419, 422
Axtell studies, 259, 268
Azam, Long and, studies, 258
Azam studies, 268
Azores, 192

B

Backhaus, Verduin and, studies, 423
Bainbridge studies, 31, 32
Baird studies, 518
Bak studies, 258, 268
Bal, Strickler and, studies, 32, 494, 503
Baldwin effect, 517–518
Baldwin studies, 517
Ballard studies, 513, 514
Baltic Sea, 32, 388

Banas studies, 161–178
Banks, *see* specific bank location
Barber studies, 329–331
Baretta studies, 388
Barlett's method, 258
Barnacle mortality, 229
Barnes, Crisp and, studies, 254
Barnes and Dick studies, 240
Barrat-Segretain, Chiarello and, studies, 392
Barstow studies, 445
Barth, and Levine, Dale, studies, 43
Barthélémy studies, 152
Bascompte and Vilà studies, 336, 350
Basset studies, 383–395
Bassler studies, 162
Bas studies, 183–196
Bathochordaeus, 331
Batschelet studies, 339
BATS data set, 257
Batty, Longley and, studies, 341
Bayliff studies, 183
Beardsley studies, 9–10
Beech forest model BEFORE, 417–419, 424
Beer studies, 544
Belew, Menczer and, studies, 510
Bell studies, 333
Bence and Nisbet studies, 230
Benfield studies, 17–28, 27
Bennett and Denman studies, 32
Benthic species, community patterns
    basics, 239–240
    CA framework, extending, 236–239
    cellular automata approach, 231–236
    empirically defined alternative models, 235–236
    Markov matrix approach, 231–236
    modeling, 231
    nonspatial (point) transition matrix models,
        233–234
    spatially explicit models, 230
    spatial scale treatments, 229–230
    spatial transition matrix models, 234–235
Benthic species, production and respiration
    basics, 97–98, 104–105
    daily potential primary production, 99–100
    gradient of exposure, 102–103
    irradiance, 102
    materials and methods, 98–99
    microphytobenthos adaptation, 102
    microscale variability, 103–104
    results, 99–104
    seasonal variations, 100–102
    temporal resolution, 102
Benthic species, spatial dynamics
    basics, 429–430
    discussion, 441–443
    intertidal zone, 432–437
    model classes, 430–432
    modeling, 432–441
    mussels, 432–437
    validation and verification, 437–441
Benzi studies, 474

Berdalet studies, 549
Berec studies, 535
Berg and Getz studies, 194
Berger studies, 411–425, 422
Bergman studies, 342
Bernard, Fiedler and, studies, 183
Bernard, Roger and, studies, 183
Bernard, Souissi and, studies, 353
Bernard and Gouze studies, 362
Bernard studies, 361–375
Bernasconi, Teranes and, studies, 128
Bernier, Kerman and, studies, 195
Beverton and Holt studies, 184, 186
Beyer and Laurence studies, 509
Beyer studies, 388, 560
Biogeochemical variability, sea surface
    basics, 215, 226
    discussion, 225–226
    model fields, 220
    modeling, 222–225
    patchiness, 222–224
    quantifying variability, 218–219
    resolution requirements, 225
    response time, 222–224
    satellite data, 220–222
    tracers, 215–218, 222–224
    vertical distribution, 224–225
Biological-physical coupling, equatorial Pacific
    analysis, 54–55
    basics, 51–52, 61–62
    cross-correlation analysis, 54, 56–57
    data, 53–54
    decorrelation timescale, 55, 57–59
    discussion, 55–61
    oceanographic setting, 52–53
    results, 55–61
    spectral analysis, 55, 59–61
Biological scale resolution, measurement comparison
    aquatic ecosystem functioning, 4–5
    basics, 3–4, 12–13
    differential structure, fluorescence signals, 10–12
    high-resolution data vs. conventional techniques, 7–12
    instrument description, 7–9
    microscale structure, 4–7
    sampling process impact, 5–7
    sensor deployment, 9–10
BIONESS casts, 28
Biophysical interactions, microscale structure, 4
BIOPMAPPER II, 27
Biotic conditions, 389–390
Bird Rock inlet, 111–113, 117–118
Biscayne Bay, 389
Bjørnsen and Nielsen studies, 31, 32, 45
Black-body source, 138
Blackwell studies, 556
Blades studies, 154
Blaine and DeAngelis studies, 392
Blakeway studies, 109–121
Blanchard studies, 103
Block studies, 196

Bluefin tuna, 196
Body weight, IBM, 528
Boehlert and Mundy studies, 183
Boersma, Mooij and, studies, 537
Boinski and Garber studies, 333
Bollens studies, 168
Boots and Getis studies, 23
Bottom-up simulation model, 411, 423
Bouma studies, 240
Bourget, Le Tourneux and, studies, 254
Bowler, Cossins and, studies, 202
Box-counting method, 341, 347–348
Boyd and Richerson studies, 517
Bradbury and Reichelt studies, 254, 336
Bradbury studies, 254, 336
Brandl and Fernando studies, 162
Brandl studies, 395
Breckling, Reuter and, studies, 391
Brewer and Coughlin studies, 352
Brewer studies, 333–358
Brightness, synchroton-based IR imaging, 138
Brooks, Mullin and, studies, 32
Broomhead, Sumpter and, studies, 535
Browman and Skiftesvik studies, 493
Browman studies, 493
Brownian motion/diffusion
    plankton dynamics, heterogeneous environment, 404
    turbulent intermittency, 460
    zooplankton swarming, 162
    zooplankton swimming behavior, 350, 353
Brown studies, 149–158
Bruce, Voorhis and, studies, 216
Bubbles
    photosynthesis effect, 84–88
    snow, 92n
    Ustica experiment, 67–68
Buckland studies, 110
Bundock, Yen and, studies, 162
Bundy and Paffenhöfer studies, 480
Bundy studies, 162, 333, 334, 337
Bunimovich studies, 196
Bunta studies, 135–145
Buoyant particles, Langmuir circulation
    basics, 445–446
    discussion, 449–450
    flow and particle patterns, 447–449
    simulation setup, 446
Burd, Jackson and, studies, 4
Burlando studies, 336
Burrows and Hawkins studies, 230
Buskey studies, 17, 163, 175
Butler studies, 31, 32

C

Caddy, Astudillo and, studies, 194
Caddy studies, 194
CA framework, extending, 236–239
Cain, McCulloch and, studies, 338

Calanus finmarchicus, 515
Calanus marcshallae, 155
Caley studies, 231
California Current, 59
Callaway and Davis studies, 232
Camera housing, zooplankton distributions, 19
Camera-net system, 18
Campbell, Mahadevan and, studies, 218
Campbell and Esaias studies, 218
Campbell studies, 215–226
Campo, Torrence and, studies, 299, 306
Canary Islands, 185, 188, 192, *see also* Skipjack tuna,
    scale similarities
Cannibalism, IBM, 530
Caparroy studies, 337, 352, 555–571
Carlton studies, 162
Carmona, Snell and, studies, 154
Carr studies, 135
Cartesian coordinates, 565
Cartwright studies, 325
Cassie studies, 31
Castellan, Siegel and, studies, 347
Castello studies, 543
Castro studies, 183–196
Casual lognormal stochastic equation, 460–462
Caswell, Maley and, studies, 510
Caswell, Pascual and, studies, 402
Caswell, Tuljapurkar and, studies, 508
Caswell and Cohen studies, 31
Caswell and John studies, 508
Caswell studies, 234
Cattaneo and Prairie studies, 208, 209
Caun studies, 149–158
CCD, *see* Charge-coupled devices (CCD)
CDF, *see* Cumulative density function (CDF)
CEH Windemere Laboratory, 203
Cellular automata (CA) approach, 231–236, 431
CENOCON software, 536
Center for Limnology, 203
Centropages adbominalis
    Mobidyc platform, 524, 526, 530, 537
    stage-structured population models, 362, 372, 373
Centropages typicus, 482
CFD, *see* Computational fluid dynamics (CFD)
    simulation
Chaetoceros muellerii, 135
Chairello and Barrat-Segretain studies, 392
Chambers studies, 509
Chaoborus sp., 516
Chapelle studies, 387
Chapman, Underwood and, studies, 229, 240
Chapman studies, 209
Characteristic functions, turbulent intermittency, 464
Chargaff's rule, 413
Charge-coupled devices, 18
Charge-coupled devices (CCD), 19
Chatfield studies, 55, 205, 258
Chavez studies, 51–62
Chervin, Semtner and, studies, 221
Chételat and Pick studies, 208
Chow-Fraser and Maly studies, 152

Cifuentes, Fogel and, studies, 128, 131
Cintrón and Schaeffer-Novelli studies, 420
Cizrók studies, 259
Claereboudt studies, 391
Clark and Levy studies, 515
Clark and Mangel studies, 512
Clark and Rose studies, 389, 391
Clark studies, 184, 193
Clements studies, 162
Clifton studies, 327
CMOS, *see* Complementary metal-oxide-silicon
    (CMOS) detectors
Coale studies, 52
Cody studies, 336
Cohen, Caswell and, studies, 31
Cohen studies, 125–131
Colbo and Li studies, 448, 449
Collins, Sundaram and, studies, 570
Command, zooplankton distributions, 21
Commercial fisheries, 195
Commito, Snover and, studies, 230, 252
Commito and Rusignuolo studies, 252
Compass method, 340–341
Compensatory dynamics, skipjack tuna, 194–195
Complementary metal-oxide-silicon (CMOS)
    detectors, 18
Computational fluid dynamics (CFD) simulation, 480
Comte-Bellot and Corrsin studies, 496
Conan studies, 194
Concentrative force, zooplankton swarming, 177
Connected habitats, plankton dynamics model
    patterns, 404–406
Conover studies, 205
Construction of model, pattern-oriented modeling,
    415
Continental shelf
    Florida, 387
    New England, 43–44
    Oregon, 33, 34–39, 43
Continuous cascades, 459–463
Continuous lognormal multifractal, 462–463
Control, zooplankton distributions, 21
Cooley, Rigler and, studies, 374
Copepods, interaction identification
    Acartia clausi, 368–372
    age-within stage model, 366–367
    basics, 361–362
    cannibalism, 368–371
    discussion, 374–375
    Euterpina acutifrons, 371–373
    eutrophic inlet location, 372–373
    extrema identification, 365–366
    mathematical proofs, 375–376
    method, 363–367
    predation, 364–365, 368–372, 376
    results, 367–373
    in situ interactions, 372–373
    stage-structured population model, 363–365
    time-series abundance data, 365–366
    transient behavior, 363–365
    transition rules, 363–365, 367–368, 375–376

Copepods, mate signaling
basics, 149–150, 157–158
discussion, 151–157
methods, 150–151
odorant levels, 155–157
pheromones, 151
results, 151–157
scent preferences, 151–154
tracking behavior, 154–155
trails, 149–157
Copepods, multiagent system IBMs
basics, 523–524, 538–539
body weight, 528
cannibalism, 530
demographic noise, 528–530
descriptions, 536
development, 524–528
discussion, 533–538
growth model, 527–528
improvements, 534–536
languages, 536
life cycle representation, 525
methodology, 534–535
Mobidyc platform, 524–525
model architecture, 525–528
molting model, 527–528
mortality simulation, 528–530
predation, 528, 530–533
reproduction, 528
results, 528–533
simulation, 528–530
spatial representation, 530
spatiotemporal dynamics, 530–533
survival model, 527–528
temperature, 528
tools, 535–536
zooplankton, 536–538
Copepods, planktonic behavior model, 544–545
Coral competition model, 394
Correlations
autocorrelations, 171–173, 229
biological-physical coupling, equatorial Pacific,
54, 55, 56–57
decorrelations, 55
multiplicative cascades, 456–457
phytoplankton, 36–37
Correll studies, 209
Corrsin, Comte-Bellot and, studies, 496
Cossins and Bowler studies, 202
Costello, Strickler and, studies, 493
Costello studies, 333, 493
Coughlin, Brewer and, studies, 352
Coughlin studies, 333, 334, 337, 350
Coullana canadensis, 162
Coulter studies, 129
County Cork, Ireland, 252
Courant-Friedrichs-Lewy (CFL) criteria, 404
Cowles and Strickler studies, 32, 482
Cowles studies, 3, 17, 31–46, 334
Craik and Leibovich studies, 446
Crisp and Barnes studies, 254

Crist studies, 343
CritterSpy, 337
Cross-correlation analysis, biological-physical
coupling scales, 54, 56–57
Crossing bank fronts, fractal characterization, 311–313
Crowder studies, 509
Cuddington and Yodzis studies, 202, 209, 210
Culex pipiens, 162
Cullen and Horrigan studies, 543, 544
Cullen and Lewis studies, 31, 43
Cullen studies, 546, 551
Cumulative density function (CDF), 10–11
Curry studies, 496
Cushing studies, 184
Cyr studies, 201–211
Cystoseira, 71
Czárán studies, 386, 413

D

Daan studies, 388
Daily potential primary production, 99–100
Dale, Barth, and Levine studies, 43
Damköhler number, 223
Damping, zooplankton swarming, 175–177
Daphnia magna, 163, 167–178
Daphnia pulex, 515–517, 537, *see also* Zooplankton
swimming behavior
Darwin studies, 507, 510
D'Asaro and Lien studies, 331
Data
biological-physical coupling scales, equatorial
Pacific, 53–54
skipjack tuna, scale similarities, 185
synchroton-based IR imaging, 141–144
zooplankton distributions, 21–22
Daubechies wavelets, 286–288
Davenport studies, 245–255, 336, 341
Davies, Reynolds and, studies, 202
Davis, Callaway and, studies, 232
Davis, Montana and, studies, 513
Davis studies, 18, 32, 162, 163
Davoult studies, 97–105
Dawkins studies, 512
DeAngelis, Blaine and, studies, 392
DeAngelis, Lee and, studies, 391
DeAngelis, Mulholland and, studies, 392
DeAngelis, Petersen and, studies, 391
DeAngelis and Mooij studies, 391, 416, 424
DeAngelis and Petersen studies, 391
DeAngelis studies, 184, 383–395, 509
De Boor studies, 471
Deboule and Dolle studies, 325
Decision-making mechanism, planktonic behavior,
547–549
Decorrelation timescale, biological-physical coupling
scales, 55, 57–59
Dekshenieks studies, 17, 32, 34
Delgado, Marin and, studies, 390
Demographic noise, 528–530

Denbo, Skyllingstad and, studies, 446
Denizot, Theodor and, studies, 321
Denman, Bennett and, studies, 32
Denman, Platt and, studies, 59, 201
Denman and Abbott studies, 31, 59, 62
Denman and Gargett studies, 43, 493
Denman and Platt studies, 32, 59, 218
Denman and Powell studies, 4
Denman studies, 31, 32, 59
Depensatory dynamics, skipjack tuna, 194–195
Deriderio studies, 3
De Rosa studies, 325
De Ross studies, 236
Derwent Water, UK, 204, 207–209
Deschamps studies, 218
Descriptions, IBM, 536
Desharnais studies, 429–443
Desharnais, Robles and, studies, 119, 392, 432
Desiderio studies, 3
Desmarestia mensiesii, 247, 254
Diatomus copepod, 152
Dick, Barnes and, studies, 240
Dickie Lake, 205
Dickson studies, 203
Diekman, Metz and, studies, 508
Diekmann, Mylius and, studies, 510
Diel vertical migration (DVM), 515
Differential structure, fluorescence signals, 10–12
Diffraction limited IR imaging, 143–144
Diffusion, zooplankton swarming, 163–164
Dill, Lima and, studies, 571
Dillion, Molot and, studies, 209
Dillion, Wainwright and, studies, 321, 322, 326
Dillion and Molot studies, 209
Dillion studies, 201–211
Dimensional representations, wavelet transform, 306–308
Dimensional swarming, 166
Dinoflagellates
    basics, 545–547
    decision-making mechanism, 547–549
    external nitrate conditions, 550–553
    simulation, 549–554
    threshold conditions, 553–554
Dioithona oculata, 162, 163
Directional characteristics, ambient noise, 91
Direct numerical simulation (DNS)
    discrete events-based Lagrangian approach, 472
    energy costs, 493, 494–496, 501–504
    lognormal theory, 472
Discrete cascades simulation, 457–459
Discrete events-based Lagrangian approach
    basics, 559–560, 571
    case study, 564–569
    discrimination, 562
    discussion, 569–571
    Gerritsen and Strickler analytical solution, 564–567
    methodology, 560–564
    numerical integration timescale, 562
    simulation, 564–567
    swimming behavior, 560–561, 567–569

vectorial estimation, 562–564
Discrete in scale, 454–455
Discrimination, discrete events-based Lagrangian approach, 562
Dissolved oxygen, 73–75, 83–84, *see also* Oxygen
Distributed Active Archive Center, 54
DNS, *see* Direct numerical simulation (DNS)
Doall studies, 149–158, 168, 177
Dodson, Krueger and, studies, 516
Dodson, Larsson and, studies, 516
Dodson studies, 334
Dolle, Deboule and, studies, 325
Donaghay, Hanson and, studies, 3, 32, 166
Donaghay studies, 3, 17, 31
Donalson and Harrison studies, 438
Donalson and Nisbet studies, 429, 431
Donalson studies, 429–443
Dorset Environmental Science Center, 203, 209
Dover studies, 131
Dowling studies, 333, 334, 336, 343, 346, 350
Dremin studies, 279–296
Duarte and Vaqué studies, 258
Duarte studies, 203
Dugatkin and Godin studies, 518
Dunne studies, 57
Dusenbery studies, 149
DVM, *see* Diel vertical migration (DVM)
Dye injections, 43–44
Dye studies, 240
Dynamic coupling, flow field numerical simulation, 481–484

E

Earn studies, 230
Eastern English Channel, 6, *see also* English Channel
East Sound, 33–34, 38, 43
Edelstein-Keshet, Ermentrout and, studies, 232, 238
Eek studies, 131
Ekholm studies, 202, 209
El Niño-La Niña cycles, 52, 54, 57, 61, 62
Eloffson studies, 162
Elsner, Tsonis and, studies, 336
Emergent patterns, spatial scale importance, 392–393
Emlet and Strathmann studies, 479
Emlet studies, 480, 487
Empirically defined alternative models, 235–236
Energy costs, planktonic organism hovering
    ambient flow field, 502
    basics, 493–494
    direct numerical simulation, 494–496
    discussion, 503
    flow fields, 494–497, 499, 502
    mean swim velocity, 501–502
    methods, 494–499
    motion equation, 497–498
    nondimensional numbers to dimensional numbers, 504
    particle rising, 502–503
    planktonic swimming models, 497–499

random flow simulation, 496–497
results, 499–503
sinking velocity, 502–503
statistics, 500–501
strain-based swimming model, 498–499
velocity-based swimming model, 498
Energy time distribution, 80–81
England, 202
English Channel, 98, 99, *see also* Eastern English
Channel
Entropy method, 295
ENVI image processing package, 118
Environmental noise, no fish, 404–405, 406
Environmental noise (none), no fish, 405–406
Environmental noise (none), one fish, 406
Environmental Sensing System (ESS), 300
Environmental sensors, zooplankton distributions, 21
Environmental structure effects, 350–351
Environmental variability, rivers and lakes
basics, 201–203, 210–211
different timescales, 210
discussion, 208–210
ecological implications, 210
methods, 203–205
results, 205–208
rivers and lakes comparison, 208–209
Epidermal cells, leaf blade gaseous exchange, 92
Epilithic communities, *see* Epiphytal and epilithnic
communities
Epiphytal and epilithnic communities
analysis and biology, 248
basics, 245–248
D estimation, 251–252
ecology, 249, 251
fractal dimensions, 249, 251–252
intertidal rocky substrates, 252–253
limitations, 254–255
value, 253–254
Equalized matched-filter processing, 73
Equatorial Pacific, *see* Biological-physical coupling,
equatorial Pacific
Equilibria, multiple (stable), 194
Erlandson, Kostylev and, studies, 230, 252
Erlandson and Kostylev studies, 251, 343, 347
Ermentrout and Edelstein-Keshet studies, 232, 238
ERSEM model, 388
Esaias, Campbell and, studies, 218
Esaias studies, 220
ESS, *see* Environmental Sensing System (ESS);
Extended self-similarity (ESS)
Estimates, zooplankton swimming behavior, 341–350
Eswaran and Pope studies, 471, 495
Euchaeta norvegica, 155
Euclidean properties, 254, 350
Euglena gracilis single cells, synchroton-based IR
imaging
basics, 135–137, 145
brightness, 138
data analysis and results, 141–144
diffraction limited IR imaging, 143–144
discussion, 144–145

experimental details, 137–141
images, 142–143
IR storage ring radiation, 137–138
microspectroscopy, 138
power, 137
sample preparation, 138–140
spectra, 141–142
spectral absorption bands, 140–141
Eulerian platform, 58
Euler studies, 508
Euterpina acutifrons, 362, 363, 368, 371–373
Evans studies, 559, 570
Excitation, zooplankton swarming, 177
Experimental Lakes Area, 209
Exposure, gradient, 102–103
Extended self-similarity (ESS), 474
External nitrate conditions, 550–553

F

Faber studies, 322
Faloutsos studies, 259
Farquhar studies, 131
Fasham studies, 44
Fast Fourier Transform function, 304, 308, 310
Fast Fourier Transform function, Finite, 55
Fast wavelet transform, 288–290
Fauchald studies, 390
Feeding efficiency, 488
Feely studies, 52
Feldman studies, 257
Feller studies, 464
Fennel and Neumann studies, 423
Fernandes studies, 229
Fernando, Brandl and, studies, 162
Fernö studies, 517, 518
Fiedler and Bernard studies, 183
Field, Moloney and, studies, 528
Fielding studies, 339
Fields, Yen and, studies, 560
Fields and Yen studies, 152, 162
Fiksen studies, 515
Film, 200 ASA Kodak Elite Chrome, 121
Finite Fast Fourier Transform function, 55
Fisher studies, 297–318, 333, 508, 517
Fish movement, spatial scale importance, 388–389
Fish-plankton dynamics model, 403–404
Fish school motion rules, 403–404
Fish stream model, 394
Fit parameters, zooplankton swarming, 171–173
Fixed genetic strategies, IBM, 510–511
Fleminger studies, 154
Florida continental shelf, 387, *see also* Biscayne Bay
Flow and particle patterns, Langmuir circulation,
447–449
Flow field, copepods
basics, 479–481, 489–490
dynamic coupling, 481–484
feeding efficiency, 488
flow geometry, 487–488

future directions, 489
  Navier-Stokes equations, 481–484
  numerical simulation, 484–488
  swimming behavior, 487–488
Flow field, energy costs, 494–497, 499, 502
Flow geometry, flow field numerical simulation, 487–488
Fluorescence fields, two-dimensional, 315–316
Fluoromap fluorometer, 272, 274
Flynn studies, 555
Fogarty studies, 184
Fogel and Cifuentes studies, 128, 131
Foley studies, 57
FONs, 420–422, 424
Fonteneau and Marcille studies, 183
Fonteneau studies, 183, 184, 195
Food web dynamics, stable isotope ecology
  basics, 125–126, 131
  direct effects, 130
  discussion, 127–129
  indirect effects, 130–131
  isotropic structure, 127–128
  methods, 126–127
  results, 127–129
  temporal fluctuations, nutrient source, 128–129
  time-sensitive factors, 129–131
Foote and Stanton studies, 27
Forrest, Mitchell and, studies, 510
Forrester studies, 389
Forstner, Riedl and, studies, 327
Foster studies, 120
Fourier and wavelet transforms, 290–291, 298
Fourier Transform function, Finite Fast, 55
Fourier Transform Infrared (FTIR) microscopy, 135, 139
Fox, Kendall and, studies, 528
Fractal analysis, benthic community patterns, 229
Fractal characterization, Georges Bank
  analysis, wavelets, 305–306
  basics, 297–299, 317–318
  crossing bank fronts, 311–313
  dimensional representations, wavelet transform, 306–308
  fluorescence fields, two-dimensional, 315–316
  future directions, 317
  hydrographic region power law behavior, 310–313
  implementation, 303–305
  interpolation, 313–317
  power law behavior, 309–313
  seasonal power law behavior, 309–310
  spectra, wavelet-based, 308
  stratified bank flanks, 313
  transform, wavelet, 301–303, 306–308
  underway oceanographic data, 300–301
  U.S. GLOBEC Broadscale Survey, 297–301, 315–316
  wavelet-based variance spectra, 301–308
  well-mixed bank crest, 310–311
Fractal dimension estimation, epiphytal and epilithnic communities
  analysis and biology, 248

basics, 245–248
D estimation, 251–252
ecology, 249, 251
fractal dimensions, 249, 251–252
intertidal rocky substrates, 252–253
limitations, 254–255
value, 253–254
Fractal dimensions, zooplankton swimming behavior, 339–350
Fractal marine system, skipjack tuna, 195–196
Fractional energy, 80
Frame, zooplankton distributions, 21
France and Peters studies, 208
FranHois, Altabet and, studies, 128
FranHois studies, 128
Frankignoul and Hasselmann studies, 210
Franks, Leising and, studies, 337
Franks and Jaffe studies, 3, 352
Franks studies, 556
Frank studies, 517
Freckleton and Watkinson studies, 240
Fredericks studies, 17–28
Freshwater models, spatial scale importance, 388
Frontier studies, 334, 336
Frost, Lam and, studies, 571
Frost, Steele and, studies, 555
Fucus patch mosaic, 238–239
Fuentes studies, 109–121
Fuhrman, Mitchell and, studies, 3, 32, 276, 334
Fung studies, 497
Fu studies, 195

G

GA, *see* Genetic algorithm (GA)
Gaertner, Pagavino and, studies, 183
Galapagos Islands, 53, 59
Gallager studies, 162
Gannes studies, 125
Garber, Boinski and, studies, 333
Garcia studies, 194
Gardner, Turner and, studies, 341
Gardner and O'Neill studies, 31
Gardner studies, 341
Gargett, Denman and, studies, 43, 493
Garrett studies, 43
Gases, Ustica experiment, 69, 91–92
Gaussian and Lévy-stable random variables, 464
Gautestad and Mysterud studies, 336, 342
Gavrikov studies, 393
Gee and Warwick studies, 249, 336
Geider studies, 555
Gelatinous organisms
  analysis, 329–330
  basics, 329, 331
  discussion, 331
Gendron studies, 162
Genetic algorithm (GA), 510–512
Gentilhomme and Lizon studies, 6
Geographical pattern control, 406

Georges Bank, fractal characterization
   analysis, wavelets, 305–306
   basics, 297–299, 317–318
   crossing bank fronts, 311–313
   dimensional representations, wavelet transform,
      306–308
   fluorescence fields, two-dimensional, 315–316
   future directions, 317
   hydrographic region power law behavior, 310–313
   implementation, 303–305
   interpolation, 313–317
   power law behavior, 309–313
   seasonal power law behavior, 309–310
   spectra, wavelet-based, 308
   stratified bank flanks, 313
   transform, wavelet, 301–303, 306–308
   underway oceanographic data, 300–301
   U.S. GLOBEC Broadscale Survey, 297–301,
      315–316
   wavelet-based variance spectra, 301–308
   well-mixed bank crest, 310–311
George studies, 209
Gerritsen and Strickler analytical solution, 564–567
Gerritsen and Strickler studies, 149, 352, 555, 560, 565,
      566, 567, 568, 570, 571
Gerritsen studies, 158, 560
Getis, Boots and, studies, 23
Getz, Berg and, studies, 194
Gilliam, Werner and, studies, 510
Gilliam's rule, 510
Ginot studies, 353, 523–540
Giordano studies, 135–145
GIS database, 113–116
Giske studies, 507–518
Glover studies, 218
Godin, Dugatkin and, studies, 518
Goldberger studies, 336
Goldberg studies, 511
Goldsmith studies, 162
Gorsky studies, 18
Gouze, Bernard and, studies, 362
Gower studies, 218
Gradient of exposure, 102–103
Graham studies, 325
Grant and Madsen studies, 326
Grant studies, 469
Grazing, 45, 352, 392
Great Lakes WATER Institute, 481
Greaves, Wilson and, studies, 338
Green Bomber tow body, 300
"Green lung of the Mediterranean," 66
Gregg studies, 177, 475
Gries studies, 155, 333
Grimm, Uchmański and, studies, 524
Grimm, V. studies, 395, 411–425, 509
Grimm and Grimm studies, 412
Grimm and Railsback studies, 422, 423
Grimm studies, 535
Grossman and Morlet studies, 280
Growth, 45, 527–528
Grünbaum studies, 32, 162

Guarini studies, 105
Gulf of Guinea, 183, 192
Gulf of Mexico, 196
Gulland studies, 184
Gunnarsson studies, 249, 348
Gunnill studies, 239
Gunther, Hardy and, studies, 32
Gurney, Nisbet and, studies, 431
Gurvich and Yaglom studies, 469, 470, 474, 475
Gyrodinium dorsum, 446

H

Haar wavelets, 284–286
Habitat structure, plankton dynamics model patterns,
      402–403
Haefner studies, 415
Haidvogel, Armi and, studies, 177
Håkanson and Peters studies, 209
Halley, Inchausti and, studies, 210
Halley studies, 201, 202, 205
Hall studies, 411
Hamilton studies, 342
Hammer studies, 162
Hamner and Robison studies, 329, 331
Hamner studies, 162
Hanski studies, 416
Hanson and Donaghay studies, 3, 32, 166
Hansson studies, 130, 131
Happ-Genzel apodization, 140
Hardy and Gunther studies, 32
Harp Lake, 205
Harrison, Donalson and, studies, 438
Harrison and Quinn studies, 230
Harris studies, 51
Hartnoll and Hawkins studies, 232, 235
Hartnoll and Wright studies, 238
Harvey, Railsback and, studies, 422, 425
Hasselmann, Frankignoul and, studies, 210
Hassel studies, 571
Hastings and Sugihara studies, 334, 339
Hattori, Wada and, studies, 128
Haury and Weihs studies, 175, 176
Haury and Yamazaki studies, 149
Haury studies, 32, 475, 550
Hausdorff parameter, 314
Havlin studies, 201
Hawkins, Burrows and, studies, 230
Hawkins, Hartnoll and, studies, 232, 235
Hawkins studies, 232
Hay studies, 375
Healy studies, 517
Heath studies, 560
Hebert studies, 162
Hecky studies, 125–131
Heino studies, 210
Henderson, Steele and, studies, 210, 317, 404
Hendry and McGlade studies, 238
Herbert and Moum studies, 43
Hermand studies, 65–94

Hermann studies, 423
Herman studies, 18, 248
Hernández-Garciá studies, 185
Heterogeneous environment, *see* Multiagent systems,
    individual-based modeling; Plankton
    dynamics, patterns in models
Hierarchy, spatial scale importance, 385–386
Higgie studies, 150
High-resolution data *vs.* conventional techniques,
    7–12
Hilborn and Mangel studies, 231
Hilborn studies, 183
Hills and Thomason studies, 254
Hills studies, 247, 249, 252, 254, 255
Hirschmugl studies, 135–145
Hiss-Purkinje cardiac conduction, 336
Hjort studies, 509
Hoeya Sound, 25, *see also* Knight Inlet
Hokkaido sardine, 194
Holland studies, 509, 510, 517, 544
Holliday studies, 3, 17, 31, 32
Holling's gradient, 429
Holling's predation cycle model, 559
Holling studies, 555, 560
Holt, Beverton and, studies, 184, 186
Hooper studies, 249
Hopcroft and Robison studies, 329
Horn studies, 232
Horrigan, Cullen and, studies, 543, 544
Horwood studies, 402
HOTS data set, 257, 265, 268
Houssin studies, 531, 536, 537
Houston and McNamara studies, 512
Howarth studies, 202, 209
Huang and Turcotte studies, 337, 350
Hubert studies, 201
Huffmann method, 295
Humes studies, 158
Humston studies, 389
Huse studies, 507–518
Hussussian studies, 167
Huston studies, 508, 509
Hutchinson studies, 258, 497
Hutchison studies, 4
Huth and Wissel studies, 390, 391
Hwang, Strickler and, studies, 150, 167
Hwang and Strickler studies, 333, 494
Hwang studies, 333, 494
Hydrographic region power law behavior, 310–313
Hypotheses, pattern-oriented modeling, 414
Hyvonen studies, 129

I

ICCAT, 183
Ichthyoplankton recorder, 18
Image-forming optics, 18
Images, 21–22, 142–143
Inchausti and Halley studies, 210
Incze studies, 18, 475

Independent predictions, pattern-oriented modeling,
    416–417
Individual-based modeling, adaptation levels
    adaptation levels, 510–514, 517
    antipredator responses, 516–517
    artificial neural networks, 513–514
    attribute vector, 509
    Baldwin effect, 517–518
    basics, 507–509, 518
    case studies, 515–517
    Daphnia, 515
    diel vertical migration, 515
    discussion, 517–518
    fixed genetic strategies, 510–511
    individual learning, 512–514
    methodology, 509–510
    overwintering diapause, 515
    phenotypic plasticity, 511–512
    reinforcement learning, 514
    salmonids, 515–516
    strategy vector, 509–510
    supervised learning, 514
Individual-based modeling, multiagent systems
    basics, 523–524, 538–539
    body weight, 528
    cannibalism, 530
    demographic noise, 528–530
    descriptions, 536
    development, 524–528
    discussion, 533–538
    growth model, 527–528
    improvements, 534–536
    languages, 536
    life cycle representation, 525
    methodology, 534–535
    Mobidyc platform, 524–525
    model architecture, 525–528
    molting model, 527–528
    mortality simulation, 528–530
    predation, 528, 530–533
    reproduction, 528
    results, 528–533
    simulation, 528–530
    spatial representation, 530
    spatiotemporal dynamics, 530–533
    survival model, 527–528
    temperature, 528
    tools, 535–536
    zooplankton, 536–538
Individual learning, 512–514
Individual organism size, spatial scale importance,
    390–392
Infinitely divisible distributions, 464–465
Infrared imaging, *see* Synchroton-based infrared
                imaging, *Euglena gracilis* single cells
Infrared storage ring (IRSR) radiation, 135–136,
    137–138
Innis, Yodzis and, studies, 390
Instrument description, biological scale resolution,
    7–9
Interpolation, 313–317

Intertidal benthic production and respiration
  basics, 97–98, 104–105
  daily potential primary production, 99–100
  gradient of exposure, 102–103
  irradiance, 102
  materials and methods, 98–99
  microphytobenthos adaptation, 102
  microscale variability, 103–104
  results, 99–104
  seasonal variations, 100–102
  temporal resolution, 102
Intertidal context, SEHRI, 109
Intertidal rocky substrates, 252–253
Intertidal zone, benthic species, 432–437
Ireland, 252
Irradiance, 92, 102
IRSR, *see* Infrared storage ring (IRSR) radiation
Ishimaru, Suzuki and, studies, 6
Isochrom Continuous Flow Stable Isotope mass
    spectrometer, 127
Isotropic structure, food web dynamics, 127–128
Itsweire studies, 331
Ivanov studies, 279–296
Ivlev studies, 571
Iwasa, Roughgarden and, studies, 386

J

Jackson and Burd studies, 4
Jaffe, Franks and, studies, 3, 352
Jaffe, McGehee and, studies, 163, 175
Jaffe studies, 3
Jakobsen studies, 560
Jamin studies, 137
Janicki and Weron studies, 464
Japan, *see* Mikame Bay; Sea of Japan; Seto Inlet
Jellyfish, *see* Gelatinous organisms
Jenkins and Watts studies, 205
Jenkins studies, 229
Jenssen, Siniff and, studies, 336
JGOFS software, 257
Jiang studies, 479–490, 481
Jiménez and Wray studies, 498
Jiménez studies, 470, 471, 477, 498
Johannes studies, 128
Johansson, Jonsson and, studies, 334
John, Caswell and, studies, 508
Johnson and Milne studies, 342
Johnson and Richardson studies, 445, 446
Johnson studies, 229–240, 338, 352, 353
Jonsson, Tiselius and, studies, 333, 337, 352
Jonsson and Johansson studies, 334
Jrrgensen studies, 556
Judson studies, 510

K

Kaitala studies, 210
Kamykowski, Yamazaki and, studies, 353, 445, 446
Kamykowski and McCollum studies, 547

Kamykowski studies, 543–556
Kareiva, Tilman and, studies, 230, 231
Kareiva and Shigesada studies, 338
Karlson studies, 391
Katona studies, 149, 162
Katsuwonus pelamis, see Skipjack tuna, scale
    similarities
Katz studies, 144
Kaufman studies, 544
Keeling studies, 230, 236, 238
Keenan studies, 17–28
Keiyu studies, 334, 544, 545
Keller studies, 512
Kelly studies, 152
Kelvin waves, 52, 61
Kendall and Fox studies, 528
Kendall studies, 131
Kenkel and Walker studies, 342
Kepler's Law, 413
Kerman and Bernier studies, 195
Kerman and Szeto studies, 195
Kerr studies, 498
Kessler studies, 271
Kierstead and Slobodkin studies, 32, 392
Kigoma, Tanzania, 125, *see also* Food web dynamics,
    stable isotope ecology
King studies, 113
Kiørboe and MacKenzie studies, 568
Kiørboe and Saiz studies, 559, 560
Kiørboe and Visser studies, 494, 503, 560, 571
Kiørboe studies, 337, 352, 560, 571
Kishi studies, 387
Klausmeier, Litchman and, studies, 210
Knap studies, 263
Knight Inlet
  camera operation, 22–24, 27
  November 21 cast, 22–23, 25–28
Koch snowflakes, 342
Kodak Elite Chrome, 200ASA film, 121
Koehl and Strickler studies, 32
Koenig studies, 230, 236
Kolmogoro-Smirnov test, 472, 473–474
Kolmogorov and Obukhov studies, 454
Kolmogorov scale
  energy costs, 495, 497, 499
  turbulence, 453, 454, 475
Kolmogorov studies, 453–454, 469
Konig studies, 322
Koscielny-Bunde studies, 201
Kosro studies, 40
Kostylev, Erlandson and, studies, 251, 343, 347
Kostylev and Erlandson studies, 230, 252
Kostylev studies, 251, 252, 253, 254
Koza studies, 511
Krause studies, 518
Krebs, Stevens and, studies, 333
Krill, 162, 390
Krueger and Dodson studies, 516
Kruskal-Wallis tests, 103, 205, 207, 349
Kullenberg studies, 43

L

Laakso studies, 202, 210
Labini, Pietronero and, studies, 336
Laborel and Vacelet studies, 321
Lagadeuc, Seuront and, studies, 12, 334, 336, 342, 344, 350, 351, 353, 537
Lagragian properties, *see also* Discrete events-based Lagrangian approach
  biological-physical coupling, 58
  Langmuir circulation, 446
  zooplankton swarming, 164
Lag times, biological-physical coupling, 57–59
Laguna de Terminos, 388
Laguncularia racemosa, 419, 422
Lake Biwa, 7, 9, 11, 12
Lake Erie, 388
Lakes, *see* Environmental variability, rivers and lakes
Lake Tanganyika, 125, 128, 129, *see also* Food web dynamics, stable isotope ecology
Lalli and Parsons studies, 546, 549
Lam and Frost studies, 571
Laminaria rodriguezi, 71
Lamirini studies, 498
Landry studies, 555
Langmuir circulation, buoyant particles
  basics, 445–446
  discussion, 449–450
  flow and particle patterns, 447–449
  simulation setup, 446
  zooplankton swarming, 177
Langmuir studies, 445
Langton studies, 543
Languages, IBM, 536
La Niña, *see* El Niño-La Niña cycles
Larson studies, 445
Larsson and Dodson studies, 516
Lasker and Sánchez studies, 325
Lasker studies, 162
Lates angustifrons, 126
Lates mariae, 126
Lates microlepis, 126
Lates stappersi, 126, 127–128
Latham, Petraitis and, studies, 392
Laurence, Beyer and, studies, 509
Lawrence studies, 512
Laws studies, 131
Law studies, 240
Lawton studies, 249
Lazier, Mann and, studies, 493
Leaf blade, gaseous interchange, 91–92
Le Crotoy, France, *see* Multiscale *in situ* measurements, intertidal benthic species
Ledwell, Sundermeyer and, studies, 43
Ledwell studies, 43
Lee and DeAngelis studies, 391
Leggett studies, 131
Le Havre, France, 98, 99
Lehman studies, 571
Leibovich, Craik and, studies, 446
Leising and Franks studies, 337

Leising and Yen studies, 162, 177
Leising studies, 337
Lenski studies, 276
Lenz studies, 18
Lesmoir-Gordon studies, 245
Lessells studies, 512
Letcher and Rice studies, 390
Letelier, Abbott and, studies, 31, 55, 58, 62
Letelier studies, 31
Leterme studies, 3–13
Le Tourneux and Bourget studies, 254
Levin, Paine and, studies, 230
Levin, Pascual and, studies, 230, 231
Levin and Pacala studies, 231, 384
Levine, Dale, Barth and, studies, 43
Levin studies, 31, 229, 385, 386, 394, 544
Levitus studies, 221
Levy, Clark and, studies, 515
Lévy index, 460
Lévy-stable random variables, 464, 465
Levy studies, 216
Lewis, Cullen and, studies, 31, 43
Lewis and Pedley studies, 560, 570
Li, Colbo and, studies, 448, 449
Liang and Uye studies, 367, 537
Liang studies, 373, 527, 528, 537
Lidén, Arheimer and, studies, 202, 209
Lien, D'Asaro and, studies, 331
Life cycle representation, 525
Lima and Dill studies, 571
Limnothrissa miodon, 126, 127–128
Limpets, 232, 234–235
Linacre studies, 208
Lindegarth studies, 229
Linuche unguiculata, 445
Li studies, 336
Litchman and Klausmeier studies, 210
Litchman studies, 210
Littman, Ackley and, studies, 514
Littorina saxatilis, 251, 253
Liu studies, 545
Lively studies, 239
Living toxicometer, zooplankton, 353
LizardTech company, 119
Lizon, Gentilhomme and, studies, 6
Lloyd's mirror effect, 81
Lobsters, 194
Loehle studies, 341
Lognormal theory, 460–463
Lognormal theory, turbulent structures
  basics, 469–471
  discussion, 474–477
  simulations, 471–474
  subsurface stratified layer, 477
  surface turbulent layer, 475–477
Long and Azam studies, 258
Long and Long studies, 248
Longley and Batty studies, 341
Loose studies, 352
Lorek and Sonnenschein studies, 535, 536
Lorenz studies, 494

Lotka studies, 508
Lotka-Volterra models, 430, 509
Louttit, Mackas and, studies, 177
Lovejoy studies, 350
Lower Atchafalaya River, 205
Lower trophic levels, spatial scale importance, 387–388
Low-pass filter, stage-structured population models, 365
Lueck, Yamazaki and, studies, 472, 474
Lumley, Tennekes and, studies, 493, 501
Lund and Rogers studies, 498
Lundberg, Ripa and, studies, 210
Lynx, 413

M

Mach studies, 336
MacIntyre studies, 546
Mackas and Louttit studies, 177
Mackas studies, 4
MacKenzie, Kiørboe and, studies, 568
MacKenzie studies, 559, 568, 570
MacNally studies, 395
Macrocystis pyrifera, 247, 251, 254
Madsen, Grant and, studies, 326
Maes studies, 544
Magnus studies, 327
Maguire and Porter studies, 391, 393
Mahadevan and Archer studies, 221
Mahadevan and Campbell studies, 218
Mahadevan studies, 215–226
Malacarne studies, 268
Malamud and Turcotte studies, 301, 303
Malamud studies, 297–318, 300
Malchow studies, 393, 401–407
Maley and Caswell studies, 510
Malik and Vassilicos studies, 570
Malkiel studies, 18
Mallorca (island), 194
Maly, Chow-Fraser and, studies, 152
Mamedov and Udalov studies, 536
Mandelbrot and Wallis studies, 202
Mandelbrot studies, 245, 300, 314, 334, 337, 339, 348, 350, 353
Mangel, Clark and, studies, 512
Mangel, Hilborn and, studies, 231
Mangrove forest model, 419–422
Mann and Lazier studies, 493
Mannini studies, 128
Mann-Whitney tests, 205, 207
Marcille, Fonteneau and, studies, 183
Marennes-Oléron Bay, 105
Marguerit studies, 353
Marin and Delgado studies, 390
Marine models and systems, spatial scale importance, 387–389
Marine Zooplankton Colloquium 2, 17
Markov models, 231–236, 240
Marmota marmota, 412

Marrasé studies, 333, 493
Marsili and Zhang studies, 259, 268
Marullo studies, 196
Mate signaling, copepods
    basics, 149–150, 157–158
    discussion, 151–157
    methods, 150–151
    odorant levels, 155–157
    pheromones, 151
    results, 151–157
    scent preferences, 151–154
    tracking behavior, 154–155
    trails, 149–157
Mate tracking behavior, 154–155
Mathis studies, 517, 518
Matlab software
    decorrelation scale analysis, 55
    rank-size analysis, 257
    ZOOVIS, 21
Matsumoto and Robison studies, 329
Matte, gases, 69
Maurolicus meulleri, 516
Maxey, Wang and, studies, 500
Maxey and Riley studies, 497
Maxwellian properties, 165, 169
May, Sugihara and, studies, 265, 336, 346
Maynard, Parker and, studies, 510
Maynou studies, 230
McCarthy studies, 547
McClain studies, 2220
McCollum, Kamykowski and, studies, 547
McCulloch and Cain studies, 338
McCulloch and Pitts studies, 513
McGehee and Jaffe studies, 163, 175
McGillicuddy studies, 313, 317
McGlade, Hendry and, studies, 238
McKee studies, 202, 209
McLachlan stability index, 105
McNamara, Houston and, studies, 512
McPhaden studies, 53
McQuinn studies, 517, 518
Mean swim velocity, 501–502
Measurements
    acoustic remote sensing, 65–94, 73–75
    biological-physical coupling, 51–62
    copepod mate signaling, 149–158
    distributions of zooplankton, 17–28
    food web dynamics, 125–131
    individual-based model validation, 161–178
    planktonic layers, 31–46
    scale resolution comparison, 3–13
    in situ, intertidal benthic species, 97–105
    spatially extensive, high resolution images, 109–121
    stable isotope ecology, 125–131
    synchrotron-based infrared imaging, 135–145
    zooplankton, 17–28, 161–178
Mediterranean Sea, 194, 196, *see also* Acoustic remote sensing
Medium response, acoustic remote sensing, 80–81
Medvinsky studies, 401–407

Menczer and Belew studies, 510
Meneguzzi, Vincent and, studies, 477, 494, 498
Mesodinium rubrum, 446
Mesopelagic fish, 516
Mesopelagic larvaceans, 331
Metz and Diekman studies, 508
Mexican Hat filter, 301–302, 305–309, 317
Mexico, 388
Michaelis-Menten equation, 547
Michener and Schell studies, 125
Microphytobenthos adaptation, 102
Microscale structure, biological scale resolution, 4–7
Microscale variability, 103–104
Microspectroscopy, *Euglena gracilis,* 138
Migné studies, 97–105
Mikame Bay, 387
MIKE21 model, 388
Millard, Wisenden and, studies, 516
Miller, Tsuda and, studies, 155
Miller and Ambrose studies, 110–111
Miller studies, 137
Milne, Johnson and, studies, 342
Milne, Wiens and, studies, 333, 353
Milne studies, 251
Minagawa and Wada studies, 125, 130
Minimum populations, skipjack tuna, 194
Missouri River, 204, 207
Mitchell, Sosik and, studies, 4
Mitchell, Waters and, studies, 274, 276, 555
Mitchell and Forrest studies, 510
Mitchell and Fuhrman studies, 3, 32, 276, 334
Mitchell studies, 257–276, 511
Mobidyc platform, 523–540
Model fields, sea surface biogeochemical variability, 220
Models
    architecture, individual-based modeling, 525–528
    benthic species, 231, 430–441
    community patterns, 231
    coral competition, 393
    dynamics model patterns, 403–404
    fish stream, 394
    marine models, 387–388
    molting, 527–528
    multiplicative cascades, 454–457
    plankton, 403–404, 497–499, 559–560
    point force model, 484
    predator-prey interaction, 559–560
    sea surface biogeochemical variability, 222–225
    spatial dynamics, 430–441
    spatial scale importance, 384, 387–388
    swimming models, 497–499
    zooplankton swarming, 165–166, 169–170, 174–177
Models, turbulent intermittency
    casual lognormal stochastic equation, 460–462
    characteristic functions, 464
    continuous cascades, 459–463
    continuous lognormal multifractal, 462–463
    discrete cascades simulation, 457–459
    infinitely divisible distributions, 464–465
    multiplicative cascades, 454–457

stable stochastic integrals, 465–466
    velocity fluctuations, 454
Moderate-resolution Imaging Spectroradiometer
    (MODIS), 220, 317
MODIS, *see* Moderate-resolution Imaging
    Spectroradiometer (MODIS)
Molofsky studies, 232
Moloney and Field studies, 528
Molot, Dillion and, studies, 209
Molot and Dillion studies, 209
Molting model, 527–528
Monin and Ozmidov studies, 469
Montana and Davis studies, 513
Monte Carlo simulations, 510
Monterey Canyon, 331
Mood studies, 471, 472
Mooij, DeAngelis and, studies, 391, 416, 424
Mooij and Boersma studies, 537
Mooij studies, 383–395
Morales studies, 210
Morel studies, 555
Morlet, Grossman and, studies, 280
Morlet wavelets, 305–306, 307
Morowitz studies, 395
Morrisey studies, 229
Morse studies, 249, 333, 341, 348
Mortality simulation, 528–530
Mosquera, Adler and, studies, 393
Mother wavelets, 285, *see also* Wavelets
Motion equation, energy costs, 497–498
Moum, Herbert and, studies, 43
Movement, spatial scale importance, 388–389
Mulholland and DeAngelis studies, 392
Mullen studies, 177
Mullin and Brooks studies, 32
Multiagent systems, individual-based modeling
    basics, 523–524, 538–539
    body weight, 528
    cannibalism, 530
    demographic noise, 528–530
    descriptions, 536
    development, 524–528
    discussion, 533–538
    growth model, 527–528
    improvements, 534–536
    languages, 536
    life cycle representation, 525
    methodology, 534–535
    Mobidyc platform, 524–525
    model architecture, 525–528
    molting model, 527–528
    mortality simulation, 528–530
    predation, 528, 530–533
    reproduction, 528
    results, 528–533
    simulation, 528–530
    spatial representation, 530
    spatiotemporal dynamics, 530–533
    survival model, 527–528
    temperature, 528
    tools, 535–536

zooplankton, 536–538

Multifractal stochastic processes
casual lognormal stochastic equation, 460–462
characteristic functions, 464
continuous cascades, 459–463
continuous lognormal multifractal, 462–463
discrete cascades simulation, 457–459
infinitely divisible distributions, 464–465
multiplicative cascades, 454–457
stable stochastic integrals, 465–466
velocity fluctuations, 454

Multipaths, photosynthesis effect, 84–88

Multiple (stable) equilibria, 194

Multiplicative cascades
correlation properties, 456–457
modeling, 454–457
scaling properties, 455–456
velocity fluctuations, 454

Multiresolution analysis, wavelets, 286–288

Multiresolution Seamless Image Database software, 119

Multiscale acoustic effects, 75–83

Multiscale *in situ* measurements, intertidal benthic species
basics, 97–98, 104–105
daily potential primary production, 99–100
gradient of exposure, 102–103
irradiance, 102
materials and methods, 98–99
microphytobenthos adaptation, 102
microscale variability, 103–104
results, 99–104
seasonal variations, 100–102
temporal resolution, 102

Mundy, Boehlert and, studies, 183

Murano, Uchima and, studies, 155

Murphy and Williams studies, 138

Mussels, 432–437

Myers and Southgate studies, 113

Mylius and Diekmann studies, 510

Mysterud, Gautestad and, studies, 336, 342

Mytilus californianus sp., 111, 117, 432

N

Narragansett Bay, 387

National Stream Water Quality Monitoring Networks, 203

Navier-Stokes equations
energy costs, 493, 494, 498
flow field numerical simulation, 481–484, 485
turbulent intermittency, 454

Nechitailo studies, 279–296

Neumann, Fennel and, studies, 423

New England continental shelf, 43–44

Newtonian terms, 164

Nicoll, Yen and, studies, 32

Nielsen, Bjørnsen and, studies, 31, 32, 45

Nielsen studies, 32, 45

Nikon SLR camera, 112

El Niño-La Niña cycles, 52, 54, 57, 61, 62

Nisbet, Bence and, studies, 230

Nisbet, Donalson and, studies, 429, 431

Nisbet, Synder and, studies, 231

Nisbet, Wilson and, studies, 230

Nisbet and Gurney studies, 431

Nisbet studies, 429–443

Niskin bottles
biological scale resolution comparison, 3, 12
microscale structure, 4, 5–7
sensor deployment, 10

No fish, environmental noise, 404–405, 406

No fish, no environmental noise, 405–406

Noise, acoustic remote sensing, 72, 88–91

Nondimensional numbers to dimensional numbers, 504

Nonnenmacher studies, 248

Non-photosynthesis-related effects, 81–83

Nonspatial (point) transition matrix models, 233–234

Normal-mode theory, 77

North Sea, 388

North Temperate Lakes Long Term Ecological Research site, 203, 209

November 19 cast, Hoeya Sound, 25

November 21 cast, Knight Inlet, 22–23, 25–28

Novikov studies, 460, 471

NTSC video format, 18

Numerical integration timescale, 562

Numerical simulation, flow field, 484–488

Numerical study, plankton dynamics model patterns, 404–406

Nutrient sources, temporal fluctuations, 128–129

Nutrient spiraling models, 392

O

Oakey studies, 475

O'Brien and Wroblewski studies, 387

Obukhov, Kolmogorov and, studies, 454

Oceanographic setting, 52–53

OceanProbe (Sontek), 33

ODE, *see* Ordinary differential equations (ODE)

Odorant levels, 155–157

Oelbermann and Sheu studies, 130

Oelleut, Poulet and, studies, 32

Oguz studies, 387

Ohman studies, 555

Oithona davisae, 155

Okubo, Powell and, studies, 308

Okubo, Yamazaki and, studies, 165

Okubo and Anderson studies, 163, 164, 165, 172, 175

Okubo studies, 32, 38, 162, 163, 165

One fish, no environmental noise, 406

O'Neill, Gardner and, studies, 31

Ontario lakes, 204, 207–209

On the scale dependence, 341–342

Oppenheim and Schafer studies, 185

Optical plankton counter, 18

Orcas Island, *see* East Sound

Ordinary differential equations (ODE), 429, 430–440, 442, 537
Oregon continental shelf, 33, 34–39, 43
O'Reilly studies, 125–131
Organism size, spatial scale importance, 390–392
Ornstein, Uhlenbeck and, studies, 166
Orr and Smith studies, 150
Ortner studies, 18
Osborn, Rothschild and, studies, 330, 493, 559, 570
Osborn, Yamazaki and, studies, 4
Osborn and Scotti studies, 493
Osborn studies, 32, 321–327, 329–331
Osteoporosis, 137
Ostrander studies, 321–327
Ouellet, Poulet and, studies, 32, 162
Ovchinnikov studies, 183
Overman and Parrish studies, 130
Overwintering diapause, *Daphnid*, 515
Owen studies, 32
Oxygen, 69, 128, *see also* Dissolved oxygen
Ozmidov, Monin and, studies, 469

P

Pacala, Levin and, studies, 231, 384
Pacific Marine Environment Laboratory, 53
Paffenhöfer, Bundy and, studies, 480
Pagavino and Gaertner studies, 183
Paine and Levin studies, 230
Paine studies, 432
Palumbi studies, 150
PAL video format, 18
Panulirus interruptus, 432
Papua New Guinea, 53
Paracalanus sp., 372
Parameterization, pattern-oriented modeling, 416
Parameters, pattern-oriented modeling, 415
Pareto, Zipf and, studies, 259
Pareto studies, 258
Parker and Maynard studies, 510
Parker studies, 201–211
Park studies, 247
Parrish, Overman and, studies, 130
Parrish, Pitcher and, studies, 193
Parsons, Lalli and, studies, 546, 549
Particle rising, 502–503
Pascual and Caswell studies, 402
Pascual and Levin studies, 230, 231
Pascual studies, 3, 232, 350, 402
Pastres studies, 387
Patchiness, *see also* Fractal characterization, Georges Bank
  basics, 32
  biogeochemical variability, 226
  microscale structure, 4, 5
  sampling process impact, 5–6
  sea surface biogeochemical variability, 222–224
Patch scale analysis, 116–118
Pattern-oriented modeling (POM)
  analysis, 416

  aquatic systems, 423–424
  assembling patterns, 414–415
  basics, 411–412
  beech forest model BEFORE, 417–419
  construction of model, 415
  discussion, 424–425
  examples, 417–423
  hypotheses, 414
  independent predictions, 416–417
  mangrove forest model, 419–422
  parameterization, 416
  parameters, 415
  patterns, 412–413
  problem formulation, 414
  protocol, 413–417
  question formulation, 414
  revisions, 416
  scales, 415
  state variables, 415
  stream trout habitat selection, 422–423
  tests, 416
Patterns, spatial scale importance, 392–393
Pauchut studies, 325
Paulik studies, 184, 193
Pavlov's dog, 514
Peak pressure-squared calculation, 80
Pedley, Lewis and, studies, 560, 570
Pegau, Zaneveld and, studies, 4
Peitgen studies, 341
Pelagic copepod, *see Temora longicornis,* mate signaling
Pelletier and Turcotte studies, 202, 209, 210
Pelletier studies, 202, 205, 210
Periáñez studies, 387
Perline studies, 268
Pershing studies, 445–446
Petchey studies, 210
Peters, France and, studies, 208
Peters, Håkanson and, studies, 209
Petersen, DeAngelis and, studies, 391
Petersen and DeAngelis studies, 391
Peterson studies, 131
Petraitis and Latham studies, 392
Petraitis studies, 109
Petrenko studies, 4
Petrovskii studies, 401–407
Phaeocystis sp., 102
Phase spaces, skipjack tuna, 193
Phenotypic plasticity, 511–512
Pheromones, 151
Phipps studies, 232
Photographic equipment and settings, 120–121
Photosynthesis influence
  bubbles, 67–68
  gases, 69
  oxygen production, 69
  Posidonia oceanica, 69
Photosynthetic activity, *see* Acoustic remote sensing
Phylum Cnidaria, 325
Physical-behavioral balances, 177
Physical pattern control, 405–406
Physics-Uspekhi Journal, 280

Phytoplankton
  Baltic concentration, 32
  light changes, 131
  microscale structure, 4–5
  physical forcing, 4
  relative nutrient availability, 131
  steepest gradient correlation, 36–37
  vertical distributions, 36–38
Phytoplankton distribution patterns, rank-size
        analysis
  basics, 257–263, 275–276
  data set size, 257–258
  initial analysis, 258
  interpretation, 259–263
  low resolution, 263–275
  methodology, 258–259
  recent use, 259
  time series fluorescence profiles, 263–275
Pick, Chételat and, studies, 208
Pietronero and Labini studies, 336
Pimm, Ariño and, studies, 210
Pink shrimp, 389
Pinnegar and Polunin studies, 130
Pisaster ochraceus, 432
Pitcher and Parrish studies, 193
Piton and Roy studies, 183
Pitts, McCulloch and, studies, 513
Pixel density, 27
Plankton, predator-prey interaction modeling
  basics, 559–560, 571
  case study, 564–569
  discrimination, 562
  discussion, 569–571
  Gerritsen and Strickler analytical solution, 564–567
  methodology, 560–564
  numerical integration timescale, 562
  simulation, 564–567
  swimming behavior, 560–561, 567–569
  vectorial estimation, 562–564
Plankton dynamics, patterns in models
  basics, 401–402, 406–407
  connected habitats, 404–406
  fish school motion rules, 403–404
  geographical pattern control, 406
  habitat structure, 402–403
  models, 403–404
  no fish, environmental noise, 404–405, 406
  no fish, no environmental noise, 405–406
  numerical study, 404–406
  one fish, no environmental noise, 406
  physical pattern control, 405–406
  plankton-fish dynamics model, 403–404
  separated habitats, 406
Plankton-fish dynamics model, 403–404
Planktonic behavior, complex adaptive system
  basics, 543–544
  copepod model, 544–545
  decision-making mechanism, 547–549
  dinoflagellate model, 545–554
  discussion, 554–556
  external nitrate conditions, 550–553

  simulation, 549–554
  threshold settings, 553–554
Planktonic layers
  ADV measurements, 39
  basics, 31–33, 46
  Continental Shelf, Oregon, 34–39
  dye injections, 43–44
  East Sound, Washington, 33–34
  future directions, 45–46
  horizontal extent, small-scale layers, 44
  issues, 45
  sharp vertical gradients, 44–45
  small scale, examples, 33–39
  vertical mixing role, 43
  vertical velocity gradients, 39–43
Planktonic swimming models, 497–499
Plankton patchiness, *see* Patchiness
Platt, Denman and, studies, 32, 59, 218
Platt and Denman studies, 59, 201
Platt and Sathyendranath studies, 31
Platt studies, 4, 59, 257, 308
Plisnier studies, 125–131
Point force model, 484
Poisson pattern, 566, 569
Polhlmann studies, 154
Pollard and Regier studies, 216
Pollicipes polymerus sp., 111, 117–118
Polunin, Pinnegar and, studies, 130
POM, *see* Pattern-oriented modeling (POM)
Pope, Eswaran and, studies, 471, 495
Pope, Yeung and, studies, 472, 495
Popple River, 204, 205
Porter, Maguire and, studies, 391, 393
Port of Mogan, 185, 188, 192, *see also* Skipjack tuna,
        scale similarities
Port River Estuary, 272
Posidonia oceanica, 66, 69, *see also* Acoustic remote
        sensing
Possingham studies, 230
Post studies, 129
Poulet and Oelleut studies, 32, 162
Powell, Denman and, studies, 4
Powell and Okubo studies, 308
Powell studies, 201, 384
Power, synchroton-based IR imaging, 137
Power law behavior, 309–313
Powers studies, 194
Power studies, 388
Power/telemetry housing, 19–20
Prairie, Cattaneo and, studies, 209
Prairie, Cattaneo studies, 208
Predation, individual-based modeling, 528, 530–533
Predator, 516–517, *see also* Plankton, predator-prey
        interaction modeling
Prey-tracking, 154
Price studies, 17, 175, 555
Pringesheim studies, 138–139
Problem formulation, 414
Process response times, *see* Biogeochemical variability,
        sea surface
Propagation channel modeling, 77–80

Proposed equations, skipjack tuna, 193
Protocol, pattern-oriented ecological modeling
  analysis, 416
  assembling patterns, 414–415
  basics, 413
  contruction of model, 415
  hypotheses, 414
  independent predictions, 416–417
  parameterization, 416
  parameters, 415
  question or problem formulation, 414
  scale, 415
  state variables, 415
Provata and Almirantis studies, 336
Prusak studies, 149–158
Pseudodiaptomus marinus, 372
Puget Sound, 33
Pyke studies, 333, 336

Q

Qiu studies, 109–121
Quantifying variability, 218–219
Question formulation, 414
QuickSCAT satellite, 53
Quinn, Harrison and, studies, 230

R

Rademacher studies, 418
Railsback, Grimm and, studies, 422, 423
Railsback and Harvey studies, 422, 425
Railsback studies, 422, 425, 508, 535, 538, 544
Raizer studies, 195
Ramos and Sangrá studies, 183
Ramos studies, 183
Rand and Wilson studies, 232, 236
Random flow simulation, 496–497
Rand studies, 231, 236
Rank-size analysis, vertical phytoplankton
    distribution patterns
  basics, 257–263, 275–276
  data set size, 257–258
  initial analysis, 258
  interpretation, 259–263
  low resolution, 263–275
  methodology, 258–259
  recent use, 259
  time series fluorescence profiles, 263–275
Rasmussen, Vander Zanden and, studies, 125
Rayleigh properties, 128–129, 322–324
Rayleigh studies, 322
Red Chalk Lake, 205
Regier, Pollard and, studies, 216
Regions of interest (ROI)
  Hoeya Sound, 25
  Knight Inlet, 25
  ZOOVIS, 21, 27–28
Reichelt, Bradbury and, studies, 254, 336
Reinforcement learning, 514

Rendell and Whitehead studies, 513
Renshaw studies, 431
Reproduction, 45, 528
Response time, sea surface biogeochemical variability,
    222–224
Retroscale analysis, 118–119
Reuter and Breckling studies, 391
Reverberation, 88
Revisions, pattern-oriented modeling, 416
Reyes studies, 388
Reynolds and Davies studies, 202
Reynolds number properties, *see also* Lognormal
    theory, turbulent structures
  energy costs, 495
  flow field numerical simulation, 482
  gelatinous organisms, 331
  mate tracking, 157
  turbulent intermittency, 454
Rhizophora mangle, 419, 421–422
Rhodomonas sp., 151
Rice, Letcher and, studies, 390
Richardson, Johnson and, studies, 445, 446
Richardson number, 39, 43
Richardson plot
    fractal dimension estimation, 251, 253
Richardson studies, 245, 247, 453, 454
Richerson, Boyd and, studies, 517
Ricker studies, 184, 186
Riedl and Forstner studies, 327
Rigler and Cooley studies, 374
Riley, Maxey and, studies, 497
Riley studies, 32
Rinaldo, Rodriguez-Iturbe and, studies, 201
Rines studies, 45
Ripa and Lundberg studies, 210
Ripa studies, 210
Riser studies, 268
Rivers, *see* Environmental variability, rivers and lakes;
    specific river
Robison, Hamner and, studies, 329, 331
Robison, Hopcroft and, studies, 329
Robison, Matsumoto and, studies, 329
Robison studies, 329, 330, 331
Robles and Desharnais studies, 119, 392, 432
Robles studies, 109–121, 111, 115, 429–443
Rocky shore communities, SEHRI
  basics, 121
  concept demonstration, 113–119
  discussion, 119–121
  GIS database, 113–116
  intertidal context, 109
  limitations of technique, 120
  optimization recommendations, 120
  patch scale analysis, 116–118
  photographic equipment, settings, and conditions,
    120–121
  retroscale analysis, 118–119
  sampling approach limitations, 110–111
  spatially extensive high resolution images, 111
  system, 111–113
  utility expansion, 119

Rodriguez-Iturbe and Rinaldo studies, 201
Roff studies, 510
Rogallo studies, 471, 494, 495, 496
Roger and Bernard studies, 183
Rogers, Lund and, studies, 498
Roger studies, 183
Rohani studies, 237
ROI, *see* Regions of interest (ROI)
Ropella studies, 415
Rose, Clark and, studies, 389, 391
Rosenblatt studies, 513, 514
Rossby properties, 52, 216
Rossi studies, 229
Rothschild and Osborn studies, 330, 493, 559, 570
Rothschild studies, 184, 194
Roughgarden, Alexander and, studies, 230
Roughgarden and Iwasa studies, 386
Roughgarden studies, 230, 231, 386, 393, 395
Roy, Piton and, studies, 183
Roy studies, 337, 350
Rummelhart studies, 513, 514
Rusak studies, 209
Rusignuolo, Commito and, studies, 252
Russell studies, 168, 202, 209
Russian River, 205, 207
Russ studies, 247
Ruxton studies, 394
R/V Elakha, 43
R/V Henderson, 33
R/V Wecoma, 43
Ryan studies, 53, 57
Rykiel studies, 432

S

Sacramento River, 205
Saffman and Turner studies, 330
Saito studies, 460
Saiz, Kiørboe and, studies, 559, 560
Saiz studies, 543, 561, 568
Sakamoto studies, 57
Salmonids, 515–516
Samorodnistky and Taqqu studies, 464
Sample preparation, *Euglena gracilis*, 138–140
Sampling approach limitations, SEHRI, 110–111
Sampling process impact, biological scale resolution, 5–7
Samson studies, 18
Sánchez, Lasker and, studies, 325
Sánchez studies, 326
Sangrá, Ramos and, studies, 183
San Joaquin River, 205, 207
San Juan Islands, Puget Sound, 33
Santa Catalina Island, 111
Sargassum, 446
Satellite data, sea surface biogeochemical variability, 220–222
Sathyendranath, Platt and, studies, 31
Scale resolution comparison
    aquatic ecosystem functioning, 4–5

basics, 3–4, 12–13
differential structure, fluorescence signals, 10–12
high-resolution data *vs.* conventional techniques, 7–12
instrument description, 7–9
microscale structure, 4–7
sampling process impact, 5–7
sensor deployment, 9–10
Scales, pattern-oriented modeling, 415
Scaling, wavelets, 293
Scaling properties, multiplicative cascades, 455–456
Scaling range, zooplankton swimming behavior, 342–346
Scanning electron microscopy (SEM), 144
Scenedesmus sp., 337
Scent preferences, 151–154
Schaeffer-Novelli, Cintrón and, studies, 420
Schafer, Oppenheim and, studies, 185
Scheffer model, 402, 403
Scheffer studies, 4, 393, 402, 509
Schell, Michener and, studies, 125
Schertzer studies, 10
Scheuring and Zeöld studies, 201
Schlieren optical system, 151, 337
Schmajuk studies, 513, 514
Schmid-Hempel, Stearns and, studies, 510
Schmitt, Seuront and, studies, 350
Schmitt and Seuront studies, 334, 337, 353
Schmitt studies, 453–466
Schneider studies, 384, 386
Schroeder studies, 334, 342, 350, 353
Schultze studies, 163
Schwehm studies, 17–28
Scilly Isles, 387
Scoglio Africa 95 experiment, 94
Scolio Africa experiment (1995), 94
Scotian Shelf, 300
Scotti, Osborn and, studies, 493
Sea-Bird 911 CTD, 33
Sea cable, 21
Sea fan orientation
    basics, 321–322, 326–327
    bottom boundary layer effects, 326
    discussion, 326–327
    growth, 325–326
    local topography impact, 326
    Rayleigh disc, 322–324
    symmetry, 324–325
Seagrass beds, acoustic remote sensing, *see* Acoustic remote sensing
Sea of Japan, 362
Seasonal power law behavior, 309–310
Seasonal variations, intertidal benthic species, 100–102
Sea-state noise, 90
Sea surface, 82, *see also* Biogeochemical variability, sea surface
Sea-Viewing Wide Field-of-View Sensor (SEAWiFS)
    biogeochemical variability, 220
    biological-physical coupling, 51, 54
    fractal characterization, 317

rank-size analysis, 257
SeaWiFS, *see* Sea-Viewing Wide Field-of-View Sensor
        (SEAWiFS)
SECAM video format, 18
Sediment, gases, 69
Seine estuary, 98–99
Sekine studies, 389
SEM, *see* Scanning electron microscopy (SEM)
Semibalanus balanoides, 251
Semtner and Chervin studies, 221
Sensor deployment, 9–10
Separated habitats, plankton dynamics model
        patterns, 406
SEPMs, *see* Spatially Explicit Population Models
        (SEPMs)
Seto Inlet
    differential structure, 11
    high resolution data *vs.* conventional techniques, 7
    rank-size analysis, 269
    sensor deployment, 9, 10
Seuront, Schmitt and, studies, 334, 337, 353
Seuront and Lagadeuc studies, 334, 336, 342, 344, 350,
        351, 353, 537
Seuront and Schmitt studies, 350
Seuront and Spilmont studies, 4, 6
Seuront studies, 3–13, 271, 333–358, 523–540
Seymour studies, 5, 258, 276
Shadowed image particle platform and evaluation
        recorder (SIPPER), 18
Shanks studies, 177
Sharp studies, 194
Sharp vertical gradients, planktonic layers, 44–45
She and Waymire studies, 460
Shenk studies, 325
Shepherd studies, 184, 186
Sheu, Oelbermann and, studies, 130
Shigesada, Kareiva and, studies, 338
Shlesinger and West studies, 336
Shorrocks studies, 249, 348
Shrimp, pink, 389
Sicily, *see* Ustica experiment (September 1999)
Siegal studies, 4
Siegel and Castellan studies, 347
Sierpinski carpet, 342
Siggia studies, 498
Signal transmission, Ustica experiment, 72
Silow studies, 388
Simulation
    adaptation levels, 507–518
    bottom-up model, 411, 423
    copepods, 479–490, 523–540
    discrete events-based Lagrangian approach,
        559–571
    energy costs, turbulence, 493–504
    flow fields, 479–490
    individual-based modeling, 507–518, 523–540
    Langmuir circulation, 445–450
    lognormal theory, 469–477
    multiagent systems, 523–540
    multifractal stochastic processes, 453–466
    multiple models approach, 429–443

pattern-oriented modeling, 411–425
plankton, 401–407, 543–556, 559–571
predator-prey interactions, 559–571
    spatial dynamics, benthic community, 429–443
    spatial scale, 383–395
    turbulent intermittency, 453–466
    turbulent structures, 469–477
Siniff and Jenssen studies, 336
Sinking velocity, energy costs, 502–503
SIPPER, *see* Shadowed image particle platform and
        evaluation recorder (SIPPER)
Size, spatial scale importance, 390–392
Skellam studies, 393
Skiftesvik, Browman and, studies, 493
Skipjack tuna, scale similarities
    basics, 183–184
    commercial fisheries, 195
    compensatory dynamics, 194–195
    data, 185
    depensatory dynamics, 194–195
    discussion, 192–196
    fractal marine system, 195–196
    methods, 185–186
    minimum populations, 194
    multiple (stable) equilibria, 194
    phase spaces, 193
    proposed equations, 193
    results, 186–192
    variable carrying capacity, 193
Skyllingstad and Denbo studies, 446
Skyllingstad studies, 445–450
Slatkin and Anderson studies, 393
Slobodkin, Kierstead and, studies, 32, 392
Small-scale, planktonic layers, 33–39, 44
Smith, Orr and, studies, 150
Smith studies, 18, 163
Snell and Carmona studies, 154
Snover and Commito studies, 230, 252
Snow, bubbles identification, 92n
Snow crabs, 194
Snowshoe hares, 413
Software
    Adobe Photoshop, 112
    ARcInfo, 113
    BATS data set, 257
    CENOCON, 536
    ENVI, 118
    HOTS data set, 257, 265, 268
    JGOFS, 257
    Matlab, 21, 55, 257
    Mobidyc platform, 523–540
    Multiresolution Seamless Image Database, 119
    SPSS, 257
    Swarm, 535
    Visual C++, 21
Sokal studies, 239
Solar heating, 83
Solari studies, 183–196
Somme estuary, *see* Multiscale *in situ* measurements,
        intertidal benthic species
Sonnenschein, Lorek and, studies, 535, 536

Sontek, 33, 40
Soranno studies, 209
Sosik and Mitchell studies, 4
Souissi and Bernard studies, 353
Souissi and Uye studies, 526, 537
Souissi studies, 361–375, 523–540, 571
Sound of Jura, 387
Sound propagation, 83–92
Sound speed profile (SSP), 83
Southgate, Myers and, studies, 113
Southward studies, 238
Spatial autocorrelation, 229
Spatial dynamics, benthic species
    basics, 429–430
    discussion, 441–443
    intertidal zone, 432–437
    model classes, 430–432
    modeling, 432–441
    mussels, 432–437
    validation and verification, 437–441
Spatial heterogeneity, *see* Multiagent systems,
        individual-based modeling; Plankton
        dynamics, patterns in models
Spatially Explicit Population Models (SEPMs), 119,
    230
Spatially extensive, high-resolution images (SEHRI)
    basics, 121
    concept demonstration, 113–119
    discussion, 119–121
    GIS database, 113–116
    intertidal context, 109
    limitations of technique, 120
    optimization recommendations, 120
    patch scale analysis, 116–118
    photographic equipment, settings, and conditions,
        120–121
    retroscale analysis, 118–119
    sampling approach limitations, 110–111
    spatially extensive high resolution images, 111
    system, 111–113
    utility expansion, 119
Spatial representation, IBM, 530
Spatial scale, importance
    abiotic conditions, 386–389
    basics, 383–384
    biotic conditions, 389–390
    discussion, 393–395
    emergent patterns, 392–393
    fish movement, 388–389
    freshwater models, 388
    hierarchy, 385–386
    individual organism size, 390–392
    lower trophic levels, 387–388
    marine models, 387–388
    marine systems, 388–389
    model considerations, 384
    origins and effects, 384–386
    streams, 389
    survey, 386–393
Spatial scale, treatments, 229–230
Spatial transition matrix models, 234–235

Spatiotemporal dynamics, 530–533
Spectra, synchroton-based IR imaging, 141–142
Spectra, wavelet-based, 308
Spectral absorption bands, 140–141
Spectral analysis, biological-physical coupling, 55,
    59–61
Spectral characteristics, ambient noise, 88–90
Spies studies, 130, 131
Spilmont, Seuront and, studies, 4, 6
Spilmont studies, 97–105
Spotted seatrout, 389
SPSS software, 257
Squires, Yamazaki and, studies, 5, 177
Squires and Yamazaki studies, 334
Squires studies, 469–477, 493–504
Squyres studies, 17–28
SSP, *see* Sound speed profile (SSP)
Stable isotope ecology, food web dynamics
    basics, 125–126, 131
    direct effects, 130
    discussion, 127–129
    indirect effects, 130–131
    isotropic structure, 127–128
    methods, 126–127
    results, 127–129
    temporal fluctuations, nutrient source, 128–129
    time-sensitive factors, 129–131
Stable stochastic integrals, 465–466
Stage-structured population models
    Acartia clausi, 368–372
    age-within stage model, 366–367
    basics, 361–362
    cannibalism, 368–371
    discussion, 374–375
    Euterpina acutifrons, 371–373
    eutrophic inlet location, 372–373
    extrema identification, 365–366
    mathematical proofs, 375–376
    method, 363–367
    predation, 364–365, 368–372, 376
    results, 367–373
    in situ interactions, 372–373
    stage-structured population model, 363–365
    time-series abundance data, 365–366
    transient behavior, 363–365
    transition rules, 363–365, 367–368, 375–376
Stanton, Foote and, studies, 27
State variables, 415
Statistical ensemble, 168–169
Statistics, energy costs, 500–501
Stavn studies, 177
Stearns and Schmid-Hempel studies, 510
Stearns studies, 168, 510
Steele and Frost studies, 555
Steele and Henderson studies, 210, 317, 404
Steele studies, 202, 205, 571
Sterner, Adams and, studies, 130
Stevens and Krebs studies, 333
Stewart studies, 472
Stigebrandt, Wulff and, studies, 388

Stochastic birth-death (SBD) model, 430–432, 434–435, 440, 442
Stochastic processes, multifractal
    casual lognormal stochastic equation, 460–462
    characteristic functions, 464
    continuous cascades, 459–463
    continuous lognormal multifractal, 462–463
    discrete cascades simulation, 457–459
    infinitely divisible distributions, 464–465
    multiplicative cascades, 454–457
    stable stochastic integrals, 465–466
    velocity fluctuations, 454
Stokes methods, 164, 446
Stolothrissa tanganicae, 126, 127–128
Stommel studies, 445, 448
Strahle studies, 314
Strain-based swimming model, 498–499
Strand studies, 510, 516
StraÓkraba studies, 208
Strategy vector, 509–510
Strathmann, Emlet and, studies, 479
Stratified bank flanks, 313
Streams, spatial scale importance, 389
Stream trout habitat selection, 422–423
Strickler, Cowles and, studies, 32, 482
Strickler, Gerritsen and, studies, 149, 352, 559, 560, 565, 566, 567, 568, 570, 571
Strickler, Hwang and, studies, 333, 494
Strickler, Koehl and, studies, 32
Strickler, Yen and, studies, 32, 333
Strickler and Bal studies, 32, 494, 503
Strickler and Costello studies, 493
Strickler and Hwang studies, 150, 167
Strickler laboratory, 481, 486
Strickler studies, 149–158, 163, 167, 330, 333–358, 479, 486, 493–504, 543
Strobed light sheet, 20–21
Strutton studies, 3, 51–62, 265
Subsurface stratified layer, 477
Sueuront and Lagadeuc studies, 12
Sugihara, Hastings and, studies, 334, 339
Sugihara and May studies, 265, 336, 346
Sumpter and Broomhead studies, 535
Sundaram and Collins studies, 570
Sundby studies, 493
Sundermeyer and Ledwell studies, 43
Supervised learning, 514
Surface turbulent layer, 475–477
Survey, spatial scale importance, 386–393
Survival model, IBM, 527–528
Suzuki and Ishimaru studies, 6
Swarming, zooplankton, *see* Zooplankton swarming
Swarm software, 535
Swimming behavior, *see also* Zooplankton swimming behavior
    discrete events-based Lagrangian approach, 560–561, 567–569
    flow field numerical simulation, 487–488
Swimming paths, zooplankton, 337–338, *see also* Zooplankton swimming behavior

Synchroton-based infrared imaging, *Euglena gracilis* single cells
    basics, 135–137, 145
    brightness, 138
    data analysis and results, 141–144
    diffraction limited IR imaging, 143–144
    discussion, 144–145
    experimental details, 137–141
    images, 142–143
    infrared imaging, 135–137
    IR storage ring radiation, 135–136, 137–138
    microspectroscopy, 138
    power, 137
    sample preparation, 138–140
    spectra, 141–142
    spectral absorption bands, 140–141
Synder and Nisbet studies, 231
Szeto, Kerman and, studies, 195

T

Talkner, Weber and, studies, 201, 205, 210
Talkner and Weber studies, 205
Tang studies, 336
Tanner studies, 232, 234, 238, 240
Tanzania, 125, *see also* Food web dynamics, stable isotope ecology
TAO, *see* Tropical Atmosphere Ocean (TAO) array
Taqqu, Samorodnistky and, studies, 464
Tarafdar studies, 339
Temora longicornis, mate signaling
    basics, 149–150, 157–158
    discussion, 151–157
    methods, 150–151
    odorant levels, 155–157
    pheromones, 151
    results, 151–157
    scent preferences, 151–154
    tracking behavior, 154–155
    trails, 149–157
Temora stylifera, 152
Temperature
    England, 202
    individual-based modeling, 528
    water, acoustic remote sensing, 83
Temporal fluctuations, nutrient source, 128–129
Temporal resolution, intertidal benthic species, 102
Temporal scaling, rivers and lakes, *see* Environmental variability, rivers and lakes
Tennekes and Lumley studies, 493, 501
Teranes and Bernasconi studies, 128
Tests, pattern-oriented modeling, 416
Test site, acoustic remote sensing, 69–71
Tett studies, 387
Thalassia testudinum, 327
Theodor and Denizot studies, 321
Thomason, Hills and, studies, 254
Thorpe studies, 43
Three-dimensional swimming behavior, 337–338
Threshold settings, planktonic behavior, 553–554

Thrush studies, 229
Thulke studies, 412
Tide, acoustic remote sensing, 81–82
Tilman and Kareiva studies, 230, 231
Time-frequency characteristics, ambient noise, 90–91
Time-integral-pressure-squared calculation, 80
Time integration, trophic levels, *see* Food web dynamics, stable isotope ecology
Timescale, numerical integration, 562
Time-sensitive factors, food web dynamics, 129–131
Time-varying medium impulse response, 75–76
Tinbergen studies, 177
Ting studies, 154
Tiselius and Jonsson studies, 333, 337, 352
Tiselius studies, 18, 31, 32, 45, 177, 333, 342, 351, 571
Titelman studies, 333, 570
TIWs, *see* Tropical instability waves (TIWs)
Topcon TotalStation survey system, 112, 113
TOPEX satellite, 53
Torrence and Campo studies, 299, 306
Toxicometer, zooplankton as, 353
Tracer distributions and transport, 215–218
Trails, 149–157
Transducer calibration, 72–73
Transform, wavelet, 301–303, 306–308
Trevorrow studies, 17–28
Tropical Atmosphere Ocean (TAO) array
  basics, 51
  biological-physical coupling, 51
  cross-correlation analysis, 54
  data, 53–54
  decorrelation analysis, 55
Tropical instability waves (TIWs), 52, 61
Tropodiaptomus simplex, 126
Tsanis studies, 388
Tsonis and Elsner studies, 336
Tsuda and Miller studies, 155
Tucker studies, 131
Tukey-Hamming method, 258
Tuljapurkar and Caswell studies, 508
Turbulence, planktonic organism hovering
  ambient flow field, 502
  basics, 493–494
  direct numerical simulation, 494–496
  discussion, 503
  flow fields, 494–497, 499, 502
  mean swim velocity, 501–502
  methods, 494–499
  motion equation, 497–498
  nondimensional numbers to dimensional numbers, 504
  particle rising, 502–503
  planktonic swimming models, 497–499
  random flow simulation, 496–497
  results, 499–503
  sinking velocity, 502–503
  statistics, 500–501
  strain-based swimming model, 498–499
  velocity-based swimming model, 498
Turbulence Ocean Microstructure Acquisition Profiler (TurboMAP), 7–9, 12

Turbulent intermittency, modeling
  basics, 453–454, 463
  casual lognormal stochastic equation, 460–462
  characteristic functions, 464
  continuous cascades, 459–463
  continuous lognormal multifractal, 462–463
  discrete cascades simulation, 457–459
  infinitely divisible distributions, 464–465
  multiplicative cascades, 454–457
  stable stochastic integrals, 465–466
  velocity fluctuations, 454
Turbulent structures, lognormal theory
  basics, 469–471
  discussion, 474–477
  simulations, 471–474
  subsurface stratified layer, 477
  surface turbulent layer, 475–477
Turchin studies, 333, 334, 336, 338, 339, 341, 342, 353
Turcotte, Huang and, studies, 337, 350
Turcotte, Malamud and, studies, 301, 303
Turcotte, Pelletier and, studies, 202, 209, 210
Turner, Saffman and, studies, 330
Turner and Gardner studies, 341
Turner studies, 117
Tuscan Arcipelago, 69
Two-dimensional *vs.* three-dimensional estimates, 348–350
Tyrrhenian Sea, *see* Ustica experiment (September 1999)
Tyutyunov studies, 194

U

Uchima and Murano studies, 155
Uchmański and Grimm studies, 524
Uchmański studies, 537
Udotea algae, 327
Ueda studies, 162
Uhlenbeck and Osnstein studies, 166
Uhlenbeck-Orstein (UO) stochastic process, 495
Underway oceanographic data, 300–301
Underwood and Chapman studies, 229, 240
Underwood studies, 229
University of Wisconsin, 203, 481
U.S. Geological Survey, 203
U.S. GLOBEC Broadscale Survey, 27, 297–301, 315–316
Usher studies, 232, 233, 234
Ustica experiment (September 1999), *see also* Acoustic remote sensing
  acoustic measurements, 72–73
  ambient noise recording, 72
  CTD and dissolved oxygen content, 73–75
  equalized matched-filter processing, 73
  experimental configuration, 71–72
  signal transmission, 72
  test site, 69–71
  transducer calibration, 72–73
  waveguiding, 94
Utility expansion, SEHRI, 119
Utne, Aksnes and, studies, 571

Utne-Palm studies, 517
Uye, Liang and, studies, 367
Uye, Souissi and, studies, 526, 537
Uye and Liang studies, 367, 537
Uye studies, 523–540

V

Vacelet, Laborel and, studies, 321
Validation and verification, 437–441
Van Atta and Yeh studies, 474
Van Ballemberghe studies, 336
Vander Zanden and Rasmussen studies, 125
Vander Zanden studies, 130
Van Duren and Videler studies, 333, 352
Van Rooij studies, 513
Van Winkle studies, 391, 394
Vaqué, Duarte and, studies, 258
Variability, 103–104, 193, *see also* Biogeochemical
　　variability, sea surface
Variable carrying capacity, 193
Vasseur and Yodzis studies, 202
Vassilicos, Malik and, studies, 570
Vectorial estimation, 562–564
Velocity, 170–171, 454
Velocity-based swimming model, 498
Venice lagoon, 387
Verburg studies, 125–131
Verduin and Backhaus studies, 423
Verity studies, 31
Vertical distribution, 224–225
Vertical gradients
　　continental shelf (Oregon), 34, 36
　　measurements and issues, 39–43
　　trophic implications, 44–45
Vertical mixing, 43
Vertical phytoplankton distribution patterns
　　basics, 257–263, 275–276
　　data set size, 257–258
　　initial analysis, 258
　　interpretation, 259–263
　　low resolution, 263–275
　　methodology, 258–259
　　recent use, 259
　　time series fluorescence profiles, 263–275
Vertical shear resolution, 39–42
Via studies, 512
Videler, van Duren and, studies, 333, 352
Video formats, 18
Video plankton recorder (VPR), 18
Vilà, Bascompte and, studies, 336, 350
Vincent and Meneguzzi studies, 477, 494, 498
Visser, Kiørboe and, studies, 494, 503, 560, 571
Visser studies, 571
Visual C++ software, 21
Volterra studies, 508
Von Bertalanffy studies, 433
Voorhis and Bruce studies, 216
Voss algorithm, 314–315
Voss studies, 314, 341

VPR, *see* Video plankton recorder (VPR)

W

Wada, Minagawa and, studies, 125, 130
Wada and Hattori studies, 128
Wainwright and Dillon studies, 321, 322, 326
Walker, Kenkel and, studies, 342
Wallis, Mandelbrot and, studies, 202
Walter studies, 423
Wang and Maxey studies, 500
Wang studies, 161–178, 471, 472
Warwick, Gee and, studies, 249, 336
Waser studies, 131
Washington, *see* East Sound; Puget Sound
Waters and Mitchell studies, 274, 276, 555
Water temperature, 83
Watkinson, Freckleton and, studies, 240
Watson studies, 18
Watts, Jenkins and, studies, 205
Wavelet-based variance spectra, 301–308
Wavelets
　　applications, 293–295
　　basics, 279–284, 296
　　Daubechies wavelets, 286–288
　　fast wavelet transform, 288–290
　　Fourier transforms, 290–291
　　fractal characterization, 298
　　Haar wavelets, 284–286
　　mother, 285
　　multiresolution analysis, 286–288
　　scaling, 293
　　technicalities, 291–292
Waymire, She and, studies, 460
Weber, Talkner and, studies, 205
Weber and Talkner studies, 201, 205, 210
Weber studies, 201
Webster studies, 209, 392
Weihs, Haury and, studies, 175, 176
Weissburg studies, 149, 155, 162, 168, 174, 333
Welch studies, 185
Weller studies, 445
Well-mixed bank crest, 310–311
Werner and Gilliam studies, 510
Weron, Janicki and, studies, 464
WESP-TOOL, 535, 536
West, Shlesinger and, studies, 336
WET Labs, Inc., 33
Whitehead, Rendell and, studies, 513
Wiebe studies, 17, 297–318
Wiegand studies, 412, 416
Wiens and Milne studies, 333, 353
Wiens studies, 336, 342, 352, 384, 386
Williams, Murphy and, studies, 138
Williamson studies, 162, 175
Wilson, Rand and, studies, 232, 236
Wilson and Adamec studies, 53, 57
Wilson and Greaves studies, 338
Wilson and Nisbet studies, 230
Wilson studies, 535, 543

Wimereux, France, *see* Multiscale *in situ*
measurements, intertidal benthic species
Winch, 21
Windemere Laboratory, 203
Wind noise, 90
Wisconsin Aquatic Technology and Environmental
Research (WATER) Institute, 481
Wisconsin lakes, 204, 207–209
Wisenden and Millard studies, 516
Wise studies, 184
Wishner studies, 162
Wissel, Huth and, studies, 390, 391
With studies, 339, 343, 353
Wolk studies, 3–13, 269, 350, 352
Woodcock studies, 445
Wood's equation, 85–86
Wootton studies, 119, 230, 231, 235, 237, 238, 239, 240
Wortmann studies, 392
Wray, Jiménez and, studies, 498
Wright, Hartnoll and, studies, 238
Wright studies, 510
Wroblewski, O'Brien and, studies, 387
Wroblewski studies, 388
Wulff and Stigebrandt studies, 388
Wunsch, Zang and, studies, 201
Wunsch studies, 210
Wu studies, 336, 342, 353

Y

Yaglom, Gurvich and, studies, 469, 470, 474, 475
Yaglom studies, 454
Yamazaki, A. studies, 543–556
Yamazaki, H. studies, 3–13, 162, 163, 469–477,
493–504, 560, 570
Yamazaki, Haury and, studies, 149
Yamazaki, Squires and, studies, 334
Yamazaki and Kamykowski studies, 353, 445, 446
Yamazaki and Lueck studies, 472, 474
Yamazaki and Okubo studies, 165
Yamazaki and Osborn studies, 4
Yamazaki and Squires studies, 5, 177
Yan studies, 202, 209
Yassen studies, 368
Yeh, van Atta and, studies, 474
Yen, Fields and, studies, 152, 162
Yen, Leising and, studies, 162, 177
Yen and Bundock studies, 162
Yen and Fields studies, 560
Yen and Nicoll studies, 32
Yen and Strickler studies, 32, 333
Yen studies, 149–158, 161–178, 333, 494
Yeung and Pope studies, 472, 495
Yoder studies, 166, 218, 276
Yodzis, Cuddington and, studies, 202, 209, 210
Yodzis, Vasseur and, studies, 202
Yodzis and Innis studies, 390
Yoshioka studies, 128
Young studies, 43, 258, 259

Z

Zaneveld and Pegau studies, 4
Zang and Wunsch studies, 201
Zaret studies, 494
Zar studies, 55, 57
Zeöld, Scheuring and, studies, 201
Zhang, Marsili and, studies, 259, 268
Zipf and Pareto studies, 259
Zipf's law, 258–259, 264
Zone of influence (ZOI), 419–420
Zooplankton, 32, 536–538
Zooplankton distributions
basics, 17–19
camera housing, 19
command and control, 21
discussion, 26–28
environmental sensors and frame, 21
Hoeya Sound, 25
image and data processing, 21–22
Knight Inlet, 22–24, 25–26
methods, 19–24
November 19 cast, 25
November 21 cast, 22–23, 25–28
power/telemetry housing, 19–20
results, 24–26
sea cable, 21
strobed light sheet, 20–21
system description, 19–22
winch, 21
Zooplankton swarming
acceleration fields, 166–167, 173–174
aggregative force field, 164–165
analysis, 168–170
autocorrelations, 171–173
basics, 161–163, 178
concentrative force, 177
damping, 175–177
diffusion, 163–164
dimensional swarming, 166
discussion, 174–177
excitation, 177
experiment, 167–168
fit parameters, 171–173
models, 165–166, 169–170, 174–177
physical-behavioral balances, 177
results, 170–174
statistical ensemble, 168–169
swarming, 163–164, 166
theory, 163–167
velocity distributions, 170–171
Zooplankton swimming behavior, *see also* Swimming
behavior
algae, 337
algorithms, robustness, 346–348
basics, 333–337, 351–353
box-counting method, 341, 347–348
characterization, 338–341
compass method, 340–341
Daphnia pulex culture, 337
environmental structure effects, 350–351

estimates, 341–350
fractal dimensions, 339–350
recording swimming paths, 337–338
on the scale dependence, 341–342
scaling range, objective, 342–346
testing estimates, 341–350
three-dimensional swimming behavior, 337–338

two-dimensional *vs.* three-dimensional estimates,
    348–350
Zooplankton Visualization and Imaging System
    (ZOOVIS)
    basics, 18–19
    Knight Inlet operations, 22–28
    methods, 19–21
Zostera, 392